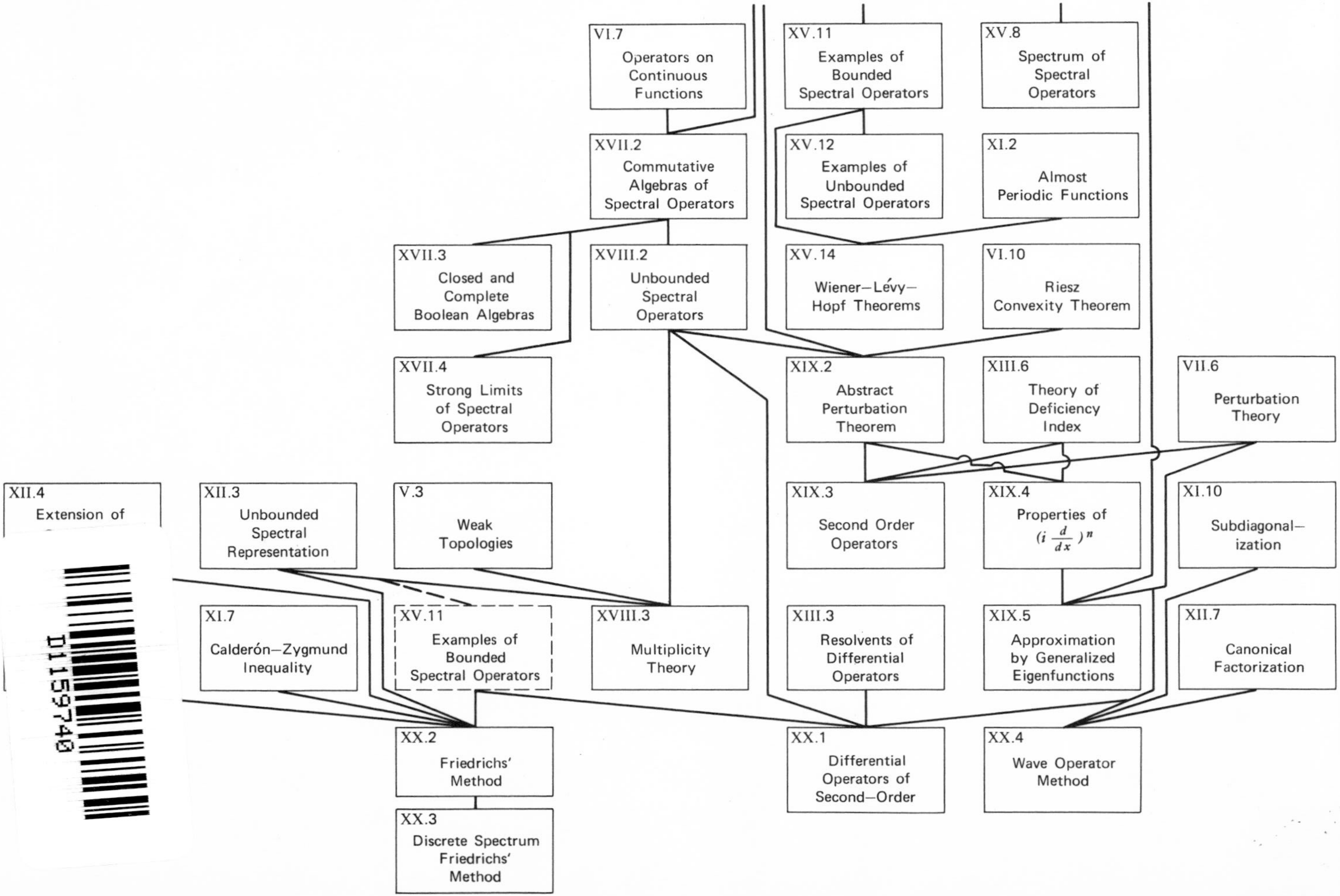

VI.7 Operators on Continuous Functions
XV.11 Examples of Bounded Spectral Operators
XV.8 Spectrum of Spectral Operators
XVII.2 Commutative Algebras of Spectral Operators
XV.12 Examples of Unbounded Spectral Operators
XI.2 Almost Periodic Functions
XVII.3 Closed and Complete Boolean Algebras
XVIII.2 Unbounded Spectral Operators
XV.14 Wiener–Lévy–Hopf Theorems
VI.10 Riesz Convexity Theorem
XVII.4 Strong Limits of Spectral Operators
XIX.2 Abstract Perturbation Theorem
XIII.6 Theory of Deficiency Index
VII.6 Perturbation Theory
XII.4 Extension of
XII.3 Unbounded Spectral Representation
V.3 Weak Topologies
XIX.3 Second Order Operators
XIX.4 Properties of $(i \frac{d}{dx})^n$
XI.10 Subdiagonal–ization
XI.7 Calderón–Zygmund Inequality
XV.11 Examples of Bounded Spectral Operators
XVIII.3 Multiplicity Theory
XIII.3 Resolvents of Differential Operators
XIX.5 Approximation by Generalized Eigenfunctions
XII.7 Canonical Factorization
XX.2 Friedrichs' Method
XX.1 Differential Operators of Second–Order
XX.4 Wave Operator Method
XX.3 Discrete Spectrum Friedrichs' Method

PURE AND APPLIED MATHEMATICS

A Series of Texts and Monographs

Vol. I:	**Supersonic Flow and Shock Waves** By R. Courant and K. O. Friedrichs
Vol. II:	**Nonlinear Vibrations in Mechanical and Electrical Systems** By J. J. Stoker
Vol. III:	**Dirichlet's Principle, Conformal Mapping and Minimal Surfaces** By R. Courant
Vol. IV:	**Water Waves** By J. J. Stoker
Vol. V:	**Integral Equations** By F. G. Tricomi
Vol. VI:	**Differential Equations: Geometric Theory** By Solomon Lefshetz
Vol. VII:	**Linear Operators—Parts I, II, and III** By Nelson Dunford and Jacob T. Schwartz
Vol. VIII:	**Modern Geometrical Optics** By Max Herzberger
Vol. IX:	**Orthogonal Functions** By G. Sansome
Vol. X:	**Lectures on Differential and Integral Equations** By K. Yosida
Vol. XI:	**Representation Theory of Finite Groups and Associative Algebras** By C. W. Curtis and I. Reiner
Vol. XII:	**Electromagnetic Theory and Geometrical Optics** By Morris Kline and Irvin W. Kay
Vol. XIII:	**Combinatorial Group Theory** By W. Magnus, A. Karrass, and D. Solitar
Vol. XIV:	**Asymptotic Expansions for Ordinary Differential Equations** By Wolfgang Wasow
Vol. XV:	**Tchebycheff Systems: With Applications in Analysis and Statistics** By Samuel Karlin and William J. Studden
Vol. XVI:	**Convex Polytopes** By Branko Grunbaum
Vol. XVII:	**Fourier Analysis in Several Complex Variables** By Leon Ehrenpreis
Vol. XVIII:	**Generalized Integral Transformations** By Armen H. Zemanian
Vol. XIX:	**Introduction to the Theory of Categories and Functors** By Ion Bucur and Aristide Deleanu
Vol. XX:	**Differential Geometry** By J. J. Stoker
Vol. XXI:	**Ordinary Differential Equations** By Jack K. Hale
Vol. XXII:	**Introduction to Potential Theory** By L. L. Helms
Vol. XXIII:	**The Functions of Mathematical Physics** By Harry Hochstadt
Vol. XXIV:	**Introduction to Mathematical Fluid Dynamics** By Richard E. Meyer
Vol. XXV:	**Differential Games** By Avner Friedman

PURE AND APPLIED MATHEMATICS

A Series of Texts and Monographs

Edited by: R. COURANT · L. BERS · J. J. STOKER

VOLUME VII

LINEAR OPERATORS

PART III

SPECTRAL OPERATORS

NELSON DUNFORD and **JACOB T. SCHWARTZ**

FORMER JAMES E. ENGLISH
PROFESSOR OF MATHEMATICS
YALE UNIVERSITY

PROFESSOR OF MATHEMATICS
COURANT INSTITUTE
NEW YORK UNIVERSITY

WITH THE ASSISTANCE OF

William G. Bade and **Robert G. Bartle**

PROFESSOR OF
MATHEMATICS
UNIVERSITY OF
CALIFORNIA, BERKELEY

PROFESSOR OF
MATHEMATICS
UNIVERSITY OF
ILLINOIS

19 71

WILEY-INTERSCIENCE

A DIVISION OF JOHN WILEY & SONS, INC.

NEW YORK · LONDON · SYDNEY · TORONTO

Library of Congress Catalog Card Number: 57-10545

ISBN 0 471 22639 4

Printed in the United States of America

10 9 8 7 6 5 4 3 2 1

Preface

As stated in the Preface to Part I of this treatise, our original intention was to include the theory of spectral operators in Part II. We thought (and still think) that this theory forms an excellent introduction to the detailed study of the much more refined and complete theory of self adjoint operators in Hilbert space. It soon became evident, however, that the natural limitations imposed by size precluded the fulfillment of this intention. The exclusion of the theory of spectral operators from Part II was not due solely to the growth of the literature in this field, but was determined also by our desire to present a number of important applications of the general theory of spectral operators, whose discussion required several hundred pages.

In presenting the theory of spectral operators in Part III, we had to cope with two independent problems of exposition not encountered in the first two parts. We briefly describe these problems here, for such a discussion gives the reader a historical perspective of Part III, as well as that of the treatise as a whole, and shows him where the existing mathematical theory, and its close relations to certain profound empirical results of contemporary physics, verges on the boundary between the known and the unknown. The applications to physics involve not only problems emerging from recent theories such as quantum mechanics, scattering theory, quantum field theory, and the quantum mechanical version of the three body problem, but also nonselfadjoint problems arising from more classical problems such as a large variety of diffusion phenomena, and particularly electromagnetic wave theory. We present the abstract operator theory in the text, and then in the Notes and Remarks, especially in Section XX.6, give the physical interpretations as well as a brief discussion of past and present related problems.

The first heretofore unencountered problem of exposition which Part III presents is that of the selection of material to be included. To comprehend the nature of this problem, let us briefly recall the corresponding situation in the spectral theory of Chapter XIII. The spectral theory for self adjoint operators had its single seed sown by B. Taylor's discussion of the vibrating string in 1713, just about a quarter of a century after the publication of Newton's *Principia*, or *Philosophiae naturalis principia mathematica*, to give this great and most influential of treatises its full title. The ensuing rapid advance in modern mathematics with a number of orthogonal expansions already well known in the eighteenth century and the beginning of a very powerful general theory of such expansions made by Sturm and Liouville in the early nineteenth century, when combined with the publication of Hermann Weyl's fundamental discoveries in 1910 and the beginnings of a most fruitful development of abstract linear analysis in the early part of the century by Volterra, F. and M. Riesz, David Hilbert and his school at Göttingen, Banach, Mazur, Schauder and many others in Poland, and the fructifying advances of the last forty years made by Gelfand, Kreĭn, and Naĭmark in the Soviet Union; Kakutani, Kato, Kuroda, and Yosida in Japan (and the United States); the Bourbaki group in France; Colojoară and Foiaş in Romania; Hille and Phillips, Friedrichs, von Neumann and Paley* and Wiener in the United States; and many others all contribute to over two and a half of the most fruitful centuries in the history of mathematics. During this period all of the foundations were laid and many of the major problems concerning self adjoint operators were solved in a satisfactory form. Thus, in presenting the spectral theory of self adjoint operators contained in Chapters X, XII, and XIII we had the benefit of not only over two hundred and fifty years of discoveries with the inevitable maturing processes, but also the historical perspective so helpful in deciding what is essential to include and what may be excluded or relegated to our various Notes and Remarks.

In contrast to this wealth of knowledge concerning self adjoint operators, we have a relatively short history of the corresponding non-selfadjoint problems. We are extremely fortunate to have the penetrating studies of G. D. Birkhoff [1–7] made at the turn of the century before the

* The English mathematician R. E. A. C. Paley (1907–1933) was working with Norbert Wiener at the Massachusetts Institute of Technology where their important discoveries were made before Paley's skiing accident in the Canadian Rockies caused his untimely death.

advent of modern operator theory. This work gives such a convincing indication that strongly convergent generalized eigenvalue expansions for a wide variety of nonselfadjoint boundary value problems exist, that it is surprising to have almost a half century pass before such generalizations were made. An exposition of Birkhoff's discoveries presented in terms of operator theory and with numerous generalizations, mostly discovered by J. T. Schwartz and extended by H. P. Kramer, are found in Chapter XIX, but a general operator theoretic method of treating the spectral theory of nonselfadjoint operators, excluding a few isolated results such as F. Riesz's elegant discussion of the theory of compact operators, is not much over a quarter of a century old. Nevertheless, the fecundity of the new theories is beyond any reasonable expectations and consequently they have proliferated in a variety of directions, presenting such an anfractuous path that it is difficult to predict which, if any, of the approaches will survive. A glance at the extent of the References which, unlike those of Part II, apply only to Part III although the earlier numbering is continued, makes it apparent that any attempt at a moderately complete covering of all relevant material would result in a heterogeneous jumble, hardly serving the worker in the field and resulting in a volume so huge that many interesting results having applications to modern physics might have to be omitted. Consequently, we are forced to relegate a large amount of interesting material to a brief reference in the Notes and Remarks. We want to stress here that such an abrogation of a topic in no way represents our opinion of its scientific value.

Our second heretofore unencountered problem is that of giving the reader a complete comprehension of the basic ideas associated with spectral operators and some of the more elementary applications to problems of mathematical physics involving nonselfadjoint operators, and to do this assuming only a minimal knowledge of elementary operator theory. Stated otherwise, we wish to make it possible to read a large portion of Part III with a knowledge of only a small amount of the contents of Parts I and II. Thus, we present in Chapter XV the concept of a spectral operator, many of its special properties, and a number of illustrative applications of the theory assuming only a little more than the knowledge that is normally acquired by a student in his first year of graduate study or earlier. Although the Interdependence Table shows parts of Chapter XV to be dependent on Chapter XIV, the only results, with one exception, used from Chapter XIV are elementary self-contained lemmas concerning approximation of integrable functions by smooth functions. The exception is Sobolev's theorem, which is used only in the proof of a corollary unessential to the main train of ideas. The reader of Chapter XV should be familiar with the elementary topics of algebra and point set topology found in Chapter I, the three basic principles of Chapter II, and the elementary properties of the theory of integration with respect to a finitely additive set function as well as with respect to a countably additive measure. Thus he need be familiar with only a relatively small part of the contents of Chapter III. He should have a good understanding of the theory of finite square matrices of complex numbers, preferably in the form presented in Section VII.1, as well as an acquaintance with finite and infinite dimensional Hilbert spaces as presented in Sections IV.3 and IV.4. Besides these topics, the reader should be familiar with the results of Section VII.3, Chapter IX on B-algebras, and the first eighteen pages of Chapter X that treat the spectral theorem for bounded self adjoint operators. This spectral theorem for self adjoint operators is not logically essential to the reading of Chapter XV but a familiarity with its meaning is helpful, since it is occasionally used for illustrative purposes. The material outlined above, with very few other references, should aid the reader in most of Part III with the exception of the important Chapter XX, where the major and deep results of Chapter XIII are used extensively with a view to making applications to some of the profound problems of contemporary physics as well as the classical theory of electromagnetic waves. One feature of Chapter XV that enables us to present the illustrations found in Sections XV.11 and XV.12 is the elementary development of the Parseval-Plancherel theory of the Fourier transform, thus avoiding any reference to Section XI.3 where this theory is developed for locally compact groups by using the general spectral theorem which involves a more complex treatment than that found in Section XV.11. The elementary treatment presented here is possible, since throughout Part III we use the Fourier transform only in finite dimensional Euclidean space. Thus the reader familiar with the usual elementary topics encountered in point set topology, algebra, analytic function theory, and real function theory (including Lebesgue integration) needs only a slight knowledge of functional analysis, most particularly Chapters II and IX, to read all but the last chapter of Part III.

In Chapter XV we try to stress most of the *basic elementary* properties of spectral operators that distinguish them from other operators. These properties assume a variety of forms not immediately apparent from the definition of a spectral operator. For example, the spectral relations, largely due to S. Foguel, found in Section XV.8 were not, for the most part, foreshadowed by any knowledge concerning the spectra of self adjoint or normal operators. One of the most interesting set of properties in the theory concerns the resolvent $R(\lambda; T) = (\lambda - T)^{-1}$ of a spectral operator T. Over twenty years ago N. Dunford and Gerhard Neubauer discovered independently that for each vector x in the complex B-space in which T operates, the analytic function $R(\lambda; T)x$ has only single valued extensions. The astute observation of S. Kakutani that this property is not enjoyed by all linear operators suggests that the proper domain for the resolution of the identity of a spectral operator should consist of the Borel sets in an infinitely many sheeted Riemann surface rather than those in the complex plane. As far as we know, this avenue of investigation has not as yet been pursued. This single valued extension property is the first of three properties enjoyed by the resolvent of a spectral operator which are explained in Chapter XV. In the following chapter it is shown just how close these three properties come to being sufficient to ensure that an operator is a spectral operator.

In order to concentrate our attention on the new concepts and avoid unnecessary digressions into unrelated problems, we limit our discussion to spectral operators in a complex B-space. However, we do not restrict ourselves to bounded operators, because the applications to contemporary atomic physics of the boundary value problems of Chapter XX require the use of unbounded spectral operators. Also, the Cauchy initial value problems solved in Chapter XV are defined in terms of certain nonself-adjoint systems of partial differential equations arising (or in some cases closely approximating the equations that arise) in a wide variety of diffusion problems. Such applications as well as the other illustrations of the general theory in Chapter XV are purposely chosen to be elementary and are usually set in Hilbert space, since the presentation is shorter than that for other B-spaces. Another reason for this restriction is that not all of the examples are valid in every B-space. We have not investigated the domain of validity of our illustrations, a fact that may tempt some readers to further investigations.

Chapter XV is an elementary introduction to the field which presents most of the distinguishing properties of spectral operators, a number of examples to illustrate phenomena not appearing in the theory of self adjoint or normal operators, and a few examples of Cauchy initial value problems represented by nonselfadjoint systems of perturbed Laplacian operators (Theorems XV.12.19, XV.12.21, and XV.12.22). In sharp contrast with some of the most innocent looking boundary value problems associated with perturbed operators such as $d^2/dt^2 + q$ with $q(t)$ complex on the infinite real interval $0 \leqq t < \infty$ whose solutions can be most scabrous, the convolution kernel giving the solution to the Cauchy initial value problem of Theorem XV.12.19 and many other similar problems may be given as a finite expression involving only elementary functions. The explicit details of calculation are in the text. Even the much more general problem of Theorem XV.12.21 has the kernel which defines its solution given by a fairly rapidly convergent series.

The Cauchy initial value problem is, in various forms, discussed briefly in all three parts of this book, thus providing an excellent opportunity to show, in Section XV.13, how it is a part of a long train of ideas crossing many borders of intellectual creation from the poetry of Homer to the work of some of our modern artists. This type of discussion is not customary in a book such as ours, perhaps even unprecedented, but is so relevant to our culture that we give a brief summary of these ideas.

In Section XV.14 there are a number of results only loosely related to spectral operators. They are concerned with different versions of the Wiener-Hopf and Wiener-Lévy theorems. The chapter ends with a large set of exercises in Section 15 and copious Notes and Remarks in Section 16. The chapter was written with the idea of forming a complete, comprehensive, and elementary unit for those who merely want an introduction to the subject and provides the essential material for a seminar or course for first or second year graduate students.

In Chapter XVI we discuss the difficult problem of determining, in terms of the properties of the resolvent of an operator, which operators are spectral operators. The fact that the illustrative examples in Chapter XV are spectral operators is proved directly from the definition of a spectral operator, whereas the type of operator discussed in Chapter XVI requires much more involved analysis, as well as the introduction of a number of new concepts. The culminating theorems of this lengthy study

(Theorems XVI.4.5., 5.15, and 5.18) we believe have some deep and interesting applications still to be discovered. Our original version of Theorem XVI.5.19 was presented as a corollary to the preceding theorem, but since it is essential for a number of the deep results in Chapter XX which have important applications to some of the most difficult problems of contemporary atomic physics, we give a proof independent of the earlier results of Chapter XVI. This was possible since it concerns a type of operator with a number of special properties including first order rate of growth of the resolvent near the spectrum. Thus the reader whose primary interest is in the applications to such topics as scattering theory, the quantum mechanical version of the three body problem, and other contemporary problems of mathematical physics to which the results of Chapter XX apply, has a direct short path to these problems; namely, Theorem XVI.5.19, its unbounded counterpart found in Theorem XVIII.2.34, and then directly to Chapter XX, thereby eliminating the involved analysis needed for most of Chapter XVI and all of the material in Chapters XVII through XIX. But we remind the reader again that the principal results of Chapter XIII are frequently used in Chapter XX.

Chapter XVII contains a study of algebras of spectral operators beginning with uniformly closed algebras, their representations, and Wedderburn type decompositions. The second part is an exposition of W. G. Bade's penetrating study of strongly closed algebras of spectral operators and complete Boolean algebras of projections. The deep and closely related multiplicity theory also due to Bade is postponed until the end of Chapter XVIII since it uses the functional calculus for unbounded spectral operators developed in the first part of that chapter.

Chapters XIX and XX contain various applications and extensions arising in connection with applications of the general theory developed in the first chapters of this volume. In Chapter XIX, operators with discrete spectrum are considered; Chapter XX focuses upon operators with continuous spectrum. The methods used in Chapter XIX are perturbation-theoretic. A spectral operator T with discrete spectrum, the points of whose spectrum are separated by sufficiently large gaps, is considered. A perturbation P, small compared to T in an appropriate technical sense, is added to T; it is then possible to estimate the resolvent of $T' = T + P$ in terms of the resolvent of T, and from this to relate the spectral projections of T' to those of T. In some cases, we may then prove that T' is spectral.

The results obtained have application to the spectral analysis of nonself-adjoint differential operators on a finite closed interval. These applications, which go back to Birkhoff and Tamarkin and extend the classical self adjoint Sturm-Liouville theory, are derived next. The work required is algebraic and establishes the validity, in the cases studied, of the asymptotic hypotheses required by the general perturbation theory. Chapter XIX ends with a section which gives results concerning completeness properties of the set of eigenvectors of a perturbed operator. These theorems make assertions less specific than those obtained in the first sections of Chapter XIX, but which can be proved under hypotheses which are considerably more general; their proofs rest ultimately on the generalized Carleman inequalities of Chapter X.

Chapter XX is devoted to applications of the general theory to the study of perturbations of operators with continuous spectra, and it is seen that many such operators either are spectral operators or have closely related properties. A number of results, with some recent generalizations, due to Naĭmark, Friedrichs, Kato, and Kuroda, are discussed. Many of the results established in Chapter XX are applicable (or, in some cases, a more involved but similar analysis is applicable) to some profound nonselfadjoint problems in contemporary as well as classical mechanics. These theorems place some empirical laws of physics, heretofore established by informal reasoning, on a rigorous mathematical foundation. The chapter brings us close to active areas of current research in mathematical physics. In the first section of Chapter XX, the general spectral theory is applied, essentially in direct fashion, to singular nonselfadjoint differential operators of second order on a half axis. These operators have the form

$$T = -\left(\frac{d}{dt}\right)^2 + q(t),$$

where q may be complex but is asymptotically very small. They are shown to be spectral, essentially by a direct asymptotic analysis which verifies the hypotheses required by the general theory of spectral operators.

The second section of Chapter XX introduces, and the third section continues, quite a different line of thought, due originally to Friedrichs and developed subsequently by him and his students and collaborators.

In this theory, one attempts to demonstrate *similarity* between an unperturbed operator T and its perturbation $T' = T + P$, that is, to show the existence of an operator U such that

$$UTU^{-1} = T'.$$

This equation may be recast to permit the application of perturbation methods; the whole approach has the fundamental advantage of being applicable in a very natural way to operators with continuous spectrum and of being independent of any assumptions of self adjointness. The analytical details necessary for the application of this method of Friedrichs are, in fact, simplest when the spectrum of the operator T fills an entire region; this case is considered first. Subsequently, the somewhat more formidable technical details belonging to the case of an operator with continuous spectrum filling an interval of the real axis are developed, and applications to Volterra integral operators and to operators derived by perturbation from Laplace's operator ∇ are studied. These latter operators, and various generalized operators of the same kind, whose considerably more formidable analysis is not given, are important in quantum theory. In several interesting subsequent studies, Friedrichs has also extended this analytic strategy to the systems of operators which occur in the quantum theory of fields.

Section 3 of Chapter XX adapts Friedrichs' method to the case of operators with discrete spectrum, deriving results which complement those given in Chapter XIX.

In Section 4 another idea, and one which has been of great significance for the application of spectral theory to physics, is discussed. This idea is as follows. If an unperturbed operator T and a perturbed operator T' derived from it are both self adjoint, then a non-rigorous formal argument can be used to buttress the expectation that the Friedrichs operator U realizing the similarity $UTU^{-1} = T'$ should have the form of a limit

$$U = \lim_{t \to \infty} e^{itT'} e^{-itT}.$$

This purely formal argument inspires us to study the limit appearing above, and indeed it develops that if the difference $T' - T$ satisfies suitable hypotheses, this limit will exist in the strong topology and will have the properties which formal reasoning ascribes to it. The rather ingenious arguments needed to prove these facts are due to Kato and Kuroda and have subsequently been generalized by various other authors, who have also established relationships between this "wave operator" or "scattering" method and various versions of the Friedrichs approach given in Section XX.2. (The formal line of reasoning underlying the rigorous results of Kato-Kuroda has a long history in physics: many elaborate calculations have been based upon it, and it has even suggested a standpoint which certain physicists have been willing to consider as an axiomatic basis for quantum theory.) It may also be mentioned that the Kato-Kuroda theory yields some of the most precise results available for the spectral analysis of partial differential operators.

For the research worker we mention that the Notes and Remarks contain a wealth of ideas with ample room for further development. Besides these ideas, the investigation of a countably additive spectral theory on a general Riemann surface has been suggested. Also, parts of Bade's multiplicity theory suggest the possibility of a spectral representation theory, analogous to the Weyl-Kodaira theory, for boundary value problems in L_1, although the technical details might be difficult. One of the most intriguing problems, about which nothing seems to be known, is that of interpreting operators, arising from natural phenomena, whose resolvent has greater than first order rate of growth at every point along a curve and to which the unbounded versions of the principal theorems of Chapter XVI apply. We see no *a priori* or philosophical reason why such phenomena should not be even more numerous than the familiar ones of the last three centuries, which involve mostly self adjoint operators. It seems possible that the dearth of such examples in the literature stems from the lack of any theory to analyze them accurately. It is common practice to slightly alter equations arising in a physical problem so as to make a rigorous analysis possible. However, such a procedure, which would change the rate of growth of the resolvent without substantially changing the problem, seems highly improbable.

We wish to mention here a slight difference in our use of the black arrow in the margin. In Parts I and II this arrow is used to call attention to ideas particularly important for the subsequent development. In Part III it continues to have this meaning but, in addition, is used to call attention to results or concepts that we consider to have special interest even though they may not be used in the further development.

We also invoke the mathematician's prerogative by using Q.E.D. in a sense slightly different from *quod erat demonstrandum.* The symbol merely signals the end of a proof. This is particularly helpful to the reader of a proof involving many pages and containing a number of proofs of preliminary statements.

We are greatly indebted to Miss R. M. Castroll for her competent editorial assistance in preparing the manuscript for the printer, checking all galley and page proof, and preparing the author index.

We also thank the staffs of the Air Force Office of Scientific Research and the Office of Naval Research for the assistance given on several occasions in the past while many of the initial results presented here were being discovered.

Sarasota, Florida NELSON DUNFORD
New York City JACOB T. SCHWARTZ
May 1971

Contents

PART III. SPECTRAL OPERATORS

La pensée n'est qu'un éclair au milieu de la nuit. Mais c'est cet éclair qui est tout.

HENRI POINCARÉ

LINEAR OPERATORS IN THREE PARTS

PART I
General Theory

PART II
Spectral Theory, Self Adjoint Operators in Hilbert Space

PART III
Spectral Operators

Nelson Dunford and Jacob T. Schwartz

CHAPTER XV

Spectral Operators

1. Introduction

The far-reaching theory of self adjoint differential boundary value problems developed in the preceding chapters illustrates once again the power and importance of the spectral reduction theory for bounded and unbounded self adjoint (or normal) operators as developed in Chapters X and XII. The problem of extending this reduction theory to non-normal operators is one of the most important unsolved problems in the theory of linear operations. Consider, for instance, the problem of finding a resolution of the identity for the formal differential operator $T = -d^2/dx^2 + q$ on the infinite internal $0 \leqq x < \infty$. If q is a real function, then, as we have seen in Chapter XIII, the properties of the operator T may be analyzed in considerable detail. But if q assumes some complex values, a situation which can easily arise in a concrete example, then the whole spectral analysis of T is invalid and it is difficult to make even the most elementary statements concerning the spectral reduction of T.

In order to give an adequate spectral analysis of this and a number of other nonselfadjoint operators we are impelled to attempt the extension of the spectral reduction theory of Chapters X and XIII to non-normal operators in Hilbert space, and to operators in B-spaces other than Hilbert space. If guided literally by the properties of normal operators, one might at first think that the appropriate operators to study were those which had the expression

$$T = \int_{\sigma(T)} \lambda E(d\lambda)$$

in terms of some projection valued measure defined on $\sigma(T)$. That such a restricted class of operators is not of sufficient generality may perhaps best be appreciated by considering the case where $\mathfrak{H}$ is a finite dimensional unitary space. If T is an operator in such a space, then an operational calculus for analytic functions may be given by the formula (cf. VII.1.8)

$$f(T) = \sum_{\lambda \in \sigma(T)} \sum_{n=0}^{\nu(\lambda)-1} \frac{(T-\lambda I)^n}{n!} f^{(n)}(\lambda) E(\lambda)$$

$$= \sum_{n=0}^{\infty} \sum_{\lambda \in \sigma(T)} \frac{(T-\lambda I)^n}{n!} f^{(n)}(\lambda) E(\lambda).$$

If T is a normal operator, then $(T-\lambda I)E(\lambda) = 0$ and this formula reduces to the formula

$$f(T) = \sum_{\lambda \in \sigma(T)} f(\lambda) E(\lambda),$$

which is not valid for an arbitrary T. To see more clearly the difference between the calculi given by these two formulas, let us rewrite them by introducing the nilpotent N and the resolution of the identity E defined by the equations

$$N = T - \sum_{\lambda \in \sigma(T)} \lambda E(\lambda), \qquad E(\delta) = \sum_{\lambda \in \delta} E(\lambda).$$

In terms of N and E the operational calculus for an arbitrary operator takes the form

$$f(T) = \sum_{n=0}^{\infty} \frac{N^n}{n!} \int_{\sigma(T)} f^{(n)}(\lambda) E(d\lambda),$$

whereas if T is normal only the first term

$$f(T) = \int_{\sigma(T)} f(\lambda) E(d\lambda)$$

of this series is needed to express $f(T)$. Since, as this example shows, the theory of normal operators is not an infallible guide even to the theory of arbitrary operators in finite dimensional spaces, we must revert to more general considerations in attempting to construct the desired general spectral reduction theory.

The most general reduction problem for an operator T in a B-space $\mathfrak{X}$ might be stated as follows: decompose $\mathfrak{X}$ into the direct sum $\mathfrak{X} = \mathfrak{X}_1 \oplus \mathfrak{X}_2$ of two proper subspaces in such a way that $T\mathfrak{X}_1 \subseteq \mathfrak{X}_1$, $T\mathfrak{X}_2 \subseteq \mathfrak{X}_2$. If E denotes the projection of $\mathfrak{X}$ onto the space $\mathfrak{X}_1$, then the two conditions $T\mathfrak{X}_1 \subseteq \mathfrak{X}_1$, $T\mathfrak{X}_2 \subseteq \mathfrak{X}_2$ are equivalent to the single algebraic condition $TE = ET$. Thus the most general reduction problem for an operator T would be: find a projection which commutes with T.

That this formulation is much too general to constitute a suitable formulation of the *spectral* reduction problem may be seen by considering the case $T = I$. Since I commutes with every projection, the above reduction problem stated for I would be: find all projections in $\mathfrak{X}$. This problem, while interesting in itself, clearly goes much further than a mere spectral analysis of the operator I. Indeed, since $\sigma(I)$ consists of one point only, so that its spectrum is geometrically irreducible, we should expect I to be irreducible from the point of view of spectral analysis also.

In order to make a spectral reduction of an operator, we must find a projection E commuting with T, such that the spectrum of the restriction $T \mid E\mathfrak{X}$ is contained in a given closed set. Stated otherwise, we desire to find, for a given subset δ of $\sigma(T)$, two subspaces $\mathfrak{X}_1$, $\mathfrak{X}_2$ such that $T\mathfrak{X}_1 \subseteq \mathfrak{X}_1$ $T\mathfrak{X}_2 \subseteq \mathfrak{X}_2$, $\mathfrak{X} = \mathfrak{X}_1 \oplus \mathfrak{X}_2$ and such that

$$\sigma(T \mid \mathfrak{X}_1) \subseteq \bar{\delta}, \qquad \sigma(T \mid \mathfrak{X}_2) \subseteq \bar{\delta}'.$$

We recall (cf. X.2.6) that the resolution of the identity E for a normal operator T in Hilbert space $\mathfrak{H}$ has the desired property $\sigma(T \mid E(\delta)\mathfrak{H}) \subseteq \bar{\delta}$ for every Borel set δ. Similarly, for an arbitrary operator T on a complex B-space we have seen (cf. VII.3.20) that for a set δ which is both open and closed in $\sigma(T)$ a projection $E(\delta)$ can be defined which commutes with T and for which the spectrum of the restriction $T \mid E(\delta)\mathfrak{X}$ is contained in δ.

The common property, that is, $\sigma(T \mid E(\delta)\mathfrak{X}) \subseteq \bar{\delta}$, in the examples just cited is not an unimportant coincidence: it lies either explicitly or implicitly at the very roots of the key results in spectral theory. Thus we are led to begin the formulation of the spectral reduction problem by requiring that for each set δ in some family Σ of sets in the plane there is a projection $E(\delta)$ which commutes with T and for which $\sigma(T \mid E(\delta)\mathfrak{X}) \subseteq \bar{\delta}$. In the examples just mentioned Σ is a field and $E(\delta)$ satisfies the algebraic laws

$$\begin{aligned} E(\delta \cap \sigma) &= E(\delta)E(\sigma), \\ E(\delta \cup \sigma) &= E(\delta) + E(\sigma) - E(\delta)E(\sigma), \\ E(\delta') &= I - E(\delta), \\ E(\sigma(T)) &= I. \end{aligned}$$

In the case of a normal operator in Hilbert space the set Σ is the σ-field of all Borel sets and E is countably additive in the strong operator topology. It is this property of countable additivity that underlies the powerful eigenvalue expansions obtained in the theory of self adjoint boundary value problems.

In the general formulation of the reduction problem we are going to require that Σ be the field of Borel sets, that $E(\delta)$ satisfy the above algebraic identities, and, in addition, that E be countably additive on Σ in the strong operator topology. Although formal definitions will be given in the next section, we shall call an operator for which the spectral reduction problem has a solution, a spectral operator. It is not the primary purpose of the present chapter to determine which operators are spectral operators but rather to discover properties of bounded spectral operators. However, in Sections 11 and 12 we do give some examples of spectral operators and in Section 13 will be found a digression into a few related topics.

In the following chapter we shall endeavor to find conditions on the resolvent of an operator which will insure an affirmative solution to the spectral reduction problem. In subsequent chapters we shall attempt to find other examples of spectral operators and to see how far the theory may be applied to non-symmetric boundary value problems.

2. Terminology and Preliminary Notions

The concepts of a Boolean algebra of projections, a spectral measure, an integral with respect to a spectral measure, and so on, which have been explained in Section X.1, are all fundamental for the present chapter. The reader may wish to review these notions in Chapter X, for they will be restated here in a concise form. There are, however, a few new concepts to be introduced here and these will be explained more fully.

Throughout the chapter the letter $\mathfrak{X}$ will be used for a complex B-space and T for a bounded linear operator in $\mathfrak{X}$. By a *projection* in $\mathfrak{X}$ is meant a bounded linear operator E in $\mathfrak{X}$ with $E^2 = E$. The *intersection* $A \wedge B$ and the *union* $A \vee B$ of two commuting projections A and B in $\mathfrak{X}$ are the projections AB and $A+B-AB$, respectively. The ranges of the intersection and union of two commuting projections are given by the equations

$$(A \wedge B)\mathfrak{X} = (A\mathfrak{X}) \cap (B\mathfrak{X}), \quad (A \vee B)(\mathfrak{X}) = (A\mathfrak{X}) + (B\mathfrak{X}) = \overline{\text{sp}}(A\mathfrak{X}, B\mathfrak{X}),$$

where $\overline{\text{sp}}(A\mathfrak{X}, B\mathfrak{X})$ means the closed linear manifold spanned by the sets $A\mathfrak{X}$ and $B\mathfrak{X}$. Thus the *natural ordering* $A \leqq B$ between two commuting projections A and B has the geometrical significance that $A \leqq B$ is equivalent to $A\mathfrak{X} \subseteqq B\mathfrak{X}$. A *Boolean algebra* of *projections* in $\mathfrak{X}$ is a set of projections in $\mathfrak{X}$ which is a Boolean algebra (cf. Section I.12) under the operations $A \wedge B$ and $A \vee B$ which has for its zero and unit elements the operators 0 and I in $\mathfrak{X}$.

1 DEFINITION. A *spectral measure* in $\mathfrak{X}$ is a homomorphic map of a Boolean algebra of sets into a Boolean algebra of projection operators in $\mathfrak{X}$ which has the additional property that it maps the unit in its domain into the identity operator I in its range. A spectral measure is said to be *bounded* if the norms of the projections in its range are bounded.

We recall that if the domain of a bounded spectral measure E is a field Σ of subsets of a set S, then the integral $\int_S f(\lambda)E(d\lambda)$ may be defined for every bounded Σ-measurable scalar function f on S. In Section X.1 it was shown that this integral is a bounded homomorphism of the B-algebra $B(S, \Sigma)$ of Σ-measurable functions on S into the B-algebra $B(\mathfrak{X})$ of bounded linear operators in $\mathfrak{X}$, that is,

$$\int_S [\alpha f(s) + \beta g(s)]E(ds) = \alpha \int_S f(s)E(ds) + \beta \int_S g(s)E(ds),$$

$$\int_S f(s)g(s)E(ds) = \left[\int_S f(s)E(ds)\right]\left[\int_S g(s)E(ds)\right],$$

$$\left|\int_S f(s)E(ds)\right| \leqq v(E) \sup_{s \in S} |f(s)|,$$

where $v(E)$ is a constant depending only upon the spectral measure E. Other properties of the integral which will be used and which were explained in Section X.1 are expressed by the equations

$$x^*\left[\int_S f(s)E(ds)\right]x = \int_S f(s)x^*E(ds)x,$$

$$\left[\int_S f(s)E(ds)\right]x = \int_S f(s)E(ds)x,$$

$$\int_S g(t) \int_S f(s)E(ds \cap dt) = \int_S g(s)f(s)E(ds),$$

and

$$\int_S f(s)E(h^{-1}(ds)) = \int_S f(h(s))E(ds),$$

where h is a map of S into itself with the property that for each δ in Σ the set $h^{-1}(\delta) = \{s \mid h(s) \in \delta\}$ is also in Σ.

It will be convenient at times to use a notion, somewhat more general than that introduced in Section X.1, of the resolution of the identity for an operator.

→ 2 DEFINITION. If Σ is a Boolean algebra of subsets of the complex plane which contains the void set and the whole plane, in short, if Σ is a field of sets in the complex plane, then a spectral measure E on Σ is called a *resolution of the identity* (or a *spectral resolution*) *for the operator* T if

$$E(\delta)T = TE(\delta), \qquad \sigma(T_\delta) \subseteq \bar{\delta}, \qquad \delta \in \Sigma.$$

Here we have used, and shall continue to use, the symbol T_δ for the restriction $T \mid \mathfrak{X}_\delta$ of T to the manifold $\mathfrak{X}_\delta = E(\delta)\mathfrak{X}$.

Let us illustrate this definition by recalling (cf. VII.3.17 and VII.3.20) that if we define a *spectral set* for the operator T to be any set δ for which $\delta \cap \sigma(T)$ is open and closed in $\sigma(T)$ and if for each δ in the field Σ of spectral sets we define the projection $E(\delta)$ by the formula

$$E(\delta) = \frac{1}{2\pi i} \int_C R(\lambda; T)\, d\lambda,$$

where C is any rectifiable Jordan curve containing $\delta \cap \sigma(T)$ but no other points of $\sigma(T)$ in its interior, then the map $\delta \to E(\delta)$ on Σ is a resolution of the identity for T. This resolution of the identity, which exists for an arbitrary bounded operator T, is not generally defined for all Borel sets in the plane, nor is it, in general, bounded or countably additive. However, in Corollary X.2.4 we have seen that a bounded normal operator in Hilbert space always has a uniquely defined bounded and countably additive resolution of the identity defined on the field of all Borel subsets of the plane.

The operators we shall study in the next few chapters will be operators which have a resolution of the identity with properties similar to those of a normal operator in Hilbert space. We shall, in fact, study resolutions of the identity which are countably additive on the field $\mathscr{B}$ of Borel sets in the plane. In this connection it should be recalled (cf. IV.10.1) that if E is a spectral measure on the σ-field $\mathscr{B}$ for which $x^*E(\cdot)x$ is countably additive for each x^* in $\mathfrak{X}^*$ and x in $\mathfrak{X}$, then E is countably additive on $\mathscr{B}$ in the strong operator topology. Since every projection $E \neq 0$ has norm $|E| \geqq 1$, a spectral measure E cannot be countably additive in the uniform operator topology unless its domain contains at most a finite number of disjoint sets. Thus we are led to the following definition and corollary.

3 DEFINITION. A spectral measure E is said to be *countably additive* if for each x^* in $\mathfrak{X}^*$ and each x in $\mathfrak{X}$ the scalar set function $x^*E(\cdot)x$ is countably additive on the domain of E.

→ 4 COROLLARY. *If the domain of a countably additive spectral measure E is a σ-field, then E is countably additive in the strong operator topology and bounded.*

The boundedness of $E(\sigma)$ follows from Corollaries IV.10.2 and II.3.21.

The spectral operators in the complex B-space $\mathfrak{X}$ which are the subject of the present chapter may now be defined as follows.

→ 5 DEFINITION. A *spectral operator* is an operator with a countably additive resolution of the identity defined on the Borel sets of the plane.

Of course it is not clear to begin with that a countably additive resolution of the identity defined on the Borel sets of the plane is uniquely determined by the spectral operator T. This is the case, but until it is proved we refer to any such spectral measure as *a* resolution of the identity for T. After uniqueness has been proved we shall refer to *the* resolution of the identity for T.

In the discussion of spectral operators we shall encounter a new notion, namely, that of an analytic extension of $R(\xi; T)x$. Here and elsewhere the symbol $R(\xi; T)$ is used for the resolvent $(\xi I - T)^{-1}$ of T at the point ξ in the resolvent set $\rho(T)$. If x is a vector in $\mathfrak{X}$, then by an *analytic extension* of $R(\xi; T)x$ will be meant an $\mathfrak{X}$-valued function f defined and analytic on an open set $D(f) \supseteq \rho(T)$ and such that

$$(\xi I - T)f(\xi) = x, \qquad \xi \in D(f).$$

It is clear that, for such an extension,

$$f(\xi) = R(\xi; T)x, \qquad \xi \in \rho(T).$$

The notion of "analytic extension" differs from that of "analytic continuation," for the domain $D(f)$ of an extension may contain points which cannot be connected with any point in $\rho(T)$ by a curve in $D(f)$.

→ 6 DEFINITION. The function $R(\xi; T)x$ is said to have the *single valued extension property* provided that for every pair f, g of analytic extensions of $R(\xi; T)x$ we have $f(\xi) = g(\xi)$ for every ξ in $D(f)\,D(g)$. The union of the sets $D(f)$ as f varies over all analytic extensions of $R(\xi; T)x$ is called the *resolvent set of* x and is denoted by $\rho(x)$. The *spectrum* $\sigma(x)$ *of* x is defined to be the complement of $\rho(x)$.

It is clear that if $R(\xi; T)x$ has the single valued extension property, then there is a maximal extension $x(\cdot)$ of $R(\xi; T)x$ whose domain is $\rho(x)$.

Throughout the rest of this section, $x(\xi)$ will denote such a maximal extension of $R(\xi; T)x$ in all cases when $R(\xi; T)x$ has the single valued extension property. In this case $x(\xi)$ is a single valued analytic function with domain $\rho(x)$ and with

$$x(\xi) = R(\xi; T)x, \qquad \xi \in \rho(T).$$

It will be shown in the next section that, if T is a spectral operator, the function $R(\xi; T)x$ has, for every x in $\mathfrak{X}$, the single valued extension property. That this is not the case for an arbitrary operator T is elegantly shown by the following example due to S. Kakutani.

Consider the space $\mathfrak{X}$ of functions f analytic in the unit circle $|z| \leqq 1$ and for which

$$f(z) = \sum_{n=0}^{\infty} c_n z^n, \quad \sum_{n=0}^{\infty} |c_n|^2 = |f|^2 < \infty.$$

In this space define T by

$$T(f, z) = \frac{f(z) - f(0)}{z}.$$

The spectrum of T is the set of λ with $|\lambda| \leqq 1$, and for $\lambda \in \rho(T)$ the function $R(\lambda; T)(g, z)$ may be calculated by solving the equation

$$(\lambda I - T)f = g$$

for f. An elementary calculation gives

$$f(z) = \frac{zg(z) - f(0)}{\lambda z - 1}.$$

Since $f(z)$ is analytic when $z = \lambda^{-1}$, we must have

$$f(0) = \lambda^{-1} g(\lambda^{-1}),$$

so that

$$R(\lambda; T)(g, z) = \frac{zg(z) - \lambda^{-1} g(\lambda^{-1})}{\lambda(z - \lambda^{-1})}.$$

Thus the vector valued analytic function $R(\lambda; T)g$, $\lambda \in \rho(T)$ will have multiple valued extensions if the function g has a multiple valued analytic continuation outside the unit circle.

3. The Resolvent of a Spectral Operator

The vector valued analytic functions $R(\xi; T)x$ associated with the resolvent of a bounded spectral operator have a number of important properties not enjoyed by functions of the form $R(\xi; T)x$ when T is not a spectral operator. In this section we shall discuss several such properties. In the following chapter it will be shown just how near three of these properties come to being sufficient for the operator T to be a spectral operator.

1 LEMMA. *Let E be a resolution of the identity for the bounded spectral operator T. Let σ be a closed set of complex numbers with $\xi_0 \notin \sigma$. If $(\xi_0 I - T)x_0 = 0$ then*

$$E(\sigma)x_0 = 0, \; E(\{\xi_0\})x_0 = x_0,$$

where $\{\xi_0\}$ is the set consisting of the single point ξ_0.

PROOF. Let T_σ be the restriction of T to the space $E(\sigma)\mathfrak{X}$, so that since $\xi_0 \notin \bar{\sigma}$, $\xi_0 \in \rho(T_\sigma)$ and

$$R(\xi_0; T_\sigma)(\xi_0 I - T)E(\sigma) = E(\sigma).$$

But since

$$(\xi_0 I - T)E(\sigma)x_0 = E(\sigma)(\xi_0 I - T)x_0 = 0,$$

we have $E(\sigma)x_0 = 0$. Now let

$$\sigma_n = \left\{\xi \,\middle|\, |\xi - \xi_0| \geqq \frac{1}{n}\right\},$$

so that by the above, $E(\sigma_n)x_0 = 0$, and since E is countably additive,

$$[I - E(\{\xi_0\})]x_0 = \lim_n E(\sigma_n)x_0 = 0,$$

and thus $x_0 = E(\{\xi_0\})x_0$. Q.E.D.

Occasionally in what follows we shall use the symbol $E(\xi_0)$ in place of $E(\{\xi_0\})$.

2 THEOREM. *If T is a bounded spectral operator in $\mathfrak{X}$, then for every x in $\mathfrak{X}$ the function $R(\xi; T)x$ has the single valued extension property.*

PROOF. Let f, g be two extensions of $R(\xi; T)x$ and define

$$h(\xi) = f(\xi) - g(\xi), \qquad \xi \in D(f)\, D(g).$$

We suppose, in order to make an indirect proof, that for some $\xi_0 \in D(f)\,D(g)$ we have $h(\xi_0) \neq 0$. Thus there is a neighborhood $N(\xi_0)$ of ξ_0 with $N(\xi_0) \subseteq D(f)\,D(g)$ and

$$\text{(i)} \qquad h(\xi) \neq 0, \qquad (\xi I - T)h(\xi) = 0, \qquad \xi \in N(\xi_0).$$

Let $\{\xi_n\}$ be a sequence of points in $N(\xi_0)$ with $\xi_n \neq \xi_0$ and $\xi_n \to \xi_0$. Then it follows from (i) and Lemma 1 that

$$0 = E(\xi_0)E(\xi_n)h(\xi_n) = E(\xi_0)h(\xi_n) \to E(\xi_0)h(\xi_0) = h(\xi_0),$$

which contradicts the fact that $h(\xi_0) \neq 0$. Q.E.D.

Thus, whenever T is spectral, and $x \in \mathfrak{X}$, $R(\xi; T)x$ has a maximal analytic extension defined in $\rho(x)$, which we shall denote as $x(\xi)$.

3 COROLLARY. *If T is a bounded spectral operator, the spectrum $\sigma(x)$ of x is void if and only if $x = 0$.*

PROOF. Using Theorem 2, we see that if $\sigma(x)$ is void, then $x(\xi)$ is everywhere defined, single valued, and hence entire. Since, by VII.3.4,

$$\lim_{\xi \to \infty} x^*x(\xi) = \lim_{\xi \to \infty} x^*R(\xi; T)x = 0,$$

it is seen that $x^*x(\xi) = 0$ for all ξ and all $x^* \in \mathfrak{X}^*$. Hence, by Corollary II.3.14, $x(\xi) = 0$ and thus $x = (\xi I - T)x(\xi) = 0$. Q.E.D.

→ 4 THEOREM. *Let T be a bounded spectral operator with resolution of the identity E, and let δ be a closed set of complex numbers. Then*

$$E(\delta)\mathfrak{X} = \{x \,|\, \sigma(x) \subseteq \delta\}.$$

PROOF. Let $E(\delta)x = x$, and let T_δ be the restriction of T to $E(\delta)\mathfrak{X}$. Since $\sigma(T_\delta) \subseteq \delta$, it is seen from the relation

$$R(\xi; T_\delta)E(\delta)x = R(\xi; T_\delta)x$$

that $R(\xi; T_\delta)E(\delta)x$ is an analytic extension of $R(\xi; T)x$ to δ', the complement of δ. Thus $\rho(x) \supseteq \delta'$ and $\sigma(x) \subseteq \delta$.

Conversely, assume that $\sigma(x) \subseteq \delta$ and let δ_1 be a closed subset of the complement δ' of δ. Then if $T_{\delta_1} = T \,|\, E(\delta_1)\mathfrak{X}$, the function $R(\xi; T_{\delta_1})E(\delta_1)x$ is an analytic extension of $R(\xi; T)E(\delta_1)x$ to δ_1'. Moreover, $E(\delta_1)x(\xi)$ is easily seen to be an extension of $R(\xi; T)E(\delta_1)x$ to $\rho(x)$. Thus, since $R(\xi; T)E(\delta_1)x$ has an analytic extension to $\rho(x) \cup \delta_1'$, that is, the whole complex plane, $\sigma(E(\delta_1)x)$ is void. By the preceding corollary, then $E(\delta_1)x = 0$. Let δ_n be an increasing sequence of closed sets whose union is δ'. Then since $E(\delta_n)x = 0$ by the above,

$$E(\delta')x = \lim_n E(\delta_n)x = 0,$$

and so $E(\delta)x = x$. Q.E.D.

5 COROLLARY. *If E is a resolution of the identity for the spectral operator T, then $E(\sigma(T)) = I$.*

6 COROLLARY. *If T is a spectral operator in $\mathfrak{X}$, then the set of all vectors whose spectrum lies in a given closed set of complex numbers is a closed linear manifold in $\mathfrak{X}$.*

→ 7 COROLLARY. *Let T be a spectral operator and A a bounded linear transformation which commutes with T. Then A commutes with every resolution of the identity for T. Moreover $\sigma(Ax) \subseteq \sigma(x)$ for every x in $\mathfrak{X}$.*

PROOF. Let δ, δ_1 be disjoint closed sets of complex numbers and let E be a resolution of the identity for T. Since

$$(\xi I - T)Ax(\xi) = A(\xi I - T)x(\xi) = Ax,$$

we see that $Ax(\xi)$ is an analytic extension of $R(\xi; T)Ax$ to $\rho(x)$. Hence $\rho(Ax) \supseteq \rho(x)$, so that $\sigma(Ax) \subseteq \sigma(x)$. Thus Theorem 4 shows that

$$AE(\delta)\mathfrak{X} \subseteq E(\delta)\mathfrak{X},$$

and hence that

$$E(\delta)AE(\delta) = AE(\delta), \qquad E(\delta)AE(\delta_1) = E(\delta)E(\delta_1)AE(\delta_1) = 0.$$

Since the complement δ' of δ is a denumerable union of closed sets, the countable additivity of E shows that $E(\delta)AE(\delta') = 0$ and hence

$$E(\delta)A = E(\delta)A[E(\delta) + E(\delta')] = E(\delta)AE(\delta) = AE(\delta).$$

Since δ is an arbitrary closed set, it follows that A commutes with $E(\sigma)$ for every Borel set σ. Q.E.D.

→ 8 COROLLARY. *Every spectral operator has a uniquely defined countably additive resolution of the identity defined on the field of Borel sets.*

PROOF. Let E, A be resolutions of the identity for T and let δ be a closed set of complex numbers. Then Theorem 4 shows that

$$A(\delta)E(\delta) = E(\delta), \qquad E(\delta)A(\delta) = A(\delta),$$

and thus Corollary 7 shows that $E(\delta) = A(\delta)$. Since E is countably additive, $E(\sigma) = A(\sigma)$ for every Borel set σ. Q.E.D.

→ 9 DEFINITION. The unique countably additive resolution of the identity defined on the Borel sets in the plane which is determined by a spectral operator T is called *the resolution of the identity for* T.

10 THEOREM. *Let* $I = E_1 + \cdots + E_n$ *where* $E_1, \ldots, E_n$ *are bounded, disjoint projections in* $\mathfrak{X}$, *each commuting with the bounded operator* T. *Then* T *is a spectral operator if and only if each restriction* $T \mid E_i \mathfrak{X}$ *is a spectral operator. If* T *is a spectral operator, then the resolution of the identity for the restriction* $T \mid E_i \mathfrak{X}$ *is the corresponding restriction of the resolution of the identity for* T.

PROOF. Let T be a spectral operator. By Corollary 7, E_i commutes with every projection $E(\sigma; T)$ in the resolution of the identity for T. It is clear that $E(\sigma; T) \mid E_i \mathfrak{X}$ is a countably additive spectral measure in $E_i \mathfrak{X}$. It is also clear that any operator A which commutes with each of the projections $E_1, \ldots, E_n$ has $\rho(A) = \bigcap_{i=1}^n \rho(A \mid E_i \mathfrak{X})$. This observation, when applied to the operator $T \mid E(\sigma; T)\mathfrak{X}$, shows that

$$\rho(T \mid E(\sigma; T)\mathfrak{X}) = \bigcap_{i=1}^{n} \rho((T \mid E(\sigma; T)\mathfrak{X}) \mid E_i \mathfrak{X})$$

$$= \bigcap_{i=1}^{n} \rho((T \mid E_i \mathfrak{X}) \mid E(\sigma; T) E_i \mathfrak{X}).$$

Thus for $\lambda \notin \bar{\sigma}$ we have λ in each of the resolvent sets

$$\rho((T \mid E_i \mathfrak{X}) \mid E(\sigma; T) E_i \mathfrak{X}),$$

which shows that

$$\sigma((T \mid E_i \mathfrak{X}) \mid E(\sigma; T) E_i \mathfrak{X}) \subseteq \bar{\sigma},$$

and proves that the restriction $E(\sigma; T) \mid E_i \mathfrak{X}$ is a resolution of the identity for the restriction $T \mid E_i \mathfrak{X}$.

Conversely, suppose that for each $i = 1, \ldots, n$ the restriction $T_i = T \mid E_i \mathfrak{X}$ is a spectral operator and let

$$E(\sigma) = \sum_{i=1}^{n} E(\sigma; T_i) E_i.$$

Then

$$E(\sigma)T = \sum_{i=1}^{n} \sum_{j=1}^{n} E(\sigma; T_i)E_i T_j E_j$$

$$= \sum_{i=1}^{n} E(\sigma; T_i)E_i T_i E_i$$

$$= \sum_{i=1}^{n} T_i E(\sigma; T_i)E_i = TE(\sigma).$$

It may similarly be verified that $E(\cdot)$ is a spectral measure and that it is countably additive. Now, if $\lambda \notin \bar{\sigma}$, then

$$\lambda \in \bigcap_{i=1}^{n} \rho(T_i \,|\, E(\sigma; T_i)E_i \mathfrak{X}) = \rho(T \,|\, E(\sigma)\mathfrak{X}).$$

This proves that $E(\cdot)$ is a resolution of the identity for T. Q.E.D.

4. The Canonical Reduction of a Spectral Operator

If $E(\lambda_i)$, $i = 1, \ldots, k$, are the projections associated with an operator T in a finite dimensional space $\mathfrak{X}$ as in Section VII.1, then (cf. VII.1.7) there are integers ν_i, $i = 1, \ldots, k$, with $(T - \lambda_i I)^{\nu_i} E(\lambda_i) = 0$. Thus the operator

$$TE(\lambda_i) = \lambda_i E(\lambda_i) + (T - \lambda_i I)E(\lambda_i)$$

is the sum of a scalar multiple of the identity operator in the space $E(\lambda_i)\mathfrak{X}$ and a nilpotent operator. Since $\mathfrak{X}$ is the direct sum (cf. VII.1.6) of the spaces $E(\lambda_i)\mathfrak{X}$, the operator T is the sum

$$T = \sum_{i=1}^{k} \lambda_i E(\lambda_i) + \sum_{i=1}^{k} (T - \lambda_i I)\, E(\lambda_i)$$

of an operator $S = \sum_{i=1}^{k} \lambda_i E(\lambda_i)$ which is equivalent to a diagonal matrix and a nilpotent operator $N = \sum_{i=1}^{k} (T - \lambda_i I)E(\lambda_i)$. Stated in other terms, this classical reduction of Jordan asserts that every finite square matrix of complex numbers is equivalent to the sum of a diagonal matrix and a nilpotent matrix.

In the present section we give an analogous canonical reduction for a bounded spectral operator in a complex B-space $\mathfrak{X}$. It will be shown that every such operator T is the sum $T = S + N$ of a quasi-nilpotent operator

N and an operator S which is of scalar type in accordance with the following definition.

→ 1 DEFINITION. A bounded operator S is said to be of *scalar type* in case it is a spectral operator which satisfies the equation

$$S = \int \lambda E(d\lambda),$$

where E is the resolution of the identity for S.

Since the spectral measure E vanishes outside the compact set $\sigma(S)$ (cf. Corollary 3.5), $f(\lambda) = \lambda$ is bounded on $\sigma(S)$ and the integral defining S exists.

It should also be observed that if one starts with a spectral measure E which is countably additive on the field of Borel sets and which vanishes outside a given compact set, then the operator S defined by the equation

$$S = \int \lambda E(d\lambda)$$

is a bounded scalar type spectral operator whose resolution of the identity is E. To see this we note first that S commutes with the projections $E(\delta)$, and second that if λ_0 is not in the closure of a set δ, then $(\lambda_0 - \lambda)^{-1}$ is bounded on δ and hence

$$\left(\int_\delta \frac{E(d\lambda)}{\lambda_0 - \lambda}\right)(\lambda_0 I - S)E(\delta) = \left(\int_\delta \frac{E(d\lambda)}{\lambda_0 - \lambda}\right)\left(\int_\delta (\lambda_0 - \lambda)E(d\lambda)\right)$$
$$= \int_\delta E(d\lambda) = E(\delta).$$

This shows that the spectrum of the restriction of S to $E(\delta)\mathfrak{X}$ is contained in $\bar{\delta}$ and proves that S is a scalar type spectral operator with resolution of the identity E.

Besides the notion of a scalar type operator the canonical decomposition to be proved for spectral operators involves the notion of a quasi-nilpotent operator, which we restate here for convenience.

2 DEFINITION. A bounded linear operator in a B-space is called a *quasi-nilpotent* if $\lim_n |T^n|^{1/n} = 0$.

The following corollary gives a useful spectral criterion which is a necessary and sufficient condition for an operator to be a quasi-nilpotent.

3 COROLLARY. *A bounded operator T is a quasi-nilpotent if and only if $\sigma(T) = \{0\}$.*

PROOF. This follows from Lemma VII.3.4. Q.E.D.

4 LEMMA. *If S and N are bounded commuting operators and if N is quasi-nilpotent, then $\sigma(S+N) = \sigma(S)$.*

PROOF. This is a corollary of Theorem VII.6.10 (or of Lemma IX.2.6 and Theorem IX.2.9). For the sake of convenience we shall reprove it here. Since N is quasi-nilpotent, it follows from Definition 2 that $|N|^k = o(\varepsilon^k)$ for every $\varepsilon > 0$ and thus for every $\lambda \in \rho(S)$ the series $\sum_{k=0}^{\infty} N^k R(\lambda; S)^k$ converges in the uniform operator topology. Since

$$\left\{\sum_{k=0}^{\infty} N^k R(\lambda; S)^k\right\}\{I - NR(\lambda; S)\} = (I - NR(\lambda; S)) \sum_{k=0}^{\infty} N^k R(\lambda; S)^k$$

$$= \sum_{k=0}^{\infty} N^k R(\lambda; S)^k - \sum_{k=1}^{\infty} N^k R(\lambda; S)^k = I,$$

the series is $(I - NR(\lambda; S))^{-1}$. This shows that the operator

$$(\lambda I - S - N)^{-1} = R(\lambda; S)(I - NR(\lambda; S))^{-1}$$

exists as an everywhere defined and bounded operator. Thus $\lambda \in \rho(S+N)$ and $\sigma(S) \supseteq \sigma(S+N)$. Similarly $\sigma(S+N) \supseteq \sigma(S)$. Q.E.D.

Using the notions of a quasi-nilpotent operator and of a scalar type operator we may state the following theorem which gives the *canonical reduction* of a spectral operator.

→ 5 THEOREM. *A bounded operator T is a spectral operator if and only if it is the sum $T = S + N$ of a bounded scalar type operator S and a quasi-nilpotent operator N commuting with S. Furthermore this decomposition is unique and T and S have the same spectrum and the same resolution of the identity.*

PROOF. It will first be shown that the sum $T = S + N$ of a scalar type operator S and a quasi-nilpotent N commuting with S is a spectral operator having the same resolution of the identity as S. Let E be the resolution of the identity for S so that, by Corollary 3.7, $NE(\delta) = E(\delta)N$ and hence $T = S + N$ commutes with $E(\delta)$. To show that E is the resolution of the identity for T it will therefore suffice to show that $\sigma(T_\delta) \subseteq \bar{\delta}$ for every Borel set δ. By Lemma 4, $\sigma(T_\delta) = \sigma(S_\delta)$, and since E is the resolution of the identity for S we have $\sigma(S_\delta) \subseteq \bar{\delta}$ and therefore $\sigma(T_\delta) \subseteq \bar{\delta}$. This shows that T and S have the same resolution of the identity. Since $S = \int \lambda E(d\lambda)$, it follows from Corollary 3.8 that S is uniquely determined by T. Hence $N = T - S$ is uniquely determined by T. It follows from Lemma 4 that T and S have the same spectrum.

Next it will be shown that every spectral operator T has the desired decomposition. The operators S and N are defined by the equations

$$S = \int \lambda E(d\lambda), \qquad N = T - S,$$

where E is the resolution of the identity for T. It is clear that S is a scalar type operator with resolution of the identity E. Also, since T commutes with $E(\delta)$ it commutes with S and thus N commutes with S. The desired conclusion will therefore be established as soon as it is shown that N is a quasi-nilpotent. To prove this we may, in view of Corollary 3, show that the spectrum $\sigma(N)$ of N is contained in the circle $C_\varepsilon = \{\lambda \mid |\lambda| \leqq \varepsilon\}$ whose radius ε is an arbitrary positive number. Now let the spectrum of T be decomposed into the union of the disjoint Borel sets $\sigma_1, \ldots, \sigma_k$, each having diameter less than a positive number $\alpha < \varepsilon$ which will be specified presently. If λ is in the resolvent set of each of the restrictions $N_{\sigma_i} = N \mid E(\sigma_i)\mathfrak{X}$, and if $R_i = R(\lambda; N_{\sigma_i})$, then putting $R = \sum_{i=1}^{k} R_i E(\sigma_i)$ we have

$$(\lambda I - N)R = \sum_{i=1}^{k} (\lambda I - N_{\sigma_i})R_i E(\sigma_i) = \sum_{i=1}^{k} E(\sigma_i) = I$$

and

$$\begin{aligned} R(\lambda I - N) &= \sum_{i=1}^{k} R(\lambda I - N)E(\sigma_i) = \sum_{i=1}^{k} R(\lambda I - N_{\sigma_i})E(\sigma_i) \\ &= \sum_{i=1}^{k} R_i(\lambda I - N_{\sigma_i})E(\sigma_i) = \sum_{i=1}^{k} E(\sigma_i) = I. \end{aligned}$$

Thus $\lambda \in \rho(N)$. Consequently the spectrum of N is contained in the union of the spectra $\sigma(N_{\sigma_i})$ of the restrictions $N \mid E(\sigma_i)\mathfrak{X}$, so that it will suffice to show that $\sigma(N_{\sigma_i}) \subseteq C_\varepsilon$ for each $i = 1, \ldots, k$. To show this we write

$$N_{\sigma_i} = (T - \lambda_i I)_{\sigma_i} + (\lambda_i I - S)_{\sigma_i},$$

where λ_i is a point in $\sigma(T_{\sigma_i})$. Since $\sigma(T_{\sigma_i}) \subseteq \bar{\sigma}_i$, we have

$$\sigma((T - \lambda_i I)_{\sigma_i}) \subseteq \bar{\sigma}_i - \lambda_i \subseteq C_\alpha \subseteq C_\varepsilon.$$

Since $(\lambda_i I - S)_{\sigma_i}$ is the restriction of $\int_{\sigma_i} (\lambda_i - \lambda)E(d\lambda)$ to σ_i we have

$$|(\lambda_i I - S)_{\sigma_i}| \leqq v(E) \max_{\lambda \in \sigma_i} |\lambda - \lambda_i| \leqq \alpha v(E)$$

and thus $(\lambda I - S)_{\sigma_i}$ is small in norm if α is small. Thus (cf. VII.6.1) $\sigma(N_{\sigma_i}) \subseteq C_\varepsilon$ for small α. By the above, this shows that $\sigma(N) \subseteq C_\varepsilon$, and since $\varepsilon > 0$ is arbitrary, it follows that $\sigma(N) = \{0\}$. It then follows from Corollary 3 that N is a quasi-nilpotent. Q.E.D.

6 Definition. The decomposition, given in Theorem 5, of a spectral operator $T = S + N$ into a sum of a scalar type operator S and a quasi-nilpotent N commuting with S is called the *canonical decomposition* of T. The operator S is called the *scalar part* of T, and N is called the *quasi-nilpotent part* or the *radical part* of T.

5. An Operational Calculus for Bounded Spectral Operators

It should be recalled that (cf. VII.3.8–10) for an arbitrary bounded operator T in a complex B-space the formula

$$f(T) = \frac{1}{2\pi i} \int_C f(\lambda)R(\lambda; T)\, d\lambda,$$

where C is an admissible contour surrounding the spectrum of T, establishes an operational calculus on the class $\mathscr{F}(T)$ of scalar functions analytic on the spectrum of T. In this section it is shown that if T is a spectral operator and if f is in $\mathscr{F}(T)$, then the operator $f(T)$ may be calculated in terms of the values which f assumes *on* the spectrum of T.

→ 1 Theorem. *Let T be a bounded spectral operator, N its radical part, and E its resolution of the identity. Then for every scalar function f analytic and single valued on the spectrum $\sigma(T)$ we have*

$$f(T) = \sum_{n=0}^{\infty} \frac{N^n}{n!} \int_{\sigma(T)} f^{(n)}(\lambda)E(d\lambda),$$

the series converging in the uniform topology of operators.

Proof. The theorem may be derived from Corollary VII.6.12 by showing that the scalar part S of T satisfies the equation $f(S) = \int f(\lambda)E(d\lambda)$ for every f in $\mathscr{F}(T)$. However, we give here a proof independent of Corollary VII.6.12. It is based on the following lemma.

2 LEMMA. *Let E be the resolution of the identity for the spectral operator T, and let N be its radical part. Then*

$$R(\xi; T) = \sum_{n=0}^{\infty} N^n \int \frac{E(d\lambda)}{(\xi-\lambda)^{n+1}},$$

the series converging in the uniform topology of operators, and uniformly with respect to ξ in any closed set ρ contained in $\rho(T)$.

PROOF. For ξ in ρ the function $(\xi-\lambda)^{-n}$ is bounded on $\sigma(T)$, so that the integral exists. Moreover,

$$\left| \int \frac{E(d\lambda)}{(\xi-\lambda)^{n+1}} \right| \leqq r^{n+1} v(E),$$

where $r = \sup |\xi - \lambda|^{-1}$, the supremum being taken over λ in $\sigma(T)$, and ξ in ρ. Since N is a generalized nilpotent,

$$\sqrt[n]{|N^n|} \to 0,$$

and hence the series

$$\sum_{n=0}^{\infty} |N^n| \, r^{n+1}$$

converges. Thus the series

$$U = \sum_{n=0}^{\infty} N^n \int \frac{E(d\lambda)}{(\xi-\lambda)^{n+1}}$$

converges in the uniform operator topology, and uniformly with respect to ξ in ρ. If S is the scalar part of T, then

$$(\xi I - S) \int \frac{E(d\lambda)}{(\xi-\lambda)^{n+1}} = \left[\int (\xi-\lambda) E(d\lambda)\right]\left[\int \frac{E(d\lambda)}{(\xi-\lambda)^{n+1}}\right] = \int \frac{E(d\lambda)}{(\xi-\lambda)^{n}},$$

and so

$$\begin{aligned}(\xi I - T) U &= (\xi I - S - N) \sum_{0}^{\infty} N^n \int \frac{E(d\lambda)}{(\xi-\lambda)^{n+1}} \\ &= \sum_{0}^{\infty} \left\{ N^n \int \frac{E(d\lambda)}{(\xi-\lambda)^{n}} - N^{n+1} \int \frac{E(d\lambda)}{(\xi-\lambda)^{n+1}} \right\} = I,\end{aligned}$$

which proves the lemma. Q.E.D.

Returning now to the proof of the theorem, let C be an admissible

Jordan curve in $\rho(T)$ containing $\sigma(T)$ in its interior and such that f is analytic within and on C. Then we have

$$f(T) = \frac{1}{2\pi i} \int_C f(\xi) R(\xi; T) \, d\xi$$

and it follows from Lemma 2 that

$$(*) \qquad f(T) = \frac{1}{2\pi i} \sum_{n=0}^{\infty} N^n \int_C f(\xi) \left[\int_{\sigma(T)} \frac{E(d\lambda)}{(\xi - \lambda)^{n+1}} \right] d\xi,$$

the series converging in the uniform topology of operators. For $x^* \in \mathfrak{X}^*$ and $x \in \mathfrak{X}$ we have, by Fubini's theorem,

$$\begin{aligned} x^* \left\{ \int_C f(\xi) \left[\int_{\sigma(T)} \frac{E(d\lambda)}{(\xi - \lambda)^{n+1}} \right] d\xi \right\} x & \\ = \int_C f(\xi) \left[\int_{\sigma(T)} \frac{x^* E(d\lambda) x}{(\xi - \lambda)^{n+1}} \right] d\xi & \\ = \int_{\sigma(T)} \left[\int_C \frac{f(\xi)}{(\xi - \lambda)^{n+1}} \, d\xi \right] x^* E(d\lambda) x & \\ = \frac{2\pi i}{n!} \int_{\sigma(T)} f^{(n)}(\lambda) x^* E(d\lambda) x. & \end{aligned}$$

Thus

$$\int_C f(\xi) \left[\int_{\sigma(T)} \frac{E(d\lambda)}{(\xi - \lambda)^{n+1}} \right] d\xi = \frac{2\pi i}{n!} \int_{\sigma(T)} f^{(n)}(\lambda) E(d\lambda),$$

so that, by $(*)$,

$$\begin{aligned} f(T) &= \sum_{n=0}^{\infty} N^n \int_{\sigma(T)} \left[\int_C \frac{f(\xi) \, d\xi}{(\xi - \lambda)^{n+1}} \right] E(d\lambda) \\ &= \sum_{n=0}^{\infty} \frac{N^n}{n!} \int_{\sigma(T)} f^{(n)}(\lambda) E(d\lambda), \end{aligned}$$

the series converging in the uniform topology of operators. Q.E.D.

→ 3 DEFINITION. An operator T is said to be of *type m* in case it is a spectral operator with resolution of the identity E and

$$f(T) = \sum_{n=0}^{m} \frac{N^n}{n!} \int f^{(n)}(\lambda) E(d\lambda), \qquad f \in \mathscr{F}(T).$$

4 THEOREM. *Let N be the radical part of the bounded spectral operator T; then T is of type m if and only if $N^{m+1} = 0$.*

PROOF. If $N^{m+1} = 0$ then clearly the formula of Theorem 1 reduces to that of Definition 3. Conversely, if T is of type m we see, by placing

$$f(\lambda) = \frac{\lambda^{m+1}}{(m+1)!}$$

in these two formulas, that

$$0 = N^{m+1} \int E(d\lambda) = N^{m+1}. \qquad \text{Q.E.D.}$$

5 COROLLARY. *A spectral operator is of scalar type if and only if it is of type 0.*

→ 6 THEOREM. *If T is a bounded spectral operator and if f is in $\mathscr{F}(T)$, then $f(T)$ is a spectral operator whose resolution of the identity is given in terms of the spectral resolution of T by the equation*

$$E(\delta; f(T)) = E(f^{-1}(\delta); T).$$

Moreover, if S is the scalar part of T, then $f(S)$ is the scalar part of $f(T)$.

PROOF. By Theorem 1,

$$f(T) = \sum_{n=0}^{\infty} f^{(n)}(S) \frac{N^n}{n!},$$

the series converging in the uniform operator topology. Since

$$f(S) = \int f(\lambda) E(d\lambda)$$

we have, from the change of measure principle, $f(S) = \int \lambda E_1(d\lambda)$ where $E_1(\delta) = E(f^{-1}(\delta))$. It follows from the remark following Definition 4.1 that $f(S)$ is a scalar type operator with resolution of the identity E_1. Thus if we can show that the operator

$$N_1 = \sum_{n=1}^{\infty} f^{(n)}(S) \frac{N^n}{n!}$$

is quasi-nilpotent, the theorem follows from Theorem 4.5. Lemma IX.2.6 shows that the radical in a commutative B-algebra is an ideal, hence that the operator $N^{(m)}$ defined by the equation

$$N^{(m)} = \sum_{n=1}^{m} f^{(n)}(S) \frac{N^n}{n!}$$

is in the radical of the closed commutative subalgebra of $B(\mathfrak{X})$ generated by I, S, and N. Lemma IX.1.12(e) shows that this radical is closed. Since $N^{(m)} \to N_1$ it follows that N_1 is also in this radical and is thus a quasi-nilpotent. Q.E.D.

7 Corollary. *Under the hypothesis of the preceding theorem, $f(T)$ is of type m if T is of type m.*

Proof. It has been shown in the course of the preceding proof that the radical part of $f(T)$ is

$$N_1 = \sum_{n=1}^{\infty} f^{(n)}(S) \frac{N^n}{n!}.$$

Since $N^{m+1} = 0$ we have

$$N_1 = \sum_{n=1}^{m} f^{(n)}(S) \frac{N^n}{n!}$$

and thus $N_1^{m+1} = 0$. Q.E.D.

6. Bounded Spectral Operators in Hilbert Space

What is the relationship between the bounded spectral operators in a Hilbert space $\mathfrak{H}$ and the bounded normal operators in $\mathfrak{H}$? The central result in this direction is a theorem of Wermer which states that every scalar type operator in Hilbert space is equivalent to a normal operator. This theorem and other special properties of spectral operators in Hilbert space are discussed in this section.

1 Lemma. *Let G be a bounded Abelian group of operators in Hilbert space $\mathfrak{H}$. Then there is a bounded self adjoint operator B in $\mathfrak{H}$ with a bounded everywhere defined inverse such that for every T in G the operator BTB^{-1} is unitary.*

Proof. Let $\mathfrak{L}$ be the linear space of all scalar functions on $\mathfrak{H} \times \mathfrak{H}$ and let $\mathfrak{P}$ consist of those f in $\mathfrak{L}$ which are bilinear, Hermitian symmetric, and have $f(x, x) \geqq 0$ for all x in $\mathfrak{H}$. Let $\mathfrak{L}$ be given its weak product topology so that, by definition, the sets

$$\{f \mid f \in \mathfrak{L}, |f(x, y) - g(x, y)| < \varepsilon\},$$

where $x, y \in \mathfrak{H}$ and $\varepsilon > 0$, form a subbase for the neighborhoods of a point g in $\mathfrak{L}$. It is easily seen that $\mathfrak{P}$ is a closed set in $\mathfrak{L}$ and that the convex set determined by functions f having the form $f(x, y) = (Tx, Ty)$ with T in G is a subset of $\mathfrak{P}$. We let $\mathfrak{K}$ be the closure of this convex set so that $\mathfrak{K}$ is a closed convex subset of $\mathfrak{P}$. If M is an upper bound for the norms of the operators in G, then $|(Tx, Ty)| \leqq M^2 |x|\,|y|$, and so

$$\text{(i)} \qquad |f(x, y)| \leqq M^2 |x|\,|y|, \qquad f \in \mathfrak{K}.$$

Also, since $|x|^2 = (T^{-1}Tx, T^{-1}Tx) \leqq M^2 |Tx|^2$, we have

$$\text{(ii)} \qquad \frac{|x|^2}{M^2} \leqq f(x, x), \qquad f \in \mathfrak{K}.$$

Since $\mathfrak{K}$ is closed in $\mathfrak{P}$ and $\mathfrak{P}$ is closed in $\mathfrak{L}$, it follows from (i) and from Tychonoff's theorem on the compactness of product spaces (I.8.5) that $\mathfrak{K}$ is a compact set. For each T in G we define the continuous linear map J_T of $\mathfrak{L}$ into itself by the formula

$$(J_T f)(x, y) = f(Tx, Ty), \qquad x, y \in \mathfrak{H}.$$

Since $J_{T_1} J_{T_2} = J_{T_1 T_2}$ and since G is an Abelian group, the collection $\{J_T\}$ is itself an Abelian group of continuous linear operators in the space $\mathfrak{L}$. Moreover, it is easily seen that $\{J_T\}$ maps $\mathfrak{K}$ into $\mathfrak{K}$. It follows from the Markov fixed point theorem (V.10.6) that there is an $f_0 \in \mathfrak{K}$ with $J_T f_0 = f_0$ for every T in G.

There is, by Lemma X.2.2, a bounded self adjoint operator A in $\mathfrak{H}$ with $f_0(x, y) = (Ax, y)$. Therefore

$$(Ax, y) = (ATx, Ty) = (T^*ATx, y), \qquad T \in G,$$

so that $A = T^*AT$ for every T in G. Since $f_0 \in P$ we have $(Ax, x) = f_0(x, x) \geqq 0$ and thus, by Theorem X.4.2, the spectrum of A is positive. If E is the resolution of the identity for A, then the operator $B = \int \lambda^{1/2} E(d\lambda)$ is a bounded self adjoint operator with $B^2 = A$. In view of (ii) we have

$$\frac{|x|^2}{M^2} \leqq (Ax, x) = (B^2 x, x) = |Bx|^2,$$

which shows that B has a bounded inverse. Thus, by Lemma XII.1.2, the range $B\mathfrak{H}$ is closed. To see that B^{-1} is everywhere defined, it will therefore suffice to show that $y = 0$ is the only vector orthogonal to $B\mathfrak{H}$. If y is orthogonal to $B\mathfrak{H}$, then $0 = (B^2 y, y) = (By, By)$ and so $By = 0$. Since B has an inverse, $y = 0$. This proves that B is a self adjoint linear homeomorphism of $\mathfrak{H}$ onto all of $\mathfrak{H}$. Now, since $T^*AT = A$ we have $B^2 T = (T^*)^{-1} B^2$, and so

$$BTB^{-1} = B^{-1}(T^*)^{-1}B = ((BTB^{-1})^*)^{-1},$$

which proves that BTB^{-1} is a unitary operator. Q.E.D.

2 Lemma. *Let $\mathfrak{B}_1, \ldots, \mathfrak{B}_k$ be a finite collection of commuting bounded Boolean algebras of projections in a Hilbert space $\mathfrak{H}$. Then there exists a bounded self adjoint operator B in $\mathfrak{H}$ with a bounded everywhere defined inverse such that BEB^{-1} is a self adjoint projection for every E in the Boolean algebra determined by the algebras $\mathfrak{B}_1, \ldots, \mathfrak{B}_k$.*

Proof. For $E \in \mathfrak{B}_i$, put $F(E) = I - 2E$. Then $F(E)^2 = I - 4E + 4E^2 = I$, and $F(E_1)F(E_2) = I - 2E_1 - 2E_2 + 4E_1E_2 = F(E_1 \wedge (I - E_2)) \vee (E_2 \wedge (I - E_1))$. Thus the collection G_i of all $F(E)$ with $E \in \mathfrak{B}_i$ forms a bounded Abelian group of operators in Hilbert space. Clearly all the elements of G_i commute with all the elements of G_j. Thus the set G of products $g_1, \ldots, g_k$ with $g_i \in G_i$ is a bounded Abelian group of operators in Hilbert space. It follows immediately from the preceding lemma that there exists a bounded self adjoint operator B with a bounded everywhere defined inverse such that $BF(E)B^{-1} = U$ is unitary for every $E \in \mathfrak{B}_i$, $i = 1, \ldots, k$. Since $F(E)^2 = I$, we have $U^2 = I$, so that $U^* = U^{-1} = U$. Thus U is self adjoint and therefore the operator

$$BEB^{-1} = \tfrac{1}{2}B(I + F(E))B^{-1} = \tfrac{1}{2}(I + U)$$

is also self adjoint for every E in $\mathfrak{B}_i$, $i = 1, \ldots, k$. From this the desired conclusion follows easily. Q.E.D.

3 Corollary. *The smallest Boolean algebra of projections in Hilbert space which contains each of a finite collection of commuting bounded Boolean algebras of projections is itself bounded.*

4 Theorem. *Let $S_1, \ldots, S_k$ be commuting scalar type operators in Hilbert space. Then there is a bounded self adjoint operator B with a bounded everywhere defined inverse such that the operators BS_iB^{-1}, $i = 1, \ldots, k$ are all normal.*

Proof. Let E_i be the resolution of the identity for S_i. By Lemma 2 there is a B of the required type such that for every Borel set δ the projections $P(\delta) = BE_i(\delta)B^{-1}$, $i = 1, \ldots, k$, are all self adjoint. Thus the operators

$$BS_iB^{-1} = \int \lambda BE_i(d\lambda)B^{-1}, \qquad i = 1, \ldots, k,$$

are normal. Q.E.D.

5 COROLLARY. *The sum and the product of two commuting bounded spectral operators in Hilbert space are also spectral operators.*

The proof of this corollary will use the following lemma.

6 LEMMA. *Let A and B be bounded operators in Hilbert space with A normal. Then if B commutes with A, it commutes with A^*.*

PROOF. This is a corollary of Corollary 3.7. Q.E.D.

PROOF OF COROLLARY 5. *Let* T_1, T_2 be commuting spectral operators in Hilbert space $\mathfrak{H}$ and let S_1+N_1, S_2+N_2 be their canonical decompositions. It follows from Corollary 3.7 that the operators S_1, S_2, N_1, and N_2 all commute with each other. Thus the sum and product of T_1 and T_2 have the forms

$$T_1+T_2=S_1+S_2+N,\ T_1T_2=S_1S_2+M,$$

where N, M are quasi-nilpotents commuting with S_1 and S_2. To see that T_1+T_2 and T_1T_2 are spectral operators it therefore suffices, in view of Theorem 4.5, to show that S_1+S_2 and S_1S_2 are scalar type operators. By Theorem 4 there is a linear homeomorphism B of $\mathfrak{H}$ onto all of $\mathfrak{H}$ such that BS_1B^{-1} and BS_2B^{-1} are commuting normal operators. Hence their sum and product are also normal, from which it is evident that S_1+S_2 and S_1S_2 are scalar type operators. Q.E.D.

We shall next endeavor to characterize operators of finite type in terms of the rate of growth of the resolvent.

7 THEOREM. *Let T be a bounded spectral operator in Hilbert space $\mathfrak{H}$, let E be its resolution of the identity, and T_σ its restriction to the manifold $E(\sigma)\mathfrak{H}$. Then T is of type $m-1$ if and only if there is a constant K independent of the Borel set σ such that*

$$(*)\qquad |R(\xi;T_\sigma)E(\sigma)|\leqq\frac{K}{\operatorname{dist}(\xi,\bar{\sigma})^m},\qquad \xi\notin\bar{\sigma},\ |\xi|\leqq|T|+1.$$

PROOF. In view of Theorem 5.4 it is sufficient to prove that the condition $(*)$ is equivalent to the condition $N^m=0$. If $N^m=0$ and $\xi\notin\bar{\sigma}$ then, by Lemma 5.2,

$$R(\xi;T_\sigma)E(\sigma)=\sum_{n=0}^{m-1}N^n\int_\sigma\frac{E(d\lambda)}{(\lambda-\xi)^{n+1}}$$

from which the condition (∗) follows.

The converse will require the following lemma.

8 LEMMA. *Let T be a spectral operator in Hilbert space $\mathfrak{H}$ and let E be its resolution of the identity. Then there is a constant M such that for any finite collection A_j, $j = 1, 2, \ldots, n$, of bounded operators in $\mathfrak{H}$ which commute with T, and any collection σ_j, $j = 1, 2, \ldots, n$, of disjoint Borel sets, we have*

$$\left|\sum_{j=1}^{n} A_j E(\sigma_j)\right| \leqq M \sup_{1 \leqq j \leqq n} |A_j|.$$

PROOF. According to Lemma 2 there is a linear one-to-one map B with $B\mathfrak{H} = \mathfrak{H}$, with B and B^{-1} both continuous and such that for each Borel set σ the projection

$$P(\sigma) = BE(\sigma)B^{-1}$$

is self adjoint. If $B_j = BA_j B^{-1}$ then

$$B\left\{\sum_{j=1}^{n} A_j E(\sigma_j)\right\}B^{-1} = \sum_{j=1}^{n} B_j P(\sigma_j).$$

By Corollary 3.7, A_j commutes with $E(\sigma)$ and hence B_j commutes with $P(\sigma)$. Thus

$$\begin{aligned}\left|\sum_{j=1}^{n} B_j P(\sigma_j)x\right|^2 = \left|\sum_{j=1}^{n} P(\sigma_j)B_j x\right|^2 &= \sum_{j=1}^{n} |P(\sigma_j)B_j x|^2 \\ &\leqq \sup_j |B_j|^2 \sum_{j=1}^{n} |P(\sigma_j)x|^2 \\ &\leqq \sup_j |B_j|^2 |x|^2,\end{aligned}$$

which proves the lemma.

Let δ be a Borel set of diameter less than ε and let ξ be a point in δ. Then

$$\begin{aligned}(T - \xi I)^m E(\delta) &= (T_\delta - \xi I_\delta)^m E(\delta) \\ &= \frac{1}{2\pi i}\int_C (\lambda - \xi)^m R(\lambda;\, T_\delta)E(\delta)\, d\lambda,\end{aligned}$$

where C is a circle with center ξ and radius 2ε. By hypothesis the integrand is bounded by $KM_1 2^m$ where $M_1 = \sup_\delta |E(\delta)|$, and so we have

$$|(T - \xi I)^m E(\delta)| \leqq 2\varepsilon K M_1 2^m.$$

Now let $\sigma(T)$ be partitioned into Borel sets σ_j, $j = 1, \ldots, n(\varepsilon)$, each of diameter less than ε, and let $\xi_j \in \sigma_j$. Then by the preceding lemma,

$$\left| \sum_{i=1}^{n(\varepsilon)} (T - \xi_i I)^m E(\sigma_i) \right| \leq 2\varepsilon K M M_1 2^m.$$

Using the binomial theorem we have

$$\sum_{i=1}^{n(\varepsilon)} (T - \xi_i I)^m E(\sigma_i) = \sum_{k=0}^{m} \binom{m}{k} (-1)^k T^{m-k} \left\{ \sum_{i=1}^{n(\varepsilon)} \xi_i^k E(\sigma_i) \right\}.$$

Since λ^k is uniformly continuous in λ as λ varies over $\sigma(T)$, $\sum_{i=1}^{n(\varepsilon)} \xi_i^k E(\sigma_i)$ tends uniformly to

$$\int_{\sigma(T)} \lambda^k E(d\lambda) = S^k$$

as $\varepsilon \to 0$. Thus, from the last two equations and the inequality preceding them, we have

$$0 = \sum_{k=0}^{m} \binom{m}{k} (-1)^k T^{m-k} S^k = (T - S)^m = N^m. \qquad \text{Q.E.D.}$$

7. Relations Between a Spectral Operator and Its Scalar Part

In this section we present the elegant results of Foguel concerning some topological and algebraic properties inherited by the scalar part of a spectral operator. The spectral relations between a spectral operator and its scalar part will be discussed in the following section. Throughout this section the letter T will stand for a spectral operator, the letters S, N for its scalar and radical parts, respectively, and E for its resolution of the identity. It should be recalled that $T = S + N$, that $\sigma(T) = \sigma(S)$, and that E is also the resolution of the identity for S. The symbol $B(\mathfrak{X})$ will be used for the algebra of all bounded linear operators in the complex B-space $\mathfrak{X}$.

1 LEMMA. *The scalar operator S is in the closed linear manifold in $B(\mathfrak{X})$ determined by those projections $E(\sigma)$ with $0 \notin \bar{\sigma}$.*

PROOF. According to the definition of the integral as given in Section X.1 there is, for each $\varepsilon > 0$, a partition $\sigma_0, \sigma_1, \ldots, \sigma_n$ of $\sigma(S)$ into sets with the point $\lambda = 0$ in at most one of the closures $\bar{\sigma}_i$ and with

$$\left| S - \sum_{i=0}^{n} \lambda_i E(\sigma_i) \right| < \varepsilon,$$

for any choice of the complex numbers λ_i in σ_i. If 0 is not in $\sigma(S)$ this proves the lemma. If $\lambda = 0$ is in $\sigma(S)$, say $0 \in \sigma_0$, then it may be supposed that $\lambda_0 = 0$ in the above inequality, which proves the lemma in this case also. Q.E.D.

2 THEOREM. *Let T belong to the right (left) ideal $\mathfrak{J}$ in $B(\mathfrak{X})$. Then every projection $E(\sigma)$ with $0 \notin \bar{\sigma}$ belongs to $\mathfrak{J}$. If $\mathfrak{J}$ is closed, then S and N also belong to $\mathfrak{J}$.*

PROOF. Let $0 \notin \bar{\sigma}$ and let $T_\sigma = TE(\sigma) \mid E(\sigma)\mathfrak{X}$, the restriction of T to the invariant subspace $E(\sigma)\mathfrak{X}$. Since $\sigma(T_\sigma) \subseteq \bar{\sigma}$, it follows that $0 \in \rho(T_\sigma)$ and hence T_σ^{-1} exists as a bounded linear operator in the space $E_\sigma(\mathfrak{X})$. Let V_σ be the bounded linear operator in $\mathfrak{X}$ defined by the equation

$$V_\sigma x = T_\sigma^{-1} E(\sigma) x, \qquad x \in \mathfrak{X}.$$

Then $TV_\sigma = E(\sigma) = V_\sigma T$, which proves that $E(\sigma)$ is in $\mathfrak{J}$. It follows from Lemma 1 that S, and hence N also, belongs to $\mathfrak{J}$ if $\mathfrak{J}$ is closed. Q.E.D

3 COROLLARY. *If T is compact, then so are S, N, and every projection $E(\sigma)$ with $0 \notin \bar{\sigma}$.*

PROOF. By Corollary VI.5.5 the compact operators form a closed two-sided ideal in $B(\mathfrak{X})$ and so the present corollary is immediate. Q.E.D.

4 COROLLARY. *If T is weakly compact, then so are S, N, and every projection $E(\sigma)$ with $0 \notin \bar{\sigma}$.*

PROOF. By Corollary VI.4.6 the weakly compact operators form a closed two-sided ideal in $B(\mathfrak{X})$, from which the present statement follows. Q.E.D.

If $\mathfrak{Y}$ is a closed linear subspace of $\mathfrak{X}$, then the set of bounded linear operators A in $\mathfrak{X}$ for which $A\mathfrak{X} \subseteq \mathfrak{Y}$ is a closed right ideal in $B(\mathfrak{X})$. Thus the following statement is an immediate corollary of Theorem 2.

5 COROLLARY. *The ranges of S, N, and $E(\sigma)$ with $0 \notin \sigma$ are contained in the closure of the range of T.*

Let A_0 be a fixed bounded linear operator in $\mathfrak{X}$. Then the set of all bounded linear operators A in $\mathfrak{X}$ for which $A_0 A = 0$ ($AA_0 = 0$) is a closed right (left) ideal in $B(\mathfrak{X})$. Thus the following statement is a direct consequence of Theorem 2.

6 COROLLARY. *If $A_0 T = 0$ (respectively $TA_0 = 0$), then $A_0 S = A_0 N = A_0 E(\sigma) = 0$ if $0 \notin \bar{\sigma}$ (respectively $SA_0 = NA_0 = E(\sigma)A_0 = 0$ if $0 \notin \bar{\sigma}$).*

7 COROLLARY. *A spectral operator is of finite type if and only if it is annihilated by some power of its radical part.*

PROOF. If T is of finite type, then $N^n = 0$ for some positive integer n (Theorem 5.4) and so $TN^n = 0$. Conversely, if some power of N annihilates T, say $TN^p = 0$, then it follows from Corollary 6 that $N^{p+1} = 0$ and so T is of finite type. Q.E.D.

8 COROLLARY. *If $Tx = 0$, then $Sx = Nx = E(\sigma)x = 0$ if $0 \notin \bar{\sigma}$.*

PROOF. For a given x in $\mathfrak{X}$, the class of bounded linear operators in $\mathfrak{X}$ for which $Ax = 0$ is a closed left ideal in $B(\mathfrak{X})$. Q.E.D.

9 COROLLARY. *Let $0 \notin \bar{\sigma}$ and let $\{x_n\}$ be a sequence in $\mathfrak{X}$ for which $\{Tx_n\}$ is convergent (convergent to zero). Then the sequence $\{E(\sigma)x_n\}$ is convergent (convergent to zero). If $\{x_n\}$ is bounded, then the sequences $\{Sx_n\}$ and $\{Nx_n\}$ are also convergent (convergent to zero).*

PROOF. The set of all A in $B(\mathfrak{X})$ for which $\{Ax_n\}$ is convergent (convergent to zero) is a left ideal, and this ideal is closed if $\{x_n\}$ is bounded. Thus the corollary follows directly from Theorem 2. Q.E.D.

10 THEOREM. *Let A be a bounded linear operator in $\mathfrak{X}$. Then $AT = 0$ if and only if $AN = 0$ and $AE(\{0\}') = 0$. Similarly $TA = 0$ if and only if $E(\{0\}')A = NA = 0$.*

PROOF. If $AT = 0$, then it follows from Corollary 6 that $AN = 0$ and

$$AE\left(\left\{\lambda \,|\, |\lambda| \geqq \frac{1}{n}\right\}\right) = 0,$$

and so $AE(\{0\}') = 0$ follows from the countable additivity of E.

Now suppose that $AN = 0$ and $AE(\{0\}') = 0$. Let $0 \notin \bar{\sigma}$ so that $\bar{\sigma} \subset \{0\}'$ and $E(\sigma) = E(\sigma)E(\{0\}')$. Then

$$AE(\sigma) = AE(\{0\}')E(\sigma) = 0.$$

Thus $E(\sigma)$ belongs to the closed ideal consisting of all C in $B\mathfrak{X}$ with $AC = 0$. It follows from Lemma 1 that $AS = 0$. Since $AN = 0$, we have $AT = 0$ also. The second part of the theorem may be proved in a similar fashion. Q.E.D

11 Corollary. *If* $E(\{0\})=0$, *then the operator* $A=0$ *is the only bounded linear operator for which either* $AT=0$ *or* $TA=0$.

Proof. If either $AT=0$ or $TA=0$ then, by the theorem, either $A=AE(\{0\})$ or $A=E(\{0\})A$. Thus $A=0$. Q.E.D.

12 Corollary. *If* $E(\{\lambda\})=0$, *then* $(\lambda I-T)\mathfrak{X}$ *is dense in* $\mathfrak{X}$.

Proof. First suppose that $\lambda=0$. If $T\mathfrak{X}$ is not dense then, by Corollary II.3.13, there is an x^* in $\mathfrak{X}^*$ with $x^*\neq 0$ and $x^*T\mathfrak{X}=0$. Let $x_1\neq 0$ and define the operator A by the equation $Ax=x^*(x)x_1$, so that $A\neq 0$. But $AT=0$, which contradicts Corollary 11. Now for an arbitrary λ, Theorem 5.6 shows that $\lambda I-T$ is a spectral operator whose spectral resolution evaluated on the set $\{0\}$ is the projection $E(\{\lambda\})$. Thus it follows by what has been proved that $(\lambda I-T)\mathfrak{X}$ is dense in $\mathfrak{X}$. Q.E.D.

13 Theorem. *If* T *has a closed range, so does* S.

Proof. The proof will be divided into two cases depending on whether the projection $E(\{0\})=0$ or not. First suppose that $E(\{0\})=0$. Then since the range of T is closed, it follows from Corollary 12 that $T\mathfrak{X}=\mathfrak{X}$. By Lemma 3.1 the operator T is one-to-one. By Theorem II.2.2, T has a bounded inverse and hence $0\in\rho(T)=\rho(S)$ and so $S\mathfrak{X}=\mathfrak{X}$.

Next suppose that $E(\{0\})\neq 0$. We note first that for any Borel set α the restriction of T to the subspace $E(\alpha)\mathfrak{X}$ is a spectral operator whose resolution of the identity F is given by the equation $F(\beta)=E(\alpha\beta)$, for each Borel set β. This observation follows immediately from Definition 2.5. Thus the restriction V of T to the subspace $E(\{0\}')\mathfrak{X}$ is a spectral operator whose resolution F is given by the equation $F(\beta)=E(\beta-\{0\})$. Hence $F(\{0\})=0$ and we may apply to V the result already proved in the first part as soon as it is shown that the range of V is closed. Let y be in the closure of the range of V. Then for some sequence $\{x_n\}$ in $E(\{0\}')\mathfrak{X}$ we have $Vx_n\to y$ and, since the range of T is closed, there is an x in $\mathfrak{X}$ with $Tx=y$. Hence

$$VE(\{0\}')x=TE(\{0\}')x=E(\{0\}')Tx=E(\{0\}')y=y.$$

Thus V satisfies the conditions assumed for T in the first part of the proof, and we may therefore conclude that the scalar part of V maps $E(\{0\}')\mathfrak{X}$ into all of itself. It follows from the uniqueness of the canonical decomposition of a spectral operator that the scalar part of a restriction is the restriction of the scalar part and thus

$$SE(\{0\}')\mathfrak{X} = E(\{0\}')\mathfrak{X}.$$

But $SE(\{0\}) = 0$ and so $S\mathfrak{X} = E(\{0\}')\mathfrak{X}$, which shows that S has a closed range. Q.E.D.

It was shown in the course of the preceding proof that for an operator T with a closed range the point $\lambda = 0$ is not in the spectrum of the operator $V = T \mid E(\{0\}')\mathfrak{X}$. Thus for all sufficiently small complex numbers $\lambda \neq 0$ the operator

$$\lambda I - T = \lambda E(\{0\}) - T_{\{0\}} + \lambda E(\{0\}') - T_{\{0\}'}$$

has a bounded everywhere defined inverse. This means that the point $\lambda = 0$, if it is in the spectrum of T at all, is an isolated point of the spectrum.

14 THEOREM. *The operator T has a closed range if and only if*

(i) *the point $\lambda = 0$ is either in the resolvent set of T or an isolated spectral point of T, and*

(ii) *the operator $TE(\{0\})$ has a closed range.*

PROOF. Let T have a closed range. We have already proved (i). To prove (ii) let y be in the closure of the range of $TE(\{0\})$ and let

$$TE(\{0\})x_n \to y.$$

Since T has a closed range, there is an x in $\mathfrak{X}$ with $Tx = y$ and hence

$$TE(\{0\})x = E(\{0\})Tx = E(\{0\})y = y,$$

which proves (ii). Conversely we assume (i) and (ii) and let y be in the closure of the range of T and suppose that $Tx_n \to y$. Then $TE(\{0\})x_n \to E(\{0\})y$, and since the range of $TE(\{0\})$ is closed, there is a vector w with $TE(\{0\})w = E(\{0\})y$. Since the point $\lambda = 0$ is isolated in $\sigma(T)$, it is in $\rho(T_{\{0\}'})$, and for some z in $E(\{0\}')\mathfrak{X}$ we have $Tz = E(\{0\}')y$. Hence

$$T(z + E(\{0\})w) = E(\{0\}')y + E(\{0\})y = y,$$

which proves (i). Q.E.D.

8. The Spectrum of a Spectral Operator

In this section the properties of the spectrum of a spectral operator T and their relationships with the corresponding properties of the scalar part of T are examined. Most of the results presented here are due to

S. R. Foguel. As before, the letter T will be used for a bounded spectral operator in the B-space $\mathfrak{X}$ and the symbols S, N, and E will be used for its scalar part, its radical part, and its resolution of the identity, respectively.

We shall be concerned with the fine structure of the spectrum, and the spectral points of an operator in $\mathfrak{X}$ will be classified, as they were in Hilbert space, according to the following definition.

→ 1 Definition. Let A be a bounded linear operator in $\mathfrak{X}$. The *point spectrum* of A is the set $\sigma_p(A)$ consisting of all complex numbers λ for which $\lambda I - A$ is not one-to-one. The *continuous spectrum* of A is the set $\sigma_c(A)$ of complex numbers λ for which $\lambda I - A$ is one-to-one and has a dense range which is not equal to $\mathfrak{X}$. The *residual spectrum* of A is the set $\sigma_r(A)$ consisting of those complex numbers λ for which $\lambda I - T$ is one-to-one and has a range not dense in $\mathfrak{X}$. The points in the point spectrum of T are sometimes called *eigenvalues* of T, and a vector $x \neq 0$ for which $(\lambda I - T)x = 0$ is called an *eigenvector* for T corresponding to the eigenvalue λ.

It is clear that the sets $\sigma_p(A)$, $\sigma_c(A)$, and $\sigma_r(A)$ are disjoint and

$$\sigma(A) = \sigma_p(A) \cup \sigma_c(A) \cup \sigma_r(A).$$

2 Theorem. *For a vector x in $\mathfrak{X}$ and a non-negative integer n we have $(\lambda I - T)^n x = 0$ if and only if $E(\{\lambda\})x = x$ and $N^n x = 0$.*

Proof. Let $(\lambda I - T)^n x = 0$ and let σ be a closed set of complex numbers not containing the number λ. Then λ is in the resolvent set of the restriction T_σ of T to $E(\sigma)\mathfrak{X}$ and

$$E(\sigma) = R(\lambda; T_\sigma)^n (\lambda I - T)^n E(\sigma),$$

which shows that

$$E(\sigma)x = R(\lambda; T_\sigma)^n E(\sigma)(\lambda I - T)^n x = 0.$$

Thus

$$E\left(\left\{\xi \,\middle|\, |\xi - \lambda| \geqq \frac{1}{n}\right\}\right) x = 0,$$

and since E is countably additive it follows that

$$E(\{\lambda\}')x = 0, \; E(\{\lambda\})x = x.$$

Hence

$$Sx = SE(\{\lambda\})x = \int_{\{\lambda\}} \mu E(d\mu)x = \lambda E(\{\lambda\})x = \lambda x,$$

which shows that

$$(\lambda I - T)x = -Nx,$$

and thus that

$$0 = (\lambda I - T)^n x = (-1)^n N^n x.$$

This proves the necessity of the conditions.

Now suppose that $E(\{\lambda\})x = x$ and $N^n x = 0$. It follows as above that $(\lambda I - S)x = 0$ and hence that $(\lambda I - T)^n x = (-1)^n N^n x$. Thus $(\lambda I - T)^n x = 0$. Q.E.D.

3 THEOREM. *If T is of finite type, its residual spectrum is void and a point λ is in its point spectrum if and only if $E(\{\lambda\}) \neq 0$.*

PROOF. Let λ be a spectral point of T. If $E(\{\lambda\}) \neq 0$ then $E(\{\lambda\})x = x$ for some $x \neq 0$ and $N^n = 0$ for some n. It follows from Theorem 2 that λ is in the point spectrum of T. If $E(\{\lambda\}) = 0$ then it follows from Theorem 2 that $\lambda I - T$ is one-to-one. By Corollary 7.12 the set $(\lambda I - T)\mathfrak{X}$ is dense in $\mathfrak{X}$ and hence λ is in the continuous spectrum of T. Q.E.D.

4 COROLLARY. *The spectrum of the scalar part S of T satisfies the relations*

$$\sigma(S) = \sigma_p(S) \cup \sigma_c(S),$$

$$\sigma_c(S) \subseteq \sigma_c(T),$$

$$\sigma_p(T) \cup \sigma_r(T) \subseteq \sigma_p(S).$$

PROOF. Since S is of finite type, $\sigma_r(S) = \phi$ and the first statement follows from the theorem. Next let λ be in $\sigma_c(S)$. Then since the resolution of the identity for S is the same as that for T, it follows from the theorem that $E(\{\lambda\}) = 0$. From this fact and Theorem 2 it follows that λ is not in $\sigma_p(T)$. The final argument of the theorem shows that $(\lambda I - T)\mathfrak{X}$ is dense in $\mathfrak{X}$ and hence, since λ is in $\sigma(T)$, it must be in $\sigma_c(T)$. This proves the second inclusion, and the final assertion follows from this by taking complements and using the fact that $\sigma_r(S) = \phi$. Q.E.D.

5 COROLLARY. *Let the complex number λ be in the complement of the Borel set σ and let T be a spectral operator. Then λ is either in the resolvent set of the restriction T_σ of T to $E(\sigma)\mathfrak{X}$ or it is in the continuous spectrum of this restriction.*

PROOF. It is clear that $T_\sigma = S_\sigma + N_\sigma$, and since the restriction of a generalized nilpotent is also a generalized nilpotent and the restriction of a scalar operator is also a scalar operator, it follows from Theorem 4.5 that S_σ is the scalar part of T_σ. Thus, by the preceding corollary, we have $\sigma_c(S_\sigma) \subseteqq \sigma_c(T_\sigma)$, and so to prove the present corollary, it suffices to prove that λ is in the continuous spectrum of S_σ. Let $(S - \lambda I)x = 0$, where x is in $E(\sigma)\mathfrak{X}$. Since S and T have the same resolution of the identity, we have, by Theorem 2, $E(\{\lambda\})x = x$, and since $\{\lambda\}$ and σ are disjoint we have $x = E(\sigma)x = 0$. Thus $S - \lambda I$ is one-to-one on $E(\sigma)\mathfrak{X}$, and so λ is either in the resolvent set of S_σ, and hence of T_σ, or else in the continuous spectrum of S_σ since, according to Theorem 3, S_σ, being of finite type, has no residual spectrum. Q.E.D.

In Theorem 3 the requirement that the spectral operator be of finite type is quite essential. The following elementary example shows that there are spectral operators with residual spectrum. Consider the operator $Tf = g$, defined in $C([0, 1])$ by the equation

$$g(t) = \int_0^t f(x)\, dx.$$

Observe first that if $g = 0$ then $f = 0$ so that T is one-to-one. Also, since every function in the range of T vanishes at the origin, this range is not dense and so the point $\lambda = 0$ is in the residual spectrum of T. We shall now show that T is a spectral operator. Indeed, it will follow from Theorem 4.5 that T is a spectral operator with scalar part $S = 0$ if it is shown that T is a quasi-nilpotent operator. To see that this is the case, observe that

$$|g(t)| \leqq t\,|f|,$$

and, inductively,

$$|(T^n f)(t)| \leqq \frac{t^n}{n!}\,|f|,$$

which shows that

$$|T^n| \leqq \frac{1}{n!}.$$

It follows from Definition 4.2 that T is a quasi-nilpotent operator.

The preceding example takes on more significance in the light of the following theorem, which shows how the existence of points in the residual spectrum of a spectral operator may be determined by the properties of its radical part.

6 THEOREM. *Let* λ *be a spectral point of the spectral operator* T. *If* $E(\{\lambda\})=0$ *then* λ *is in the continuous spectrum of* T. *If* $E(\{\lambda\})\neq 0$ *let* N_λ *be the restriction of* N *to* $E(\{\lambda\})\mathfrak{X}$. *Then*

(a) $\lambda \in \sigma_p(T)$ *if and only if* $0 \in \sigma_p(N_\lambda)$;
(b) $\lambda \in \sigma_r(T)$ *if and only if* $0 \in \sigma_r(N_\lambda)$;
(c) $\lambda \in \sigma_c(T)$ *if and only if* $0 \in \sigma_c(N_\lambda)$.

PROOF. If $E(\{\lambda\})=0$ it follows from Theorem 2 that λ is not in the point spectrum of T, and it is seen from Corollary 7.12 that λ is indeed in the continuous spectrum of T. Now suppose that $E(\{\lambda\})\neq 0$. Since $SE(\{\lambda\})=\lambda E(\{\lambda\})$ we have

$$(T-\lambda I)E(\{\lambda\})=NE(\{\lambda\}). \tag{i}$$

Now

$$(T-\lambda I)\mathfrak{X}=(T-\lambda I)E(\{\lambda\})\mathfrak{X}\oplus(T-\lambda I)E(\{\lambda\}')\mathfrak{X}. \tag{ii}$$

Since Corollary 5 shows that $(T-\lambda I)E(\{\lambda\}')\mathfrak{X}$ is dense in $E(\{\lambda\}')\mathfrak{X}$, it follows from (ii) that $(T-\lambda I)\mathfrak{X}$ is dense in $\mathfrak{X}$ if and only if $(T-\lambda I)E(\{\lambda\})\mathfrak{X}$ is dense in $E(\{\lambda\})\mathfrak{X}$. Since, by Theorem 2, any eigenvector for T corresponding to λ must be in the space $E(\{\lambda\})\mathfrak{X}$, the three statements (a), (b), (c) now follow from (i). Q.E.D.

7 THEOREM. *If the space* $\mathfrak{X}$ *is separable, then the point and residual spectra of a spectral operator are countable.*

PROOF. Let T be a spectral operator with scalar part S and resolution of the identity E. It follows from Theorem 6 that

$$\sigma_p(T)\cup\sigma_r(T)\subseteq\{\lambda\,|\,E(\{\lambda\})\neq 0\}.$$

Now if M is a bound for the resolution E and if x_λ is a vector with $|x_\lambda|=1$ and $E(\{\lambda\})x_\lambda=x_\lambda$, then for two distinct such points λ and μ we have

$$|x_\lambda-x_\mu|\geqq\frac{1}{M}\,|E(\{\lambda\})(x_\lambda-x_\mu)|=\frac{|x_\lambda|}{M}=\frac{1}{M}.$$

Since the set on the right side of the above inclusion relation is separable, the preceding inequality shows that it is countable. Q.E.D.

8 THEOREM. *The spectrum of a spectral operator T consists of those complex numbers λ for which there is a sequence $\{x_n\}$ of vectors with*

(i) $$|x_n| = 1, \ (T - \lambda I)x_n \to 0.$$

PROOF. It is clear that such a point λ belongs to the spectrum of T. Conversely, let λ be in the spectrum of T. If λ is an eigenvalue, then we may take $x_n = x$, where x is an eigenvector for T corresponding to λ and normalized so that $|x| = 1$. Thus we may and shall suppose that $T - \lambda I$ is one-to-one. First suppose that $E(\{\lambda\}) = 0$. It follows from Corollary 7.12 that λ is in the continuous spectrum and thus that the inverse $(\lambda I - T)^{-1}$ is discontinuous. Thus the existence of a sequence satisfying (i) is assured. Finally, suppose that $E(\{\lambda\}) \neq 0$. Since $SE(\{\lambda\}) = \lambda E(\{\lambda\})$ we have

(ii) $$(T - \lambda I)^n E(\{\lambda\}) = N^n E(\{\lambda\}), \qquad n \geqq 0.$$

Now let $x \neq 0$ be an arbitrary vector in $E(\{\lambda\})\mathfrak{X}$. Since λ is not in the point spectrum of T, it follows from (ii) that $N^n x \neq 0$, for $n \geqq 0$, and we define $x_n = (N^n x)/|N^n x|$. It will now be shown that, for some subsequence $\{x_{n_i}\}$, $Nx_{n_i} \to 0$. If this is not the case, then for some positive δ,

$$\delta < \frac{|N^{n+1}x|}{|N^n x|}, \qquad n \geqq 0,$$

and hence

$$|x| < \frac{|Nx|}{\delta} < \frac{|N^2 x|}{\delta^2} < \cdots < \frac{|N^n x|}{\delta^n} < \cdots.$$

Thus

$$\begin{aligned} \limsup |N^n|^{1/n} &= \limsup |N^n|^{1/n} |x|^{1/n} \\ &\geqq \limsup |N^n x|^{1/n} \\ &\geqq \limsup \delta |x|^{1/n} = \delta > 0, \end{aligned}$$

which is impossible since N is a generalized nilpotent. Thus a subsequence $\{x_{n_i}\}$ has $Nx_{n_i} \to 0$ and, since $x_{n_i} = E(\{\lambda\})x_{n_i}$, it follows from (ii) that $(T - \lambda I)x_{n_i} \to 0$. Q.E.D.

9. The Algebras $\mathfrak{A}^p$ and $\hat{\mathfrak{A}}^p$

In this section a certain elementary type of non-commutative B^*-algebra of operators on a direct sum of Hilbert spaces is introduced. Stated briefly, the algebra $\mathfrak{A}^p$ is the algebra of all operators A in the direct sum $\mathfrak{H}^p$ of a Hilbert space $\mathfrak{H}$ with itself p times whose matrix representation (a_{ij}) has, for its elements, the operators a_{ij} which all belong to a commutative B^*-algebra $\mathfrak{A}$ of operators on $\mathfrak{H}$. Only a few basic properties of the algebra $\mathfrak{A}^p$ and its elements are discussed here. In the following section the problem of determining all spectral operators in $\mathfrak{A}^p$ will be solved, and these results are illustrated in Section 11.

We begin with a brief discussion of operators on a direct sum

$$\mathfrak{H}^p = \mathfrak{H} \oplus \cdots \oplus \mathfrak{H} \tag{1}$$

of a Hilbert space $\mathfrak{H}$ with itself p times. Here p is a positive integer and $\mathfrak{H}$ is assumed to have positive dimension. The elements of $\mathfrak{H}^p$ are, by definition, the finite ordered sets $x = [x_1, \ldots, x_p]$ of p elements in $\mathfrak{H}$, and addition, scalar multiplication, and scalar products are defined in $\mathfrak{H}^p$ in terms of the corresponding operations in $\mathfrak{H}$ by the equations

$$\begin{aligned} [x_1, \ldots, x_p] + [y_1, \ldots, y_p] &= [x_1 + y_1, \ldots, x_p + y_p], \\ \alpha[x_1, \ldots, x_p] &= [\alpha x_1, \ldots, \alpha x_p], \\ ([x_1, \ldots, x_p], [y_1, \ldots, y_p]) &= \sum_{i=1}^{p} (x_i, y_i). \end{aligned} \tag{2}$$

With these definitions the space $\mathfrak{H}^p$ is indeed a Hilbert space (IV.4.19) with norm

$$|[x_1, \ldots, x_p]| = \left(\sum_{i=1}^{p} |x_i|^2 \right)^{1/2}. \tag{3}$$

It is clear that if a_{ij}, $i, j = 1, \ldots, p$, is any set of p^2 bounded linear operators in $\mathfrak{H}$, the equations

$$y_i = \sum_{j=1}^{p} a_{ij} x_j, \qquad i = 1, \ldots, p, \tag{4}$$

define a bounded linear map $A : [x_1, \ldots, x_p] \to [y_1, \ldots, y_p]$ of $\mathfrak{H}^p$ into itself. If we use the symbols x and y for the vectors $[x_1, \ldots, x_p]$ and $[y_1, \ldots, y_p]$ respectively, the equations (4) then may be stated simply as $y = Ax$. Conversely, if one starts with an arbitrary bounded linear map $A : x \to y$ of $\mathfrak{H}^p$ into itself, it is easy to define a set a_{ij} of p^2 bounded linear maps in $\mathfrak{H}$ such that the equation $y = Ax$ is equivalent to the system (4). To see this, let A be a bounded linear map in $\mathfrak{H}^p$ and let $\mathfrak{H}_i$ be the set of vectors $[x_1, \ldots, x_p]$ in $\mathfrak{H}^p$ with $x_j = 0$ for $j \neq i$. Thus the map H_i defined on $\mathfrak{H}_i$ by the equation $H_i x = x_i$ is clearly a linear isometric map of $\mathfrak{H}_i$ onto all of $\mathfrak{H}$, and the map

$$P_i[x_1, \ldots, x_p] = H_i^{-1} x_i \tag{5}$$

is the orthogonal projection of $\mathfrak{H}^p$ onto $\mathfrak{H}_i$. Thus, for vectors $x = [x_1, \ldots, x_p]$ and $Ax = y = [y_1, \ldots, y_p]$, we have

$$x_j = H_j P_j x,$$

$$x = \sum_{j=1}^{p} P_j x = \sum_{j=1}^{p} H_j^{-1} x_j,$$

and so

$$y_i = H_i P_i y = H_i P_i Ax = \sum_{j=1}^{p} H_i P_i A H_j^{-1} x_j.$$

Therefore if we let $a_{ij} = H_i P_i A H_j^{-1}$ then these elements a_{ij}, $i, j = 1, \ldots, p$ are bounded linear operators in $\mathfrak{H}$ and the equations (4) state the same relationship between the vectors x and y as does the equation $y = Ax$. This representation of A by means of the matrix (a_{ij}) is unique, for if there were another set b_{ij} of p^2 operators in $\mathfrak{H}$ which so represented A, the difference $c_{ij} = a_{ij} - b_{ij}$ would have the property that $\sum_{j=1}^{p} c_{ij} x_j = 0$ for each $i = 1, \ldots, p$ and each $x = [x_1, \ldots, x_p]$ in $\mathfrak{H}^p$. If one replaces x by $P_k x$, it follows that $c_{ik} = 0$ for $i, k = 1, \ldots, p$. Thus the mapping $A \leftrightarrow (a_{ij})$ just established is a one-to-one correspondence between the algebra $B(\mathfrak{H}^p)$ of bounded linear operators in $\mathfrak{H}^p$ and the matrix algebra $\mathfrak{M}_p(B(\mathfrak{H}))$ of order p over the algebra $B(\mathfrak{H})$ of bounded linear maps in $\mathfrak{H}$. This algebra $\mathfrak{M}_p(B(\mathfrak{H}))$ is, by definition, the set of $p \times p$ matrices (a_{ij}) of elements a_{ij} in $B(\mathfrak{H})$, and addition, multiplication, and scalar multiplication are defined by the equations

$$\begin{aligned} (a_{ij}) + (b_{ij}) &= (a_{ij} + b_{ij}), \\ (a_{ij})(b_{ij}) &= \left(\sum_{k=1}^{p} a_{ik} b_{kj} \right), \\ \alpha(a_{ij}) &= (\alpha a_{ij}). \end{aligned} \tag{6}$$

The unit in $\mathfrak{M}_p(B(\mathfrak{H}))$ is the matrix (e_{ij}) defined as

$$\begin{aligned} e_{ij} &= 0, \qquad i \neq j, \\ &= e, \qquad i = j, \end{aligned} \tag{7}$$

where e is the unit in $B(\mathfrak{H})$, that is, the identity operator in $\mathfrak{H}$.

The symbol I will be used for the unit in $B(\mathfrak{H}^p)$, that is, the identity operator in $\mathfrak{H}^p$. Thus $I \leftrightarrow (e_{ij})$, which is an exception to the general notational practice of using corresponding upper and lower case letters for corresponding elements. It is clear that the mapping $A \leftrightarrow (a_{ij})$ is linear. It is, in fact, an algebraic isomorphism between the algebras $B(\mathfrak{H}^p)$ and $\mathfrak{M}_p(B(\mathfrak{H}))$, for if $A \leftrightarrow (a_{ij})$ and $B \leftrightarrow (b_{ij})$ and $y = ABx$, then

$$y_i = \sum_{k=1}^{p} a_{ik}\left(\sum_{j=1}^{p} b_{kj} x_j\right) = \sum_{j=1}^{p}\left(\sum_{k=1}^{p} a_{ik} b_{kj}\right) x_j,$$

which shows that $AB \leftrightarrow (a_{ij})(b_{ij})$.

The algebra $\mathfrak{M}_p(B(\mathfrak{H}))$ is normed by defining

$$|(a_{ij})| = |A|, \tag{8}$$

where A and (a_{ij}) are corresponding elements so that the mapping $A \leftrightarrow (a_{ij})$ is an isometric isomorphism between the B-algebras $B(\mathfrak{H}^p)$ and $\mathfrak{M}_p(B(\mathfrak{H}))$. It should be noted that convergence in the topology defined by (8) is equivalent to convergence defined by the norm

$$|(a_{ij})|_0 = \sup_{1 \leq i,\, j \leq p} |a_{ij}|.$$

To see this, note first that since $|H_i| = |P_i| = |H_j^{-1}| = 1$, we have $|a_{ij}| \leqq |A|$ and so $|(a_{ij})|_0 \leqq |(a_{ij})|$. On the other hand, if $|x| \leqq 1$, then $|x_i| \leqq 1$ and equation (4) shows that $|y_i| \leqq p\,|(a_{ij})|_0$, $|y_i|^2 \leqq p^2\,|(a_{ij})|_0^2$, $|y|^2 \leqq p^3\,|(a_{ij})|_0^2$, and $|y| = |Ax| \leqq p^{3/2}\,|(a_{ij})|_0$. Thus

$$|(a_{ij})|_0 \leqq |(a_{ij})| \leqq p^{3/2}\,|(a_{ij})|_0.$$

We next examine the nature of the conjugate space $(\mathfrak{H}^p)^*$. A linear functional x^* on $\mathfrak{H}^p$ determines p linear functionals $x_1^*, \ldots, x_p^*$ on $\mathfrak{H}$ according to the relations

$$x_i^* x_i = x^* H_i^{-1} x_i, \qquad x_i \in \mathfrak{H}, \tag{9}$$

and, conversely, any set $x_1^*, \ldots, x_p^*$ of p linear functionals uniquely determines a point x^* in $(\mathfrak{H}^p)^*$ related to them by the equation (9). This functional x^* may be defined by the equation

$$x^*[x_1, \ldots, x_p] = x_1^* x_1 + \cdots + x_p^* x_p, \qquad [x_1, \ldots, x_p] \in \mathfrak{H}^p. \tag{10}$$

It is clear that this correspondence $x^* \leftrightarrow [x_1^*, \ldots, x_p^*]$ is a one-to-one linear map between the two spaces $(\mathfrak{H}^p)^*$ and $(\mathfrak{H}^*)^p$. We next point out that it is also an isometric map, that is,

$$\sup_{|x|\leqq 1} |x^*x| = \left(\sum_{i=1}^{p} |x_i^*|^2\right)^{1/2}, \tag{11}$$

provided the functionals x_i^* are related to x^* by equation (9). Since $\mathfrak{H}^p$ is a Hilbert space, there is (IV.4.5) a uniquely defined point $y = [y_1, \ldots, y_p]$ in $\mathfrak{H}^p$ such that $x^*x = (x, y)$ for every x in $\mathfrak{H}^p$, and this unique y also has the property that $|x^*| = |y|$. Since $(x, y) = (x_1, y_1) + \cdots + (x_p, y_p)$ for every x in $\mathfrak{H}^p$, equation (10) shows that $x_i^*x_i = (x_i, y_i)$ for every x_i in $\mathfrak{H}$. Thus this same theorem (IV.4.5) shows that $|y_i| = |x_i^*|$ and hence, since $|x^*| = |y|$ that $|x^*| = |[x_1^*, \ldots, x_p^*]|$. This last equation is merely a restatement of (11), which is thus established. The correspondence $x^* \leftrightarrow [x_1^*, \ldots, x_p^*]$ given by equation (9) is therefore an isometric isomorphism between the spaces $(\mathfrak{H}^p)^*$, and $(\mathfrak{H}^*)^p$ and we may, when it appears that no confusion should arise, write $\mathfrak{H}^{*p} = (\mathfrak{H}^*)^p = (\mathfrak{H}^p)^*$ and $x^* = [x_1^*, \ldots, x_p^*]$.

Let A be a bounded linear operator in $\mathfrak{H}^p$ and let A^* be its Hilbert space adjoint so that

$$(Ax, y) = (x, A^*y), \qquad x, y \in \mathfrak{H}^p. \tag{12}$$

Let (a_{ij}) and (b_{ij}) be the matrices in $\mathfrak{M}_p(B(\mathfrak{H}))$ corresponding to A and A^* respectively. Then (12) gives

$$\sum_{i=1}^{p}\left(\sum_{j=1}^{p} a_{ij}x_j, y_i\right) = \sum_{i=1}^{p}\left(x_i, \sum_{k=1}^{p} b_{ik}y_k\right), \tag{13}$$

which is an identity in the $2p$ elements $x_i, y_i, i = 1, \ldots, p$ in $\mathfrak{H}$. Fix two elements x, y in $\mathfrak{H}$ and two integers μ, ν with $1 \leqq \mu, \nu \leqq p$. Let $x_\mu = x$, $y_\nu = y$, and $x_i = y_j = 0$ if $i \neq \mu$ and $j \neq \nu$. Then (13) reduces to the equation $(a_{\mu\nu}x, y) = (x, b_{\nu\mu}y)$, which, since x and y are arbitrary, shows that $b_{\nu\mu} = a_{\mu\nu}^*$. Since $B(\mathfrak{H}^p)$ is a B^*-algebra (IX.3.2) under the involution $A \to A^*$, it follows that the map $(a_{ij}) \to (b_{ij})$ is an involution in the algebra $\mathfrak{M}_p(B(\mathfrak{H}))$ and that, with this involution, $\mathfrak{M}_p(B(\mathfrak{H}))$ is a B^*-algebra. The mapping $A \leftrightarrow (a_{ij})$ between the B^*-algebras $B(\mathfrak{H}^p)$ and $\mathfrak{M}_p(B(\mathfrak{H}))$ is then an isometric $*$-isomorphism, and these algebras are $*$-equivalent. We shall henceforth identify them and simply write $A = (a_{ij})$ instead of $A \leftrightarrow (a_{ij})$.

The operators which we shall be discussing belong to a certain non-commutative B^*-subalgebra $\mathfrak{A}^p$ of $B(\mathfrak{H}^p)$ for which all of the corresponding matrix elements a_{ij} belong to a commutative B^*-subalgebra of $B(\mathfrak{H})$. The algebra $\mathfrak{A}^p$ will now be defined explicitly in terms of the notion of a spectral measure space. Let $\mathfrak{S}$ be an arbitrary set and Σ a σ-field of sets in $\mathfrak{S}$ with

$\mathfrak{S}$ in Σ. Let $e(\cdot)$ be a countably additive spectral measure on Σ whose values are self adjoint projections in $\mathfrak{H}$. Corollary 2.4 shows that $e(\cdot)$ is bounded and countably additive in the strong operator topology. We shall assume that Σ is complete in the sense that it contains all subsets of any set σ in Σ for which $e(\sigma) = 0$. Thus the triple $(\mathfrak{S}, \Sigma, e)$ constitutes what we shall call a *complete and countably additive self adjoint measure space of projections in* $\mathfrak{H}$.

Associated with the spectral measure space $(\mathfrak{S}, \Sigma, e)$ is the B^*-algebra $eB(\mathfrak{S}, \Sigma)$ of e-essentially bounded Σ-measurable complex valued functions $\hat{a}$ on $\mathfrak{S}$. The norm in $eB(\mathfrak{S}, \Sigma)$ is its e-essential supremum, which is defined by the equation

$$(14) \qquad e\text{-ess}\sup_{s\in\mathfrak{S}} |\hat{a}(s)| = \inf_{e(\delta)=e} \sup_{s\in\delta} |\hat{a}(s)|,$$

and the involution $\hat{a} \to \hat{a}^*$ is the complex conjugate $\hat{a}^*(s) = \overline{\hat{a}(s)}$. Although we speak of the elements of $eB(\mathfrak{S}, \Sigma)$ as functions, they are, more accurately, equivalence classes of functions where the two functions $\hat{a}$ and $\hat{b}$ on $\mathfrak{S}$ are equivalent if $\hat{a}(s) = \hat{b}(s)$ for e-almost all s in $\mathfrak{S}$. The space $eB(\mathfrak{S}, \Sigma)$, with the natural algebraic operations, is clearly a commutative B^*-algebra with unit $\hat{e}(s) = 1$ for all s in $\mathfrak{S}$.

The symbol $\hat{\mathfrak{A}}$ will be used for a B^*-subalgebra of $eB(\mathfrak{S}, \Sigma)$ with the same unit $\hat{e}$. The integral

$$(15) \qquad a = \int_{\mathfrak{S}} \hat{a}(s)e(ds), \qquad \hat{a} \in \hat{\mathfrak{A}},$$

maps the algebra $\hat{\mathfrak{A}}$ into a commutative B^*-subalgebra $\mathfrak{A}$ of $B(\mathfrak{H})$, and this mapping $\hat{a} \to a$ is an isometric $*$-isomorphism between $\hat{\mathfrak{A}}$ and $\mathfrak{A}$ (X.2.9). The symbols $\hat{\mathfrak{A}}$ and $\mathfrak{A}$ will be used throughout our discussion for the B^*-algebras introduced here which are $*$-equivalent under the isomorphism given by the equation (15). Usually we have in mind the case where $\hat{\mathfrak{A}} = eB(\mathfrak{S}, \Sigma)$, but we prefer to formulate the development for the case of an arbitrary B^*-subalgebra $\hat{\mathfrak{A}}$ whose unit is also $\hat{e}$. Since $eB(\mathfrak{S}, \Sigma)$ and $\hat{\mathfrak{A}}$ have the same unit, Corollary IX.3.10 shows that an element in $\hat{\mathfrak{A}}$ has an inverse in $\hat{\mathfrak{A}}$ if and only if it has an inverse in $eB(\mathfrak{S}, \Sigma)$. We shall need this fact also for non-commutative B^*-algebras, and it is proved in the following Lemma 2.

In case $\mathfrak{S}$ is a topological space, we shall assume that Σ contains the Borel subsets of $\mathfrak{S}$ and that $e(\sigma) \neq 0$ if σ is a non-void set open; that is,

$\mathfrak{S}$ is the support of the spectral measure $e(\cdot)$. This enables us to consider the B^*-algebra $C(\mathfrak{S})$ of all bounded continuous complex functions on $\mathfrak{S}$ as a B^*-subalgebra of $eB(\mathfrak{S}, \Sigma)$, for if $e(\delta) = e$, then δ is dense in $\mathfrak{S}$ and hence

$$\sup_{s \in \mathfrak{S}} |\hat{a}(s)| = e\text{-ess}\sup_{s \in \mathfrak{S}} |\hat{a}(s)|, \qquad \hat{a} \in C(\mathfrak{S}). \tag{16}$$

Thus any B^*-subalgebra $\hat{\mathfrak{C}}$ of $C(\mathfrak{S})$ whose unit is the same as that of $C(\mathfrak{S})$ is a permissible choice for the algebra $\hat{\mathfrak{A}}$. In case $\hat{\mathfrak{A}}$ is such an algebra of continuous functions, we shall sometimes write $\mathfrak{C}$ instead of $\mathfrak{A}$ for the equivalent algebra of operators. In this case we may also write $\hat{\mathfrak{C}}^p$ and $\mathfrak{C}^p$ for the two algebras $\hat{\mathfrak{A}}^p$ and $\mathfrak{A}^p$ defined below.

An arbitrary commutative B^*-subalgebra $\mathfrak{A}$ of $B(\mathfrak{H})$ may be represented in the manner described by formula (15). In fact, according to the general spectral theorem for commutative B^*-algebras of operators (X.2.1), every such algebra $\mathfrak{A}$ determines a compact Hausdorff space $\mathfrak{S}$, a regular countably additive self adjoint spectral measure $e(\cdot)$ defined on the Borel sets of $\mathfrak{S}$ such that the equation (15) gives a *-equivalence between $\mathfrak{A}$ and the B^*-algebra $\hat{\mathfrak{C}} = C(\mathfrak{S})$ of all continuous complex valued functions on $\mathfrak{S}$. This spectral measure $e(\cdot)$, whose existence is asserted by the spectral theorem, has the property just mentioned; that is, $e(\sigma) \neq 0$ if σ is non-void and open. To see this we note that, since $\mathfrak{S}$ is a normal space (I.5.9), Urysohn's theorem (I.5.2) assures the existence of a non-zero real continuous function $\hat{a}$ on $\mathfrak{S}$ which vanishes on the complement of σ. Thus, by the general spectral theorem (X.2.1) the operator a of equation (15) is not zero. However, since $\hat{a}$ vanishes on σ', we have $\hat{a}(s) = \hat{a}(s)\chi_\sigma(s)$, which, because of the homomorphic nature of the map (15), shows that $a = ae(\sigma)$ and thus that $e(\sigma) \neq 0$.

We associate with $\mathfrak{A}$ and $\hat{\mathfrak{A}}$ the two algebras $\mathfrak{A}^p$ and $\hat{\mathfrak{A}}^p$ as follows. The algebra $\mathfrak{A}^p$ is that B^*-subalgebra of $B(\mathfrak{H}^p)$ consisting of all operators $A = (a_{ij})$ with a_{ij} in $\mathfrak{A}$. The algebra $\hat{\mathfrak{A}}^p$ is the matrix algebra of order p over $\hat{\mathfrak{A}}$, that is, the algebra of all matrices $\hat{A} = (\hat{a}_{ij})$ whose elements $\hat{a}_{ij}$ are in $\hat{\mathfrak{A}}$. The elements $\hat{A}$ in $\hat{\mathfrak{A}}^p$ are to be regarded as e-essentially bounded maps $s \to \hat{A}(s) = (\hat{a}_{ij}(s))$ of $\mathfrak{S}$ into $B(E^p)$, the B^*-algebra of linear operators in p-dimensional unitary space E^p, and are normed as such. Thus, for ξ in E^p,

$$|\hat{A}(s)| = \sup_{|\xi|=1} |\hat{A}(s)\xi|,$$

and

$$|\hat{A}| = e\text{-ess}\sup_{s\in\mathfrak{S}} |\hat{A}(s)|. \tag{17}$$

The algebraic operations, the involution, and the unit $\hat{I}$ in $\hat{\mathfrak{A}}^p$ are defined by the equations

$$(\alpha\hat{A})(s) = \alpha\hat{A}(s),\ (\hat{A}+\hat{B})(s) = \hat{A}(s)+\hat{B}(s),\ (\hat{A}\hat{B})(s) = \hat{A}(s)\hat{B}(s)$$
$$(\hat{A}^*)(s) = \hat{A}(s)^*,\ \hat{I}(s) = (\hat{e}_{ij}(s)),$$

where, for all s in $\mathfrak{S}$, $\hat{e}_{ij}(s) = 1$ if $i = j$ and $\hat{e}_{ij}(s) = 0$ if $i \neq j$. It is clear that, under these definitions, the algebra $\hat{\mathfrak{A}}^p$ is a B-algebra with involution. It is also a B^*-algebra for

$$|\hat{A}^*\hat{A}| = e\text{-ess}\sup_{s\in\mathfrak{S}} |\hat{A}^*(s)\hat{A}(s)| = e\text{-ess}\sup_{s\in\mathfrak{S}} |\hat{A}(s)|^2 = |\hat{A}|^2,$$

where here, when we write $|\hat{A}^*(s)\hat{A}(s)| = |\hat{A}(s)|^2$, we use the fact that $B(E^p)$ is a B^*-algebra (IX.3.2). Just as was the case with $eB(\mathfrak{S}, \Sigma)$ the elements of the algebra $\hat{\mathfrak{A}}^p$ are equivalence classes where two matrices $\hat{A}$ and $\hat{B}$ are equivalent if $\hat{A}(s) = (s)\,\hat{B}$ for e-almost all s in $\mathfrak{S}$.

For a given matrix $\hat{A} = (\hat{a}_{ij})$ in $\hat{\mathfrak{A}}^p$ we shall use the symbol $\int_{\mathfrak{S}}\hat{A}(s)e(ds)$ for the matrix $A = (a_{ij})$ whose elements are $a_{ij} = \int_{\mathfrak{S}}\hat{a}_i(s)e(ds)$. Thus the matrix form of equation (15) is

$$A = \int_{\mathfrak{S}} \hat{A}(s)e(ds). \tag{18}$$

We shall see presently that this equation establishes an isometric *-isomorphism between the two algebras $\mathfrak{A}^p$ and $\hat{\mathfrak{A}}^p$.

Since the algebras $\mathfrak{A}^p$ and $\hat{\mathfrak{A}}^p$ are non-commutative (if $p > 1$), we now give two elementary lemmas concerning B^*-algebras which are not necessarily commutative. The first of these will be used to see that equation (18) establishes an isometric *-isomorphism between the algebras $\mathfrak{A}^p$ and $\hat{\mathfrak{A}}^p$.

1 LEMMA. *Every *-isomorphism between B^*-algebras is an isometry.*

PROOF. Let x and y be corresponding elements in the *-isomorphic B^*-algebras $\mathfrak{X}$ and $\mathfrak{Y}$. Since $|x^*|^2 = |x^*x|$ it suffices to prove that the corresponding self adjoint elements $u = x^*x$, $v = y^*y$ have the same norm. Since the algebras are isomorphic, the spectra $\sigma(u)$ and $\sigma(v)$ are the same, and this shows that u and v have the same spectral radius. Thus (IX.1.8) we have $\lim_n |u^n|^{1/n} = \lim_n |v^n|^{1/n}$. But, since u and v are self adjoint, they generate commutative B^*-subalgebras in $\mathfrak{X}$ and $\mathfrak{Y}$ respectively; and so (IX.3.3) we have $|u^n| = |u|^n$ and $|v^n| = |v|^n$ if n is a power of 2. Hence

$$|u| = \lim_n |u^n|^{1/n} = \lim_n |v^n|^{1/n} = |v|. \qquad \text{Q.E.D.}$$

2 Lemma. *If a B^*-subalgebra $\mathfrak{X}$ of a B^*-algebra $\mathfrak{Y}$ has the same unit e as $\mathfrak{Y}$, then an element in $\mathfrak{X}$ with an inverse in $\mathfrak{Y}$ has this inverse also in $\mathfrak{X}$.*

Proof. We first show that $e = e^*$. Since e is the unit, $e^* = ee^*$, and so $e = ee^* = (ee^*)^* = e^{**}e^* = ee^* = e^*$. Now let x in $\mathfrak{X}$ have the inverse x^{-1} in $\mathfrak{Y}$. Then $(x^{-1})^*x^* = (xx^{-1})^* = e^* = e$ so that x^* also has an inverse in $\mathfrak{Y}$. Since $\mathfrak{X}$ is a B^*-subalgebra of $\mathfrak{Y}$, we have x^* in $\mathfrak{X}$ and hence the element $z = xx^*$ is in $\mathfrak{X}$. If z^{-1} is in $\mathfrak{X}$, then $e = xx^*z^{-1}$ and so x has its inverse x^*z^{-1} in $\mathfrak{X}$. This reduces the problem to showing that a self adjoint element z in $\mathfrak{X}$ with an inverse z^{-1} in $\mathfrak{Y}$ has z^{-1} in $\mathfrak{X}$. Since $z = z^*$, the resolvent sets of z when considered as an element of $\mathfrak{X}$ and of $\mathfrak{Y}$ both contain the imaginary axis with the possible exception of the point $\lambda = 0$. Thus for a purely imaginary $\lambda \neq 0$ the element $\lambda e - z$ has inverses in both $\mathfrak{X}$ and $\mathfrak{Y}$. Since e is the unit for both of these algebras, these inverses are the same element, which we shall call $(\lambda e - z)^{-1}$. Now since z^{-1} exists in $\mathfrak{Y}$ and the resolvent is a continuous function of λ, we have

$$-z^{-1} = \lim_{\lambda \to 0} (\lambda e - z)^{-1},$$

which, since $\mathfrak{X}$ is closed, shows that z^{-1} is in $\mathfrak{X}$. Q.E.D.

This lemma shows that if an operator $A = (a_{ij})$ in $\mathfrak{A}^p$ has an inverse A^{-1} in the algebra $B(\mathfrak{H}^p)$ of all bounded linear operators on $\mathfrak{H}^p$, then this inverse A^{-1} is also in $\mathfrak{A}^p$.

→ 3 Theorem. *The algebras $\mathfrak{A}^p$ and $\hat{\mathfrak{A}}^p$ are isometrically $*$-isomorphic under the correspondence $A \leftrightarrow \hat{A}$ given by the equation*

$$A = \int_{\mathfrak{S}} \hat{A}(s)e(ds).$$

Proof. The correspondence is one-to-one, for if $\int \hat{A}(s)e(ds) = 0$ then $\int \hat{a}_{ij}(s)e(ds) = 0$, and since equation (15) gives a $*$-equivalence between $\mathfrak{A}$ and $\hat{\mathfrak{A}}$ we see that $\hat{a}_{ij} = 0$; which means that $\hat{A} = 0$. Thus $\mathfrak{A}^p$ and $\hat{\mathfrak{A}}^p$ are $*$-isomorphic B^*-algebras, and the desired conclusion follows from Lemma 1. Q.E.D.

4 COROLLARY. *If N is a generalized nilpotent in $\mathfrak{A}^p$, then $N^p = 0$.*

PROOF. If $|N^n|^{1/n} \to 0$ then, for e-almost all s in $\mathfrak{S}$, $|\hat{N}(s)^n|^{1/n} \to 0$, which shows that for these s the $p \times p$ matrix $\hat{N}(s)$ of complex numbers is a generalized nilpotent in $B(E^p)$. Hence $\hat{N}^p(s) = 0$ for e-almost all s in $\mathfrak{S}$ and therefore $N^p = 0$. Q.E.D.

5 COROLLARY. *For every A in $\mathfrak{A}^p$,*

$$\sup_{|x|=1} |Ax| = e\text{-ess}\sup_{s\in\mathfrak{S}} |\hat{A}(s)| .$$

PROOF. This is just a restatement of the isometric property of the map $A \leftrightarrow \hat{A}$. Q.E.D.

For an operator $A = (a_{ij})$ in $\mathfrak{A}^p$ the notion of the determinant, $\det(a_{ij})$, may be defined in the usual manner (I.13), since the elements of the matrix belong to the commutative algebra $\mathfrak{A}$. The determinant $\delta = \det(a_{ij})$ is an element of $\mathfrak{A}$ also, and since the map $\hat{a} \to a$ given by (15) is a homomorphism, it follows that (with the notation of I.13)

$$\hat{\delta} = \sum_{i_1=1}^{p} \cdots \sum_{i_p=1}^{p} \delta_{i_1, \ldots, i_p} \hat{a}_{i_1 1} \cdots \hat{a}_{i_p p} = \det(\hat{a}_{ij}),$$

and thus that

$$\det(a_{ij}) = \int_{\mathfrak{S}} \det(\hat{a}_{ij}(s)) e(ds). \tag{19}$$

The following corollary enables us to determine the spectrum of an operator in $\mathfrak{A}^p$ in terms of the spectra $\sigma(\hat{A}(s))$ of the finite $p \times p$ matrices $\hat{A}(s)$ of complex numbers.

→ 6 COROLLARY. *For an operator A in $\mathfrak{A}^p$ the following statements are equivalent.*

(i) *A has an inverse in $B(\mathfrak{H}^p)$;*
(ii) *A has an inverse in $\mathfrak{A}^p$;*
(iii) *$\det(a_{ij})$ has an inverse in $\mathfrak{A}$;*
(iv) *e-ess sup $|\det(\hat{a}_{ij}(s))|^{-1} < \infty$.*

PROOF. The equivalence of (i) and (ii) is a consequence of Lemma 2. It follows from the theorem that (ii) is equivalent to the existence of $\hat{A}^{-1}$ in $\hat{\mathfrak{A}}^p$, which in turn is equivalent to the e-essential boundedness of the numerical function $\det(\hat{A}^{-1}(s)) = [\det(\hat{a}_{ij}(s))]^{-1}$. This proves the equivalence of (i), (ii), and (iv). Equation (19), together with the fact that the map $a \leftrightarrow \hat{a}$ given by equation (15) is an isomorphism between the algebras $\mathfrak{A}$ and $\hat{\mathfrak{A}}$, shows that (iii) and (iv) are equivalent. Q.E.D.

7 Corollary. *If $\mathfrak{S}$ is a compact space and $\hat{\mathfrak{A}} = \hat{\mathfrak{C}}$, then the condition* (iv) *of the preceding corollary may be replaced by the*

$$\det \hat{A}(s) \neq 0, \qquad s \in \mathfrak{S}.$$

8 Corollary. *Under the hypothesis of the preceding corollary the spectrum of an operator A in $\mathfrak{C}^p$ is*

$$\sigma(A) = \bigcup_{s \in \mathfrak{S}} \sigma(\hat{A}(s)).$$

In general, without any topology in $\mathfrak{S}$, the spectrum of an operator A in $\mathfrak{A}^p$ is determined as follows.

→ 9 Corollary. *The spectrum of an operator A in $\mathfrak{A}^p$ is given by the formula*

$$\sigma(A) = \bigcap_{e(\delta)=e} \overline{\bigcup_{s \in \delta} \sigma(\hat{A}(s))}.$$

Proof. Let δ be an arbitrary set in Σ with $e(\delta) = e$ and let λ_0 be a complex number with

$$\lambda_0 \notin \overline{\bigcup_{s \in \delta} \sigma(\hat{A}(s))}.$$

Then $(\lambda_0 \hat{I}(s) - \hat{A}(s))^{-1}$ exists and, by Cramer's rule, is bounded on δ since the elements of $\lambda_0 \hat{I}(s) - \hat{A}(s)$ are bounded on δ. This shows that $(\lambda_0 \hat{I} - \hat{A})^{-1}$ exists as an element of $\hat{\mathfrak{A}}^p$. It follows from Corollary 6 that λ_0 is in the resolvent set $\rho(A)$. This shows that if $e(\delta) = e$, then

$$\sigma(A) \subseteq \overline{\bigcup_{s \in \delta} \sigma(\hat{A}(s))}$$

and so

$$(*) \qquad \sigma(A) \subseteq \bigcap_{e(\delta)=e} \overline{\bigcup_{s \in \delta} \sigma(\hat{A}(s))}.$$

Now let λ_0 be an arbitrary point in $\rho(A)$. By Corollary 6, λ_0 is in $\rho(\hat{A})$, and $(\lambda_0 \hat{I} - \hat{A})^{-1}(s) = (\lambda_0 \hat{I} - \hat{A}(s))^{-1}$ exists for e-almost all s in $\mathfrak{S}$ and is e-essentially bounded on $\mathfrak{S}$. Thus, for some δ in Σ with $e(\delta) = e$, λ_0 is in $\rho(\hat{A}(s))$ for every s in δ and

$$\sup_{s \in \delta} |(\lambda_0 \hat{I} - \hat{A}(s))^{-1}| = K < \infty.$$

This inequality shows not only that λ_0 is in the set $R = \bigcap_{s\in\delta} \rho(\hat{A}(s))$, but also that it is in its interior R^0. For if not, there is a sequence $\{s_n\}$ of points in δ and a sequence $\{\lambda_n\}$ with λ_n in $\sigma(\hat{A}(s_n))$ and $\lambda_n \to \lambda_0$. We may choose the sequence $\{\xi_n\}$ in E^p such that $\hat{A}(s)\xi_n = \lambda_n \xi_n$ and $|\xi_n| = 1$. Thus the vectors

$$\eta_n = (\lambda_0 \hat{I} - \hat{A}(s_n))\xi_n = (\lambda_0 - \lambda_n)\xi_n \to 0,$$

and

$$1 = |\xi_n| = |(\lambda_0 \hat{I} - \hat{A}(s_n))^{-1}\eta_n| \leqq K|\eta_n| \to 0,$$

a contradiction which shows that λ_0 is in R^0.

Since λ_0 is an arbitrary number in $\rho(A)$, this proves that

$$\rho(A) \subseteq \bigcup_{e(\delta)=e} \Big(\bigcap_{s\in\delta} \rho(\hat{A}(s))\Big)^0,$$

and, upon taking complements, that

$$\sigma(A) \supseteq \bigcap_{e(\delta)=e} \overline{\bigcup_{s\in\delta} \sigma(\hat{A}(s))},$$

which, when combined with the inequality (∗), gives the desired expression for the spectrum $\sigma(A)$. Q.E.D.

10. The Spectral Analysis of Operators in $\mathfrak{A}^p$

Throughout this section the algebra $\hat{\mathfrak{A}}$ will be $eB(\mathfrak{S}, \Sigma)$. We study the individual operators in $\mathfrak{A}^p$ with a view to discovering their spectral properties and, in particular, we give a method for determining whether or not a given operator in $\mathfrak{A}^p$ has a countably additive resolution of the identity. We begin the study of this problem by examining some relations between the roots and the coefficients of a polynomial.

1 LEMMA. *There is a Borel measurable map $\Gamma \to [\lambda_1(\Gamma), \ldots, \lambda_p(\Gamma)]$ of $B(E^p)$ into E^p with $\sigma(\Gamma) = \{\lambda_1(\Gamma), \ldots, \lambda_p(\Gamma)\}$.*

PROOF. Let the complex number system be ordered by defining $u_1 \prec u_2$ to mean that $|u_1| \leqq |u_2|$ and if $|u_1| = |u_2|$, then $\arg u_1 < \arg u_2$. By $\arg u$ is understood the smallest non-negative real number for which $u = |u| \exp(i \arg u)$. In terms of this ordering, let $\lambda_1(\Gamma)$ be the smallest root of the polynomial $d(\lambda; \Gamma) = \det(\lambda I - \Gamma)$. It will first be shown that $|\lambda_1(\Gamma)|$ is a continuous function of Γ. Let $\Gamma_m \to \Gamma_0$, and since $|\lambda_1(\Gamma)| \leqq |\Gamma|$

(VII.3.4), we may choose a subsequence $\{\Gamma'_m\}$ for which $\lambda_1(\Gamma'_m)$ converges to a complex number u. Since $0 = d(\lambda_1(\Gamma'_m); \Gamma_m) \to d(u; \Gamma_0)$, it follows that either $\lambda_1(\Gamma_0) \prec u$ or $\lambda_1(\Gamma_0) = u$. Thus if $|\lambda_1(\Gamma_m)|$ does not converge to $|\lambda_1(\Gamma_0)|$, there is a subsequence $\{\Gamma''_m\}$ for which

$$|\lambda_1(\Gamma''_m)| > |\lambda_1(\Gamma_0)| + \varepsilon$$

for some positive ε. Thus the number $\lambda_1 = \lambda_1(\Gamma_0)$ has distance of at least ε from each of the spectra $\sigma(\Gamma''_m)$, and so the sequence $\{R(\lambda_1; \Gamma''_m)\}$ of resolvents is bounded. We may and shall assume that it has been replaced by a convergent subsequence, and thus that $R(\lambda_1; \Gamma''_m) \to B$. Hence

$$I = (\lambda_1 I - \Gamma''_m)R(\lambda_1; \Gamma''_m) = R(\lambda_1; \Gamma''_m)(\lambda_1 I - \Gamma''_m),$$

and, by passing to the limit, it is seen that B is an inverse to the singular matrix $\lambda_1(\Gamma_0)I - \Gamma_0$, which is impossible. This proves that $|\lambda_1(\Gamma)|$ is a continuous function of Γ. Now let the sequence $\{\Gamma_m\}$ approaching Γ_0 be chosen so that

$$\arg \lambda_1(\Gamma_m) \to \theta_0,$$

where $\theta_0 = \liminf_{\Gamma \to \Gamma_0} \arg \lambda_1(\Gamma)$. Thus, since $|\lambda_1(\Gamma_m)| \to |\lambda_1(\Gamma_0)|$, we have

$$\lim_m \lambda_1(\Gamma_m) = |\lambda_1(\Gamma_0)|\, e^{i\theta_0}.$$

It follows, as was shown above, that either $\lambda_1(\Gamma_0) \prec |\lambda_1(\Gamma_0)| \exp(i\theta_0)$ or $\arg \lambda_1(\Gamma_0) = \theta_0$. In either case,

$$\arg \lambda_1(\Gamma_0) \leqq \liminf_{\Gamma \to \Gamma_0} \arg \lambda_1(\Gamma).$$

Now, for non-negative real numbers r and θ, let

$$S(r, \theta) = \{u \,|\, |u| \leqq r, \arg u \leqq \theta\}.$$

It follows from the properties of $|\lambda_1(\Gamma)|$ and $\arg \lambda_1(\Gamma)$ which have just been established, that the set $\lambda_1^{-1}(S(r, \theta))$ is closed. Since the sets $S(r, \theta)$ generate the Borel field of complex numbers, the function λ_1 is a Borel measurable function of Γ. The proof may be completed by choosing $\lambda_2(\Gamma)$ to be the smallest root of the polynomial $d(\lambda; \Gamma)(\lambda - \lambda_1(\Gamma))^{-1}$ and repeating this process to define all the roots $\lambda_1(\Gamma), \ldots, \lambda_p(\Gamma)$. Q.E.D.

2 Corollary. *Let $s \to \Gamma(s)$ be a Σ-measurable map from $\mathfrak{S}$ into $B(E^p)$. Then there are Σ-measurable scalar functions $\hat\lambda_1, \ldots, \hat\lambda_p$ on $\mathfrak{S}$ for which*

$$\sigma(\Gamma(s)) = \{\hat\lambda_1(s), \ldots, \hat\lambda_p(s)\}, \qquad s \in \mathfrak{S}.$$

PROOF. Let $\hat{\lambda}_i(s) = \lambda_i(\Gamma(s))$, where λ_i is the function of Lemma 1, and let G be an open set of complex numbers. Then

$$\hat{\lambda}_i^{-1}(G) = \Gamma^{-1}(\lambda_i^{-1}(G)),$$

which shows that $\hat{\lambda}_i$ is Σ-measurable. Q.E.D.

3 LEMMA. *Let $s \to \Gamma(s)$ be a Σ-measurable map from $\mathfrak{S}$ into $B(E^p)$. Then there are disjoint sets $\mathfrak{S}_1, \ldots, \mathfrak{S}_p$ in Σ, whose union is $\mathfrak{S}$, such that*

$$\text{(i)} \qquad \det(\lambda I - \Gamma(s)) = \prod_{j=1}^{i} (\lambda - \hat{\lambda}_{ij}(s))^{\nu_{ij}(s)}, \qquad s \in \mathfrak{S}_i,$$

where $\hat{\lambda}_{ij}$, $j = 1, \ldots, i$ are Σ-measurable functions on $\mathfrak{S}_i$ and, for $1 \leq i \leq p$, we have

$$\text{(ii)} \qquad \hat{\lambda}_{i1}(s) \prec \hat{\lambda}_{i2}(s) \prec \cdots \prec \hat{\lambda}_{ii}(s), \qquad s \in \mathfrak{S}_i.$$

The resolution of the identity for $\Gamma(s)$ is given by the formula

$$\text{(iii)} \qquad E(\sigma; \Gamma(s)) = \sum_{j=1}^{i} \chi_\sigma(\hat{\lambda}_{ij}(s)) E(\hat{\lambda}_{ij}(s); \Gamma(s)), \qquad s \in \mathfrak{S}_i,\ \sigma \in \mathscr{B}.$$

The projection $E(\hat{\lambda}_{ij}(s); \Gamma(s))$ corresponding to the single eigenvalue $\hat{\lambda}_{ij}(s)$ of $\Gamma(s)$ is a Σ-measurable function of s and has the form

$$\text{(iv)} \qquad E(\hat{\lambda}_{ij}(s); \Gamma(s)) = \frac{R_{ij}}{m_{ij}(\hat{\lambda}_{ij}(s))^{\nu_{ij}}}, \qquad s \in \mathfrak{S}_i,$$

where R_{ij} is a polynomial in $\Gamma(s)$ and $\hat{\lambda}_{ij}(s)$,

$$\text{(v)} \qquad m_{ij}(\lambda) = \frac{\prod_{k=1}^{i} (\lambda - \hat{\lambda}_{ik}(s))^{\nu_{ik}}}{(\lambda - \hat{\lambda}_{ij}(s))^{\nu_{ij}}}$$

and ν_{ij} is any set of non-negative integers with

$$\text{(vi)} \qquad \sup_{s \in \mathfrak{S}_i} \nu_{ij}(s) \leq \nu_{ij}.$$

PROOF. Let $\mathfrak{S}_i$ be the set of all s in $\mathfrak{S}$ for which the spectrum $\sigma(\Gamma(s))$ has i points. Then, clearly, the sets $\mathfrak{S}_i$ are disjoint and their union is $\mathfrak{S}$. Corollary 2 and the equations

$$\mathfrak{S}_1 = \{s \mid \hat{\lambda}_1(s) = \hat{\lambda}_2(s) = \cdots = \hat{\lambda}_p(s)\},$$

$$\mathfrak{S}_2 = \bigcup_{k=1}^{p-1} \{s \mid \hat{\lambda}_1(s) = \cdots = \hat{\lambda}_k(s) \neq \hat{\lambda}_{k+1}(s) = \cdots = \hat{\lambda}_p(s)\},$$

$$\mathfrak{S}_3 = \bigcup_{1 \leq k \leq j < p} \{s \mid \hat{\lambda}_1(s) = \cdots = \hat{\lambda}_k(s) \neq \hat{\lambda}_{k+1}(s) = \cdots = \hat{\lambda}_j(s) \neq \hat{\lambda}_{j+1}(s) = \cdots = \hat{\lambda}_p(s)\},$$

and so on, show that the sets $\mathfrak{S}_1, \ldots, \mathfrak{S}_p$ are in Σ. For each s in $\mathfrak{S}_i$ we let $\hat{\lambda}_{ij}(s)$, $j = 1, \ldots, i$ be the distinct characteristic numbers of $\Gamma(s)$ arranged in the same increasing order as they were in Corollary 2, so that the functions $\hat{\lambda}_{ij}$ are Σ-measurable functions on $\mathfrak{S}_i$ satisfying the inequalities (ii). Thus the determinant $\det(\lambda I - \Gamma(s))$ has the form stated in (i). The remaining conclusions concern the form of the projections in the resolution of the identity for a finite matrix.

Let us recall that, for a $p \times p$ matrix Γ of complex numbers with

$$\det(\lambda I - \Gamma) = \prod_{k=1}^{i} (\lambda - \lambda_k)^{m_k},$$

and $\lambda_1, \ldots, \lambda_i$ distinct, the projection $E(\lambda_j; \Gamma)$ associated with the single eigenvalue λ_j is the matrix $e_j(\Gamma)$ where $e_j(\lambda)$ is an analytic function which is identically 1 in a neighborhood of λ_j and identically 0 in a neighborhood of every spectral point λ_k with $k \neq j$. It follows (VII.1.5) that the projection $E(\lambda_j; \Gamma)$ is a polynomial $P_j(\Gamma)$ in Γ and that any polynomial P_j with the properties

$$P_j^{(\nu)}(\lambda_k) = 0, \qquad 0 \leqq \nu < \nu_k, \qquad k \neq j,$$

$$P_j(\lambda_j) = 1, \qquad P_j^{(\nu)}(\lambda_j) = 0, \qquad 0 < \nu < \nu_j,$$

where the numbers ν_k are arbitrary integers with $\nu_k \geqq m_k$, $k = 1, \ldots, i$, will have the property that $P_j(\Gamma) = E(\lambda_j; \Gamma)$. To construct such an interpolating polynomial we let

$$m(\lambda) = \prod_{k=1}^{i} (\lambda - \lambda_k)^{\nu_k}, \quad m_j(\lambda) = \frac{m(\lambda)}{(\lambda - \lambda_j)^{\nu_j}}, \quad 1 \leqq j \leqq i,$$

and note that, for an arbitrary choice of the scalars $c_0, \ldots, c_{\nu_j - 1}$, the polynomial

$$P_j(\lambda) = m_j(\lambda) \sum_{k=0}^{\nu_j - 1} c_k (\lambda - \lambda_j)^k$$

and its first $\nu_k - 1$ derivatives vanish at the point λ_k, provided that $k \neq j$. We wish to choose the ν_j constants $c_0, \ldots, c_{\nu_j - 1}$ so that P_j satisfies the ν_j conditions required at the point λ_j. Using the Leibniz rule for the derivative of a product, it is seen that these ν_j conditions are

$$P_j(\lambda_j) = m_j(\lambda_j) c_0 = 1,$$

$$\vdots$$

$$P_j^{(\nu)}(\lambda_j) = \sum_{r=0}^{\nu} \binom{\nu}{r} r!\, m_j^{(\nu - r)}(\lambda_j) c_r = 0, \qquad 0 < \nu < \nu_j.$$

These equations may be solved recursively, and it is found that their unique solution $c_0, \ldots, c_{\nu_j - 1}$ has the form

$$c_\nu = \frac{Q_j}{m_j(\lambda_j)^{\nu+1}}, \qquad 0 \leqq \nu < \nu_j,$$

where Q_j is a polynomial in the roots $\lambda_1, \ldots, \lambda_i$. Thus the polynomial P_j has the form

$$P_j(\lambda) = \frac{R_j(\lambda; \lambda_1, \ldots, \lambda_i)}{m_j(\lambda_j)^{\nu_j}}, \qquad 1 \leqq j \leqq i,$$

where R_j is a polynomial in λ and the roots $\lambda_1, \ldots, \lambda_i$, and the elementary projections are given by the equation $E(\lambda_j; \Gamma) = P_j(\Gamma)$. Thus, for an arbitrary Borel set σ of complex numbers,

$$E(\sigma; \Gamma) = \sum_{j=1}^{i} \chi_\sigma(\lambda_j) E(\lambda_j; \Gamma).$$

The preceding remarks, when applied to the matrix $\Gamma(s)$, complete the proof. Q.E.D.

REMARK. In applying this lemma as well as other theorems involving the functions $\hat{\lambda}_{ij}$, it is not essential that they be ordered as in (ii). The numbering of these functions was merely defined by this ordering to insure their measurability. *What is essential is that for each* $i = 1, \ldots, p$ *and* $j = 1, \ldots, i$ *the functions* $\hat{\lambda}_{ij}$ *be defined and* Σ*-measurable on* $\mathfrak{S}_i$ *and for each* s *in* $\mathfrak{S}_i$ *the numbers* $\hat{\lambda}_{i1}(s), \ldots, \hat{\lambda}_{ii}(s)$ *be the eigenvalues of the matrix* $\Gamma(s)$.

→ 4 LEMMA. *Let* $s \to \Gamma(s)$ *be a* Σ*-measurable map from* $\mathfrak{S}$ *into* $B(E^p)$. *Then the following statements are equivalent:*

(i) $$\sup_{\sigma \in \mathscr{B}} e\text{-}\operatorname{ess\,sup}_{s \in \mathfrak{S}} |E(\sigma; \Gamma(s))| < \infty;$$

(ii) $$e\text{-}\operatorname{ess\,sup}_{s \in \mathfrak{S}_i} |E(\hat{\lambda}_{ij}(s); \Gamma(s)) < \infty, \qquad 1 \leqq j \leqq i \leqq p;$$

(iii) $$e\text{-}\operatorname{ess\,sup}_{s \in \mathfrak{S}} \sup_{\sigma \in \mathscr{B}} |E(\sigma; \Gamma(s)| < \infty.$$

PROOF. The formula (iii) of Lemma 3 shows that (ii) implies (iii). It is clear that (iii) implies (i), and so to prove the lemma it will suffice to prove that (i) implies (ii). Let

(a) $$\sup_{\sigma \in \mathscr{B}} e\text{-}\operatorname{ess\,sup}_{s \in \mathfrak{S}} |E(\sigma; \Gamma(s))| = K < \infty,$$

and suppose that for some $i = 1, \ldots, p$, some $j = 1, \ldots, i$, and some set $G \subseteq \mathfrak{S}_i$ with $e(G) \neq 0$ we have

$$\text{(b)} \qquad |E(\hat{\lambda}_{ij}(s); \Gamma(s))| > K, \qquad s \in G.$$

Let $\delta(s)$ be defined on $\mathfrak{S}_i$ by the equation

$$\delta(s) = \inf_{k \neq j} |\hat{\lambda}_{ik}(s) - \hat{\lambda}_{ij}(s)|, \qquad s \in \mathfrak{S}_i.$$

It follows from Lemma 3(ii) that $\delta(s) > 0$. We may therefore choose an $\varepsilon > 0$ so that the set $G(\varepsilon) = \{s \mid s \in G, \delta(s) > 2\varepsilon\}$ has e-measure $e(G(\varepsilon)) \neq 0$. Now choose λ_0 so that

$$|\hat{\lambda}_{ij}(s) - \lambda_0| < \varepsilon,$$

for every s in a set $G_1(\varepsilon) \subseteq G(\varepsilon)$ with $e(G_1(\varepsilon)) \neq 0$. Let $\sigma = \{\lambda \mid |\lambda - \lambda_0| < \varepsilon\}$ so that if $s \in \hat{\lambda}_{ik}^{-1}(\sigma) \cap \hat{\lambda}_{ij}^{-1}(\sigma)$ for some $k \neq j$, we have $|\hat{\lambda}_{ij}(s) - \hat{\lambda}_{ik}(s)| < 2\varepsilon$, and therefore s is not in $G(\varepsilon)$. Hence for every $k \neq j$ we have

$$\text{(c)} \qquad \hat{\lambda}_{ik}^{-1}(\sigma) \cap \hat{\lambda}_{ij}^{-1}(\sigma) \cap G(\varepsilon) = \phi,$$

the void set. From Lemma 3(iii),

$$\text{(d)} \qquad \chi_{G(\varepsilon)}(s)E(\sigma; \Gamma(s)) = \sum_{k=1}^{i} \chi_\sigma(\hat{\lambda}_{ik}(s))E(\hat{\lambda}_{ik}(s); \Gamma(s))\chi_{G(\varepsilon)}(s).$$

Now, if s is in $G_1(\varepsilon)$ then $\hat{\lambda}_{ij}(s) \in \sigma$ and $s \in G(\varepsilon)$ and thus (c) shows that $\hat{\lambda}_{ik}(s) \notin \sigma$ for $k \neq j$. Thus, for s in $G_1(\varepsilon)$, the equation (d) gives

$$\text{(e)} \qquad E(\sigma; \Gamma(s)) = E(\hat{\lambda}_{ij}(s); \Gamma(s)), \qquad s \in G_1(\varepsilon),$$

and since $G_1(\varepsilon) \subseteq G(\varepsilon)$, equations (b) and (e) show that

$$|E(\sigma; \Gamma(s))| = |E(\hat{\lambda}_{ij}(s); \Gamma(s))| > K, \qquad s \in G_1(\varepsilon).$$

Since $e(G_1(\varepsilon)) \neq 0$, this contradicts the equation (a). Q.E.D.

We now return our attention to the B^*-algebras $\mathfrak{A}^p$ and $\hat{\mathfrak{A}}^p$ which, according to Theorem 9.3, are $*$-equivalent under the map $A \leftrightarrow \hat{A}$ given by the formula

$$(1) \qquad A = \int_{\mathfrak{S}} \hat{A}(s)e(ds), \qquad \hat{A} \in \hat{\mathfrak{A}}^p.$$

We first focus our attention on the problem of determining those operators in $\mathfrak{A}^p$ which are spectral operators. To this end the following Fubini type theorem is fundamental.

5 THEOREM. *Let the operator A in $\mathfrak{A}^p$ have the property*

$$\text{(i)} \qquad \sup_{\sigma \in \mathscr{B}} e\text{-ess} \sup_{s \in \mathfrak{S}} |E(\sigma; \hat{A}(s))| < \infty.$$

Then for every bounded Borel scalar function φ defined on the spectrum $\sigma(A)$, the integral

$$\text{(ii)} \qquad \int_{\sigma(A)} \varphi(\lambda) E(d\lambda; \hat{A}(s))$$

is an e-essentially bounded Σ-measurable function of s. The integral

$$\text{(iii)} \qquad \int_{\mathfrak{S}} E(\sigma; \hat{A}(s)) e(ds), \qquad \sigma \in \mathscr{B},$$

is a bounded, countably additive spectral measure in $\mathfrak{H}^p$ and

$$\text{(iv)} \quad \int_{\mathfrak{S}} \left[\int_{\sigma(A)} \varphi(\lambda) E(d\lambda; \hat{A}(s)) \right] e(ds) = \int_{\sigma(A)} \varphi(\lambda) \left[\int_{\mathfrak{S}} E(d\lambda; \hat{A}(s)) e(ds) \right].$$

PROOF. For s in $\mathfrak{S}_i$ the integral (ii) is

$$\text{(v)} \quad \int_{\sigma(A)} \varphi(\lambda) E(d\lambda; \hat{A}(s)) = \sum_{j=1}^{i} \varphi(\hat{\lambda}_{ij}(s)) E(\hat{\lambda}_{ij}(s); \hat{A}(s)), \qquad s \in \mathfrak{S}_i,$$

and Lemma 3 shows that this is a Σ-measurable function of s. The inequality (ii) of Lemma 4 shows that it is an e-essentially bounded function of s. This proves the first conclusion. Now let $e_{ij}(\sigma; \hat{A}(s))$ be the element in the ith row and jth column of $E(\sigma; \hat{A}(s))$ and let x and y be arbitrary vectors in $\mathfrak{H}$. Then the countable additivity of the integral (iii) is equivalent to that of the integral

$$\int_{\mathfrak{S}} e_{ij}(\sigma; \hat{A}(s))(e(ds)x, y).$$

If $\{\sigma_m\}$ is an increasing sequence of Borel sets with union σ, it follows from Lemma 4(iii) that the functions $e_{ij}(\sigma_m; \hat{A}(s))$ are uniformly bounded in m and s except for s in an e-null set. Thus, since the measure $(e(\cdot)x, y)$ is bounded,

$$\lim_{m \to \infty} \int_{\mathfrak{S}} e_{ij}(\sigma_m; \hat{A}(s))(e(ds)x, y) = \int_{\mathfrak{S}} e_{ij}(\sigma; \hat{A}(s))(e(ds)x, y).$$

This proves the countable additivity of the integral (iii). The fact that this integral is a spectral measure in $\mathfrak{H}^p$ follows from Theorem 9.3, and its boundedness follows from Corollary 2.4. This proves the second conclusion.

Since the integral (iii) is bounded in σ, the integral on the right side of (iv) is a continuous function of φ regarded as an element of the B-space of bounded Borel functions on $\sigma(A)$. From equation (v) and Lemma 3 it follows that the left side of (iv) is also continuous in φ. Since both sides are linear in φ, to prove their equality it will suffice to prove equality for every φ in some fundamental set. Such a set consists of the characteristic functions of Borel sets. If φ is the characteristic function of the Borel set σ, then both sides of this equation reduce to the integral (iii). Q.E.D.

6 THEOREM. *An operator A in $\mathfrak{A}^p$ is a spectral operator if and only if*

(i) $$\sup_{\sigma\in\mathscr{B}} e\text{-ess}\sup_{s\in\mathfrak{S}} |E(\sigma; \hat{A}(s))| < \infty.$$

When this condition is satisfied, A is a spectral operator of type $p-1$ whose resolution of the identity is given by the formula

(ii) $$E(\sigma; A) = \int_{\mathfrak{S}} E(\sigma; \hat{A}(s))e(ds), \qquad \sigma \in \mathscr{B}.$$

PROOF. The preceding theorem shows that the operator valued measure $E(\sigma; A)$ given by equation (ii) is a countably additive spectral measure in $\mathfrak{H}^p$ defined on the Borel sets $\mathscr{B}$ in the complex plane. Since $E(\sigma; \hat{A}(s))$ and $\hat{A}(s)$ commute for each s in $\mathfrak{S}$, and since equation (1) represents a *-isomorphism between $\hat{\mathfrak{A}}^p$ and $\mathfrak{A}^p$, the operators $E(\sigma; A)$ and A also commute. We now apply the preceding theorem with $\varphi(\lambda) = \lambda$ so that the scalar part of the matrix $\hat{A}(s)$, that is, the matrix

$$\hat{S}(s) = \int_{\sigma(A)} \lambda E(d\lambda; \hat{A}(s)), \qquad s \in \mathfrak{S},$$

defines a point $\hat{S}$ in $\hat{\mathfrak{A}}^p$. Hence, if $\hat{N}(s)$ is the radical part of $\hat{A}(s)$, the function $\hat{N}$ is in $\hat{\mathfrak{A}}^p$ and $\hat{N}^p = 0$. Thus if

$$S = \int_{\mathfrak{S}} \hat{S}(s)e(ds), \qquad N = \int_{\mathfrak{S}} \hat{N}(s)e(ds),$$

we have $A = S + N$, $SN = NS$, $N^p = 0$, and, by equation (iv) of the preceding theorem,

$$S = \int_{\sigma(A)} \lambda E(d\lambda; A).$$

This shows that A is a spectral operator of type $p-1$ whose scalar part is S and whose radical part is N.

Now, conversely, suppose that for some $\hat{A}$ in $\hat{\mathfrak{A}}^p$ the operator A on $\mathfrak{H}^p$ which is given by equation (1) is a spectral operator. For every positive integer k let

$$\mathfrak{S}^k = \{s \mid \sup_{1 \leqq j \leqq i \leqq p} |E(\hat{\lambda}_{ij}(s); \hat{A}(s))| \leqq k\}.$$

Let E_k be the projection in $\mathfrak{H}^p$ defined by the matrix $E_k = Ie(\mathfrak{S}^k)$ with $e(\mathfrak{S}^k)$ on the principal diagonal and zeros elsewhere, so that $E_k A = AE_k$. Theorem 3.10 shows that the restriction $A \mid E_k \mathfrak{H}^p$ is a spectral operator whose resolution of the identity is $E(\sigma; A) \mid E_k \mathfrak{H}^p$. Since $E(\sigma; A)$ is bounded in σ,

$$|E(\sigma; A)\, E_k \mathfrak{H}^p| \leqq |E(\sigma; A)| \leqq K, \qquad \sigma \in \mathscr{B},$$

for some constant K. The restriction $A \mid E_k \mathfrak{H}^p$ satisfies the condition (i) on the set $\mathfrak{S}^k$, and so it follows from what has already been proved that

$$E(\sigma; A) \mid E_k \mathfrak{H}^p = \int_{\mathfrak{S}^k} E(\sigma; \hat{A}(s)) e(ds) \mid E_k \mathfrak{H}^p.$$

Thus, since the map (1) is isometric, we have

$$e\text{-ess}\sup_{s \in \mathfrak{S}^k} |E(\sigma; \hat{A}(s))| = |E(\sigma; A) \mid E_k \mathfrak{H}^p| \leqq K.$$

Since $\{\mathfrak{S}^k\}$ is an increasing sequence with union $\mathfrak{S}$, it follows that

$$e\text{-ess}\sup_{s \in \mathfrak{S}} |E(\sigma; \hat{A}(s))| \leqq K.$$

The desired inequality (i) follows, since the constant K is independent of the arbitrary Borel set σ. Q.E.D.

REMARK. To illustrate more clearly the relationship between Theorem 6 and the well known result for self adjoint operators, we observe that a self adjoint operator or, more generally, *a normal operator in* $\mathfrak{A}^p$ *is a spectral operator*. For if A is a normal operator in $\mathfrak{A}^p$, it follows from Theorem 9.3 that for e-almost all s in $\mathfrak{S}$ the $p \times p$ matrix $\hat{A}(s)$ is normal. Since, for a Borel set σ, the projection $E = E(\sigma; \hat{A}(s))$ is a polynomial in $\hat{A}(s)$, it follows that, for e-almost all s in $\mathfrak{S}$, E is normal. Thus

$$|E|^2 = |E^2|^2 = |(E^2)^* E^2| = |(E^*E)^* E^*E| = |E^*E|^2 = |E|^4,$$

which shows that $|E|$ is either 0 or 1. This proves that, for e-almost all s in $\mathfrak{S}$, $|E(\sigma; \hat{A}(s)| \leqq 1$ and shows that the condition of the theorem is satisfied.

7 Corollary. *Let A be in $\mathfrak{A}^p$. For some constant K and e-almost all s in $\mathfrak{S}$ let*

$$\frac{1}{|\lambda-\mu|} \leqq K$$

for every pair of distinct eigenvalues of $\hat{A}(s)$. Then A is a spectral operator of type $p-1$.

Proof. It follows from equations (iv) and (v) of Lemma 3 that $E(\hat{\lambda}_{ij}(s);\ \hat{A}(s))$ is e-essentially bounded on $\mathfrak{S}$. Lemma 4 then shows that condition (i) of the theorem is satisfied. Q.E.D.

8 Corollary. *Every operator A in $\mathfrak{A}^p$ is the strong limit of a sequence of spectral operators.*

Proof. Let $\mathfrak{S}^k$ and E_k be as in the proof of the theorem and let

$$\begin{aligned} \hat{A}_k(s) &= \hat{A}(s), & s &\in \mathfrak{S}^k, \\ &= 0, & s &\notin \mathfrak{S}^k, \end{aligned}$$

so that $\hat{A}_k$ is in $\hat{\mathfrak{A}}^p$. The corresponding operator $A_k = \int_{\mathfrak{S}} \hat{A}_k(s)e(ds)$ has its restrictions $A_k \,|\, E_k \mathfrak{H}^p = A \,|\, E_k \mathfrak{H}^p$, and $A_k \,|\, (I-E_k)\mathfrak{H}^p = 0$, both spectral operators. Hence Theorem 3.10 shows that A_k is a spectral operator. Now since $\mathfrak{S}^k \to \mathfrak{S}$ it follows that $e(\mathfrak{S}^k)x \to x$ for every x in $\mathfrak{H}$ and thus that $E_k x \to x$ for every x in $\mathfrak{H}^p$. This shows that $A_k x = AE_k x \to Ax$ for every x in $\mathfrak{H}^p$. Q.E.D.

9 Corollary. *Let A be a spectral operator in $\mathfrak{A}^p$. Then the scalar and radical parts S and N of A are also in $\mathfrak{A}^p$. If $\hat{A}$, $\hat{S}$, $\hat{N}$ are the elements in $\hat{\mathfrak{A}}^p$ corresponding to A, S, N under the isomorphism* (1) *then, for e-almost all s in $\mathfrak{S}$, the matrices $\hat{S}(s)$ and $\hat{N}(s)$ are the scalar and radical parts of $\hat{A}(s)$. If φ is a bounded Borel scalar function on $\sigma(A)$, then $\varphi(S)$ is in $\mathfrak{A}^p$ and*

$$\widehat{\varphi(S)}(s) = \varphi(\hat{S}(s)),$$

for e-almost all s in $\mathfrak{S}$.

Proof. In the proof of the theorem it was shown that the scalar and radical parts of A are obtained by integrating, over $\mathfrak{S}$ with respect to $e(\cdot)$, the scalar and radical parts of $\hat{A}(s)$. This shows that S and N are in $\mathfrak{A}^p$ and, since the equation (1) establishes an isomorphism between $\mathfrak{A}^p$ and $\hat{\mathfrak{A}}^p$, it also shows that the scalar and radical parts of $\hat{A}(s)$ are functions in the equivalence classes in $\hat{\mathfrak{A}}^p$ corresponding to S and N respectively. Now equation (iv) of Theorem 5 shows that

$$\int_{\mathfrak{S}} \varphi(\hat{S}(s))e(ds) = \int_{\sigma(A)} \varphi(\lambda)E(d\lambda; A),$$

and so

$$\int_{\mathfrak{S}} \varphi(\hat{S}(s))e(ds) = \varphi(S) = \int_{\mathfrak{S}} \widehat{\varphi(S)}(s)e(ds),$$

from which it follows that $\varphi(\hat{S}(s)) = \widehat{\varphi(S)}(s)$ for e-almost all s in $\mathfrak{S}$. Q.E.D.

It will at times be convenient to consider $\mathfrak{A}$ as a subalgebra of $\mathfrak{A}^p$. This mapping of $\mathfrak{A}$ into $\mathfrak{A}^p$ is done as follows.

10 DEFINITION. For each operator b on $\mathfrak{H}$ and vector $x = [x_1, \ldots, x_p]$ in $\mathfrak{H}^p$ we define

$$bx = [bx_1, \ldots, bx_p].$$

Under this mapping the algebra $\mathfrak{A}$ is clearly *-isomorphic to the B^*-subalgebra of $\mathfrak{A}^p$ consisting of all operators having the diagonal form

$$A = \begin{pmatrix} a & \cdots & 0 \\ \vdots & & \vdots \\ 0 & \cdots & a \end{pmatrix}$$

with all diagonal elements the same element in A. It is also clear (without necessarily referring to Lemma 9.1) that this mapping of $\mathfrak{A}$ into $\mathfrak{A}^p$ is an isometry. When no confusion seems likely we shall use the same symbol a for an element in $\mathfrak{A}$ as well as for its correspondent in $\mathfrak{A}^p$. It is in this sense that we interpret expressions such as aA and Aa with a in $\mathfrak{A}$ and A in $\mathfrak{A}^p$. If $A = (a_{ij})$ and if a is in $\mathfrak{A}$, then $aA = (aa_{ij})$, $aA = Aa$, and $\widehat{Aa} = \hat{A}\hat{a}$, that is,

$$\left(\int_{\mathfrak{S}} \hat{a}(s)e(ds)\right)\left(\int_{\mathfrak{S}} \hat{A}(s)e(ds)\right) = \int_{\mathfrak{S}} \hat{a}(s)\hat{A}(s)e(ds).$$

11 COROLLARY. *Let S and N be the scalar and radical parts, respectively, of the spectral operator A in $\mathfrak{A}^p$. For each $i = 1, \ldots, p$ let $\mathfrak{S}_i$ be the set of all s in $\mathfrak{S}$ where $\hat{A}(s)$ has i distinct eigenvalues. For those i for which $\mathfrak{S}_i$ is not void, let $\lambda_{i1}(s), \ldots, \lambda_{ii}(s)$ be the i distinct eigenvalues of $\hat{A}(s)$ chosen in*

such a way that the functions $\hat{\lambda}_{ij}$ are Σ-measurable on $\mathfrak{S}_i$ (see the remark after Lemma 3). For other values of i let $\hat{\lambda}_{ij}(s) = 0$. Let the operators λ_{ij} and E_{ij} be defined, for $1 \leqq j \leqq i \leqq p$, by the equations

$$\text{(i)} \quad \lambda_{ij} = \int_{\mathfrak{S}_i} \hat{\lambda}_{ij}(s) e(ds), \qquad E_{ij} = \int_{\mathfrak{S}_i} E(\hat{\lambda}_{ij}(s); \hat{A}(s)) e(ds).$$

Then $\lambda_{ij} \in \mathfrak{A}$, $E_{ij} \in \mathfrak{A}^p$, $E_{ij}^2 = E_{ij}$, and $E_{ij} E_{mn} = 0$ unless $i = m$ and $j = n$. Furthermore, these operators λ_{ij} and E_{ij} commute with A, S, N, and the resolution of the identity $E(\sigma; A)$. For every bounded Borel function φ on $\sigma(A)$ we have

$$\text{(ii)} \qquad \varphi(S) = \sum_{i=1}^{p} \sum_{j=1}^{i} \varphi(\lambda_{ij}) E_{ij}.$$

If φ is analytic and single valued in a neighborhood of the spectrum, then

$$\text{(iii)} \qquad \varphi(A) = \sum_{\nu=1}^{p-1} \frac{N^\nu}{\nu!} \sum_{i=1}^{p} \sum_{j=1}^{i} \varphi^{(\nu)}(\lambda_{ij}) E_{ij}.$$

Proof. We observe first that for e-almost all s in $\mathfrak{S}$, $|\hat{\lambda}_{ij}(s)| \leqq |\hat{A}(s)| \leqq |\hat{A}| = |A|$, so that the functions $\hat{\lambda}_{ij}$ are bounded e-almost everywhere on $\mathfrak{S}_i$. Lemma 3 shows that they are Σ-measurable and thus λ_{ij} is in $\mathfrak{A}$. Since A is a spectral operator, Theorem 6 and Lemma 4 show that $E(\hat{\lambda}_{ij}(s); \hat{A}(s))$ is e-essentially bounded and Lemma 3 shows that it is Σ-measurable. Thus the operator E_{ij} is in $\mathfrak{A}^p$. Now Theorem 6(ii) and Lemma 3(iii) show that

$$\begin{aligned}
E(\sigma; A) &= \int_{\mathfrak{S}} E(\sigma; \hat{A}(s)) e(ds) = \sum_{i=1}^{p} \int_{\mathfrak{S}_i} E(\sigma; \hat{A}(s)) e(ds) \\
&= \sum_{i=1}^{p} \int_{\mathfrak{S}_i} \sum_{j=1}^{i} \chi_\sigma(\hat{\lambda}_{ij}(s)) E(\hat{\lambda}_{ij}(s); \hat{A}(s)) e(ds) \\
&= \sum_{i=1}^{p} \sum_{j=1}^{i} \left[\int_{\mathfrak{S}_i} \chi_\sigma(\hat{\lambda}_{ij}(s)) e(ds)\right] \left[\int_{\mathfrak{S}_i} E(\hat{\lambda}_{ij}(s); \hat{A}(s)) e(ds)\right] \\
&= \sum_{i=1}^{p} \sum_{j=1}^{i} e(\hat{\lambda}_{ij}^{-1}(\sigma)) \int_{\mathfrak{S}_i} E(\hat{\lambda}_{ij}(s); \hat{A}(s)) e(ds) \\
&= \sum_{i=1}^{p} \sum_{j=1}^{i} E(\sigma; \lambda_{ij}) E_{ij},
\end{aligned}$$

where, in arriving at this last equation, we have used Corollary X.2.10.

Thus we have

$$\begin{aligned}\varphi(S) &= \int_{\sigma(A)} \varphi(\xi)E(d\xi; A) \\ &= \sum_{i=1}^{p} \sum_{j=1}^{i} \int_{\sigma(A)} \varphi(\xi)E(d\xi; \lambda_{ij})E_{ij} \\ &= \sum_{i=1}^{p} \sum_{j=1}^{i} \varphi(\lambda_{ij})E_{ij}.\end{aligned}$$

By Corollary 9, $\hat{S}(s)$ and $\hat{N}(s)$ are, for e-almost all s in $\mathfrak{S}$, the scalar and radical parts of $\hat{A}(s)$, and so $E(\hat{\lambda}_{ij}(s); \hat{A}(s))$ commutes with $\hat{A}(s)$, $\hat{S}(s)$, $\hat{N}(s)$, and $E(\sigma; \hat{A}(s))$. Thus E_{ij} commutes with A, S, N, $E(\sigma; A)$. The element λ_{ij} also commutes with these operators, since it is in $\mathfrak{A}$. Now, using the homomorphic property of the map $A \leftrightarrow \hat{A}$, it is seen that

$$E_{ij}E_{mk} = \int_{\mathfrak{S}_i \cap \mathfrak{S}_m} E(\hat{\lambda}_{ij}(s); \hat{A}(s))E(\hat{\lambda}_{mk}(s); \hat{A}(s))e(ds),$$

which shows that $E_{ij}E_{mk} = 0$ if $i \neq m$, that $E_{ij}E_{ik} = 0$ if $j \neq k$, and that $E_{ij}^2 = E_{ij}$.

Theorem 6 shows that $N^p = 0$, and thus the final conclusion follows from Theorem 5.1 and equation (ii) of the present corollary. Q.E.D.

12 COROLLARY. *Every spectral operator A in $\mathfrak{A}^p$ determines a decomposition of $\mathfrak{H}^p$ into a finite number of subspaces, each invariant under A, and such that the restriction of A to each subspace consists of multiplication by an element of $\mathfrak{A}$ plus a nilpotent operator of order at most p.*

PROOF. Using the notations and results of the preceding corollary, it is seen, by taking $\varphi(\lambda) = 1$ in (ii), that $\mathfrak{H}^p$ is the direct sum of its subspaces $E_{ij}\mathfrak{H}^p$ and that each of these subspaces is invariant under A. On $E_{ij}\mathfrak{H}^p$ we see, by taking $\varphi(\lambda) = \lambda$ in (ii), that

$$Ax = \lambda_{ij}x + Nx, \qquad x \in E_{ij}\mathfrak{H}^p,$$

where $N^p = 0$ by Theorem 6. Q.E.D.

13 COROLLARY. *Let $\mathfrak{S}$ be a compact space and, in the notation of Corollary* 11, *let the sets $\mathfrak{S}_i$ be closed and the roots $\hat{\lambda}_{ij}(s)$, $j = 1, \ldots, i$, be continuous on $\mathfrak{S}_i$. Then A is a spectral operator of type $p-1$.*

PROOF. If $\mathfrak{S}_i$ is not void, the function $|\hat{\lambda}_{ij}(s) - \hat{\lambda}_{ik}(s)|^{-1}$ is, for $1 \leqq j \leqq k \leqq i$, defined and continuous on $\mathfrak{S}_i$. Since $\mathfrak{S}_i$ is closed, it is compact and this function is also bounded on $\mathfrak{S}_i$ by some constant K which we take to be independent of the integer $i = 1, \ldots, p$. Since every s belongs to one of the sets $\mathfrak{S}_i$, we have $|\lambda - \mu|^{-1} \leqq K$ for every pair λ, μ of distinct eigenvalues of $\hat{A}(s)$, and Corollary 7 shows that A is a spectral operator of type $p-1$. Q.E.D.

14 COROLLARY. *Let A be in $\mathfrak{A}^p$ and let $\mathfrak{S}$ be a compact space. Suppose that, for each s in $\mathfrak{S}$, there are p distinct eigenvalues $\hat{\lambda}_1(s), \ldots, \hat{\lambda}_p(s)$ of the matrix and that these functions $\hat{\lambda}_i(s)$ are continuous. Then A is a spectral operator of scalar type.*

PROOF. The argument of the preceding corollary shows that A is a spectral operator. Since $\hat{A}(s)$ has distinct eigenvalues, it is a scalar operator, that is, its radical part is zero. Thus Corollary 9 shows that A is also a scalar type operator. Q.E.D.

11. Some Examples of Bounded Spectral Operators

In this section some of the preceding results will be illustrated by considering the algebra $\mathfrak{A}$ of all convolution operators in $\mathfrak{H} = L_2(R^N)$, the Hilbert space of square integrable functions on real Euclidean space R^N of N dimensions. We shall show that this algebra of *all* convolutions defined on $\mathfrak{H}$ regardless of whether they be defined in terms of bounded additive set functions, countably additive set functions, ordinary Lebesgue integrals, proper value integrals or any other improper integral, is a B^*-algebra which is *-equivalent to the B^*-algebra $\hat{\mathfrak{A}} = L_\infty(R^N, \Sigma, ds)$ where Σ is the σ-field of Lebesgue measurable sets and ds is Lebesgue measure. The operators in the non-commutative B^*-algebra $\mathfrak{A}^p$ are then the operators A in $\mathfrak{H}^p = L_2(R^N) + \cdots + L_2(R^N)$ whose matrix representation $A = (a_{jk})$ consists of convolution operators a_{jk} in $\mathfrak{H}$. We sometimes find it convenient to consider the basic space $\mathfrak{S}$ of the preceding section as the one point compactification of R^N by the addition of the single point ∞. Lebesgue measure is extended to $\mathfrak{S}$ by letting the set $\{\infty\}$ have measure zero so that the measure spaces (R^N, Σ, ds) and $(\mathfrak{S}, \Sigma, ds)$ are the same. The algebra $\hat{\mathfrak{C}}$ consists of all $\hat{a}$ which are continuous on $\mathfrak{S}$, that is, all continuous complex functions on R^N for which the limit $\hat{a}(\infty) = \lim_{|s|\to\infty} \hat{a}(s)$ exists. The set $\hat{\mathfrak{C}}_0$ is the subalgebra (without unit) of $\hat{\mathfrak{C}}$ for which $\hat{a}(\infty) = 0$.

The examples to be discussed here and in the following section are of two types: bounded spectral operators in $\mathfrak{A}^p$ and a closely related type of unbounded spectral operator associated with certain linear systems of partial differential equations. A few properties of the Fourier transform will be essential in the discussion of both types of problems, and we begin by developing these properties. The discussion will be independent of the theory of the Fourier transform in locally compact groups as developed in XI.3, for in the case of N-dimensional Euclidean space, a more elementary approach and one better suited to our present needs is possible.

For an ordered set $\alpha = [\alpha_1, \ldots, \alpha_N]$ of N non-negative integers, points $s = [s_1, \ldots, s_N]$ and $t = [t_1, \ldots, t_N]$ in R^N, a polynomial $P(s_1, \ldots, s_N)$ in N variables, and a function φ in $C^\infty = C^\infty(R^N)$, we use the notation

$$|\alpha|_1 = \alpha_1 + \cdots + \alpha_N, \qquad |s| = (|s_1|^2 + \cdots + |s_2|^2)^{1/2},$$
$$s^\alpha = s_1^{\alpha 1} s_2^{\alpha 2} \cdots s_N^{\alpha N}, \qquad P(s) = P(s_1, \ldots, s_N),$$
$$st = (s, t) = s_1 t_1 + \cdots + s_N t_N,$$

$$\varphi^{(\alpha)} = \partial_s^\alpha \varphi = \left(\frac{\partial}{\partial s}\right)^\alpha \varphi = \left(\frac{\partial}{\partial s_1}\right)^{\alpha 1} \cdots \left(\frac{\partial}{\partial s_N}\right)^{\alpha N} \varphi, \qquad \left(\frac{\partial}{\partial s_j}\right)^0 \varphi = \varphi,$$

$$P(\partial) = P(\partial_s) = P\left(\frac{\partial}{\partial s}\right) = P\left(\frac{\partial}{\partial s_1}, \ldots, \frac{\partial}{\partial s_N}\right).$$

The symbol Φ will be used for the set of those functions in C^∞ which, together with all derivatives of all orders, approach zero, as $|s| \to \infty$, faster than any power of $1/|s|$. The functions in Φ are called *rapidly decreasing* functions on R^N. The set Φ is clearly a linear subset of all the Lebesgue spaces $L_p = L_p(R^N)$, $1 \leqq p \leqq \infty$, and, for φ in Φ, the symbol $|\varphi|_p$ is used for the norm of φ as an element of L_p. Besides having topologies as a subset of various B-spaces, the set Φ will be given a topology of its own by prescribing a set of basic neighborhoods of the origin as follows: for every pair p, q of positive integers and every real $\varepsilon > 0$, let $U(p, q, \varepsilon)$ consist of all φ in Φ for which

$$(1) \qquad (1 + |s|^2)^p |\varphi^{(\alpha)}(s)| < \varepsilon, \qquad s \in R^N,$$

for every α with $|\alpha|_1 \leqq q$. Elementary considerations show that Φ is locally convex and complete in the sense that a sequence $\{\varphi_m\}$ for which $\varphi_p - \varphi_q \to 0$ as $p \to \infty$, $q \to \infty$ has a limit in Φ. Besides being a linear space,

Φ clearly has the property that it contains every derivative of every one of its elements. Also, if f is a complex C^∞ function on R^N which is *slowly increasing* in the sense that every $\alpha = [\alpha_1, \ldots, \alpha_N]$ determines an integer $m > 0$ and a constant C such that

$$|f^{(\alpha)}(s)| \leqq C(1 + |s|^2)^m, \qquad s \in R^N, \tag{2}$$

then for every φ in Φ we have also $f\varphi$ in Φ. Thus, in particular, Φ contains, together with φ, every product $P\varphi$ where P is a polynomial in $s_1, \ldots, s_N$. It is clear that convergence $\varphi_m \to \varphi$ in Φ is equivalent to the assertion that for every pair P, Q of polynomials in N variables we have

$$P(s)Q\left(\frac{\partial}{\partial s}\right)\varphi_m(s) \to P(s)Q\left(\frac{\partial}{\partial s}\right)\varphi(s), \tag{3}$$

uniformly for s in R^N.

The *Fourier transform* $F\varphi$ of a function φ in Φ is given by the integral

$$(F\varphi)(s) = \frac{1}{(2\pi)^{N/2}} \int_{R^N} e^{-ist}\varphi(t)\, dt, \qquad s \in R^N. \tag{4}$$

The basic *inversion formula* for this integral is due to Fourier and is contained in the following theorem.

→ 1 THEOREM. *The Fourier transform is a homeomorphic isomorphism of Φ onto all of itself whose inverse is*

$$\varphi(s) = \frac{1}{(2\pi)^{N/2}} \int_{R^N} e^{ist}(F\varphi)(t)\, dt, \qquad s \in R^N. \tag{5}$$

PROOF. Since φ is rapidly decreasing, the integral (4) may be differentiated under the integral sign to give

$$(F\varphi)^{(\alpha)}(s) = \frac{1}{(2\pi)^{N/2}} \int_{R^N} e^{-ist}(-it)^\alpha\varphi(t)\, dt,$$

which shows that $F\varphi$ is in C^∞. Thus, for any polynomial P in N variables,

$$(P(\partial)F\varphi)(s) = \frac{1}{(2\pi)^{N/2}} \int_{R^N} e^{-ist}P(-it)\varphi(t)\, dt. \tag{6}$$

If we use the fact that $\varphi^{(\alpha)}(s) \to 0$ as $|s| \to \infty$, then an integration by parts of the Fourier transform of $\varphi^{(\alpha)}$ gives

$$F(\varphi^{(\alpha)})(s) = (is)^\alpha(F\varphi)(s),$$

and thus, for a general polynomial Q,

$$F(Q(\partial)\varphi)(s) = Q(is)F(\varphi)(s). \tag{7}$$

If P and Q are two polynomials, then (6) and (7) may be combined and written as

$$Q(is)P\left(\frac{\partial}{\partial s}\right)(F\varphi)(s) = \frac{1}{(2\pi)^{N/2}}\int_{R^N} e^{-ist}Q\left(\frac{\partial}{\partial t}\right)\{P(-it)\varphi(t)\}\,dt. \tag{8}$$

Since $Q(\partial/\partial t)\{P(-it)\varphi(t)\}$ is Lebesgue integrable on R^N, equation (8) shows that $F\varphi$ is rapidly decreasing and thus that F maps Φ into Φ. Equation (8) also shows that F is a continuous map of Φ into Φ, for if $\varphi_n \to 0$ in Φ then the sequence $Q(\partial/\partial t)\{P(-it)\varphi_n(t)\}$ also approaches zero in Φ, and for every $\varepsilon > 0$

$$\left|Q\left(\frac{\partial}{\partial t}\right) P(-it)\varphi_n(t)\right| \leqq \frac{\varepsilon}{(1+|t|^2)^N}, \qquad t \in R^N,$$

for all sufficiently large integers n. It thus follows from (8) that $Q(is)P(\partial/\partial s)F(\varphi_n)(s)$ approaches zero uniformly on R^N, which proves that $F\varphi(_n) \to 0$ in Φ. Hence, F is a continuous map of Φ into Φ.

We next establish the inversion formula (5). Once this is done, it will follow that F is one-to-one on Φ and that $\varphi(s) = (F^2\varphi)(-s)$, which proves that $F\Phi = \Phi$ and thus that the inverse $(F^{-1}\varphi)(s) = (F^2\varphi)(-s)$ is everywhere defined and continuous on Φ. Hence, to complete the proof of the theorem it suffices to establish the inversion formula (5).

By considering the integral as an iterated integral, it is seen that it suffices to establish (5) in the case $N = 1$. In this case the right side of (5) is

$$\begin{aligned}(F^2\varphi)(-s) &= \frac{1}{2\pi}\int_{-\infty}^{\infty} e^{ist}\left\{\int_{-\infty}^{\infty} e^{-itu}\varphi(u)\,du\right\}dt \\ &= \lim_{\alpha\to\infty}\frac{1}{2\pi}\int_{-\infty}^{\infty} e^{ist}\left\{\int_{-\alpha}^{\alpha} e^{-itu}\varphi(u)\,du\right\}dt \\ &= \lim_{\alpha\to\infty}\frac{1}{\pi}\int_{-\infty}^{\infty}\frac{\sin\alpha(s-u)}{s-u}\varphi(u)\,du \\ &= \lim_{\alpha\to\infty}\frac{1}{\pi}\int_{-\infty}^{\infty}\varphi(s-t)\frac{\sin\alpha t}{t}\,dt,\end{aligned}$$

and so equation (5) may be established by proving the following lemma.

2 LEMMA. *Let φ be a complex valued function in $L_1(-\infty, \infty)$, continuous at a point s, and of bounded variation in some neighborhood of s. Then*

$$\varphi(s) = \lim_{\alpha\to\infty} \frac{1}{\pi}\int_{-\infty}^{\infty} \varphi(s-t)\,\frac{\sin \alpha t}{t}\,dt. \tag{9}$$

PROOF. To prove the lemma we first observe that for every $\delta > 0$ the function $\varphi(s-t)/t$ is integrable on the infinite intervals $[-\infty, -\delta]$, $[\delta, \infty]$ and so

$$\lim_{\alpha\to\infty} \int_{-\infty}^{-\delta} \varphi(s-t)\,\frac{\sin \alpha t}{t}\,dt = 0 = \lim_{\alpha\to\infty} \int_{\delta}^{\infty} \varphi(s-t)\,\frac{\sin \alpha t}{t}\,dt.$$

It is therefore sufficient to show that, for some $\delta > 0$,

$$\varphi(s) = \lim_{\alpha\to\infty} \frac{1}{\pi}\int_{-\delta}^{\delta} \varphi(s-t)\,\frac{\sin \alpha t}{t}\,dt, \tag{10}$$

and, by using the real and imaginary parts of φ separately, we may assume that φ is real valued. Now, by assumption, $\varphi(u)$ is continuous at s and of bounded variation for u in some closed interval $s-\delta \leqq u \leqq s+\delta$ about s, and we may choose $\delta > 0$ so small that

$$|\varphi(s-t) - \varphi(s)| < \varepsilon, \qquad |t| \leqq \delta, \tag{11}$$

where ε is an arbitrary positive number. Since φ is of bounded variation. it is the difference of two non-decreasing functions, and so it suffices to prove (10) under further restriction that φ is non-decreasing. We have, by the second theorem of the mean for Riemann integrals, a number β in $[-\delta, \delta]$ with

$$\begin{aligned}
&\frac{1}{\pi}\int_{-\delta}^{\delta} \varphi(s-t)\,\frac{\sin \alpha t}{t}\,dt \\
&= \varphi(s)\,\frac{1}{\pi}\int_{-\delta}^{\delta} \frac{\sin \alpha t}{t}\,dt - \frac{1}{\pi}\int_{-\delta}^{\delta} \{\varphi(s) - \varphi(s-t)\}\,\frac{\sin \alpha t}{t}\,dt \\
&= \varphi(s)\,\frac{1}{\pi}\int_{-\delta\alpha}^{\delta\alpha} \frac{\sin u}{u}\,du - \{\varphi(s) - \varphi(s-\delta)\}\int_{\beta}^{\delta} \frac{\sin \alpha t}{t}\,dt \\
&= \varphi(s)\,\frac{1}{\pi}\int_{-\delta\alpha}^{\delta\alpha} \frac{\sin u}{u}\,du - \{\varphi(s) - \varphi(s-\delta)\}\int_{\alpha\beta}^{\alpha\delta} \frac{\sin u}{u}\,du.
\end{aligned} \tag{12}$$

Now

$$\lim_{\alpha\to\infty} \frac{1}{\pi}\int_{-\delta\alpha}^{\delta\alpha} \frac{\sin u}{u}\,du = 1 \tag{13}$$

and

$$\left|\int_{\alpha\beta}^{\alpha\delta} \frac{\sin u}{u}\,du\right| \leqq C, \tag{14}$$

where the constant C is independent of α and β. Thus (11), ..., (14) show that

$$\limsup_{\alpha\to\infty} \left|\varphi(s) - \frac{1}{\pi}\int_{-\delta}^{\delta} \varphi(s-t)\,\frac{\sin \alpha t}{t}\,dt\right| \leqq \varepsilon C,$$

which, since $\varepsilon > 0$ is arbitrary, proves (10) and completes the proof of Lemma 2 as well as Theorem 1. Q.E.D.

→ 3 COROLLARY (*Parseval-Plancherel*). *The set Φ of rapidly decreasing functions is dense in the Hilbert space $\mathfrak{H} = L_2(R^N)$, and the Fourier transform as defined on Φ by the equation* (4) *has a unique extension to a unitary operator F in $\mathfrak{H}$ with the properties*

(i) $(F\varphi, \psi) = (\varphi, F^{-1}\psi), \qquad \varphi, \psi \in \mathfrak{H};$

(ii) $|F\varphi| = |\varphi|, \qquad \varphi \in \mathfrak{H};$

(iii) $(F\varphi)(s) = \underset{r\to\infty}{\text{l.i.m.}}\ \dfrac{1}{(2\pi)^{N/2}}\displaystyle\int_{|t|\leqq r} e^{-ist}\varphi(t)\,dt, \qquad \varphi \in \mathfrak{H};$

(iv) $(F^{-1}\varphi)(s) = \underset{r\to\infty}{\text{l.i.m.}}\ \dfrac{1}{(2\pi)^{N/2}}\displaystyle\int_{|t|\leqq r} e^{ist}\varphi(t)\,dt, \qquad \varphi \in \mathfrak{H};$

where the notation l.i.m. *is used for the limit in the mean of order* 2, *that is, the limit in the norm of* $\mathfrak{H}$.

PROOF. The density of Φ in $\mathfrak{H}$ is a corollary of XIV.2.3. It follows from Theorem 1 and the Fubini theorem that, for φ and ψ in Φ,

$$\begin{aligned}(F\varphi, \psi) &= \frac{1}{(2\pi)^{N/2}}\int_{R^N}\left(\int_{R^N} e^{-ist}\varphi(t)\,dt\right)\overline{\psi(s)}\,ds \\ &= \frac{1}{(2\pi)^{N/2}}\int_{R^N}\varphi(t)\,\overline{\left(\int_{R^N} e^{ist}\psi(s)\,ds\right)}\,dt \\ &= (\varphi, F^{-1}\psi),\end{aligned}$$

and, by placing $\psi = F\varphi$, that $|F\varphi| = |\varphi|$, which proves that F is continuous in the $\mathfrak{H}$ topology of Φ. Since Φ is dense in $\mathfrak{H}$, F has a continuous extension to $\mathfrak{H}$ which clearly satisfies (i) and (ii) and is thus a unitary operator. To prove (iii) we let $\varphi_r(t) = \varphi(t)$ for $|t| \leqq r$ and $\varphi_r(t) = 0$ for $|t| > r$. Then, since Φ is dense in $\mathfrak{H}$ and convergence in the mean of order 2 on the set of t with $|t| \leqq r$ implies convergence in L_1 on this set, we see that $F\varphi_r$ is given by equation (4), that is,

$$(F\varphi_r)(s) = \frac{1}{(2\pi)^{N/2}} \int_{|t| \leqq r} e^{-ist}\varphi(t)\, dt.$$

Since $\varphi_r \to \varphi$ as $r \to \infty$, equation (iii) follows from the continuity of F and (iv) is proved similarly. Q.E.D.

→ REMARK. As the proof shows, we also have

$$F\varphi = \underset{n\to\infty}{\text{l.i.m.}} \frac{1}{(2\pi)^{N/2}} \int_{K_n} e^{-ist}\varphi(t)\, dt,$$

where $\{K_n\}$ is any increasing sequence of sets of finite measure whose union differs from $\mathfrak{S}$ by a set of measure zero.

The basic spectral measure e on Σ which is used to define the algebras $\mathfrak{A}$ and $\mathfrak{A}^p$ is defined in terms of the Fourier transform F on $\mathfrak{H}$ by the equation

$$(15) \qquad e(\sigma) = F^{-1}\mu(\sigma)F, \qquad \sigma \in \Sigma,$$

where $\mu(\sigma)$ is the multiplication projection defined as

$$(16) \qquad \begin{aligned} (\mu(\sigma)\varphi)(s) &= \varphi(s), & s &\in \sigma, \\ &= 0, & s &\notin \sigma. \end{aligned}$$

It is clear that μ is a countably additive self adjoint spectral measure in $\mathfrak{H}$, and thus, since F is unitary, the measure $e(\sigma)$, $\sigma \in \Sigma$, is a spectral measure with these same properties. We also have $e(\sigma) = 0$ if and only if σ has Lebesgue measure zero so that the notions of e-almost everywhere, e-essential boundedness, and so on, are the same as the corresponding notions referring to Lebesgue measure. We shall therefore usually omit the reference to e in such expressions. Thus the algebra $\mathfrak{A}$ consists of all operators a in $B(\mathfrak{H})$ which have the form

$$(17) \qquad a = \int_{\mathfrak{S}} \hat{a}(s)e(ds),$$

for some Σ-measurable and essentially bounded complex valued function $\hat{a}$ on R^N.

Before illustrating the results of the preceding section, we shall examine the structure of these operators in $\mathfrak{A}$ and in particular show that many of the convolution and singular convolution operators, which are of frequent occurrence in analysis, belong to $\mathfrak{A}$. Convolutions have, in recent years, enjoyed a renewed interest largely because the distribution theory of L. Schwartz [5] has shown their increased usefulness in the theory of partial differential equations. Here, we shall first be concerned with certain special examples of convolutions which map $\mathfrak{H}$ into $\mathfrak{H}$, which belong to the algebra $\mathfrak{A}$, and which have an integral representation in one of the two forms

$$(f * \varphi)(s) = \int_{R^N} \varphi(s-t) f(t)\, dt, \qquad \varphi \in \mathfrak{H}, \tag{18}$$

or

$$(\lambda * \varphi)(s) = \int_{R^N} \varphi(s-t) \lambda(dt), \qquad \varphi \in \mathfrak{H}, \tag{19}$$

determined by a complex valued function f on R^N or a complex valued set function λ defined on a family $\Sigma(\lambda)$ of sets in R^N. The representation (18) or (19) of a given convolution operator depends upon the interpretation of the integral, that is, whether the integral is an ordinary Lebesgue integral defined for almost all s, a Cauchy type principal value integral defined as a certain limit of Lebesgue integrals, an integral of a vector valued function, or a principal value integral of a vector valued function. For a given φ in $\mathfrak{H}$ and a point t in R^N the translate φ_t in $\mathfrak{H}$, defined as $\varphi_t(s) = \varphi(s-t)$, when regarded as a map $t \rightarrow \varphi_t$ of R^N into $\mathfrak{H}$, is everywhere continuous (XI.3.1(f)) and, since $|\varphi_t| = |\varphi|$, bounded on R^N. This elementary observation suggests that it might be easier to give certain convolutions an integral representation if we use the integral of vector valued functions, and this we shall do.

We begin by defining the convolution integral (19) determined by a complex valued finitely additive and bounded set function λ defined on a field $\Sigma(\lambda)$ which contains the open sets in R^N. Since such a function λ has bounded variation, the integral

$$\int_{R^N} f(t) \lambda(dt) \tag{20}$$

exists for every vector valued function f on R^N which is bounded and λ-measurable. In particular, if f is continuous and has its values in a compact set, it is λ-measurable and thus integrable. For, in this case, the range of f may be covered with a finite number of open sets $G_1, \ldots, G_n$, each having diameter less than a given positive number ε and, if we let

$$A_1 = G_1, \qquad A_j = G_j - \bigcup_{k=1}^{j-1} G_k, \qquad j = 2, \ldots, n,$$

then the sets $B_j = f^{-1}(A_j)$ are in $\Sigma(\lambda)$. Thus the finitely valued function

$$f_\varepsilon = \sum_{j=1}^{n} x_j \chi_{B_j},$$

where x_j is any point of A_j if A_j is not void, and otherwise $x_j = 0$, is λ-measurable and

$$|f_\varepsilon(s) - f(s)| < \varepsilon, \qquad s \in R^N,$$

which proves much more than the λ-measurability of f. In particular, a bounded continuous complex valued function on R^N is λ-integrable. This observation enables us to define the integral

$$(21) \qquad \tilde{\lambda}(s) = \int_{R^N} e^{-ist}\lambda(dt), \qquad s \in R^N,$$

which will be used presently. The translation φ_t, however, is not compact valued, and we proceed as follows. For an $\mathfrak{H}$ valued function f, bounded and continuous on R^N, and a functional x^* in $\mathfrak{H}^*$, the preceding remarks show that x^*f is λ-integrable and since

$$\left|\int_{R^N} x^*f(t)\lambda(dt)\right| \leqq |x^*| \sup_t |f(t)|\, v(\lambda; R^N),$$

the integral $\int x^*f(t)\lambda(dt)$ is continuous in x^*. Since $\mathfrak{H}$ is reflexive, there is a unique point in $\mathfrak{H}$, which we designate by the symbol $\int_{R^N} f(t)\lambda(dt)$, such that

$$(22) \qquad x^*\int_{R^N} f(t)\lambda(dt) = \int_{R^N} x^*f(t)\lambda(dt).$$

The integral (20) is thus defined for bounded continuous $\mathfrak{H}$ valued functions on R^N and, in particular, if $f(t) = \varphi_t$, for some φ in $\mathfrak{H}$. The integral in (19) is defined to be the integral (20). It is clear that if for almost all s

the function $\varphi(s-t)$ is λ-integrable in t, then the two meanings assigned to the integral in (19) coincide almost everywhere and thus define the same point in $\mathfrak{H}$. Thus every bounded finitely additive complex valued set function λ defined on a field containing the open sets in R^N defines a convolution operator in $\mathfrak{H}$ by means of equation (19).

Let T be any continuous linear map in $\mathfrak{H}$, let f be a bounded continuous $\mathfrak{H}$ valued function on R^N, and let x^* be in $\mathfrak{H}^*$. Then (22) shows that

$$x^*T\int_{R^N} f(t)\lambda(dt) = \int_{R^N} x^*Tf(t)\lambda(dt)$$

$$= x^*\int_{R^N} Tf(t)\lambda(dt),$$

and thus that

$$T\int_{R^N} f(t)\lambda(dt) = \int_{R^N} Tf(t)\lambda(dt). \tag{23}$$

If we let $\tau = (2\pi)^{N/2}F$, where F is the Fourier transform in $\mathfrak{H}$, and set $\tau\varphi = \tilde{\varphi}$, then (23) gives

$$\tau(\lambda * \varphi) = \int_{R^N} \tau\varphi_t\,\lambda(dt),$$

and, for almost all s in R^N, this function has the value

$$\begin{aligned} \tau(\lambda * \varphi)(s) &= \int_{R^N}\left\{\underset{r\to\infty}{\text{l.i.m.}}\int_{|u|\leq r} e^{-isu}\varphi(u-t)\,du\right\}\lambda(dt) \\ &= \int_{R^N}\left\{\underset{r\to\infty}{\text{l.i.m.}}\int_{|u|\leq r} e^{-is(u-t)}\varphi(u-t)\,du\right\}e^{-ist}\lambda(dt) \\ &= \tilde{\lambda}(s)\tilde{\varphi}(s), \end{aligned} \tag{24}$$

where in this last step we have used the remark following the proof of Corollary 3. Since $\tau(\lambda * \varphi)$ and $\tilde{\varphi}$ are both Lebesgue measurable functions and $\tilde{\varphi}$ is an arbitrary function in $\mathfrak{H}$, this equation (24) shows that $\tilde{\lambda}$ is Lebesgue measurable also. If we write $\tilde{\lambda}$ for the operation in $\mathfrak{H}$ of multiplication by $\tilde{\lambda}$, that is, $(\tilde{\lambda}\varphi)(s) = \tilde{\lambda}(s)\varphi(s)$, then (24) may be written as

$$\lambda * \varphi = \tau^{-1}\tilde{\lambda}\tau\varphi, \qquad \varphi \in \mathfrak{H}. \tag{25}$$

We shall sometimes write this operation of convolution in still a different form. Note that the operation $\varphi \to \hat{a}\varphi$ of multiplication by a function $\hat{a}$

in $\hat{\mathfrak{A}}$ may be written in terms of the measure μ of equation (16) as

$$\hat{a}\varphi = \int_{\mathfrak{S}} \hat{a}(s)\mu(ds)\varphi, \qquad \varphi \in \mathfrak{H}. \tag{26}$$

This is clear if $\hat{a}$ is the characteristic function of a set in Σ and, since linear combinations of such functions are dense in $\hat{\mathfrak{A}}$, it holds for all $\hat{a}$ in $\hat{\mathfrak{A}}$. Thus in view of (26) the equation (24) may also be written as

$$\tau(\lambda * \varphi) = \int_{\mathfrak{S}} \tilde{\lambda}(s)\mu(ds)\tau\varphi.$$

By the Plancherel theorem (Corollary 3) the operator τ^{-1} exists and is continuous on $\mathfrak{H}$, and since $e(\sigma) = F^{-1}\mu(\sigma)F = \tau^{-1}\mu(\sigma)\tau$, the preceding equation gives

$$\lambda * \varphi = \int_{\mathfrak{S}} \tilde{\lambda}(s)e(ds)\varphi, \qquad \varphi \in \mathfrak{H}. \tag{27}$$

This equation shows that the integral $\int \varphi_t \lambda(dt)$ of the vector valued function φ_t with respect to the *finitely additive* set function λ is the same as the integral of the numerical function $\tilde{\lambda}$ with respect to the *countably additive* vector valued measure $e(\sigma)\varphi$. To summarize, we have the following.

4 THEOREM. *Every bounded additive complex valued set function λ defined on a field containing the open sets in R^N determines a convolution integral*

$$\lambda * \varphi = \int_{R^N} \varphi_t \lambda(dt), \qquad \varphi \in \mathfrak{H},$$

where $\varphi_t(s) = \varphi(s-t)$, and the map $\boldsymbol{\lambda}: \varphi \to \lambda * \varphi$ maps $\mathfrak{H}$ into $\mathfrak{H}$, belongs to $\mathfrak{A}$, and has

$$\boldsymbol{\lambda} = \int_{\mathfrak{S}} \tilde{\lambda}(s)e(ds), \qquad |\boldsymbol{\lambda}| = \operatorname*{ess\,sup}_{s \in \mathfrak{S}} |\tilde{\lambda}(s)|,$$

where $\tilde{\lambda}$ is given by (21). *Stated otherwise, we have $\boldsymbol{\lambda} = F^{-1}\tilde{\lambda}F$ where $\tilde{\lambda}$ is the operation of multiplication by $\tilde{\lambda}$ and F is the Fourier transform in $\mathfrak{H}$.*

Many convolution operators occur as a limit

$$\boldsymbol{\lambda}\varphi = \lim_n \lambda_n * \varphi, \qquad \varphi \in \mathfrak{H}, \tag{28}$$

of a sequence of convolutions determined by bounded finitely additive set functions. The Cauchy principal value integrals used in defining the

Hilbert transform and the singular convolutions of Calderón-Zygmund type are of this variety. The following theorem gives a useful criterion for the existence of the limit (28).

→ 5 THEOREM. *Let* $\{\lambda_n\}$ *be a sequence of bounded finitely additive complex valued set functions defined on the field determined by the open sets in* R^N. *Then the convolution operator* $\boldsymbol{\lambda}$ *given by equation* (28) *is everywhere defined on* $\mathfrak{H}$ *if and only if*

$$\sup_n |\tilde{\lambda}_n|_\infty < \infty, \tag{29}$$

and the limit

$$\tilde{\lambda}(s) = \lim_n \tilde{\lambda}_n(s) \tag{30}$$

exists in measure on every bounded measurable set in R^N. *When it exists,* $\boldsymbol{\lambda}$ *is in* $\mathfrak{A}$ *and*

$$\boldsymbol{\lambda} = \int_{\mathfrak{S}} \tilde{\lambda}(s)e(ds), \qquad |\boldsymbol{\lambda}| = \operatorname{ess\,sup}_s |\tilde{\lambda}(s)|. \tag{31}$$

Stated otherwise, $\boldsymbol{\lambda} = F^{-1}\tilde{\lambda}F$ *where* F *is the Fourier transform in* $\mathfrak{H}$ *and* $\tilde{\lambda}$ *is the operation of multiplication by the function* $\tilde{\lambda}$.

PROOF. Let $\boldsymbol{\lambda}_n\varphi$ converge for each φ in $\mathfrak{H}$. Theorem 4 shows that $|\boldsymbol{\lambda}_n| = |\tilde{\lambda}_n|_\infty$, and Theorem II.3.6 proves (29). Let $\varphi = F^{-1}\chi$, where χ is the characteristic function of the bounded measurable set σ. Then, by Theorem 4, $\boldsymbol{\lambda}_n\varphi = F^{-1}\tilde{\lambda}_n\chi$, and since $\boldsymbol{\lambda}_n\varphi$ converges, we see from the continuity of F that

$$\lim_{m,n\to\infty} \int_\sigma |\tilde{\lambda}_n(s) - \tilde{\lambda}_m(s)|^2\, ds = 0,$$

which implies (III.3.6) that $\tilde{\lambda}_n$ converges in measure on σ. Conversely, if (29) and (30) hold, then, since the norms $|\boldsymbol{\lambda}_n| = |\tilde{\lambda}_n|_\infty$ are bounded, to prove the convergence of $\boldsymbol{\lambda}_n\varphi$ for every φ in $\mathfrak{H}$ it suffices to prove convergence for each φ in a fundamental set. Since F is a homeomorphism in $\mathfrak{H}$, such a fundamental set consists of all $\varphi = F^{-1}\psi$ where ψ vanishes outside a bounded measurable set σ. For this φ we see from Theorem 4 that $\boldsymbol{\lambda}_n\varphi = F^{-1}\tilde{\lambda}_n\psi$ and hence $\boldsymbol{\lambda}_n\varphi$ converges in $\mathfrak{H}$ provided that

$$\lim_{m,n\to\infty} \int_\sigma |\tilde{\lambda}_n(s) - \tilde{\lambda}_m(s)|^2\, |\psi(s)|^2\, ds = 0.$$

This equation follows from (29), (30), and Theorem III.3.6 and proves that the limit (28) exists for every φ in $\mathfrak{H}$.

Now let (29) and (30) hold, let ψ be an arbitrary vector in $\mathfrak{H}$, and let $\varepsilon > 0$. Then there is a bounded measurable set σ with

$$2 \sup_n |\tilde{\lambda}_n|_\infty \int_{\sigma'} |\psi(s)|^2 \, ds < \varepsilon,$$

and so

$$\int_{R^N} |\tilde{\lambda}_n(s) - \tilde{\lambda}(s)|^2 |\psi(s)|^2 \, ds \leqq \int_{\sigma} |\tilde{\lambda}_n(s) - \tilde{\lambda}(s)|^2 \, |\psi(s)|^2 \, ds + \varepsilon,$$

and (30) proves that $\tilde{\lambda}_n \psi \to \tilde{\lambda}\psi$ in $\mathfrak{H}$. Thus, for every φ in $\mathfrak{H}$,

$$\boldsymbol{\lambda}\varphi = \lim_n \boldsymbol{\lambda}_n \varphi = \lim_n F^{-1}\tilde{\lambda}_n F\varphi = F^{-1}\tilde{\lambda} F\varphi,$$

from which the equations (31) follow. Thus, when it exists as an everywhere defined operator, $\boldsymbol{\lambda}$ belongs to $\mathfrak{A}$. Q.E.D.

The reader will observe that the argument used in proving Theorem 5 may likewise be used to prove the following.

6 THEOREM. *The algebra* $\mathfrak{A}$ *is complete in the strong topology of operators in the sense that if* a_n *is in* $\mathfrak{A}$ *for each* $n = 1, 2, \ldots$, *and if* $a_n \varphi \to a\varphi$ *for each* φ *in* $\mathfrak{H}$, *then* a *is in* $\mathfrak{A}$.

We next observe that every operator a in $\mathfrak{A}$ is a convolution operator of the type described in Theorem 5 and that the approximating set functions may be taken as countably additive and absolutely continuous with respect to Lebesgue measure. To see this we use the lemma.

7 LEMMA. *For each* ψ *in* L_∞ *there is a sequence* $\{\varphi_n\}$ *of* C^∞ *functions with* $\varphi_n(s) = 0$ *for* $|s| > n$, *with* $|\varphi_n|_\infty \leqq \sqrt{2}\,|\psi|$, *and such that* $\varphi_n \to \psi$ *in measure on every bounded measurable set in* R^N.

PROOF. We first suppose that ψ is real, measurable, and bounded on R^N. Let $0 < \varepsilon_n \to 0$ and $\mathfrak{S}_n = \{s \in R^N \mid |s| < n\}$. Since ψ is measurable, there is a closed set $\delta_n \subset \mathfrak{S}_n$ such that the restriction of ψ to δ_n is continuous on δ_n and

$$\text{(i)} \qquad \text{meas } \delta_n > \text{meas } \mathfrak{S}_n - \varepsilon_n .$$

Let $\psi_n(s) = \psi(s)$ on δ_n and $\psi_n(s) = 0$ on the complement of $\mathfrak{S}_n$. Then it is seen from the Tietze extension theorem (I.5.3) that there is a function

Ψ_n continuous on R^N with $\Psi_n(s) = \psi_n(s) = \psi(s)$ for s in δ_n, $\Psi_n(s) = \psi_n(s) = 0$ for $|s| \geqq n$, and

$$\text{(ii)} \qquad |\Psi_n|_\infty = \sup_{s\in\delta_n} |\psi_n(s)| = \sup_{s\in\delta_n} |\psi(s)| \leqq |\psi|_\infty .$$

Since Ψ_n is uniformly continuous, there is an $\eta_n > 0$ such that $|\Psi_n(s) - \Psi_n(s')| < \varepsilon_n$ if $|s - s'| < \eta_n$. Now choose disjoint closed subsets δ_n^k, $k = 1, \ldots, m = m(n)$, of δ_n each having diameter less than η_n and such that

$$\text{(iii)} \qquad \operatorname{meas}\left(\bigcup_{k=1}^{m} \delta_n^k\right) > \operatorname{meas} \delta_n - \varepsilon_n ,$$

and disjoint open subsets ω_n^k of $\mathfrak{S}_n$ with $\omega_n^k \supset \delta_n^k$ and diam $\omega_n^k < \eta_n$. From Lemma XIV.2.1 it is seen that there are functions φ_n^k on R^N with the properties

$$\text{(iv)} \quad \varphi_n^k \in C_0^\infty ;\ 0 \leqq \varphi_n^k(s) \leqq 1;\ \varphi_n^k(s) = 1,\ s \in \delta_n^k;\ \varphi_n^k(s) = 0,\ s \notin \omega_n^k .$$

Let $\alpha_n^k = \Psi_n(s_n^k)$ where s_n^k is some point in δ_n^k and let

$$\zeta_n(s) = \sum_{k=1}^{m} \alpha_n^k \varphi_n^k(s).$$

If s belongs to one of the sets ω_n^k then

$$|\zeta_n(s)| = |\alpha_n^k \varphi_n^k(s)| \leqq |\alpha_n^k| \leqq \sup_{s\in\omega_n^k} |\Psi_n(s)| \leqq |\Psi_n|_\infty ,$$

and thus (iii) shows that

$$\sup_{s\in\bigcup_{k=1}^{m}\omega_n^k} |\zeta_n(s)| \leqq |\psi|_\infty .$$

If s is not in $\bigcup_{k=1}^m \omega_n^k$ then $\zeta_n(s) = 0$ which, together with the preceding inequality, proves

$$\text{(v)} \qquad |\zeta_n|_\infty \leqq |\psi|_\infty .$$

The statements (iv) show that ζ_n is in C^∞ and vanishes outside $\mathfrak{S}_n$. Thus in view of (v) all that remains to prove is that $\zeta_n \to \psi$ in measure on every bounded set. For s in δ_n^k we have $\zeta_n(s) = \alpha_n^k$, and since δ_n^k has diameter less than η_n, $|\zeta_n(s) - \psi(s)| = |\zeta_n(s) - \Psi_n(s)| < \varepsilon_n$ which shows that

$$\text{(vi)} \qquad |\zeta_n(s) - \psi(s)| < \varepsilon_n , \qquad s \in \bigcup_{k=1}^{m} \delta_n^k .$$

Now let B be any bounded measurable set and $\varepsilon > 0$. Choose n so large that $B \subset \mathfrak{S}_n$ and $\varepsilon_n < \varepsilon$. Then from (i), (iii) and (vi) it is seen that

$$\begin{aligned}\text{meas}\{s \in B \,|\, |\zeta_n(s) - \psi(s)| > \varepsilon\} &\leqq \text{meas}\{s \in \mathfrak{S}_n \,|\, |\zeta_n(s) - \psi(s)| > \varepsilon_n\} \\ &\leqq 2\varepsilon_n \to 0,\end{aligned}$$

which proves that $\zeta_n \to \psi$ in measure on B. For a complex valued function ψ in $L_\infty(R^N)$ the stated conclusions may be obtained by applying the results just proved to the real and imaginary parts of ψ. Q.E.D.

8 THEOREM. *Every operator a in $\mathfrak{A}$ is a convolution*

$$a\varphi = \lim_n \lambda_n * \varphi, \qquad \varphi \in \mathfrak{H},$$

with the set functions λ_n defined and absolutely continuous on the field of Lebesgue measurable sets in R^N.

PROOF. According to Lemma 7 there is a sequence $\{\varphi_n\}$ in Φ with $|\varphi_n| \leqq \sqrt{2}\,|\hat{a}|$ and $\varphi_n \to \hat{a}$ in measure on every bounded measurable set in R^N. Theorem 1 shows that the functions $\psi_n = \tau^{-1}\varphi_n$ are in Φ and thus in L_1. If we let

$$\lambda_n(\sigma) = \int_\sigma \psi_n(s)\, ds, \qquad \sigma \in \Sigma,$$

then λ_n is an absolutely continuous measure defined on Σ and

$$\tilde{\lambda}_n(s) = \int_{R^N} e^{-ist}\lambda_n(dt) = \int_{R^N} e^{-ist}\psi_n(t)\, dt = \tilde{\psi}_n(s) = \varphi_n(s).$$

Thus, by Theorem 5, for every φ in $\mathfrak{H}$,

$$\begin{aligned}\lim_n \lambda_n * \varphi &= \lim_n \int_{R^N} \varphi_n(s)e(ds)\varphi \\ &= \int_{\mathfrak{S}} \hat{a}(s)e(ds)\varphi = a\varphi. \qquad \text{Q.E.D.}\end{aligned}$$

9 COROLLARY. *An operator a in the Hilbert space $\mathfrak{H}$ belongs to $\mathfrak{A}$ if and only if it has the form*

$$(a\varphi)(s) = \underset{n\to\infty}{\text{l.i.m.}} \int_{R^N} f_n(s-t)\varphi(t)\, dt, \qquad \varphi \in \mathfrak{H},$$

with f_n in $L_1(R^N)$. For each φ in $\mathfrak{H}$ the integral exists in the sense of Lebesgue for almost all s in R^N.

PROOF. Since

$$\int_{R^N} \varphi_t \lambda_n(dt) = \int_{R^N} \varphi_t f_n(t)\, dt,$$

it follows from III.11.17 that

$$(\lambda_n * \varphi)(s) = \int_{R^N} \varphi(s-t) f_n(t)\, dt = \int_{R^N} f_n(s-t)\varphi(t)\, dt,$$

and that the integral exists in the sense of Lebesgue for almost all s in R^N. The corollary thus follows from Theorems 5 and 8. Q.E.D.

10 COROLLARY. *The convolution operators which are defined everywhere on* $\mathfrak{H}$ *by equation* (28) *form a commutative* B^**-algebra of operators in* $\mathfrak{H}$ *which is* $*$*-equivalent to the* B^**-algebra* $L_\infty(R^N)$.

The following examples are among the most familiar convolutions in $\mathfrak{A}$.

11 EXAMPLE (Translation). Let $\lambda_t(\sigma) = 1$ if t is in σ; otherwise let $\lambda_t(\sigma) = 0$. Then $(\lambda_t * \varphi)(s) = \varphi(s-t)$.

12 EXAMPLE (Convolution by an L_1 function). Let f be in L_1 and

$$\lambda(\sigma) = \int_\sigma f(t)\, dt, \qquad \sigma \in \Sigma.$$

Then

$$\begin{aligned}(\lambda * \varphi)(s) &= \int_{R^N} \varphi(s-t) f(t)\, dt \\ &= \int_{R^N} f(s-t)\varphi(t)\, dt, \qquad \varphi \in \mathfrak{H},\end{aligned}$$

and the integral exists as a Lebesgue integral for almost all s in R^N. This convolution is also written as $f * \varphi$. We shall sometimes use the symbol $\mathbf{f}$ for the operator in $\mathfrak{H}$ which maps φ into $f * \varphi$. Thus, for f in L_1, we have

$$\mathbf{f} = \int_{\mathfrak{S}} \tilde{f}(s) e(ds), \qquad |\mathbf{f}| = |\tilde{f}|_\infty .$$

13 EXAMPLE (Convolution by an L_2 function). An L_2 function f whose Fourier transform $\tilde{f}$ is in L_∞ determines a convolution operator $\mathbf{f}$ in $\mathfrak{A}$. To see this we note that for any pair f, φ in L_2 the convolution integral

$$(32) \qquad (f * \varphi)(s) = \int_{R^N} f(s-t)\varphi(t)\, dt$$

exists, as a Lebesgue integral, for each s in R^N. Consider first the case where φ is in $L_1 \cap L_2$. The preceding example shows that

$$f * \varphi = \varphi * f = \int_{\mathfrak{S}} \tilde{\varphi}(s) e(ds) f$$

is in L_2 and thus

$$\tau(f * \varphi) = \int_{\mathfrak{S}} \tilde{\varphi}(s) \tau e(ds) f = \int_{\mathfrak{S}} \tilde{\varphi}(s) \mu(ds) \tilde{f}$$

$$= \tilde{f}\, \tilde{\varphi} = \int_{\mathfrak{S}} \tilde{f}(s) \mu(ds) \tilde{\varphi}$$

which shows that

$$f * \varphi = \int_{\mathfrak{S}} \tilde{f}(s) e(ds) \varphi.$$

By using a sequence $\{\varphi_n\}$ in $L_1 \cap L_2$ to approach an arbitrary φ in L_2, this equation, together with (32), shows that the operator

$$\mathbf{f} = \int_{\mathfrak{S}} \tilde{f}(s) e(ds)$$

in $\mathfrak{A}$ maps φ into $f * \varphi$.

14 Example (Principal value convolutions). A point s_0 in $\mathfrak{S}$ is called a *singular* point for the measurable function f defined almost everywhere on $\mathfrak{S}$ if f is not integrable in the sense of Lebesgue on any neighborhood of s_0. The singular points clearly form a closed set in $\mathfrak{S}$ which we assume to have measure zero. Let the ε neighborhood of ∞ be the set of all s with $|s| > 1/\varepsilon$ and the ε neighborhood of a point s_0 in R^N consist of s with $|s - s_0| < \varepsilon$. Let $f_\varepsilon(s) = f(s)$ if s is not in the ε neighborhood of any singular point, and otherwise let $f_\varepsilon(s) = 0$. Then for each $\varepsilon > 0$ the function f_ε is in L_1 and, according to Theorem 5, the convolution

$$(33) \qquad f * \varphi = \lim_{\varepsilon \to 0} f_\varepsilon * \varphi, \qquad \varphi \in \mathfrak{H},$$

exists for every φ in $\mathfrak{H}$ if and only if $|\tau f_\varepsilon|_\infty$ is bounded on the interval $0 < \varepsilon < 1$ and the limit, $\lim_{\varepsilon \to 0} f_\varepsilon$, exists in measure on every bounded measurable set. If f is not in L_1, the convolution (33) is often called a *singular convolution*. The following two examples illustrate this concept.

15 EXAMPLE (The Hilbert transform). Here $N = 1$ and $f(s) = 1/s$ so that $s = 0$ and $s = \infty$ are the only singular points of f. Let $r = 1/\varepsilon$ so that

$$\tilde{f}_\varepsilon(s) = \left[\int_{-r}^{-\varepsilon} + \int_{\varepsilon}^{r}\right] \frac{e^{-ist}}{t}\, dt$$

$$= -2i \int_{\varepsilon}^{r} \frac{\sin st}{t}\, dt$$

$$= 2i \operatorname{sgn} s \int_{|s|\varepsilon}^{|s|r} \frac{\sin t}{t}\, dt,$$

which shows that $|\tilde{f}_\varepsilon|_\infty$ is bounded and that $\tilde{f}_\varepsilon(s) \to -\pi i \operatorname{sgn} s$ for every $s \neq 0$. Thus the Hilbert transform

$$(34) \qquad (\mathbf{h}\varphi)(s) = \underset{\varepsilon \to 0}{\text{l.i.m.}} \int_{|s-t|>\varepsilon} \frac{\varphi(t)}{s-t}\, dt, \qquad \varphi \in \mathfrak{H},$$

is in $\mathfrak{A}$ and has

$$(35) \qquad \mathbf{h} = -\pi i \int_{\mathfrak{S}} \operatorname{sgn} s\, e(ds), \qquad |\mathbf{h}| = \pi.$$

16 EXAMPLE (Calderón and Zygmund). An analogue of the Hilbert transform in case the dimension N is greater than 1 is given by an *odd kernel of Calderón-Zygmund* type. Let f be a measurable function on R^N which is odd and homogeneous of degree $-N$, that is,

$$(36) \qquad f(-s) = -f(s), \qquad f(rs) = r^{-N}f(s), \qquad r > 0, \qquad 0 \neq s \in R^N,$$

and for which

$$(37) \qquad \int_{\mathfrak{M}} |f(\omega)|\, m(d\omega) < \infty,$$

where m is the measure on the hypersurface $\mathfrak{M}$ of the unit sphere in R^N. Then, by changing to spherical polar coordinates (r, ω) where $s = r\omega$ with $r \geqq 0$, $|\omega| = 1$, we have, for $0 < \varepsilon < r$,

$$\int_{\varepsilon \leqq |s| \leqq r} |f(s)|\, ds = \int_{\varepsilon}^{r} \left\{\int_{\mathfrak{M}} |f(\rho\omega)|\, \rho^{N-1}\right\} m(d\omega)\, d\rho$$

$$= \int_{\varepsilon}^{r} \frac{d\rho}{\rho} \left\{\int_{\mathfrak{M}} |f(\omega)|\, m(d\omega)\right\} < \infty,$$

which shows that f has no singular points other than perhaps 0 and ∞.

Since $m(\sigma) = m(-\sigma)$, we have, upon setting $r = 1/\varepsilon$,

$$\begin{aligned}\tilde{f}_\varepsilon(s) &= \int_{R^N} e^{-ist} f_\varepsilon(t)\, dt \\ &= \int_{\mathfrak{M}} f(\omega) \left\{ \int_\varepsilon^r \frac{e^{-is\omega\rho}}{\rho}\, d\rho \right\} m(d\omega) \\ &= \int_{\mathfrak{M}} f(\omega) \left\{ -i \int_\varepsilon^r \frac{\sin s\omega\rho}{\rho}\, d\rho \right\} m(d\omega) \\ &= \int_{\mathfrak{M}} f(\omega) \left\{ -\operatorname{sgn} s\omega \int_{|s\omega|\varepsilon}^{|s\omega| r} \frac{\sin \rho}{\rho} \right\} m(d\omega),\end{aligned}$$

and since the function in the braces is bounded in all variables, the condition (37) shows that $|\tilde{f}_\varepsilon|_\infty$ is bounded in ε. Also, if $s \neq 0$, the set of ω in $\mathfrak{M}$ for which $s\omega = 0$ has hypersurface measure zero, which shows that $\lim_{\varepsilon \to 0} \tilde{f}_\varepsilon(s)$ exists if $s \neq 0$ and is

$$\tilde{f}(s) = \frac{-\pi i}{2} \int_{\mathfrak{M}} f(\omega) \operatorname{sgn} s\omega m(d\omega), \qquad s \neq 0. \tag{38}$$

Thus the singular convolution operator

$$\mathbf{f}\varphi = f * \varphi = \underset{\varepsilon \to 0}{\text{l.i.m.}} \int_{|s-t| > \varepsilon} f(s-t)\varphi(t)\, dt, \qquad \varphi \in \mathfrak{H}, \tag{39}$$

is in $\mathfrak{A}$ and has the representation

$$\mathbf{f} = \frac{-\pi i}{2} \int_{\mathfrak{S}} \left\{ \int_{\mathfrak{M}} f(\omega) \operatorname{sgn} s\omega\ m(d\omega) \right\} e(ds) \tag{40}$$

and norm

$$|\mathbf{f}| = \frac{\pi}{2} \operatorname*{ess\,sup}_{s \in \mathfrak{S}} \left| \int_{\mathfrak{M}} f(\omega) \operatorname{sgn} s\omega\ m(d\omega) \right|. \tag{41}$$

Having seen some examples of the operators in $\mathfrak{A}$, we now give some examples of spectral operators in $\mathfrak{A}^p$. In view of Lemma 10.4 and Theorem 10.6, the central result for determining such operators is as follows.

17 Theorem. *An operator A in $\mathfrak{A}^p$ is a spectral operator if and only if*

$$\operatorname*{ess\,sup}_{s \in \mathfrak{S}_i} |E(\hat{\lambda}_{ij}(s); \hat{A}(s))| < \infty, \qquad 1 \leqq j \leqq i \leqq p. \tag{42}$$

In applying this criterion, one does not have to calculate the norms of the projections $E(\hat{\lambda}_{ij}(s); \hat{A}(s))$ for, since they are finite matrices, their norms are essentially bounded if and only if their elements are essentially bounded. One does need to know the eigenvalues $\hat{\lambda}_{ij}(s)$ of the matrix $\hat{A}(s)$ which makes it difficult to state any general rule, other than (42), for determining the spectral operators in $\mathfrak{A}^p$. In the case $p = 2$, however, the condition (42) may be restated in a more applicable form. For s in $\mathfrak{S}_1$ the spectrum $\sigma(\hat{A}(s))$ has only one point and so $E(\hat{\lambda}_{11}(s); \hat{A}(s)) = I$; consequently, we need only consider the case where s is in $\mathfrak{S}_2$. For brevity we let $\hat{a}_{ij} = \hat{a}_{ij}(s)$ and $\hat{\lambda}_j = \hat{\lambda}_{2j}(s)$ so that $\det(\lambda I - \hat{A}(s)) = (\lambda - \hat{\lambda}_1)(\lambda - \hat{\lambda}_2)$ with

$$\begin{aligned}\hat{\lambda}_1 &= \tfrac{1}{2}(\hat{a}_{11} + \hat{a}_{22} + \hat{\delta}),\\ \hat{\lambda}_2 &= \tfrac{1}{2}(\hat{a}_{11} + \hat{a}_{22} - \hat{\delta}),\end{aligned} \tag{43}$$

where $\hat{\delta}$ is chosen as any Σ-measurable square root of

$$\hat{\delta}^2 = (\hat{a}_{11} - \hat{a}_{22})^2 + 4\hat{a}_{12}\hat{a}_{21}. \tag{44}$$

The sets $\mathfrak{S}_1$ and $\mathfrak{S}_2$ may be expressed in terms of $\hat{\delta}$ and are

$$\mathfrak{S}_1 = \{s \mid \hat{\delta}(s) = 0\}, \qquad \mathfrak{S}_2 = \{s \mid \hat{\delta}(s) \neq 0\}. \tag{45}$$

Since the polynomial $P(\lambda) = (\hat{\lambda}_1 - \hat{\lambda}_2)^{-1}(\lambda - \hat{\lambda}_2)$ has the properties that $P(\hat{\lambda}_1) = 1$ and $P(\hat{\lambda}_2) = 0$, it is seen that

$$\begin{aligned}E(\hat{\lambda}_{21}(s); \hat{A}(s)) &= \frac{1}{\hat{\lambda}_1 - \hat{\lambda}_2}(\hat{A}(s) - \hat{\lambda}_2 I)\\ &= \frac{1}{2}\begin{pmatrix} 1 + \dfrac{\hat{a}_{11} - \hat{a}_{22}}{\hat{\delta}} & \dfrac{2\hat{a}_{12}}{\hat{\delta}} \\ \dfrac{2\hat{a}_{21}}{\hat{\delta}} & 1 - \dfrac{\hat{a}_{11} - \hat{a}_{22}}{\hat{\delta}} \end{pmatrix},\end{aligned} \tag{46}$$

and that this matrix is essentially bounded on $\mathfrak{S}_2$ if and only if each of the three functions

$$\frac{\hat{a}_{11} - \hat{a}_{22}}{\hat{\delta}}, \qquad \frac{\hat{a}_{12}}{\hat{\delta}}, \qquad \frac{\hat{a}_{21}}{\hat{\delta}}$$

is essentially bounded on $\mathfrak{S}_2$. Thus the projections $E(\hat{\lambda}_{21}(s); \hat{A}(s))$ and $E(\hat{\lambda}_{22}(s); \hat{A}(s)) = I - E(\hat{\lambda}_{21}(s); \hat{A}(s))$ are both essentially bounded on $\mathfrak{S}_2$, that is, the condition (42) is satisfied, if and only if

$$\operatorname*{ess\,sup}_{s \in \mathfrak{S}_2} \frac{|\hat{a}_{11} - \hat{a}_{22}|^2 + |\hat{a}_{12}|^2 + |\hat{a}_{21}|^2}{|\hat{\delta}^2|} < \infty, \tag{47}$$

and we have proved the following corollary to Theorem 17.

→ 18 THEOREM. *An operator* $A = (a_{ij})$ *in* $\mathfrak{A}^2$ *has a resolution of the identity if and only if the inequality* (47) *is satisfied.*

19 COROLLARY. *An operator* $A = (a_{ij})$ *in* $\mathfrak{A}^2$ *is a spectral operator if* $\mathfrak{S}_2$ *has measure zero.*

20 COROLLARY. *An operator* A *in* $\mathfrak{C}^2$ *is a spectral operator if* $\mathfrak{S}_2 = \mathfrak{S}$.

The operational calculus of Corollary 10.11 for a spectral operator A in $\mathfrak{A}^2$ requires the calculation of the projection E_{ij}, and this we proceed to do. It will be convenient to introduce an operator δ^{-1}, in general unbounded, which is defined by the equations

$$\mathfrak{D}(\delta^{-1}) = \left\{\varphi \,\middle|\, \varphi \in \mathfrak{H},\ \int_{\mathfrak{S}_2} \left|\frac{\tilde{\varphi}(s)}{\hat{\delta}(s)}\right|^2 ds < \infty\right\}, \tag{48}$$

and

$$\delta^{-1}\varphi = \tau^{-1}\left(\frac{\chi_{\mathfrak{S}_2}}{\hat{\delta}}\right)\tau\varphi, \qquad \varphi \in \mathfrak{D}(\delta^{-1}). \tag{49}$$

Let φ be an arbitrary point of $\mathfrak{H}$ and let

$$\mathfrak{S}_n = \mathfrak{S}_1 \cup \left\{s \,\middle|\, s \in \mathfrak{S}_2, \qquad \frac{1}{|\hat{\delta}(s)|} \leqq n\right\}.$$

Then the function $\tau^{-1}\chi_{\mathfrak{S}_n}\tilde{\varphi}$ is in $\mathfrak{D}(\delta^{-1})$ and since $\tau^{-1}\chi_{\mathfrak{S}_n}\tilde{\varphi} \to \varphi$, we see that $\mathfrak{D}(\delta^{-1})$ is dense so that the operator δ^{-1} is densely defined. It is also closed, for suppose that $\varphi_n \in \mathfrak{D}(\delta^{-1})$, $\varphi_n \to \varphi$, and $\delta^{-1}\varphi_n \to \psi$. Then by using subsequences, we may suppose that $\tau\varphi_n \to \tau\varphi$ and $\tau\delta^{-1}\varphi_n \to \tau\psi$ almost everywhere as well as in $\mathfrak{H}$. Then we have, for almost all s,

$$(\tau\psi)(s) = \lim_n (\tau\delta^{-1}\varphi_n)(s) = \lim_n \left(\frac{\chi_{\mathfrak{S}_2}(s)}{\hat{\delta}(s)}\right)(\tau\varphi_n)(s)$$

$$= \left(\frac{\chi_{\mathfrak{S}_2}(s)}{\hat{\delta}(s)}\right)\tilde{\varphi}(s),$$

which shows that φ is in $\mathfrak{D}(\delta^{-1})$ and that $\psi=\delta^{-1}\varphi$. Thus δ^{-1} is a closed densely defined operator. If a is in $\mathfrak{A}$ and $|\hat{a}(s)\hat{\delta}^{-1}(s)|$ is bounded on $\mathfrak{S}_2$, then it is clear that the range of a is in the domain of δ^{-1} and that $\delta^{-1}a$ is a bounded operator and in $\mathfrak{A}$. Hence for a spectral operator A in $\mathfrak{A}^2$ we see from Theorem 18 that the operators $\delta^{-1}(a_{11}-a_{22})$, $\delta^{-1}a_{21}$, $\delta^{-1}a_{12}$, and $\delta^{-1}\delta=e(\mathfrak{S}_2)$ are all operators in the algebra $\mathfrak{A}$. It follows from (46) that the projections E_{21} and E_{22} have the matrix representations

$$(50)\qquad E_{21}=\frac{\delta^{-1}}{2}\begin{pmatrix}\delta+(a_{11}-a_{22}) & 2a_{12}\\ 2a_{21} & \delta-(a_{11}-a_{22})\end{pmatrix}e(\mathfrak{S}_2),$$

$$E_{22}=\frac{\delta^{-1}}{2}\begin{pmatrix}\delta-(a_{11}-a_{22}) & -2a_{12}\\ -2a_{21} & \delta+(a_{11}-a_{22})\end{pmatrix}e(\mathfrak{S}_2),$$

and that the right sides of these equations represent bounded everywhere defined operators on $\mathfrak{H}^2$ and, in fact, operators in $\mathfrak{A}^2$. As observed earlier $E(\hat{\lambda}_{11}(s);\hat{A}(s))=I$ on $\mathfrak{S}_1$ so that (50), (51), and

$$(52)\qquad E_{11}=\begin{pmatrix}e & 0\\ 0 & e\end{pmatrix}e(\mathfrak{S}_1)$$

give the projections needed for the operational calculus. For example, a bounded Borel function φ of the scalar part S of A is

$$(53)\qquad \varphi(\mathfrak{S})=\varphi(\lambda_{11})E_{11}+\varphi(\lambda_{21})E_{21}+\varphi(\lambda_{22})E_{22},$$

and the resolution of the identity $E(\sigma;A)=E(\sigma;S)$ may be obtained from (53) by placing $\varphi=\chi_\sigma$, the characteristic function of σ. Since $\chi_\sigma(\lambda_{ij})=E(\sigma;\lambda_{ij})=e(\hat{\lambda}_{ij}^{-1}(\sigma))$, we have

$$(54)\qquad E(\sigma;A)=e(\hat{\lambda}_{11}^{-1}(\sigma))E_{11}+e(\hat{\lambda}_{21}^{-1}(\sigma))E_{21}+e(\hat{\lambda}_{22}^{-1}(\sigma))E_{22}.$$

It should be mentioned that this development of the operational calculus for a spectral operator A in $\mathfrak{A}^2$ may be carried out quite analogously for an operator A in $\mathfrak{A}^p$ with $p>2$. This is seen from equation (iv) of Lemma 10.3, which shows that the general matrix element of the bounded projection E_{ij} may be written as the product $\delta_{ij}^{-1}R_{ij}$, where R_{ij} is a polynomial in the roots and the matrix elements a_{ij}, and δ_{ij}^{-1} is a closed densely defined operator.

21 EXAMPLE. Let $N=1$. The operator

$$(55)\qquad A=\begin{pmatrix}a & \mathbf{h}\\ \mathbf{h} & b\end{pmatrix},$$

where **h** is Hilbert's singular integral (34), is an operator in $\mathfrak{A}^2$ which is not self adjoint, and not normal unless $i(a-b)$ is self adjoint but is, for many choices of a and b, a spectral operator. To see this, equation (35) shows that the condition (47) is satisfied if and only if

$$\operatorname*{ess\,sup}_{s \in \mathfrak{S}_2} |(\hat{a}(s) - \hat{b}(s))^2 - 4\pi^2|^{-1} < \infty, \tag{56}$$

which is certainly true, for example, if a and b both have norms less than π.

Other examples that are clearly spectral operators are those operators whose matrix representation has one of the forms

$$\begin{pmatrix} \mathbf{h}+a & 0 \\ b & a \end{pmatrix}, \quad \begin{pmatrix} e+a & 0 \\ b & a \end{pmatrix}, \tag{57}$$

while the operator

$$\begin{pmatrix} e-a & 0 \\ b & a \end{pmatrix} \tag{58}$$

does not have a resolution of the identity unless

$$|\hat{b}(s)| \leqq M\,|1 - 2\hat{a}(s)|, \tag{59}$$

for some constant M and almost all s in $\mathfrak{S}_2$, which, in this case, is the set where $\hat{a}(s) \neq \frac{1}{2}$.

As the following theorem shows, the nilpotent part of a spectral operator in $\mathfrak{A}^2$ is

$$N = e(\mathfrak{S}_1)\begin{pmatrix} \frac{1}{2}(a_{11}-a_{22}) & a_{12} \\ a_{21} & \frac{1}{2}(a_{22}-a_{11}) \end{pmatrix},$$

from which it follows that a spectral operator A in $\mathfrak{A}^2$, not of the form $A = \lambda I$ with λ in $\mathfrak{A}$, has a non-zero radical part if $e(\mathfrak{S}_1) \neq 0$. By Theorem 6.4 the spectral operators with non-zero radical parts are the only ones not similar to normal operators. Thus both of the operators in (57) are scalar type operators, and for any a and b in $\mathfrak{A}$, they are similar to normal operators. In the case of the operators (55) and (58) the situation is somewhat different. If, in (55), the norms $|a|$ and $|b|$ are both less than π, it follows that $\mathfrak{S}_1$ is void and thus that $N = 0$, but since $\hat{a}$ and $\hat{b}$ are arbitrary functions in L_∞, the sets $\mathfrak{S}_1$ and $\mathfrak{S}_2$ may both have positive measure, in which case the operator (55) is not similar to a normal operator. If $\hat{a}$ and $\hat{b}$ are both continuous, as they would be if a and b are convolutions by L_1 functions, then the inequality (56) cannot hold unless either $\mathfrak{S}_1$ or $\mathfrak{S}_2$ is void and in either of these cases the operator (55) is a scalar type operator. It is different with the operator in (58), where $\mathfrak{S}_1$ and $\mathfrak{S}_2$ may both have positive measure, in which case the operator is not similar to a normal operator, even though a and b are convolutions by L_1 functions.

Every operator A in $\mathfrak{A}^2$ has a decomposition similar to the canonical decomposition of a spectral operator. This decomposition, which is given in the following theorem, will enable us to establish an operational calculus for A even though it may not have a resolution of the identity.

22 THEOREM. *Every operator A in $\mathfrak{A}^2$ uniquely determines two operators S and N in $\mathfrak{A}^2$ with the following properties:*

(i) $$A = S + N, \qquad SN = NS;$$

(ii) $$N^2 = 0;$$

(iii) *for almost all s in $\mathfrak{S}$, the minimal polynomial of the matrix $\hat{S}(s)$ of complex numbers has only simple roots.*

PROOF. Let

$$S = \begin{pmatrix} \frac{1}{2}(a_{11}+a_{22})e(\mathfrak{S}_1)+a_{11}e(\mathfrak{S}_2) & a_{12}e(\mathfrak{S}_2) \\ a_{21}e(\mathfrak{S}_2) & \frac{1}{2}(a_{11}+a_{22})e(\mathfrak{S}_1)+a_{22}e(\mathfrak{S}_2) \end{pmatrix} \tag{60}$$

and

$$N = e(\mathfrak{S}_1)\begin{pmatrix} \frac{1}{2}(a_{11}-a_{22}) & a_{12} \\ a_{21} & \frac{1}{2}(a_{22}-a_{11}) \end{pmatrix}, \tag{61}$$

where $\mathfrak{S}_1$ and $\mathfrak{S}_2$ are given by (45). The equations (i) follow from an elementary calculation, and if

$$\delta = \int_{\mathfrak{S}} \delta(s)e(ds), \tag{62}$$

then (45) shows that $e(\mathfrak{S}_1)\delta = 0$ and a direct multiplication gives $N^2 = 0$, proving (ii). It follows from (60) that, for s in $\mathfrak{S}_1$, $\hat{S}(s) = \hat{\lambda}_{11}(s)I$ and hence its minimal polynomial is $\lambda - \hat{\lambda}_{11}(s)$ whereas, for s in $\mathfrak{S}_2$, $\hat{S}(s) = \hat{A}(s)$ and, for such s, $\hat{S}(s)$ has two distinct roots. Hence for every s in $\mathfrak{S}$ the minimal polynomial of the matrix $\hat{S}(s)$ has only simple roots. It remains to be shown that this decomposition is unique. This becomes clear when one observes that a $p \times p$ matrix of complex numbers whose minimal polynomial has only simple roots is a scalar type spectral operator in p-dimensional unitary space and conversely. Thus if S_1, N_1 is another pair of operators in $\mathfrak{A}^2$ satisfying (i), (ii), (iii), then for almost all s in $\mathfrak{S}$ we have

$\hat{N}_1^2(s)=0$, and $\hat{A}(s)=\hat{S}_1(s)+\hat{N}_1(s)$ is therefore the canonical reduction of the spectral operator $\hat{A}(s)$ in $B(E^2)$. Since such a reduction is unique, we must have $\hat{S}_1(s)=\hat{S}(s)$ and $\hat{N}_1(s)=\hat{N}(s)$, which proves that $S_1=S$ and $N_1=N$. Q.E.D.

The next corollary follows from Corollary 9.9 and equation (61).

23 COROLLARY. *The operators A and S have the same spectrum and the operator $N=0$ if and only if, for some $\hat{\lambda}$ in $\hat{\mathfrak{A}}$, $\hat{A}(s)=\hat{\lambda}(s)I$ for almost all s in $\mathfrak{S}_1$.*

It is clear that if A is a spectral operator, the decomposition of Theorem 22 is its canonical decomposition. The important fact for our present purposes is that the decomposition exists for every A in $\mathfrak{A}^2$ even though A, and therefore S, may have no resolution of the identity. In what follows S and N will be the operators of Theorem 22 determined by the operator A in $\mathfrak{A}^2$.

Let A be an arbitrary element of $\mathfrak{A}^2$ and let φ be a complex valued function analytic and single valued on the spectrum $\sigma(A)$ so that the operator $\varphi(A)$ is defined as

$$\varphi(A)=\frac{1}{2\pi i}\int_C \varphi(\lambda)(\lambda I-A)^{-1}\,d\lambda, \tag{63}$$

where C consists of a finite number of positively oriented rectifiable Jordan curves forming the boundary of an open set containing $\sigma(A)$ and upon whose closure φ is analytic and single valued. By Corollary 9.6(ii), $(\lambda I-A)^{-1}$ is in $\mathfrak{A}^2$, and it follows from (63) that $\varphi(A)$ is also in $\mathfrak{A}^2$. It will now be shown that $\widehat{\varphi(A)}(s)=\varphi(\hat{A}(s))$. By Corollary 9.9, φ is analytic on $\hat{A}(s)$ for almost all s, and since, by Theorem 9.3,

$$(\lambda I-A)^{-1}=\int_{\mathfrak{S}}(\lambda I-\hat{A}(s))^{-1}e(ds),$$

we have

$$\varphi(A)=\frac{1}{2\pi i}\int_C \varphi(\lambda)\left\{\int_{\mathfrak{S}}(\lambda I-\hat{A}(s))^{-1}e(ds)\right\}d\lambda$$
$$=\int_{\mathfrak{S}}\left\{\frac{1}{2\pi i}\int_C \varphi(\lambda)(\lambda I-\hat{A}(s))^{-1}d\lambda\right\}e(ds),$$

where the change in the order of integration is justified since it is valid if the measure $e(\sigma)$ is replaced by $(e(\sigma)x, y)$ for an arbitrary pair x, y in $\mathfrak{H}$. This proves that

$$\varphi(A) = \int_{\mathfrak{S}} \varphi(\hat{A}(s))e(ds), \tag{64}$$

a formula with some advantages over (63) but one which can be improved for the purpose of calculation. The derivation of equation (64) made no use of the assumption that $p=2$, and thus this equation holds for an operator A in $\mathfrak{A}^p$ for any $p \geqq 1$, a fact that will be used later. Since $\sigma(A)=\sigma(S)$, it follows from VII.3.10 that

$$\varphi(A) = \sum_{k=0}^{\infty} N^k \frac{\varphi^{(k)}(S)}{k!}$$

and thus, since $N^2=0$,

$$\varphi(A) = \varphi(S) + N\varphi'(S), \tag{65}$$

where $\varphi'(S)$ is the operator (63) or (64) with A replaced by S and $\varphi(\lambda)$ by $\varphi'(\lambda)$. Now it follows from (60) that $\hat{S}(s) = \hat{\lambda}_{11}(s)I$ for s in $\mathfrak{S}_1$ and so

$$\begin{aligned} \varphi(S)e(\mathfrak{S}_1) &= \left\{\int_{\mathfrak{S}_1} \varphi(\hat{\lambda}_{11}(s))e(ds)\right\}I = \varphi(\lambda_{11})I, \\ \varphi'(S)e(\mathfrak{S}_1) &= \left\{\int_{\mathfrak{S}_1} \varphi'(\hat{\lambda}_{11}(s))e(ds)\right\}I = \varphi'(\lambda_{11})I. \end{aligned} \tag{66}$$

On the other hand, for s in $\mathfrak{S}_2$, it is seen from (46) that

$$\begin{aligned} \varphi(\hat{A}(s)) &= \varphi(\hat{\lambda}_{21}(s)) \frac{\hat{A}(s) - \hat{\lambda}_{22}(s)I}{\hat{\lambda}_{21}(s) - \hat{\lambda}_{22}(s)} + \varphi(\hat{\lambda}_{22}(s)) \frac{\hat{A}(s) - \hat{\lambda}_{21}(s)I}{\hat{\lambda}_{22}(s) - \hat{\lambda}_{21}(s)} \\ &= \frac{\varphi(\hat{\lambda}_{21}(s)) - \varphi(\hat{\lambda}_{22}(s))}{\hat{\lambda}_{21}(s) - \hat{\lambda}_{22}(s)} (\hat{A}(s) - \hat{\lambda}_{21}(s))I + \varphi(\hat{\lambda}_{21}(s))I. \end{aligned} \tag{67}$$

From (64) and (65) we have

$$\begin{aligned} \varphi(A) &= \varphi(A)e(\mathfrak{S}_1) + \varphi(A)e(\mathfrak{S}_2) \\ &= \varphi(S)e(\mathfrak{S}_1) + N\varphi'(S)e(\mathfrak{S}_1) + \int_{\mathfrak{S}_2} \varphi(\hat{A}(s))e(ds), \end{aligned}$$

and thus (66) and (67) give

$$\varphi(A) = \varphi(\lambda_{11})I + \varphi'(\lambda_{11})N + \varphi(\lambda_{21})I + (\Delta\varphi)(A - \lambda_{21}I), \tag{68}$$

where $\Delta\varphi$ is the operator in $\mathfrak{A}$ given by

$$\Delta\varphi = \int_{\mathfrak{S}_2} \frac{\varphi(\hat{\lambda}_{21}(s)) - \varphi(\hat{\lambda}_{22}(s))}{\hat{\lambda}_{21}(s) - \hat{\lambda}_{22}(s)}\, e(ds). \tag{69}$$

For computational purposes equation (68) has an advantage over (64), for it expresses $\varphi(A)$ as a linear function of A and N with coefficients from $\mathfrak{A}$. We summarize by stating

24 Theorem. *For every operator A in $\mathfrak{A}^2$ and every complex function φ, analytic and single valued on its spectrum, the operators $\varphi(A)$ of equations* (63) *and* (68) *are the same.*

Equation (67) suggests that for operators A in $\mathfrak{A}^2$ with $e(\mathfrak{S}_1) = 0$, there is an operational calculus on an algebra larger than that consisting of analytic functions. To see that this is indeed the case, let A be in $\mathfrak{A}^2$ and have $e(\mathfrak{S}_1) = 0$ and let $\mathfrak{L}(A)$ be the algebra of all complex functions φ on $\sigma(A)$ which satisfy the Lipschitz condition

$$|\varphi|_\Delta = \sup_{\lambda \neq \mu} \frac{|\varphi(\lambda) - \varphi(\mu)|}{|\lambda - \mu|} < \infty. \tag{70}$$

It is readily proved that, with the natural operations of addition, multiplication, multiplication by scalars, and norm

$$|\varphi| = |\varphi|_\Delta + |\varphi|_\infty, \qquad \varphi \in \mathfrak{L}(A), \tag{71}$$

the set $\mathfrak{L}(A)$ is a B-algebra whose unit is the constant function 1. Now, by Corollary 9.9, $\hat{\lambda}_{21}(s)$ and $\hat{\lambda}_{22}(s)$ are in $\sigma(A)$ for almost all s. Thus, equation (67) shows that $\varphi(\hat{A}(s))$ is essentially bounded on $\mathfrak{S}$, and hence for φ in $\mathfrak{L}(A)$ the operator

$$\varphi(A) = \int_{\mathfrak{S}} \varphi(\hat{A}(s))e(ds), \qquad \varphi \in \mathfrak{L}(A), \tag{72}$$

exists as an element of $\mathfrak{A}^2$. We have

25 Theorem. *If the operator A in $\mathfrak{A}^2$ has $e(\mathfrak{S}_1) = 0$, then the map $\varphi \to \varphi(A)$ given by* (72) *is a continuous operational calculus for A which maps $\mathfrak{L}(A)$ into $\mathfrak{A}^2$.*

Proof. The conclusion of the theorem means that

$$\varphi(A) \in \mathfrak{A}^2, \qquad \varphi \in \mathfrak{L}(A); \tag{73}$$

that the map $\varphi \to \varphi(A)$ is an algebraic homomorphism, that is,

$$(74)\quad (\alpha\varphi + \beta\psi)(A) = \alpha\varphi(A) + \beta\psi(A),\ (\varphi\psi)(A) = \varphi(A)\psi(A), \qquad \varphi, \psi \in \mathfrak{L}(A),$$

which is also an isomorphism, that is,

$$(75)\qquad \varphi(A) = 0 \quad \text{implies} \quad \varphi = 0,$$

and continuous, that is,

$$(76)\qquad |\varphi(A)| \leqq K|\varphi|, \qquad \varphi \in \mathfrak{L}(A),$$

for some constant K independent of φ; and finally that the map $\varphi \to \varphi(A)$ is consistent with the meaning assigned to $\varphi(A)$ for analytic functions φ, that is, the operators (63) and (72) are the same if φ is analytic and single valued on $\sigma(A)$.

This last statement was already established in (64). Also, we have already observed that $\varphi(A)$ is in $\mathfrak{A}^2$ if φ is in $\mathfrak{L}(A)$. Statement (74) follows at once from the corresponding identities for the finite matrices $\varphi(\hat{A}(s))$ and $\psi(\hat{A}(s))$.

To prove (75), let φ be in $\mathfrak{L}(A)$ and $\varphi(A) = 0$. It follows from Theorem 9.3 that $\varphi((\hat{A}(s)) = 0$ almost everywhere on $\mathfrak{S}$. Thus, for almost all s, φ vanishes on the spectrum $\sigma(\hat{A}(s))$. So, for some set σ_0 in Σ with $e(\sigma_0) = e$, the function φ vanishes on $\bigcup_{s\in\sigma_0} \sigma(\hat{A}(s))$, and since φ is continuous, it also vanishes on the closure of this set. Thus Corollary 9.9 shows that φ vanishes on $\sigma(A)$, which means that $\varphi = 0$.

To prove (76) it is seen from Theorem 9.3 and equation (67) that

$$\begin{aligned} |\varphi(A)| &= \operatorname*{ess\,sup}_{s\in\mathfrak{S}} |\varphi(\hat{A}(s))| \\ &\leqq 2|\varphi|_\Delta |A| + |\varphi|_\infty \leqq K|\varphi|, \end{aligned}$$

where K is the larger of $2|A|$ and 1. Q.E.D.

12. Some Examples of Unbounded Spectral Operators

Although the topic of unbounded spectral operators will be treated in some detail in Chapter XVIII and many illustrations of such operators will be found in Chapters XIX and XX, we introduce the subject here briefly in order to show how some of the criteria for the determination of the spectral operators in the algebra $\mathfrak{A}^p$ may be carried over the case of certain related unbounded operators arising in the study of linear systems of partial differential equations with constant coefficients.

The notation will be that of the preceding section, but we shall now be concerned with $p \times p$ matrices $\hat{A}(s) = (\hat{a}_{jk}(s))$ whose elements are measurable complex valued functions defined almost everywhere on $\mathfrak{S}$ but not necessarily bounded. For every set σ in Σ and every such matrix $\hat{A}(s)$ we define the matrix

$$\hat{A}_\sigma(s) = \hat{A}(s), \qquad s \in \sigma, \\ = 0, \qquad s \notin \sigma, \tag{1}$$

and the operator $\hat{A}_\sigma$ in $\mathfrak{H}^p$ according to the equations

$$\begin{aligned} \mathfrak{D}(\hat{A}_\sigma) &= \{\psi \mid \psi \in \mathfrak{H}^p, \ \int_\sigma |\hat{A}(s)\psi(s)|^2 \, ds < \infty\}, \\ (\hat{A}_\sigma \psi)(s) &= \hat{A}_\sigma(s)\psi(s), \qquad \psi \in \mathfrak{D}(\hat{A}_\sigma). \end{aligned} \tag{2}$$

Throughout the present discussion the matrix $\hat{A}(s)$ will be fixed, and hence notations for various concepts dependent upon $\hat{A}(s)$ may not show such dependence. For example, we shall use the symbol Σ_0 for the family of all sets σ in Σ upon which $\hat{A}(s)$ is essentially bounded. It is clear that for such sets σ the operator $\hat{A}_\sigma$ is a bounded everywhere defined operator. In general, however, $\hat{A}_\sigma$ need not be bounded but it is always a closed and densely defined operator. To see this let $\psi_m \in \mathfrak{D}(\hat{A}_\sigma)$, $\psi_m \to \psi$, and $\hat{A}_\sigma \psi_m \to \varphi$. We may, by using subsequences, assume that $\hat{A}_\sigma \psi_m$ and ψ_m both converge almost everywhere as well as in $\mathfrak{H}^p$. But then it follows that $\hat{A}_\sigma(s)\psi(s) = \varphi(s)$ almost everywhere, which proves that ψ is in $\mathfrak{D}(\hat{A}_\sigma)$, that $\hat{A}_\sigma \psi = \varphi$, and thus that $\hat{A}_\sigma$ is closed. To see that it is densely defined we choose a sequence $\{\sigma_m\}$ with

$$\sigma_m \in \Sigma_0, \qquad \sigma_m \subseteq \sigma_{m+1}, \qquad \text{meas}\{(\bigcup_m \sigma_m)'\} = 0, \tag{3}$$

an arbitrary vector ψ in $\mathfrak{H}^p$, and let $\psi_m(s) = \psi(s)$ for s in σ_m and $\psi_m(s) = 0$ elsewhere. Then, since σ_m is in Σ_0, ψ_m is in $\mathfrak{D}(\hat{A}_\sigma)$ and, in view of (3), $\psi_m \to \psi$ in $\mathfrak{H}^p$, which shows that $\hat{A}_\sigma$ is densely defined.

The Fourier transform F in $\mathfrak{H}^p$ is defined as $F[\psi_1, \ldots, \psi_p] = [F\psi_1, \ldots, F\psi_p]$. The matrix $\hat{A}(s)$ and an arbitrary set σ in Σ determine an operator A_σ in $\mathfrak{H}^p$, which is defined by the equations

$$\mathfrak{D}(A_\sigma) = F^{-1}\mathfrak{D}(\hat{A}_\sigma), \qquad A_\sigma \varphi = F^{-1}\hat{A}_\sigma F\varphi, \qquad \varphi \in \mathfrak{D}(A_\sigma). \tag{4}$$

Since F is a homeomorphism in $\mathfrak{H}^p$ and $\hat{A}_\sigma$ is closed and densely defined, it follows that A_σ is also closed and densely defined. Similarly, it is seen that A_σ is bounded and everywhere defined for each σ in Σ_0. For an arbitrary σ in Σ the operator A_σ of equation (4) is determined by the bounded operators A_σ with σ in Σ_0, as is evident from the following lemma.

1 LEMMA. *Let A_σ be the operator* (4) *and let $\{\sigma_m\}$ be a sequence satisfying* (3). *Then*

(i) $$\mathfrak{D}(A_\sigma) = \{\varphi \mid \lim_{m\to\infty} A_{\sigma\sigma_m}\varphi \text{ exists}\},$$

(ii) $$A_\sigma\varphi = \lim_{m\to\infty} A_{\sigma\sigma_m}\varphi, \qquad \varphi \in \mathfrak{D}(A_\sigma).$$

PROOF. Let $\varphi = F^{-1}\psi \in \mathfrak{D}(A_\sigma)$ so that ψ is in $\mathfrak{D}(\hat{A}_\sigma)$ and hence

$$FA_{\sigma\sigma_m}\varphi = \hat{A}_{\sigma\sigma_m}\psi \to \hat{A}_\sigma\psi.$$

This shows that

$$A_{\sigma\sigma_m} \to F^{-1}\hat{A}_\sigma F\varphi = A_\sigma\varphi.$$

Now, conversely, let φ be an arbitrary vector in $\mathfrak{H}^p$ for which $A_{\sigma\sigma_m}\varphi$ converges in $\mathfrak{H}^p$ to some vector ξ. Then, upon letting $\psi = F\varphi$ and $\psi_m(s) = \chi_{\sigma_m}(s)\psi(s)$, we have, in the norm of $\mathfrak{H}^p$,

$$\hat{A}_\sigma\psi_m = \hat{A}_{\sigma\sigma_m}\psi = FA_{\sigma\sigma_m}\varphi \to F\xi,$$

as well as $\psi_m \to \psi$. Since $\hat{A}_\sigma$ is closed and ψ_m is in $\mathfrak{D}(\hat{A}_\sigma)$, it follows that the vector ψ is in $\mathfrak{D}(\hat{A}_\sigma)$ and that $\hat{A}_\sigma\psi = F\xi$. Thus $F^{-1}\psi = \varphi$ is in $\mathfrak{D}(A_\sigma)$ and

$$\lim_m A_{\sigma\sigma_m}\varphi = \xi = F^{-1}\hat{A}_\sigma F\varphi = A_\sigma\varphi,$$

which completes the proof of the lemma. Q.E.D.

It is clear that, for every pair σ_1, σ_2 in Σ we have $\mu(\sigma_1)\mathfrak{D}(\hat{A}_{\sigma_2}) \subseteq \mathfrak{D}(\hat{A}_{\sigma_2})$, and thus it follows that

(5) $$e(\sigma_1)\mathfrak{D}(A_{\sigma_2}) \subseteq \mathfrak{D}(A_{\sigma_2}), \qquad \sigma_1, \sigma_2 \in \Sigma.$$

If the set σ differs from $\mathfrak{S}$ by a null set, we shall usually write A and $\hat{A}$ for A_σ and $\hat{A}_\sigma$, respectively, so that from (5) we have

(6) $$e(\sigma)\mathfrak{D}(A) \subseteq \mathfrak{D}(A), \qquad \sigma \in \Sigma.$$

Thus we may speak of the restrictions of A and $\hat{A}$ to the subspaces $e(\sigma)\mathfrak{D}(A)$ and $\mu(\sigma)\mathfrak{D}(\hat{A})$, respectively. It follows that the restriction of A_σ to

$e(\sigma)\mathfrak{D}(A)$ coincides with the restriction of A to $e(\sigma)\mathfrak{D}(A)$, that is,

$$A_\sigma e(\sigma)\varphi = Ae(\sigma)\varphi, \qquad \sigma \in \Sigma,\ \varphi \in \mathfrak{D}(A). \tag{7}$$

For σ in Σ_0, $\hat{A}_\sigma$ is in $\hat{\mathfrak{A}}$ and thus A_σ is in $\mathfrak{A}$ so that, as in the preceding section,

$$A_\sigma = \int_\sigma \hat{A}(s)e(ds), \qquad \sigma \in \Sigma_0. \tag{8}$$

If $\{\sigma_m\}$ is a sequence of sets in Σ satisfying (3), then

$$A\varphi = \lim_m \int_{\sigma_m} \hat{A}(s)e(ds)\varphi, \qquad \varphi \in \mathfrak{D}(A), \tag{9}$$

by Lemma 1, and so the operator A is a type of unbounded convolution. The preceding discussion may be summarized as follows.

2 THEOREM. *For each measurable $p \times p$ matrix $\hat{A}(s)$ of complex functions defined almost everywhere on $\mathfrak{S}$ and each measurable set σ in $\mathfrak{S}$ the operators $\hat{A}_\sigma$ and A_σ defined in* (2) *and* (4) *are closed and densely defined in $\mathfrak{H}^p$. For any set σ in Σ we have the relations* (6) *and* (7) *and, for σ in Σ_0, A_σ is in $\mathfrak{A}$ and* (8) *holds. For a sequence $\{\sigma_m\}$ satisfying* (3), *A is given by* (9) *and the domain of A consists precisely of those φ for which the limit in* (9) *exists.*

Our next concern will be with the question of the existence of a resolution of the identity for A. This leads us to formulate the definition of an unbounded spectral operator in terms of the previously defined notion of a bounded spectral operator.

→ 3 DEFINITION. The operator A given in (4) will be called a *spectral operator* if for each σ in Σ_0 the bounded operator A_σ is a spectral operator and if, in addition, the resolutions of the identity $E(\delta; A_\sigma)$ are uniformly bounded, that is,

$$\sup_{\sigma \in \Sigma_0} \sup_{\delta \in \mathscr{B}} |E(\delta; A_\sigma)| < \infty. \tag{10}$$

The operator A is called a *scalar type* operator if it is a spectral operator for which each of the operators A_σ with σ in Σ_0 is a scalar type operator.

The existence of a resolution of the identity for such unbounded spectral operators is given in the following theorem.

→ 4 THEOREM. *The closed operator A given in (4) is a spectral operator if and only if*

$$\text{(i)} \qquad \operatorname*{ess\,sup}_{s\in\mathfrak{S}_j} |E(\hat{\lambda}_{jk}(s); \hat{A}(s))| < \infty, \qquad 1 \leqq k \leqq j \leqq p.$$

When this condition is satisfied, the set function

$$\text{(ii)} \qquad E(\delta; A) = \int_{\mathfrak{S}} E(\delta; \hat{A}(s))e(ds), \qquad \delta \in \mathscr{B},$$

is a bounded countably additive spectral measure defined on the Borel sets in the complex plane and whose values are projection operators in $\mathfrak{H}^p$. Furthermore, for every set σ in Σ_0, we have

$$\text{(iii)} \qquad E(\delta; A)A_\sigma = A_\sigma E(\delta; A), \qquad \delta \in \mathscr{B},$$

and

$$\text{(iv)} \qquad E(\delta; A)e(\sigma) = E(\delta; A_\sigma)e(\sigma), \qquad \delta \in \mathscr{B}.$$

This last property uniquely determines the spectral measure $E(\delta; A)$.

PROOF. The first statement is a corollary of Lemma 10.4 and Theorem 10.6 and the second follows from Theorem 10.5 by observing that its proof does not use the boundedness of $\hat{A}(s)$. Since $E(\delta; \hat{A}(s))$ and $\hat{A}(s)$ commute for each s in $\mathfrak{S}$, the equation (iii) follows from Theorem 9.3. To prove (iv) we see from (1) that

$$\begin{aligned} E(\delta; A)e(\sigma) &= \int_\sigma E(\delta; \hat{A}(s))e(ds) \\ &= \int_\sigma E(\delta; \hat{A}_\sigma(s))e(ds) \\ &= \left\{\int_{\mathfrak{S}} E(\delta; \hat{A}_\sigma(s))e(ds)\right\}e(\sigma) \\ &= E(\delta; A_\sigma)e(\sigma). \end{aligned}$$

The final assertion is clear since $E(\delta; A_\sigma)$ is uniquely determined by A, $\sigma \in \Sigma_0$, and δ in $\mathscr{B}$. Q.E.D.

A scalar type operator may be expressed in terms of its resolution of the identity, as is shown in the following general type eigenvalue expansion theorem.

5 THEOREM. *If A is a scalar type operator, then*

$$A\varphi = \lim_{r\to\infty} \int_{|\lambda|\leqq r} \lambda E(d\lambda; A)\varphi, \qquad \varphi \in \mathfrak{D}(A).$$

PROOF. Let $\{\sigma_m\}$ be a sequence satisfying (3) and let δ_r be the set of complex numbers λ with $|\lambda| \leqq r$. Then, for φ in $\mathfrak{D}(A)$, $e(\sigma_m)\varphi$ is in $\mathfrak{D}(A)$ by (6). Also it follows from (7) and (iv) of Theorem 4 that

$$\begin{aligned} E(\delta_r; A)Ae(\sigma_m)\varphi &= E(\delta_r; A)A_{\sigma_m}e(\sigma_m)\varphi \\ &= E(\delta_r; A_{\sigma_m})A_{\sigma_m}e(\sigma_m)\varphi \\ &= \int_{|\lambda|\leqq r} \lambda E(d\lambda; A_{\sigma_m})e(\sigma_m)\varphi \\ &= \int_{|\lambda|\leqq r} \lambda E(d\lambda; A)e(\sigma_m)\varphi. \end{aligned}$$

It is seen from (9) that $Ae(\sigma_m)\varphi \to A\varphi$, so that

$$E(\delta_r; A)A\varphi = \int_{|\lambda|\leqq r} \lambda E(d\lambda; A)\varphi,$$

and the desired conclusion follows, from the countable additivity of E, by letting $r \to \infty$. Q.E.D.

As a corollary to Theorem 4 we see from Theorem 11.18 and Corollary 11.23 that in case $p = 2$, the criteria may be stated as follows.

6 THEOREM. *Let $p = 2$, $\hat{\delta}^2 = (\hat{a}_{11} - \hat{a}_{22})^2 + 4\hat{a}_{12}\hat{a}_{21}$, and define the sets $\mathfrak{S}_1 = \{s \mid \hat{\delta}(s) = 0\}$, and $\mathfrak{S}_2 = \{s \mid \hat{\delta}(s) \neq 0\}$. Then the operator A of equation* (4) *is a spectral operator if and only if*

$$\text{(i)} \quad \operatorname*{ess\,sup}_{s\in\mathfrak{S}_2} \frac{|\hat{a}_{11}(s) - \hat{a}_{22}(s)|^2 + |\hat{a}_{21}(s)|^2 + |\hat{a}_{12}(s)|^2}{|\hat{\delta}^2(s)|} < \infty.$$

It is a scalar type spectral operator if and only if in addition to (i) *we have*

$$\text{(ii)} \qquad \hat{A}(s) = \hat{\lambda}(s)I$$

for some measurable function $\hat{\lambda}$ on R^N and almost all s in $\mathfrak{S}_1$.

The preceding results may be readily applied to a formal differential operator $A = (a_{jk})$ where a_{jk} is a polynomial $a_{jk}(\partial/\partial s_1, \ldots, \partial/\partial s_N)$, with constant coefficients, in the partial derivatives $\partial/\partial s_1, \ldots, \partial/\partial s_N$. We let $\Phi^p = \Phi \oplus \cdots \oplus \Phi$ be the dense linear subspace of $\mathfrak{H}^p$ determined by the set Φ of rapidly decreasing functions on R^N. For $\varphi = \{\varphi_1, \ldots, \varphi_p\}$ in Φ^p, let $A\varphi$ be that vector in $\mathfrak{H}^p$ whose jth component is

$$(A\varphi)_j = \sum_{k=1}^{p} a_{jk}(\partial)\varphi_k, \qquad \varphi \in \Phi^p. \tag{11}$$

Using the inversion formula

$$\varphi_k(s) = \frac{1}{(2\pi)^{N/2}} \int_{R^N} e^{ist}(F\varphi_k)(t)\, dt,$$

we have

$$\begin{aligned} a_{jk}(\partial_s)\varphi_k(s) &= \frac{1}{(2\pi)^{N/2}} \int_{R^N} e^{ist} a_{jk}(it)(F\varphi_k)(t)\, dt \\ &= (F^{-1}a_{jk}(i\,\cdot)F\varphi_k)(s). \end{aligned}$$

Here we have used the symbol $a_{jk}(i\,\cdot)$ for the operation of multiplication by the function $a_{jk}(it)$. Thus if $\hat{A} = (\hat{a}_{jk}(s))$, where $\hat{a}_{jk}(s) = a_{jk}(is)$, the equations (11) have the form

$$A\varphi = F^{-1}\hat{A}F(\varphi), \qquad \varphi \in \Phi^p. \tag{12}$$

This equation, together with Theorem 2, shows that the differential operator of (11) has a closed extension given by equation (4) with $\sigma = \mathfrak{S}$. We shall call this closed densely defined operator $A_{\mathfrak{S}}$ *the natural closed extension of* A.

The presence of the constant $i = \sqrt{-1}$ in the formal differential operator

$$\begin{pmatrix} \dfrac{\partial^2}{\partial s_1^2} & i\dfrac{\partial^2}{\partial s_1\, \partial s_2} \\ \dfrac{\partial^2}{\partial s_1\, \partial s_2} & \dfrac{\partial^2}{\partial s_2^2} \end{pmatrix} \tag{13}$$

prevents its natural closed extension to $\mathfrak{D}(A_{\mathfrak{S}})$ from being self adjoint, but nevertheless it has a resolution of the identity, for we have

$$\hat{A}(s) = \begin{pmatrix} -s_1^2 & -is_1s_2 \\ -s_1s_2 & -s_2^2 \end{pmatrix},$$

$$\delta^2(s) = (s_1^2 - s_2^2)^2 + 4is_1^2s_2^2,$$

and the fraction in the inequality (i) of Theorem 6 is

$$\frac{|s_1^2 - s_2^2|^2 + 2\,|s_1 s_2|^2}{\{(s_1^2 - s_2^2)^4 + 16 s_1^4 s_2^4\}^{1/2}}$$

which is bounded on all of R^n. Thus $A_{\mathfrak{S}}$ has a resolution of the identity. Furthermore, the set $\mathfrak{S}_1$ is the null set $\{s \mid s_1 = s_2 = 0\}$, so that $A_{\mathfrak{S}}$ is a scalar type operator to which Theorem 5 applies.

As another example, consider the formal differential operator

$$\begin{pmatrix} \dfrac{\partial^2}{\partial s_1^2} & -\alpha i \\ \beta i & \dfrac{\partial^2}{\partial s_2^2} \end{pmatrix}, \tag{14}$$

where α, β are positive real numbers. If $\alpha \neq \beta$, the corresponding closed operator $A_{\mathfrak{S}}$ cannot be self adjoint, but it always has a resolution of the identity for

$$\hat{A}(s) = \begin{pmatrix} -s_1^2 & -\alpha i \\ \beta i & -s_2^2 \end{pmatrix},$$

$$\hat{\delta}^2(s) = (s_1^2 - s_2^2)^2 + 4\alpha\beta,$$

so that the inequality (i) of Theorem 6 holds. Here the set $\mathfrak{S}_1$ is void so that $A_{\mathfrak{S}}$ is a scalar type spectral operator to which Theorem 5 applies.

Another type of differential operator whose extension is a spectral operator which is not a scalar operator is illustrated by the formal differential operator:

$$\begin{pmatrix} 2\dfrac{\partial^2}{\partial s_1\,\partial s_2} & i\dfrac{\partial^2}{\partial s_2^2} \\ i\dfrac{\partial^2}{\partial s_1^2} & 0 \end{pmatrix}. \tag{15}$$

Here we have $\mathfrak{S}_1 = R^N$ and $\mathfrak{S}_2$ is void, so that inequality (i) of Theorem 6 is satisfied but equation (ii) is not. Thus the natural closed extension $A_{\mathfrak{S}}$ of the formal operator (15) is a spectral operator which is neither a scalar type operator nor a nilpotent operator.

Occasionally in what follows we shall simply write $\hat{A}$ for the operator $\hat{A}_{\mathfrak{S}}$, since no confusion should arise. This simplification will not be made in the notation for the natural closed extension $A_{\mathfrak{S}}$, for in this case the symbol A is used for the restriction $A_{\mathfrak{S}}$ to Φ, that is, the formal differential operator which defines $A_{\mathfrak{S}}$.

The spectra of the unbounded operators we have been discussing in this section are not always as readily obtained as they were for the bounded operators in $\mathfrak{A}^p$. The expression given in Corollary 9.9 for the spectrum of an operator A in $\mathfrak{A}^p$ is not necessarily correct when the elements of $\hat{A}$ are unbounded functions of s. For a closed operator $A_{\mathfrak{S}}$ the resolvent set $\rho(A_{\mathfrak{S}})$ has been defined (see Section VII.9) as the set of all complex numbers λ for which $(\lambda I - A_{\mathfrak{S}})^{-1}$ exists as a bounded everywhere defined operator. The spectrum $\sigma(A_{\mathfrak{S}})$ of $A_{\mathfrak{S}}$ is defined to be the complement $\rho(A_{\mathfrak{S}})$. It is clear from the equation (4) that $\rho(A_{\mathfrak{S}}) = \rho(\hat{A}_{\mathfrak{S}})$ and thus $\sigma(A_{\mathfrak{S}}) = \sigma(\hat{A}_{\mathfrak{S}})$. It follows that $\rho(A_{\mathfrak{S}})$ consists of those λ for which $(\lambda I - \hat{A}(s))^{-1}$ exists for almost all s in $\mathfrak{S}$ and is essentially bounded on $\mathfrak{S}$. The last part of the proof of Corollary 9.9 does not use the boundedness of the elements of the matrix $\hat{A}(s)$ and shows that even for the unbounded operators $A_{\mathfrak{S}}$ considered in this section the spectrum $\sigma(A_{\mathfrak{S}})$ contains the set $\sigma_0(A_{\mathfrak{S}})$ defined by the equation

$$\sigma_0(A_{\mathfrak{S}}) = \bigcap_{e(\delta)=e} \overline{\bigcup_{s\in\delta} \sigma(\hat{A}(s))}. \tag{16}$$

In general, all we can say here is contained in the following theorem.

→ 7 THEOREM. *Let $\hat{A}_{\mathfrak{S}}$, $A_{\mathfrak{S}}$ be the closed densely defined operators defined in equations* (2) *and* (4) *respectively. Then $\sigma(A_{\mathfrak{S}}) = \sigma(\hat{A}_{\mathfrak{S}})$, $\rho(A_{\mathfrak{S}}) = \rho(\hat{A}_{\mathfrak{S}})$ and $\sigma(A_{\mathfrak{S}}) \supseteq \sigma_0(A_{\mathfrak{S}})$. If $A_{\mathfrak{S}}$ is a scalar type spectral operator, then $\sigma(A_{\mathfrak{S}}) = \sigma_0(A_{\mathfrak{S}})$. For any $A_{\mathfrak{S}}$, not necessarily a spectral operator, the spectrum $\sigma(A_{\mathfrak{S}})$ consists of $\sigma_0(A)$ and all complex numbers λ in its complement for which*

$$\operatorname*{ess\,sup}_{s\in\mathfrak{S}} |(\lambda I - \hat{A}(s))^{-1}| = \infty. \tag{i}$$

The resolvent set consists of all complex numbers $\lambda \notin \sigma_0(A_{\mathfrak{S}})$ for which

$$\operatorname*{ess\,sup}_{s\in\mathfrak{S}} |(\lambda I - \hat{A}(s))^{-1}| < \infty. \tag{ii}$$

PROOF. It should be noted that from the definition of $\sigma_0(A_{\mathfrak{S}})$ it follows that $(\lambda I - \hat{A}(s))^{-1}$ exists for almost all s in $\mathfrak{S}$ provided that λ is not in $\sigma_0(A)$ and so the expressions occurring in (i) and (ii) are well defined. In view of the observations made preceding the statement of the theorem, the last two conclusions of the theorem are clear and we shall only show that $\sigma(A_{\mathfrak{S}}) = \sigma_0(A_{\mathfrak{S}})$ for scalar type spectral operators $A_{\mathfrak{S}}$. If $A_{\mathfrak{S}}$ is such an operator then it follows from Definition 3 and Corollary 10.9 that $\hat{A}_{\mathfrak{S}}$ is a scalar type spectral operator and we see from Theorem VII.1.8 that for $\lambda \notin \sigma_0(A_{\mathfrak{S}})$

$$(\lambda I - \hat{A}_{\mathfrak{S}}(s))^{-1} = \sum_{j=1}^{k} \frac{E(\hat{\lambda}_{kj}(s); \hat{A}(s))}{\lambda - \hat{\lambda}_{kj}(s)}, \qquad s \in \mathfrak{S}_k, \, k = 1, \ldots, p.$$

Since $\lambda \notin \sigma_0(A_{\mathfrak{S}})$ the functions $(\lambda - \hat{\lambda}_{kj}(s))^{-1}$ are essentially bounded on $\mathfrak{S}_k$ and since $A_{\mathfrak{S}}$ is a spectral operator it follows from Theorem 4 that the functions $|E(\hat{\lambda}_{kj}(s); \hat{A}(s))|$ are essentially bounded on $\mathfrak{S}_k$, $k = 1, \ldots, p$. Thus (ii) holds and λ is therefore in the resolvent set $\rho(\hat{A}_{\mathfrak{S}}) = \rho(A_{\mathfrak{S}})$. This proves that $\sigma_0(A_{\mathfrak{S}}) = \sigma(A_{\mathfrak{S}})$ if $A_{\mathfrak{S}}$ is a scalar type spectral operator. Q.E.D.

The equation $\sigma_0(A_{\mathfrak{S}}) = \sigma(A_{\mathfrak{S}})$ is sometimes true for spectral operators other than those of scalar type. Also the spectrum of certain nilpotent operators can be the whole complex plane. Some illustrative examples will be illuminating. Let $p = 2$ and consider the formal differential operator

$$A = \begin{pmatrix} 0 & \partial/\partial s_1 \\ 0 & 0 \end{pmatrix}. \tag{17}$$

Then $\sigma_0(\hat{A}) = \{0\}$, $\hat{A}(s)^2 = 0$, and for every $\lambda \notin \sigma_0(A_{\mathfrak{S}})$, that is for every $\lambda \neq 0$, we have

$$(\lambda - \hat{A}(s))^{-1} = \frac{I}{\lambda} + \frac{1}{\lambda^2} \begin{pmatrix} 0 & is_1 \\ 0 & 0 \end{pmatrix},$$

which is not essentially bounded. Hence the spectrum $\sigma(A_{\mathfrak{S}})$ is the whole plane. In fact, if p and N are arbitrary positive integers and the formal differential operator

$$A = (a_{jk}), \qquad a_{jk} = a_{jk}(\partial/\partial s_1, \ldots, \partial/\partial s_N), \tag{18}$$

where a_{jk} is a polynomial in $s_1, \ldots, s_N$ with constant coefficients, is nilpotent on Φ^p, that is, $A^n \Phi^p = 0$ for some positive integer n, then $\sigma(A_{\mathfrak{S}})$ is the whole complex plane unless all the polynomials a_{jk}, $1 \leqq j, k \leqq p$ are constants (that is, independent of $s_1, \ldots, s_N$), in which case $\sigma(A_{\mathfrak{S}}) = \{0\}$. For, since $\hat{N}^n(s) = 0$ almost everywhere, we have $\sigma_0(A_{\mathfrak{S}}) = \{0\}$ and for any $\lambda \neq 0$

$$(\lambda I - \hat{A}(s))^{-1} = \frac{I}{\lambda} + \frac{\hat{A}(s)}{\lambda^2} + \cdots + \frac{\hat{A}^{n-1}(s)}{\lambda^n},$$

which shows that the elements of $(\lambda I - \hat{A}(s))^{-1}$ are polynomials in $s_1, \ldots, s_N$ and so $|(\lambda I - \hat{A}(s))^{-1}|$ cannot be essentially bounded unless all these polynomials are constants. If this is the case, then $\lambda I - \hat{A}(s) = \{(\lambda I - \hat{A}(s))^{-1}\}^{-1}$ has constant elements and so does $\hat{A}(s)$. We may therefore conclude from Theorem 7 that the natural closed extension $A_{\mathfrak{S}}$ of the formal differential operator (18) with $A^n\Phi^p = 0$ has its spectrum $\sigma(A_{\mathfrak{S}})$ the whole complex plane unless A is of order zero, that is, none of its elements a_{jk} contain any derivatives, in which case $\sigma(A_{\mathfrak{S}}) = \{0\}$.

Another illuminating example, with $p = 2$, is the perturbed Laplacian

$$A = \begin{pmatrix} \nabla^2 & a \\ 0 & \nabla^2 \end{pmatrix}, \qquad \nabla^2 = \frac{\partial^2}{\partial s_1^2} + \cdots + \frac{\partial^2}{\partial s_N^2}, \tag{19}$$

and $a = a(\partial/\partial s_1, \ldots, \partial/\partial s_N)$ where a is a polynomial of degree at most 4. Here

$$\hat{A}(s) = \begin{pmatrix} -|s|^2 & \hat{a}(s) \\ 0 & -|s|^2 \end{pmatrix} \tag{20}$$

and so $\sigma_0(A_{\mathfrak{S}}) = (-\infty, 0]$. If $\hat{a}(s) \equiv 0$ then Theorem 6 shows that $\hat{A}$ is a scalar type operator and the preceding theorem shows that $\sigma(A_{\mathfrak{S}}) = (-\infty, 0]$. So we shall suppose that $\hat{a}(s)$ is not identically zero. We have, for $\lambda \notin (-\infty, 0]$

$$(\lambda I - \hat{A}(s))^{-1} = \frac{I}{\lambda + |s|^2} + \frac{1}{(\lambda + |s|^2)^2} \begin{pmatrix} 0 & \hat{a}(s) \\ 0 & 0 \end{pmatrix},$$

and since $|\lambda + |s|^2|^{-1}$ and $|\hat{a}(s)|\,|\lambda + |s|^2|^{-2}$ are bounded on R^N we have λ in $\rho(A_{\mathfrak{S}})$. Thus $\sigma(A_{\mathfrak{S}}) = (-\infty, 0] = \sigma_0(A_{\mathfrak{S}})$ even though $A_{\mathfrak{S}}$ is not a scalar type spectral operator.

Other examples are afforded by the formal differential operators (13), (14), and (15) and the spectra of their natural closed extensions are readily obtained.

An amusing example which has features not seen in the preceding examples is given by the formal differential operator

$$A = \begin{pmatrix} \dfrac{\partial}{\partial s_1} & -\dfrac{\partial}{\partial s_2} \\[2ex] \dfrac{\partial}{\partial s_2} & \dfrac{\partial}{\partial s_1} \end{pmatrix}. \tag{21}$$

Here we have

$$\hat{A}(s) = +i \begin{pmatrix} s_1 & -s_2 \\ s_2 & s_1 \end{pmatrix}, \qquad s = (s_1, s_2) \in R^2. \tag{22}$$

For each s in R^2 this matrix is normal and

$$\hat{A}(s)\hat{A}^*(s) = |s|^2 I = \hat{A}^*(s)\hat{A}(s), \qquad s \in R^2, \tag{23}$$

so that

$$\frac{\hat{A}(s)}{|s|}\left(\frac{\hat{A}(s)}{|s|}\right)^* = I, \qquad 0 \neq s \in R^2. \tag{24}$$

Equation (24) shows that $|s|^{-1}\hat{A}(s)$ is unitary for $s \neq 0$. Thus

$$\hat{A}^{-1}(s) = \frac{-i}{|s|^2}\begin{pmatrix} s_1 & s_2 \\ -s_2 & s_1 \end{pmatrix}, \qquad 0 \neq s \in R^2, \tag{25}$$

and

$$|\hat{A}(s)\psi(s)| = |s|\,|\psi(s)|, \qquad s \in \mathfrak{S},\ \psi \in \mathfrak{H}^2. \tag{26}$$

The equation $A\varphi = 0$, being equivalent to the Cauchy-Riemann equation for the real and imaginary parts of a holomorphic function, has no non-zero solution φ in Φ^2. For, by classical function theory, any solution $\varphi = (\varphi_1, \varphi_2)$ in Φ^2 of the equation $A\varphi = 0$ would define an everywhere analytic function $f(z) = \varphi_1(s) + i\varphi_2(s)$ of the variable $z = s_1 + is_2$, and since f vanishes at infinity, it is identically zero.

Equation (26) shows more, for it shows that the only ψ in $\mathfrak{D}(\hat{A}_1)$ with $\hat{A}_1\psi = 0$ is $\psi = 0$ and thus the only φ in $\mathfrak{D}(A_{\mathfrak{S}})$ with $A_{\mathfrak{S}}\varphi = 0$ is $\varphi = 0$.

The set $\mathfrak{S}_1$ for the operator (21) is a null set and

$$\hat{\lambda}_{21}(s) = s_2 + is_1, \qquad \hat{\lambda}_{22}(s) = -s_2 + is_1, \tag{27}$$

and the corresponding projections $\hat{E}_{2j}(s) = E(\hat{\lambda}_{2j}(s); \hat{A}(s))$, $j = 1, 2$, which may be obtained from equation (46) in Section 11, are

$$\hat{E}_{21}(s) = \frac{1}{2}\begin{pmatrix} 1 & i \\ -i & 1 \end{pmatrix}, \qquad \hat{E}_{22}(s) = \frac{1}{2}\begin{pmatrix} 1 & -i \\ i & 1 \end{pmatrix} \tag{28}$$

and thus seen to be independent of s. It is clear from Theorem 7 and equations (27) that the spectrum $\sigma(A_{\mathfrak{S}})$ is the whole complex plane and elementary considerations show that every spectral point is in the continuous spectrum. The fact that the projections $\hat{E}_{jk}$ are independent of s

simplifies the operational calculus for $A_{\mathfrak{S}}$. To understand what we mean by this statement the reader need not be familiar with Section VII.9 which develops an operational calculus for unbounded closed operators, nor does he have to know the contents of Chapter XVIII which contain the development of an operational calculus for unbounded spectral operators. For here we are dealing with a type of operator $A_{\mathfrak{S}}$ with special properties which make the definition of functions $f(A_{\mathfrak{S}})$ of $A_{\mathfrak{S}}$ very easy, as we shall now show.

Suppose that the formal differential operator A on Φ^p has the property that for almost all s in R^N all eigenvalues of the matrices $\hat{A}(s)$ are simple roots of the minimal polynomial for $\hat{A}(s)$. This is the case if $A_{\mathfrak{S}}$ is a scalar type operator. Let f be any measurable function defined on the spectrum $\sigma(A_{\mathfrak{S}}) = \sigma(\hat{A})$. We simply define the $p \times p$ matrix $f(\hat{A})(s)$ whose elements are measurable functions on R^N by the equation $f(\hat{A})(s) = f(\hat{A}(s))$. (If the roots of the minimal polynomials of $\hat{A}(s)$ are not all simple, f would have to have derivatives at certain points of the spectrum and in general if we have no special information about A, then this equation still defines $f(\hat{A})(s)$ provided f has $p-1$ derivatives on $\sigma(A_{\mathfrak{S}})$.) Thus the operator $f(\hat{A})$ is defined by equation (2) and $f(A_{\mathfrak{S}})$ by equation (4). These operators are, as we have observed, bounded if f is essentially bounded on the spectrum. For example, if $f(\lambda) = \exp t\lambda$ and if the spectrum $\sigma(A_{\mathfrak{S}})$ lies in a left half plane, then $\exp tA_{\mathfrak{S}}$ is a bounded operator for $t \geqq 0$. This is the function one forms to solve the Cauchy problem $\varphi'(t) = A_{\mathfrak{S}}\varphi(t)$, $\varphi(0) = \varphi_0$; which we shall briefly illustrate in Theorems 19 and 21 that follow. But now let us return to elucidate our earlier statement: the fact that the projections $\hat{E}_{jk}$ given in (28) are independent of s simplifies the operational calculus for the natural closed extension $A_{\mathfrak{S}}$ of the operator (21). Suppose f is a bounded measurable function on the complex plane (which is the spectrum of $A_{\mathfrak{S}}$) and is in $L_1(R^2)$. Then, using (27) and setting

$$f_{21}(s) = f(s_2 + is_1), \qquad f_{22}(s) = f(-s_2 + is_1)$$

we have

$$f(\hat{A}(s)) = \hat{E}_{21}f_{21}(s) + \hat{E}_{22}f_{22}(s)$$

and equation (4) which defines $f(A_{\mathfrak{S}})$ gives, since $\hat{E}_{ij}$ is independent of s,

$$f(A_{\mathfrak{S}}) = \hat{E}_{21}F^{-1}\mathbf{f}_{21}F + \hat{E}_{22}\,F^{-1}\mathbf{f}_{22}\,F,$$

where $\mathbf{f}_{ij}$ is the operation of multiplication by f_{ij}. Thus it follows that

$$f(A_{\mathfrak{S}})\varphi = \frac{1}{2}\begin{pmatrix} 1 & i \\ -i & 1 \end{pmatrix}(\tau^{-1}f_{21}) * \varphi + \frac{1}{2}\begin{pmatrix} 1 & -i \\ i & 1 \end{pmatrix}(\tau^{-1}f_{22}) * \varphi, \qquad \varphi \in \mathfrak{H}^2.$$

That the convolution $(\tau^{-1}f_{jk}) * \varphi$ exists for every φ in $\mathfrak{H}^2$ is readily seen from Theorem 11.5 by using the sequence

$$\lambda_n(\sigma) = \int_{|s| \leqq n} f_{jk}(s)\chi_\sigma(s)\, ds.$$

There are two more illustrative examples we should like to discuss. They are both perturbed Laplacian operators, the first being (here again we take $p = N = 2$)

$$A = \begin{pmatrix} \nabla^2 & a \\ b & \nabla^2 \end{pmatrix}, \tag{29}$$

where

$$\nabla^2 = \frac{\partial^2}{\partial s_1^2} + \frac{\partial^2}{\partial s_2^2}, \tag{30}$$

$$a = \alpha_0 + \alpha_1 \frac{\partial}{\partial s_1} + \alpha_2 \frac{\partial}{\partial s_2}, \quad b = \beta_0 + \beta_1 \frac{\partial}{\partial s_1} + \beta_2 \frac{\partial}{\partial s_2}.$$

We have

$$\hat{A}(s) = \begin{pmatrix} -|s|^2 & \hat{a}(s) \\ \hat{b}(s) & -|s|^2 \end{pmatrix}, \tag{31}$$

and $\delta^2(s) = 4\hat{a}(s)\hat{b}(s)$. Thus Theorem 6 shows that if either a or b is zero, the natural closed extension $A_{\mathfrak{S}}$ of A is a spectral operator (since in this case $\mathfrak{S}_1 = R^2$) which is of scalar type if and only if both a and b are zero. In case $\hat{a}(s)\hat{b}(s)$ is not identically zero, the set $\mathfrak{S}_1$ consists of the two lines defined by the equation $\hat{a}(s)\hat{b}(s) = 0$ and is thus a null set. In this case the fraction appearing in (i) of Theorem 6 is

$$\frac{|\hat{a}(s)|^2 + |\hat{b}(s)|^2}{4\,|\hat{a}(s)\hat{b}(s)|}, \qquad s \in \mathfrak{S}_2, \tag{32}$$

which is clearly essentially bounded on $\mathfrak{S}_2$ if and only if for some constant $\alpha \neq 0$, $\hat{b} = \alpha\hat{a}$ and hence the expression (32) is the constant $(1 + |\alpha|^2)/4\,|\alpha|$. We therefore conclude from Theorem 6 that if $\hat{a}\hat{b} \neq 0$, the operator $A_{\mathfrak{S}}$ is a spectral operator if and only if $\hat{a}$ and $\hat{b}$ are proportional, in which case $A_{\mathfrak{S}}$ is a scalar type spectral operator.

It may therefore be concluded that the natural closed extension of the formal differential operator (29) is a spectral operator if and only if it has the form

$$A = \begin{pmatrix} \nabla^2 & \beta^2 a \\ \alpha^2 a & \nabla^2 \end{pmatrix} \tag{33}$$

where α, β are any complex constants. The natural closed extension $A_{\mathfrak{S}}$ of the operator (33) is of scalar type if and only if either $\alpha\beta a \neq 0$ or $a = 0$. Let us consider the case where $\alpha\beta a \neq 0$. Then $\mathfrak{S}_1$ is a null set and we may take

$$\hat{\lambda}_{21}(s) = -|s|^2 + \alpha\beta\hat{a}(s), \qquad \hat{\lambda}_{22}(s) = -|s|^2 - \alpha\beta\hat{a}(s), \qquad s \in \mathfrak{S}_2. \tag{34}$$

Since $\hat{a}(s)$ is linear in s_1, s_2, Theorem 7 shows that the spectrum $\sigma(A_{\mathfrak{S}})$ lies in some left half plane $\mathscr{R}(\lambda) \leqq \omega$. This suggests the possibility that $A_{\mathfrak{S}}$ may be the infinitesimal generator of a strongly continuous semi-group $T(t)$, $t \geqq 0$ of bounded linear operators in $\mathfrak{H}^2$ and that we may be able to solve the abstract Cauchy problem, $\varphi' = A_{\mathfrak{S}}\varphi$, $\varphi(0) = \varphi_0$, which will be precisely formulated presently. As the following theorem shows, the Hille-Yosida-Phillips theorem (VIII.1.13) is readily applied to this operator.

8 THEOREM. *If the natural closed extension $A_{\mathfrak{S}}$ of the formal differential operator $A = (a_{ij})$, where $a_{ij} = a_{ij}(\partial/\partial s_1, \ldots, \partial/\partial s_N)$, $1 \leqq i, j \leqq p$, are polynomials with constant coefficients, is a scalar type spectral operator, then it is the infinitesimal generator of a strongly continuous semi-group (defined on $[0, \infty)$) of bounded linear operators in $\mathfrak{H}^p$ if and only if its spectrum lies in some left half plane.*

PROOF. Suppose that $A_{\mathfrak{S}}$ is a scalar type operator whose spectral points satisfy the inequality $\mathscr{R}(\lambda) \leqq \omega$. Then it follows from Corollary 10.9 that for almost all s in R^N, $\hat{A}(s)$ is a scalar type operator in E^p. Thus there is a set R_1^N in R^N whose complement is a null set such that for every s in R_1^N we have $\nu(\hat{\lambda}(s)) = 1$ for every eigenvalue $\hat{\lambda}(s)$ of the matrix $\hat{A}(s)$, where, as usual, the number $\nu(\hat{\lambda}(s))$ is the multiplicity of $\hat{\lambda}(s)$ as a root of the minimal polynomial for $\hat{A}(s)$. Then for every real $\lambda > \omega$ and every positive integer n it follows from Theorem VII.1.8 that

$$R(\lambda; \hat{A}(s))^n = \sum_{j=1}^{k} \frac{\hat{E}_{kj}(s)}{(\lambda - \hat{\lambda}_{kj}(s))^n}, \qquad s \in \mathfrak{S}_k \cap R_1^N, \tag{35}$$

where $\hat{E}_{kj}(s) = E(\hat{\lambda}_{kj}(s);\ \hat{A}(s))$. Since $\lambda - \omega < |\lambda - \hat{\lambda}_{kj}(s)|$, Theorem 4 and equation (35) show that

$$\text{(36)} \qquad \operatorname*{ess\,sup}_{s \in \mathfrak{S}} |R(\lambda; \hat{A}(s))^n| \leqq \frac{M}{(\lambda - \omega)^n}, \qquad n = 1, 2, \ldots .$$

It follows from Corollaries 9.5 and 10.9 that

$$\text{(37)} \qquad |R(\lambda; A_{\mathfrak{S}})^n| = \operatorname*{ess\,sup}_{s \in \mathfrak{S}} |R(\lambda; \hat{A}_{\mathfrak{S}})^n(s)| = \operatorname*{ess\,sup}_{s \in \mathfrak{S}} |R(\lambda; \hat{A}(s))^n|$$

and we thus conclude that

$$\text{(38)} \qquad |R(\lambda; A_{\mathfrak{S}})^n| \leqq \frac{M}{(\lambda - \omega)^n}, \qquad \lambda > \omega, \qquad n = 1, 2, \ldots .$$

It follows from the Hille-Yosida-Phillips theorem (VIII.1.13) that $A_{\mathfrak{S}}$ is the infinitesimal generator of a strongly continuous semi-group (defined on $[0, \infty)$) of bounded linear operators in $\mathfrak{H}^p$. This theorem shows also that if, conversely, $A_{\mathfrak{S}}$ is such an infinitesimal generator then (38) holds for some real constants M and ω. To complete the proof it thus suffices to show that (38) implies that every spectral point λ of $A_{\mathfrak{S}}$ has $\mathscr{R}(\lambda) \leqq \omega$. Now (37) and (38) show that there is a set $\mathfrak{S}_0$ whose complement in $\mathfrak{S}$ is a null set and for which

$$\text{(39)} \qquad |R(\lambda; \hat{A}(s))^n| \leqq \frac{M}{(\lambda - \omega)^n}, \qquad \lambda > \omega, \qquad s \in \mathfrak{S}_0 .$$

According to Theorem 7 it suffices to prove that for every s in $\mathfrak{S}_0$ the real parts of all the eigenvalues of $\hat{A}(s)$ are at most equal to ω. If this is not true, there are integers j, k with $\mathscr{R}(\hat{\lambda}_{kj}(s)) > \omega$ for some s in $\mathfrak{S}_k \cap \mathfrak{S}_0$. Since $\hat{E}_{kj}(s) \neq 0$ there is a vector $\psi(s)$ in E^p with $|\psi(s)| = 1$ and $\psi(s) = \hat{E}_{kj}(s)\psi(s)$. Thus, since $\hat{E}_{kj}(s)$ and $\hat{E}_{kq}(s)$ are disjoint projections if $q \neq j$, it follows from (35) that $R(\lambda; \hat{A}(s))^n\psi(s) = \psi(s)(\lambda - \hat{\lambda}_{kj}(s))^{-n}$ and (39) gives

$$\text{(40)} \qquad 1 = |\psi(s)| \leqq M \frac{|\lambda - \hat{\lambda}_{kj}(s)|^n}{(\lambda - \omega)^n}, \qquad n = 1, 2, \ldots .$$

Since $\mathscr{R}(\hat{\lambda}_{kj}(s)) > \omega$ we may fix λ so large that $|\lambda - \hat{\lambda}_{kj}(s)| < \lambda - \omega$ and thus the fraction appearing in (40) approaches zero as $n \to \infty$. Q.E.D.

It would be erroneous to conjecture that the only spectral operators which are infinitesimal generators of strongly continuous semi-groups (on $[0, \infty)$) are of scalar type.

To see this we consider another perturbed Laplacian operator for arbitrary positive integers p and N. The unperturbed operator is one that gives rise to a set $\varphi_j'(t) = \alpha^2 \nabla^2 \varphi_j(t)$, $j = 1, \ldots, p$, of diffusion equations which occur in the theory of heat flow and in other similar problems such as the diffusion of light or the theory of neutron diffusion. Actually these diffusion equations are only approximations to the true equations in the case of light or neutron diffusion. In these cases the true equations correspond to a perturbed Laplacian operator. What we shall endeavor to see here is how much our theory so far developed will allow us to perturb the operator $\alpha^2\nabla^2 I$ and still permit us to solve the Cauchy problem for the perturbed diffusion equations. The perturbation will be made in two steps. We first use a nilpotent perturbation consisting of differential operators of orders at most 2. After that we perturb further by allowing the addition of any bounded operator in $\mathfrak{H}^p$. Similar examples of higher order, some having unbounded perturbations, will be found in Exercises 65, ..., 69, inclusive, of Section 14.

We begin here by considering the operator

$$A = \alpha^2\nabla^2 I + (a_{jk}) = \begin{pmatrix} \alpha^2\nabla^2, & a_{12}, & \ldots, & a_{1p} \\ 0, & \alpha^2\nabla^2 & & \cdot \\ \cdot & & & \cdot \\ \cdot & & & a_{(p-1)p} \\ 0, & 0, & \ldots, & \alpha^2\nabla^2 \end{pmatrix} \tag{41}$$

where $\alpha \neq 0$ is a constant either real or purely imaginary,

$$\nabla^2 = \frac{\partial^2}{\partial s_1^2} + \cdots + \frac{\partial^2}{\partial s_N^2}, \qquad a_{jk} = a_{jk}(\partial/\partial s_1, \ \ldots, \ \partial/\partial s_N), \tag{42}$$

and a_{jk} is a polynomial in $s_1, \ldots, s_N$ with constant coefficients, of degree at most 2, and with $a_{jk} = 0$ for $j \geqq k$. In this example $\mathfrak{S}_1 = R^N$, $\hat{\lambda}_{11}(s) = -\alpha^2|s|^2 = -\alpha^2(s_1^2 + \cdots + s_N^2)$, $\sigma_0(A_{\mathfrak{S}}) = (-\infty, 0]$, if α is real, $\sigma_0(A_{\mathfrak{S}}) = [0, \infty)$ if α is purely imaginary, and the matrix $\hat{N}(s) = \hat{A}(s) - \hat{\lambda}_{11}(s)I$ is a nilpotent; $\hat{N}^p(s) = 0$ for every s in R^N. We shall from here on assume that $\alpha^2 > 0$. The case where $\alpha^2 < 0$ is entirely analogous and the reader will have no difficulty in supplying the changes necessary for that case. An alternate way of treating the case $\alpha^2 < 0$ is to apply the results obtained for the case $\alpha^2 > 0$ to the operator $-A$.

We fix $\omega > 0$ which makes the constant

$$K = \sup_{s \in R^N} \frac{|\hat{N}(s)|}{\omega + \alpha^2|s|^2} \tag{43}$$

finite since the elements of $\hat{N}(s)$ are polynomials in $s_1, \ldots, s_N$ of degree at most 2. For $n = 1, 2, \ldots,$ and every real $\lambda > \omega$ it follows from Theorem VII.1.8 that

$$(\lambda - \omega)^n R(\lambda; \hat{A}(s))^n \tag{44}$$

$$= \sum_{\nu=0}^{p-1} \frac{n(n+1)\cdots(n+\nu-1)}{\nu\,!} \frac{(\lambda-\omega)^n}{(\lambda+\alpha^2|s|^2)^{n+\nu}} \hat{N}^\nu(s),$$

where the term corresponding to $\nu = 0$ in this sum is $(\lambda-\omega)^n(\lambda+\alpha^2|s|^2)^{-n}I$ and thus has norm at most 1 for all $\lambda > \omega$, $s \in R^N$ and $n = 1, 2, \ldots$. If $\nu > 0$ the function of λ given by the expression $(\lambda-\omega)^n(\lambda+\alpha^2|s|^2)^{-n-\nu}$ on the interval $\omega \leqq \lambda < \infty$ has its maximum value when $\lambda = \omega + n(\omega+\alpha^2|s|^2)\nu^{-1}$ from which it follows that

$$\sup_{\omega \leqq \lambda < \infty} \frac{(\lambda-\omega)^n}{(\lambda+\alpha^2|s|^2)^{n+\nu}} = \nu^\nu \frac{n^n}{(n+\nu)^{n+\nu}} \frac{1}{(\omega+\alpha^2|s|^2)^\nu},$$

and from (43) that

$$\left| \frac{n(n+1)\cdots(n+\nu-1)}{\nu\,!} \frac{(\lambda-\omega)^n}{(\lambda+\alpha^2|s|^2)^{n+\nu}} \hat{N}^\nu(s) \right|$$

$$\leqq \frac{\nu^\nu}{\nu!} \frac{n(n+1)\cdots(n+\nu-1)}{n^\nu} \left(\frac{n}{n+\nu}\right)^{n+\nu} K^\nu,$$

which is bounded in n since, as $n \to \infty$, it converges to $e^{-\nu}\nu^\nu K^\nu/\nu!$. Thus equations (44) and (37) establish the inequality (38) and the Hille-Yosida-Phillips theorem (VIII.1.13) shows that the natural closed extension $A_{\mathfrak{S}}$ of the formal partial differential operator (41) is the infinitesimal generator of a semi-group (on $[0, \infty)$) of bounded linear operators in $\mathfrak{H}^p$.

We shall next show that the spectrum $\sigma(A_{\mathfrak{S}}) = (-\infty, 0]$. To do this we appeal to Theorem 7 and fix any $\lambda \notin \sigma_0(A_{\mathfrak{S}}) = (-\infty, 0]$. The matrix $(\lambda I - \hat{A}(s))^{-1}$ is given by equation (44) with $n = 1$ and so

$$(\lambda I - \hat{A}(s))^{-1} = \sum_{\nu=0}^{p-1} \frac{\hat{N}^\nu(s)}{(\lambda+\alpha^2|s|^2)^{\nu+1}}. \tag{45}$$

Since $\lambda \notin (-\infty, 0]$ and the elements of $\hat{N}(s)$ are polynomials in $s_1, \ldots, s_N$ of degree at most 2, both $|\lambda+\alpha^2|s|^2|^{-1}$ and $|\hat{N}(s)(\lambda+\alpha^2|s|^2)^{-1}|$ are bounded on R^N and so equation (45) shows that the condition (ii) of Theorem 7 is satisfied, which proves that λ is in $\rho(A_{\mathfrak{S}})$ and thus that $\sigma(A_{\mathfrak{S}}) = (-\infty, 0]$.

Since $\mathfrak{S}_1 = R^N$ we have $E(\hat{\lambda}_{11}(s); \hat{A}(s)) = I$ for every s in R^N and so it is clear from Theorem 4 that $A_{\mathfrak{S}}$ is a spectral operator.

The fact that $A_{\mathfrak{S}}$ is a spectral operator is, as the argument shows, independent of any restrictions on the order of the operators a_{jk}.

We next give characterizations of the domains $\mathfrak{D}(\hat{A}_{\mathfrak{S}})$ and $\mathfrak{D}(A_{\mathfrak{S}})$ which are more convenient than those used in (2) and (4) which define them. From (2) we have

$$\mathfrak{D}(\hat{A}_{\mathfrak{S}}) = \{\psi \in \mathfrak{H}^p \,|\, \int_{\mathfrak{S}} |\hat{A}(s)\psi(s)|^2 \, ds < \infty\}.$$

If ψ is in $\mathfrak{D}(\hat{A}_{\mathfrak{S}})$ and $\zeta = \hat{A}\psi$ then

$$\zeta_j(s) = -\alpha^2 |s|^2 \psi_j(s) + \sum_{k=j+1}^{p} \hat{a}_{jk}(s)\psi_k(s), \qquad j = 1, \ldots, p. \tag{46}$$

The requirement that a given ψ in $\mathfrak{H}^p$ be in $\mathfrak{D}(\hat{A}_{\mathfrak{S}})$ is the same as requiring $|\zeta(s)|^2$ to be integrable on $\mathfrak{S}$, which in turn is equivalent to the requirement that each $|\zeta_j(s)|^2$, $j = 1, \ldots, p$ be integrable on $\mathfrak{S}$. We suppose ζ to be such a vector. For $j = p$ equation (46) is $\zeta_p(s) = -\alpha^2 |s|^2 \psi_p(s)$ so that $|\zeta_p(s)|^2$ is integrable on $\mathfrak{S}$ if and only if $|s|^4 |\psi_p(s)|^2$ is integrable on $\mathfrak{S}$. For $j = p-1$ equation (46) becomes

$$\zeta_{p-1}(s) = -\alpha^2 |s|^2 \psi_{p-1}(s) + \hat{a}_{(p-1)p}(s)\psi_p(s),$$

and since $\hat{a}_{(p-1)p}(s)$ is a polynomial of degree at most 2 in $s_1, \ldots, s_N$ we have $|\hat{a}_{(p-1)p}(s)| = O(|s|^2)$ as $|s| \to \infty$, and since $|\zeta_{p-1}(s)|^2$ is integrable on $\mathfrak{S}$ it follows that $|s|^4 |\psi_{p-1}(s)|^2$ is integrable on $\mathfrak{S}$. Thus, by downward induction, the equations (46) show that if ψ is in $\mathfrak{D}(\hat{A}_{\mathfrak{S}})$ we have $\int_{R^N} |s|^4 |\psi(s)|^2 \, ds < \infty$. Conversely, if a vector ψ in $\mathfrak{H}^p$ satisfies this condition, it clearly belongs to $\mathfrak{D}(\hat{A}_{\mathfrak{S}})$. Thus we may say that if the polynomials $\hat{a}_{jk}(s)$ in the operator (41) are of degree at most 2, then

$$\mathfrak{D}(\hat{A}_{\mathfrak{S}}) = \{\psi \in \mathfrak{H}^p \,|\, \int_{R^N} |s|^4 |\psi(s)|^2 \, ds < \infty\}. \tag{47}$$

In order to characterize $\mathfrak{D}(A_{\mathfrak{S}})$ in a form more convenient than its definition $\mathfrak{D}(A_{\mathfrak{S}}) = F^{-1}\mathfrak{D}(\hat{A}_{\mathfrak{S}})$ as given in equation (4), we shall find it convenient to use the notion of a tempered distribution.

9 DEFINITION. A linear functional on the space Φ of rapidly decreasing functions on R^N which is continuous in the topology of Φ defined by the neighborhoods given by the expression (1) of Section 11 is called a *tempered distribution* in R^N. The set of all tempered distributions

in R^N will be denoted by $T(R^N)$. It is clear that if T, U are in $T(R^N)$ and α, β are constants, then the function $\alpha T + \beta U$ defined on Φ by

$$\text{(i)} \qquad (\alpha T + \beta U)(\varphi) = \alpha T(\varphi) + \beta U(\varphi), \qquad \varphi \in \Phi,$$

is in $T(R^N)$ and $T(R^N)$ is thus a linear space. The *complex conjugate* $\overline{T}$, the *Fourier transform* FT, and the *inverse Fourier transform* $F^{-1}T$ of a tempered distribution T are defined by the equations

$$\text{(ii)} \qquad \overline{T}(\varphi) = \overline{T(\overline{\varphi})}, \qquad \varphi \in \Phi,$$

and

$$\text{(iii)} \quad (FT)(\varphi) = T(F\varphi), \qquad (F^{-1}T)(\varphi) = T(F^{-1}\varphi), \qquad \varphi \in \Phi.$$

Theorem 11.1 shows that FT and $F^{-1}T$ are tempered distributions. If T, T_n are in $T(R^N)$ where (n) is a directed set, then we define $T = \lim_n T_n$ to mean that

$$\text{(iv)} \qquad T\varphi = \lim_n T_n \varphi, \qquad \varphi \in \Phi.$$

As an example of a tempered distribution we might mention the function T_ν determined by a bounded finitely additive set function ν (defined on some field of sets in R^N which includes the open sets) by the equation

$$\text{(48)} \qquad T_\nu(\varphi) = \int_{R^N} \varphi(s)\nu(ds), \qquad \varphi \in \Phi.$$

Similarly, functions on R^N may determine tempered distributions just as they sometimes determine distributions. For example, since convergence $\varphi_n \to \varphi$ in Φ is equivalent to the statement that for any two polynomials P, Q in N variables we have $P(s)Q(\partial/\partial s)\varphi_n(s) \to P(s)Q(\partial/\partial s)\varphi(s)$ uniformly for s in R^N, it is clear that a function ψ in any one of the Lebesgue spaces $L_p(R^N)$, $1 \leqq p \leqq \infty$, determines a tempered distribution by the equation

$$T(\varphi) = \int_{R^N} \varphi(s)\psi(s)\, ds, \qquad \varphi \in \Phi.$$

To be precise we formulate this relationship in the following definition.

10 Definition. A scalar function ψ on R^N is said to *determine the tempered distribution* T_ψ in R^N if for every φ in Φ, $\varphi(s)\psi(s)$ is Lebesgue integrable on R^N and the functional T_ψ defined by the equation

$$\text{(49)} \qquad T_\psi(\varphi) = \int_{R^N} \varphi(s)\psi(s)\, ds, \qquad \varphi \in \Phi,$$

is in $T(R^N)$. The set of such scalar functions ψ which determine tempered distributions T_ψ will be denoted by $S(R^N)$.

Since $C_0^\infty(R^N) \subset \Phi$ and the convergence $\varphi_n \rightrightarrows \varphi$ in $C_0^\infty(R^N)$ (XIV.3.1) implies that $\varphi_n \to \varphi$ it is clear that the restriction of a tempered distribution to $C_0^\infty(R^N)$ is a distribution in R^N. This observation enables us to deduce the following:

11 LEMMA. *If, in the sense of the preceding definition, a tempered distribution corresponds to two functions, then these two functions coincide almost everywhere on* R^N.

PROOF. For any given compact set K there is a function φ in Φ for which $\varphi(s) = 1$ on K (XIV.2.1). It follows that any function which determines a tempered distribution is Lebesgue integrable over every compact set. Thus Lemma XIV.3.3 applies to show that two functions which determine the same tempered distribution can differ only on a null set. Q.E.D.

This lemma gives a linear one-to-one correspondence between $S(R^N)$ and a linear subset of $T(R^N)$ and enables us to make the following definition.

12 DEFINITION. A tempered distribution T which corresponds to a function ψ in the sense of Definition 10 will be said to *be a function*. If ψ is continuous, or differentiable, or belongs to $L_p(R^N)$ or $C^\infty(R^N)$, etc., then T will be said to be continuous, or differentiable, or belong to $L_p(R^N)$ or $C^\infty(R^N)$, etc. In other words *we shall simply identify a tempered distribution which is a function with the function to which it corresponds.* If $\psi \in S(R^N)$ and $T = T_\psi$ we may sometimes write $T(s)$ instead of $\psi(s)$.

We summarize in the following lemma some of the elementary properties of the map $\psi \to T_\psi$.

13 LEMMA. *The map* $\psi \to T_\psi$ *of* $S(R^N)$ *into* $T(R^N)$ *is a linear map with the additional properties:*

(i) $\overline{T}_\psi = T_{\overline{\psi}}$,

(ii) *If* $1 \leqq p \leqq \infty$, $p^{-1} + q^{-1} = 1$, *and* $\psi \in L_p(R^N)$, *then*

$$|T_\psi \varphi| \leqq |\varphi|_q |\psi|_p, \qquad \varphi \in \Phi.$$

(iii) *If* $1 < p < \infty$, $p^{-1} + q^{-1} = 1$ *and* $\psi \in L_p(R^N)$, *then*

$$\sup_{\varphi \in \Phi, |\varphi|_q = 1} |T_\psi \varphi| = |\psi|_p .$$

(iv) *For ψ in $L_2(R^N)$*

$$FT_\psi = T_{F\psi}, \qquad F^{-1}T_\psi = T_{F^{-1}\psi}.$$

PROOF. The linearity of the map $\psi \to T_\psi$ and (i) are readily verified. Statement (ii) follows from Hölder's inequality (III.3.2). Since $C_0^\infty(R^N) \subset \Phi \subset L_p(R^N)$, it follows from Lemma XIV.2.2 that, for $p < \infty$, the subset Φ of $L_p(R^N)$ is dense in $L_p(R^N)$ and so (iii) follows from Theorem IV.8.1. To prove (iv) suppose first that ψ is in Φ in which case a permissible interchange of integration gives

$$(FT_\psi)(\varphi) = \int_{R^N} (F\varphi)(s)\psi(s)\, ds = \int_{R^N} \varphi(s)(F\psi)(s)\, ds = T_{F\psi}(\varphi), \qquad \varphi \in \Phi.$$

Now Φ is dense in $\mathfrak{H} = L_2(R^N)$ and thus there is for an arbitrary ψ in $\mathfrak{H}$ a sequence $\{\psi_n\} \subset \Phi$ with $\psi_n \to \psi$ in $\mathfrak{H}$. Since $F\Phi = \Phi$ (XI.1) and F is continuous on $\mathfrak{H}$, the inequality (ii) shows that $FT_{\psi_n} \to FT_\psi$. On the other hand, the preceding equation shows that $FT_{\psi_n} = T_{F\psi_n}$, which by (ii), converges to $T_{F\psi}$. This proves the first part of (iv) and the second part is proved similarly. Q.E.D.

We define derivatives of tempered distributions as we did for distributions. The motivating factor for the definition is that if a tempered distribution T_{φ_0} is a sufficiently smooth differentiable function φ_0 then we would like $\partial^\alpha T_{\varphi_0}$ to be the tempered distribution corresponding to $\partial^\alpha \varphi_0$, provided, of course, that the derivative $\partial^\alpha \varphi_0$ determines a tempered distribution. Suppose, for example, that φ_0, φ are both in Φ. Then an integration by parts in the variable s_j gives

$$\int_{R^N} \varphi(s) \frac{\partial}{\partial s_j} \varphi_0(s)\, ds = -\int_{R^N} \frac{\partial \varphi(s)}{\partial s_j} \varphi_0(s)\, ds,$$

which leads us to define the derivative $(\partial/\partial s_j)T$ of an arbitrary tempered distribution by the equation

$$\frac{\partial}{\partial s_j} T(\varphi) = -T\left(\frac{\partial \varphi}{\partial s_j}\right), \qquad \varphi \in \Phi. \tag{50}$$

It is clear that $\partial T/\partial s_j$ is a tempered distribution and repeated differentiation for any set $\alpha = (\alpha_1, \ldots, \alpha_N)$ of N non-negative integers shows that $\partial^\alpha T$ or $(\partial/\partial s)^\alpha T$ is given by the equation

$$(\partial^\alpha T)(\varphi) = (-1)^{|\alpha|_1} T(\partial^\alpha \varphi), \qquad \varphi \in \Phi. \tag{51}$$

For a formal differential operator

$$a = \sum_{|\alpha|_1 \leqq m} a_\alpha \, \partial^\alpha \tag{52}$$

with constant coefficients a_α the tempered distribution aT is thus given by the formula

$$(aT)(\varphi) = T(a^+\varphi), \qquad \varphi \in \Phi, \tag{53}$$

where

$$a^+ = \sum_{|\alpha|_1 \leqq m} (-1)^{|\alpha|_1} a_\alpha \, \partial^\alpha. \tag{54}$$

In connection with the derivatives of tempered distributions the following lemma will be needed.

14 LEMMA. *If both* T *and* $\partial^\alpha T$ *are functions for some* T *in* $T(R^N)$ *and some* α, *then so are* $\bar{T}$ *and* $\partial^\alpha \bar{T}$ *functions, and*

$$\bar{T}(s) = \overline{T(s)}, \qquad (\partial^\alpha \bar{T})(s) = \overline{(\partial^\alpha T)(s)}.$$

PROOF. The statement about $\bar{T}$ is already contained in part (i) of Lemma 13. Since

$$\bar{T}(\varphi) = \int_{R^N} \varphi(s) \, \overline{T(s)} \, ds, \qquad \varphi \in \Phi,$$

we have, by definition,

$$\begin{aligned} (\partial^\alpha \bar{T})(\varphi) &= \overline{\int_{R^N} (-1)^{|\alpha|_1} \, \partial^\alpha \overline{\varphi(s)} T(s) \, ds} \\ &= \overline{(\partial^\alpha T)(\bar{\varphi})} \\ &= \overline{\int_{R^N} \bar{\varphi}(s) (\partial^\alpha T)(s) \, ds} = \int_{R^N} \varphi(s) \overline{(\partial^\alpha T)(s)} \, ds, \end{aligned}$$

which shows that $\partial^\alpha \bar{T}$ is a function and that $(\partial^\alpha \bar{T})(s) = \overline{(\partial^\alpha T)(s)}$. Q.E.D.

The following definition and lemma are useful in determining the domains of certain natural closed extensions of formal differential operators. Also, it should be noted that if the tempered distribution T belongs to the space $T^{(k)}(R^N)$ of the following definition, then the preceding lemma shows that the complex conjugate tempered distribution $\bar{T}$ is also in $T^{(k)}(R^N)$.

15 DEFINITION. For $k = 1, 2, \ldots,$ let $T^{(k)}(R^N)$ be the set of all tempered distributions T in R^N for which $\partial^\alpha T$ is in $\mathfrak{H} = L_2(R^N)$ for every α with $|\alpha|_1 \leqq k$. For each pair T, U in $T^{(k)}(R^N)$ we define

$$\text{(i)} \qquad (T, U)^{(k)} = \sum_{0 \leqq |\alpha|_1 \leqq k} \int_{R^N} (\partial^\alpha T)(s)(\partial^\alpha \bar{U})(s)\, ds,$$

$$\text{(ii)} \qquad |T|_{(k)} = \{(T, T)_{(k)}\}^{1/2}.$$

Lemma 14 shows that the scalar product $(T, U)_{(k)}$ satisfies all the requirements demanded in the definition of a Hilbert space (IV.2.26) except that of completeness. This property will now be proved.

16 LEMMA. *The space $T^{(k)}(R^N)$ of the preceding definition is a complete Hilbert space.*

PROOF. As just observed, we need only prove completeness. Let $\{T_m\}$ be a Cauchy sequence in $T^{(k)}(R^N)$. Since $|T|_{(k)} \geqq |T|$, the norm of T as a function in $\mathfrak{H} = L_2(R^N)$, then T_n converges in $\mathfrak{H}$ to some T. Similarly $|T_n - T_m|_{(k)} \geqq |\partial^\alpha T_n - \partial^\alpha T_m|$ for $0 \leqq |\alpha|_1 \leqq k$ and so the sequences $\{\partial^\alpha T_m\}$ all converge in $\mathfrak{H}$ to functions T_α in $\mathfrak{H}$. Let φ be in Φ, then

$$\begin{aligned}\int_{R^N} \varphi(s) T_\alpha(s)\, ds &= \lim_{m\to\infty} (\partial^\alpha T_m)(\varphi),\\ &= \lim_{m\to\infty} (-1)^{|\alpha|_1} T_m(\partial^\alpha \varphi),\\ &= (-1)^{|\alpha|_1} T(\partial^\alpha \varphi),\\ &= (\partial^\alpha T)(\varphi),\end{aligned}$$

which shows that $\partial^\alpha T = T_\alpha$ and is thus in $\mathfrak{H}$. Q.E.D.

As we shall see presently, the following space is useful in characterizing the domains $\mathfrak{D}(A_{\mathfrak{S}})$ of the natural closed extensions of certain formal differential operators A in Φ^p.

17 DEFINITION. For every integer $p \geqq 1$ the space $(p)T^{(k)}(R^N)$ is defined to be the direct sum

$$(p)T^{(k)}(R^N) = T^{(k)}(R^N) \oplus \cdots \oplus T^{(k)}(R^N),$$

where there are p summands. For $T = [T_1, \ldots, T_p]$ and $U = [U_1, \ldots, U_p]$ in $(p)T^{(k)}(R^N)$ the scalar product and norm in $(p)T^{(k)}(R^N)$ are defined, as usual, by the equations

$$(T, U)_{(p,k)} = \sum_{j=1}^{p} (T_j, U_j)_{(k)},$$

$$|T|_{(p,k)} = \{(T, T)\}^{1/2}.$$

Since the direct sum of an arbitrary number of Hilbert spaces is always a complete Hilbert space (IV.4.19), the space $(p)T^{(k)}(R^N)$ is a Hilbert space.

We are now prepared to return to the study of our illustrative example of the perturbed Laplacian as given in equation (41) and to determine the domain $\mathfrak{D}(A_{\mathfrak{S}})$ in case the disturbing operators are of order at most 2. The characterization of $\mathfrak{D}(A_{\mathfrak{S}})$ is incorporated in the following theorem which also summarizes what has already been proved about this operator.

18 THEOREM. *Let the formal differential operator* $P = (a_{jk})$, *where* $a_{jk} = (\partial/\partial s_1, \ldots, \partial/\partial s_N)$, j, $k = 1, \ldots, p$, *are polynomials with constant coefficients and degrees at most* 2 *and let* $a_{jk} = 0$ *for* $j \geqq k$. *Let the formal operator* A *be defined on* Φ^p *by the matrix*

$$A = \alpha^2\nabla^2 I + P = \begin{pmatrix} \alpha^2\nabla^2, & a_{12}, & \cdots, & a_{1p} \\ 0, & \alpha^2\nabla^2, & \ldots, & \cdot \\ \cdot & \cdot & & \cdot \\ \cdot & \cdot & & a_{(p-1)p} \\ 0, & 0, & \ldots, & \alpha^2\nabla^2 \end{pmatrix}$$

where $\alpha \neq 0$ *is either a real constant or a purely imaginary one, and*

$$\nabla^2 = \frac{\partial^2}{\partial s_1^2} + \cdots + \frac{\partial^2}{\partial s_N^2}.$$

Then the natural closed extension $A_{\mathfrak{S}}$ *of* A *is a spectral operator with spectrum* $\sigma(A_{\mathfrak{S}}) = (-\infty, 0]$ *in case* $\alpha^2 > 0$ *and* $\sigma(A_{\mathfrak{S}}) = [0, \infty)$ *in case* $\alpha^2 < 0$. *In either case the domain* $\mathfrak{D}(A_{\mathfrak{S}})$ *is*

$$\mathfrak{D}(A_{\mathfrak{S}}) = (p)T^{(2)}(R^N).$$

The operator $A_{\mathfrak{S}}$ *is also the infinitesimal generator of a strongly continuous semi-group (on* $[0, \infty)$ *in case* $\alpha^2 > 0$ *and on* $(-\infty, 0]$ *in case* $\alpha^2 < 0$*) of bounded linear operators in* $\mathfrak{H}^p$.

REMARK. As the reader will readily observe, Theorems 18, 19 and 21 may be stated in a manner that allows α to be an arbitrary complex number.

PROOF. All of the conclusions except the formula for $\mathfrak{D}(A_{\mathfrak{S}})$ have been established. Since this formula is a special case of the conclusion (vii) of Theorem 19 that follows we shall omit its proof here.

19 THEOREM. *Let A be the operator of Theorem 18 with $\alpha^2 > 0$. Then for every $\varphi^{(0)}$ in $\mathfrak{H}^p$ there is one and only one continuous map $t \to \varphi(t)$ of $[0, \infty)$ into $\mathfrak{H}^p$ which is differentiable for $t > 0$ and has the properties*

$$\text{(i)} \qquad \varphi(t) \in \mathfrak{D}(A_{\mathfrak{S}}), \qquad 0 < t < \infty,$$

$$\text{(ii)} \qquad \varphi'(t) = A_{\mathfrak{S}}\varphi(t), \qquad 0 < t < \infty,$$

$$\text{(iii)} \qquad \varphi(0) = \varphi^{(0)}.$$

This unique function φ is given by the equation $\varphi(t) = T(t)\varphi^{(0)}$ where $T(t)$, $0 \leqq t < \infty$ is the strongly continuous semi-group for which $A_{\mathfrak{S}}$ is the infinitesimal generator. The semi-group $T(t)$ and the function φ have the further properties:

(iv) *The derivative $\varphi'(0)$ exists and $\varphi'(0) = A_{\mathfrak{S}}\varphi(0)$ if and only if $\varphi^{(0)} \in \mathfrak{D}(A_{\mathfrak{S}})$.*

(v) *The semi-group $T(t)$ has a strongly analytic extension to a semi-group $T_1(\lambda)$ defined for λ in the half plane $\mathscr{R}(\lambda) > 0$ and thus the unique solution φ of the Cauchy problem presented by equations* (i), (ii), (iii) *is analytic and has the analytic extension $\varphi_1(\lambda) = T_1(\lambda)\varphi^{(0)}$ to the half plane $\mathscr{R}(\lambda) > 0$.*

(vi) *The operator $A^k_{\mathfrak{S}}$ is the natural closed extension of the formal differential operator A^k on Φ^p and for $\mathscr{R}(\lambda) > 0$,*

$$\varphi_1(\lambda) \in \bigcap_{k=1}^{\infty} \mathfrak{D}(A^k_{\mathfrak{S}}), \qquad \varphi_1^{(k)}(\lambda) = A^k_{\mathfrak{S}}\varphi_1(\lambda), \qquad k = 1, 2, \ldots,$$

and

$$\text{(vii)} \qquad \mathfrak{D}(A^k_{\mathfrak{S}}) = (p)T^{(2k)}(R^N), \qquad k = 1, 2, \ldots, .$$

(viii) *For $\mathscr{R}(\lambda) > 0$ the function $\varphi_1(\lambda) = \varphi_1(s_1, \ldots, s_N; \lambda)$ has continuous derivatives with respect to $s = [s_1, \ldots, s_N]$ of all orders and $\partial_s^\beta \varphi_1$ belongs to $\mathfrak{H}^p$ for every $\beta = [\beta_1, \ldots, \beta_N]$.*

If $\alpha^2 < 0$ then the analogous conclusions obtained by reflecting the complex plane in the imaginary axis also hold.

PROOF. Let $\hat{N}(s) = \hat{A}(s) + \alpha^2 |s|^2 I$. It follows from Theorem VII.1.8 that the matrix $\hat{T}(s; t) = \exp(t\hat{A}(s))$ is given by the formula

$$(55) \qquad \hat{T}(s; t) = e^{-t\alpha^2|s|^2} e^{tN(s)}$$

$$= e^{-t\alpha^2|s|^2} \sum_{\nu=0}^{p-1} \frac{t^\nu}{\nu!} \hat{N}^\nu(s)$$

and has the properties

$$(56) \qquad \hat{T}(s; t+u) = \hat{T}(s; t)\hat{T}(s; u), \qquad \hat{T}(0) = I, \qquad s \in R^N.$$

The equation (54) shows that $T(s; t)$ is bounded in s if $t \geqq 0$ and thus the operator $T(t)$ in $\mathfrak{H}^p$ defined by the equation

$$(57) \qquad (\hat{T}(t)\psi)(s) \equiv \hat{T}(s; t)\psi(s), \qquad \psi \in \mathfrak{H}^p, \qquad 0 \leqq t < \infty,$$

is a bounded linear operator in $\mathfrak{H}^p$ and it follows from (56) that $\hat{T}(t)$, $0 \leqq t < \infty$, is a semi-group, that is,

$$(58) \qquad \hat{T}(t+u) = \hat{T}(t)\hat{T}(u), \qquad \hat{T}(0) = I, \qquad 0 \leqq t, u < \infty.$$

It is clear from (55) that, for any $\psi \in \mathfrak{H}^p$, $T(t)\psi$ is continuous on the open interval $(0, \infty)$. We wish to show now that it is also continuous at $t = 0$. To show this we shall establish the fact that $|\hat{T}(t)|$ is bounded on $0 < t < \infty$. Since $\hat{N}(s)$ has no elements of degree more than 2 in $s_1, \ldots, s_N$ we have $|\hat{N}(s)| \leqq K|s|^2$ and thus, in view of (55), to establish the boundedness of $|\hat{T}(t)|$ it suffices to show that for each $j = 0, \ldots, p$,

$$(59) \qquad K(t) = \sup_{s \in R^N} t^j e^{-t\alpha^2|s|^2} |s|^{2j}$$

is finite and bounded in t. An elementary calculation shows that

$$(60) \qquad K(t) = \left(\frac{j}{\alpha^2 e}\right)^j, \qquad 0 < t < \infty,$$

and is thus independent of t, which proves that

$$(61) \qquad |\hat{T}(t)| \leqq M, \qquad 0 \leqq t < \infty.$$

Thus for ψ in $\mathfrak{H}^p$ it follows from (55) that

$$\lim_{t \to 0} \hat{T}(s; t)\psi(s) = \psi(s), \qquad s \in R^N,$$

and from (61) that

$$|\hat{T}(s; t)\psi(s)| \leqq M|\psi(s)|, \qquad s \in R^N,$$

and so it follows from the dominated convergence theorem (III.6.16) that $\lim_{t\to 0} \hat{T}(t)\psi = \psi$ in $\mathfrak{H}^p$. This proves that $\hat{T}(t)$ is a strongly continuous semi-group on $[0, \infty)$ of bounded linear operators in $\mathfrak{H}^p$. Let $\hat{B}$ be the infinitesimal generator of the semi-group $\hat{T}(t)$, $0 \leqq t < \infty$. It will now be shown that $\hat{B} = \hat{A}$. For $\psi \in \mathfrak{D}(\hat{B})$, we have, by definition,

$$\lim_{h\to 0+} \frac{\hat{T}(h) - \hat{I}}{h} \psi = \hat{B}\psi.$$

On the other hand, it follows from (55) that for each s in R^N

$$\lim_{h\to 0+} \left(\frac{\hat{T}(h) - \hat{I}}{h} \psi\right)(s) = \hat{A}(s)\psi(s),$$

which shows that $\hat{A}(s)\psi(s)$ is square integrable on R^N and that $\hat{B}\psi = \hat{A}\psi$. Thus $\hat{A} \supseteqq \hat{B}$. Now every positive λ is in $\rho(\hat{A})$ and thus the Hille-Yosida theorem (VIII.1.13) shows that all large real λ are in $\rho(\hat{A}) \cap \rho(\hat{B})$. For such λ we have, since $\hat{A} \supseteqq \hat{B}$, $(\lambda\hat{I} - \hat{A})\mathfrak{D}(\hat{B}) = (\lambda\hat{I} - \hat{B})\mathfrak{D}(\hat{B}) = \mathfrak{H}^p$. However we also have $(\lambda\hat{I} - \hat{A})\mathfrak{D}(\hat{A}) = \mathfrak{H}^p$, which proves that $\mathfrak{D}(\hat{A}) = \mathfrak{D}(\hat{B})$ and thus that $\hat{A} = \hat{B}$. We now define

$$T(t) = F^{-1}\hat{T}(t)F, \qquad 0 \leqq t < \infty, \tag{62}$$

so that $T(t)$ is a strongly continuous semi-group on $[0, \infty)$ of bounded linear operators in $\mathfrak{H}^p$. Its infinitesimal generator is, by definition, the operator B whose domain consists of all φ in $\mathfrak{H}^p$ for which the limit

$$\begin{aligned} B\varphi &= \lim_{h\to 0+} \frac{T(h) - I}{h} \varphi \\ &= \lim_{h\to 0+} F^{-1} \frac{\hat{T}(h) - \hat{I}}{h} F\varphi, \end{aligned}$$

exists. Thus φ is in $\mathfrak{D}(B)$ if and only if $F\varphi$ is in $\mathfrak{D}(\hat{A})$, that is, φ is in $F^{-1}\mathfrak{D}(\hat{A}) = \mathfrak{D}(A_{\mathfrak{S}})$. Since the preceding equation shows that $B\varphi = F^{-1}\hat{A}F\varphi$, $\varphi \in \mathfrak{D}(A_{\mathfrak{S}})$, it follows that $B = A_{\mathfrak{S}}$ and that $A_{\mathfrak{S}}$ is the infinitesimal generator of the semi-group (62). Let $\varphi(t) = T(t)\varphi^{(0)}$. It is clear from equation (55) that the function $\psi(t) = \hat{T}(t)\psi^{(0)}$, where $\psi^{(0)} = F\varphi^{(0)}$, is in $\mathfrak{D}(\hat{A})$ for $t > 0$ and that $\psi'(t) = \hat{A}\psi(t)$ for $t > 0$. Thus (i) and (ii) follow since F and F^{-1} are isometries in $\mathfrak{H}^p$. Equation (iii) is an immediate consequence of the definition of $\varphi(t)$.

We shall next prove that the map $t \to \varphi(t)$ of $[0, \infty)$ into $\mathfrak{H}^p$ is the only one with the properties (i), (ii), and (iii). Let φ_t be another function on $[0, \infty)$ to $\mathfrak{H}^p$ with the properties (i), (ii), and (iii). Then for $u > t > 0$

$$\begin{aligned}\frac{\partial}{\partial t} T(u-t)\varphi_t &= \lim_{h \to 0} \frac{1}{h}[T(u-t-h)\varphi_{t+h} - T(u-t)\varphi_t] \\ &= \lim_{h \to 0} T(u-t-h)\frac{\varphi_{t+h}-\varphi_t}{h} \\ &\quad + \lim_{h \to 0} \frac{T(u-t-h)-T(u-t)}{h}\varphi_t .\end{aligned}$$

Since $|T(t)|$ is bounded and φ_t satisfies (ii),

$$\lim_{h \to 0} T(u-t-h)\frac{\varphi_{t+h}-\varphi_t}{h} = T(u-t)A_{\mathfrak{S}}\varphi_t .$$

Since $d(T(t)\varphi^{(0)})/dt = A_{\mathfrak{S}} T(t)\varphi^{(0)}$ for every $\varphi^{(0)}$ in $\mathfrak{H}^p$ and every $t > 0$ we see by placing φ_t for $\varphi^{(0)}$ and $u-t$ for t that

$$\lim_{h \to 0} \frac{T(u-t-h)-T(u-t)}{h}\varphi_t = -T(u-t)A_{\mathfrak{S}}\varphi_t .$$

Thus

$$\frac{\partial}{\partial t} T(u-t)\varphi_t = 0, \qquad 0 < t < u < \infty,$$

and

$$\frac{\partial}{\partial t} T(u-t)[\varphi_t - T(t)\varphi^{(0)}] = \frac{\partial}{\partial t}[T(u-t)\varphi_t - T(u)\varphi^{(0)}] = 0,$$

which shows that the vector

$$\zeta = T(u-t)[\varphi_t - T(t)\varphi^{(0)}]$$

is independent of t and thus also independent of u. By letting $u \to t+$ we have

$$\zeta = \varphi_t - T(t)\varphi^{(0)}, \qquad 0 < t < \infty,$$

and by letting $t \to 0+$ we see that $\zeta = \varphi^{(0)} - \varphi^{(0)} = 0$. This shows that $\varphi_t = T(t)\varphi^{(0)} = \varphi(t)$ and proves the uniqueness of $\varphi(t)$.

Since $A_{\mathfrak{S}}$ is the infinitesimal generator of the semi-group $T(t)$ the conclusion (iv) follows from Definition VIII.1.6 and Lemma VIII.1.7(b).

For every complex number λ with $\mathscr{R}(\lambda) > 0$ let $T_1(\lambda) = F^{-1}\hat{T}_1(\lambda)F$, $\varphi_1(\lambda) = T_1(\lambda)\varphi^{(0)}$, where

$$\hat{T}_1(s;\lambda) = e^{-\lambda\alpha^2|s|^2} \sum_{\nu=0}^{p-1} \frac{\lambda^\nu}{\nu!} \hat{N}^\nu(s), \tag{63}$$

and where, as before, the operator $\hat{T}_1(\lambda)$ is defined on $\mathfrak{H}^p$ by the equation

$$(\hat{T}_1(\lambda)\psi)(s) = \hat{T}_1(s;\lambda)\psi(s), \qquad \psi \in \mathfrak{H}^p. \tag{64}$$

Since $|\hat{N}(s)| = O(|s|^2)$ as $|s| \to \infty$, it is clear from (63) that $\hat{T}_1(\lambda)$ is strongly analytic on the half plane $\mathscr{R}(\lambda) > 0$ and since F and F^{-1} are isometries in $\mathfrak{H}^p$, it follows that $T_1(\lambda)$ is also strongly analytic when $\mathscr{R}(\lambda) > 0$ which proves (v).

To prove (vi) we observe first that while the powers $\hat{A}^k$ of an unbounded operator $\hat{A}$ are defined inductively by taking $\mathfrak{D}(\hat{A}^k) = \{\psi \in \mathfrak{H}^p \mid \hat{A}^{k-1}\psi \in \mathfrak{D}(\hat{A})\}$ and $\hat{A}^k\psi = \hat{A}(\hat{A}^{k-1}\psi)$ it follows from the particular form of our operator $\hat{A}$ that the power $\hat{A}^k$ is given by the formulas

$$\mathfrak{D}(\hat{A}^k) = \{\psi \in \mathfrak{H}^p \mid \int_{R^N} |s|^{4k} |\psi(s)|^2 \, ds < \infty\} \tag{65}$$

and

$$(\hat{A}^k\psi)(s) = (\hat{A}(s))^k\psi(s), \qquad \psi \in \mathfrak{D}(\hat{A}^k). \tag{66}$$

The formula (65) results from the fact that the matrix $(\hat{A}(s))^k$ has $(-1)^k\alpha^{2k}|s|^{2k}$ for its elements on the principal diagonal, zeros below the diagonal, polynomials of degree at most $2k$ above it, and may be derived by the argument used to prove equation (47). Since the map $A \to \hat{A}(s)$ of the algebra of formal differential operators $A = (a_{jk})$ on Φ^p is a homomorphism, it follows from equation (66) that $A^k_{\mathfrak{S}}$ is the natural closed extension of A^k.

It follows from (63) that for the vector $\psi_1(\lambda) = \hat{T}(\lambda)\psi^{(0)}$, we have

$$\frac{\partial}{\partial\lambda} \psi_1(\lambda)(s) = \hat{A}(s)\psi_1(\lambda)(s), \qquad \mathscr{R}(\lambda) > 0,\ s \in R^N, \tag{67}$$

and, since $|\hat{N}(s)| = O(|s|^2)$ as $|s| \to \infty$, it follows from (63) and (65) that

$$\psi_1(\lambda) \in \mathfrak{D}(\hat{A}^k), \qquad \mathscr{R}(\lambda) > 0,\ k = 1, 2, \ldots. \tag{68}$$

In view of (67) and the fact that the derivative $\psi_1'(\lambda)$ exists in the norm of $\mathfrak{H}^p$ we have

$$\psi_1'(\lambda) = \hat{A}\psi_1(\lambda), \qquad \mathscr{R}(\lambda) > 0. \tag{69}$$

By using the Fourier transform, (68) and (69) become

(70) $$\varphi_1(\lambda) \in \mathfrak{D}(A^k_{\mathfrak{S}}), \qquad \mathscr{R}(\lambda) > 0,\ k = 1, 2, \ldots,$$

and

(71) $$\varphi_1'(\lambda) = A_{\mathfrak{S}}\, \varphi_1(\lambda), \qquad \mathscr{R}(\lambda) > 0.$$

Suppose that we have established the equation

(72) $$\varphi_1^{(k)}(\lambda) = A^k_{\mathfrak{S}}\, \varphi_1(\lambda), \qquad \mathscr{R}(\lambda) > 0.$$

Since φ_1 is analytic, the derivative $\varphi_1^{(k+1)}$ exists and equation (72) shows that

$$\varphi_1^{(k+1)}(\lambda) = \lim_{h\to 0} \frac{\varphi_1^{(k)}(\lambda+h) - \varphi_1^{(k)}(\lambda)}{h} = \lim_{h\to 0} A^k_{\mathfrak{S}} \frac{\varphi_1(\lambda+h) - \varphi_1(\lambda)}{h}.$$

Since $A^k_{\mathfrak{S}}$ is closed, (71) and (73) show that $\varphi_1^{(k+1)}(\lambda) = A^{(k+1)}_{\mathfrak{S}}\varphi_1(\lambda)$ which completes the proof of (vi).

To prove (vii) we recall that, by definition, (since $A^k_{\mathfrak{S}}$ is the natural closed extension of A^k) we have $\mathfrak{D}(A^k_{\mathfrak{S}}) = F^{-1}\mathfrak{D}(\hat{A}^k)$ and so the statement

(73) $$(p)T^{(2k)}(R^N) \subseteq \mathfrak{D}(A^k_{\mathfrak{S}})$$

is equivalent to the statement

(74) $$FT = [FT_1, \ldots, FT_p] \in \mathfrak{D}(\hat{A}^k), \qquad T \in (p)T^{(2k)}(R^N).$$

It is apparent from (65) that a vector $\psi = [\psi_1, \ldots, \psi_p]$ is in $\mathfrak{D}(\hat{A}^k)$ if and only if each component ψ_j has $\int_{R^N} |s|^{4k} |\psi_j(s)|^2\, ds < \infty$. Hence the statement (73) is equivalent to

(75) $$\int_{R^N} |s|^{4k} |(FT)(s)|^2\, ds < \infty, \qquad T \in T^{(2k)}(R^N).$$

In other words, this just amounts to observing that (73) is true for every positive integer p if it holds for $p = 1$. Consequently, we let T be an arbitrary element of $T^{(2k)}(R^N)$. Lemma 14 shows that $\overline{T} \in T^{(2k)}(R^N)$ and so

$$\nabla^{2k}\overline{T} = \left(\frac{\partial^2}{\partial s_1^2} + \cdots + \frac{\partial^2}{\partial s^2}\right)^k \overline{T} \in \mathfrak{H}$$

and

(76) $$(\nabla^{2k}\overline{T})(\varphi) = \int_{R^N} \varphi(s)(\nabla^{2k}\overline{T})(s)\, ds, \qquad \varphi \in \Phi.$$

Also, for $\psi = F\varphi$ we have

$$(77) \qquad (\nabla^{2k}\overline{T})(\varphi) = (\nabla^{2k}\overline{T})(F^{-1}\psi) = \int_{R^N} (\nabla^{2k}F^{-1}\psi)(s)\overline{T}(s)\,ds.$$

Since ψ is in Φ (Theorem 11.1), we may differentiate under the integral sign to see that

$$(\nabla^{2k}F^{-1}\psi)(s) = \frac{1}{(2\pi)^{N/2}} \int_{R^N} e^{isu}(-1)^k |u|^{2k}\psi(u)\,du = F^{-1}((-1)^k |\cdot|^{2k}\psi(\cdot))(s)$$

and so from (77) we obtain

$$(78) \qquad (\nabla^{2k}\overline{T})(\varphi) = \int_{R^N} F^{-1}((-1)^k |\cdot|^{2k}\psi(\cdot))(s)\overline{T}(s)\,ds = \int_{R^N} \psi(s)(-1)^k |s|^{2k}(F^{-1}\overline{T})(s)\,ds = \int_{R^N} \psi(s)(-1)^k |s|^{2k}\overline{(FT)(s)}\,ds.$$

Since F is an isometry in $\mathfrak{H}$, $|\psi|_2 = |\varphi|_2$ and since $F\Phi = \Phi$ is dense in $\mathfrak{H}$ it follows from (76) and (78) that

$$|\nabla^{2k}\overline{T}|_2 = \sup_{|\varphi|_2=1} |(\nabla^{2k}\overline{T})(\varphi)| = \sup_{|\psi|_2=1} \left| \int_{R^N} \psi(s)(-1)^k|s|^{2k}\,\overline{(FT)(s)}\,ds \right| = \left\{ \int_{R^N} |s|^{4k}|(FT)(s)|^2\,ds \right\}^{1/2}$$

which establishes (75) and proves (73).

Now conversely let $\varphi^{(0)} \in \mathfrak{D}(A^k_{\mathfrak{S}})$ so that $\psi^{(0)} = F\varphi^{(0)}$ is in $\mathfrak{D}(\hat{A}^k)$ which is equivalent to the assertion that each component $\psi_j^{(0)}$ of $\psi^{(0)}$ is in $\mathfrak{H}$ and satisfies the inequality

$$(79) \qquad \int_{R^N} |s|^{4k} |\psi_j^{(0)}(s)|^2\,ds < \infty.$$

We wish to show that the tempered distribution $T_{\varphi_j^{(0)}}$ corresponding to $\varphi_j^{(0)}$ is in $T^{(2k)}(R^N)$. In view of Lemma 14 it suffices to show that $\overline{T}_{\varphi_j^{(0)}}$ is in $T^{(2k)}(R^N)$. Let $|\alpha|_1 \leqq 2k$, $\varphi \in \Phi$, $\psi = F\varphi$. Since $\psi \in \Phi$ (Theorem 11.1) we have, by differentiating under the integral sign as before,

$$\begin{aligned}(\partial^\alpha \overline{T}_{\varphi_j^{(0)}})(\varphi) &= \int_{R^N} (-1)^{|\alpha|_1}\, \partial^\alpha(F^{-1}\psi)(s)\overline{\varphi_j^{(0)}(s)}\, ds \\ &= \int_{R^N} (-1)^{|\alpha|_1} F^{-1}((i\,\cdot)^\alpha \psi(\cdot))(s)\overline{\varphi_j^{(0)}(s)}\, ds \\ &= \int_{R^N} \psi(s)(-1)^{|\alpha|_1}(is)^\alpha(F^{-1}\overline{\varphi}_j^{(0)})(s)\, ds \\ &= \int_{R^N} \psi(s)(-1)^{|\alpha|_1}(is)^\alpha\overline{(F\varphi_j^{(0)})(s)}\, ds \\ &= \int_{R^N} \psi(s)(-1)^{|\alpha|_1}(is)^\alpha\overline{\psi_j^{(0)}(s)}\, ds.\end{aligned}$$

Since $\psi_j^{(0)}$ satisfies (79), it follows that the coefficient of $\psi(s)$ in the preceding integrand is square integrable on R^N and the same is true of its Fourier transform $F((-1)^{|\alpha|_1}(i\,\cdot)^\alpha\overline{\psi_j^{(0)}(\cdot)})$. The preceding equation thus gives

$$(\partial^\alpha \overline{T}_{\varphi_j^{(0)}})(\varphi) = \int_{R^N} \varphi(s)F((-1)^{|\alpha|_1}(i\,\cdot)^\alpha\overline{\psi_j^{(0)}}(\cdot))(s)\, ds, \qquad \varphi \in \Phi,$$

and proves that $\overline{T}_{\varphi_j^{(0)}}$ is in $T^{(2k)}(R^N)$ and that

$$[T_{\varphi_j^{(0)}}, \ldots, T_{\varphi_p^{(0)}}] \in (p)T^{(2k)}(R^N).$$

This shows that

$$\mathfrak{D}(A_{\mathfrak{S}}^k) \subseteq (p)T^{(2k)}(R^N) \tag{80}$$

which, when combined with (73), establishes conclusion (vii).

According to conclusions (vi) and (vii), for each λ with $\mathscr{R}(\lambda) > 0$, $\varphi_1(\lambda)$ is a vector $T = [T_1, \ldots, T_p]$ with T_j, $j = 1, \ldots, p$, a tempered distribution in R^N whose derivatives of all orders are functions in $\mathfrak{H} = L_2(R^N)$. The restriction $F_j = T_j \,|\, C_0^\infty(R^N)$ of T_j to $C_0^\infty(R^N)$ is a distribution in R^N whose restriction $F_j \,|\, \mathfrak{S}_r$ to the open ball $\mathfrak{S}_r = \{s \in R^N \,|\, |s| < r\}$ has for its derivatives functions which are the restrictions of the corresponding derivatives of T_j. Thus conclusion (ii) of Theorem XIV.4.5 applies with $n = N$, $p = 2$, and k arbitrarily large, to show that all derivatives of T_j are continuous on the closure of $\mathfrak{S}_r$. Since $r > 0$ is arbitrary, $\partial_s^\beta \varphi_1(s_1, \ldots, s_N; \lambda)$ exists for each β as a continuous vector function of s on R^N and

$$\int_{\mathfrak{S}_r} |\partial_s^\beta \varphi_1(s; \lambda)|^2\, ds = \int_{\mathfrak{S}_r} |(\partial_s^\beta T)(s)|^2\, ds \leqq \int_{R^N} |(\partial_s^\beta T)(s)|^2\, ds < \infty$$

shows that these derivatives $\partial_s^\beta \varphi_1$ all belong to $\mathfrak{H}^p$. Q.E.D.

20 COROLLARY. *Let k be a positive integer and let $\varphi(s) = [\varphi_1(s)_1, \ldots, \varphi_p(s)]$ be defined on R^N. If, for all sets $\beta = [\beta_1, \ldots, \beta_N]$ of N non-negative integers, $(\partial_s^\beta \varphi)(s)$ exists, is continuous on R^N and square integrable on R^N, then φ belongs to $\mathfrak{D}(A_{\mathfrak{S}}^k)$.*

21 THEOREM. *Using the notation of Theorem 19 and letting B be an arbitrary bounded linear operator in $\mathfrak{H}^p$, we have:*

(i) *The operator $A_{\mathfrak{S}} + B$ with domain $\mathfrak{D}(A_{\mathfrak{S}})$ is the infinitesimal generator of a strongly continuous semi-group $S(t)$, $0 \leqq t < \infty$.*

For every $\varphi^{(0)}$ in $\mathfrak{H}^p$ there is one and only one continuous map $t \to \varphi(t)$ of $[0, \infty)$ into $\mathfrak{H}^p$ which is differentiable for $t > 0$ and has the properties

(ii) $\varphi(t) \in \mathfrak{D}(A_{\mathfrak{S}}), \quad 0 < t < \infty,$

(iii) $\varphi'(t) = (A_{\mathfrak{S}} + B)\varphi(t), \quad 0 < t < \infty,$

(iv) $\varphi(0) = \varphi^{(0)}.$

(v) *This unique function φ is given by the equation $\varphi(t) = S(t)\varphi^{(0)}$. The semi-group $S(t)$ has a strongly analytic extension to a semi-group $S(\zeta)$ defined for ζ in the half plane $\mathscr{R}(\zeta) > 0$. The unique solution to the Cauchy problem presented by the equations* (ii), (iii), (iv) *is analytic in t and has an analytic extension $\varphi_1(\zeta) = S(\zeta)\varphi^{(0)}$ to the half plane $\mathscr{R}(\zeta) > 0$ which satisfies the equations* (ii) *and* (iii) *for complex values of t with $\mathscr{R}(t) > 0$.*

(vi) $\mathfrak{D}(A_{\mathfrak{S}} + B) = (p)T^{(2)}(R^N).$

(vii) *The derivative $\varphi'(0)$ exists and $\varphi'(0) = (A_{\mathfrak{S}} + B)\varphi(0)$ if and only if $\varphi^{(0)}$ is in $(p)T^{(2)}(R^N)$.*

If $\alpha^2 < 0$ then the analogous conclusions obtained by reflecting the complex plane in the imaginary axis also hold.

PROOF. We shall first need a bound for the norm of the analytic extension $T_1(\zeta)$ of the semi-group $T(t)$ to the half plane $\mathscr{R}(\zeta) > 0$. Let $\mathscr{R}(\zeta) > 0$ so that $\zeta = |\zeta| e^{i\theta}$ with $-\pi/2 < \theta < \pi/2$. Then since

$$\sup_{s \in R^N} e^{-\mathscr{R}(\zeta)\alpha^2 |s|^2} |s|^{2\nu} = \frac{1}{\mathscr{R}(\zeta)^\nu} \left(\frac{\nu}{e\alpha^2}\right)^\nu$$

it follows from (63) and the fact that $|\hat{N}(s)| = O(|s|^2)$ as $|s| \to \infty$ that $|\hat{T}_1(\zeta)| \leqq K(\sec\theta)^{p-1}$ and since F is an isometry in $\mathfrak{H}^p$ that

$$|T_1(\zeta)| \leqq K(\sec\theta)^{p-1}, \qquad \zeta = |\zeta| e^{i\theta}, \qquad \frac{-\pi}{2} < \theta < \frac{\pi}{2}. \tag{81}$$

On the real axis $\theta = 0$ and hence, since $T(t)$ is strongly continuous at $t = 0$,

$$|T(t)| \leqq K, \qquad 0 \leqq t < \infty, \tag{82}$$

a fact which also follows from (61). The inequality (81) shows that $T_1(\zeta)$ is strongly continuous at $\zeta = 0$ provided ζ remains in a sector $-\varphi \leqq \theta \leqq \varphi$ with $0 \leqq \varphi < \pi/2$. We now define inductively operators $S_n(\zeta)$ for $\mathscr{R}(\zeta) > 0$, $n = 0, 1, \ldots$, and $\varphi \in \mathfrak{H}^p$, by the equations

$$S_0(\zeta) = T_1(\zeta), \qquad S_n(\zeta)\varphi = \int_0^\zeta T_1(\zeta - u) B S_{n-1}(u)\varphi \, du \tag{83}$$

where the path of integration is the straight line segment from 0 to ζ. Since $T_1(\zeta)$ is strongly analytic for $\mathscr{R}(\zeta) > 0$, it is clear from (83) that $S_n(\zeta)$ is also. Using (81) and (83) we obtain, inductively, the estimates

$$|S_n(\zeta)| \leqq [K(\sec\,\theta)^{p-1}]^{n+1} |B|^n \frac{|\zeta|^n}{n!}. \tag{84}$$

Thus the series

$$S(\zeta) = \sum_{n=0}^{\infty} S_n(\zeta), \qquad \mathscr{R}(\zeta) > 0, \tag{85}$$

converges absolutely, showing that $S(\zeta)$ is strongly analytic for $\mathscr{R}(\zeta) > 0$ and that

$$|S(\zeta)| \leqq K(\sec\,\theta)^{p-1} \exp\{K(\sec\,\theta)^{p-1} |B| \, |\zeta|\} \tag{86}$$

and along the positive real axis,

$$|S(t)| \leqq K e^{K|B|t}, \qquad 0 < t < \infty. \tag{87}$$

Thus for any real number $\omega > K|B|$,

$$\int_0^\infty e^{-\lambda t} S(t)\varphi \, dt = \sum_{n=0}^{\infty} \int_0^\infty e^{-\lambda t} S_n(t)\varphi \, dt, \qquad \mathscr{R}(\lambda) > \omega, \ \varphi \in \mathfrak{H}^p. \tag{88}$$

Now, using Corollary VIII.1.16, we obtain, for $\mathscr{R}(\lambda) > \omega$, the formula

$$\begin{aligned} \int_0^\infty e^{-\lambda t} S_n(t)\varphi \, dt &= \int_0^\infty e^{-\lambda t} \left\{ \int_0^t T(t-u) B S_{n-1}(u)\varphi \, du \right\} dt \\ &= \int_0^\infty e^{-\lambda u} \left\{ \int_u^\infty e^{-\lambda(t-u)} T(t-u) B S_{n-1}(u)\varphi \, dt \right\} du \\ &= \int_0^\infty e^{-\lambda u} \left\{ \int_0^\infty e^{-\lambda t} T(t) B S_{n-1}(u)\varphi \, dt \right\} du \end{aligned}$$

$$= \int_0^\infty e^{-\lambda t}\, T(t)\left\{\int_0^\infty e^{-\lambda u} BS_{n-1}(u)\varphi\, du\right\} dt$$

$$= R(\lambda; A_{\mathfrak{S}})B \int_0^\infty e^{-\lambda u} S_{n-1}(u)\varphi\, du,$$

whose repeated application gives, since $S_0(t) = T(t)$,

$$\begin{aligned}\int_0^\infty e^{-\lambda t} S_n(\lambda)\varphi\, dt &= [R(\lambda); A_{\mathfrak{S}})B]^n \int_0^\infty e^{-\lambda u} T(u)\, du \\ &= [R(\lambda; A_{\mathfrak{S}})B]^n R(\lambda; A_{\mathfrak{S}}) \\ &= R(\lambda; A_{\mathfrak{S}})[BR(\lambda; A_{\mathfrak{S}})]^n,\end{aligned}$$

which, together with (88) shows that for every φ in $\mathfrak{H}^p$

(89) $$\int_0^\infty e^{-\lambda t} S(t)\varphi\, dt = R(\lambda; A_{\mathfrak{S}}) \sum_{n=0}^{\infty} [BR(\lambda; A_{\mathfrak{S}})]^n \varphi, \qquad \mathscr{R}(\lambda) > \omega.$$

Now let $\varphi \in \mathfrak{H}^p$, $\mathscr{R}(\lambda) > \omega$, $\gamma = K|B|\omega^{-1}$, so that from (82) it follows that

$$\begin{aligned}|BR(\lambda; A_{\mathfrak{S}})\varphi| &= \left|\int_0^\infty e^{-\lambda t} BT(t)\varphi\, dt\right| \\ &\leqq K|B|\,|\varphi|\,\mathscr{R}(\lambda)^{-1} < \frac{K|B|}{\omega}|\varphi| = \gamma|\varphi|,\end{aligned}$$

and thus that $|BR(\lambda; A_{\mathfrak{S}})| \leqq \gamma < 1$ for $\mathscr{R}(\lambda) > \omega$, which shows that the series (89) converges absolutely. It will next be shown that the series (89) is the resolvent $R(\lambda; A_{\mathfrak{S}} + B)$. For brevity we let $R = R(\lambda; A_{\mathfrak{S}})$, $B_0 = \sum_{n=0}^\infty [BR(\lambda; A_{\mathfrak{S}})]^n$, $B_1 = B_0 - I = \sum_{n=1}^\infty [BR(\lambda; A_{\mathfrak{S}})]^n$. Then

$$(\lambda I - A_{\mathfrak{S}} - B)RB_0 = (I - BR)B_0 = B_0 - BRB_0 = B_0 - B_1 = I,$$

and, for $\varphi \in \mathfrak{D}(A_{\mathfrak{S}})$,

$$\begin{aligned}RB_0(\lambda I - A_{\mathfrak{S}} - B)\varphi &= RB_0(\lambda I - A_{\mathfrak{S}})\varphi - RB_0 B\varphi \\ &= R(I + B_1)(\lambda I - A_{\mathfrak{S}})\varphi - RB_0 B\varphi \\ &= \varphi + RB_1(\lambda I - A_{\mathfrak{S}})\varphi - RB_0 B\varphi,\end{aligned}$$

and since $(\lambda I - A_{\mathfrak{S}})\varphi = R^{-1}\varphi = (BR)^{-1}B\varphi$, the above equation becomes

$$\begin{aligned}RB_0(\lambda I - A_{\mathfrak{S}} - B)\varphi &= \varphi + R\{B_1(BR)^{-1}B\varphi - B_0 B\varphi\} \\ &= \varphi + R\{B_0 B\varphi - B_0 B\varphi\} = \varphi,\end{aligned}$$

which proves that $RB_0 = R(\lambda; A_{\mathfrak{S}} + B)$ and, together with (89), gives

$$R(\lambda; A_{\mathfrak{S}} + B)\varphi = \int_0^\infty e^{-\lambda t} S(t)\varphi\, dt, \qquad \mathscr{R}(\lambda) > \omega,\ \varphi \in \mathfrak{H}^p. \tag{90}$$

Corollary VIII.1.16 thus shows that $S(t)$, $0 \leqq t < \infty$ is a strongly continuous semi-group and that $A_{\mathfrak{S}} + B$ with domain $\mathfrak{D}(A_{\mathfrak{S}})$ is its infinitesimal generator. We have already observed that $S(\zeta)$ is strongly analytic for $\mathscr{R}(\zeta) > 0$. We shall now show that it is a semi-group, that is, $S(\zeta_1 + \zeta_2) = S(\zeta_1)S(\zeta_2)$ if $\mathscr{R}(\zeta_1) > 0$ and $\mathscr{R}(\zeta_2) > 0$. For any φ in $\mathfrak{H}^p$ and $t > 0$ the function

$$f_1(\zeta_1) = [S(\zeta_1 + t) - S(\zeta_1)S(t)]\varphi$$

is analytic and vanishes for ζ_1 on the positive real axis. Thus $f_1(\zeta_1) = 0$ for $\mathscr{R}(\zeta_1) > 0$. Thus for $\mathscr{R}(\zeta_1) > 0$ the function

$$f_2(\zeta_2) = [S(\zeta_1 + \zeta_2) - S(\zeta_1)S(\zeta_2)]\varphi$$

vanishes for ζ_2 on the positive real axis and hence $f_2(\zeta_2) = 0$ if $\mathscr{R}(\zeta_2) > 0$, which proves that $S(\zeta)$ is a strongly analytic semi-group on the half plane $\mathscr{R}(\zeta) > 0$.

We shall next prove that $S(\zeta)\varphi \in \mathfrak{D}(A_{\mathfrak{S}})$ for $\varphi \in \mathfrak{H}^2$ and $\mathscr{R}(\zeta) > 0$. At one place in the proof we have $A_{\mathfrak{S}} T(\zeta - u)BS_{n-1}(u)$ as an integrand which has a singularity of magnitude $(\zeta - u)^{-2}$ when $u = \zeta$. To avoid this we use a most elementary type of summability by introducing the operators defined on $\mathfrak{H}^p$ for every integer $n \geqq 1$, for $0 < r < 1$, and $\mathscr{R}(\zeta) > 0$ by the equation,

$$S_{n,r}(\zeta)\varphi = \int_0^{r\zeta} T(\zeta - u)BS_{n-1}(u)\varphi\, du, \qquad \varphi \in \mathfrak{H}^p, \tag{91}$$

where, as before, the path of integration is the straight line segment joining 0 with $r\zeta$.

By Theorem 19 (vi)

$$S_0(\zeta)\varphi \in \mathfrak{D}(A_{\mathfrak{S}}), \qquad S_0'(\zeta)\varphi = A_{\mathfrak{S}} S_0(\zeta)\varphi, \qquad \mathscr{R}(\zeta) > 0. \tag{92}$$

We wish now to calculate the derivative $S_n'(\zeta)\varphi$ for $n \geqq 1$. We have

$$\frac{S_{n,r}(\zeta + h) - S_{n,r}(\zeta)}{h}\varphi$$

$$= T_1(h)\frac{1}{h}\int_{r\zeta}^{r\zeta + rh} T_1(\zeta - u)BS_{n-1}(u)\varphi\, du$$

(continued)

$$+\int_0^{r\zeta} \frac{T_1(\zeta+h-u)-T_1(\zeta-u)}{h} BS_{n-1}(u)\varphi\, du$$

which shows that

$$\frac{d}{d\zeta} S_{n,r}(\zeta)\varphi = rT_1(\zeta - r\zeta)BS_{n-1}(r\zeta)\varphi \tag{93}$$

$$-\int_0^{r\zeta} A_{\mathfrak{S}}\, T_1(\zeta-u)BS_{n-1}(u)\varphi\, du.$$

We know that the last integral exists, since for each u in the segment of integration, $T_1(\zeta-u)BS_{n-1}(u)\varphi$ is in $\mathfrak{D}(A_{\mathfrak{S}})$ (Theorem 19(vi)) and the integrability with respect to u of the function $A_{\mathfrak{S}}\, T_1(\zeta-u)BS_{n-1}(u)\varphi$ follows immediately from the form of the function $\hat{T}_1(s;\lambda)$ given in equation (63) for, since we are integrating only from 0 to $r\zeta$ with $r<1$, the argument $\zeta-u$ remains away from zero. In view of Theorem III.6.20 equation (93) may be written as

$$\frac{d}{d\zeta} S_{n,r}(\zeta)\varphi = rT_1(\zeta - r\zeta)BS_{n-1}(r\zeta)\varphi \tag{94}$$

$$+A_{\mathfrak{S}}\int_0^{r\zeta} T_1(\zeta-u)BS_{n-1}(u)\varphi\, du.$$

It follows from (91) (or from the analyticity of $S_{n,r}(\zeta)$) that

$$\lim_{\to 1}\frac{d}{d\zeta} S_{n,r}(\zeta)\varphi = \frac{d}{d\zeta} S_n(\zeta)\varphi$$

and, since

$$\lim_{r\to 1} rT_1(\zeta - r\zeta)BS_{n-1}(r\zeta)\varphi = BS_{n-1}(\zeta)\varphi,$$

the last term in equation (94) has a limit as $r\to 1$. Since $A_{\mathfrak{S}}$ is a closed operator and

$$\lim_{r\to 1}\int_0^{r\zeta} T_1(\zeta-u)BS_{n-1}(u)\varphi\, du = S_n(\zeta)\varphi,$$

it follows that $S_n(\zeta)\varphi$ is in $\mathfrak{D}(A_{\mathfrak{S}})$ and that

$$\begin{aligned} S_n'(\zeta)\varphi &= BS_{n-1}(\zeta)\varphi + A_{\mathfrak{S}}S_n(\zeta)\varphi \\ &= (A_{\mathfrak{S}}+B)S_n(\zeta)\varphi + B(S_{n-1}(\zeta)-S_n(\zeta))\varphi. \end{aligned} \tag{95}$$

Since $S(\zeta)$ and $S_n(\zeta)$, $n \geqq 0$, are analytic on the half plane $\mathscr{R}(\zeta) > 0$, it follows from (85) that

$$\text{(96)} \qquad S'(\zeta)\varphi = \sum_{n=0}^{\infty} S_n'(\zeta)\varphi, \qquad \mathscr{R}(\zeta) > 0,\ \varphi \in \mathfrak{H}^p,$$

and from (92), (95) and (84) that

$$\begin{aligned}\text{(97)} \qquad S'(\zeta)\varphi &= \lim_{m\to\infty} \sum_{n=0}^{m} S_n'(\zeta)\varphi \\ &= \lim_{m\to\infty} (A_{\mathfrak{S}} + B) \sum_{n=0}^{m} S_n(\zeta)\varphi - \lim_{m\to\infty} BS_m(\zeta)\varphi \\ &= \lim_{m\to\infty} (A_{\mathfrak{S}} + B) \sum_{n=0}^{m} S_n(\zeta)\varphi.\end{aligned}$$

Since $A_{\mathfrak{S}} + B$ is a closed operator, it follows from (85) and (97) that

$$\text{(98)} \quad S(\zeta)\varphi \in \mathfrak{D}(A_{\mathfrak{S}} + B), \qquad S'(\zeta)\varphi = (A_{\mathfrak{S}} + B)S(\zeta)\varphi, \qquad \mathscr{R}(\zeta) > 0,\ \varphi \in \mathfrak{H}^p.$$

Thus if we set $\varphi(t) = S(t)\varphi^{(0)}$, $0 \leqq t < \infty$, the function φ satisfies the conclusions (i), (ii) and (iii) and its uniqueness may be established as it was in Theorem 19. We have already seen that $S(t)$ has a strongly analytic extension to a semi-group $S(\zeta)$ defined for $\mathscr{R}(\zeta) > 0$ and (98) shows that the conclusions (ii) and (iii) are satisfied by complex values of t with $\mathscr{R}(t) > 0$. This completes the proof of conclusions (i), . . . , (v). The conclusion (vi) follows from Theorem 19 (vii). Since $A_{\mathfrak{S}} + B$ is the infinitesimal generator of the semi-group $S(t)$, $0 \leqq t < \infty$, conclusion (vii) follows from (vi), Definition VIII.1.6, and Lemma VIII.1.7(b). Q.E.D.

A few comments on a method of solving some of the Cauchy problems we have encountered may be in order. It happens that a number of the examples we have discussed and, as the reader will readily observe, many that we have not discussed, only require elementary functions for the explicit representation of the solution of the initial value problem. We shall illustrate this by giving the specific elementary details of how the calculations are made in a few examples. According to Corollary 11.10, the operator algebra of all convolutions (proper and improper) defined on $\mathfrak{H}$ is *-equivalent to the B^*-algebra $L_\infty(R^N)$. It follows that the algebra $\mathfrak{A}^p$ of operators A in $\mathfrak{H}^p$ defined by the equation

(99) $$A = \int_{R^N} \hat{A}(s)e(ds),$$

where the elements of $\hat{A}(s) = (a_{jk}(s))$ are bounded measurable functions on R^N, is a convolution algebra and

(100) $$A\varphi = (\tau^{-1}\hat{A}) * \varphi, \qquad \varphi \in \mathfrak{H}^p,$$

where

(101) $$\tau = (2\pi)^{N/2}F, \qquad \tau^{-1} = \frac{1}{(2\pi)^{N/2}} F^{-1},$$

and F is the Fourier transform in $\mathfrak{H}^p$. The natural closed extensions $A_{\mathfrak{S}}$ of the formal differential operators A illustrating the Cauchy initial value problems that we have discussed, except that of Theorem 21 , are all infinitesimal generators of semi-groups in this algebra $\mathfrak{A}^p$. For such an infinitesimal generator $A_{\mathfrak{S}}$, the strongly continuous semi-group $T(t), t \geqq 0$ (or $t \leqq 0$) of operators in $\mathfrak{H}^p$ that it generates is given by the formula

(102) $$T(t) = e^{tA_{\mathfrak{S}}} = \int_{R^N} (e^{t\hat{A}})(s)e(ds).$$

We may therefore, according to (100), express the solution $\varphi(t) = T(t)\varphi^{(0)}$ of the initial value problem

(103) $$\varphi'(t) = A_{\mathfrak{S}}\varphi(t), \qquad \varphi(0) = \varphi^{(0)},$$

as a convolution integral

(104) $$\varphi(t) = (\tau^{-1} \exp t\hat{A}) * \varphi^{(0)}.$$

The formula (104) solves, in the sense that term is used in discussing differential equations, the Cauchy problem (103) for it reduces the problem to that of a quadrature.

We shall first illustrate the use of (104) by solving the diffusion equation with $p = 1$, N arbitrary, and the initial value a prescribed function $\varphi^{(0)}$ in $L_2(R^N)$, that is, the problem of solving the following equations for $\varphi(s; t)$:

(105) $$\frac{\partial}{\partial t} \varphi(s; t) = \alpha^2 \nabla^2 \varphi(s; t), \qquad \varphi(s; 0) = \varphi(s)^{(0)},$$

where $\nabla^2 = \partial^2/\partial s_1^2 + \cdots + \partial^2/\partial s_N^2$. Since $\sin u$ is an odd function, we see by iterated integration that

$$(106)\qquad (\tau^{-1}e^{t\hat{A}})(s) = \frac{1}{(2\pi)^N}\int_{R^N} e^{-\alpha^2 t|u|^2}\cos su\,du$$

$$= \prod_{j=1}^{N}\frac{1}{2\pi}\int_{-\infty}^{\infty} e^{-\alpha^2 tr^2}\cos s_j r\,dr$$

$$= \tfrac{1}{2}(\pi\alpha^2 t)^{-N/2}e^{-|s|^2/4\alpha^2 t},$$

where in the last step we have used the well-known (cf. Bartle [8], p. 369) definite integral

$$(107)\qquad \int_{-\infty}^{\infty} e^{-cr^2}\cos(tr)\,dr = \sqrt{\frac{\pi}{c}}\,e^{-t^2/4c}, \qquad c > 0.$$

Thus (104) shows that the solution of (105) is

$$(108)\qquad \varphi(s;t) = \tfrac{1}{2}(\pi\alpha^2 t)^{-N/2}\int_{R^N} e^{-|s-u|^2/4\alpha^2 t}\varphi^{(0)}(u)\,du.$$

The kernel here is the N-dimensional form of the Gauss-Weierstrass transformation. If $\alpha^2 > 0$ and $N = 3$, equation (108) gives Laplace's solution to the problem of heat flow in an infinite solid. Here $\varphi^{(0)}(u)$ is the initial temperature at the point $u = (u_1, u_2, u_3)$, $\varphi(s;t)$ is the temperature at the point $s = (s_1, s_2, s_3)$ and time t, and $\alpha^2 = K/C$ where K is the heat conductivity of the material and C is the heat capacity of the material per unit volume. The equation (105) also occurs in the study of a solute diffusing through a solvent in which case $\varphi(s;t)$ is the density of the solute at the point s and time t. The solute as well as the solvent may both be liquids, and so equation (105) arises in connection with the diffusion of one liquid through another and also in the study of neutron diffusion through matter. It is for such reasons that (105) is known as the *diffusion equation.* The fact that diffusion is an irreversible process reflects itself, in the mathematical solutions of such problems as the preceding, by yielding semi-groups instead of full groups. Thus a proof of the unique existence of a solution to a Cauchy problem corresponding to a diffusion equation is another way of asserting that the present determines the future (but not the past, as is the case with certain conservative mechanical systems encountered in celestial mechanics and wave propagation where there is no energy lost or entropy gained).

We next consider the problem of finding the explicit form of the solution to the Cauchy problem whose existence is established in Theorem 19. As before, we assume that $\alpha^2 t > 0$. It follows from equation (55) which gives the analytical expression for $(\exp t\hat{A})(s)$ that the calculation of the operator $\boldsymbol{\tau}^{-1} \exp t\hat{A}$ involves the calculation of integrals of the form

$$\int_{R^N} e^{isu} e^{-\alpha^2 t|u|^2} u_1^{j_1} \cdots u_N^{j_N} \, du$$

$$= \prod_{k=1}^{N} \int_{-\infty}^{\infty} e^{-\alpha^2 t u_k^2} (\cos s_k u_k + i \sin s_k u_k) u_k^{j_k} \, du_k .$$

Thus the calculation of $\tau^{-1} \exp t\hat{A}$ reduces to the evaluation of *only a finite number* of functions defined on the real axis by the equations

$$\text{(109)} \qquad \begin{cases} F_j(y) = \displaystyle\int_{-\infty}^{\infty} e^{-\alpha^2 t x^2} (\cos xy) x^j \, dx, \\ G_j(y) = \displaystyle\int_{-\infty}^{\infty} e^{-\alpha^2 t x^2} (\sin xy) x^j \, dx. \end{cases}$$

It is clear that $F_j(y) = 0 = G_{j+1}(y)$ if j is odd. Equation (107) shows that

$$\text{(110)} \qquad F_0(y) = \sqrt{\frac{\pi}{\alpha^2 t}} \, e^{-y^2/4\alpha^2 t}$$

and we may calculate the remaining functions $F_{2n}(y)$ and $G_{2n+1}(y)$ inductively from (110) and the relations

$$\text{(111)} \qquad G_{j+1}(y) = -F_j'(y), \qquad F_{j+1}(y) = G_j'(y).$$

We next consider a different type of perturbed Laplacian with $p = 2$, N arbitrary, and given by the formal differential operator defined on Φ^2 by the matrix

$$\text{(112)} \qquad A = \begin{pmatrix} \alpha^2 \nabla^2 & a \\ \beta^2 a & \alpha^2 \nabla^2 \end{pmatrix},$$

where $\beta \neq 0$ is a complex constant,

$$a = \alpha_0 + \alpha_1 \frac{\partial}{\partial s_1} + \cdots + \alpha_N \frac{\partial}{\partial s_N},$$

and the complex constants $\alpha_0, \ldots, \alpha_N$ are not all zero. As before α^2 is a positive or negative real number and $\nabla^2 = \partial^2/\partial s_1^2 + \cdots + \partial^2/\partial s_N^2$. We assume that $\beta \neq 0$ and that not all of the constants $\gamma_0, \ldots, \gamma_N$ are zero, for otherwise the operator (112) reduces to the case covered by the preceding example, whereas with our assumptions a slightly different type of integral arises. We have already observed in our discussion of the operator (33) (which only differs from (112) by a multiplicative constant) that the natural closed extension $A_{\mathfrak{S}}$ of the operator (112) is a scalar type spectral operator. We have

$$\hat{A}(s) = \begin{pmatrix} -\alpha^2 |s|^2 & \hat{a}(s) \\ \beta^2 \hat{a}(s) & -\alpha^2 |s|^2 \end{pmatrix}, \tag{113}$$

whose characteristic roots are

$$\hat{\lambda}_{21}(s) = -\alpha^2 |s|^2 + \beta \hat{a}(s), \qquad \hat{\lambda}_{22}(s) = -\alpha^2 |s|^2 - \beta \hat{a}(s), \tag{114}$$

and have the associated projections $\hat{E}_j(s) = E(\hat{\lambda}_{2j}(s); \hat{A}(s))$, $j = 1, 2$, given, according to equation (11.41), by

$$\hat{E}_1(s) = \frac{1}{2}\begin{pmatrix} 1 & \beta^{-1} \\ \beta & 1 \end{pmatrix}, \qquad \hat{E}_2(s) = \frac{1}{2}\begin{pmatrix} 1 & -\beta^{-1} \\ -\beta & 1 \end{pmatrix}. \tag{115}$$

Thus (VII.1.8)

$$(e^{t\hat{A}})(s) = e^{t\hat{\lambda}_{21}(s)} \hat{E}_1 + e^{t\hat{\lambda}_{22}(s)} \hat{E}_2, \tag{116}$$

where we have omitted the variable s in the symbols $\hat{E}_j(s)$ to emphasize, as (115) shows, that they are independent of s. Since $A_{\mathfrak{S}}$ is a scalar type spectral operator, Theorem 7 shows that $\sigma(A_{\mathfrak{S}}) = \sigma_0(A_{\mathfrak{S}})$, and since $\hat{a}(s)$ is a linear form, the equations (114) show that $\sigma(A_{\mathfrak{S}})$ lies in a half plane. It follows from Theorem 8 that $A_{\mathfrak{S}}$ is the infinitesimal generator of the strongly continuous semi-group $T(t) = F^{-1} \exp(t\hat{A})F$. For $u = (u_1, \ldots, u_N) \in R^N$ we have $\beta\hat{a}(u) = \beta\alpha_0 + \delta u + i\varepsilon u$ where the linear forms δu and εu are real and homogeneous, that is, δ_j and ε_j, $j = 1, \ldots, N$ are real constants. It is thus apparent from (114), (115) and (116) that the calculation of the operator $\tau^{-1}\exp(t\hat{A})$ only involves evaluation of the integrals

$$\int_{R^N} e^{isu}\, e^{-\alpha^2 t|u|^2 \pm t\beta\alpha_0 \pm t\delta u \pm it\varepsilon u}\, du$$

$$= e^{\pm t\beta\alpha_0} \prod_{j=1}^{N} \int_{-\infty}^{\infty} e^{-\alpha^2 t u_j^2 \pm t\delta_j u_j} \{\cos(s_j \pm t\varepsilon_j)u_j + i \sin(s_j \pm t\varepsilon_j)u_j\}\, du_j.$$

The integrals here are of the following two types, with $c = \alpha^2 t > 0$, b real and $y = s_j + t\varepsilon_j$.

$$\int_{-\infty}^{\infty} e^{-c(u-b)^2} \cos(uy)\, du$$

$$= \cos(by) \int_{-\infty}^{\infty} e^{-cr^2} \cos(ry)\, dr = \sqrt{\frac{\pi}{c}}\,(e^{-y^2/4c})\cos(by),$$

$$\int_{-\infty}^{\infty} e^{-c(u-b)^2} \sin(uy)\, du$$

$$= \sin(by) \int_{-\infty}^{\infty} e^{-cr^2}\cos(ry)dr = \sqrt{\frac{\pi}{c}}\,(e^{-y^2/4c})\sin(by),$$

where again we have used formula (107). This shows that an explicit form for the matrix representing the convolution kernel operator $\tau^{-1} \exp(t\hat{A})$ in $\mathfrak{H}^2$ may be given in terms of a finite number of elementary functions.

The reader may have observed that, in the general theory of strongly continuous semi-groups, the differentiability of $T(t)\varphi$ is only assured if φ is in the domain of the infinitesimal generator, whereas in the examples of Theorems 19 and 21 the function $\varphi(t) = T(t)\varphi^{(0)}$ was differentiable (even analytic) for every $t > 0$ and every $\varphi^{(0)}$ in $\mathfrak{H}^p$, thus allowing the initial value $\varphi^{(0)}$ in the corresponding Cauchy problem to be an arbitrary vector in $\mathfrak{H}^p$. From the abstract theory of semi-groups this phenomenon may be explained by the fact that, in these examples, we have $T(t)\mathfrak{H}^p \subseteqq \mathfrak{D}(A_{\mathfrak{S}})$ if $t > 0$ (Hille and Phillips [1, Theorem 10.3.5, p. 310]). In the examples of Theorems 19 and 21 we did not appeal to the general theory, as it was clear from the form of exp $t\hat{A}$ that $T(t)\varphi$ was differentiable for $t > 0$ and an arbitrary φ in $\mathfrak{H}^p$. It is even clear that the nth order derivative $T^{(n)}(t)$ exists in the uniform operator topology for $t > 0$ and every positive integer n.

The same observations are true of the natural closed extension of the formal differential operator (112). This is apparent from the formulas (114), (115), (116) and the fact that $\hat{a}(s)$ is a linear form. It is also clear from (113) that

$$\mathfrak{D}(\hat{A}) = \{\psi \in \mathfrak{H}^2 \mid \int_{R^N} |s|^4\, |\psi(s)|^2\, ds < \infty\} \tag{117}$$

and thus if follows, as the proof of Theorem 19 shows, that

$$\mathfrak{D}(A_{\mathfrak{S}}) = (2)T^{(2)}(R^N).$$

We summarize a few of the properties of the perturbed Laplacian operator (112) in the following theorem.

22 THEOREM. *The natural closed extension $A_{\mathfrak{S}}$ of the formal differential operator* (112) *is a scalar type spectral operator which is the infinitesimal generator of a strongly continuous semi-group $T(t)$ of linear operators in $\mathfrak{H}^2$. The domain of $A_{\mathfrak{S}}$ is*

(i) $$\mathfrak{D}(A_{\mathfrak{S}}) = (2)T^{(2)}(R^N),$$

and

(ii) $$T(t)\mathfrak{H}^2 \subseteq (2)T^{(2)}(R^N), \qquad t > 0.$$

For every $\varphi^{(0)}$ in $\mathfrak{H}^2$ there is a uniquely determined continuous map $t \to \varphi(t)$ of $[0, \infty)$ into $\mathfrak{H}^2$ (of $(-\infty, 0]$ into $\mathfrak{H}^2$ if $\alpha^2 < 0$) which is differentiable for $t > 0$ and has the properties

(iii) $\varphi(t) \in (2)T^{(2)}(R^N), \qquad t > 0,$

(iv) $\varphi'(t) = A_{\mathfrak{S}}\varphi(t), \qquad t > 0,$

(v) $\varphi(0) = \varphi^{(0)}.$

(vi) *This uniquely determined function $\varphi(t)$ is given by the equation*

$$\varphi(t) = T(t)\varphi^{(0)} = (\tau^{-1} \exp(t\hat{A})) * \varphi^{(0)}$$

and the elements of the matrix representing the convolution operator $\tau^{-1}\exp(t\hat{A})$ are expressible in terms of a finite number of elementary functions.

(vii) *For $t > 0$ the derivatives $T^{(n)}(t)$, $n = 1, 2, \ldots$, all exist in the uniform operator topology.*

PROOF. Statement (ii) follows (i) and (114) ... (117); (vi) shows that $(\tau\varphi)(t) = (\exp\, t\hat{A})\tau\varphi^{(0)}$, which proves (vii) and the existence of $\varphi'(t)$, $t > 0$. The uniqueness of $\varphi(t)$ may be proved as it was in Theorem 19. The other conclusions have already been proved. Q.E.D.

13. Determinism

Besides the few examples mentioned in the present section we have met the linear Cauchy initial value problem in Sections VIII.2.14, XIV.7 and XIV.8, but nowhere have we mentioned the position it holds in a long train of philosophic thought concerning the dichotomy of free will and determinism in human destiny. In this general sense it has, perhaps, a history older than any other topic discussed in these volumes and, since mathematics, physics, astronomy, and philosophy have never been isolated from the other creative developments of civilizations in the past, we would feel a bit remiss if we did not give a short account of this history.

The fact that an initial state of a system may determine all of its future states, as is the case with some of the Cauchy problems that we have discussed, is an example of a many-faceted proposition that has captured the inquisitorial imagination of scientists, mathematicians, philosophers, and poets for some twenty-eight centuries. The idea of determinism or predestination shows itself as early as the 8th century B.C. in the poems of Homer and a group of Ionian bards known as the Homeridae (literally, the sons of Homer) and several centuries later in the writings of the tragedians Aeschylus (525–456 B.C.), Sophocles (495?–406? B.C.), and Euripides (480–406? B.C.). With all of these poets and dramatists it took form in the *fate* of one or more individuals and was not a universal doctrine as was the case with the various philosophical theories of determinism. It sometimes occurred merely as a short *deus ex machina* to bring a play to a timely conclusion but, more importantly, fate appeared either as a curse on a certain family, usually to last for at most a couple of generations, and pronounced by an oracle of a god or by direct intervention of the gods themselves as, for example, in the machinations of Athene who was so prone to help in the plight of Achilles in the Trojan war or in that of Odysseus in his great journey home. Likewise Euripides, in the protasis of his *Helen*, using a version of the legend made famous by the poet Stesichorus (sixth century B.C.) and documented by the historian Herodotus (484?–425? B.C.), relates how Helen spent the years of the Trojan war in Egypt under the protection of its virtuous king Proteus and never did go to Troy with Paris. There are several versions of how this fortunate bit of fate was achieved by the gods. Judging from many such examples in their poetry and drama, one concludes that the Greeks of antiquity believed that a man's fate was completely in the hands of the gods and the most he could do was to make supplications and sacrificial offerings to keep in their good graces.

Somewhat later, around the fourth and third centuries B.C., a school of philosophers, sometimes referred to as the *Logical Determinists* believed that men's wills are fettered and therefore nothing can be altered. Such views were formulated by Zeno (340–265 B.C.) and his school of Stoics and, even earlier, by Diodorus Cronus (fourth century B.C.) who was a member of the school known as the Megarians, founded by Euclid (450–368 B.C., not the geometer). Their views, like those of the poets, were associated with the idea of fate, but were backed up by their principle of logic, *tertium non datur*, which means in modern terms that they admitted only a two-valued logic—every statement was either true or false. (This principle does not contradict Gödel's Theorem but does presuppose the exclusion of self-contradictory statements, the existence of which they were certainly aware. The Megarian Eubulides was famous for his logical paradoxes—the familar paradox "If a man says he is a liar, is he telling the truth?" is due to him.) This proposition was assumed to apply even to statements that had never been formulated as well as to statements regarding future events—ergo, quite simply, everything is predetermined. However, the Stoic logician Chrysippus (280–206 B.C.) found it convenient to modify their basic premise to allow escape from such unfortunate disturbances as a sick man would suffer from the statement "I shall not recover from this disease." For, whether it be true or false, he would have no need to call a doctor. Chrysippus introduced the notion of "condestinate" facts which allows the man to recover only if he calls a physician. These two facts, that of recovering and that of calling a physician, are "condestinate." Even though the logical determinists took their logic quite seriously, they had more faith in portents if confirmed by an oracle of a god.

The *Ethical Determinists*, who were more or less contemporary with the tragedians, assumed that man's fate was determined by himself but not by an act of free will. Plato (427–347 B.C.) and Socrates (469–399? B.C.) assumed that one's decisions were made for one's own good and thus in any situation a man's actions were predetermined by his reasoned decision as to what was best for him. Descartes' (1596–1650) and Leibniz' (1646–1716) ideas on determinism were very similar to those of Plato. But Plato's pupil Aristotle (384–322 B.C.) as well as John Locke (1632–1704), and many philosophers between these two, rejected Plato's reasoning on the grounds that it contradicts the fact that man is an incontinent being. Thus man's decisions, which affect his actions and thus determine his future, are made from desire and appetite and not from a reasoned judgment as to what is best for him. A gourmet chooses his food for his pleasure, not for its benefit to him. Aristotle's views are probably those held by most of his successors until the dawn of modern science with Newton in the seventeenth century, when many philosophers began to switch to the idea that human behavior and future events were determined by the laws of nature.

Actually, the laws of nature as an explanation of human behavior and as a basis for rejecting free will in favor of determinism did not originate in modern times but predates the Christian era. As early as the fifth century B.C., Democritus and Leucippus were among the founders of the atomic theory of matter which held that all matter was composed of indivisible, indestructible, and undiminishable atoms in constant motion. Such a scientific view which, strangely enough, was also held by the Epicureans (who even believed that souls were composed of atoms), made everything ultimately depend on the motions and collisions of atoms which, at any instant, have positions and velocities determining their future states or configurations. It seems probable that the Epicureans could not tolerate the thought of absolutely no free will, for they found it convenient to alter the atomic theory slightly by introducing the concept of a special motion of the atoms which they called "swerve" and which was assumed to occasionally take place and allow a deviation from the otherwise preordained state of affairs.

This idea of swerve is today trenchantly replaced by the empirical fact of the motion of the perihelion of Mercury (observed in 1845 by Urbain Leverrier and satisfactorily explained much later by Einstein's general theory of relativity), by the knowledge that the internal energy of an atom can be converted into radiated energy or light, that the mass of a particle varies with its velocity, that mass may be converted into energy, that light is affected by a gravitational field, that no observations or measurements can tell precisely the simultaneous position and momentum of certain elementary particles; any one of these provides a deviation from the classical theories of Newtonian mechanics, cosmology or light. Thus Niels Bohr (1885–1962), Lorentz (1853–1928), Einstein (1875–1955), Heisenberg (1901–   ), and other modern physicists have given ample evidence to require a modification of Laplace's most incisive and universal of all formulations of the doctrine of *scientific determinism*.

Laplace (1749–1827), having the benefit of the rapid growth of theoretical mechanics since the time of Newton (1642–1727), was able to explain all the anomalies, such as the changing speeds of Saturn and Jupiter, in Newtonian mechanics that were known at his time and thus demonstrated the stability of the solar system, a law of nature which Newton had doubted. In fact, Newton believed that stability could only be achieved with the help of God. Laplace's unswerving confidence in Newton's laws is responsible for his famous reply to Napoleon's query concerning the lack of God in his theories. Laplace merely said that he had no need of that hypothesis. Laplace was greatly impressed by the irrefutable implications of Newton's laws to the doctrine of determinism. In explaining these implications in nonmathematical terms he envisaged, in his *Philosophical Essay on Probabilities* (1814), a kind of superhuman calculator who could, upon having complete knowledge of the present state of the universe, tell with absolute certainty its exact state at any time, future or past. The obvious impossibility of the existence of an intelligence so superior that it could ascertain the initial conditions (the present state), let alone perform the calculations necessary for prognostications, is what led Laplace to his studies in the theory of probabilities.

The aberrations in Newtonian mechanics discovered around the turn of the century do, of course, refute Laplace's conclusions when applied on the universal scale which Newton's laws permitted. However, when applied to the solar system and certain other systems of macroscopic as opposed to microscopic nature, these laws allow the scientist to forecast the future with great accuracy simply because the recently discovered anomalies from the classical laws are quite small. For example, the more recent revisions of Leverrier's original measurement of the change in the perihelion of Mercury show it to be about 43 seconds of arc per century.

Great accuracy in cosmological predictions has been a characteristic of mathematical astronomy since ancient times. Perhaps the most familiar and commonly used result of these quite abstract, and by no means trivial, mathematical deductions of antiquity is the calendar, which is, in essence, a sort of ephemeris. There were at least nine ancient civilizations that developed calendars, our own western calendar being a descendant, with a few variations, from the Roman, but perhaps the most interesting one is the lunar calendar of the Babylonians, whose scientific basis has only recently been discovered. Our present knowledge of Babylonian cosmology, which is one of the most fascinating historical discoveries of modern times, is largely due to the scholarly researches of O. Neugebauer and his predecessors Fathers Epping, Strassmaier, and Kugler whose work, published between 1881 and 1924, made it clear that Babylonian lunar theory was of a truly sound mathematical character and used a minimal amount of empirical data. Prior to their discoveries it was believed that Babylonian astronomy was based on mysticism and astrology. The basic problem of Babylonian lunar theory was their calendar, which defined the beginning of each "month" as the instant when the new crescent moon was first visible after sunset; the specific problem was the determination of which months had 30 days and which 29. The construction of a calendar involved deductions from a combination of many *independent* problems of which the Babylonians had an accurate grasp; problems such as the effect upon its visibility of the periodic deviations of the moon from the ecliptic, the variable inclination between the ecliptic and horizon, the latitude of the moon, and the determination of the shortest distance between the moon and the sun at sunset that will permit visibility, which is a problem involving the determination of the variations in lunar and solar velocities. The Babylonians not only realized the independence of these problems but developed ephemerides to solve them, and they also devised a theory which allowed them to make the very accurate predictions of the combined effect of these phenomena which were needed to develop their calendar. Neugebauer, a leading authority on mathematical astronomy in the ancient world, regards this achievement as one of the greatest to be made in the exact sciences of antiquity. We should add that the Babylonians accomplished this without the aid of any kind of geometric model of the solar system or even a model of just the sun, the earth, and the moon. The tools that allowed them to become their own oracles were the ephemerides. These lunar ephemerides were rectangular matrices whose rows represent consecutive months and whose columns contain specific lunar and solar data such as velocities, longitudes, and magnitude of eclipses, such data depending on the particular problem to be solved by the given ephemeris. The general theory developed by the Babylonians led to elementary algorithms, based on arithmetic progressions, for the calculation of subsequent rows from the preceding ones.

As Neugebauer has discovered, the Babylonians also had planetary ephemerides capable of giving very accurate predictions of future planetary phenomena.

The reader who desires more knowledge of the Babylonian civilization than that in Neugebauer's magnum opus [1] of three volumes which contains all texts concerning mathematical astronomy known up to 1955, will be pleased to know that there are still many (perhaps *hundreds of thousands*) tablets in the British Museum and elsewhere which remain unpublished!

In this extremely brief capsule summary of the theory of determinism we have mentioned but a few of the many theories that have been formulated but we do, in conclusion, wish to note a few dissenting voices. We have already mentioned how some recent scientific discoveries have partially undermined the universal scientific determinism as formulated by Laplace. We also briefly note that the existentialist novelists, who are among the modern proponents of the existence of man as an individual with a completely free will, have for the most part taken a hands-off attitude towards the laws of nature and normally attack other restrictions upon their freedom which perhaps do not have such a formidable foundation.

However, some bold intellectuals such as Diderot (1713–1784) are exceptions for he questioned, in his *Letter on the Blind* (1749), the existence of order in the universe. This was all the more daring because of his close association with the mathematician d'Alembert, an authority on theoretical dynamics, with whom he edited the most prodigious of all 18th century publications, the great *Encyclopédie ou Dictionnaire raisonné des Sciences, des Arts et des Métiers*. We also find in Diderot's *Rameau's Nephew* a foreshadowing of existentialism, although here there is no questioning of the laws of nature or design in the universe. In his famous dialogue with Diderot (his shocked interlocutor) the self-imposed moral debasement of the nephew, achieved by making a parasitic attachment to 18th century French society, was his form of revolt against corruption and ignorance in that society which was not only a deterrrent to the *Encyclopédie* but to the arts in general.

However, Dostoevsky (1821–1881), in his *Notes from Underground* (1864), has his underground man deliver a masterful diatribe against the scientific determinism of Laplace as well as the ethical determinism of Plato. He does not try to refute scientific determinism but regards "twice two makes four" and the laws of nature as facts which he will not accept or tolerate, as they fetter him like a surrounding wall which he cannot break down.

In Kafka (1883–1924), most particularly in *The Trial* and *The Castle*, there is no revolt—no bitterness. Here one finds humility in the articulate and sensitive prose of a devout and gifted author who, in these humorous and at the same time pathetic fantasies, portrays the futility of any communication with the divine, which is symbolized by a bureaucratic establishment replete with a hierarchy of illogical, obstinate, shadowy and illusory officials; files jammed full of documents in complete disarray; telephones of questionable existence and always with unreliable connections; etc. The humorous and reverent satire in these novels is the very antithesis of the cynical revolts found in *Rameau's Nephew* and Dostoevsky's *Notes from Underground*. What fettered Kafka was that which separates humans from any knowledge of divine justice.

Sartre (1905–), like Kafka, appears to live in peaceful coexistence with science, possibly because of the amusing fact that his metaphysical rationale seems to lead to a belief in the unreliability of reason. Thus, it is not surprising that that enigmatically gifted novelist Jean Genet (1910–), Sartre's *Saint Genet* and his prototype of the perfect existentialist, is not only undisturbed by the laws of nature but seems to be completely unaware of them. However, regardless of such presumed unawareness, he is certainly capable of deriving much pleasure from these laws, for it is inconceivable that at many times in his life he would not have been happy to hear the prison warden say "You're due to be released next *month*"—a statement which, as we have indicated, requires a considerable amount of abstract and non-trivial mathematics together with some accurate astronomical observations to give it precise meaning.

Thus, while there are many stimulating ideas in the various theories of existentialism, we believe that these doctrines or those of nihilism or any other doctrine cannot regard mathematics and the exact sciences as isolated intellectual disciplines, for they are inextricably fused into our life and culture and will become more so in the future. This fusion to which we refer is vividly illustrated by Neugebauer [2]. Although his primary purpose in this book is the detailed presentation of the mathematical astronomy of ancient Egypt and Babylonia, he allows himself some fascinating digressions into the Middle Ages. From the many examples of sculpture, paintings, illuminated manuscripts, and so on, of the Middle Ages and the Renaissance that incorporate astronomical phenomena he singles out one, namely, the *Book of Hours* of Jean de France, Duc de

Berry, and by reproducing one of the twelve most famous pages (those representing the twelve months), he describes in detail the interesting observation that the artists accurately portray the anomalies in the sun's velocity, a fact known to the ancient Greeks as well as to the Babylonians. This is something hardly to be expected by the uninitiated in a prayerbook executed in the first quarter of the 15th century. This remarkable manuscript, now the property of the Musée Condé in Chantilly, France, has in 1969, after almost five and a half centuries, been published under the title of *The Très Riches Heures of Jean, Duke of Berry*. This same rapport between artists and the exact sciences of antiquity continues to exist today. Marianne Moore's *Icosasphere* describes a recent engineering application of the icosahedron, one of the five Platonic solids; Frank Lloyd Wright's many uses of the cantilever and Alexander Calder's mobiles, both using Archimedes' principle of the lever, are familiar examples of applications of ancient principles which have been recorded by modern artists. Every reader can easily supply many such examples.

However, the massive but still incomplete fusion of modern science into our civilization is much more complex. Only the educated know the meaning of the equation $E = mc^2$ and the resulting uses of atomic energy, and only the educated know how Newton's laws are used to determine orbits in space travel. It is the small minority which knows that the 19th century's greatest mathematical physicist, James Clerk Maxwell, proposed, in 1864, four exact laws which govern electromagnetic waves (the existence of such waves was not established until almost a quarter of a century later by the German physicist Hertz), and that these laws have given us radio, television, television via satellite, radio control of space travel, and radar; phenomena that are familiar to practically every inhabitant of every civilized part of this Earth whether he be educated or not. Despite the complexity and abstract character of these modern laws of nature, we still find an appreciation of their beauty by some artists and the resulting friendly rapport with the mathematical physicist. This is documented, for example, by the correspondence between one of our greatest poets, Paul Valéry, and André Gide [1] which makes it evident that Valéry (who studied mathematics before entering the legal profession) appreciated, as an artist, the works of such men as Laplace and Maxwell. Maxwell's genius has given us a most powerful means of instantaneous visual communication and mass education, both of which may well be prerequisite to world peace. It is to be hoped that they will be used intelligently, for, if not, we may end up with an incredible number of so called "educated" intellectual robots. However, television or any other engineering application of a scientific principle cannot replace a good book or a work of art, so that as long as we continue to produce the great creators that we have in the past, the pleasure of our increased cultural development is assured.

14. Wiener-Lévy-Hopf Theorems

As in Sections 11 and 12, $\mathfrak{H}$ will denote the Hilbert space $L_2(R^N)$, $\hat{\mathfrak{A}}$ the B^*-algebra $L_\infty(R^N)$,and $\mathfrak{A}$ the B^*-subalgebra of $B(\mathfrak{H})$ consisting of all operators having the form

$$(1) \qquad a = \int_{\mathfrak{S}} \hat{a}(s)e(ds), \qquad \hat{a} \in \hat{\mathfrak{A}},$$

where $\mathfrak{S}$ is the one point compactification of N-dimensional Euclidean space R^N and e is the self adjoint spectral measure given in equation (11.15). We know from Lemma 9.2 that if the elements a, A in $\mathfrak{A}$, $\mathfrak{A}^p$, respectively, possess bounded inverses a^{-1}, A^{-1} existing as operators in $\mathfrak{H}$, $\mathfrak{H}^p$, respectively, then these inverses are in $\mathfrak{A}$, $\mathfrak{A}^p$, respectively. The proof given in Lemma 9.2 for these statements used, in an essential manner, the fact that $\mathfrak{A}$ and $\mathfrak{A}^p$ are B^*-algebras. This important property of an operator algebra containing all inverses that exist as bounded everywhere defined operators is, of course, not restricted to B^*-algebras. We shall discuss here certain *incomplete* subalgebras of $\mathfrak{A}$ and the corresponding subalgebras of $\mathfrak{A}^p$ which also enjoy this property.

We first discuss the subalgebra $\mathfrak{A}_1$ of $\mathfrak{A}$ which consists of all operators having the form

$$(2) \qquad a = \alpha e + \mathbf{f}, \qquad f \in L_1,$$

where $\mathbf{f}$ is the convolution operator $\mathbf{f}g = f * g$, $g \in \mathfrak{H}$. The element a in $\mathfrak{A}_1$ uniquely determines the complex number α and the function f in L_1. To see this we first observe that the element $\hat{a}$ in $\hat{\mathfrak{A}}$ corresponding to the operator (2) is, by Theorem 11.4,

$$(3) \qquad \hat{a}(s) = \alpha + \tilde{f}(s), \qquad s \in \mathfrak{S},\ a \in \mathfrak{A}_1,$$

where $\tilde{f} = \tau f$ and $\tau = (2\pi)^{N/2}F$. Since $a = 0$ if and only if $\hat{a} = 0$, to prove that a determines α and f uniquely it suffices to show that $\alpha = 0$ and $f = 0$

if $\hat{a} = 0$. To prove this we shall use the facts that $\tilde{f}$ is in C_0 and that $\tilde{f}$ uniquely determines f, which is the well known *uniqueness theorem for the Fourier transform*. Both of these statements are contained in

1 THEOREM. *For f in L_1 the function $\tilde{f}$ is continuous on $\mathfrak{S}$ and $\tilde{f}(\infty) = 0$. Also $f = 0$ only if $\tilde{f}(s) = 0$ for every s in $\mathfrak{S}$.*

PROOF. Theorem 11.1 shows that for f in the space Φ of rapidly decreasing functions the function $\tilde{f}$ is continuous on $\mathfrak{S}$ and $\tilde{f}(\infty) = 0$. Lemma XIV.2.3 shows that Φ is dense in L_1. Thus for f in L_1 we have a sequence $\{f_n\}$ in Φ with $|f_n - f|_1 \to 0$, and so $|\tilde{f}_n - \tilde{f}|_\infty \leqq |f_n - f|_1 \to 0$, which proves that $\tilde{f}$ is continuous and that $\tilde{f}(\infty) = 0$. Now let $\tilde{f} = 0$. Since $|\mathbf{f}| = |\tilde{f}|_\infty$ we have $f * g = 0$ for every g in L_2. Let g be the characteristic function of a compact set σ so that

$$f * g = \int_\sigma f_t \, dt = 0.$$

Thus, since f_t is continuous in t (with values in L_1), it vanishes identically and $0 = f_0 = f$. Q.E.D.

This theorem shows that if the function (3) vanishes on $\mathfrak{S}$ then $\alpha = \hat{a}(\infty) = 0$, $\tilde{f}(s) = 0$, and $f = 0$. Hence the operator a in $\mathfrak{A}_1$ given by equation (2) determines α and f uniquely.

It is readily seen that $\mathfrak{A}_1$ is a subalgebra of $\mathfrak{A}$ which contains the adjoint of each of its elements and that the adjoint of the operator (2) is

$$a^* = \bar{\alpha}e + \mathbf{f}^*, \qquad f^*(t) = \overline{f(-t)}. \tag{4}$$

It follows from Theorem 11.4 that

$$|a| \leqq |\alpha| + |f|_1, \tag{5}$$

where $|f|_1$ is the L_1 norm of f. The algebra $\mathfrak{A}_1$ is not complete as a subalgebra of $\mathfrak{A}$, but if we renorm it by placing

$$|a|_1 = |\alpha| + |f|_1, \tag{6}$$

then the completeness of L_1 assures the completeness of $\mathfrak{A}_1$ under the norm (6) and the inequality (5) gives

$$|a| \leqq |a|_1, \tag{7}$$

which shows that convergence in $\mathfrak{A}_1$ implies convergence in $\mathfrak{A}$. The algebra $\mathfrak{A}_1$ with the norm (6) is not a B^*-algebra, for it is not true that $|aa^*| = |a|^2$, but $\mathfrak{A}_1$ with the norm (6) satisfies all the axioms defining a commutative B-algebra with an involution and an identity. These elementary properties are easily verified and have been established in Section XI.3. It follows from Theorem 1 that $\mathfrak{A}_1$ is isometrically isomorphic to the algebra obtained from the L_1 group algebra by adjoining the unit e, a fact already observed in a more general setting (XI.3.3).

Since, for a in $\mathfrak{A}_1$, the function $\hat{a}$ is continuous on the compact space $\mathfrak{S}$, it follows that an operator a in $\mathfrak{A}_1$ has an inverse in $\mathfrak{A}$ if $\hat{a}(s)$ does not vanish on $\mathfrak{S}$. A celebrated theorem of N. Wiener gives more by asserting that the inverse a^{-1} is in $\mathfrak{A}_1$.

The basic notions underlying the proof of Wiener's theorem as it will be presented here are those to be found in I. M. Gelfand's theory of commutative normed rings, or B-algebras as we have called them. It should be recalled (IX.2.1) that an element a and a maximal ideal $\mathfrak{M}$ in a commutative B-algebra $\mathfrak{A}_1$ uniquely determine a complex number $a(\mathfrak{M})$ for which $a+\mathfrak{M}=a(\mathfrak{M})e+\mathfrak{M}$. A topology (IX.2.7) may be defined in the set of maximal ideals so that the resulting space $\sigma(\mathfrak{A}_1)$, called the *structure space* or the *spectrum* of the algebra $\mathfrak{A}_1$, is a compact Hausdorff space (IX.2.8) and the complex functions $a(\mathfrak{M})$, $\mathfrak{M} \in \sigma(\mathfrak{A}_1)$, are continuous on $\sigma(\mathfrak{A}_1)$ with $|a(\mathfrak{M})| \leqq |a|$ (IX.2.9). An element a in $\mathfrak{A}_1$ will have an inverse in $\mathfrak{A}_1$ if and only if it belongs to no maximal ideal (IX.1.12(e)), which is just another way of saying that a has an inverse in $\mathfrak{A}_1$ if and only if the function $a(\mathfrak{M})$ has no zeros on $\sigma(\mathfrak{A}_1)$. Thus the Gelfand theory gives a general procedure for determining whether or not an element possesses an inverse and so embeds the Wiener theorem just cited in a more general theory which unifies many similar phenomena. In order to apply this procedure to a given algebra, it is sufficient to give a satisfactory representation of its spectrum and the functions $a(\mathfrak{M})$ on it. For the algebra $\mathfrak{A}_1$ this problem is completely solved in the following theorem.

2 THEOREM (*Gelfand-Raĭkov*). *The spectrum of the algebra* $\mathfrak{A}_1$ *is homeomorphic to the space* $\mathfrak{S}$, *and for corresponding elements* s *and* $\mathfrak{M}_s$, *we have*

$$a(\mathfrak{M}_s)=\hat{a}(s), \qquad a \in \mathfrak{A}_1,\ s \in \mathfrak{S}.$$

PROOF. A one-to-one correspondence between $\sigma(\mathfrak{A}_1)$ and $\mathfrak{S}$ will first be established. To do this it suffices, in view of Lemma IX.2.2, to show that the set of non-zero complex valued homomorphisms h on $\mathfrak{A}_1$ is in a one-to-one correspondence with the points s in $\mathfrak{S}$ in such a way that for corresponding elements h and s we have $h(a) = \hat{a}(s)$ for every a in $\mathfrak{A}_1$. Since, for a fixed s in $\mathfrak{S}$, the map $a \to \hat{a}(s)$ is clearly a non-zero homomorphism in $\mathfrak{A}_1$, it suffices to show that every such homomorphism h on $\mathfrak{A}_1$ uniquely determines a point s in $\mathfrak{S}$ for which $h(a) = \hat{a}(s)$ for every a in $\mathfrak{A}_1$. If $h(\mathbf{f}) = 0$ for every f in L_1, then we have $h(a) = \hat{a}(\infty)$ for every a in $\mathfrak{A}_1$ and so the point ∞ in $\mathfrak{S}$ corresponds to this particular h. For other h we fix a function f in L_1 for which $h(\mathbf{f}) \neq 0$ and, in terms of the translation $f_t(s) = f(s-t)$, define the complex valued function

$$c(t) = \frac{h(\mathbf{f}_t)}{h(\mathbf{f})}, \qquad t \in R^N. \tag{8}$$

Since $f_t * g = g_t * f$ for every pair of functions f, g in L_1, it is seen that $c(t)$ depends only on t and not on the particular function f that is used in (8). Since $t \to f_t$ is a continuous map of R^N into L_1 (XI.3.1(f)) and since h is continuous (IX.2.3), the inequality (7) shows that $c(t)$ is continuous in t. Since $f_s * f_t = f_{s+t} * f$ we have $h(\mathbf{f}_s)h(\mathbf{f}_t) = h(\mathbf{f}_{s+t})h(\mathbf{f})$, which shows that $c(s+t) = c(s)c(t)$, that is, c is a homomorphism of the additive group R^N into the multiplicative group of complex numbers and thus c is a continuous character of R^N. Since $|\mathbf{f}_t| \leqq |f_t|_1 = |f|_1$, the equation (8) shows that c is bounded on R^N and thus $c(mt) = c(t)^m$ and $c(-mt) = c(t)^{-m}$ are bounded in t and m, which proves that $|c(t)| = 1$ for all t in R^N. Thus c is a continuous homomorphism of the additive group R^N onto the multiplicative group of the unit circle in the complex plane. A well known elementary result then gives the existence of a uniquely determined point $s = s(h)$ in R^N with

$$c(t) = e^{-ist}, \qquad t \in R^N. \tag{9}$$

Now let x^* be the continuous linear functional on L_1 determined by the relation $h(\mathbf{g}) = x^*g$ for every g in L_1. Then, since the vector valued function f_t is continuous and bounded as a map from R^N into L_1, the vector valued function $f_t g(t)$ is integrable in t and

$$\begin{aligned} h(\mathbf{f})h(\mathbf{g}) &= h(\mathbf{fg}) = x^*(f * g) = x^* \int_{R^N} f_t g(t)\, dt \\ &= \int_{R^N} x^* f_t g(t)\, dt = \int_{R^N} h(\mathbf{f}_t) g(t)\, dt \\ &= h(\mathbf{f}) \int_{R^N} c(t) g(t)\, dt = h(\mathbf{f}) \int_{R^N} e^{-ist} g(t)\, dt. \end{aligned}$$

Thus, since $h(\mathbf{f}) \neq 0$, this proves that the homomorphism h uniquely determines a point s in $\mathfrak{S}$ such that $h(\mathbf{g}) = (\tau g)(s)$ for every g in L_1, and hence $h(a) = \hat{a}(s)$ for every a in $\mathfrak{A}_1$. Lemma IX.2.2 shows that the identity $h(a) = \hat{a}(s)$, $a \in \mathfrak{A}_1$, may be written as $a(\mathfrak{M}_s) = \hat{a}(s)$, $a \in \mathfrak{A}_1$, where $\mathfrak{M}_s$ is the maximal ideal

$$\mathfrak{M}_s = \{a \,|\, a \in \mathfrak{A}_1, \hat{a}(s) = 0\}.$$

Thus we have proved that this equation establishes a one-to-one correspondence between $\mathfrak{S}$ and the spectrum $\sigma(\mathfrak{A}_1)$. Since the spectrum of $\mathfrak{A}_1$ and $\mathfrak{S}$ are both compact Hausdorff spaces, to see that the map $s \to \mathfrak{M}_s$ is a homeomorphism it suffices to observe that it is continuous. Let $\varepsilon > 0$ and let $a_1, \ldots, a_m$ be elements of $\mathfrak{A}_1$. These determine (IX.2.3) a basic neighborhood N of the point $\mathfrak{M}_{s_0}$ in the spectrum of $\mathfrak{A}_1$ which is given by the formula

$$N = \{\mathfrak{M}_s \,|\, |a_i(\mathfrak{M}_s) - a_i(\mathfrak{M}_{s_0})| < \varepsilon, i = 1, \ldots, m)\}.$$

Since $a(\mathfrak{M}_s) = \hat{a}(s)$, the set of all s in $\mathfrak{S}$ which map into N under the map $s \to \mathfrak{M}_s$ is the set

$$\{s \,|\, s \in \mathfrak{S}, |\hat{a}_i(s) - \hat{a}_i(s_0)| < \varepsilon, i = 1, \ldots, m\},$$

which is an open set in $\mathfrak{S}$, since the function $\hat{a}(s) = \alpha + (\tau f)(s)$ is continuous for every f in L_1. This shows that the map $s \to \mathfrak{M}_s$ is continuous and is, therefore, also a homomorphism. Q.E.D.

As explained earlier, an important result of N. Wiener is a corollary of the Gelfand-Raĭkov theorem and may be stated as follows.

3 Corollary (*N. Wiener*). *If an element a in $\mathfrak{A}_1$ has $\hat{a}(s) \neq 0$ for every s in $\mathfrak{S}$, then there is an element b in $\mathfrak{A}_1$ with $\hat{b}(s) = \hat{a}(s)^{-1}$ for every s in $\mathfrak{S}$.*

We shall find it convenient to state this theorem of Wiener and other similar results in terms of the following concept.

4 Definition. An algebra $\mathfrak{A}_0$ of operators in a B-space $\mathfrak{X}$ is said to *contains all inverses* if $\mathfrak{A}_0$ contains the identity operator in $\mathfrak{X}$ and the inverse of any one of its elements which possesses an inverse in the algebra $B(\mathfrak{X})$ of bounded linear operators in $\mathfrak{X}$.

This notion of containing all inverses is, of course, relative to the basic B-space in which the operators a in $\mathfrak{A}_0$ act. This should cause no confusion, since we are dealing with *operator* algebras rather than abstract algebras.

For an operator a in $\mathfrak{A}_1$ the function $\hat{a}$ is continuous on the compact space $\mathfrak{S}$, and so its non-vanishing is equivalent to the existence of the inverse a^{-1} in $\mathfrak{A}$ and, in view of Lemma 9.2, this is equivalent to the existence of an inverse in $B(\mathfrak{H})$. Thus Wiener's result may be stated as follows.

5 COROLLARY (*N. Wiener*). *The operator algebra* $\mathfrak{A}_1$ *contains all inverses.*

There are a number of immediate corollaries to this theorem of Wiener which will become apparent from the following lemma.

6 LEMMA. *Let* $\mathfrak{J}$ *be an ideal in the subalgebra* $\mathfrak{A}_0$ *of* $\mathfrak{A}$ *and let* $\mathfrak{A}(\mathfrak{J})$ *be the algebra of all operators having the form*

$$a = \alpha e + b, \qquad b \in \mathfrak{J}. \tag{10}$$

If $\mathfrak{A}_0$ *contains all inverses, so does* $\mathfrak{A}(\mathfrak{J})$.

PROOF. Let the operator (10) have an inverse a^{-1} in $B(\mathfrak{H})$. Then a^{-1} is in $\mathfrak{A}_0$, and since $\mathfrak{J}\mathfrak{A}_0 = \mathfrak{J}$, the elementary identity $a^{-1} = \alpha^{-1}(e - ba^{-1})$ shows that a^{-1} is in $\mathfrak{A}(\mathfrak{J})$. Q.E.D.

Let L_0 be a linear subset of L_1 with the property that

$$f * g \in L_0, \qquad f \in L_1,\ g \in L_0. \tag{11}$$

Then the set of all convolution operators $\mathbf{f}$ with f in L_0 forms an ideal in $\mathfrak{A}_1$, and in view of Lemma 6, Wiener's theorem shows that the algebra $\mathfrak{A}_0$ of operators having the form

$$a = \alpha e + \mathbf{f}, \qquad f \in L_0, \tag{12}$$

contains all inverses. We shall give some examples of such algebras $\mathfrak{A}_0$ by specifying various choices for the linear space L_0. It is clear that the linear space $L_0 = L_1 \cap L_\infty$ satisfies (11). Also the space $L_0 = L_1 \cap C$ of all integrable functions which coincide almost everywhere with a continuous function clearly satisfies (11). Example 11.12 shows that the space $L_0 = L_1 \cap L_1$ satisfies (11). More generally, for any p with $1 \leqq p \leqq \infty$, the space $L_0 = L_1 \cap L_p$ satisfies (11). To see this, let f be in L_1 and g be in $L_1 \cap L_p$. If p is either 1 or ∞, it is clear that $|f * g|_p \leqq |f|_1 \cdot |g|_p$ and so this same inequality, for arbitrary p, follows from the Riesz convexity theorem (VI.10.11). This proves that the space $L_0 = L_1 \cap L_p$ satisfies (11). The spaces $L_0 = L_1 \cap C^{(k)}$ and $L_0 = L_1 \cap C^\infty$ are other examples of linear subsets of L_1 which satisfy (11). For $k < \infty$ the symbol $C^{(k)}$ is used for the set of all complex functions on R^N having all derivatives of orders $\leqq k$ existing everywhere on R^N as bounded continuous functions.

These results are summarized in the following definition and corollary.

7 Definition. Let L_0 be a set of functions on R^N each having an essentially bounded Fourier transform. Then for f in L_0 the convolution operator $\mathbf{f}g = f * g = \int \tilde{f}(s)e(ds)$ is defined as a bounded linear map $g \to f * g$ of $\mathfrak{H}$ into itself. We shall say that an operator a in $\mathfrak{A}$ *is of type* L_0 if it has the form $a = \alpha e + \mathbf{f}$ with f in L_0. An operator $A = (a_{ij})$ in $\mathfrak{A}^p$ is said to *be of type* L_0 if each a_{ij} is of type L_0.

8 Corollary. *Let the operator a in $\mathfrak{A}$ have an inverse in $B(\mathfrak{H})$. If a is of type $L_1 \cap L_q$ for some q in the range $1 \leqq q \leqq \infty$ or of type $L_1 \cap C^{(k)}$ for some integer k in the range $0 \leqq k \leqq \infty$, then the inverse a^{-1} has this same property.*

For a subalgebra $\mathfrak{A}_0$ of $\mathfrak{A}$ we shall use the symbol $\mathfrak{A}_0^p$ for the subalgebra of $\mathfrak{A}^p$ consisting of those operators $A = (a_{ij})$ with a_{ij} in $\mathfrak{A}_0$.

9 Theorem. *Let $\mathfrak{A}_0$ be a subalgebra of $\mathfrak{A}$ which contains all inverses. Then $\mathfrak{A}_0^p$ contains all inverses.*

Proof. If the operator A in $\mathfrak{A}_0^p$ has an inverse in $B(\mathfrak{H}^p)$ then, by Corollary 9.6, A^{-1} is in $\mathfrak{A}^p$ and the determinant $\delta = \det(a_{ij})$ has an inverse in $\mathfrak{A}$. Since $\mathfrak{A}_0$ contains all inverses, δ^{-1} is in $\mathfrak{A}_0$. It follows from the definition of the determinant of a matrix of operators that $\hat{\delta}(s) = \det(\hat{a}_{ij}(s))$, a fact already observed in equation (9.19), and thus if A_{ij} is the cofactor of a_{ij} then $\hat{A}_{ij}(s)$ is the cofactor of $\hat{a}_{ij}(s)$ as an element of $\hat{A}(s) = (\hat{a}_{ij}(s))$. This elementary observation shows that Cramer's rule applies to the matrices in $\mathfrak{A}^p$, that is, $A^{-1} = (b_{ij})$ where $b_{ij} = \delta^{-1}A_{ji}$, which proves that A^{-1} is in $\mathfrak{A}_0^p$. Q.E.D.

10 Corollary. *Let the operator A in $\mathfrak{A}^p$ have an inverse in $B(\mathfrak{H}^p)$. If A is of type $L_1 \cap L_q$ for some q in the range $1 \leqq q \leqq \infty$ or of type $L_1 \cap C^{(k)}$ for some integer k in the range $0 \leqq k \leqq \infty$, then the inverse A^{-1} has this same property.*

P. Lévy has generalized Wiener's result by showing that, for an operator a in $\mathfrak{A}_1$ and a function φ which is analytic on the spectrum $\sigma(a)$, the operator $\varphi(a)$ is also in $\mathfrak{A}_1$ and $\widehat{\varphi(a)}(s) = \varphi(\hat{a}(s))$ for every s in $\mathfrak{S}$. The

Wiener theorem is the case where $\varphi(\xi) = \xi^{-1}$. The following result extends the Lévy theorem to various subalgebras $\mathfrak{A}_0^p$ of $\mathfrak{A}_1^p$.

11 COROLLARY. *Let φ be a complex function, analytic on the spectrum of the operator A in $\mathfrak{A}^p$. Then*

$$\widehat{\varphi(A)}(s) = \varphi(\hat{A}(s)), \tag{13}$$

for almost all s in $\mathfrak{S}$. If A is of type $L_1 \cap L_q$ for some q in the range $1 \leqq q \leqq \infty$ or of type $L_1 \cap C^{(k)}$ for some non-negative integer k or for $k = \infty$, then $\varphi(A)$ has this same property and the equation (13) *holds for every s in $\mathfrak{S}$.*

PROOF. The first conclusion has already been established in equation (11.64). Let $\mathfrak{X}$ be a B-space of functions on R^N (or equivalence classes of such functions) and let $L_0 = L_1 \cap \mathfrak{X}$. We assume that

(14) if a sequence $\{f_n\}$ in L_0 converges to f in L_1 and g in $\mathfrak{X}$, then $f = g$; and

$$L_0 * L_0 \subseteqq L_0 . \tag{15}$$

In view of (14) the linear set L_0 is a B-space under the norm

$$|f|_0 = |f|_1 + \|f\|, \qquad f \in L_0, \tag{16}$$

where $\|f\|$ is the norm in $\mathfrak{X}$. We let $\mathfrak{A}_0$ consist of all operators in $\mathfrak{H}$ which have the form

$$a = \alpha e + \mathbf{f}, \qquad f \in L_0 . \tag{17}$$

Theorem 1 shows that the operator a determines α and f uniquely, and so we may define a norm in $\mathfrak{A}_0$ by the equation

$$|a|_0 = |\alpha| + |f|_0, \qquad a \in \mathfrak{A}_0 . \tag{18}$$

It is clear that $\mathfrak{A}_0$ is a B-space under this norm. It is also an algebra, for the product of two operators $a = \alpha e + \mathbf{f}$ and $b = \beta e + \mathbf{g}$ in $\mathfrak{A}_0$ is

$$ab = \alpha\beta e + \alpha\mathbf{g} + \beta\mathbf{f} + \mathbf{h}, \tag{19}$$

where $h = f * g$ is in L_0 by (15). We next observe that for a fixed a in $\mathfrak{A}_0$ the product ab is a continuous function of b in $\mathfrak{A}_0$. To see this it suffices, in view of (19), to show that the map $g \to f * g$ is a continuous map of L_0 into itself. This fact follows immediately from the closed graph theorem (II.2.4), for if $g_n \to g$ and $f * g_n \to k$ in L_0, then $f * g_n \to f * g$ in L_1, which shows that $k = f * g$. Thus, multiplication in $\mathfrak{A}_0$ is continuous in each variable separately and the unit e has norm $|e|_0 = 1$. The elementary process discussed in IX.1 shows that $\mathfrak{A}_0$ may be renormed with a norm $\|a\|_0$ under which $\mathfrak{A}_0$ becomes a B-algebra with $\|a\|_0 \leqq M|a|_0 \leqq K\|a\|_0$ for some positive numbers M and K. This shows that $\mathfrak{A}_0$ is metrically and algebraically equivalent to a B-algebra.

Now let $\mathfrak{X}$ be one of the B-spaces L_q with $1 \leqq q \leqq \infty$ or $C^{(k)}$ with $1 \leqq k \leqq \infty$. Then the conditions (14) and (15) are clearly satisfied. Corollary 10 shows that if A is of type L_0, then for every λ in the resolvent set of A, $R(\lambda; A)$ is also of type L_0 and belongs to $\mathfrak{A}_0^p$. It is clear from the remarks in IX.1 that $\mathfrak{A}_0^p$ is a B-algebra under a norm in which convergence of a sequence $A^{(n)} = (a_{ij}^{(n)})$ is equivalent to the convergence in L_0 of the elements $a_{ij}^{(n)}$ for each $i, j = 1, \ldots, p$. Thus Lemma IX.1.5 shows that $R(\lambda; A)$ is a continuous (even analytic) function of λ with values in the normed B-algebra $\mathfrak{A}_0^p$. Hence the Riemann integral (11.63), which defines $\varphi(A)$ in terms of $\varphi(\lambda)$ and $R(\lambda; A)$, is also in $\mathfrak{A}_0^p$. It follows from Theorem 1 that $\widehat{\varphi(A)}(s)$ and $\hat{A}(s)$ are continuous in s, which proves that (13) holds for every s in $\mathfrak{S}$. Finally, if A is of type $L_1 \cap C^{(\infty)}$, it follows from what has been proved that $\varphi(A)$ is of type $L_1 \cap C^{(k)}$ for every positive integer $k < \infty$ and thus $\varphi(A)$ is also of type $L_1 \cap C^{(\infty)}$. Q.E.D.

We conclude this section by mentioning very briefly an interesting application of the Wiener-Lévy result, made by M. G. Kreĭn, to the problem of solving a non-homogeneous integral equation of the Wiener-Hopf type. Kreĭn gives a thorough and general treatment of this equation, which is of importance in physics and astrophysics in connection with problems involving the transfer of radiant energy. I. C. Gohberg and M. G. Kreĭn have jointly made a similar study of systems of such equations. The Wiener-Hopf equation is

$$x(s) + \int_0^\infty f(s-t)x(t) = y(s), \qquad 0 \leqq s < \infty, \tag{20}$$

where s and t are real variables. Here f and y are given functions and x is to be determined.

We shall discuss the case where $s = [s_1, \ldots, s_N]$ and $t = [t_1, \ldots, t_N]$ are points in N-dimensional Euclidean space R^N and give the simplification over the original work of Wiener and Hopf, introduced by Kreĭn, for solving this equation under special hypotheses. Let $\alpha_1, \ldots, \alpha_N$ be real numbers, let

$$\begin{aligned} \mathfrak{S}_+ &= \{s \mid s \in R^N, \alpha_1 s_1 + \cdots + \alpha_N s_N \geqq 0\}, \\ \mathfrak{S}_- &= \{s \mid s \in R^N, \alpha_1 s_1 + \cdots + \alpha_N s_N < 0\}, \end{aligned} \tag{21}$$

and consider the equation

$$\alpha x(s) + \int_{\mathfrak{S}_+} f(s-t)x(t) = y(s), \qquad s \in \mathfrak{S}_+, \tag{22}$$

where α is a given complex number. We shall assume that f is in L_1 and that the operator a in $\mathfrak{A}_1$ given by the equation

$$a = \alpha e + \mathbf{f} \tag{23}$$

has the property that

(24) the points 0 and ∞ in the complex plane both belong to the same connected component of the resolvent set of a.

Under this assumption it will be shown that for every y in $L_2(\mathfrak{S}_+)$ there is one and only one x in $L_2(\mathfrak{S}_+)$ which satisfies the equation (22) for almost all s in $\mathfrak{S}_+$. For f in $L_1 \cup L_2$ we shall let $f_\pm = P_\pm f$, where $P_\pm$ are the projections defined as

$$\begin{aligned} (P_+ f)(s) &= f(s), \qquad & s \in \mathfrak{S}_+, \\ &= 0, & s \in \mathfrak{S}_-, \end{aligned} \tag{25}$$

and $P_- f = f - f_+$. We let $L_\pm = P_\pm L_1$ and $\mathfrak{H}_\pm = P_\pm \mathfrak{H}$ so that we may identify $L_2(\mathfrak{S}_+)$ with $\mathfrak{H}_+$. It is readily seen that

$$L_\pm * L_\pm \subseteqq L_\pm, \qquad L_\pm * \mathfrak{H}_\pm \subseteqq \mathfrak{H}_\pm, \tag{26}$$

for if f and g are both in L_+ then, for s in $\mathfrak{S}_-$,

$$(f * g)(s) = \int_{\mathfrak{S}_+} f(s-t)g(t)\, dt = 0$$

since $s - t$ is in $\mathfrak{S}_-$ for every t in $\mathfrak{S}_+$. This shows that $L_+ * L_+ \subseteqq L_+$, and an elementary continuity argument shows that $L_+ * \mathfrak{H}_+ \subseteqq \mathfrak{H}_+$. The other inclusions in (26) are proved similarly.

In order to avoid ambiguity in notation, we shall, in the discussion immediately following, sometimes write $\zeta(f)$ for the convolution operator $\mathbf{f}$ in $\mathfrak{A}_1$ corresponding to the function f in L_1. There is an important elementary decomposition of the operators in $\zeta(L_1)$ which is defined as follows. An operator $\zeta(f)$ in $\zeta(L_1)$, according to Theorem 1, uniquely determines the function f in L_1, and so the operators $\zeta(f_+)$ and $\zeta(f_-)$ are uniquely determined by $\zeta(f)$ and have the properties

$$\zeta(f) = \zeta(f_+) + \zeta(f_-), \qquad \zeta(f_+)\zeta(f_-) = \zeta(f_-)\zeta(f_+), \tag{27}$$

and

$$\zeta(f_\pm)\mathfrak{H}_\pm \subseteq \mathfrak{H}_\pm . \tag{28}$$

The first equation in (27) follows from the definitions of the operators $\zeta(f_\pm)$, the second from the fact that they belong to the commutative algebra $\mathfrak{A}_1$, and equation (28) follows from (26).

With each operator a in $\mathfrak{A}_1$ we associate an operator a_+ in the algebra $B(\mathfrak{H}_+)$ of bounded linear operators on $\mathfrak{H}_+$ defined by the equation

$$a_+ x = P_+ ax, \qquad x \in \mathfrak{H}_+ . \tag{29}$$

We shall show that the operator a_+ has a bounded everywhere defined inverse on $\mathfrak{H}_+$, that is, a_+^{-1} exists as an element of $B(\mathfrak{H}_+)$. Under the assumption (24) the branch, $\ln \lambda$, of the natural logarithm for which $\ln 1 = 0$ is single valued and analytic on a neighborhood of the spectrum of a; thus, according to Corollary 11, the operator $b = -\ln a = \ln a^{-1}$ is in $\mathfrak{A}_1$ and hence has the form $b = \beta e + \zeta(g)$ for some g in L_1. By applying the decomposition (27) to the operator $\zeta(g)$, we have the factorization of a^{-1} as given.by

$$a^{-1} = e^b = e^\beta e^{\zeta(g)} = e^\beta e^{\zeta(g+)} e^{\zeta(g-)}, \tag{30}$$

and it follows from (28) that

$$e^{\zeta(g\pm)}\mathfrak{H}_\pm \subseteq \mathfrak{H}_\pm , \qquad e^{-\zeta(g\pm)} \subseteq \mathfrak{H}_\pm . \tag{31}$$

Now suppose that the vectors x, y in $\mathfrak{H}_+$ satisfy the equation (22), that is,

$$y = a_+ x \tag{32}$$

and let us write this equation as

$$y = ax - P_- ax. \tag{33}$$

Equations (30)and (33) give

$$e^{\zeta(g-)}y = e^{-\beta} e^{-\zeta(g+)}x - e^{\zeta(g-)}P_- ax. \tag{34}$$

Since x is in $\mathfrak{H}_+$ and $P_- ax$ is in $\mathfrak{H}_-$, it follows from (31) and (34) that

$$P_+ e^{\zeta(g-)}y = e^{-\beta}e^{-\zeta(g+)}x, \tag{35}$$

$$x = e^{\beta}e^{\zeta(g+)}P_+ e^{\zeta(g-)}y, \tag{36}$$

which shows that the vector x in $\mathfrak{H}_+$ is uniquely determined by y. In other words, the operator a_+ is one-to-one on $\mathfrak{H}_+$. Now let y be an arbitrary vector in $\mathfrak{H}_+$ and define the vector x by the equation (36). Then (31) shows that x is in $\mathfrak{H}_+$ and equation (35) holds. This means that for some vector z in $\mathfrak{H}_-$ we have

$$e^{\zeta(g-)}y = e^{-\beta}e^{-\zeta(g+)}x + z,$$

and, using (30), it is seen that

$$y = ax + e^{-\zeta(g-)}z.$$

The second inclusion in (31) shows that the last term in this equation is in $\mathfrak{H}_-$. Thus $P_+ y = P_+ ax = a_+ x$, and since y is in $\mathfrak{H}_+$, we have $y = P_+ y = a_+ x$. This proves that a_+^{-1} exists as an everywhere defined operator in $\mathfrak{H}_+$, and (36) gives the formula

$$(a_+)^{-1} = e^{\beta}e^{\zeta(g+)}P_+ e^{\zeta(g-)}, \tag{37}$$

where β and g are defined by the equation $\beta e + \zeta(g) = -\ln a$.

In summary we may state

12 THEOREM. *Let the operator $a = \alpha e + \mathbf{f}$ in $\mathfrak{A}_1$ have the property* (24). *Then for every y in $L_2(\mathfrak{S}_+)$ there is one and only one x in $L_2(\mathfrak{S}_+)$ which satisfies equation* (22) *for almost all s in $\mathfrak{S}_+$. This unique solution x is given by equation* (36).

15. Exercises

1 Find a spectral operator T which is not weakly compact but whose scalar part S is compact.

2 Find a spectral operator T with $\sigma_p(S) = \sigma_r(T) = \sigma(T)$.

3 Find a spectral operator T such that $\sigma_p(S) = \sigma_c(T) = \sigma(T)$.

4 Find a spectral operator T such that S and T have closed ranges but N does not.

5 Find a spectral operator T such that S has a closed range but T does not.

6 Show that the operator $g = Tf$ in $C[0, 1]$ defined by the equation

$$g(s) = f(s) + \int_0^s K(s, t) f(t)\, dt,$$

where K is a bounded, measurable function on $[0, 1] \times [0, 1]$, is a spectral operator.

7 Every scalar operator S is the sum $S = S_1 + iS_2$ where S_1 and S_2 are commuting scalar operators with real spectra. Furthermore, the Boolean algebra of projections generated by the resolutions of the identity of S_1 and S_2 is bounded.

8 Let T be a spectral operator. There are two operators R and J such that:

(i) $T = R + iJ, \qquad RJ = JR.$

(ii) The sets $\sigma(R)$ and $\sigma(J)$ are real.

(iii) R is a scalar operator and J is a spectral operator.

(iv) The Boolean algebra of projections generated by the resolutions of the identity of R and J is bounded.

9 If T is a spectral operator and R_1 and J_1 are operators satisfying the relations (i) and (ii) of the preceding exercise, then $R_1 = R + M$ and $J_1 = J + iM$ where M is a generalized nilpotent. Thus prove that (i), (ii), and (iii) in Exercise 8 insure the uniqueness of the operators R and J.

10 Every scalar operator S is the product of two commuting scalar operators S_1, S_2 where S_1 has a non-negative spectrum and $\sigma(S_2)$ is a subset of the unit circle. Furthermore, the Boolean algebra of projections generated by the resolutions of the identity of S_1 and S_2 is bounded.

11 For each spectral operator T there are two operators P and U with the properties:

(i) $T = PU = UP$.

(ii) The spectrum $\sigma(P)$ is a set of non-negative numbers and $\sigma(U)$ is a subset of the unit circle.

(iii) The operator U is a scalar operator and P is a spectral operator.

(iv) The Boolean algebra of projections generated by the resolutions of the identity for P and U is bounded.

12 Let P_1, U_1 satisfy the conditions (i) and (ii) of the preceding exercise. Then there are generalized nilpotents N_1 and N_2 commuting with T and such that

$$U_1 = U + N_1, \qquad P_1 = P + N_2,$$

$$N_2 = \sum_{n=0}^{\infty} (-N_1 U^{-1})^{n+1} P.$$

13 If for some λ in the spectrum of the spectral operator T in $\mathfrak{X}$ the manifold $(\lambda I - T)\mathfrak{X}$ is closed, then λ is in the point spectrum of T.

14 Let T be an operator (not necessarily spectral) on a complex B-space $\mathfrak{X}$ and let T^* be its adjoint. Let $\sigma(x^*)$ be the spectrum of a point x^* in $\mathfrak{X}^*$ relative to T^*. If for some x in $\mathfrak{X}$ and x^* in $\mathfrak{X}^*$ we have $\sigma(x)$ and $\sigma(x^*)$ disjoint, then $x^*Sx = 0$ for every bounded linear operator S in $\mathfrak{X}$ which commutes with T.

15 Let T be a spectral operator with spectral resolution E. Let δ be a non-void subset of the spectrum $\sigma(T)$ of T which is open in the relative topology of $\sigma(T)$. Then $E(\delta) \neq 0$.

16 Let f be a bounded Borel function defined on the spectrum of the scalar operator S. Then the operator $f(S) = \int_{\sigma(S)} f(\lambda)E(d\lambda)$ has a bounded everywhere defined inverse if and only if, for some Borel set δ with $E(\delta) = I$, we have

$$\sup_{\lambda \in \delta} \frac{1}{|f(\lambda)|} < \infty.$$

17 For a compact spectral operator, derive the properties given in Theorem VII.4.5 directly from the theory in Section XV.7 without recourse to the theory in Section VII.4.

18 (McGarvey) Let $\mathfrak{X}$ be a linear space, Σ a field of sets, and $E(\cdot)$ a function defined on Σ and having its values in the set of all everywhere defined projections in $\mathfrak{X}$. It is assumed that for every σ and δ in Σ the projections $E(\sigma)$ and $E(\delta)$ commute and, if σ and δ are disjoint, that $E(\delta \cup \sigma) = E(\delta) + E(\sigma)$. Then, for every δ, σ in Σ,

$$E(\delta \cap \sigma) = E(\delta)E(\sigma), \qquad E(\delta \cup \sigma) = E(\delta) \cup E(\sigma).$$

19 Let A be in $\mathfrak{A}^p$ and s in $\mathfrak{S}$. Show that the scalar part of $\hat{A}(s)$ is

$$\hat{S}(s) = \sum_{i=1}^{p} \sum_{j=1}^{i} \hat{\lambda}_{ij}(s)E(\hat{\lambda}_{ij}(s); \hat{A}(s)).$$

(Hint: Use Theorems VII.1.7 and VII.1.8). From this, without recourse to Theorem 10.6, prove that A is a spectral operator if

$$e\text{-}\operatorname*{ess\,sup}_{s \in \mathfrak{S}_i} |E(\hat{\lambda}_{ij}(s); \hat{A}(s))| < \infty, \qquad 1 \leqq j \leqq i \leqq p.$$

20 Let $\{A^{(m)}\}$ be a sequence of operators in $\mathfrak{A}^p$ for which the limit

$$\text{(i)} \qquad a_{ij}(s) = \lim_m a_{ij}^{(m)}(s)$$

exists for e-almost all s in $\mathfrak{S}$. Then $A^{(m)}$ converges strongly to the operator $A = (a_{ij})$ if and only if, for some K,

$$\text{(ii)} \quad e\text{-}\operatorname*{ess\,sup}_{s \in \mathfrak{S}} |a_{ij}^{(m)}(s)| \leqq K, \qquad 1 \leqq j \leqq i \leqq p, \qquad 1 \leqq m < \infty.$$

21 Let p be a polynomial with complex coefficients and not identically zero. Show that the closed extension A, given by equation (12.4), of the formal differential operator

$$\begin{pmatrix} 2p\left(\dfrac{\partial}{\partial s}\right) & ip\left(\dfrac{\partial}{\partial s}\right) \\ ip\left(\dfrac{\partial}{\partial s}\right) & 0 \end{pmatrix}$$

is a spectral operator which is not of scalar type. Show that $\sigma(A) = p(iR^N)$.

22 Let α, β be positive numbers and let A be the closed extension, given by (12.4), of the formal differential operator (12.14) in the Hilbert space $L_2(R^N) \oplus L_2(R^N)$. Show that the spectrum of A is given by equation (12.17).

The next few results, due to F. Wolf, use the notations introduced in the following definition.

23 Definition. Let $\mathfrak{X}$ be a Banach space, F a bounded countably additive set function whose values are in $B(\mathfrak{X})$, and $\mathfrak{A}_1^k$ the set of all operators in $\mathfrak{X}$ with spectrum lying in the unit circle and such that for every function f analytic in a neighborhood of the unit circle, the operator $f(A)$ has the form

$$\text{(i)} \qquad f(A) = \sum_{j=0}^{k-1} f^{(j)}(1)A_j + \int_0^{2\pi} \frac{d^k}{d\theta^k} f(e^{i\theta})F(d\theta).$$

Let $\mathfrak{A}_2^k$ consist of all A in $B(\mathfrak{X})$ with spectrum in the unit circle and such that

$$\text{(ii)} \qquad |f(A)| \leqq M \sum_{j=0}^{k} \sup_{\theta} |f^{(j)}(e^{i\theta})|$$

for every function f analytic in a neighborhood of the unit circle. Let $\mathfrak{A}_3^k$ be the set of A in $B(\mathfrak{X})$ which satisfy the inequalities

$$\text{(iii)} \qquad |A^n| \leqq M|n|^k, \qquad n = \pm 1, \pm 2, \ldots,$$

and $\mathfrak{A}_4^k$ the set of A in $B(\mathfrak{X})$ for which

$$|R(\lambda; A)| \leqq \frac{M}{|1-|\lambda||^k}, \qquad |\lambda| \neq 1.$$

24 Show that $\mathfrak{A}_1^k \subseteqq \mathfrak{A}_2^k$.

25 If $\mathfrak{X}$ is weakly complete, then $\mathfrak{A}_2^k \subseteqq \mathfrak{A}_1^k$.

26 Prove that $\mathfrak{A}_2^k \subseteqq \mathfrak{A}_3^k$ and that $\mathfrak{A}_2^k \subseteqq \mathfrak{A}_4^k$.

27 Prove that $\mathfrak{A}_3^k \subseteqq \mathfrak{A}_4^{k+1}$.

28 Show that $\mathfrak{A}_4^k \subseteqq \mathfrak{A}_3^k$.

29 Let $f(z) = \sum_{-\infty}^{+\infty} a_n z^n$ be analytic in a neighborhood of the unit circle. Then

$$\int_0^{2\pi} |f^{(k+1)}(e^{i\theta})|^2 \, d\theta = 2\pi \sum_{-\infty}^{+\infty} |a_n n(n-1) \cdots (n-k)|^2,$$

and

$$\left\{ \sum_{-\infty}^{\infty} |a_n n(n-1) \cdots (n-k+1)| \right\}^2$$
$$\leqq \left(\sum_{-\infty}^{\infty} |a_n n(n-1) \cdots (n-k)|^2 \right) \left(\sum_{n \neq k} \frac{1}{|n-k|^2} \right).$$

Hence, prove that

$$\sup_{|z|=1} |f^{(k+1)}(z)| \geqq M \sum_{-\infty}^{\infty} |a_n n(n-1) \cdots (n-k+1)|.$$

30 If A is in $\mathfrak{A}_3^k$, then

$$\sup_{|z|=1} |f^{(k+1)}(z)| \geqq M|f(A)|,$$

and thus $\mathfrak{A}_3^k \subseteqq \mathfrak{A}_2^{k+1}$.

31 If the space $\mathfrak{X}$ is weakly complete, then

$$\mathfrak{A}_1^k = \mathfrak{A}_2^k \subseteqq \mathfrak{A}_4^k \subseteqq \mathfrak{A}_3^k \begin{cases} \subseteqq \mathfrak{A}_4^{k+1} \\ \subseteqq \mathfrak{A}_2^{k+1} = \mathfrak{A}_1^{k+1}. \end{cases}$$

32 Let $\mathfrak{X} = L_1(-\infty, \infty)$ and let A be the mapping in $\mathfrak{X}$ which maps $f(t)$ into $f(t+s)$ where $s \neq 0$ is a fixed constant. Prove that 1 is in the residual spectrum of A and that A is not a spectral operator.

33 Let A be defined in the B-space $C[0, 1]$ by the map

$$A: f(t) \to tf(t).$$

Then the algebra generated by A and the identity operator is equivalent to the algebra of continuous functions on $[0, 1]$ but A is not a scalar operator. Hence Theorem XVII.2.5 is not valid if the space $\mathfrak{X}$ is not assumed to be weakly complete.

34 Let Ω denote the solid unit sphere in the complex plane and let μ denote Lebesgue measure on Ω normalized so that $\mu(\Omega) = 1$. If $1 \leqq p \leqq \infty$, and if S is the operator in $L_p(\Omega)$ defined by

$$(Sf)(t) = tf(t), \qquad t \in \Omega,$$

then S is a scalar type spectral operator with $E(\sigma)f = \chi_\sigma f$, $\sigma(S) = \Omega$, and

$$R(\lambda; S)f(t) = \frac{1}{\lambda - t} f(t).$$

35 Let $T \in B(\mathfrak{X})$ and let E be a spectral measure on the Borel sets of the complex plane which commutes with T. Show that in order to prove that E is a resolution of the identity for T it is enough to prove that $\sigma(T_c) \subseteqq c$ for each closed set c.

36 Let $\mathfrak{X}$ be one of the B-spaces c_0, l_1, l_∞ and let T be defined by

$$T(\xi_1, \xi_2, \xi_1, \ldots) = (\tfrac{1}{1}\xi_1, \tfrac{1}{2}\xi_2, \tfrac{1}{3}\xi_1, \ldots).$$

Show that in each of these spaces T is compact. Calculate $\sigma(T)$ and $R(\lambda; T)$. Show that T is a spectral operator in c_0 or l_1 and exhibit the resolution of the identity for T. Show that T is not a spectral operator in l_∞. [Hence the adjoint of a compact spectral operator does not need to be a spectral operator.]

37 Let S be the right shift in l_1; that is,

$$S(\xi_1, \xi_2, \xi_3, \ldots) = (0, \xi_1, \xi_2, \ldots).$$

Show that $\sigma(S) = \sigma_r(S) = \{\lambda \mid |\lambda| \leqq 1\}$. Prove that if $x \neq 0$, then $\sigma(x) = \sigma(S)$.

38 Let l_1 be the B-space of all absolutely convergent complex sequences $x = (\xi_n)$, $n = 1, 2, \ldots$ and let L be the left shift operator

$$L(\xi_1, \xi_2, \ldots) = (\xi_2, \xi_3, \ldots).$$

Show that $\sigma(L) = \{\lambda \mid |\lambda| \leqq 1\}$, in fact, $\sigma_p(L) = \{\lambda \mid |\lambda| < 1\}$ and $\sigma_c(L) = \{\lambda \mid |\lambda| = 1\}$. Calculate the maximal extension of $R(\lambda; L)x_m$, where $x_m = (\delta_{mn})$.

39 Let $\mathfrak{X}$ be the B-space of all bounded complex sequences $x = \{\xi_n \mid -\infty < n < +\infty\}$ and let S be the right shift operator such that the nth coordinate of Sx is the $(n-1)$st coordinate of x. Show that $\sigma(S) = \sigma_p(S) = \{\lambda\colon |\lambda| = 1\}$. If x_1 is the sequence with $\xi_1 = 1$ and $\xi_n = 0$ for $n \neq 1$, determine $R(\lambda; S)x_1$ and show that $\sigma(x_1) = \sigma(S)$.

40 (Saffern) Let $A \in B(\mathfrak{X})$ have the single-valued extension property and let $B = SAS^{-1}$, where S is a regular operator. Show that B has the single-valued extension property and that if $x \in \mathfrak{X}$, then

$$\sigma_A(x) = \sigma_B(Sx), \qquad S[x_A(\lambda)] = (Sx)_B(\lambda).$$

41 Let $\mathscr{B}$ be a bounded Boolean algebra of projections in $B(\mathfrak{X})$ and let $\|x\|$ be defined by

$$\|x\| = \sup\{|Ex| \mid E \in \mathscr{B}\}.$$

Show that $\|x\|$ is a norm on $\mathfrak{X}$ and is equivalent to the original norm; in fact, $|x| \leqq \|x\| \leqq K|x|$. If $E \in \mathscr{B}$ and $E \neq 0$, then $\|E\| = 1$ and if $x \in E\mathfrak{X}$ and $y \in F\mathfrak{X}$ with $EF = 0$, then x and y are mutually orthogonal in the sense that

$$\|x\| \leqq \|x + ty\|, \qquad \|y\| \leqq \|y + tx\|$$

for all scalars t.

42 Let E be a bounded Boolean algebra of projections in $B(\mathfrak{X})$ and let

$$v(E, x, x^*) = \sup \sum |x^*E_j x|,$$

where the supremum is taken over all finite sets $\{E_j\}$ which are mutually disjoint in the sense that $E_j E_k = 0, j \neq k$. Show that there exists a constant K such that

$$v(E, x, x^*) \leqq K |x| |x^*| .$$

[Hint: Examine the proof of Lemma III.1.5.]

43 (Berkson) Let E be a bounded Boolean algebra of projections in $B(\mathfrak{X})$ and let $\|x\|_E$ be defined by

$$\|x\|_E = \sup\{v(E, x, x^*) \big| |x^*| = 1,\, x^* \in \mathfrak{X}^*\}.$$

Show that $\|x\|_E$ is a norm on $\mathfrak{X}$ and that

$$|x| \leqq \|x\|_E \leqq K |x| .$$

44 (Berkson) With the norm introduced in the preceding exercise, each E_j satisfies

$$\|I + itE_j\|_E = |1 + it| = \sqrt{1 + t^2}$$

for t real. [Hint. If $\{F_k\}$ is a mutually disjoint set of elements in E, then the set $\{F_k E_j,\, F_k(I - E_j)\}$ is mutually disjoint.]

45 Using the preceding exercise, show that $\|E_j\|_E = 1$ and that if $E_j E_k = 0$ and $x = E_j x$, $y = E_k y$, then x and y are mutually orthogonal in the sense that

$$\|x\|_E \leqq \|x + ty\|_E , \qquad \|y\|_E \leqq \|x + ty\|_E$$

for all real t.

46 (Fixman) Let T be a spectral operator in $\mathfrak{X}$ with resolution of the identity E. Suppose that $\mathfrak{Y}$ is a closed subspace of $\mathfrak{X}$ which is invariant under $E(\sigma)$ for each Borel set σ. Then $\mathfrak{Y}$ is invariant under T if and only if it is invariant under $R(\lambda; T)$ for each $\lambda \in \rho(T)$.

47 Let U be the right shift operator on $\mathfrak{X} = l_2(-\infty, \infty)$, so that U is a unitary operator. Let $\mathfrak{Y}$ be those elements (ξ_n) of $\mathfrak{X}$ such that $\xi_n = 0$ for $n \leqq 0$. Show that $\mathfrak{Y}$ is invariant under U but it is not invariant under $R(\lambda; U)$, $|\lambda| < 1$, and hence $\mathfrak{Y}$ is not invariant under $E(\sigma)$ for every Borel set σ.

48 Let Ω be a compact Hausdorff space and let S be a scalar spectral operator in $C(\Omega)$. Show that S is weakly compact if and only if it is compact. (The same result holds for S in L_1 over a positive measure space.)

49 (McCarthy) Let E be a spectral measure and let μ be a positive measure defined on the Borel field Σ of a set S of complex numbers. Suppose that there exists a constant M such that $|E(\sigma)| \leqq M\mu(\sigma)$ for all $\sigma \in \Sigma$. Let $x_0 \in \mathfrak{X}$ and $x_0^* \in \mathfrak{X}^*$, and let ψ be a bounded measurable function. Define Φ: $L_1(S, \Sigma, \mu) \to \mathfrak{X}^*$ and Ψ: $L_\infty(S, \Sigma, \mu) \to \mathfrak{X}$ by

$$\Phi(f) = \int_S f(s)E^*(ds)x_0^*, \qquad f \in L_1,$$

$$\Psi(h) = \int_S h(s)\psi(s)E(ds)x_0, \qquad h \in L_\infty.$$

Show that Φ and Ψ are bounded linear operators and that

$$\Phi(f)\Psi(h) = \int_S f(s)h(s)\psi(s)x_0^*E(ds)x_0.$$

50 (McCarthy) Continuing the preceding exercise, let x_0^*, x_0, and σ_0 be chosen so that $x_0^*E(\sigma_0)x_0 \neq 0$ and let g be the Radon-Nikodým derivative of $x_0^*E(\cdot)x_0$ with respect to μ. There is a subset σ_1 of σ_0 on which g is bounded away from zero and we let $\psi = 1/g$ on σ_1. Letting Φ and Ψ be operators defined on $L_1(\sigma_1, \mu)$ and $L_\infty(\sigma_1, \mu)$ as in the preceding exercise, show that Ψ is a one-to-one and bicontinuous mapping of the non-separable space $L_\infty(\sigma_1, \mu)$ onto a closed separable subspace $\overline{\text{sp}}\{E(\sigma)x_0 \mid \sigma \in \Sigma\}$ of $\mathfrak{X}$. This contradiction shows that a spectral measure cannot satisfy such a Lipschitz condition.

51 (McCarthy) Let Ω, μ be as in the preceding exercise and let $\mathfrak{X}$ be the direct sum

$$\mathfrak{X} = L_\infty(\Omega) \oplus L_2(\Omega) \oplus L_1(\Omega).$$

Let S and N be defined in $\mathfrak{X}$ by

$$S(f(t), g(t), h(t)) = (tf(t), tg(t), th(t)), \qquad t \in \Omega,$$
$$N(f, g, h) = (0, f, g).$$

Show that S is a scalar type spectral operator and that $N^2 \neq 0$, $N^3 = 0$. Show that the spectral operator $T = S + N$ satisfies the first order rate of growth condition

$$|R(\lambda; T_\sigma)E(\sigma)| \leqq K/\delta,$$

where $\delta = \text{dist}(\lambda, \bar{\sigma})$. [Hint. Show that if $p \geqq 1$, then

$$\int_\sigma |\lambda - \xi|^{-(p+2)}\mu(d\xi) \leqq \int_{|\lambda-\xi|\geqq\delta} |\lambda - \xi|^{-(p+2)}\mu(d\xi) \leqq \left(\frac{2}{p}\right)\delta^{-p}.]$$

52 (McCarthy) By taking m copies of $L_2(\Omega)$ in the preceding exercise, obtain a spectral operator T which has mth order rate of growth and such that the radical part N of T satisfies $N^{m+1} \neq 0$.

53 Let $\{\lambda_n\}$ be a sequence of distinct non-zero complex numbers, let $\{P_n\}$ be a sequence of mutually orthogonal non-zero projections in $B(\mathfrak{X})$, and suppose that

$$\sum_{n=1}^{\infty} |\lambda_n P_n|$$

is convergent. Prove that $\lambda_n \to 0$ and the series

$$B = \sum_{n=1}^{\infty} \lambda_n P_n$$

converges in the uniform topology of $B(\mathfrak{X})$. Show $\sigma(B) = \{0, \lambda_1, \lambda_2, \ldots\}$ and that

$$R(\lambda; B) = \frac{I}{\lambda} + \sum_{n=1}^{\infty} \frac{\lambda_n}{\lambda(\lambda - \lambda_n)} P_n,$$

where the series converges uniformly for λ in a compact subset of $\rho(B)$. Moreover, each of the points λ_n is a simple pole of $R(\lambda; B)$ and the residue of $R(\lambda; B)$ at λ_n is P_n.

54 (Taylor) Let $A \in B(\mathfrak{X})$ have a spectrum consisting of a sequence $\{\lambda_n\}$ converging to 0 such that $R(\lambda; A)$ has a simple pole at λ_n with residue P_n and suppose that $\Sigma |\lambda_n P_n|$ is convergent. Let B, C be defined by

$$B = \sum_{n=1}^{\infty} \lambda_n P_n, \qquad C = A - B.$$

Since $AP_n = \lambda_n P_n$, show that A, B, C commute and that $AB = B^2$ and $BC = 0$. Show that, at least for sufficiently large λ,

$$R(\lambda; A) = R(\lambda; B) \sum_{n=0}^{\infty} (CR(\lambda; B))^n = R(\lambda; B) \sum_{n=0}^{\infty} \frac{C^n}{\lambda^n}.$$

Using the fact that the difference $R(\lambda; A) - R(\lambda; B)$ is analytic for $\lambda \neq 0$, prove that C is a quasi-nilpotent operator and that

$$R(\lambda; A) = R(\lambda; B) + R(\lambda; C) - \frac{I}{\lambda}.$$

55 (McCarthy) Let T be a spectral operator in a complex B-space $\mathfrak{X}$ which satisfies the growth condition (*) in Theorem XV.6.7, namely

$$(*) \qquad |R(\xi; T_\sigma)E(\sigma)| \leqq \frac{K}{\text{dist}\,(\xi, \bar{\sigma})^m}$$

for $\xi \notin \bar{\sigma}$, $|\xi| \leqq |T| + 1$. If k is a natural number then there exists a constant M_k such that if $0 \leqq \varepsilon \leqq 1$ and σ is a Borel set with diameter at most ε, then $|N^k E(\sigma)| \leqq M_k \varepsilon^{k+1-m}$ (where N is the radical part of T.)

56 (McCarthy) Let T be a spectral operator in a complex B-space which satisfies condition (*) of the preceding exercise. Let σ be a Borel subset such that for every $\varepsilon > 0$ there is a covering $\{\sigma_j\}$ of σ by disjoint sets σ_j of diameter ε_j such that $\Sigma \varepsilon_j^p \leqq \varepsilon$ for some positive number p. Show that $N^k E(\sigma) = 0$ when $k \geqq p + m - 1$. [This condition is satisfied if σ has Hausdorff p-measure equal to zero.]

57 (McCarthy) Let T be a spectral operator in a complex B-space $\mathfrak{X}$ satisfying the mth order condition (*). If N is the radical part of T, then $N^{m+2} = 0$.

58 (McCarthy) Let T be a spectral operator in a complex B-space $\mathfrak{X}$ which satisfies the mth order condition (*) and suppose that $\sigma(T)$ has plane measure zero. If N is the radical part of T, then $N^{m+1} = 0$.

59 (Foguel, Fixman) If $\mathfrak{X}$ is a B-space, then an element $U \in B(\mathfrak{X})$ which is isometric and maps onto $\mathfrak{X}$ is called a *unitary operator*. Prove that $\sigma(U) \subseteqq \{\lambda \mid |\lambda| = 1\}$, and that

$$|R(\lambda; U)| \leqq \frac{1}{|1 - |\lambda||}, \qquad |\lambda| \neq 1.$$

If U is a spectral operator, then it is of scalar type.

60 (Bishop) Let $\mathscr{B}$ denote the Borel subsets of the complex plane C and let $T \in B(\mathfrak{X})$. A vector valued measure (cf. IV.10) m on $\mathscr{B}$ to $\mathfrak{X}$ is said to be a *T-measure* in case

$$Tm(e) = \int_e \lambda m(d\lambda), \qquad e \in \mathscr{B}.$$

If $x = m(C)$, we say that *x has a T-measure m*.

(a) Every eigenvector for T has a T-measure.

(b) If T is a scalar type spectral operator, then every vector in $\mathfrak{X}$ has a T-measure.

(c) If m is a T-measure and if $y = m(e)$, then $R(\lambda; T)y$ has an analytic extension for $\lambda \notin \bar{e}$.

(d) If m is a T-measure, then $m(\sigma(T)) = m(C)$.

(e) If m is a T-measure and $\lambda_0 \in e$, then

$$|(T - \lambda_0 I)m(e)| \leqq \operatorname{diam}(e)\, \|m\|\,(e),$$

where the symbol $\|m\|\,(e)$ denotes the semi-variation of m over e.

61 Let $T \in B(\mathfrak{X})$ and let $\mathfrak{Y}$ be a closed subspace of $\mathfrak{X}$ which is invariant under T. Let σ be a closed subset of the complex plane which has a connected complement and suppose that $\mathfrak{Y} \subseteqq \{x \in \mathfrak{X} \mid \sigma(x) \subseteqq \sigma\}$. Prove that $\sigma(T \mid \mathfrak{Y}) \subseteqq \sigma$. Show that this may fail if the complement is not connected.

62 For the operator A of equation (21) in Section 12 establish the equations (27), (28), $\sigma(A_{\mathfrak{S}}) = \sigma_c(A_{\mathfrak{S}})$, and

$$\hat{E}_{21}\mathfrak{H}^2 = \{(\zeta, -i\zeta), \zeta \in \mathfrak{H}\}, \qquad \hat{E}_{22}\mathfrak{H}^2 = \{(\eta, i\eta), \eta \in \mathfrak{H}\}.$$

63 Let C be a closed densely defined operator in $\mathfrak{H}^p$ and introduce the scalar product and norm into $\mathfrak{D}(C)$ by the equations

$$\begin{aligned} (\varphi, \zeta)_1 &= (\varphi, \zeta) + (C\varphi, C\zeta). \qquad && \varphi, \zeta \in \mathfrak{D}(C), \\ |\varphi|_1 &= \{(\varphi, \varphi)\}^{\frac{1}{2}}, && \varphi \in \mathfrak{D}(C). \end{aligned}$$

Show that $\mathfrak{D}(C)$ becomes a complete Hilbert space and C is a continuous map of $\mathfrak{D}(C)$ into $\mathfrak{H}^p$.

64 Let μ be a finitely additive real or complex valued set function defined on a field of sets containing the open sets in R^N. Let

$$\int_{R^N} \frac{v(\mu; ds)}{(1 + |s|^2)^k} < \infty$$

for some positive integer k. Prove that the function T defined by the equation

$$T(\varphi) = \int_{R^N} \varphi(s)\mu(ds), \qquad \varphi \in \Phi$$

is a tempered distribution.

65 Let $A = (a_{jk})$ be the formal differential operator of equation (12.11) whose elements have orders at most $2q$ and let its natural closed extension $A_{\mathfrak{S}}$ be a spectral operator. If

$$\operatorname*{ess\,sup}_{s \in R^N} \sup_{\hat{\lambda}(s) \in \sigma(\hat{A}(s))} |s|^{pq} e^{t\mathscr{R}(\hat{\lambda}(s))} < \infty,$$

for some real t, then $A_{\mathfrak{S}} \exp(tA_{\mathfrak{S}})$ is a bounded operator.

66 Let q be a positive integer, $(\nabla^2)^q = (\partial^2/\partial s_1^2 + \cdots + \partial^2/\partial s_N^2)^q$, $a_{jk} = a_{jk}(\partial/\partial s_1, \ldots, \partial/\partial s_N)$ be polynomials with constant coefficients and having degree $\leqq 2q$ if $j < k$ and $a_{jk} = 0$ if $j \geqq k$. Let $A_{\mathfrak{S}}$ be the natural closed extension of the formal differential operator

$$A = \alpha^2(\nabla^2)^q I + (a_{jk})$$

on Φ^p, where α^2 is a positive or negative real number. Prove that:

(i) The natural closed extension $A_{\mathfrak{S}}$ of A is a spectral operator with spectrum $\sigma(A_{\mathfrak{S}}) = (-\infty, 0]$ if $(-1)^q\alpha^2 < 0$ and $\sigma(A_{\mathfrak{S}}) = [0, \infty)$ if $(-1)^q \alpha^2 > 0$.

(ii) The operator $A_{\mathfrak{S}}$ is the infinitesimal generator of a strongly continuous semi-group $T(t)$ (of bounded linear operators in $\mathfrak{H}^p = L_2(R^N) + \cdots + L_2(R^N)$) on $[0, \infty)$ if $(-1)^q\alpha^2 < 0$ and on $(-\infty, 0]$ if $(-1)^q\alpha^2 > 0$.

(iii) The semi-group $T(t)$ has a strongly analytic extension to the half plane $\mathscr{R}(\lambda) > 0$ in case $(-1)^q\alpha^2 < 0$, and to the half plane $\mathscr{R}(\lambda) < 0$ in case $(-1)^q\alpha^2 > 0$.

(iv) $\mathfrak{D}(A_{\mathfrak{S}}) = (p)T^{(2q)}(R^N)$.

67 Let $A_{\mathfrak{S}}$ and $T(t)$ be the operators of the preceding exercise with $(-1)^q\alpha^2 < 0$. Then for every $\varphi^{(0)}$ in $\mathfrak{H}^p$ there is one and only one continuous map $t \to \varphi(t)$ of $[0, \infty)$ into $\mathfrak{H}^p$ which is differentiable for $t > 0$ and has the properties

(i) $$\varphi(t) \in (p)T^{(2q)}(R^N), \qquad 0 < t < \infty,$$

(ii) $$\varphi'(t) = A_{\mathfrak{S}} \varphi(t), \qquad 0 < t < \infty,$$

(iii) $$\varphi(0) = \varphi^{(0)}.$$

This unique function φ is given by the equation $\varphi(t) = T(t)\varphi^{(0)}$ and has an analytic extension to the half plane $\mathscr{R}(\lambda) > 0$.

(iv) The derivative $\varphi'(0)$ exists and $\varphi'(0) = A_{\mathfrak{S}}\varphi(0)$ if and only if $\varphi^{(0)} \in (p)T^{(2q)}(R^N)$.

(v) For $\mathscr{R}(\lambda) > 0$ the analytic extension $\varphi_1(\lambda) = \varphi_1(s_1, \ldots, s_N; \lambda)$ of $\varphi(t)$ as a function of $s = (s_1, \ldots, s_N)$, has each of its p components belonging to $C^\infty(R^N)$ and all partial derivatives $\partial_s^\beta \varphi_1$ belonging to $\mathfrak{H}^p$.

Analogous conclusions hold if $(-1)^q\alpha^2 > 0$.

68 Let $A_{\mathfrak{S}}$ be the operator of the preceding exercise and let B be an arbitrary bounded linear operator in $\mathfrak{H}^p$. Then

(i) The operator $A_{\mathfrak{S}} + B$ with domain $(p)T^{(2q)}(R^N)$ is the infinitesimal generator of a strongly continuous semi-group $S(t)$, $t \geqq 0$, of bounded linear operators in $\mathfrak{H}^p$ and $S(t)$ has a strongly analytic extension to a semi-group $S(\zeta)$ defined for ζ in the half plane $\mathscr{R}(\zeta) > 0$.

For every $\varphi^{(0)}$ in $\mathfrak{H}^p$ there is one and only one continuous map $t \to \varphi(t)$ of $[0, \infty)$ into $\mathfrak{H}^p$ which is differentiable for $t > 0$ and has the properties

(ii) $\qquad \varphi(t) \in (p)T^{(2q)}(R^N), \qquad 0 < t < \infty,$

(iii) $\qquad \varphi'(t) = (A_{\mathfrak{S}} + B)\varphi(t), \qquad 0 < t < \infty,$

(iv) $\qquad \varphi(0) = \varphi^{(0)}.$

(v) This unique function $\varphi(t)$ is given by the equation $\varphi(t) = S(t)\varphi^{(0)}$ and thus has an analytic extension $\varphi_1(\zeta) = S(\zeta)\varphi^{(0)}$ to the half plane $\mathscr{R}(\zeta) > 0$. This extension φ_1 satisfies (ii) and (iii) for complex values of t with $\mathscr{R}(t) > 0$.

(vi) The derivative $\varphi'(0)$ exists and $\varphi'(0) = (A_{\mathfrak{S}} + B)\varphi(0)$ if and only if $\varphi^{(0)} \in (p)T^{(2q)}(R^N)$.

Analogous conclusions hold if $(-1)^q\alpha^2 > 0$.

(69) Let $A_{\mathfrak{S}}$ and B be as in the preceding exercise and let $D = (d_{jk})$ be a formal differential operator on Φ^p whose elements are polynomials with constant coefficients and of degrees at most $2q - 2$. Then

(i) The domain $\mathfrak{D}(D_{\mathfrak{S}})$ of the natural closed extension of D contains $(p)T^{(2q)}(R^N)$ and the operator $A_{\mathfrak{S}} + D_{\mathfrak{S}} + B$ with domain $(p)T^{(2q)}(R^N)$ is closed.

(ii) The closed operator $A_{\mathfrak{S}} + D_{\mathfrak{S}} + B$ on $(p)T^{(2q)}(R^N)$ is the infinitesimal generator of a strongly continuous semi-group $Q(t)$, $t \geqq 0$, of bounded linear operators in $\mathfrak{H}^p$ and $Q(t)$ has a strongly analytic extension to a semi-group $Q_1(\zeta)$ defined for ζ in the half plane $\mathscr{R}(\zeta) > 0$.

For every $\varphi^{(0)}$ in $\mathfrak{H}^p$ there is one and only one continuous map $t \to \varphi(t)$ of $[0, \infty)$ into $\mathfrak{H}^p$ which is differentiable for $t > 0$ and has the properties

(iii) $\qquad \varphi(t) \in (p)T^{(2q)}(R^N), \qquad 0 < t < \infty,$

(iv) $\qquad \varphi'(t) = (A_{\mathfrak{S}} + D_{\mathfrak{S}} + B)\varphi(t), \qquad 0 < t < \infty,$

(v) $\qquad \varphi(0) = \varphi^{(0)}.$

(vi) This unique function $\varphi(t)$ is given by the equation $\varphi(t) = Q(t)\varphi^{(0)}$ and thus has an analytic extension $\varphi_1(\zeta) = Q_1(\zeta)\varphi^{(0)}$ to the half-plane

$\mathscr{R}(\zeta) > 0$. This extension φ_1 satisfies (iii) and (iv) for complex values of t with $\mathscr{R}(t) > 0$.

(vii) The derivative $\varphi'(0)$ exists and $\varphi'(0) = (A_{\mathfrak{S}} + D_{\mathfrak{S}} + B)\varphi(0)$ if and only if $\varphi^{(0)} \in (p)T^{(2q)}(R^N)$.

Analogous results hold in case $(-1)^q\alpha^2 < 0$.

70 Let $\hat{A} = (\hat{a}_{jk}(s))$ be a $p \times p$ matrix of real or complex measurable functions on R^N and let $A_{\mathfrak{S}}$ be the densely defined closed operator given by equations 12.4. Show that if $A_{\mathfrak{S}}$ is a scalar type spectral operator whose spectrum lies in a left half plane then it is the infinitesimal generator of the strongly continuous semi-group $T(t)$, $t \geqq 0$, given by the equation

$$T(t)\varphi = \int_{\sigma(A_{\mathfrak{S}})} e^{t\lambda} E(d\lambda)\varphi, \qquad \varphi \in \mathfrak{H}^p,$$

where $E(\cdot)$ is the resolution of the identity for $A_{\mathfrak{S}}$. The function $T(t)\varphi$ is also given by the convolution

$$T(t)\varphi = (\tau^{-1}e^{t\hat{A}}) * \varphi, \qquad \varphi \in \mathfrak{H}^p.$$

71 Let $\varphi_j^{(0)} \in L_1(R^3)$, $j = 1, 2, 3$, $s = (s_1, s_2, s_3) \in R^3$, $t > 0$ and $\nabla^2 = \partial^2/\partial s_1^2 + \partial^2/\partial s_2^2 + \partial^2/\partial s_3^2$. Give an explicit form for the solution of the initial value problem

$$\frac{\partial}{\partial t}\varphi_1(s; t) = \tfrac{1}{4}\nabla^2\varphi_1(s; t) + \left(\frac{\partial}{\partial s_2} + \frac{\partial}{\partial s_3}\right)\varphi_2(s; t) + \frac{\partial^2}{\partial s_3^2}\varphi_3(s; t)$$

$$\frac{\partial}{\partial t}\varphi_2(s; t) = \tfrac{1}{4}\nabla^2\varphi_2(s; t) + \frac{\partial^2}{\partial s_1^2}\varphi_3(s; t)$$

$$\frac{\partial}{\partial t}\varphi_3(s; t) = \tfrac{1}{4}\nabla^2\varphi_3(s; t)$$

$$\lim_{t\to 0+} \varphi_j(\cdot; t) = \varphi_j^{(0)} \text{ in } L_2(R^3), \qquad j = 1, 2, 3.$$

72 Let $\varphi_j^{(0)} \in L_2(R^2)$, $j = 1, 2$, $s = (s_1, s_2) \in R^2$, $t > 0$ and $\nabla^2 = \partial^2/\partial s_1^2 + \partial^2/\partial s_2^2$. Give an explicit representation for the convolution matrix kernel $K(s; t)$ such that the vector $\varphi(t) = (\varphi_1(t), \varphi_2(t))$ given by the formula

$$\varphi(t) = \int_{-\infty}^{\infty} K(s - u; t)\varphi^{(0)}(u)\, du$$

satisfies the equations

$$\varphi'(t) = \begin{pmatrix} \frac{1}{4}\nabla^2 & 2\left(\frac{\partial}{\partial s_1} + \frac{\partial}{\partial s_2}\right) \\ \frac{\partial}{\partial s_1} + \frac{\partial}{\partial s_2} & \frac{1}{4}\nabla^2 \end{pmatrix} \varphi(t), \qquad t > 0,$$

$$\lim_{t \to 0+} \varphi_j(t) = \varphi_j^{(0)} \text{ in } L_2(R^2), \qquad j = 1, 2.$$

16. Notes and Remarks

Since the theory of spectral operators has been developed rather recently, there is a comparatively small bibliography associated with it; this bibliography is steadily growing, however. We shall indicate here the sources of most of the material presented in the text, and attempt to direct the reader to additional papers where results pertaining to spectral operators can be found that are not included here.

In addition to the theory of spectral operators, there are several other aspects of functional analysis which border on the topics presented in this volume. Although it is not possible to go into much detail about these other theories, we shall try to provide a brief sketch of their main aspects, and their relation to the theory of spectral operators, as well as give references for further reading. These references are not intended to provide complete coverage, but emphasize general expository accounts and recent literature.

For additional work on aspects of the theory of spectral operators (as presented here) and spectral measures, we cite the following.

Apostol [6, 7, 8, 10], Bade [2, 3, 4, 5], Berkson [2, 3, 4, 5], Berkson and Dowson [1], Colojoară [1, 2, 3, 4], Colojoară and Foiaş [4], Dean [1], Dowson [1, 2, 3, 4, 5, 6], Dunford [17, 18, 19, 20, 21], Edwards and Ionescu Tulcea [1], Feldman [1], Feldzamen [1, 2], Fixman [1], Foguel [1, 2, 3, 4], Foiaş [3, 10, 11, 13, 15, 16], Gray [1], Heyn [1], Ionescu Tulcea [3], Kakutani [15], Kantorovitz [1, 2, 3, 4, 5, 6, 7, 8, 9], Keldyš and Lidskiĭ [1], Kesel'man [1, 2], Kluvánek [1, 2], Kluvánek and Kovářiková [1], Kovářiková [1], Krabbe [3, 6], Ljance [1, 2], Lumer [2], McCarthy [1, 2], McCarthy and Schwartz [1], McCarthy and Stampfli [1], McCarthy and Tzafriri [1], McGarvey [1], Maeda [1, 2, 3, 4, 5, 6, 7, 8], Moyal [1],

Nel [1], Neubauer [2], Oberai [1, 2, 3, 4], Panchapagesan [1], Pedersen [1], Plafker [1], Saffern [1], Salehi [1], Schaefer [7, 10, 11], Schaefer and Walsh [1], J. Schwartz [2, 6, 7], Simpson [1, 2, 3], Smart [3], Stampfli [1, 2, 7, 11], Suzuki [1], Turner [1, 2], Tzafriri [1, 2, 3, 4, 5, 6], Walsh [1, 2, 3], Wermer [3].

A list of references dealing with algebras of operators and multiplicity theory will be given later.

Recently there has been considerable work done on the theory of "generalized spectral operators" and "decomposable operators." We shall describe the general outlines of this work and cite pertinent references later.

In addition, numerous papers have been written which deal with the general topic of spectral theory for operators in B-spaces, but which are not specifically (or intimately) related to spectral operators. The following list includes papers of this nature.

Allan [1, 2], Apostol [10, 11], Bartle [6, 7], Berkson [1, 2], Bishop [1, 2], Brown [1], Deal [1, 2], Deprit [1, 2], Derr and Taylor [1], Dollinger [1, 2], Dolph [1, 2], Dolph and Penzlin [1], Dunford [6, 7, 14, 15, 16, 20, 21], Edwards and Ionescu Tulcea [1], Foguel [12], Foiaş [1, 3, 4, 9, 12, 13], Gindler [1], Gindler and Taylor [1], Gohberg and Kreĭn [2, 7, 8], Harazov [4], Kantorovitz [1, 2, 3, 4, 5, 6, 7, 8, 9], Kariotis [1], Kocan [1], Krabbe [4, 5, 6, 7, 8, 9, 12, 13], Kultze [1], Leaf [1, 2], Lebow [1], Ljubič [1, 2, 3], Ljubič and Macaev [1, 2], Lorch [7], Macaev [1, 2, 3], Maeda [1, 2, 3, 4, 5, 6, 7, 8], Neubauer [1], Pietsch [1, 2, 3, 4, 5, 6], Plafker [1], Ringrose [3], Saphar [1, 2, 6], Schaefer [7, 10, 11, 15], J. Schwartz [4], Sebastião e Silva [4, 5], Sills [1], Sine [1], Smart [1, 2], Taylor [15, 16, 17, 18], Tillmann [2], Trampus [1], Vasilescu [1, 4], Waelbroeck [1, 2, 3], Wolf [4, 5].

Some of the work presented in these papers will be briefly described in the following material, but it will not be possible to do more than suggest some directions of research, and the reader is urged to examine the original sources for fuller details.

Spectral operators. For a general survey of the theory of spectral operators up to 1958, the reader is referred to the article of Dunford [19], which gives the historical background and a general expository account of much of the material in this volume. Colojoară [5] has presented the theory of spectral operators from a different point of view, by starting with decomposable operators, then generalized spectral operators, and finally spectral operators.

Most of the theorems presented in Sections 1-5 are taken from Dunford [17, 18], although a number of modifications have been made in the exposition presented here. In 1953–1954 Neubauer communicated some valuable ideas to us concerning the work in Dunford [17]. He suggested a number of improvements and observed that the resolvent of a spectral operator has the single valued extension property.

In addition to considerable elaboration of the material presented in this volume, which will be detailed below, there are three types of extensions of the theory of spectral operators that have been made. The first extension, initiated by C. Ionescu Tulcea [3], is a generalization of the theory to certain locally convex spaces. The generalization, however, is not done in a routine fashion but is substantially recast. Let E be a locally convex linear Hausdorff space which is "quasi-complete and barreled." Ionescu Tulcea [3] introduces a family $F = \{m_{x, x^*}\}$ of bounded complex Radon measures on the complex numbers C, which he calls a *spectral family* in case there exists a homomorphism U of the algebra of all bounded Borel measurable functions into the algebra of all continuous linear mappings in E such that $U(1) = I$ and

$$x^*U(f)x = \int_C f(s)m_{x,x^*}(ds)$$

for all $x \in E$, $x^* \in E^*$, and all bounded Borel functions f. The notion of a spectral operator is introduced in such a way as to include operators which are not everywhere defined; hence it unifies Chapters XV and XVIII in spirit. Unfortunately, full details of this work have not been published; however, further research along this line has been done by F.-Y. Maeda (see [1, 2]), who also generalized some work of Bishop on weak T-measures as well as some results of J. Schwartz [2]. Additional research in this direction has been made by Oberai [1, 4], Plafker [1], and Simpson [1, 2, 3].

A second direction in which the theory of spectral operators has been extended is represented by work of Schaefer [7, 10, 11]. Although he considered spectral measures and scalar operators in locally convex spaces, the noteworthy feature of his work is that it exploits the connection between spectral properties and *order* properties. It also develops the theory for operators which may not be defined on the entire space. A brief sketch of this research, and some closely related work, is given below under the heading "Spectral measures, locally convex spaces, and order."

The third (but not entirely different) direction that has been developed is represented by the incisive research of I. Colojoară and C. Foiaş on *decom-*

posable and on *generalized spectral operators*. Fortunately, their work, as well as that of their collaborators, and its relation to the work of Maeda and Kantorovitz are well summarized in the splendid book of Colojoară and Foiaş [4]. However, because of its interest, we shall sketch the main outlines of their research below under the two headings "Decomposable operators" and "Operational calculi and spectral theory."

The *single valued extension property*. The example of an operator which does not have the single valued extension property that is given in Section 2 is due to S. Kakutani (see Dunford [18]). Kesel'man [1] gave necessary conditions for an operator to have this property. Colojoară and Foiaş [4; p.5] showed that if f is analytic and non-constant on every component of an open set containing $\sigma(T)$, then $f(T)$ has the single valued extension property if and only if T does. In Colojoară and Foiaş [1] [4; Chap. I] the following idea is introduced: Let T, $U \in B(\mathfrak{X})$ and let

$$(T-U)^{[n]} = \sum_{k=0}^{n} (-1)^{n-k} \binom{n}{k} T^k U^{n-k};$$

then we say that T and U are *quasi-nilpotent equivalent* in case

$$\lim_{n\to\infty} |(T-U)^{[n]}|^{1/n} = \lim_{n\to\infty} |(U-T)^{[n]}|^{1/n} = 0.$$

(It is clear that if $U = T + N$ where $TN = NT$ is a quasi-nilpotent operator, then $(T-U)^{[n]} = (T-U)^n$, and T and U are quasi-nilpotent equivalent; the general notion extends this case.) The relation of being quasi-nilpotent equivalent is indeed an equivalence relation and, when T and U are quasi-nilpotent equivalent, then (i) $\sigma(T) = \sigma(U)$, (ii) T has the single valued extension property if and only if U does, and (iii) if T (hence U) has the single valued extension property and $x \in \mathfrak{X}$, then $\sigma_T(x) = \sigma_U(x)$.

The set of operators in $B(\mathfrak{X})$ with the single valued extension property is not closed. However, Vasilescu [2] has shown that if $\{T_k\}$ is a sequence in $B(\mathfrak{X})$ with the single valued extension property and if $T \in B(\mathfrak{X})$ is such that

$$\lim_{k\to\infty} (\limsup_{n\to\infty} |(T_k - T)^{[n]}|^{1/n}) = 0,$$

then T has the single valued extension property. In particular it follows that if $\{T_k\}$ is a sequence of commuting operators with the single valued extension property, and if $T = \lim T_k$, then T has the single valued extension property.

If $T \in B(\mathfrak{X})$ has the single valued extension property and $x \in \mathfrak{X}$, then we have $\sigma(x) \subseteqq \sigma(T)$. Conversely, Sine [1] observed that if $\lambda \in \rho(x)$ for all $x \in \mathfrak{X}$ and if $S : \mathfrak{X} \to \mathfrak{X}$ is defined by $Sx = x(\lambda)$, then $S = (\lambda I - T)^{-1}$ from which we deduce that

$$\sigma(T) = \bigcup \{\sigma(x) \,|\, x \in \mathfrak{X}\}.$$

This result has been considered by Gray in a study of the analytic extensions of the resolvent of vectors. Dollinger [1] has also considered the spectrum $\sigma_T(x)$ both as a function of $x \in \mathfrak{X}$ and of $T \in B(\mathfrak{X})$. Colojoară and Foiaş [1] (see also [4; p.1]) show that $\sigma(x(\lambda)) = \sigma(x)$ for every $x \in \mathfrak{X}$, $\lambda \in \rho(x)$.

If $T \in B(\mathfrak{X})$ has the single valued extension property and F is a closed subset of the complex plane C, let

$$\mathfrak{X}_T(F) = \mathfrak{X}(F) = \{x \in \mathfrak{X} \,|\, \sigma(x) \subseteqq F\}.$$

It is readily seen that $\mathfrak{X}(F)$ is a linear manifold in $\mathfrak{X}$, but it need not be closed even when F is closed; an elementary counterexample is given in Colojoară and Foiaş [4; p.25]. If T is a spectral operator with resolution of the identity E and if δ is a closed set, then Theorem 3.4 asserts that the set $\mathfrak{X}(\delta)$ is closed; indeed, it coincides with the set $E(\delta)\mathfrak{X}$. In the case of a normal operator in Hilbert space, Corollary 3.7 was conjectured by von Neumann [24] and proved by Fuglede [1] and Halmos [3] [6; p.68]. The general case was given by Dunford [18; p.329].

If Γ is a total manifold in $\mathfrak{X}^*$ and $T \in B(\mathfrak{X})$ has a bounded resolution of the identity E such that $x^*E(\cdot)x$ is countably additive for all $x^* \in \Gamma$, $x \in \mathfrak{X}$, we say that T is a *spectral operator of class* (Γ); if there exists some total set Γ for which T is spectral of class (Γ), we say that T is *prespectral.* Although many of the properties of spectral operators extend to prespectral operators, not all do. In particular, Theorems 3.2 and 3.4 remain valid for prespectral operators, but Corollary 3.7 does not. Indeed, Fixman [1] constructed an example where this corollary fails and showed that it is possible for an operator in l_∞ to be a spectral operator of two different classes, and to have two (non-commuting) resolutions of the identity. It can be seen that if T is prespectral with two resolutions of the identity E_1, E_2 of class (Γ), and if E_1 and E_2 commute, then $E_1 = E_2$. However, it is not known whether an operator can have distinct (hence non-commuting) resolutions of the identity *of the same class.*

It follows from some results of Bade [4] (see also XVII.2.1 and XVII.2.12) that if $\mathfrak{X}$ is a weakly complete B-space, then any prespectral operator is automatically spectral, and so has a unique resolution of the identity. Berkson and Dowson [1] have considered prespectral operators in some detail and have obtained a number of results concerning them. They show that if T is spectral, then T^* has a unique resolution of the identity of class $(\mathfrak{X})$; moreover, if T^* is spectral of class (Γ) where $\Gamma \subseteq \mathfrak{X}^{**}$, then it has a unique resolution of the identity of class (Γ). If $\mathfrak{X}^*$ is weakly complete and T is spectral of class (Γ), then T has a unique resolution of the identity of class (Γ). Moreover, (Berkson and Dowson [1]), if T is spectral of class (Γ) and the spectrum of T is either totally disconnected or an R-set (that is, the rational functions which are analytic on $\sigma(T)$ are dense in $C(\sigma(T))$), then T has a unique resolution of the identity of class (Γ).

Restrictions and quotients. Theorem 3.10 shows that if a spectral operator $T \in B(\mathfrak{X})$ is reduced by a closed subspace $\mathfrak{Y} \subseteq \mathfrak{X}$ and one of its complements (that is, if T commutes with some projection of $\mathfrak{X}$ onto $\mathfrak{Y}$), then the restriction $T \mid \mathfrak{Y}$ of T to $\mathfrak{Y}$ is spectral. The situation corresponding to an invariant closed subspace of T is not so simple. However, Fixman [1] proved that the restriction of a spectral operator T to an invariant closed subspace $\mathfrak{Y}$ of $\mathfrak{X}$ is spectral if and only if the resolution of the identity E of T also leaves $\mathfrak{Y}$ invariant; in this case the resolution of the identity of $T \mid \mathfrak{Y}$ is given by $E \mid \mathfrak{Y}$. Moreover, Dowson [1] proved that $\sigma(T \mid \mathfrak{Y}) \subseteq \sigma(\mathfrak{Y})$; he further showed that if S is a scalar type operator whose spectrum is nowhere dense and does not separate the plane, then the restriction of S to *any* invariant closed subspace is spectral. Similarly if S is a scalar type operator whose spectrum is an R-set and if $\mathfrak{Y}$ is an invariant closed subspace, then $S \mid \mathfrak{Y}$ is spectral if and only if $\sigma(S \mid \mathfrak{Y}) \subseteq \sigma(S)$. An example (due to J. R. Ringrose) shows that these two results do not generalize to spectral operators that are not scalar type. However, if T is a scalar type operator whose spectrum is totally disconnected, then the restriction of T to any invariant closed subspace is spectral. In [5], Dowson considered the restriction of prespectral operators; it was seen that the three results just stated for spectral operators do not hold for prespectral operators.

The analogous problem of operators induced by spectral operators on quotient spaces was considered by Dowson [3]. He showed that if $T \in B(\mathfrak{X})$ is a spectral operator and $\mathfrak{Y}$ is a closed subspace invariant under T, then the operator $T_{\mathfrak{Y}}$ (defined on $\mathfrak{X}/\mathfrak{Y}$ by $T_{\mathfrak{Y}}[x] = [Tx]$) is a spectral operator if and only if $\mathfrak{Y}$ is invariant under the resolution of the identity of T. In this case $\sigma(T_{\mathfrak{Y}}) \subseteq \sigma(T)$ and the resolution of the identity of $T_{\mathfrak{Y}}$ is obtained by taking the quotient under $\mathfrak{Y}$ of the resolution of the identity of T. Thus we infer that *if T is spectral and $\mathfrak{Y}$ is an invariant closed subspace, then the quotient operator $T_{\mathfrak{Y}}$ is spectral if and only if the restriction $T \mid \mathfrak{Y}$ is spectral.* Consequently, the analogs of the results stated in the preceding paragraph for restrictions of spectral and scalar type operators also hold for their quotients.

Although the restrictions of operators with the single valued extension property have this property (see Dowson [1]), the corresponding result for quotients is not true. Indeed, Dowson [3] notes that the unitary shift operator has quotients which do not have the single valued extension property. On the other hand, Vasilescu [8] proved that if $T \in B(\mathfrak{X})$ and M is a closed subset of the complex plane C such that: (i) if f is an analytic function on an open set $D_f \subseteq M'$ to $\mathfrak{X}$ such that $(\lambda I - T)f(\lambda) = 0$ for all $\lambda \in D_f$, then $f(\lambda) = 0$ for $\lambda \in D_f$, and (ii) $\mathfrak{X}(M) = \{x \in \mathfrak{X} \mid \sigma(x) \subseteq M\}$ is a closed subspace of $\mathfrak{X}$; then the quotient operator induced by T in $\mathfrak{X}/\mathfrak{X}(M)$ has the single valued extension property.

Another aspect of the restrictions of scalar type operators in Hilbert space to invariant closed subspaces was made by Saffern [1], who defined an operator in a Hilbert space to be *subscalar* if it is the restriction of a scalar type operator to an invariant closed subspace. (This generalizes the related notion of a subnormal operator that was introduced by Halmos.) As would be expected, many properties of subnormal operators can be extended to subscalar operators; for example, there exist minimal scalar extensions of subscalar operators. Although minimal normal extensions of a subnormal operator are unitarily equivalent, minimal scalar extensions of a subscalar operator need not even be similar. Hence the natural definition of a resolution of the identity of a subscalar operator (as the restriction of the resolution of the identity of a minimal scalar extension) is not uniquely defined. However, relations between the various parts of the spectrum of a subscalar operator and a minimal scalar extension are obtained and a functional calculus for a subscalar operator can be defined. It is proved that a subscalar operator has the single valued extension property and that the resulting set of vector valued analytic extensions of the resolvent operator forms a set of similarity invariants for the subscalar operator.

The notion of subscalar operators and scalar extensions in B-spaces has been pursued by Ionescu Tulcea [4, 5] and Ionescu Tulcea and Plafker [1]. An operator $T \in B(\mathfrak{X})$ is said to be *subscalar* if there exists a B-space $\tilde{\mathfrak{X}} \supseteq \mathfrak{X}$, a continuous projection of $\tilde{\mathfrak{X}}$ onto $\mathfrak{X}$, and a scalar type operator $\tilde{T} \in B(\tilde{\mathfrak{X}})$ such that $\tilde{T} \mid \mathfrak{X} = T$. It is proved that T is subscalar if and only if there exists a compact set $Z \subseteq C$ and a continuous linear mapping $U: f \rightarrow U(f)$ of the space of bounded Borel functions measurable on Z into $B(\mathfrak{X})$ such that (i) $I = U(f_0)$ for $f_0(\lambda) \equiv 1$ and $T = U(f_1)$ for $f_1(\lambda) \equiv \lambda$, (ii) $U(ff_1) = U(f)T$ for all f, (iii) if $\{f_n\}$ is a bounded sequence which converges pointwise to zero, then $\{U(f_n)\}$ converges strongly to zero. Another characterization of subscalar operators is given, based on a theorem which asserts that, under certain hypotheses, a continuous linear mapping U of a Banach algebra $\mathfrak{A}$ with unit e into $B(\mathfrak{X})$ is the projection of a continuous homomorphism of $\mathfrak{A}$ into $B(\tilde{\mathfrak{X}})$ for some B-space $\tilde{\mathfrak{X}} \supseteq \mathfrak{X}$ which has a continuous projection onto $\mathfrak{X}$. A second proof of this latter result is given in Ionescu Tulcea and Plafker [1].

The Canonical Reduction. Theorem 4.5, which gives the reduction of a spectral operator into the sum of scalar type and quasi-nilpotent operators may be regarded as a generalization of the Jordan Decomposition Theorem; it was proved by Dunford [18; p. 333], although the proof there is based on results from the theory of B-algebras. A proof similar to the one given here was communicated to the authors by Foiaş. A version of the canonical reduction for everywhere defined spectral operators in a locally convex linear space was given by Ionescu Tulcea [3] when the space is quasicomplete and barreled. See below for remarks concerning the extension of this result to generalized spectral operators.

In the case of prespectral operators, Theorem 4.5 has the following formulation, noted by Berkson and Dowson [1]:

THEOREM. *Let $T \in B(\mathfrak{X})$ be spectral of class (Γ) with a resolution of the identity E, and define S and N by*

$$S = \int_{\sigma(T)} \lambda E(d\lambda), \qquad N = T - S.$$

Then S is spectral of class (Γ) with E as a resolution of the identity, $\sigma(S) = \sigma(T)$, and N is a quasi-nilpotent operator commuting with E.

Conversely, let S be spectral of class (Γ) with a resolution of the identity E such that $S = \int_{\sigma(T)} \lambda E(d\lambda)$, and let N be a quasi-nilpotent operator commuting *with E. Then $S + N$ is spectral of class (Γ) with E as a resolution of the identity and $\sigma(S+N) = \sigma(S)$.*

A number of results under which a unique decomposition can be obtained for prespectral operators are given in Berkson and Dowson [1]. In addition, they modify an example due to Fixman [1; p. 1035] to construct bounded operators S and A on l_∞ such that S is a scalar type operator on l_∞ of class (l_1) (and the adjoint of a scalar type operator), that $\sigma(S)$ consists of $\{1\} \cup \{(n-2)/(n-1) \mid n = 3, 4, \ldots\}$, and that $A^2 = 0$, $AS = SA$. However A does not commute with any resolution of the identity for S; moreover, the operator $S + A$ is not spectral of *any* class.

The operational calculus for spectral operators that is presented in Section 5 is due to Dunford [17]. One of the interesting developments in some of the recent work on spectral theory has been an approach "opposite" to that here; namely, one *starts* with an operational calculus for an appropriate class of functions defined on a set containing the spectrum of an operator. We shall discuss some of this work below under a separate heading.

One class of operators (that has an operational calculus) for which an interesting generalization of Theorem 4.5 holds is the collection of $T \in B(\mathfrak{X})$ such that for some $k \geqq 0$ we have $|e^{itT}| = O(|t|^k)$ for $t \in R$, $|t| \to \infty$. (Such operators are also characterized by the condition that $\sigma(T)$ is real and $|R(\lambda; T)| = O(|\mathscr{I}(\lambda)|^{-\beta})$ as $\mathscr{I}(\lambda) \to 0$ for some $\beta \geqq 1$.) S. Kantorovitz [6, 7] showed that if this condition is satisfied for an operator in a reflexive space $\mathfrak{X}$ and if $\sigma(T)$ has one-dimensional Lebesgue measure zero, then there exist a linear manifold $\mathfrak{D} \subseteq \mathfrak{X}$ (which he calls the *Jordan manifold* for T), linear transformations S and N on $\mathfrak{D}$ into $\mathfrak{D}$, and a function E defined for each Borel set in R with values which are linear transformations in $\mathfrak{D}$ such that (i) $E(R)x = x$ for all $x \in \mathfrak{D}$, and (ii) $E(\cdot)x$ is a bounded regular strongly countably additive vector measure for each $x \in \mathfrak{D}$, such that:

(a) $T \mid \mathfrak{D} = S + N$;

(b) $SN = NS$;

(c) $N^{k+1} = 0$;

(d) $p(S)x = \int_{\sigma(T)} p(t)E(dt)x,$

for each $x \in \mathfrak{D}$ and each polynomial p. Moreover, in a certain sense the manifold $\mathfrak{D}$ is maximal and unique. Further hypotheses imply that E is multiplicative. If $\mathfrak{Y}$ is the completion of $\mathfrak{D}$ under a certain norm, then the adjoint of the extension of T to $\mathfrak{Y}$ yields a spectral operator of class $\mathfrak{Y}$.

By examining the operator T in C [0, 1] defined by $Tf(x) = xf(x)$, Deal [1] arrived at some conditions which are sufficient to guarantee that there exists a homomorphism E of the Boolean algebra of finite unions of intervals in [0, 1] into an algebra of projections such that an operator has the representation

$$T = \int_0^1 \lambda E(d\lambda) + N,$$

when N is a quasi-nilpotent which may not commute with T. The conditions imply that $\sigma(T) = \sigma_r(T)$. See also Deal [2].

The following result was proved by Sine [1], using techniques similar to those in Smart [2]. Let $T \in B(\mathfrak{X})$ with $\sigma(T) = \sigma_c(T) \subseteqq [0, 1]$ with a bounded commuting strongly continuous family $E(t)$, $t \in [0, 1]$, of projections such that (i) $E(0) = 0$, $E(1) = I$, (ii) $E(t)T = TE(t)$, (iii) if $s \leqq t$ then $E(s)E(t) = E(s)$, and (iv) $\sigma(T \mid E(t))\mathfrak{X} \subseteqq [0, t]$. Then $T = S + N$, where

$$S = \int_0^1 t\, dE(t),$$

and S is well-bounded (in a sense defined below) and N is a commuting quasi-nilpotent.

Spectral operators in Hilbert space. The fact that a finite number of commuting spectral operators in a Hilbert space can be simultaneously transformed into normal operators by passing to an equivalent inner product is due to Wermer [3]. He based his argument on a result for spectral measures due to Lorch [1] and Mackey [4]. The proof given here depends on the result, due to Sz.-Nagy [7], that a bounded Abelian group of operators in a Hilbert space is equivalent to a unitary group (cf. 6.1). (See also Day [8] and Dixmier [1].)

Sums and products of spectral operators. One of the most important applications of Wermer's theorem is to prove that the sum and product of commuting spectral operators in Hilbert space are also spectral operators. Unfortunately, this result does not extend to an arbitrary B-space; indeed, Kakutani [15] gave an example of two operators whose sum is not spectral. Let S and T be the Cantor set in [0, 1] and consider the linear manifold $C(S) \otimes C(T)$ of $C(S \times T)$ consisting of all finite sums of the form

$$(*) \qquad z(s, t) = \sum_{i=1}^{n} x_i(s) y_i(t),$$

where $x_i \in C(S)$, $y_i \in C(T)$ for $i = 1, \ldots, n$. We define a new norm for the element $z \in C(S) \otimes C(T)$ by

$$(**) \qquad \|z\| = \inf \sum_{i=1}^{n} |x_i|\,|y_i|,$$

where the infinum is taken over all representations of z in the form $(*)$. Then $|z| \leqq \|z\|$, where $|z|$ denotes the norm in $C(S \times T)$. If we complete $C(S) \otimes C(T)$ relative to the norm defined in $(**)$, we obtain a B-space $C(S) \hat{\otimes} C(T)$ which is a subset of $C(S \times T)$. Kakutani showed that, for some continuous function f defined on the Cantor set, the operators A and B defined on $C(S) \hat{\otimes} C(T)$ by

$$Az(s, t) = f(s) z(s, t), \qquad Bz(s, t) = f(t) z(s, t),$$

are spectral operators, but that $A + B$ is not a spectral operator since the appropriate Boolean algebra of projection operators is not bounded.

By modifying Kakutani's construction, McCarthy [2, I] showed that the sum of two commuting scalar type operators on a separable reflexive B-space need not be a spectral operator.

In the affirmative direction Foguel [2] noted that in any space the sum (or product) of two commuting spectral operators is spectral if and only if the sum (or product) of their scalar parts is spectral. In addition, Dunford [18] and Foguel [1] proved that if $\mathfrak{X}$ is weakly complete, then the sum and product of spectral operators is spectral provided the Boolean algebra generated by their spectral measures is bounded. McCarthy [2, I] showed that if one of the operators has finite multiplicity, then the sum of spectral operators is spectral; moreover, for certain separable reflexive spaces this condition is necessary. Subsequently, McCarthy [2, II] proved that if $\mathscr{E}$ and $\mathscr{F}$ are commuting bounded Boolean algebras of projections in L_p with $1 < p < +\infty$, then the Boolean algebra generated by $\mathscr{E}$ and $\mathscr{F}$ is bounded. (See also Littman, McCarthy, and Riviere [1] where the case $0 < p \leqq 2$ is treated.) As a consequence of this remarkable result it follows that the sum and product of commuting spectral operators in L_p, $1 \leqq p < +\infty$, are spectral operators; there is no restriction on the multiplicity of the operators or the separability of the space. This work was extended to certain locally convex spaces resembling L_p by Oberai [1].

It follows from some deep work of Lindenstrauss and Pełczyński [1] that if $\mathfrak{X}$ is a complemented subspace of an L_1-space and $\mathscr{E}$ is a bounded Boolean algebra of projections, then there exists a constant M such that for every finite family $\{E_k\}$ of disjoint projections in $\mathscr{E}$ one has

$$\sum_{k=1}^{n} |E_k x| \leqq M \left| \left(\sum_{k=1}^{n} E_k \right) x \right|, \qquad x \in \mathfrak{X}.$$

Similarly, if $\mathfrak{X}^*$ is (isometric to) a complemented subspace of an L_1-space (in particular, if $\mathfrak{X} = C(K)$ for K compact Hausdorff), then the above conditions imply that

$$\left| \left(\sum_{k=1}^{n} E_k \right) x \right| \leqq M \sup_k |E_k x|, \qquad x \in \mathfrak{X}.$$

From this McCarthy and Tzafriri [1] show that if $\mathfrak{X}$ is a complemented subspace of an L_p-space, with $1 \leqq p \leqq +\infty$, then the Boolean algebra generated by two commuting bounded Boolean algebras of projections is bounded. *Thus the sum and product of commuting spectral operators in a complemented subspace* $\mathfrak{X}$ *of an* L_p*-space with* $1 \leqq p \leqq +\infty$, *are also spectral operators*; as noted above, this holds when $\mathfrak{X} = C(K)$. (This result does not contradict Kakatuni's example since $C(S) \hat{\otimes} C(T)$ is not a complemented subspace of an L_∞-space.) On the other hand it is proved that scalar type operators in a L_∞-space are of a surprisingly special form in that their spectral measures are finite. Thus if S is a scalar type operator in such a space, then $S = \sum_{i=1}^{n} \lambda_i E_i$. Moreover, every scalar type operator in $L_\infty(0, 1)$ is similar to an operator of the form $Tf = gf$ where g is a simple function; this was proved by McCarthy and Tzafriri [1]. (See also Dean [1].) In the space $C[0, 1]$, however, there are complete Boolean algebras of projections which are not finite. Even so, any scalar type operator S in $C(K)$ has a representation of the form $Sx = \Sigma_\gamma \lambda_\gamma E(\{\lambda_\gamma\})x$ where $\{\lambda_\gamma\}$ is the set of eigenvalues of S (cf. McCarthy and Tzafriri [1; p. 539]).

As we shall see below, the sum and product of two commuting generalized spectral operators is a generalized spectral operator. This was proved by Foiaş [9]; see also Colojoară and Foiaş [4] and Kantorovitz [5].

In case the sum of two commuting spectral operators is a spectral operator, it is to be expected that its resolution of the identity should be given by a "convolution integral" such as

$$E(\sigma) = \int E_1(\sigma - \lambda) E_2(d\lambda),$$

where E_1, E_2 are the resolutions of the identity of the given operators. Such a formula has been established by Foguel [1] under the hypotheses that (i) the Boolean algebra generated by E_1 and E_2 is bounded, (ii) $\mathfrak{X}$ is weakly complete, and (iii) the E-measure of the boundary of σ is zero. Further work along this line was done by Kantorovitz [1, 2, 4], who showed that the above formula holds in the strong operator topology without condition (iii). The same result was established earlier by Pedersen [1] for the case of a reflexive space, and his argument extended by Oberai [1] to the case of certain locally convex spaces. Similar results also hold for the resolution of the identity of the product of two commuting spectral operators. Kluvánek and Kováříková [1] studied conditions under which there exists a spectral measure G on the Borel subsets of $C \times C$ such that $G(\sigma \times \delta) = E_1(\sigma)E_2(\delta)$ for all Borel sets σ, δ. A necessary and sufficient condition for this is that for each $x \in \mathfrak{X}$, the set

$$\left\{\sum_{j=1}^{n} E_1(\sigma_j)E_2(\delta_j)x\right\},$$

as $\sigma_j \times \delta_j$, $j = 1, \ldots, n$, varies over all finite sets of mutually disjoint rectangles, is weakly sequentially compact. If $\mathfrak{X}$ is weakly complete, this condition is satisfied. More generally, it is always possible to find a G with values in $B(\mathfrak{X}^{**})$ such that $G(\sigma \times \delta) = E_1(\sigma)^{**}E_2(\delta)^{**}$ for all Borel sets σ, δ.

The scalar and nilpotent parts of a spectral operator. Theorem 6.7 shows that if $T \in B(\mathfrak{X})$ is a spectral operator in a Hilbert space and if the resolvent operator $\lambda \to R(\lambda; T)$ satisfies a mth order rate of growth condition, then the nilpotent part N of T is of type $m - 1$; that is, $N^m = 0$. This was proved by Dunford [18]; however, McCarthy [1, I] showed that this assertion does not hold in an arbitrary B-space. In fact, all that can be inferred in an arbitrary B-space is that $N^{m+2} = 0$; however, if $\sigma(T)$ has plane measure zero, or if $\mathfrak{X}$ is weakly complete or separable, then $N^{m+1} = 0$. Additional results, as well as examples to show that these indices of nilpotency cannot be reduced, are given in McCarthy [1]. Some of these results were extended to spectral operators in certain locally convex spaces by Simpson [1, 2]. Neubauer [2] gave a necessary and sufficient condition that the nilpotent part of a spectral operator be of order m. The condition is that there exists $M > 0$ such that if $\sigma_1, \ldots, \sigma_n$ are disjoint Borel sets in $\sigma(T)$ and $\xi_1, \ldots, \xi_n$ are complex numbers such that $|\xi_j| \leq |T| + 1$ and $\xi_j \notin \bar{\sigma}_j$, then

$$\left| \sum_{j=1}^{n} R(\xi_j; T \mid E(\sigma_j)\mathfrak{X})E(\sigma_j) \right| \leqq M/d^m,$$

where $d = \min\{\text{dist}(\xi_j, \bar{\sigma}_j) \mid j = 1, \ldots, n\}$.

The results of Section 7 relating properties of a spectral operator and its scalar part are due to Foguel [2], as are most of the results in Section 8 concerning the spectral properties of T, S, and N. Oberai [2] discussed some of these results for operators on certain locally convex spaces.

The sections 9–11 inclusive constitute a slight expansion and elaboration of results of Dunford [21] some of which are intimately related to and inspired by the work of Foguel [3]. Just recently T. R. Chow [1] has, by using the reduction theory of von Neumann, generalized Theorem 10.6 by proving that any closed spectral operator in Hilbert space can be decomposed into a direct integral of closed irreducible spectral operators. The examples in Section 12 are not intended to be complete—they do not by any means illustrate all types of phenomena which occur. Other types will be found in the exercises in Section 15 and the inquisitive reader will no doubt observe many others. He will also observe that some of the results stated for Hilbert space in this and the preceding sections can be appropriately formulated for the spaces $L_p(R^N)$, $1 \leqq p < \infty$, as well. The bibliography and history of the Cauchy problem, even the linear initial value form of it that we have illustrated, are too large for us to give here, especially since we have not attempted to discuss the theory of this problem but have only illustrated it with a few examples related to the theory of spectral operators. We mention some pertinent references. For the early literature on the Cauchy problem, Hadamard [2] contains many references in the preface and the footnotes. I. M. Gelfand and G. E. Šilov [2] have a modern treatment and a good bibliography. Finally, E. Hille [1, 5, 7] and E. Hille and R. S. Phillips [1] have an interesting discussion containing a number of types that we have not mentioned.

We are indebted to the fascinating lectures of O. Neugebauer [2], delivered at Cornell, for the few remarks we have made about Babylonian astronomy. They contain a detailed and lucid explanation of all we have mentioned about Babylonian astronomy and much more. They also have a bibliography and illuminating remarks at the end of each chapter.

The results of Section 14 are rather loosely related to those of the preceding sections. Theorem 2 may be found in Gelfand, I. M., and Raĭkov, D. A. [1] and its Corollaries 3 and 5 in Wiener, N. [4] and [5].

This Wiener theorem and P. Lévy's generalization obtained by replacing $1/\lambda$ by any function $f(\lambda)$ analytic for λ in a neighborhood of the spectrum of the element a in $\mathfrak{A}_1$ are extensions of the results to be found in IX.4.10 and IX.4.11. The elementary Lemma 6 which permits these Wiener-Lévy results to be extended to such non-closed ideals as $L_1 \cap L_q$ and $L_1 \cap C^{(k)}$ does not seem to have been previously observed in connection with results of the type given in Corollaries 8, 10, and 11.

The equation (20) was first solved (without our restriction (24)) in the homogeneous case $y = 0$ in Wiener, N., and Hopf, E. [1] with certain growth restrictions on the kernel f. In his well known monograph on the subject, E. Hopf [4] gives analytic formulas for solutions of the Wiener-Hopf equation in the non-homogeneous as well as the homogeneous cases. M. G. Kreĭn [26] has given a deep and comprehensive study of non-homogeneous equations of the Wiener-Hopf type. Oddly enough, it appears that Kreĭn was the first to obtain the factorization given in equation (30) by using the Wiener-Lévy theorem, although essentially the same factorization appears in the early work of Wiener and Hopf [1]. The paper of Kreĭn has an excellent bibliography on the subject and contains an account of its historical development.

In recent years there has been an increasing amount of interest in linear operators satisfying algebraic identities obtained by abstract Wiener-Hopf methods. Along these lines we refer the reader to the papers by Andersen [1, 2], Atkinson [6], Baxter [2], Rota [6, 7] and Spitzer [1].

Finite linear systems of equations of the Wiener-Hopf type have been extensively studied by Gohberg, I. C., and Kreĭn, M. G. [3]. In this situation the factorization corresponding to our equation (30) is more difficult but may be accomplished by means of the solution of the homogeneous Hilbert problem for matrix functions. For a solution to this problem we refer to Gohberg-Kreĭn [3] and to the excellent treatise of N. I. Muskhelishvili [1].

Spectral and Hermitian operators. If T is a scalar type operator in $\mathfrak{X}$ with resolution of the identity E defined on the Borel sets $\mathscr{B}$ of C and if we define A and B by

$$A = \int_C \mathscr{R}(\lambda)E(d\lambda), \qquad B = \int_C \mathscr{I}(\lambda)E(d\lambda),$$

then A and B are commuting scalar type operators whose resolutions of the identity can be found by a change of measure, and $T = A + iB$.

Alternatively (cf. Kantorovitz [4]), we can define E_R and E_J on the Borel sets of R by

$$E_R(\sigma) = E(\delta \times R), \qquad E_J(\sigma) = E(R \times \delta)$$

and obtain

$$A = \int_R \lambda E_R(d\lambda), \qquad B = \int_R \lambda E_J(d\lambda).$$

Thus a scalar type (and hence also a spectral) operator can be broken into "real" and "imaginary parts." Analogously, a scalar type (and hence a spectral) operator has a "polar decomposition" similar to that given in Section XII.7 for a normal operator. (See Foguel [2].) This polar decomposition is also to be found in Kovářiková [1] who was concerned with expressing the spectral measure of the given operator as a product of the spectral measures of the factors.

In a sense, a scalar type operator with real spectrum can be thought of as a generalization of a Hermitian or self adjoint operator in a Hilbert space. However, there is another relevant notion which is of interest and importance. Lumer [1] has observed that in any complex B-space $\mathfrak{X}$ one can introduce (at least one) *semi-inner product* consistent with the original norm. (By a semi-inner product we mean a mapping $[\cdot, \cdot]$ of $\mathfrak{X} \times \mathfrak{X}$ into C such that:

(i) for each $y \in \mathfrak{X}$, the map $x \to [x, y]$ is linear on $\mathfrak{X}$;
(ii) for each $x \in \mathfrak{X}$, $[x, x] \geqq 0$;
(iii) $[x, x] = 0$ if and only if $x = 0$;
(iv) $|[x, y]|^2 \leqq [x, x][y, y]$ for $x, y \in \mathfrak{X}$.

The semi-inner product induces a norm $\|x\| = [x, x]^{1/2}$ and we say that the semi-inner product is consistent with the original norm in case the norms $\|\cdot\|$ and $|\cdot|$ are equivalent.) If $T \in B(\mathfrak{X})$ we can define its *numerical range* with respect to the semi-inner product to be

$$W(T) = \{[Tx, x] \mid [x, x] = 1, x \in \mathfrak{X}\}.$$

Although different semi-inner products yield different numerical ranges for T, they all have the same convex hull. Hence if $T \in B(\mathfrak{X})$ has a real numerical range with respect to one consistent semi-inner product, it has a real numerical range with respect to all. Lumer called such operators

Hermitian operators on $\mathfrak{X}$; he further showed that this notion coincides with one previously introduced by Vidav [2], namely, that

$$|I - itT| = 1 + o(t) \quad \text{as} \quad t \to 0 \qquad (t \in R).$$

By now there is a substantial literature on semi-inner product spaces, numerical ranges, and Hermitian operators. See Bonsall and Duncan [1].

Berkson [2] showed that if E is a bounded spectral measure and if one defines

$$\| x \| = \sup\{\text{var } x^*E(\cdot)x \,|\, |x^*| = 1\},$$

then $\|\cdot\|$ is a norm equivalent to $|\cdot|$ and relative to which all the operators $E(\delta)$ become Hermitian. It follows from this and Berkson [5; p.3] that if f is continuous on the compact support K of E, then

$$\| \int f(s)E(ds) \| = \sup_{s \in K} |f(s)| .$$

Berkson [2] further showed that if T is a scalar type operator then there exist unique operators A, B such that (i) $T = A + iB$, (ii) $AB = BA$, and (iii) $A^m B^n$ is Hermitian for $m, n = 0, 1, 2, \ldots$ for some equivalent norm. Conversely, if $\mathfrak{X}$ is reflexive (or even weakly complete), then these three conditions characterize scalar type operators.

Lumer [2] showed that if T_1 and T_2 are commuting scalar type operators, then there is an equivalent renorming of $\mathfrak{X}$ under which $T_j = A_j + iB_j$ where A_j, B_j are commuting Hermitian operators. Hence $T_1 + T_2 = A + iB$ where A, B are commuting Hermitian operators. Among other things, he showed that if $\mathcal{E}$ is a Boolean algebra of projection operators in $B(\mathfrak{X})$, then the following assertions are equivalent: (a) $\mathcal{E}$ is uniformly bounded; (b) the uniform closure A of the real linear span of $\mathcal{E}$ consists of operators which are Hermitian under some equivalent norm; (c) this closure A consists of operators with spectral adjoints.

Panchapagesan [1] considered the relation between the polar decomposition of scalar type and Hermitian operators and proved some results analogous for those of Berkson cited above.

Recently, Stampfli [11] has introduced an interesting class of operators. If $x \in \mathfrak{X}$, by the Hahn-Banach Theorem there exists an $x^* \in \mathfrak{X}^*$ such that $|x^*| = |x|$ and $x^*x = |x|^2$. Let φ: $\mathfrak{X} \to \mathfrak{X}^*$ be any mapping such that $\varphi(x) = x^*$ and $\varphi(\lambda x) = \bar{\lambda}\varphi(x)$. In general φ is not unique, linear, or continuous, but it induces a semi-inner product $[x, y] = (\varphi(y))x$ on $\mathfrak{X}$. Given such a mapping φ, we say that $A \in B(\mathfrak{X})$ is *adjoint Abelian* if $A^*\varphi = \varphi A$ (equivalently, if $(Ax)^* = A^*x^*$ for all $x \in \mathfrak{X}$). If A is adjoint Abelian, then $\sigma(A)$ is real and A^{2n} is Hermitian for $n = 1, 2, \ldots$; however; A need not be Hermitian, and $cI + A$ need not be adjoint Abelian. It is proved that if A is adjoint Abelian, then (i) $(A^2)^*$ is a scalar type operator of class $(\mathfrak{X})$, and (ii) if $\mathfrak{X}$ is weakly complete, then A^2 is a scalar type operator. Moreover, if A is adjoint Abelian and $\mathfrak{X}$ is weakly complete, then A is scalar type if and only if either (a) A is invertible, (b) 0 is an isolated point of $\sigma(A)$, or (c) $\sigma(A)$ is non-negative (or non-positive). It is not known, however, whether every adjoint Abelian operator in a weakly complete space is scalar.

Spectral operators and unconditional convergence. If T is a scalar type operator whose eigenvectors are a fundamental set, then the expansion in terms of these eigenvectors is unconditionally convergent. Conversely, Smart [3] showed that if $T \in B(\mathfrak{X})$ is such that its eigenvectors $\{x_k \mid k = 1, 2, \ldots\}$ have the property that no x_k is in the closed subspace spanned by $\{x_k \mid k \neq n\}$, then there exist $\{f_k \mid k = 1, 2, \ldots\}$ in $\mathfrak{X}^*$ such that $f_j(x_k) = \delta_{jk}$. If $\mathfrak{Y}$ is the set of all $x \in \mathfrak{X}$ such that $\Sigma f_k(x)x_k$ is unconditionally convergent to x, then $\mathfrak{Y}$ is a B-space under the norm:

$$\|x\| = \sup\{|\sum_{k \in I} f_k(x)x_k| \,\big|\, I \text{ finite}\}.$$

Moreover, $\mathfrak{Y}$ is invariant under T, and $T \mid \mathfrak{Y}$ is a bounded scalar type operator on $\mathfrak{Y}$ with resolution of the identity given by

$$E(\sigma)x = \sum_{\lambda_k \in \sigma} f_k(x)x_k,$$

where $\{\lambda_k\}$ denotes the set of eigenvalues of T.

Roots, logarithms, and semi-groups of spectral operators. If S is an invertible scalar type (respectively, spectral) operator and T satisfies $T^n = S$ for some positive integer n, then Stampfli [1] showed that T is a scalar type (respectively, spectral) operator. Results dealing with the existence (and non-existence) of roots and logarithms of operators have been established by Apostol [5, 6, 7, 8, 9, 10, 11], Colojoară [4], Deckard and Pearcy [2], Halmos and Lumer [1], Halmos, Lumer and Schäffer [1], Hille [6], Krabbe [1], Kurepa [1, 2, 3, 4, 6], Lanier [1], Lumer [3], Putnam [19, 21, 22], Schäffer [3], and Stampfli [1], although some of these authors dealt with operators that are more (or less) general than spectral operators in B-spaces. To cite one result, Apostol [6] proved that if f is analytic on a neighborhood of $\sigma(T)$, if $f'(\lambda) \neq 0$ for $\lambda \in \sigma(T)$, and if $f(T)$ is a spectral (respectively, scalar type) operator, then T is spectral (respectively, scalar type).

A number of papers dealing with groups or semi-groups of spectral operators have been published. See Berkson [5], Foiaş [3, 8], Ionescu Tulcea [2], Lanier [1], Lumer [2, 3], McCarthy and Stampfli [1] and Panchapagesan [1] for results along this line.

The following papers are less closely related to spectral operators, but consider similar problems: Adamjan and Arov [2], Foiaş and Gehér [1], Gustafson [1], Hasegawa [1], Heyn [1], Ito [1], Kleinecke [4], Komatsu [1], Maltese [1, 2], Miyadera [2], Mlak [5], Singbal-Vedak [1], Yosida [13].

Spectral measures, locally convex spaces and order. Spectral measures have been studied in connection with (partially) ordered spaces by Schaefer [7, 10, 11], Schaefer and Walsh [1], and Walsh [1, 2, 3]. We shall give a condensed description of some of this work, but, in order to be brief, it is convenient to specialize some of their results. Much of this work is arranged so as to treat bounded and unbounded operators simultaneously; in this sense it presents a unification of Chapters XV and XVIII. On the other hand, some of their results indicate that spectral measures and spectral operators in various non-normable spaces that are of interest in analysis are often of a more special character than might be expected.

Let X be a compact Hausdorff space and let A be a locally convex Hausdorff space which is also an algebra over C with identity e and in which multiplication is separately continuous. We shall also assume that A is "semi-complete" in the sense that Cauchy sequences are convergent in A (note that A is not necessarily metrizable; hence the qualifier "semi"). By a *spectral measure* μ on X to A we mean a mapping $\delta \to \mu(\delta)$ from the Baire field $\mathscr{B}_0$ (the σ-field generated by the compact G_δ sets in X) of X into A which is weakly countably additive and satisfies $\mu(X) = e$, $\mu(\delta_1 \cap \delta_2) = \mu(\delta_1)\mu(\delta_2)$ for δ_1, δ_2 in $\mathscr{B}_0$. (Under suitable hypotheses one can associate such spectral measures with continuous homomorphisms of $C(X)$ into A, thus affording an alternative approach—see Schaefer [11; p. 470].)

It is proved that if μ is a spectral measure, then the intersection K of all convex cones containing $\{\mu(\delta) \mid \delta \in \mathscr{B}_0\}$ is a "weakly normal" cone and defines a (partial) order in A with respect to which μ is a positive measure. *Thus every spectral measure on X to A is positive with respect to a suitable order on A.*

Let μ (respectively, ν) be a spectral measure on X (respectively, Y) to A. *Then in order for there to exist a necessarily unique spectral measure* λ *on the Baire sets in* $X \times Y$ *to* A *such that* $\lambda(\delta \times \sigma) = \mu(\delta)\nu(\sigma)$ *for all* δ, σ, *it is necessary and sufficient that the values of* μ *and* ν *are contained in a commutative subalgebra of* A *and that there exists a partial order of* A *for which both* μ *and* ν *are positive.* This result throws light on when the sum and product of scalar type operators are scalar type operators.

We shall say that an element $a \in A$ is *scalar* (although Schaefer [10; p. 143] used the word "spectral") if there exists a compact space X, a spectral measure μ on X to A and a bounded Baire measurable function $f\colon X \to C$ such that $a = \int f(x)\mu(dx)$. It follows that there exists a spectral measure ν_a defined on a compact subset of C to A such that $a = \int z\nu_a(dz)$; moreover, the measure ν_a is unique and its support is the spectrum of a in A (that is, the complement of the largest open set in C in which the map $\lambda \to (\lambda e - a)^{-1}$ is locally holomorphic in A). The elements in the range of ν_a are, in a natural sense, "functions of a" and the mapping $g \to g(a) = \nu_a(g)$ is an operational calculus for a.

We say that a scalar element $a \in A$ is *real* (respectively, *positive*) in case its spectrum $\sigma(a)$ is contained in R (respectively, $[0, +\infty)$). It can be proved that an $a \in A$ is a real scalar element if and only if there exists an ordering of A with respect to which a is in the real linear hull of the order interval $[0, e]$, and that $a \in A$ is positive if and only if there exists an ordering of A and a non-negative number γ such that $0 \leqq a \leqq \gamma e$. Algebras in which every element is a scalar element are considered in Schaefer [10, 11] and Walsh [1]. Basically they turn out to be commutative and equivalent to an algebra of the form $C(\Omega)$. Results of this character may be compared with those in Section XVII.2.

These (and other) results are applied to study the case where A is the algebra of all continuous linear mappings on a locally convex space E with the strong operator topology; here we assume that E is semi-complete and barreled. In this setting, scalar operators can be introduced; in fact, it is possible to study unbounded operators. We state only one result in this direction (see Schaefer [10; p. 169]). Let T be a closed linear operator with dense domain in E; then T is a "real scalar operator" if and only if $(I + T^2)^{-1}$ exists and there is an ordering in A for which $0 \leqq (I + T^2)^{-1} \leqq I$ and $-I \leqq T(I + T^2)^{-1} \leqq I$.

In Schaefer [11], similar results are obtained in a more "algebraic" setting; here, spectral measures are considered to be continuous homomorphisms of the algebra $C(X)$ into A.

Schaefer and Walsh [1] gave some examples of operators which are scalar (in the sense defined above) in certain non-normable locally convex spaces. For instance: (a) Let $f \in D(T^n)$ be a distribution on the n-dimensional torus T^n; then the convolution mappings $u \to f * u$ on the spaces $C_0^\infty(T^n)$ or $D(T^n)$ are scalar. Thus differential operators with constant coefficients are scalar on $C_0^\infty(T^n)$ and $D(T^n)$. (b) Let p and q be real-valued C_0^∞ functions on the one-dimensional torus T^1. Then the mappings $u \to (d/dt)(pdu/dt) + qu$ on $C_0^\infty(T^1)$ and $D(T^1)$ are scalar. (c) The mappings $u \to -d^2u/dt^2 + t^2u$ on the spaces Φ (respectively, T^1) of rapidly decreasing C^∞ functions (respectively, tempered distributions) on the real line are scalar. Thus the Fourier transform $u(t) \to \int_R e^{-2\pi its}u(s)\,ds$ in Φ and its adjoint in T^1 are scalar. The proofs that these operators are scalar follow from the fact that the eigenfunctions of certain boundary value problems form absolute bases in $C_0^\infty(T^n)$ or Φ and from results in Schaefer [10].

Walsh [2] considers weakly countably additive spectral measures μ defined on a σ-field Σ of subsets of a set X whose values are an equicontinuous family of continuous linear operators on a locally convex space E. If E is complete it is seen (Walsh [2; p. 308]) that each such spectral measure is the projective limit of a generalized sequence of spectral measures with values in $B(\mathfrak{X}_\alpha)$ where $\mathfrak{X}_\alpha$ is a B-space.

Let μ be such a measure, let E be boundedly complete and, for each $x \in E$, let $\mathfrak{M}(x)$ (respectively, $\mathfrak{M}_R(x)$) be the closed complex (respectively, real) linear span of $\{\mu(\delta)x \mid \delta \in \Sigma\}$; these spaces are called the *cyclic* (respectively, *real cyclic*) *subspaces generated by* x. It is proved that the spectral measure induces on $\mathfrak{M}(x)$ and $\mathfrak{M}_R(x)$ an order structure with strong properties. In particular, if $\mathfrak{M}(x)$ is complete, then it is isomorphic as a vector space with a space of equivalence classes of integrable functions on X modulo certain "null functions;" similarly, $\mathfrak{M}_R(x)$ is isomorphic both as a vector space and a vector lattice with a space of equivalence classes of real-valued integrable functions. This enables him to extend certain theorems of Bade [4] to suitable metrizable locally convex spaces. For instance, if E is a complete metrizable locally convex space then there exists a *countable* set of continuous linear functionals $\{x_n^*\}$ such that if $x_n^*\mu(\delta)x = 0$ for all n, then $\mu(\delta)x = 0$ (and hence the restriction of $\mu(\delta)$ to $\mathfrak{M}(x)$ is 0). In the case where E is a B-space this countable set can be replaced by a *single* functional, but in general this cannot be done.

One of the most surprising results due to Walsh [2] is the result that if μ is an equicontinuous Borel spectral measure into the space of continuous operators in a space E in which closed bounded sets are compact (for example, a Montel space), then μ is purely atomic. In Walsh [3] this was extended to the case where E has the property that weakly compact subsets are compact (for example, l_1). Thus the scalar operators in such spaces are limits in the strong operator topology of finite dimensional operators. Since some interesting non-normable locally convex spaces are Montel spaces, this means that scalar operators in these spaces are of a very restricted type.

Interpolation of spectral operators. Questions concerning the interpolation of spectral operators and their resolutions of the identity were considered by Krabbe [13] and Oberai [3]. Krabbe's paper considers a rather general setting; Oberai's refers to the case of the L_p-spaces and is easier to describe. Let (S, Σ, μ) be a finite measure space and denote $L_p = L_p(S, \Sigma, \mu)$. Let $1 \leqq r \leqq s \leqq +\infty$ and let T be a spectral operator on L_p with resolution of the identity E. If T leaves L_s invariant and $T_s = T \mid L_s$ is spectral, then its resolution of the identity is $E \mid L_s$. Moreover, if these hypotheses are satisfied and $r \leqq p \leqq s$, then T leaves L_p invariant, and $T_p = T \mid L_p$ is a spectral operator with resolution of the identity $E_p = E \mid L_p$. Similarly, if S generates a continuous operator on both L_r and L_s, and if S is spectral in one of these spaces, then a necessary and sufficient condition that it be spectral in the other spaces is that the restriction (or extension) of the spectral measure be bounded in the second space.

T-measures. The notion of a scalar type spectral operator was generalized in an interesting way by E. Bishop [1]. Although he considered closed operators, we shall limit our attention to operators $T \in B(\mathfrak{X})$, where $\mathfrak{X}$ is a reflexive B-space.

A vector valued measure m on the Borel sets $\mathscr{B}$ of the complex plane C to $\mathfrak{X}$ is said to be a *T-measure* if

$$Tm(\delta) = \int_\delta \lambda m(d\lambda), \qquad \delta \in \mathscr{B}.$$

If m is a T-measure and $x = m(C)$, we say that *x has a T-measure m*. In a sense, a T-measure is a generalization of an eigenvector of T. It is easy to see that if T is a scalar type spectral operator with resolution of the identity E, then $E(\cdot)x$ is a T-measure of x. However, not every operator has a non-trivial T-measure; consider the right shift operator in l_2, or a quasi-nilpotent operator with empty point spectrum.

It is proved that if m is a T-measure, then m vanishes outside $\sigma(T)$. Moreover, *T is a scalar type spectral operator if and only if every vector has a unique T-measure.*

A functional calculus is defined for operators which have sufficiently many T-measures; this calculus is quite analogous to that for a self adjoint operator; however, it is possible for $f(T)$ to be unbounded even when f and T are bounded.

In order to handle quasi-nilpotent operators, Bishop introduces the notion of a *weak T-measure.* He shows that *an operator T in a reflexive B-space $\mathfrak{X}$ is a spectral operator if and only if each $x \in \mathfrak{X}$ has a weak T-measure and each $x^* \in \mathfrak{X}^*$ has a weak T^*-measure.*

The main results in Bishop [1] are extended to a locally convex space which is barreled and quasi-complete by Maeda [1, 2].

Bishop's "duality theories." In a very incisive paper, Errett Bishop [2] established some interesting results relating certain manifolds pertaining to an operator T in a reflexive complex B-space $\mathfrak{X}$ with those manifolds pertaining to its adjoint T^*. He referred to these as "duality theories," but they very clearly belong to spectral theory.

Let F be a closed subset of the complex plane C. The *strong spectral manifold* $\mathfrak{M}(F, T)$ is defined to be the closure of the set of all x in $\mathfrak{X}$ which have the property that there exists an analytic function f on the complement F' to $\mathfrak{X}$ such that $(\lambda I - T)f(\lambda) = x$ for all $\lambda \in F'$. The *weak spectral manifold* $\mathfrak{N}(F, T)$ is defined to be the set of all vectors $x \in \mathfrak{X}$ which have the property that for every $\varepsilon > 0$ there exists an analytic function f on F' to $\mathfrak{X}$ such that $|(\lambda I - T)f(\lambda) - x| < \varepsilon$ for all $\lambda \in F'$.

Evidently these spectral manifolds are closed and $\mathfrak{M}(F, T) \subseteq \mathfrak{N}(F, T)$; examples are given to show that this inclusion can be proper.

Bishop introduces four types of duality theory for an operator T in a reflexive space $\mathfrak{X}$. (i) The operator T is said to admit a *duality theory of type* 1 if $\mathfrak{M}(F_1, T)^\perp \subseteq \mathfrak{M}(F_2, T^*)$, whenever F_1 and F_2 are disjoint compact sets, and if $\mathfrak{M}(\bar{G}_1, T)^\perp \subseteq \mathfrak{M}(\bar{G}_2, T^*)$ whenever G_1 and G_2 are open sets which cover C. (ii) The operator T admits a *duality theory of type* 2 if $\mathfrak{M}(\bar{G}_1, T), \ldots, \mathfrak{M}(\bar{G}_n, T)$ span $\mathfrak{X}$ whenever $G_1, \ldots, G_n$ are open sets which cover C. (iii) The operator T admits a *duality theory of type* 3 if for arbitrary open sets $G_1, \ldots, G_n$ which cover the complex plane, there exist closed linear subspaces $\mathfrak{M}_1, \ldots, \mathfrak{M}_n$ which span $\mathfrak{X}$, are invariant under T, and such that $\sigma(T \mid \mathfrak{M}_i) \subseteq \bar{G}_i$, $i = 1, \ldots, n$. (iv) The operator T admits a *duality theory of type* 4 if

$$\mathfrak{M}(F_1, T)^\perp \supseteq \mathfrak{N}(F_2, T^*), \qquad \mathfrak{N}(F_1, T)^\perp \supseteq \mathfrak{M}(F_2, T^*),$$
$$\mathfrak{M}(\bar{G}_1, T)^\perp \subseteq \mathfrak{N}(\bar{G}_2, T^*), \qquad \mathfrak{N}(\bar{G}_1, T)^\perp \subseteq \mathfrak{M}(\bar{G}_2, T^*),$$

whenever F_1 and F_2 are disjoint compact sets, and G_1 and G_2 are open sets which cover C. (It should be noted that T and T^* play symmetric roles in (i) and (iv).)

An example is given to show that not every operator admits a duality theory of type 1; however, if T satisfies the condition:

$$(\alpha) \qquad \mathfrak{N}(F_1, T) \subseteq \mathfrak{M}(F_2, T), \quad \text{if} \quad F_1 \subseteq \text{the interior of} \quad F_2,$$

then T does admit such a duality theory.

It is a distinctly non-trivial fact that every bounded linear operator T in a reflexive B-space $\mathfrak{X}$ admits a duality theory of type 4. The proof of this theorem is based on an analysis of certain B-spaces of vector valued analytic functions.

We say that T satisfies condition (β) if for every open set U and every sequence $\{f_n\}$ of analytic functions from U to $\mathfrak{X}$ and $x \in \mathfrak{X}$ such that $(\lambda I - T)f_n(\lambda) \to x$ uniformly in compact subsets of U, it follows that $\{f_n\}$ is uniformly bounded on compact subsets of U.

It is proved that if T satisfies condition (β) and F is closed in C, then (1) $\mathfrak{M}(F, T) = \mathfrak{N}(F, T)$, (2) for $x \in \mathfrak{M}(F, T)$ there exists an analytic function $f: F' \to \mathfrak{X}$ such that $(\lambda I - T)f(\lambda) = x$ for all $\lambda \in F'$, and (3) T satisfies condition (α), and (4) T admits a duality theory of type 1.

It is proved that if T^* satisfies condition (β), then T admits a duality theory of type 2. Moreover, if T and T^* satisfy condition (β), then T admits a duality theory of type 3.

In particular, let T be an operator in a reflexive B-space $\mathfrak{X}$ with the property that for every open set U and every sequence $\{f_n\}$ of complex valued analytic functions on U such that $|f_n(\lambda)| \leqq |R(\lambda; T)|$ for all $\lambda \in U \cap \rho(T)$, then $\{f_n\}$ is uniformly bounded on compact subsets of U. It then follows that T and T^* satisfy condition (β) and hence admit duality theories of types 1, 2, 3, and 4.

Applying a theorem due to Wolf [3], Bishop concludes that if $\mathfrak{X}$ is reflexive, $\sigma(T)$ is real, and

$$\int \log^+ \log^+ \sup\{|R(\lambda;\, T)| \mid \mathscr{I}(\lambda) = y\}\, dy < +\infty,$$

then T admits duality theories of types 1, 2, 3, and 4.

Decomposable operators. Foiaş [12] introduced a large and important class of operators which he called "decomposable." This class appears to contain all operators that have a reasonably rich spectral theory; indeed, it will be seen that decomposable operators admit a duality theory of type 3 in the sense of E. Bishop. We shall give a brief description of the theory of decomposable operators; for more details the reader should consult Foiaş [12] and Colojoară and Foiaş [1, 4]. See also Apostol [4, 5, 11, 15], Colojoară and Foiaş [2, 5], Kariotis [1], and Vasilescu [3, 4, 5, 6, 7, 8] for other results concerning decomposable operators.

Let $\mathfrak{H}$ be a Hilbert space with an orthonormal basis $\{x_n \mid n = 0, \pm 1, \pm 2, \ldots\}$, let T be the "right shift" operator $Tx_n = x_{n+1}$ $(n = 0, \pm 1, \pm 2, \ldots)$, and let $\mathfrak{Y} = \overline{\text{sp}}\{x_n \mid n \geqq 0\}$. Then $\mathfrak{Y}$ is invariant under T, T is unitary and $\sigma(T) = \{\lambda \in C \mid |\lambda| = 1\}$, while $\sigma(T \mid \mathfrak{Y}) = \{\lambda \in C \mid |\lambda| \leqq 1\}$. Thus the restriction of an operator T to an arbitrary invariant closed subspace may have spectrum larger than $\sigma(T)$.

Foiaş [12] defined a closed linear subspace $\mathfrak{Y}$ of a B-space $\mathfrak{X}$ to be a *spectral maximal subspace* of $T \in B(\mathfrak{X})$ if (i) $\mathfrak{Y}$ is invariant under T, and (ii) if $\mathfrak{Z}$ is a closed linear subspace of $\mathfrak{X}$ which is invariant under T and $\sigma(T \mid \mathfrak{Z}) \subseteq \sigma(T \mid \mathfrak{Y})$, then $\mathfrak{Z} \subseteq \mathfrak{Y}$. (See also Ljubič and Macaev [2; §4, IV].)

It can be shown that if $\mathfrak{Y}$ is a spectral maximal subspace of T, then $\sigma(T \mid \mathfrak{Y}) \subseteq \sigma(T)$; moreover, $\mathfrak{Y}$ is invariant under any $S \in B(\mathfrak{X})$ which commutes with T. More generally, if $\mathfrak{Y}_1$ and $\mathfrak{Y}_2$ are spectral maximal subspaces of T, then $\mathfrak{Y}_1 \subseteq \mathfrak{Y}_2$ if and only if $\sigma(T \mid \mathfrak{Y}_1) \subseteq \sigma(T \mid \mathfrak{Y}_2)$.

Ljubič and Macaev [2] give an example of an operator T in Hilbert space with $\sigma(T) = [0, 1]$ and such that the restriction of T to any non-trivial invariant closed linear subspace $\mathfrak{Y}$ has the property that $\sigma(T \mid \mathfrak{Y}) = [0, 1]$. To contrast with this, Colojoară and Foiaş [1; p. 526] [4; p. 23] showed that if T has the single valued extension property and if

$$\mathfrak{X}_T(F) = \{x \in \mathfrak{X} \mid \sigma_T(x) \subseteq F\}$$

is closed, then $\mathfrak{X}_T(F)$ is a spectral maximal subspace of T and

$$(*) \qquad \sigma(T \mid \mathfrak{X}_T(F)) \subseteq \sigma(T) \cap F.$$

The conclusion (*) was obtained earlier by Ljubič and Macaev [2] and Bartle [6] for operators with real spectrum and whose resolvent satisfies a growth condition and by Foiaş [9] for "generalized scalar operators."

It is not difficult to see that if T is a spectral operator with resolution of the identity E, then for F closed in $\sigma(T)$, the subspace $E(F)\mathfrak{X}$ is a spectral maximal subspace for T. Similarly if σ is a spectral set (in the sense of VII.3.17) and if E_σ is the corresponding projection operator, then $E_\sigma\mathfrak{X}$ is a spectral maximal subspace of T. Hence both spectral and compact operators have "many" spectral maximal subspaces. It can be seen that they are decomposable in the following sense.

DEFINITION. An operator $T \in B(\mathfrak{X})$ is said to be *decomposable* if for every finite open cover $G_1, \ldots, G_n$ of $\sigma(T)$ there is a family $\mathfrak{Y}_1, \ldots, \mathfrak{Y}_n$ of spectral maximal subspaces of T such that (i) $\sigma(T \mid \mathfrak{Y}_j) \subseteq G_j, j = 1, \ldots, n$, and (ii) $\mathfrak{X} = \sum_{j=1}^n \mathfrak{Y}_j$.

(It is stressed that the subspaces $\mathfrak{Y}_j$ are not uniquely determined, nor do we assume that $\mathfrak{Y}_j \cap \mathfrak{Y}_k = \{0\}$ for $j \neq k$; hence the sum in (ii) is not necessarily direct. All that is intended is that every $x \in \mathfrak{X}$ can be represented in the form $x = \sum_{j=1}^n y_j$, $y_j \in \mathfrak{Y}_j$.)

Foiaş [12] showed that every decomposable operator T has the single valued extension property; hence it makes sense to define $\mathfrak{X}_T(F)$ where F is a closed set in C. Moreover $\mathfrak{X}_T(F)$ is a closed linear subspace and even a spectral maximal subspace of T, whence it follows that relation (*) holds. Conversely, if $\mathfrak{Y}$ is a spectral maximal subspace of the decomposable operator T, then $\mathfrak{Y} = \mathfrak{X}_T(\sigma(T \mid \mathfrak{Y}))$. Thus a closed linear subspace of $\mathfrak{X}$ is a spectral maximal subspace of a decomposable operator T if and only if it is of the form $\mathfrak{X}_T(F)$ for some closed set F.

The perturbation of decomposable operators by quasi-nilpotent operators and related questions was considered in Colojoară and Foiaş [1, 4]. Recall that T, $U \in B(\mathfrak{X})$ are said to be *quasi-nilpotent equivalent* in case

$$\lim_{n\to\infty} |(T - U)^{[n]}|^{1/n} = \lim_{n\to\infty} |(U - T)^{[n]}|^{1/n} = 0;$$

here we have used the notation:

$$(T - U)^{[n]} = \sum_{k=0}^{n} (-1)^{n-k}\binom{n}{k} T^k U^{n-k}.$$

(If $TU = UT$, then T and U are quasi-nilpotent equivalent if and only if $T - U$ is quasi-nilpotent.) It is proved that if T is decomposable and T and U are quasi-nilpotent equivalent, then U is decomposable. Moreover, if T and U are decomposable, then $\mathfrak{X}_T(F) = \mathfrak{X}_U(F)$ for all closed sets F if and only if T and U are quasi-nilpotent equivalent.

If T is a spectral operator and T and U are quasi-nilpotent equivalent, then U is a spectral operator. Moreover, if T and U are spectral, then they are quasi-nilpotent equivalent if and only if they have the *same* resolution of the identity, and if and only if the single condition: $\lim_n |(T - U)^{[n]}|^{1/n} = 0$ is satisfied.

If T is decomposable and f is analytic on an open set containing $\sigma(T)$, then $f(T)$ is decomposable (see Colojoară and Foiaş [2, 4]). Conversely, if f is analytic on a neighborhood of $\sigma(U)$, if f is one-to-one on $\sigma(U)$, and if $f(U)$ is decomposable, then U is decomposable. Similarly, Apostol [5, 11] has shown that if f is analytic on a neighborhood of $\sigma(U)$, if the zeros of f' have no accumulation point in $\sigma(U)$, and if $f(U)$ is decomposable, then U is decomposable.

In Apostol [4, 11] the restrictions of a decomposable operator to subspaces, and the operators induced by a decomposable operator in quotient spaces are considered in detail.

Apostol [11] introduced the notion of a "spectral capacity," which is seen to be intimately related to decomposable operators. A *spectral capacity* $\mathscr{E}$ is a mapping of the collection $\mathscr{F}$ of all closed subsets of C into the collection $\mathscr{S}(\mathfrak{X})$ of closed linear subspaces of $\mathfrak{X}$ which satisfies the following conditions:

(i) $\mathscr{E}(\phi) = \{0\}, \qquad \mathscr{E}(C) = \mathfrak{X};$

(ii) $\displaystyle\bigcap_{n=1}^{\infty} \mathscr{E}(F_n) = \mathscr{E}\left(\bigcap_{n=1}^{\infty} F_n\right), \qquad F_n \in \mathscr{F};$

(iii) if $\{G_1, \ldots, G_n\}$ is an open covering of C, then

$$\mathfrak{X} = \sum_{j=1}^{n} \mathscr{E}(\bar{G}_j).$$

We say that an operator $T \in B(\mathfrak{X})$ has a spectral capacity $\mathscr{E}$, if for every $F \in \mathscr{F}$:

(iv) $T\mathscr{E}(F) \subseteq \mathscr{E}(F);$

(v) $\sigma(T \mid \mathscr{E}(F)) \subseteq F.$

It should be noted that if E is a spectral measure on the Borel sets $\mathscr{B}$ of C, then the function $\mathscr{E}$ defined for $F \in \mathscr{F}$ by $\mathscr{E}(F) = E(F)\mathfrak{X}$ yields a spectral capacity.

It was proved by Apostol [11] that if $T \in B(\mathfrak{X})$ is a decomposable operator, then the mapping $\mathscr{E} : \mathscr{F} \to \mathscr{S}(\mathfrak{X})$ defined by

$$(**) \qquad \mathscr{E}(F) = \mathfrak{X}_T(F), \qquad F \in \mathscr{F},$$

is a spectral capacity for T. Conversely, Foiaş [17] showed that if $T \in B(\mathfrak{X})$ has a spectral capacity $\mathscr{E}$, then T is decomposable and (**) holds. More generally, he showed that if T is an operator for which there exists a function $\mathscr{E}$ satisfying conditions (i), (iii), (iv), (v) and

$$(\text{ii}') \qquad \mathscr{E}(F_1) \cap \mathscr{E}(F_2) = \mathscr{E}(F_1 \cap F_2), \qquad F_j \in \mathscr{F};$$

then T is a decomposable operator and

$$\mathfrak{X}_T(F) = \bigcap\{\mathscr{E}(\bar{G}) \,|\, G \text{ open}, \ F \subseteq G\}.$$

Operational calculi and spectral theory. The existence of the operational calculus for analytic functions of an arbitrary operator in $B(\mathfrak{X})$, as presented in Chapter VII, and of the operational calculus for continuous (or even essentially bounded Borel) functions of a self adjoint operator in Hilbert space, as presented in Chapter X, have been known for a long time. Similarly, the operational calculus of bounded Borel functions of a spectral operator was developed in Dunford [17]; in a sense spectral operators were introduced as a class of operators in $B(\mathfrak{X})$ for which a rich operational calculus was available. Thus the operational calculus was considered to be a consequence of the spectral properties of the operator.

In recent years a substantial amount of work has been done by considering the spectral theory as a consequence of a suitable operational calculus. To some extent, the work of Lorch [7], for an operator T in a reflexive B-space satisfying $|T^n| \leq K$ for $n = 0, \pm 1, \pm 2, \ldots$, was based on the possibility of defining $f(T) = \sum_{n=-\infty}^{+\infty} c_n T^n$, where $f(\theta) = \sum_{n=-\infty}^{+\infty} c_n e^{in\theta}$ is an absolutely convergent Fourier series. Similarly, a scalar type spectral operator T can be defined to be an operator for which there exists a continuous homomorphism of the B-algebra $B(C, \mathscr{B})$ of bounded Borel functions on the complex plane into $B(\mathfrak{X})$ which maps the function $f_0(\lambda) \equiv 1$ into I and the function $f_1(\lambda) \equiv \lambda$ into T. Once this point of view has been taken it is very natural to replace the B-algebra $B(C, \mathscr{B})$ by smaller algebras of functions. Thus we can attempt to classify operators $T \in B(\mathfrak{X})$ by the algebras $\mathfrak{A}$ of functions, defined on a suitable subset of

C, for which there exists a continuous homomorphism of $\mathfrak{A}$ into $B(\mathfrak{X})$ which maps f_0 into I and f_1 into T. Such an approach was rather explicitly suggested in the papers of Wolf [4, 5]; for example, in [4] the algebra was C^n on the unit circle or the real line, and in [5] more general algebras were proposed. Similarly, Smart [2] worked with absolutely continuous functions of a "well-bounded" operator to obtain a spectral decomposition (see below).

Despite these references, it seems fair to say that the first work which systematically adopts the functional calculus as an approach to obtain the spectral theory of the operator is the paper of Foiaş [7] which appeared in 1960. This research was followed, in 1962, by papers of Colojoară [1, 2] and Maeda [5] and, in 1964, by Kantorovitz [3] and Sine [1]. Since then a number of contributions have been made along this line; see especially Apostol [3, 4, 5, 9, 11, 15], Colojoară [1, 2, 3, 4, 5], Colojoară and Foiaş [1, 2, 3], Foiaş [9, 10, 11, 13], Ionescu Tulcea [5], Kantorovitz [3, 5, 6, 7, 8, 9], Maeda [4, 5, 6, 7, 8], Ringrose [3], Smart [2], Sine [1], Sills [1], Suciu [1], Tillmann [1, 2], and Vasilescu [1, 4]. Fortunately, the splendid treatise of Colojoară and Foiaş [4] (see also Colojoară [5]) deals precisely with this aspect of spectral theory, and is both extensive and up-to-date. Hence we refer the reader to this reference for details and will content ourselves with a bird's-eye view of the subject.

Let Ω be an open set in C. Then Foiaş [19] defines a *spectral distribution* to be a linear map U of the algebra $C^\infty(\Omega)$ of complex valued infinitely differentiable functions on Ω into the algebra $B(\mathfrak{X})$ such that (i) U is continuous with the topology of uniform convergence of all derivatives on compact subsets of Ω, (ii) U has compact support in Ω, (iii) $U(\varphi\psi) = U(\varphi)U(\psi)$ for φ, ψ in $C^\infty(\Omega)$, and (iv) $U(f_0) = I$ for $f_0(\lambda) \equiv 1$, $\lambda \in \Omega$. An operator $T \in B(\mathfrak{X})$ is said to be a *generalized scalar operator* if there exists a spectral distribution $U : C^\infty(\Omega) \to B(\mathfrak{X})$ such that $U(f_1) = T$ for $f_1(\lambda) \equiv \lambda$, $\lambda \in \Omega$; in this case U is said to be a *spectral distribution* for T.

Unfortunately there is not a unique spectral distribution for a generalized scalar operator T; indeed, if U is one such distribution and Q is a nilpotent operator (say $Q^{n+1} = 0$) commuting with $U(f)$ for all $f \in C^\infty$, and if $Df = \frac{1}{2}[(\partial f/\partial x) + i(\partial f/\partial y)]$, then

$$V(f) = \sum_{k=0}^{n} \frac{Q^k}{k!} U(D^k f), \qquad f \in C^\infty(\Omega)$$

is a spectral distribution for T. (Note that $D^k f_1 = 0$ for all $k \geqq 1$; hence $V(f_1) = T$.)

A spectral distribution $U : C^\infty(\Omega) \to B(\mathfrak{X})$ is said to be *regular* if $A \in B(\mathfrak{X})$, and $AU(f_1) = U(f_1)A$ implies that $AU(\varphi) = U(\varphi)A$ for all $\varphi \in C^\infty(\Omega)$. A generalized scalar operator $T \in B(\mathfrak{X})$ is said to be regular if it has a regular spectral distribution. Although it is not known whether or not every generalized scalar operator is regular (unless the spectrum is sufficiently "thin"), given any two *regular* spectral distributions U, V for a generalized scalar operator, there exists an integer $p > 0$ such that $(U(\varphi) - V(\varphi))^p = 0$ for all $\varphi \in C^\infty(\Omega)$. More generally, if U and V are arbitrary (not necessarily commuting) spectral distributions of a generalized scalar operator, then there exists an integer $p > 0$ such that $(U(\varphi) - V(\varphi))^{[p]} = 0$ for all $\varphi \in C^\infty(\Omega)$; thus the spectral distributions are unique up to quasi-nilpotence equivalence, at least. Similarly, if a generalized spectral operator has multiplicity 1 (in the sense that the only quasi-nilpotent operator commuting with T is 0), then the operator is a regular generalized spectral operator with unique spectral distribution. (See Colojoară and Foiaş [4; p. 103].)

The class of generalized scalar operators contains the class of scalar type operators, for we can define

$$U(\varphi) = \int \varphi(\lambda) E(d\lambda), \qquad \varphi \in C^\infty(C),$$

if E is a resolution of the identity for S. More generally, if $T = S + N$ is the canonical decomposition of a spectral operator of *finite type* (say $N^n = 0$), we can define U as above and

$$V(\varphi) = \sum_{k=0}^{n-1} \frac{N^k}{k!} U(D^k \varphi)$$

to show that such a spectral operator is a generalized scalar operator. Conversely, if a spectral operator is a generalized scalar operator, then it is of finite type (see Foiaş [11], or Colojoară and Foiaş [4]). For examples of generalized scalar operators that are not spectral operators, let $\mathfrak{X} = C^r[0, 1]$, $(r = 0, 1, 2, \ldots)$ with norm

$$|f| = \sup_{t \in [0,1]} \{|f(t)|, |f'(t)|, \ldots, |f^{(r)}(t)|\};$$

then the operator T defined by $(Tf)(t) = tf(t)$, $t \in [0, 1]$, is a generalized scalar operator with spectral distribution $(U(\varphi)f)(t) = \varphi(t)f(t)$ for $\varphi \in C^\infty$.

It is proved (see Foiaş [9] or Colojoară and Foiaş [4]) that if U is a spectral distribution on $C^\infty(\Omega)$ to $B(\mathfrak{X})$ and if $\varphi \in C^\infty(\Omega)$, then $U(\varphi)$ is a generalized scalar operator with spectrum contained in the image, under φ, of the support of U. In particular, if T is a generalized scalar operator and U is a spectral distribution for T, then $\sigma(T)$ coincides with the support of U. In addition, every generalized scalar operator T is decomposable (in the sense described above) and hence has the single valued extension property. Moreover, the spectral maximal subspace $\mathfrak{X}_T(F)$, for F closed, can be characterized as the intersection

$$\bigcap \{\mathfrak{X}^G \mid G \text{ open}, F \subseteqq G\},$$

where $\mathfrak{X}^G$ is the subspace consisting of $U(\varphi)x$ where $x \in \mathfrak{X}$ and $\varphi \in C^\infty(\Omega)$ has support contained in G.

Using results concerning the tensor product of commuting spectral distributions, Foiaş proved that the sum (and product) of two generalized scalar operators having commuting spectral distributions is a generalized scalar operator. Hence the sum (and product) of two commuting *regular* generalized scalar operators is a generalized scalar operator. In particular, the sum (and product) of two commuting spectral operators of finite type is a generalized scalar operator.

We have already noted that the notion of a generalized scalar operator has been extended even further by Colojoară [1, 2], Maeda [4, 5, 6], Kantorovitz [3] and Sine [1]. In particular, Maeda [4] replaced the algebra C^∞ by a topological algebra $\mathfrak{A}$ of complex valued locally bounded Borel functions satisfying certain properties, and studied continuous homomorphisms of $\mathfrak{A}$ into $B(\mathfrak{X})$.

In their monograph [4], Colojoară and Foiaş introduce a still more general operational calculus. Let Ω be a subset of the complex plane; then an algebra $\mathfrak{A}$ of complex valued functions on Ω is said to be *admissible* if it (i) contains the functions $f_0(\lambda) \equiv 1$ and $f_1(\lambda) \equiv \lambda$ for all $\lambda \in \Omega$, (ii) for every open cover $\{G_1, \ldots, G_n\}$ of $\overline{\Omega}$ there exist non-negative functions $\{\varphi_1, \ldots, \varphi_n\}$ with the support of φ_i in G_i and $\sum_{i=1}^n \varphi_i = 1$ on Ω, and (iii) for every $f \in \mathfrak{A}$ and ξ not in the support of f, the function f_ξ defined by

$$\begin{aligned} f_\xi(\lambda) &= f(\lambda)/(\xi - \lambda), \qquad && \lambda \in \Omega - \{\xi\}, \\ &= 0, && \lambda \in \Omega \cap \{\xi\}, \end{aligned}$$

belongs to $\mathfrak{A}$. If $\mathfrak{A}$ is an admissible algebra, then a mapping $f \to U_f$ of $\mathfrak{A}$

into $B(\mathfrak{X})$ is called an $\mathfrak{A}$-*spectral function* if (i) the map $f \to U_f$ is an algebraic homomorphism with $U_{f_0} = I$, and (ii) the map $\xi \to U_{f_\xi}$ of Ω into $B(\mathfrak{X})$ is analytic on the complement of the support of f. An operator $S \in B(\mathfrak{X})$ is called $\mathfrak{A}$-*scalar* if there exists an $\mathfrak{A}$-spectral function $U : \mathfrak{A} \to B(\mathfrak{X})$ such that $S = U_{f_1}$ (where $f_1(\lambda) \equiv \lambda$). Every scalar type spectral operator is $\mathfrak{A}$-*scalar*, with $\mathfrak{A} =$ the algebra of bounded Borel functions. Similarly, if $\mathfrak{X} = L_p(\Omega)$, $1 \leqq p \leqq \infty$, and $\mathfrak{A} = L_\infty(\Omega)$, then the multiplication operator $Sg(t) = tg(t)$, for $g \in L_p(\Omega)$, $t \in \Omega$, is $\mathfrak{A}$-scalar. In addition, if S is a compact operator and $\mathfrak{A}$ is the algebra of all Borel functions f defined on $\Omega = \{\lambda \mid |\lambda| \leqq |S| + 1\}$ and which are analytic on some open neighborhood G_f of $\sigma(S)$, then S is $\mathfrak{A}$-scalar.

It is proved that if S is an $\mathfrak{A}$-scalar operator with $\mathfrak{A}$-spectral function U, then S is decomposable and hence has the single valued extension property; moreover, $\sigma(S)$ is the support of U.

Suppose now that Ω is a closed set in C and that $\mathfrak{A}$ is an admissible algebra of continuous functions on Ω which is inverse closed (in the sense that if $f \in \mathfrak{A}$ and $1/f \in C(\Omega)$, then $1/f \in \mathfrak{A}$). If U is an $\mathfrak{A}$-spectral function and if $\mathfrak{A}_1$ is the B-algebra generated in $B(\mathfrak{X})$ by $\{U_f \mid f \in \mathfrak{A}\}$, then the space of maximal ideals in $\mathfrak{A}_1$ can be identified with the spectrum of the $\mathfrak{A}$-scalar operator $S = U_{f_1}$ and that with this identification the Gelfand map $B \to \hat{B}$ on $\mathfrak{A}_1$ to $C(\sigma(S))$ has the properties that

$$\hat{U}_f = f \mid \sigma(S), \qquad \sigma(U_f) = f(\sigma(S))$$

for all $f \in \mathfrak{A}$. This extends a result established for spectral distributions by Vasilescu [1]. It follows from this that if $\mathfrak{A}$ is as above and U and V are two $\mathfrak{A}$-spectral functions for the same $\mathfrak{A}$-scalar operator, then U_f and V_f are quasi-nilpotent equivalent for each $f \in \mathfrak{A}$.

The notion of a spectral operator was generalized by Colojoară [1] and Maeda [4] and these ideas have been even further extended to that of an $\mathfrak{A}$-spectral operator by Colojoară and Foiaş [4; p. 76]. It is shown that $T \in B(\mathfrak{X})$ has the property that there exists an $\mathfrak{A}$-spectral function U such that T is quasi-nilpotent equivalent to $S = U_{f_1}$ if and only if we have

$$T\mathfrak{X}_S(F) \subseteqq \mathfrak{X}_S(F) \qquad \text{and} \qquad \sigma(T \mid \mathfrak{X}_S(F)) \subseteqq F$$

for every closed set F in C. If an operator T has this property and commutes with the $\mathfrak{A}$-spectral function, it is called an $\mathfrak{A}$-*spectral operator*. It is proved that T is an $\mathfrak{A}$-spectral operator if and only if $T = S + N$ where S

is an $\mathfrak{A}$-scalar operator and N is a quasi-nilpotent operator commuting with an $\mathfrak{A}$-spectral function of S. This, of course, is a generalization of the canonical decomposition theorem.

Similarly, one can introduce $\mathfrak{A}$-*unitary* and $\mathfrak{A}$-*self adjoint* operators as $\mathfrak{A}$-scalar operators for an admissible algebra of functions defined on the unit circle C_1 or on the real line R, respectively. If $S \in B(\mathfrak{X})$ satisfies $|S^n| = O(|n|^\alpha)$ as $|n| \to \infty$ for some $\alpha \geqq 0$, then S is a C^m-unitary operator for $m > \alpha + 1$. Likewise, if the resolvent satisfies

$$|R(\lambda; S)| = O(|1 - |\lambda||^{-\beta}), \qquad |\lambda| \neq 1,$$

as $|\lambda| \to 1$ for some $\beta \geqq 1$, then S is a C^m-unitary operator for $m > \beta + 1$. More generally, let $S \in B(\mathfrak{X})$ have $\sigma(S) \subseteqq C_1$, let $\rho_n = |S^n|$, and let $\mathfrak{A}_S$ be the algebra of all functions f on $C_1 \to C$ such that

$$f(e^{it}) = \sum_{n=-\infty}^{+\infty} a_n e^{int}, \qquad \sum_{n=-\infty}^{+\infty} |a_n| \rho_n < +\infty.$$

Then $\mathfrak{A}_S$ is a B-algebra under pointwise operations and norm $|f|_S = \sum |a_n| \rho_n$. If S satisfies the condition

$$\sum_{n=-\infty}^{+\infty} \frac{\log \rho_n}{1 + n^2} < +\infty,$$

due to A. Beurling, then $\mathfrak{A}_S$ is a regular B-algebra and S is $\mathfrak{A}_S$-unitary. Indeed, an $\mathfrak{A}$-spectral function for S can be defined by

$$U_f = \sum_{n=-\infty}^{+\infty} a_n T^n$$

when f has the above form. This result extends some theorems of Wermer [1, 7]. Similarly, if $\sigma(S) \subseteqq C_1$ and

$$|R(\lambda; S)| \leqq M \exp(K||\lambda| - 1|^{-\beta})$$

for some $\beta > 0$, then S is an $\mathfrak{A}_S$-unitary operator.

By a change of variable (the "Cayley transform"), the study of $\mathfrak{A}$-self adjoint operators can be referred to that of $\mathfrak{A}$-unitary operators. Thus it can be shown that if $S \in B(\mathfrak{X})$ has $\sigma(S) \subseteqq R$ and if

$$|R(\lambda; S)| \leqq M \exp(K|\mathscr{I}\lambda|^{-\beta}), \qquad \mathscr{I}\lambda \neq 0,$$

for some $\beta > 0$, then S is an $\mathfrak{A}$-self adjoint operator for an appropriate algebra $\mathfrak{A}$. More specifically, if

$$|R(\lambda; S)| = O(|\mathscr{I}\lambda|^{-\beta}), \qquad \mathscr{I}\lambda \neq 0,$$

as $\mathscr{I}\lambda \to 0$ for some $\beta > 0$, then S is C^m-self adjoint for $m > [\beta] + 1$. Similarly, if

$$|e^{itS}| = O(|t|^\gamma)$$

as $|t| \to \infty$, then S is C^m-self adjoint for $m > [\gamma] + 2$. (See also Kantorovitz [5] and Tillmann [1].)

Subdiagonalization and Volterra operators. When applied to operators in Hilbert space, the above results yield some interesting information. Let $T \in B(\mathfrak{H})$ be such that $T - T^*$ belongs to the Carleman class C_p for some $1 \leqq p < \infty$ (XI.9.1). Then there exists an algebra $\mathfrak{A}$ such that T is $\mathfrak{A}$-self adjoint. (See Colojoară and Foiaş [4; p. 166].) Hence, if $\sigma(T)$ is not a single point, then T has non-trivial spectral maximal subspaces. This result improves one due to Sahnovič [5] and J. Schwartz [6]. These ideas have further application to the theory of "subdiagonalization" (or triangularization) of operators T for which $T - T^* \in C_p$ that was discussed in Section IX.10. For other papers on this and related topics, see: Brodskiĭ [1, 2, 5, 6], Cekanovskiĭ [1], Duren [1], Gohberg and Kreĭn [4, 7, 8, 9], Keldyš and Lidskiĭ [1], Kreĭn [23], Livšic [6], Ringrose [4, 5], Sahnovič [4, 5], J. Schwartz [6, 7], Sz.-Nagy and Foiaş [18].

In the theory of subdiagonalization, *Volterra operators* (i.e., compact quasi-nilpotent operators) play an important role. Such operators have been discussed in: Brodskiĭ [3, 4, 5, 6], Brodskiĭ and Kisilevskiĭ [1], J. M. Freeman [1, 3], Gohberg and Kreĭn [1, 5, 7, 8], Gol'dengeršel [1, 2], Kalisch [1, 2, 3, 4, 6], Kal'muševskiĭ [1], Kisilevskiĭ [1, 2], Macaev [1], Osher [1], Ringrose [2], Sahnovič [1, 6], Sarason [2], Suzuki [2].

Well-bounded operators. Let $\mathfrak{X}$ be a reflexive (or even weakly complete) B-space and let $T \in B(\mathfrak{X})$ be such that $\sigma(T) \subseteqq [0, 1]$; then it is seen that T is a scalar type spectral operator if and only if there exists a constant $K > 0$ such that

$$|p(T)| \leqq K \sup_{t \in [0,1]} |p(t)|$$

for each polynomial p. Analogously, Smart [2] defined an operator T to be *well-bounded* if there exists a constant $K > 0$ such that $|p(T)| \leqq K \|p\|$, whence $\|p\|$ denotes the norm on the space $BV[0, 1]$ (i.e.,

$$\|p\| = |p(0+)| + v(p, [0, 1]),$$

where $v(p, [0, 1])$ denotes the total variation of p on the interval $[0, 1]$).

Smart proved that if $\mathfrak{X}$ is reflexive, then for any real number t there exists a unique projection $E(t)$ in $B(\mathfrak{X})$ such that:

(i) $E(t)$ commutes with any bounded operator commuting with T.
(ii) $|E(t)| \leqq 2K$.
(iii) $E(t) = 0$ for $t < 0$, and $E(t) = I$ for $t \geqq 1$.
(iv) $E(s) = E(s)E(t) = E(t)E(s)$ for $s \leqq t$.
(v) $\lim_{t \downarrow s} E(t)x = E(s)x$ for $x \in \mathfrak{X}$.
(vi) $\sigma(T \mid E(t)\mathfrak{X}) \subseteqq (-\infty, t] \cap \sigma(T)$, and
$\sigma(T \mid (I - E(t))\mathfrak{X}) \subseteqq [t, +\infty) \cap \sigma(T)$.

Improving a result of Smart, Ringrose [3, I] showed that

$$T = \int_0^1 t \, dE(t),$$

where the integral exists as a Riemann integral in $B(\mathfrak{X})$. (See also Sine [1] for related results.)

The approach used by Smart and Ringrose was based in part on the fact that a well-bounded operator admits a functional calculus for absolutely continuous functions, but was basically "constructive" in character. Subsequently, Sills [1] presented a different method of attack, which we shall describe. Let AC_0 denote the B-algebra of all absolutely continuous functions on [0, 1] which vanish at 0; Sills introduced the "Arens multiplication" in AC_0^{**} to obtain a B-algebra which is neither commutative nor semi-simple. However, he was able to identify a collection of idempotents in AC_0^{**} corresponding to the non-zero multiplicative linear functionals on $L_\infty(0, 1) \cong AC_0^*$; these can be associated with the points of [0, 1]. If $T \in B(\mathfrak{X})$ is well-bounded, it generates an operational calculus $f \to f(T)$ of $AC_0 \to B(\mathfrak{X})$, and if $\mathfrak{X}$ is reflexive, this homomorphism can be extended to a homomorphism of the algebra AC_0^{**} into $B(\mathfrak{X})$. The extended homomorphism maps the idempotents of AC_0^{**} into projection operators from which the integral representation of T can be derived.

Ringrose [3, II] discussed well-bounded operators in a non-reflexive B-space $\mathfrak{X}$. It turns out that well-boundedness of $T \in B(\mathfrak{X})$ is equivalent to the existence of a family of projections $\{F(t) \mid t \in R\}$ in $B(\mathfrak{X}^*)$ (called the "decomposition of the identity for T") satisfying certain natural properties, and such that the equation

$$T^* = I - \int_0^1 F(t) \, dt$$

holds in the weak operator topology. (Note that this equation is obtained from $\int_0^1 t\,dF(t)$ by "integrating by parts.") In this case, the family $\{F(t)\}$ is not necessarily unique; however, if each $F(t)$ is the adjoint of an operator in $\mathfrak{X}$, then uniqueness does hold.

Berkson and Dowson [2] consider well-bounded operators possessing a family $\{E(t)\}$ of projections in $\mathfrak{X}$ such that $F(t) = E(t)^*$ for each $t \in [0, 1]$. Suppose also that (i) E is strongly continuous on the right, and (ii) the strong limit $\lim_{\lambda \uparrow \mu} E(\lambda)$ exists and is denoted by $E(\mu-)$. (These conditions are automatic if $\mathfrak{X}$ is reflexive.) In this case, for $f \in AC[0, 1]$ we have

$$f(T) = \int_{0-}^{1} f(\lambda)\, dE(\lambda)$$

where this Riemann integral exists strongly; moreover, $E(\mu) - E(\mu-)$ is a projection of $\mathfrak{X}$ onto $\{x \mid Tx = \mu x\}$, and $\sigma_r(T) = \phi$. They further show that an operator $T \in B(\mathfrak{X})$ with $\sigma(T) \subseteq R$ has adjoint of scalar type of class $(\mathfrak{X})$ if and only if T is well-bounded and the function $xF(\cdot)x^*$ is in $BV[0, 1]$ for every $x \in \mathfrak{X}$, $x^* \in \mathfrak{X}^*$. Also, a well-bounded spectral operator is of scalar type and properties (i), (ii) above are satisfied.

Unbounded spectral measures. In Section X.1 we introduced spectral measures E, defined on a field Σ of subsets of a set, which are bounded in the sense that for some constant K we have $|E(\sigma)| \leqq K$ for all $\sigma \in \Sigma$. In the present chapter we have dealt almost exclusively with countably additive spectral measures defined on a σ-field Σ, in which case the boundedness condition is automatically satisfied. Since this boundedness and countable additivity is bound up with unconditional convergence of eigenfunction expansions, considerations from the theory of boundary-value problems for differential equations suggest that a somewhat more general theory might be useful. Such a theory has been developed by Ljance [2] in Hilbert space. He defines a *generalized spectral measure* (g.s.m.) P in a Hilbert space $\mathfrak{H}$ to be a mapping of a set $\mathfrak{D}(P)$ into $B(\mathfrak{H})$ satisfying:

(i) $\mathfrak{D}(P)$ is a collection of Borel subsets of the complex plane C which contains every Borel subset of any set in $\mathfrak{D}(P)$, and the union of any two sets in $\mathfrak{D}(P)$.

(ii) For each $\delta_1, \delta_2 \in \mathfrak{D}(P)$ we have $P(\delta_1)P(\delta_2) = P(\delta_1 \cap \delta_2)$.

(iii) If $\{\delta_1, \delta_2, \ldots\}$ is a partition of $\delta \in \mathfrak{D}(P)$ into mutually disjoint

Borel sets and if $x, y \in \mathfrak{H}$, then

$$(P(\delta)x, y) = \sum_{k=1}^{+\infty} (P(\delta_k)x, y).$$

(iv) The sets $\{P(\delta) \mid \delta \in \mathfrak{D}(P)\}$ and $\{P(\delta)^* \mid \delta \in \mathfrak{D}(P)\}$ are total on $\mathfrak{H}$.

It is proved that every g.s.m. can be extended to a class $\mathfrak{D}_0(P)$ of Borel subsets of C which is maximal (in a certain sense) and that such an extension is uniquely determined. In fact, let $\mathfrak{D}_0(P)$ consist of the set of all Borel sets $\delta \subseteq C$ such that $\sup\{|P(\sigma)| \,|\sigma \in \mathfrak{D}(P),\ \sigma \subseteq \delta\} < +\infty$. Moreover, if $\delta \in \mathfrak{D}_0(P)$, then the generalized sequence $\{P(\sigma) \mid \sigma \in \mathfrak{D}(P),\ \sigma \subseteq \delta\}$ converges strongly in $\mathfrak{H}$ to an operator. If we define $P_0(\delta)$ to be this strong limit, we obtain a g.s.m. P_0 on $\mathfrak{D}_0(P)$ which extends P and which is maximal in a certain technical sense.

Let $\tilde{\mathfrak{H}}$ be the set of elements $x \in \mathfrak{H}$ such that $x = P(\delta)x$ for some $\delta \in \mathfrak{D}(P)$; Ljance calls $\tilde{\mathfrak{H}}$ the *space of basic elements* corresponding to P. If one regards $\mathfrak{D}(P)$ as a directed set under inclusion, then we can regard $\tilde{\mathfrak{H}}$ as the "inductive limit" of the spaces $P(\delta)\mathfrak{H}$ with an appropriate topology. Similarly one defines $\hat{\mathfrak{H}}$, the *space of generalized elements* corresponding to P, to be the collection of all generalized sequences $\hat{x} = \{x_\sigma \mid \sigma \in \mathfrak{D}(P)\}$ with $x_\sigma \in P(\sigma)\mathfrak{H}$ and such that $x_\sigma = P(\sigma)x_\delta$ if $\sigma \subseteq \delta$. The space $\hat{\mathfrak{H}}$ is a vector space under the obvious pointwise operations and can be considered to be the "projective limit" of the spaces $P(\delta)\mathfrak{H}$ with an appropriate topology. The space $\tilde{\mathfrak{H}}$ is dense in $\mathfrak{H}$, and the mapping $x \to \{P(\delta)x\}$ embeds $\mathfrak{H}$ densely in $\hat{\mathfrak{H}}$. Each operator $P(\delta)$, $\delta \in \mathfrak{D}(P)$, can be extended by continuity to a projection $\hat{P}(\delta)$ in $\mathfrak{H}$; more generally, for any Borel set $\sigma \subseteq C$ one can define $\hat{P}(\sigma)$ for $\hat{x} = \{x_\sigma\}$ in $\hat{\mathfrak{H}}$ by

$$\hat{P}(\sigma)\hat{x} = \{x_{\sigma \cap \delta} \mid \sigma \in \mathfrak{D}(P)\}.$$

Then the function $\hat{P}$ is a countably additive spectral measure in $B(\mathfrak{H})$ defined on the Borel sets of C and such that $\hat{P}(C)$ is the identity operator in $\hat{\mathfrak{H}}$.

Under a certain countability condition, Ljance gives a generalization of the Lorch-Mackey theorem relating generalized spectral measures P with a self adjoint spectral measure E. One has a relation of the form $P(\delta) = M^{-1}E(\delta)M$, although the one-to-one operator M is unbounded on $\mathfrak{H}$, and is continuous only in the topologies of the spaces $\tilde{\mathfrak{H}}$ and $\hat{\mathfrak{H}}$.

Ljance introduces the collection $\mathfrak{A}(P)$ of all closed linear operators of $\mathfrak{H}$ into $\mathfrak{H}$ which "commute" in an appropriate sense with the g.s.m. P. One can introduce algebraic and topological structures on $\mathfrak{A}(P)$ which make it isomorphic and homeomorphic with the topological algebra $\check{\mathfrak{A}}(P)$ (consisting of the continuous linear operators in $\check{\mathfrak{H}}$ which commute with the restriction of P to $\check{\mathfrak{H}}$) and with the topological algebra $\hat{\mathfrak{A}}(P)$ (consisting of the continuous linear operators in $\hat{\mathfrak{H}}$ which commute with the extension of P to $\hat{\mathfrak{H}}$). An operational calculus can also be introduced: Let $\mathscr{P}$ be the class of all Borel functions $f: C \to C$ which are bounded on each set $\delta \in \mathfrak{D}(P)$. If $f \in \mathscr{P}$ and $\delta \in \mathfrak{D}(P)$ define

$$f_\delta = \int_\delta f(\lambda) P(d\lambda),$$

whence it follows that

$$|f_\delta| \leqq 4\{\sup_{\substack{\sigma \subseteqq \delta \\ \sigma \in \mathfrak{D}(P)}} |P(\sigma)|\}\{\sup_{\lambda \in \sigma} |f(\lambda)|\}.$$

If we define T_f to be the closure in $\mathfrak{H}$ of the operator in $\check{\mathfrak{H}}$ which has the form $x \to f_\delta x$ for $x = P(\delta)x$, then the map $f \to T_f$ is a homomorphism of the algebra $\mathscr{P}$ into $\mathfrak{A}(P)$, and is continuous with a reasonable topology on $\mathscr{P}$.

A generalization of the canonical reduction theorem can be obtained. We say that a closed operator N commuting with P is *P-quasi-nilpotent* if, for each $\delta \in \mathfrak{D}(P)$, the restriction of N to $P(\delta)\mathfrak{H}$ is a quasi-nilpotent operator. We say that $S \in \mathfrak{A}(P)$ is *P-scalar* if it is obtained via the above operational calculus corresponding to the function $f_1(\lambda) \equiv \lambda$, $\lambda \in C$. Then, an element $T \in \mathfrak{A}(P)$ admits a representation $T = S + N$ where S is P-scalar and N is P-quasi-nilpotent if and only if the spectrum of $T \mid P(\delta)\mathfrak{H}$ is contained in $\bar{\delta}$ for every $\delta \in \mathfrak{D}(P)$.

Finally, a theory of spectral representation is developed for generalized spectral measures.

The ideas described above have applications to differential operators. Let R^+ be the semi-axis $[0, +\infty)$, let $p: R^+ \to C$ be integrable on R^+, and let $\theta \in C$. Let $\mathfrak{D}(L)$ be the set of all $f \in L_2(R^+)$ whose derivatives f' are absolutely continuous on every finite interval in R^+ and satisfying the boundary condition

$$f'(0) - \theta f(0) = 0$$

and such that

$$l(f) = -f'' + pf \in L_2(R^+).$$

We define L to be the operator with domain $\mathfrak{D}(L)$ such that

$$L(f) = -f'' + pf, \qquad f \in \mathfrak{D}(L).$$

Naĭmark [12] showed that there exists a function A which is analytic in some half-plane $\mathscr{I}(s) > -\eta (\eta > 0)$ such that the spectrum of L consists of the interval $[0, +\infty)$ together with a finite number of points $\lambda_1, \ldots, \lambda_n$ such that $\lambda_k = s_k^2$, where $\mathscr{I}(s_k) > 0$ and $A(s_k) = 0$. The points $\lambda_1, \ldots, \lambda_n$ are eigenvalues of finite multiplicity of L, while the points $\lambda \in [0, +\infty)$ are in the continuous spectrum of L. If A does not have any non-zero real roots, then we can write $L_2(R^+) = \mathfrak{M} \oplus \mathfrak{N}$ where $\mathfrak{M}$ corresponds to the continuous spectrum of L, and $\mathfrak{N}$ corresponds to the point spectrum of L. If p satisfies the condition

$$\int_0^{+\infty} e^{\varepsilon x} |p(x)| \, dx < +\infty$$

for some $\varepsilon > 0$, then $\mathfrak{N}$ is finite dimensional. In this case it follows from the work of Naĭmark [12] and Levin [1] that L is similar to a self adjoint operator on $\mathfrak{M}$; hence L is a spectral operator. (See Chapter XX where results of this character will be proved.)

However, if the function A does have non-zero real roots, $\sigma_1, \ldots, \sigma_m$, then the numbers $\tilde{\lambda}_k = \sigma_k^2$, $k = 1, \ldots, m$, are called the *spectral singularities* of L; they play an important role in the discussion of L and were studied by Ljance [4]. He showed that if L has spectral singularities, then $L_2(R^+)$ *cannot* be represented as $\mathfrak{M} \oplus \mathfrak{N}$, where L is similar to a self adjoint operator on $\mathfrak{M}$, and $\mathfrak{N}$ is finite dimensional. Nevertheless, he shows that one can define a *generalized* spectral measure P on the collection of all Borel sets which have a positive distance from the spectral singularities, but that $|P(\delta)| \to +\infty$ when the distance between δ and some spectral singularity tends to zero. Although not every function in $L_2(R^+)$ is the strong limit of its eigenfunction expansion with respect to L, the set of functions for which such expansions exist is dense in $L_2(R^+)$. For further properties and a detailed study of the operator L see Ljance [4]. Spectral singularities of this character are also considered in Ljance [8], Pavlov [1, 2], and J. Schwartz [4].

Operators in Hilbert space. Some interesting expositions of the spectral theorem for self adjoint or normal operators have been given recently.

For example, see Berberian [2, 4], Bernau [1], Bernau and Smithies [1], Bonsall [8], Halmos [12], Halperin [11], and Whitley [2]. We remark only that Bernau [1], Bernau and Smithies [1], and Whitley [2] have given "elementary" proofs of the fact that if T is a normal operator and p is a polynomial in two variables, then

$$|p(T, T^*)| = \sup\{|p(\lambda, \bar{\lambda})| \mid \lambda \in \sigma(T)\}.$$

This result provides a key to a rapid proof of the spectral theorem for normal operators.

There are a number of other aspects of the spectral theory of operators in Hilbert space, such as the notion of the numerical range, Hilbert-Schmidt operators, the spectral sets of von Neumann, and so on, that have received attention. Some of these have been extended in various ways to B-spaces (see the remarks above on Hermitian operators in B-spaces and the numerical range defined by means of a semi-inner product); we refer the reader to a forthcoming monograph by Bonsall and Duncan [1] for an account of some of this work. But there still are many results which are meaningful only in Hilbert space.

The reader may refer to the following articles and books: Andô [3], Aleksandrjan and Mkrtčjan [1], Apostol [2], Berberian [1, 3], Berberian and Orland [1], Bernau [2, 3], Biriuk and Coddington [1], Bonsall [6], Bos [1], Broido [1], Brown and Pearcy [1], Cartan [2], Coddington [5], Davis and Rider [1], Deckard and Pearcy [1], Dolph [1, 2], Dolph and Penzlin [1], Donoghue [2], Durszt [1], Foiaş [1, 2, 3, 4, 5, 7], George [1], Ghika [1], Gohberg and Markus [1], Gonshor [1], Hadeler [1], Halmos [11, 14], Halmos and McLaughlin [1], Hempel [1], Hestenes [1], S. Hildebrandt [1, 2], Inoue [1, 2, 3, 4, 5, 6], Istrăteşcu [1], Jakubov [1], G. I. Kac [1], Kacnel'son and Macaev [1], Kalisch [5], Kamowitz [1], Kaniel [1], MacCluer [1], McKelvey [1, 2], Maurin [4, 5], Maurin and Maurin [2], Mitjagin and Pełczyński [1], Nevanlinna and Nieminen [1], Nieminen [1], Olagunju and West [1], Orland [1], Putnam [30, 31], Saitô and Yoshino [2], Sarason [1], Schaefer [8], Schreiber [6, 7], Sheth [1], Stampfli [2, 4, 6, 7, 8, 9], Suzuki [1], Sz.-Nagy [17], Sz.-Nagy and Foiaş [1, 2, 3, 4, 5, 6, 7, 8, 9, 10, 12], Tillmann [1], Williams [1], Yoshino [1, 2].

Invariant subspaces. On pages 929–930 we discussed the question of whether an operator in $B(\mathfrak{X})$ has a non-trivial invariant closed subspace. This question has still not been answered for $\mathfrak{X} = \mathfrak{H}$. However, a considerable amount of work has been done on it and related theories. The reader may refer to: Andô [4], Apostol [12], Aronszajn and Smith [1], Arveson and Feldman [1], Bernstein and Robinson [1], de Branges [3], de Branges and Rovnyak [1, 2], Brodskiĭ and Smulyan [1], Ciorănescu [1], Crimmins and Rosenthal [1], Donoghue [1], Duren [1], Fan [6], Ginzburg [1], Godič [1], Gol'dman and Levič [1], Halmos [11, 16], Hasumi and Srinivasan [1, 2], Helson [1], Helson and Lowdenslager [1], Kreĭn [28], Nikol'skiĭ [1], Rota [1, 4], Saphar [3], Sarason [3], Schaefer [13], Scroggs [1], Srinivasan [1, 2], Sz.-Nagy and Foiaş [1, 6, 9, 10], Volk [1], Wermer [1, 2, 4].

Contractions and dilations. On pages 931–932 we have briefly discussed contractions and dilations of operators in Hilbert space. This theory, which was sparked by some fundamental discoveries of Sz.-Nagy, has been very fully developed on recent years. It has been explored from many points of view: prediction theory, stationary stochastic processes, Fourier analysis, subdiagonalization, operational calculus, invariant subspaces, operator valued analytic functions, and so on. To describe these researches is far beyond our scope. Fortunately, the recent book of Sz.-Nagy and Foiaş [10] presents an authoritative and excellent exposition of many of these developments.

Other work can be found in the following papers: Adamjan and Arov [1, 2], Andô [2, 3], Berberian [5], Biriuk and Coddington [1], Bram [1], Brehmer [1], de Bruijn [1], Coddington [5], Coddington and Gilbert [1], Čumakin [1], Durszt [1, 2], Foguel [9, 10, 11], Foiaş [4, 5, 14], Foiaş and Gehér [1], Foiaş and Mlak [1], Gilbert [1], Gohberg and Kreĭn [6], Halmos [14, 15], Halperin [6, 7, 8, 9, 10], Ionescu Tulcea [4], Ionescu Tulcea and Plafker [1], Itô [1], Lebow [1, 2], McKelvey [1], Mlak [1, 2, 3, 4, 5], Nakano [19], Orland [2], Pflüger [1], Saffern [1], Saitô and Yoshino [1], Sarason [1], Schreiber [2, 3, 4, 5], Suciu [1], Sz.-Nagy [17, 18, 19, 20, 21, 22], Sz.-Nagy and Foiaş [1, 2, 3, 4, 5, 6, 7, 8, 9, 10].

Positive operators. One general class of operators that is of importance and has received considerable attention recently is the class of *positive operators* on an ordered vector space. We shall make a few comments about these operators here, after some preliminary ideas are clarified.

A vector space $\mathfrak{B}$ is said to be *ordered* by a (reflexive, transitive, and anti-symmetric binary) relation $\leqq$ if (a) $x \leqq y$ implies that $x + z \leqq y + z$ for all x, y, z in $\mathfrak{B}$, and (b) $x \leqq y$ implies that $\lambda x \leqq \lambda y$ for all x, y in $\mathfrak{B}$ and $\lambda \in R$, $\lambda \geqq 0$. If $\mathfrak{B}$ is an ordered vector space under $\leqq$, then the set

$K = \{x \in V \mid 0 \leqq x\}$ is called the *positive cone* of $\mathfrak{V}$ (with respect to $\leqq$); it is easy to see that K satisfies (i) $K + K \subseteqq K$, (ii) $\lambda K \subseteqq K$ for all $\lambda \in R$, $\lambda \geqq 0$, and (iii) $K \cap (-K) = \{0\}$. Conversely, if K is a subset of $\mathfrak{V}$ satisfying (i), (ii), (iii), and if we define $x \leqq y$ to mean that $y - x \in K$, then we obtain a relation ordering $\mathfrak{V}$. A linear topological space which is also ordered is said to be an *ordered topological vector space* in case its positive cone is closed. For an exposition of ordered topological vector spaces, we refer the reader to the books of Day [12], Kelley and Namioka [1], Schaefer [18] and Peressini [1].

The notions of order are most closely tied to real coefficients, while spectral theoretic notions are simpler for complex coefficients. To bridge this gap, we employ the following construction. If $\mathfrak{X}$ is a real B-space which is ordered by $\leqq$, let $\mathfrak{X}_1 = \mathfrak{X} \oplus \mathfrak{X}$ and define:

$$(\alpha + i\beta)[x, y] = [\alpha x - \beta y, \alpha y + \beta x], \qquad \alpha, \beta \in R,\ x, y \in \mathfrak{X};$$

$$|[x, y]| = \sup_{0 \leqq \theta \leqq 2\pi} |(\cos\theta)x + (\sin\theta)y|;$$

$$[x_1, y_1] \leqq [x_2, y_2] \qquad \text{whenever} \quad x_1 \leqq x_2 \quad \text{and} \quad y_1 \leqq y_2;$$

then $\mathfrak{X}_1$ is a complex B-space, the map $x \to [x, 0]$ is an isometric isomorphism of $\mathfrak{X}$ into $\mathfrak{X}_1$, and $\mathfrak{X}_1$ becomes an ordered B-space, called the *complexification of* $\mathfrak{X}$. Finally, if $T \in B(\mathfrak{X})$, we define $T_1[x, y] = [Tx, Ty]$; then the spectral properties of T can be defined and studied by examining the corresponding properties of T_1.

If $\mathfrak{X}$ is an ordered real B-space with positive cone K and if $T \in B(\mathfrak{X})$, then T is said to be *positive* in case $T(K) \subseteqq K$ (equivalently, if $Tx \geqq 0$ for all $x \geqq 0$, $x \in \mathfrak{X}$).

One of the most important and celebrated results concerning positive operators is:

THEOREM. (*Kreĭn-Rutman* [1]). *Let $\mathfrak{X}$ be an ordered real B-space such that $\overline{\text{sp}}(K) = \mathfrak{X}$. If $T \in B(\mathfrak{X})$ is a compact positive operator with positive spectral radius $r(T)$, then $r(T)$ is an eigenvalue and has a corresponding positive eigenvector.*

This theorem should be regarded as a generalization of a classical theorem concerning positive matrices due to Frobenius and Perron. In turn, the theorem of Kreĭn-Rutman has been extended in various ways. For instance (see Schaefer [18; p. 264]), if we replace the compactness of T by the hypothesis that the resolvent operator $R(\lambda; T_1)$ has a pole on the circle $|\lambda| = r(T)$, then we infer that $r(T) \in \sigma(T)$; moreover, if $r(T)$ is a pole of this resolvent, it is of maximal order on this circle.

In the case of "irreducible" positive operators with $r(T) = 1$ on the real B-space $C(\Omega)$, Ω compact, it can be shown (see Schaefer [18; p. 272]) that the eigenvalues on the unit circle are cyclically located and each has multiplicity one; if this point spectrum contains an isolated point, then these eigenvalues are the nth roots of unity for some n. If one such eigenvalue is a pole of the resolvent, then all are poles of order 1. The number 1 is the only eigenvalue with a positive eigenfunction. If Ω is connected, then 1 is the only root of unity which can belong to the point spectrum of T.

The reader should consult the article of Kreĭn and Rutman [1] for remarks concerning the history of the study of positive operators. For a survey of the recent work, see the forthcoming volume of Schaefer [20] and the following references: Andô [1], Bahtin [1, 2], Bahtin, Krasnosel'skiĭ, and Stečenko [1], Birkhoff [9], Bonsall [2, 3, 4, 5, 7], Bonsall, Lindenstrauss, and Phelps [1], V. M. Brodskiĭ [1], Eberly [1], A. J. Ellis [1], Esajan and Stečenko [1], Karlin [3], Krasnosel'skiĭ [5], Lotz [1], Lotz and Schaefer [1], Marek [1, 2, 3], Mewborn [1], Niiro [1], Niiro and Sawashima [1], Peressini [1], Peressini and Sherbert [1, 2], Phelps [1], Putnam [20, 23, 25], Riedl [1], Rota [5], Sasser [1], Sawashima [1, 2, 3], Schaefer [1, 3, 4, 6, 11, 12, 14, 15, 16, 17, 18, 19], J. Schwartz [5], and Thompson [1].

Generalizations of compact operators. Since the class of compact operators in $B(\mathfrak{X})$ has such well-behaved spectral properties and frequently arises in problems of analysis, very many papers have been written which either examine special types of compact operators (such as Hilbert-Schmidt operators), explore the determinant theory approach to certain compact operators, or propose generalizations or extensions of the classical results of Fredholm and Riesz.

The following list of papers are cited in this connection: Altman [6], Andô [4], Balslev and Gamelin [1], Bonsall [9], Bonsall and Tomiuk [1], Breuer [1], Breuer and Cordes [1], Buraczewski [1, 2], Caradus [1, 2, 3], Chung [1], Coburn [1], Coburn and Lebow [1, 2], Cordes [4], Deckard and Pearcy [1], Deprit [1, 2], De Wilde [1, 2], Donoghue [1], Eberly [1], R. J. Ellis [1], Fišman and Valickiĭ [1], Gamelin [1], Gil'derman and Korotkov [1], Gillespie and West [1], Gohberg and Kreĭn [1, 7, 8], Gohberg and Markus [1], Goldberg [2], Goldberg and Thorp [1, 2], Gramsch [1, 2], Graves [6], Grothendieck [6], Haahti [1], Halmos [11], Harazov [6, 7, 8], Heuser [1, 2, 3], Hukuhara and Sibuya [1, 2], Iohvidov [2], Kaashoek [1], Kaashoek and Lay [1], Kaniel and Schechter [1], Kleinecke [5], Kultze [1], Leżański [1], Lindenstrauss [1], Lindenstrauss and Pełczyński [1], Luxemburg and Zaanen [1], Macaev [1], Olagunju and West [1], Pełczyński [1], Pettineo [1, 2], Pietsch [1, 2, 3, 4, 5, 6, 7, 9, 10, 12], Przeworska-Rolewicz [3], Ringrose [1, 4, 6], Ruston [2, 3, 5, 6], Saphar [7, 8], Schaefer [2, 5], Schatten [2], J. Schwartz [5], Sikorski [1, 2, 3, 4, 5, 6, 7, 8, 9], Silverman and Yen [1], Smithies [1], Stečenko [1], Weidmann [1], West [1, 3, 4], Williamson [3], Whitley [1], Zabreĭko, Krasnosel'skiĭ and Stečenko [1].

The index theory. In recent years the notion of the index of an operator has played a significant role in diverse areas of analysis and geometry. We shall introduce this idea for certain operators in $B(\mathfrak{X})$.

Let $T \in B(\mathfrak{X})$, let $\mathfrak{N}(T) = \{x \in \mathfrak{X} \mid Tx = 0\}$ be the null space of T, and let $\alpha(T)$ (called the *nullity* of T) be the dimension of $\mathfrak{N}(T)$ if it is finite dimensional and $+\infty$ otherwise. If the range $\mathfrak{R}(T) = \{Tx \mid x \in \mathfrak{X}\}$ is closed in $\mathfrak{X}$, we let $\beta(T)$ (called the *deficiency* of T) be the dimension of $\mathfrak{X}/\mathfrak{R}(T)$ if this space is finite dimensional and $+\infty$ otherwise. An operator T will be called a *Fredholm operator* in case $\mathfrak{R}(T)$ is closed and both $\alpha(T)$ and $\beta(T)$ are finite; it will be called a *semi-Fredholm operator* in case $\mathfrak{R}(T)$ is closed and at least one of $\alpha(T)$ and $\beta(T)$ is finite. If T is a semi-Fredholm operator, we define the *index* of T to be

$$\kappa(T) = \alpha(T) - \beta(T)$$

(although the negative of this quantity is sometimes taken by other authors).

It can be proved that if $T \in B(\mathfrak{X})$ is a semi-Fredholm operator, then $\mathfrak{R}(T^*)$ is closed and $\mathfrak{N}(T^*) = \mathfrak{R}(T)^\perp$ and $\mathfrak{R}(T^*) = \mathfrak{N}(T)^\perp$, whence it follows that $\alpha(T^*) = \beta(T)$ and $\beta(T^*) = \alpha(T)$. Therefore $T^* \in B(\mathfrak{X}^*)$ is a semi-Fredholm operator and $\kappa(T^*) = \kappa(T)$. Similarly, if T and S are Fredholm operators, then TS is a Fredholm operator and

$$\kappa(TS) = \kappa(T) + \kappa(S).$$

Probably the most important theorems concerning the index are the following two results concerning the effect of perturbing a semi-Fredholm operator.

First Stability Theorem. *Let $T \in B(\mathfrak{X})$ be a semi-Fredholm operator. Then there exists $\delta > 0$ such that if $S \in B(\mathfrak{X})$ is such that $|S - T| < \delta$, then S is a semi-Fredholm operator and $\kappa(S) = \kappa(T)$. In addition, we have*

$$\alpha(S) \leqq \alpha(T), \qquad \beta(S) \leqq \beta(T).$$

Second Stability Theorem. *Let $T \in B(\mathfrak{X})$ be a semi-Fredholm operator. If $K \in B(\mathfrak{X})$ is compact, then $S = T + K$ is a semi-Fredholm operator with $\kappa(S) = \kappa(T)$.*

It was observed by Gol'dman [1] that both of these theorems fail if T is not a semi-Fredholm operator (i.e., if $\alpha(T) = \beta(T) = \infty$).

For proofs of these theorems we refer the reader to Kato [13], and to Gohberg and Kreĭn [2] and Goldberg [2], where there are other references, historical remarks, and applications. We remark only that the notion of the index arose in 1921 in connection with the study of certain singular integral equations by F. Noether. In essence, the first stability theorem was proved by Dieudonné [22] for Fredholm operators, although he did not specifically define the index. In 1951 Atkinson proved both stability theorems. In that same year both Yood [2] and Gohberg [7] also proved the second theorem. Since that time there have been a large number of papers generalizing and using the index. For example, extensions are available for unbounded closed operators, for perturbations K more general than compact operators, for operators in $B(\mathfrak{X}, \mathfrak{Y})$, or in linear topological spaces. Applications are made to equations in linear spaces, to integral equations, and to differential operators on manifolds. We refer the reader to the following books or papers: Atkinson [2, 4], Breuer [1], Breuer and Cordes [1], Caradus [1, 2], Coburn and Lebow [1, 2], Cordes [3, 4], Cordes and Labrousse [1], Dieudonné [22], Gamelin [1], Gohberg [4, 7], Gohberg and Kreĭn [2], Goldberg [2], Gol'dman [1], Gol'dman and Kračkovskiĭ [1], Gramsch [1, 2], Kaashoek [1, 2], Kato [11, 13], Kreĭn and Krasnosel'skiĭ [1], Martirosjan [1], Neubauer [3], Newberger [1], Paraska [1], Pettineo [1], Przeworska-Rolewicz and Rolewicz [1, 2, 3, 4], Saphar [3, 4, 5], Schaefer [2], Schechter [1, 2], Seeley [1], Švarc [1], Tôgô and Shiraishi [1], and Yood [2].

CHAPTER XVI

Spectral Operators: Sufficient Conditions

1. Statement of the Problem

In the preceding chapter we studied the properties of bounded spectral operators, that is, operators which have a countably additive resolution of the identity defined on the field of Borel sets. These operators were found to have a number of interesting properties which generalize those of a bounded normal operator in Hilbert space, and which also generalize those of an arbitrary linear operator in a complex finite dimensional space.

In this chapter we shall give conditions that are sufficient for a bounded operator T to be a spectral operator. The conditions given will be expressed in terms of the resolvent $R(\xi; T)$ and the analytic extensions $x(\xi)$ of $R(\xi; T)x$. We have already established a number of key properties of the resolvents of spectral operators. First of all, by Theorem XV.3.2, we have

(A) *For each x in $\mathfrak{X}$ the function $R(\xi; T)x$ has the single valued extension property.*

Secondly, we have

(B) *There is a constant K, depending only upon T, such that for every pair x, y of vectors with $\sigma(x)$, $\sigma(y)$ disjoint we have*

$$|x| \leqq K|x+y|.$$

To see this, note that it follows from Theorem XV.3.4 and Corollary XV.3.7 that $E(\sigma(x))x = x$ and that the spectrum of $E(\sigma(x))y$ is void. Thus, by Corollary XV.3.3, $E(\sigma(x))y = 0$. Hence

$$|x| = |E(\sigma(x))(x+y)| \leqq K|x+y|,$$

where K is a bound for E.

Finally, it follows from Corollary XV.3.6 that spectral operators have the following property.

(C) *For every closed set* δ *of complex numbers the set of all vectors* x *with* $\sigma(x) \subseteq \delta$ *is also closed.*

In this chapter we shall see that the properties (A), (B), and (C) come near to being sufficient for an arbitrary operator T to be a spectral operator.

The present chapter is divided into two main parts, Sections 2, 3, 4 being the first part and Section 5 the second. In the first part it is shown that an operator T (in a weakly complete space) which has the properties (A), (B), and (C) has a uniquely determined countably additive resolution of the identity defined on a certain σ-field $\mathscr{M}(T)$ of sets. In general, this field need not contain all Borel sets and may not even contain enough sets to be particularly useful. Thus our analysis is only perfected in the second part (Section 5), where conditions are given which guarantee that $\mathscr{M}(T)$ contains the field of all Borel sets.

The conditions of Section 5 require first of all that the spectrum of T lie in a rectifiable Jordan curve and that the resolvent $R(\lambda; T)$ of T have a finite rate of growth as λ approaches $\sigma(T)$. Operators satisfying these conditions automatically satisfy conditions (A) and (C). Thus the rate of growth conditions of Section 5, and the boundedness assumption (B), are together almost sufficient to guarantee that T is a spectral operator.

For this reason the problem of verifying the boundedness condition (B) must be regarded as a key problem in applying the theory of spectral operators. The reader will readily perceive that (B) is not the sort of condition that is likely to be easy to verify. This is not surprising, for what (B) amounts to in the final analysis is the assertion that the resolution of the identity is countably additive. Thus (B) is the condition which gives rise to the phenomenon of unconditional convergence of the eigenvalue expansions.

What we mean, in more detail, is this. It is because of the countable additivity of the resolution of the identity that spectral operators have important expansion theorems associated with them. For example, if T is a spectral operator and its spectrum is denumerable, then every x in $\mathfrak{X}$ has an unconditionally convergent expansion of the type $x = \sum x_n$ $(= \sum_{\lambda \in \sigma(T)} E(\lambda)x)$ where the spectrum of x_n consists of just one point λ_n, so that x_n is a kind of "generalized eigenvector" associated with λ_n. If T is a spectral operator of scalar type, then the generalized eigenvectors are simply eigenvectors in the ordinary sense. If T is a spectral operator of type m, then the generalized eigenvectors x_n satisfy the equations $(\lambda_n I - T)^{m+1}x_n = 0$, $n = 1, 2, \ldots$. Thus, as long as T has a countably additive resolution of the identity, we are not far from the simple situation characteristic of normal operators in Hilbert space. If the countable additivity of the spectral resolution fails, so do many other convenient eigenvalue expansion properties.

It should be noted that it is by no means the case that all the familiar eigenvalue expansions of classical analysis are unconditionally convergent. Indeed, there are many examples, such as Fourier series expansions in $L_p(0, 2\pi)$ with $1 < p < \infty, p \neq 2$, where the expansion converges, but only conditionally. Other examples will be encountered later, in Chapter XIX. This seems to indicate that further developments in spectral theory will include a theory of conditionally convergent expansions associated with discrete and continuous spectra. Nevertheless, the cases where one does have unconditionally convergent eigenvalue expansions are of sufficient importance to justify studying them for their own sake. It is this fact that lends importance to the problem of discovering which operators are spectral operators.

2. Consequences of the Condition (A).

In this section we shall be concerned with a bounded linear transformation T in a complex B-space $\mathfrak{X}$. We shall establish a few properties of T on the basis of the single assumption (A), which is repeated here for convenience of reference.

(A) *For each x in $\mathfrak{X}$ the function $R(\xi; T)x$ has the single valued extension property.*

Since T satisfies condition (A) we recall (cf. Definition XV.2.6) that the *resolvent set of x*, which is denoted by $\rho(x)$, may be defined as the union of all the domains $D(f)$, the union being taken as f varies over all analytic extensions of $R(\xi; T)x$. Thus $\rho(x)$ is an open set containing $\rho(T)$, and its complement $\sigma(x)$ is a closed subset of $\sigma(T)$. The set $\sigma(x)$ is called the *spectrum of x*. It is clear that there is a unique maximal analytic extension of $R(\xi; T)x$. This extension, which is a single valued analytic function defined on $\rho(x)$, will be denoted by $x(\cdot)$. Thus the function $x(\cdot)$ has, by definition, the properties

$$(\xi I - T)x(\xi) = x, \qquad \xi \in \rho(x),$$
$$x(\xi) = R(\xi; T)x, \qquad \xi \in \rho(T).$$

Even though (A) is taken as a standing assumption throughout this section, it will be indicated parenthetically in the statement of each lemma in the proof of which it is used.

1 Lemma (A). *If α, β are complex numbers and x, y are vectors in $\mathfrak{X}$, then*

$$\sigma(x+y) \subseteq \sigma(x) \cup \sigma(y),$$
$$\alpha x(\xi) + \beta y(\xi) = (\alpha x + \beta y)(\xi), \qquad \xi \in \rho(x)\rho(y).$$

Proof. The function $\alpha x(\xi) + \beta y(\xi)$ is an analytic extension of

$$R(\xi; T)\alpha x + R(\xi; T)\beta y = R(\xi; T)(\alpha x + \beta y), \qquad \xi \in \rho(T),$$

defined on the open set $\rho(x)\rho(y)$. Thus $\rho(\alpha x + \beta y) \supseteq \rho(x)\rho(y)$. For $\xi \in \rho(x)\rho(y)$ we have

$$(\alpha x + \beta y)(\xi) = \alpha x(\xi) + \beta y(\xi),$$

by (A). Q.E.D.

2 Lemma (A). *The spectrum $\sigma(x)$ is void if and only if $x = 0$.*

Proof. The proof of Corollary XV.3.3 may be used to prove the present lemma. Q.E.D.

3 Lemma (A). *Let σ be a set of complex numbers, and σ' its complement. If $x + y = x_1 + y_1$, where $\sigma(x), \sigma(x_1) \subseteq \sigma$ and $\sigma(y), \sigma(y_1) \subseteq \sigma'$, then $x = x_1$, $y = y_1$.*

Proof. By Lemma 1

$$\sigma(x - x_1) \subseteq \sigma(x) \cup \sigma(x_1) \subseteq \sigma,$$
$$\sigma(y_1 - y) \subseteq \sigma(y) \cup \sigma(y_1) \subseteq \sigma',$$

and so the vector $x - x_1 = y_1 - y$ has a void spectrum. Thus Lemma 2 shows that $x = x_1$, $y = y_1$. Q.E.D.

4 Lemma (A). *If P is a bounded linear operator in $\mathfrak{X}$ which commutes with T, then*

$$\sigma(Px) \subseteq \sigma(x), \qquad x \in \mathfrak{X}.$$

PROOF. Since P commutes with T, it commutes with the resolvent $R(\xi; T)$ for every ξ in $\rho(T)$. From the equation $R(\xi; T)Px = PR(\xi; T)x$ it is clear that $Px(\xi)$ is an analytic extension of $R(\xi; T)Px$ to the domain $\rho(x)$. Thus $\rho(Px) \supseteqq \rho(x)$ and hence $\sigma(Px) \subseteqq \sigma(x)$. Q.E.D.

3. Consequences of the Conditions (A) and (B)

Throughout this section it will be assumed that the operator T satisfies condition (A) of Section 2 as well as the fundamental boundedness condition (B), which is stated as follows.

(B) *There is a constant K, depending only on T, such that for every pair x, y of vectors with $\sigma(x)$, $\sigma(y)$ disjoint we have*

$$|x| \leqq K|x+y|.$$

Although the assumptions (A) and (B) will be standing assumptions throughout the section, they will be indicated in the statement of each lemma where they are used.

Assumption (B) allows us to associate projections $E(\delta)$ with certain sets δ of complex numbers.

1 DEFINITION. The symbol $\mathscr{S}_1(T)$ will be used for the family of all sets σ with the property that vectors of the form $x+y$ with $\sigma(x) \subseteqq \sigma$, $\sigma(y) \subseteqq \sigma'$ are dense in $\mathfrak{X}$.

It is clear that if σ is in $\mathscr{S}_1(T)$, then the complement σ' is also in $\mathscr{S}_1(T)$.

2 LEMMA (A, B). *If σ is in $\mathscr{S}_1(T)$, there is one and only one bounded projection $E(\sigma)$ on $\mathfrak{X}$ with the properties $E(\sigma)x = x$ if $\sigma(x) \subseteqq \sigma$ and $E(\sigma)x = 0$ if $\sigma(x) \subseteqq \sigma'$. Moreover,*

$$E(\sigma) + E(\sigma') = I, \qquad E(\sigma)E(\sigma') = 0, \qquad |E(\sigma)| \leqq K.$$

PROOF. The properties

$$(*) \qquad E(\sigma)x = x \quad \text{if} \quad \sigma(x) \subseteqq \sigma, \qquad E(\sigma)x = 0 \quad \text{if} \quad \sigma(x) \subseteqq \sigma'$$

define the projection $E(\sigma)$ on the dense set $\mathfrak{D} = \{x+y \mid \sigma(x) \subseteqq \sigma, \sigma(y) \subseteqq \sigma'\}$. Thus the uniqueness of $E(\sigma)$ is assured by the requirement that it be bounded.

To prove that an $E(\sigma)$ with the properties (*) exists, note that by Lemma 2.3 the properties (*) define a single valued projection on $\mathfrak{D}$. Assumption (B) merely states that this projection is bounded, with bound at most K. Thus it has a unique extension by continuity to a projection, with bound at most K, defined on $\mathfrak{X}$. Since it is clear that $(E(\sigma)+E(\sigma'))x = x$ and $E(\sigma)E(\sigma')x = 0$ for x in $\mathfrak{D}$, it follows by continuity that these properties hold for all x in $\mathfrak{X}$. Q.E.D.

3 LEMMA (A, B). *If P is a bounded linear operator which commutes with T, then*

$$PE(\sigma) = E(\sigma)P, \qquad \sigma \in \mathscr{S}_1(T).$$

PROOF. For σ in $\mathscr{S}_1$, vectors of the form $z = x + y$ with $\sigma(x) \subseteqq \sigma$ and $\sigma(y) \subseteqq \sigma'$ are dense in $\mathfrak{X}$. For such a vector z we have $Pz = Px + Py$ and, by Lemma 2.4, $\sigma(Px) \subseteqq \sigma$ and $\sigma(Py) \subseteqq \sigma'$. Thus, by Lemma 2, $E(\sigma)Px = Px$ and $E(\sigma)Py = 0$ and so

$$E(\sigma)Pz = Px = PE(\sigma)z.$$

Since the vectors z are dense in $\mathfrak{X}$, we have $E(\sigma)P = PE(\sigma)$. Q.E.D.

We now introduce a subclass of $\mathscr{S}_1(T)$, which will be shown to be a Boolean algebra.

4 DEFINITION (A, B). The symbol $\mathscr{S}_2(T)$ will be used for the family of all sets σ having the property that for every x in $\mathfrak{X}$ and every $\varepsilon > 0$, there are vectors x_1, x_1' with $\sigma(x_1) \subseteqq \sigma(x)\sigma$, $\sigma(x_1') \subseteqq \sigma(x)\sigma'$, and $|x_1 + x_1' - x| < \varepsilon$.

It is clear that $\mathscr{S}_2(T)$ is closed under complementation, and contains the void set and the whole plane.

5 LEMMA (A, B). *The family $\mathscr{S}_2(T)$ is a Boolean algebra.*

PROOF. Since $\mathscr{S}_2(T)$ is closed under complementation, to prove the lemma it is sufficient to show that it contains the union of every pair of its elements.

Let σ_1 and σ_2 be sets in $\mathscr{S}_2(T)$ and, for every set μ of complex numbers, let

$$\mathfrak{M}(\mu) = \{x \mid \sigma(x) \subseteqq \mu\}.$$

If x is in $\mathfrak{X}$, then, since σ_1 is in $\mathscr{S}_2(T)$, x is in the closure of $\mathfrak{M}(\sigma_1\sigma(x)) + \mathfrak{M}(\sigma_1'\sigma(x))$. On the other hand, it follows, since σ_2 is in $\mathscr{S}_2(T)$, that

$\mathfrak{M}(\sigma_2\sigma_1'\sigma(x))+\mathfrak{M}(\sigma_2'\sigma_1'\sigma(x))$ is dense in $\mathfrak{M}(\sigma_1'\sigma(x))$. Since, by Lemma 2.1, $\mathfrak{M}(\sigma_1\sigma(x))+\mathfrak{M}(\sigma_2\sigma_1'\sigma(x))$ is contained in $\mathfrak{M}((\sigma_1\cup\sigma_2)\sigma(x))$, it follows immediately that x is in the closure of $\mathfrak{M}((\sigma_1\cup\sigma_2)\sigma(x))+\mathfrak{M}((\sigma_1\cup\sigma_2)'\sigma(x))$. But this means that $\sigma_1\cup\sigma_2$ is in $\mathscr{S}_2(T)$. Q.E.D.

6 LEMMA (A, B). *The restriction of the projection valued function E from $\mathscr{S}_1(T)$ to the Boolean algebra $\mathscr{S}_2(T)$ is a spectral measure.*

PROOF. We use the notations of the proof of the preceding lemma and let σ be in $\mathscr{S}_2(T)$. Since an arbitrary vector is in the closure of $\mathfrak{M}(\sigma\sigma(x))+\mathfrak{M}(\sigma'\sigma(x))$ and since, by Lemma 2, $E(\sigma)(z+y)=z$ for z in $\mathfrak{M}(\sigma\sigma(x))$ and y in $\mathfrak{M}(\sigma'\sigma(x))$, it follows that $E(\sigma)x$ is in $\overline{\mathfrak{M}(\sigma\sigma(x))}$, and that $E(\sigma)x=x$ for x in $\overline{\mathfrak{M}(\sigma\sigma(x))}$. Thus, if σ_1, σ_2 are in $\mathscr{S}_2(T)$, then $E(\sigma_1)E(\sigma_2)x$ is in $\overline{\mathfrak{M}(\sigma_1\sigma_2\sigma(x))}$. Hence

$$E(\sigma_1\sigma_2)E(\sigma_1)E(\sigma_2)x=E(\sigma_1)E(\sigma_2)x.$$

Since

$$E(\sigma_1\sigma_2)x\subseteq\overline{\mathfrak{M}(\sigma_1\sigma_2\sigma(x))}\subseteq\overline{\mathfrak{M}(\sigma_1\sigma(x))\cap\mathfrak{M}(\sigma_2\sigma(x))},$$

we have

$$E(\sigma_1)E(\sigma_2)E(\sigma_1\sigma_2)x=E(\sigma_1\sigma_2)x.$$

Since all the projections $E(\cdot)$ commute with T and hence with each other, by Lemma 3 it follows that $E(\sigma_1)E(\sigma_2)=E(\sigma_1\sigma_2)$. Then we have

$$\begin{aligned}E(\sigma_1)\vee E(\sigma_2)&=E(\sigma_1)+E(\sigma_2)-E(\sigma_1\sigma_2)\\&=I-(I-E(\sigma_1))(I-E(\sigma_2))\\&=I-(E(\sigma_1')E(\sigma_2'))=I-E(\sigma_1'\sigma_2')\\&=E((\sigma_1'\sigma_2')')=E(\sigma_1\cup\sigma_2).\end{aligned}$$ Q.E.D.

7 DEFINITION (A, B). The symbol $\mathscr{S}(T)$ will be used for the collection of those sets $\sigma\in\mathscr{S}_2(T)$ for which there exist closed sets $\mu_n, \nu_n\in\mathscr{S}_2(T)$ with $\mu_n\subseteq\sigma$, $\nu_n\subseteq\sigma'$, $n=1, 2, \ldots$, and

$$x=\lim_{n\to\infty}[E(\nu_n)+E(\mu_n)]x, \qquad x\in\mathfrak{X}.$$

8 LEMMA (A, B). *The family $\mathscr{S}(T)$ is a Boolean algebra.*

PROOF. It is clear that $\mathscr{S}(T)$ is closed under complementation. Hence, in order to show that $\mathscr{S}(T)$ is a Boolean algebra, it will suffice to show that it is closed under the operation of forming unions.

Let σ, $\tilde{\sigma} \in \mathscr{S}(T)$. Let $\{\mu_n\}$ and $\{\nu_n\}$ be as in Definition 7, and let $\{\tilde{\mu}_n\}$ and $\{\tilde{\nu}_n\}$ be sequences of closed sets in $\mathscr{S}_2(T)$ such that $\tilde{\mu}_n \subseteq \tilde{\sigma}$, $\tilde{\nu}_n \subseteq \tilde{\sigma}'$, and

$$x = \lim_{n \to \infty} \{E(\tilde{\nu}_n)x + E(\tilde{\mu}_n)x\}, \qquad x \in \mathfrak{X}.$$

Since the sequence $\{E(\nu_n) + E(\mu_n)\}$ is strongly convergent, it is bounded (cf. II.3.6) and the operators $E(\nu_n) + E(\mu_n)$ are therefore equi-continuous. Thus we have

$$\begin{aligned} x &= \lim_{n \to \infty} [E(\nu_n) + E(\mu_n)][E(\tilde{\mu}_n) + E(\tilde{\nu}_n)]x \\ &= \lim_{n \to \infty} \{E(\mu_n \tilde{\mu}_n \cup \mu_n \tilde{\nu}_n \cup \nu_n \tilde{\mu}_n)x + E(\nu_n \tilde{\nu}_n)x\}, \qquad x \in \mathfrak{X}. \end{aligned}$$

Since $\{\mu_n \tilde{\mu}_n \cup \mu_n \tilde{\nu}_n \cup \nu_n \tilde{\mu}_n\}$ and $\{\nu_n \tilde{\nu}_n\}$ are sequences of closed sets in $\mathscr{S}_2(T)$ contained in $\sigma \cup \tilde{\sigma}$ and $(\sigma \cup \tilde{\sigma})'$, respectively, it follows that $\sigma \cup \tilde{\sigma}$ is in $\mathscr{S}(T)$. Q.E.D.

9 LEMMA (A, B). *The set $\sigma(T)$ belongs to $\mathscr{S}(T)$, and $E(\sigma(T)) = I$. Furthermore, every subset δ of the resolvent set $\rho(T)$ is in $\mathscr{S}(T)$ and has $E(\delta) = 0$.*

PROOF. Since $\sigma(x) \subseteq \sigma(T)$ for all $x \in \mathfrak{X}$, it is clear from Definition 4 and Lemma 2 that $\sigma(T)$ is in $\mathscr{S}_2(T)$ and that $E(\sigma(T))x = x$ for all $x \in \mathfrak{X}$. Since $\sigma(T)$ is closed (cf. VII.3.2), it is clear from Definition 7 that $\sigma(T)$ is in $\mathscr{S}(T)$. If $\delta \subseteq \rho(T)$, then the void set ϕ and the spectrum $\sigma(T)$ are closed subsets of δ, δ' respectively and $E(\phi)x + E(\sigma(T))x = x$, which proves that δ is in $\mathscr{S}(T)$. Since $E(\rho(T)) = 0$, we have $E(\delta) = E(\delta)E(\rho(T)) = 0$. Q.E.D.

10 LEMMA (A, B). *Let $\{\sigma_m\}$ be a decreasing sequence of sets in $\mathscr{S}(T)$ whose limit σ is also in $\mathscr{S}(T)$. Then*

$$E(\sigma)x = \lim_{m \to \infty} E(\sigma_m)x, \qquad x \in \mathfrak{X}.$$

PROOF. We wish to show that $\lim_{m \to \infty} E(\sigma_m - \sigma)x = 0$ for all x. Thus we may pass without loss of generality from consideration of the sequence $\{\sigma_m\}$ to consideration of the sequence $\{\sigma_m - \sigma\}$; that is, we may and shall assume without loss of generality that σ is void. Since $E(\sigma_m) = E(\sigma_m \sigma(T))$, by the preceding lemma, we may also assume that $\sigma_m \subseteq \sigma(T)$.

Suppose that our assertion is false, so that there is a $p > 0$ and a vector x such that $|E(\sigma_m)x| \geqq p$ for arbitrarily large m. Passing without loss of generality to a subsequence, we may assume that $|E(\sigma_m)x| \geqq p$ for all m.

For σ in $\mathscr{S}_2(T)$ let $M(\sigma) = \sup_{\mu \subseteqq \sigma} |E(\mu)x|$, $\mu \in \mathscr{S}_2(T)$. It is clear that if $\nu_1 \subseteqq \nu_2$ then $M(\nu_1) \leqq M(\nu_2)$. Let $\mu \subseteqq \nu_1 \cup \nu_2$. Since $|E(\mu)x| = |E(\mu\nu_1)x + E(\mu\nu_1'\nu_2)x| \leqq M(\nu_1) + M(\nu_2)$, it follows immediately that

$$M(\nu_1 \cup \nu_2) \leqq M(\nu_1) + (M\nu_2).$$

Since

$$|E(\mu)x| = |E(\mu)E(\sigma)x| \leqq K\,|E(\sigma)x|$$

for $\mu \subseteqq \sigma$, it is seen that

$$M(\sigma) \leqq K\,|E(\sigma)x|.$$

Since σ_m is in $\mathscr{S}(T)$, we can find closed sets μ_m and ν_m in $\mathscr{S}_2(T)$ such that $\mu_m \subseteqq \sigma_m$, $\nu_m \subseteqq \sigma_m'$, and

$$|E((\mu_m \cup \nu_m)')x| \leqq pK^{-1}2^{-m-1}.$$

Then

$$M((\mu_m \cup \nu_m)') \leqq p2^{-m-1},$$

so that, putting $\delta_m = \sigma_m - \mu_m$, we have $\delta_m = \sigma_m \mu_m' \subseteqq \mu' \cap \nu_m' = (\mu_m \cup \nu_m)'$ and so

$$M(\delta_m) \leqq p2^{-m-1}.$$

It follows that no finite sum $\delta_1 \cup \cdots \cup \delta_n$ can cover σ_n. Indeed,

$$M(\delta_1 \cup \cdots \cup \delta_n) \leqq \sum_{i=1}^{n} p2^{-i-1} \leqq \tfrac{1}{2}p,$$

while $|E(\sigma_n)x| \geqq p$. Hence

$$\sigma_n \mu_1\mu_2 \cdots \mu_n = \sigma_n - \bigcup_{i=1}^{n} \sigma_n \delta_i$$

is non-void. Since $\bigcap_{i=1}^{n} \mu_i$ is a decreasing sequence of non-void closed subsets of the compact set $\sigma(T)$, we have $\bigcap_{i=1}^{\infty} \mu_i \neq \phi$. Thus, since $\mu_i \subseteqq \sigma_i$ it follows that $\bigcap_{i=1}^{\infty} \sigma_i \neq \phi$, contrary to assumption. Q.E.D.

The following three theorems summarize the results of this section and at the same time lay the foundation for the studies to be made in Sections 4 and 5 that follow.

11 THEOREM (A, B). *Let T be a bounded linear operator in the complex B-space $\mathfrak{X}$. Then there is a unique spectral measure on the field $\mathscr{S}(T)$ with the properties*

$$\begin{aligned} E(\delta)x &= x, \qquad \delta \in \mathscr{S}(T),\ \sigma(x) \subseteq \delta, \\ &= 0, \qquad \delta \in \mathscr{S}(T),\ \sigma(x) \subseteq \delta'. \end{aligned}$$

This spectral measure is bounded, is countably additive on $\mathscr{S}(T)$, and commutes with T.

PROOF. From Definitions 1, 4, and 7 it is seen that $\mathscr{S}(T) \subseteq \mathscr{S}_1(T)$, and thus for each δ in $\mathscr{S}(T)$ there is, by Lemma 2, one and only one projection $E(\delta)$ with $E(\delta)x = x$ if $\sigma(x) \subseteq \delta$ and $E(\delta)x = 0$ if $\sigma(x) \subseteq \delta'$. Lemma 2 also shows that $|E(\delta)|$ is bounded in δ. Lemma 3 shows that $E(\delta)$ commutes with T and Lemmas 6, 8, and 10 show that E is a countably additive spectral measure on $\mathscr{S}(T)$. Q.E.D.

Since the field $\mathscr{S}(T)$ is not necessarily a σ-field, it is natural to ask whether or not the spectral measure E may be extended to the σ-field generated by $\mathscr{S}(T)$. The following definition and theorems are concerned with this question.

12 DEFINITION. The symbol $\mathscr{M}(T)$ will be used for the σ-complete Boolean algebra (or σ-field) determined by the Boolean algebra $\mathscr{S}(T)$. The sets in $\mathscr{M}(T)$ are called *sets measurable* T or T*-measurable sets.*

13 THEOREM (A, B). *Let T be a bounded linear operator in the complex B-space $\mathfrak{X}$ and let E be the associated spectral measure whose existence was established in Theorem* 11. *Then, in the conjugate space $\mathfrak{X}^*$, there is a unique extension of the adjoint E^* to a spectral measure on the σ-field $\mathscr{M}(T)$ of sets measurable T which is countably additive on $\mathscr{M}(T)$ in the $\mathfrak{X}$ topology of $\mathfrak{X}^*$. This unique extension is bounded and commutes with T^*.*

PROOF. For every x in $\mathfrak{X}$ and x^* in $\mathfrak{X}^*$ there is, according to the Hahn extension theorem (cf. Corollary III.5.9), a unique countably additive extension $m(e, x, x^*)$ of $x^*E(e)x$ from $\mathscr{S}(T)$ to $\mathscr{M}(T)$. From its uniqueness it is seen that $m(e, x, x^*)$ is bilinear in x, x^*, and from the boundedness of $|E(e)|$ follows the boundedness of $m(e, x, x^*)$. Thus for each e in $\mathscr{M}(T)$ there is a uniquely defined bounded linear operator $A(e)$ in $\mathfrak{X}^*$ for which

$$xA(e)x^* = m(e, x, x^*).$$

It will next be shown that the mapping $e \to A(e)$ of $\mathcal{M}(T) \to B(\mathfrak{X}^*)$ is a spectral measure. It clearly preserves finite disjoint unions, takes complements into complements, is countably additive in the $\mathfrak{X}$ topology of $\mathfrak{X}^*$, and is bounded. It remains only to show that

$$A(\sigma)A(\delta) = A(\sigma\delta).$$

It is seen, by using the above remarks, that for a fixed σ the family of δ for which the equation is valid is a σ-field. Thus if σ is in $\mathcal{S}(T)$, the equation holds for all δ in $\mathcal{M}(T)$. Analogously, if δ is fixed in $\mathcal{M}(T)$, then since the equation holds for σ in a σ-field containing $\mathcal{S}(T)$, it must hold for all σ in $\mathcal{M}(T)$.

Since T and $E(\delta)$ commute and since $A(\delta) = E(\delta)^*$ for δ in $\mathcal{S}(T)$, we have

$$xT^*A(\delta)x^* = xA(\delta)T^*x^*, \qquad x \in \mathfrak{X},\ x^* \in \mathfrak{X}^*,\ \delta \in \mathcal{S}(T).$$

Since A is countably additive in the $\mathfrak{X}$ topology of $\mathfrak{X}^*$, this identity holds for every δ in the σ-field determined by $\mathcal{S}(T)$ and this proves that $A(\delta)$ commutes with T^* for every T-measurable set δ. Since $m(e, x, x^*) = x^*A(e)x$ is bounded in e, it follows from the principle of uniform boundedness (cf. II.3.21) that $|A(e)|$ is bounded for e in $\mathcal{M}(T)$. Q.E.D.

→ 14 THEOREM (A, B). *Let T be a bounded linear operator in the weakly complete complex B-space $\mathfrak{X}$ and let E be the associated spectral measure whose existence was established in Theorem* 11. *Then there is a uniquely determined extension of E to a spectral measure on the σ-field $\mathcal{M}(T)$ of sets measurable T which is countably additive on $\mathcal{M}(T)$ in the strong operator topology. This extension is bounded and commutes with T.*

PROOF. Let A be the spectral measure in the adjoint space $\mathfrak{X}^*$ which is associated with T^* as in the preceding theorem. If $\mathfrak{X}$ is weakly complete, $A(\delta)$ is the adjoint of an operator $E(\delta)$ in $\mathfrak{X}$. To see this, note that the family of all sets δ for which there exists an operator $E(\delta)$ in $\mathfrak{X}$ with $A(\delta) = E(\delta)^*$ contains $\mathcal{S}(T)$. This family is also a Boolean algebra since A is a spectral measure. Since $\mathfrak{X}$ is weakly complete, it is a σ-complete Boolean algebra and hence coincides with $\mathcal{M}(T)$. Since $E(\delta)^*$ commutes with T^*, it follows that $E(\delta)$ commutes with T for every δ in $\mathcal{M}(T)$. Theorem 13 shows that E is countably additive on $\mathcal{M}(T)$, and Corollary XV.2.4 shows that E is countably additive on $\mathcal{M}(T)$ in the strong operator topology. The boundedness of E follows from that of A. Q.E.D.

4. Consequences of the Conditions (A, B, C): Necessary and Sufficient Conditions for Spectral Operators

Theorem 3.14 falls short of proving that T is a spectral operator in two ways. First of all the spectral measure E is not necessarily a resolution of the identity for T, for, even though it commutes with T, it may not satisfy the inclusion relation $\sigma(T_\delta) \subseteq \bar{\delta}$ (cf. Definition XV.2.2, or Definition 1 below). Second, the field $\mathscr{S}(T)$ or the σ-field $\mathscr{M}(T)$ may not contain all Borel sets. The first of these difficulties will be eliminated by the hypothesis (C), to be made presently. The second of these difficulties leads us to consider operators which are spectral relative to a field other than the field of Borel sets. Such operators are described as follows.

1 Definition. Let Σ be a field of sets in the complex plane and let T be a linear operator in the complex B-space $\mathfrak{X}$. Then a spectral measure E on Σ is said to be a *resolution of the identity for* T if it commutes with T and satisfies

$$\sigma(T_\delta) \subseteq \bar{\delta}, \qquad \delta \in \Sigma,$$

where T_δ is the restriction of T to $E(\delta)\mathfrak{X}$. The operator T is said to be a *spectral operator of class* $(\Sigma, \mathfrak{X}^*)$ if it has a bounded resolution of the identity on Σ for which the set functions $x^*E(\cdot)x$, with x in $\mathfrak{X}$ and x^* in $\mathfrak{X}^*$, are all countably additive on Σ. An operator U in $\mathfrak{X}^*$ is said to be a *spectral operator of class* $(\Sigma, \mathfrak{X})$ if it has a bounded resolution of the identity A on Σ for which the set functions $xA(\cdot)x^*$, with x in $\mathfrak{X}$ and x^* in $\mathfrak{X}^*$, are all countably additive on Σ.

Thus T is a spectral operator if and only if it is a spectral operator of class $(\mathscr{B}, \mathfrak{X}^*)$, where $\mathscr{B}$ is the field of Borel sets in the plane.

Besides conditions (A) and (B) of Sections 2 and 3, the following condition (C) will be assumed in most of what follows. However, when any of the assumptions (A), (B), (C) are made in a lemma or theorem, they will be indicated parenthetically.

(C) *For every closed set δ of complex numbers the set of all vectors x with $\sigma(x) \subseteq \delta$ is also closed.*

2 Lemma (A, B, C). *For every set δ in $\mathscr{S}_1(T)$ and every vector z in $\mathfrak{X}$ we have $\sigma(E(\delta)z) \subseteq \bar{\delta}\sigma(z)$.*

PROOF. Since δ is in $\mathscr{S}_1(T)$, an arbitrary vector z in $\mathfrak{X}$ is the limit of a sequence $z_n = x_n + y_n$ with $\sigma(x_n) \subseteq \delta$ and $\sigma(y_n) \subseteq \delta'$. Thus

$$\sigma(E(\delta)z_n) = \sigma(x_n) \subseteq \bar{\delta},$$

and since $E(\delta)z_n \to E(\delta)z$, it follows from (C) that

$$\sigma(E(\delta)z) \subseteq \bar{\delta}.$$

Since $E(\delta)$ commutes with T (cf. Lemma 3.3), it is seen from Lemma 2.4 that $\sigma(E(\delta)z) \subseteq \sigma(z)$. Thus $\sigma(E(\delta)z) \subseteq \bar{\delta}\sigma(z)$. Q.E.D.

3 LEMMA (A, B, C). *For δ in $\mathscr{S}_1(T)$ let T_δ be the restriction of T to $E(\delta)\mathfrak{X}$. Then*

$$\sigma(T_\delta) \subseteq \bar{\delta}, \qquad \delta \in \mathscr{S}_1(T).$$

PROOF. It follows from Lemma 3.3 that T commutes with $E(\delta)$ and so T maps $E(\delta)\mathfrak{X}$ into itself. It is therefore meaningful to speak of the spectrum of the restriction of T to $E(\delta)\mathfrak{X}$.

Let $\xi \notin \bar{\delta}$. It will first be shown that $\xi I - T$ is one-to-one on $E(\delta)\mathfrak{X}$. If x is in $E(\delta)\mathfrak{X}$ and $(\xi I - T)x = 0$, then, since

$$R(\lambda; T) = \sum_{n=0}^{\infty} \frac{(T - \xi I)^n}{(\lambda - \xi)^{n+1}},$$

for all large λ, it is seen that $x(\lambda) = x/(\lambda - \xi)$ for $\lambda \neq \xi$. Thus the spectrum $\sigma(x)$ contains at most the point ξ, and therefore $\bar{\delta}\sigma(x)$ is void. Since $x = E(\delta)x$ it follows from Lemma 2 that $\sigma(x)$ is void and from Lemma 2.2 that $x = 0$. This shows that $\xi I - T$ is one-to-one on $E(\delta)\mathfrak{X}$.

It will now be shown that $(\xi I - T)E(\delta)\mathfrak{X} = E(\delta)\mathfrak{X}$. Let $x = E(\delta)x$ be an arbitrary point in $E(\delta)\mathfrak{X}$. Then, by Lemma 2, $\sigma(x) \subseteq \bar{\delta}$ and so $\xi \in \rho(x)$. Thus $(\xi I - T)x(\xi) = x$ and hence $(\xi I - T)E(\delta)x(\xi) = E(\delta)x = x$, which shows that $(\xi I - T)E(\delta)\mathfrak{X} = E(\delta)\mathfrak{X}$. The operator $\xi I - T$ therefore maps $E(\delta)\mathfrak{X}$, in a one-to-one manner, onto all of itself. This means that ξ is in $\rho(T_\delta)$ and thus $\sigma(T_\delta) \subseteq \bar{\delta}$. Q.E.D.

4 THEOREM. *A spectral operator T has the properties* (A), (B), *and* (C). *Conversely, if the bounded linear operator T has these properties, it is a spectral operator of class $(\mathscr{S}(T), \mathfrak{X}^*)$. Moreover, T has a resolution of the identity which is countably additive in the strong operator topology.*

PROOF. It was shown in Theorem XV.3.2 and Corollary XV.3.6 that the spectral operator T has the properties (A) and (C). Let E be the resolution of the identity for T and let $|E(\delta)| \leqq K$ for all Borel sets δ. Let $\sigma(x)$ and $\sigma(y)$ be disjoint. Then by Theorem XV.3.4 we have $E(\sigma(x))x = x$, $E(\sigma(y))y = y$. Hence

$$0 = E(\sigma(x)\sigma(y))y = E(\sigma(x))E(\sigma(y))y = E(\sigma(x))y.$$

Thus

$$|x| = |E(\sigma(x))x| = |E(\sigma(x))(x+y)| \leqq K\,|x+y|.$$

Conversely, let the bounded linear operator T satisfy conditions (A), (B), and (C). Then, by Theorem 3.11 and Lemma 3, T is a spectral operator of class $(\mathscr{S}(T), \mathfrak{X}^*)$ with a resolution of the identity which is countably additive in the strong operator topology. Q.E.D.

→ 5 THEOREM. *Let T be a bounded linear operator in a weakly complete space. Then T is a spectral operator if and only if T satisfies conditions* (A), (B), (C), *and the following condition* (D):

(D) *Every complex number is interior to a set of arbitrarily small diameter belonging to $\mathscr{S}(T)$.*

PROOF. It was noted in the preceding theorem that a spectral operator has properties (A) through (C). To show that a spectral operator T has property (D), let δ be a closed set in the complex plane, and let $\{\delta_n\}$ be an increasing sequence of closed sets whose union is the complement δ' of δ. Let E be the spectral resolution of T. Then

$$x = \lim_n \{E(\delta)x + E(\delta_n)x\}.$$

By Theorem XV.3.4, $\sigma(E(\delta)x) \subseteq \delta$ and $\sigma(E(\delta_n)x) \subseteq \delta_n$. This shows that δ is in $\mathscr{S}_1(T)$. Lemma 2 shows that δ is in $\mathscr{S}_2(T)$, and since δ_n is closed, the above equation proves that δ is in $\mathscr{S}(T)$. Thus $\mathscr{S}(T)$ contains every closed set, and property (D) is evident.

Conversely, if the operator T satisfies conditions (A) through (D), then, by Theorem 4, it is a spectral operator of class $(\mathscr{S}(T), \mathfrak{X}^*)$. According to Theorem 3.14, the resolution of the identity for T has a unique extension to a countably additive spectral measure E on $\mathscr{M}(T)$. It will next be shown that $\mathscr{M}(T)$ contains all Borel sets.

To do this, let U be an open set of the complex plane and let K be a compact subset of U. Then, by (D), each point p in K is interior to a certain set σ_p in $\mathscr{S}(T)$ with $\sigma_p \subseteq U$. Since K is compact, it is contained in the union σ of a finite collection of the sets σ_p. Thus we have shown that if K is a compact subset of U, there exists a $\sigma \in \mathscr{S}(T)$ such that $K \subseteq \sigma \subseteq U$. Since U is the union of a countable infinity of its own compact subsets, it follows that U is in $\mathscr{M}(T)$. Since $\mathscr{M}(T)$ contains all open sets, it contains the family $\mathscr{B}$ of all Borel sets.

To complete the proof it will suffice to show that $\sigma(T_\delta) \subseteq \bar{\delta}$ for every Borel set δ. If $\lambda \notin \bar{\delta}$, then, using (D), the compact set $\bar{\delta}\sigma(T)$ may be covered by a set σ in the field $\mathscr{S}(T)$ with $\lambda \notin \bar{\sigma}$. Since T is a spectral operator of class $(\mathscr{S}(T), \mathfrak{X}^*)$, we have $\sigma(T_\sigma) \subseteq \bar{\sigma}$ and consequently λ is in $\rho(T_\sigma)$, which means that $\lambda I - T$ is a one-to-one map of $E(\sigma)\mathfrak{X}$ into all of itself. Since $\sigma \supseteq \delta\sigma(T)$, we have $E(\sigma) \supseteq E(\delta\sigma(T)) = E(\delta)$ and consequently $E(\delta)\mathfrak{X}$ is an invariant subspace of $E(\sigma)\mathfrak{X}$. Thus $\lambda I - T$ is a one-to-one map of $E(\delta)\mathfrak{X}$ into all of itself. This proves that λ is in $\rho(T_\delta)$ and thus that $\sigma(T_\delta) \subseteq \bar{\delta}$. Q.E.D.

We conclude this section with two results on adjoint operators.

6 LEMMA. *Let Σ be a field of sets in the complex plane and let T be a spectral operator of class $(\Sigma, \mathfrak{X}^*)$. Then its adjoint T^* is a spectral operator of class $(\Sigma, \mathfrak{X})$.*

PROOF. Let E be a resolution of the identity for T. Then the mapping $\sigma \to E(\sigma)^*$ of Σ into $B(\mathfrak{X}^*)$ is a spectral measure in $\mathfrak{X}^*$. Moreover, $xE(\sigma)^*x^*$ is evidently countably additive on Σ for each $x \in \mathfrak{X}$ and $x^* \in \mathfrak{X}^*$. Let $\lambda \notin \bar{\sigma}$. Then the restriction of $\lambda I - T$ to $E(\sigma)\mathfrak{X}$ has an inverse R_σ. Define the operator P_σ in $\mathfrak{X}$ by putting $P_\sigma = R_\sigma E(\sigma)$. Then clearly $E(\sigma)P_\sigma = P_\sigma = P_\sigma E(\sigma)$. Hence

$$P_\sigma^* E(\sigma)^* = E(\sigma)^* P_\sigma^*,$$

so that P_σ^* maps $E(\sigma)^*\mathfrak{X}^*$ into itself. Also $(\lambda I - T)P_\sigma = E(\sigma)$, and $P_\sigma(\lambda I - T) = P_\sigma E(\sigma)(\lambda I - T) = P_\sigma(\lambda I - T)E(\sigma) = E(\sigma)$. Thus

$$P_\sigma^*(\lambda I^* - T^*) = (\lambda I^* - T^*)T_\sigma^* = E(\sigma)^*.$$

Consequently the restriction of P_σ^* to $E(\sigma)^*\mathfrak{X}^*$ is the inverse of the restriction of $\lambda I^* - T^*$ to $E(\sigma)^*\mathfrak{X}^*$. Hence λ is in the resolvent of the restriction $(T^*)_\sigma$ of T^* to $E(\sigma)^*\mathfrak{X}^*$. This shows that $\sigma((T^*)_\sigma) \subseteq \bar{\sigma}$ and completes the proof. Q.E.D.

→ 7 THEOREM (A, B, C, D). *Let T be a bounded linear operator in the complex B-space $\mathfrak{X}$ and let $\mathscr{B}$ be the field of Borel sets in the plane. Then T^* is a spectral operator of class $(\mathscr{B}, \mathfrak{X})$.*

PROOF. In view of condition (D), we have $\mathscr{B} \subseteq \mathscr{M}(T)$. By Theorem 3.13 the spectral measure E^* of the preceding lemma may be extended from $\mathscr{S}(T)$ to a spectral measure defined on $\mathscr{M}(T)$. Then, as in the proof of Theorem 5, it may be shown that $\sigma(T \mid E(\delta)\mathfrak{X}) \subseteq \bar{\delta}$ for each δ in $\mathscr{M}(T)$. Q.E.D.

5. Operators Whose Spectra Lie in a Jordan Curve

In the preceding section it was seen (4.5 and 4.7) that operators satisfying the conditions (A), ..., (D) are spectral operators. In this section it is shown that, in certain important special cases, all of these conditions, except possibly the boundedness condition (B), are automatically satisfied. Thus, for the special types of operators, condition (B) becomes the condition which is necessary as well as sufficient for the operator to be a spectral operator.

1 LEMMA. *Condition* (A) *of Section* 2 *is satisfied if the spectrum of* T *is nowhere dense in the complex plane.*

PROOF. If the resolvent set is dense, then any two analytic, or even continuous, extensions of $R(\lambda; T)x$ must coincide on their common domain of continuity. Q.E.D.

All of the special type operators to be considered in the present section will have nowhere dense spectra so that, according to Lemma 1, condition (A) will be satisfied by all of the operators that will be studied here.

The following theorem, which applies in particular to compact operators, gives a topological restriction on the spectrum of T which guarantees that (A), (C), and (D) are all satisfied.

→ 2 THEOREM. *If the spectrum of an operator in a weakly complete space is totally disconnected, then it is a spectral operator if and only if the boundedness condition* (B) *of Section* 3 *is satisfied.*

PROOF. Let T be a bounded linear operator in the weakly complete B-space $\mathfrak{X}$. To prove the theorem it will, in view of Theorem 4.5, suffice to show that T has properties (A), (C), and (D). Since the spectrum $\sigma(T)$ of T is totally disconnected, it is nowhere dense and, according to Lemma 1, condition (A) is satisfied.

It is clear from Theorem VII.3.20 that every spectral set belongs to $\mathscr{S}_2(T)$, and hence it follows from Definition 3.7 that every spectral set belongs to $\mathscr{S}(T)$. Since the spectrum is totally disconnected, every spectral point is contained in a spectral set of arbitrarily small diameter and thus in an $\mathscr{S}(T)$ set of arbitrarily small diameter. Since it is clear that every subset of the resolvent set is an $\mathscr{S}(T)$ set, condition (D) is immediate.

To verify condition (C), let δ be a closed set of complex numbers and let

$$\mathfrak{M}(\delta) = \{x \,|\, \sigma(x) \subseteq \delta\}.$$

Condition (C) will be proved by showing that $\mathfrak{M}(\delta)$ is closed. Since $\sigma(x) \subseteq \sigma(T)$ we have $\mathfrak{M}(\delta) = \mathfrak{M}(\delta\sigma(T))$, and it may therefore be assumed, without loss of generality, that $\delta \subseteq \sigma(T)$. Since $\sigma(T)$ is totally disconnected, the closed set δ is an intersection $\bigcap_\alpha \delta_\alpha$ of spectral sets δ_α. Now clearly

$$\mathfrak{M}(\delta) = \mathfrak{M}(\bigcap_\alpha \delta_\alpha) = \bigcap_\alpha \mathfrak{M}(\delta_\alpha),$$

and so to see that $\mathfrak{M}(\delta)$ is closed it will suffice to see that $\mathfrak{M}(\delta_\alpha)$ is closed. Since δ_α is a spectral set, this follows from Theorem VII.3.20. Q.E.D.

Theorem 2 suggests that the difficulties which may be encountered in verifying conditions (C) and (D) are, in some way, related to the presence of connected components of the spectrum. The remainder of the present section will be devoted to a study of the case where the spectrum is contained in a finite disjoint union of connected sets each one of which is a Jordan arc. Before passing to the details of this study, the following result will be introduced to allow us, without any loss of generality, to study the case where the whole spectrum is contained in one Jordan arc.

3 THEOREM. *Let T be an operator in the B-space $\mathfrak{X}$. If $\mathfrak{X}$ is the direct sum of two of its closed subspaces $\mathfrak{X}_1$ and $\mathfrak{X}_2$, each invariant under T, and if the restrictions of T to $\mathfrak{X}_1$ and $\mathfrak{X}_2$ are both spectral operators, then T is a spectral operator.*

PROOF. This is a corollary of the case $n = 2$ of Theorem XV.3.10. Q.E.D.

In most of the remainder of the present section it will be assumed that the spectrum $\sigma(T)$ of T is contained in a closed Jordan curve Γ_0. In order to avoid technical complications it will be convenient to assume also that Γ_0 is smoothly embedded in a one parameter family of closed rectifiable Jordan curves. More specifically, and as a basis for the analytical discussion that follows, it will be assumed that there is a function $\xi = \xi(t,\delta)$ which is twice continuously differentiable on its domain $-1 \leqq t,\ \delta \leqq 1$ of definition and which has the following properties. The equation $\xi(-1, \delta) = \xi(+1, \delta)$ holds for all δ in the interval $-1 \leqq \delta \leqq 1$, whereas $\xi(s, \delta) \neq \xi(t, \delta)$ unless $s = t$ or the pair s, t is the pair $-1, 1$. Thus $\xi(\cdot, \delta)$ is the parametric representation of a simple closed rectifiable Jordan curve Γ_δ. It is assumed that the curves Γ_δ are mutually disjoint, that Γ_{δ_1} lies inside Γ_{δ_2} if $-1 \leqq \delta_1 < \delta_2 \leqq 1$, and that Γ_0 contains the spectrum $\sigma(T)$. There will be occasion to integrate around the curve Γ_δ with respect to its arc length, and for this reason it is supposed that the curves Γ_δ are oriented in the positive sense customary in the theory of complex variables. The simple Jordan arc Δ_{λ_0} which is parametrized by the function $\xi(t_0, \cdot)$ is called the *transversal through the point* $\lambda_0 = \xi(t_0, 0)$. The principal assumption that will be made throughout most of this section is that the resolvent $R(\lambda; T)$ has a finite rate of growth as λ approaches a spectral point λ_0 along the transversal Δ_{λ_0} through λ_0. This rate-of-growth hypothesis is stated formally as follows.

(G) *The spectrum of T is contained in the rectifiable Jordan curve Γ_0 described above. Moreover, for each spectral point λ_0 there are two positive integers $\nu = \nu(\lambda_0)$ and $M = M(\lambda_0)$, depending upon λ_0, and such that*

$$|(\lambda - \lambda_0)^\nu R(\lambda; T)| \leqq M, \qquad \lambda \neq \lambda_0, \quad \lambda \in \Delta_{\lambda_0}.$$

Although the rate-of-growth condition (G) will be assumed in most of what follows, it will be stated either explicitly or parenthetically (as was done with conditions (A), ..., (D)) in any theorem where it is used. It will be seen that this growth condition implies conditions (A) and (C) and that the boundedness and growth conditions (B) and (G) together come very near to insuring that the operator T is a spectral operator.

4 LEMMA. *An operator with property* (G) *also has properties* (A) *and* (C).

PROOF. If the operator T in the B-space $\mathfrak{X}$ satisfies the growth condition (G), its spectrum lies in the rectifiable Jordan curve Γ_0. Thus the spectrum is nowhere dense, and condition (A) follows from Lemma 1.

To prove (C), let δ be a closed subset of the complex plane and let

$$\mathfrak{M}(\delta) = \{x \,|\, x \in \mathfrak{X},\ \sigma(x) \subseteq \delta\}.$$

It will be shown that $\mathfrak{M}(\delta)$ is closed. For every x we have $\sigma(x) \subseteq \sigma(T) \subseteq \Gamma_0$ and thus $\mathfrak{M}(\delta) = \mathfrak{M}(\delta\Gamma_0)$, which allows us to assume, with no loss of generality, that δ is a subset of the curve Γ_0. The set δ is therefore an intersection $\delta = \bigcap_\alpha \delta$ of sets δ_α each one of which is the complement in Γ_0 of an open subinterval of Γ_0. Since

$$\mathfrak{M}(\delta) = \mathfrak{M}(\bigcap_\alpha \delta_\alpha) = \bigcap_\alpha \mathfrak{M}(\delta_\alpha),$$

in order to see that $\mathfrak{M}(\delta)$ is closed it will suffice to prove that $\mathfrak{M}(\delta_\alpha)$ is closed. In other words, we may and shall assume that δ is the complement of an open subinterval γ of Γ_0. Let $\{x_n\}$ be a sequence in $\mathfrak{X}$, convergent to the point x, and with $\rho(x_n) \supseteq \gamma$. To prove (C) it will be shown that $\rho(x) \supseteq \gamma$. To do this it is evidently sufficient to show that $\rho(x)$ contains an arbitrary open subinterval γ_0 of γ. Let a and b be the end points of γ_0 and let C be a simple Jordan curve composed of the transversals Δ_a, Δ_b and arcs connecting their end points in such a way that C includes γ_0 in its interior, intersects Γ_0 only at the points a, b, and includes the rest of Γ_0 in its exterior.

Condition (G) shows that there is an integer N such that

$$\lim_{\substack{\lambda\to a\\ \lambda\in C}} (\lambda-a)^N(\lambda-b)^N x_n(\lambda) = \lim_{\substack{\lambda\to b\\ \lambda\in C}} (\lambda-a)^N(\lambda-b)^N x_n(\lambda) = 0$$

uniformly in $n = 1, 2, \ldots$. Thus there are open subarcs N_a, N_b of C containing a, b, respectively, and such that the vector $y_n(\lambda) = (\lambda-a)^N(\lambda-b)^N x_n(\lambda)$ has norm

$$|y_n(\lambda)| < \frac{\varepsilon}{2}, \qquad n \geqq 1, \quad \lambda \in N_a \cup N_b.$$

Since $x_n \to x$ we have

$$\lim_{n\to\infty} y_n(\lambda) = (\lambda-a)^N(\lambda-b)^N R(\lambda; T)x$$

uniformly for λ in $C - N_a - N_b$. This fact, together with the preceding inequality, shows that for some integer n_0 depending upon ε we have

$$|y_n(\lambda) - y_m(\lambda)| < \varepsilon, \qquad \lambda \in C, \qquad n, m \geqq n_0,$$

and thus proves that the sequence $\{y_n(\lambda)\}$ converges uniformly for λ in C. By the maximum modulus principle this sequence converges uniformly on the union of C and its interior to an analytic function $y(\lambda)$. Since

$$y(\lambda) = \lim_n y_n(\lambda) = (\lambda - a)^N (\lambda - b)^N x(\lambda)$$

for $\lambda \notin \Gamma_0$, the vector

$$X(\lambda) = \frac{y(\lambda)}{(\lambda - a)^N (\lambda - b)^N}$$

is an analytic continuation of $x(\lambda)$ into the interior of C. Since $(\lambda I - T)x(\lambda) = x$ for $\lambda \notin \Gamma_0$, it follows that $(\lambda I - T)X(\lambda) = x$ for all λ interior to C. Thus $\rho(x)$ includes the interior of C and, since the interior of C includes γ_0, the proof is complete. Q.E.D.

According to the preceding lemma and Theorem 4.5, an operator T in a weakly complete space will be spectral if it satisfies (B), (G), and (D). We shall now study condition (D) more carefully and see that conditions (B) and (G) come very near to implying condition (D). This will allow us to replace condition (D), and in a variety of ways, by more satisfactory conditions. For example, it will be seen that an operator T in a reflexive space which satisfies (G) and whose adjoint satisfies (B) is a spectral operator.

Before starting this study, it will be convenient to restate condition (G) in a form more suitable to the ana'ysis that follows. In the first place, it is clear that the integers $\nu = \nu(\lambda_0)$ and $M = M(\lambda_0)$, with the required properties, exist for every λ_0 in Γ_0 even if λ_0 is not in the spectrum. Also, it may be assumed that for each λ_0 in Γ_0 the transversal $\Delta_{\lambda 0}$ lies within the circle of radius $\frac{1}{2}$ and center λ_0. This shortening of the transversal $\Delta_{\lambda 0}$ may be achieved by replacing $\xi(\lambda, \delta)$ by $\xi_1(\lambda, \delta_1)$ where $\delta_1 = K\delta$ with K sufficiently large. Now, if every point of $\Delta_{\lambda 0}$ is within a distance of $\frac{1}{2}$ from λ_0, it follows from (G) that

$$\lim_{N \to \infty} (\lambda - \lambda_0)^N R(\lambda; T) = 0$$

uniformly for λ in $\Delta(\lambda_0)$. Thus there is an integer valued function $\nu = \nu(\lambda_0)$ defined for every λ_0 in Γ_0 and such that $|(\lambda - \lambda_0)^{\nu(\lambda_0)} R(\lambda; T)| \leqq 1$ for every λ in $\Delta_{\lambda 0}$ except $\lambda = \lambda_0$. In other words the function $M = M(\lambda_0)$ of condition (G) may, without loss of generality, be assumed to be identically one. Thus condition (G) may be restated in the following equivalent form.

(G) *The spectrum of T is contained in the rectifiable Jordan curve Γ_0. Moreover, there is an integer valued function ν defined on Γ_0 such that for every λ_0 in Γ_0,*

$$|(\lambda-\lambda_0)^{\nu(\lambda_0)}R(\lambda; T)| \leqq 1, \qquad \lambda \neq \lambda_0, \qquad \lambda \in \Delta_{\lambda_0}.$$

In the analysis to follow, it is this inequality that will be used rather than the one in the earlier formulation of growth condition (G).

5 DEFINITION. An integer valued function ν defined on Γ_0 and satisfying the preceding inequality is called an *index function* for T. An *interval of constancy* relative to T is a non-void open subinterval of Γ_0 upon which some index function for T is constant. A point λ in Γ_0 is said to be *regular relative* to T if it belongs to an interval of constancy and if, in addition, there is an integer n such that the manifold

$$(\lambda I - T)^n\mathfrak{X} + \{x \mid (\lambda I - T)^n x = 0\}$$

is dense in $\mathfrak{X}$.

It should be noted that if

$$\mathfrak{X} = \overline{(T-\lambda I)^n\mathfrak{X} + \{x \mid (T-\lambda I)^n x = 0\}},$$

then, by applying the operator $(T-\lambda I)^n$ to both sides of this equation, one obtains the inclusion relation

$$(T-\lambda I)^n\mathfrak{X} \subseteq \overline{(T-\lambda I)^{2n}\mathfrak{X}}.$$

Thus, since the manifolds $(T-\lambda I)^m\mathfrak{X}$ decrease as m increases, it follows that

$$(T-\lambda I)^n\mathfrak{X} \subseteq \overline{(T-\lambda I)^{n+1}\mathfrak{X}}.$$

Since the manifolds $\{x \mid (T-\lambda I)^m x = 0\}$ increase with m, it follows that

$$(T-\lambda I)^{n+1}\mathfrak{X} + \{x \mid (T-\lambda I)^{n+1} x = 0\}$$

is dense in $\mathfrak{X}$. By induction, it is seen that the manifold

$$(T-\lambda I)^{n+k}\mathfrak{X} + \{x \mid (T-\lambda I)^{n+k} x = 0\}$$

is dense in $\mathfrak{X}$ for all $k \geqq 0$. This fact will be stated in the following lemma for future reference.

6 LEMMA. *A complex number λ is regular relative to T if and only if it is contained in an interval of constancy relative to T and, for all sufficiently*

large integers n, the manifold

$$(T - \lambda I)^n \mathfrak{X} + \{x \,|\, (T - \lambda I)^n x = 0\}$$

is dense in $\mathfrak{X}$.

It should also be noted that every point λ of Γ_0 which is in the resolvent set of T is regular relative to T. This follows since the resolvent set is open and $R(\lambda; T)$ is continuous so that λ is interior to an interval where some index function is constant. Also, for λ in the resolvent set, the density requirement of Definition 5 is satisfied, since for such λ, $(T - \lambda I)\mathfrak{X} = \mathfrak{X}$.

7 Lemma (A). *If* $(\lambda_0 I - T)^n x = 0$ *for some integer n and some* $x \neq 0$, *then* $\sigma(x) = \{\lambda_0\}$.

Proof. Since it is finite, the series

$$\begin{aligned} X(\lambda) &= \sum_{j=0}^{\infty} \frac{(-1)^j}{(\lambda - \lambda_0)^{j+1}} (\lambda_0 I - T)^j x \\ &= \sum_{j=0}^{n-1} \frac{(-1)^j}{(\lambda - \lambda_0)^{j+1}} (\lambda_0 I - T)^j x \end{aligned}$$

converges for every $\lambda \neq \lambda_0$ and satisfies the equation $(\lambda I - T)X(\lambda) = x$. Thus $X(\lambda)$ is an analytic extension of $R(\lambda; T)x$ to the complement of $\{\lambda_0\}$. This means that $\sigma(x) \subseteqq \{\lambda_0\}$. Since, by Lemma 2.2, $\sigma(x)$ is not void, we have $\sigma(x) = \{\lambda_0\}$. Q.E.D.

8 Lemma (B, G). *Every closed subinterval of* Γ_0 *whose end points are regular relative to T belongs to* $\mathscr{S}_1(T)$.

Proof. Let γ be a closed subinterval of Γ_0 whose end points λ_1, λ_2 are regular relative to T. It is clear that, by making a suitable change in the parameter s in the function $\xi(s, \delta)$, it may be assumed that $\lambda_1 = \xi(-\frac{1}{2}, 0)$ and $\lambda_2 = \xi(\frac{1}{2}, 0)$. Since λ_1 and λ_2 are interior to intervals of constancy relative to T, there is an $\varepsilon > 0$ such that for $|\lambda_0 - \lambda_1| < \varepsilon$ or $|\lambda_0 - \lambda_2| < \varepsilon$ the inequality

$$\text{(i)} \qquad |\lambda - \lambda_0|^N |R(\lambda; T)| \leqq 1, \qquad \lambda_0 \neq \lambda \in \Delta_{\lambda_0},$$

holds for all sufficiently large values of N. In view of Lemma 6 the integer N may be fixed so that the inequality (i) holds and also so that the manifolds

$$\mathfrak{M}_i = (\lambda_i I - T)^N \mathfrak{X} + \{x \,|\, (\lambda_i I - T)^N x = 0\}, \qquad i = 1, 2,$$

are both dense in $\mathfrak{X}$. Since $\mathfrak{M}_2$ is dense in $\mathfrak{X}$, the manifold

$$(\lambda_1 I - T)^N \mathfrak{M}_2 + \{x \mid (\lambda_1 I - T)^N x = 0\}$$

is dense in $\mathfrak{X}$, so that

$$(\lambda_1 I - T)^N (\lambda_2 I - T)^N \mathfrak{X} + \{x \mid (\lambda_1 I - T)^N x = 0\} \\ + \{x \mid (\lambda_2 I - T)^N x = 0\}$$

is also dense in $\mathfrak{X}$. By Lemma 7, $\sigma(x) \subset \gamma$ if $(\lambda_i I - T)^N x = 0$ for $i = 1$ or $i = 2$, so that to prove the present lemma it will be sufficient to show that every element of the form $y = (\lambda_1 I - T)^N (\lambda_2 I - T)^N x$ may be approximated arbitrarily closely by a sum $z_1 + z_2$ where $\sigma(z_1)$ is contained in γ and $\sigma(z_2)$ is contained in the complement γ'.

Actually more will be proved, for it will be shown that z_1 and z_2 may be chosen so that $z_1 + z_2$ is arbitrarily close to y and the spectra $\sigma(z_1)$, $\sigma(z_2)$ are contained in the *interior* of γ, γ', respectively. Using the operational calculus of Chapter VII, we have

$$\text{(ii)} \quad y = (\lambda_1 I - T)^N (\lambda_2 I - T)^N x$$
$$= \frac{1}{2\pi i} \left\{ \int_{\Gamma_1} - \int_{\Gamma_{-1}} \right\} (\lambda_1 - \lambda)^N (\lambda_2 - \lambda)^N R(\lambda; T)\, x \, d\lambda.$$

The transversals Δ_{λ_1} and Δ_{λ_2} divide the annular region between Γ_1 and Γ_{-1} into two simply connected areas, each bounded by a curve consisting of the arcs Δ_{λ_1}, Δ_{λ_2} and portions of Γ_1, Γ_{-1}. Let the area containing γ be called A_1, and its positively oriented bounding curve C_1. Let the area

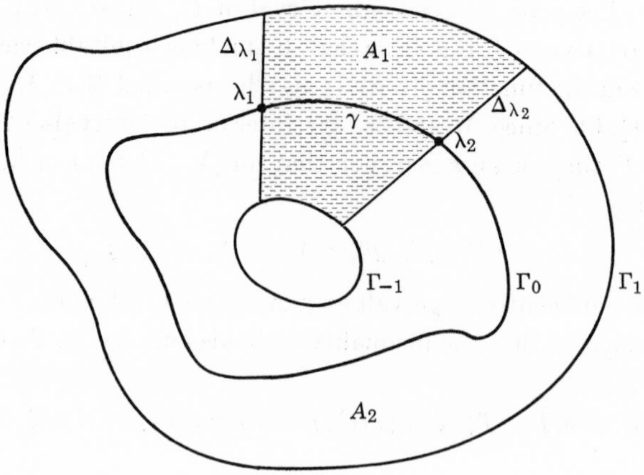

containing the complement γ' be called A_2, and its positively oriented bounding curve C_2. Then, since the integrand in the expression (ii) for y is bounded on C_1 and C_2, we have

$$y = y_{C_1} + y_{C_2},$$

where

$$\text{(iii)} \qquad y_{C_i} = \frac{1}{2\pi i}\int_{C_i} (\lambda_1 - \lambda)^N(\lambda_2 - \lambda)^N R(\lambda; T)x\, d\lambda, \qquad i = 1, 2.$$

It will be shown that y_{C_1} can be approximated arbitrarily closely by vectors whose spectra lie interior to C_1, and that y_{C_2} can be approximated arbitrarily closely by elements whose spectra lie interior to C_2. The details will be given only for y_{C_2}, but the proof for y_{C_1} is quite similar.

Let Δ_ε^-, Δ_ε^+ be the transversals through the points $\xi((-\frac{1}{2} - \varepsilon), 0)$, $\xi((\frac{1}{2} + \varepsilon), 0)$, respectively. Let A_1^ε, A_2^ε be the two simply connected regions into which the annular region between Γ_1 and Γ_{-1} is divided by the transversals Δ_ε^- and Δ_ε^+. Let A_2^ε be that one of these areas contained in A_2, and let C_2^ε be its bounding curve. In view of (i), the integrand in (iii) is uniformly bounded, and it follows from an elementary calculation using Lebesgue's dominated convergence theorem that

$$y_{C_2} = \lim_{\varepsilon \to 0} y_{C_2^\varepsilon},$$

where

$$y_{C_2^\varepsilon} = \frac{1}{2\pi i}\int_{C_2^\varepsilon} (\lambda_1 - \lambda)^N(\lambda_2 - \lambda)^N R(\lambda; T)x\, d\lambda.$$

Thus, to see that y_{C_2} is the limit of a sequence of vectors whose spectra are interior to C_2, it will suffice to show that the spectrum $\sigma(y_{C_2^\varepsilon})$ is contained in A_2^ε. It is evident that the integral

$$I(\xi) = \frac{1}{2\pi i}\int_{C_2^\varepsilon} (\lambda_1 - \lambda)^N(\lambda_2 - \lambda)^N(\xi - \lambda)^{-1} R(\lambda; T)x\, d\lambda$$

is analytic for ξ outside A_2^ε. Since

$$\begin{aligned}(\xi I - T)I(\xi) &= \frac{1}{2\pi i}\int_{C_2^\varepsilon} (\lambda_1 - \lambda)^N(\lambda_2 - \lambda)^N(\xi - \lambda)^{-1}(\xi I - T)R(\lambda; T)x\, d\lambda \\ &= \frac{1}{2\pi i}\int_{C_2^\varepsilon} (\lambda_1 - \lambda)^N(\lambda_2 - \lambda)^N(\xi - \lambda)^{-1}(\xi - \lambda)R(\lambda; T)x\, d\lambda \\ &= y_{C_2^\varepsilon},\end{aligned}$$

it follows that $I(\xi)$ is an analytic extension of $R(\xi; T)y_{C_2^\varepsilon}$. Thus $\rho(y_{C_2^\varepsilon})$ includes the entire complement of A_2^ε, and so the spectrum $\sigma(y_{C_2^\varepsilon})$ is contained in A_2^ε. Q.E.D.

9 COROLLARY (B, G). *Let γ be a closed subinterval of Γ_0 whose end points λ_1, λ_2 belong to intervals of constancy relative to T. Then, for sufficiently large N, the manifold $(\lambda_1 I - T)^N(\lambda_2 I - T)^N\mathfrak{X}$ is contained in the closed manifold determined by vectors of the form $z_1 + z_2$ where $\sigma(z_1)$ is interior to γ and $\sigma(z_2)$ is interior to the complementary arc γ'.*

PROOF. The above statement is what was actually proved in the preceding proof. Q.E.D.

10 LEMMA (B, G). *If the set of points regular relative to T is dense on Γ_0, then every closed subinterval of Γ_0 whose end points are regular relative to T is in $\mathscr{S}(T)$ and every Borel subset of the plane is measurable T.*

PROOF. Let γ be a closed subinterval of Γ_0 whose end points are regular relative to T. Since the points regular relative to T are dense in Γ_0, there is an increasing sequence $\{\gamma_n\}$ of open subintervals of Γ_0, each member of which has regular end points in the arc γ' complementary to γ, and whose union is the whole arc γ'. By Lemma 8 the intervals γ and $\bar{\gamma}_n$, $n = 1, 2, \ldots$, are all in $\mathscr{S}_1(T)$. Since γ is in $\mathscr{S}_1(T)$ there are, for each x in $\mathfrak{X}$ and $\varepsilon > 0$, vectors y and z such that

$$|y + z - x| < \varepsilon, \qquad \sigma(y) \subseteq \gamma, \qquad \sigma(z) \subseteq \gamma'.$$

Since $\sigma(z)$ is compact, we have $\bar{\gamma}_n \supseteq \gamma_n \supseteq \sigma(z)$ for all sufficiently large n. Thus, by Lemma 3.2, $[E(\gamma) + E(\bar{\gamma}_n)](y + z) = y + z$ for all large n. By the boundedness assumption (B) the norms of the projections $E(\gamma) + E(\bar{\gamma}_n)$ are bounded in n, and since $\varepsilon > 0$ is arbitrary, we have

$$\text{(i)} \qquad x = \lim_n [E(\gamma)x + E(\bar{\gamma}_n)]x, \qquad x \in \mathfrak{X}.$$

Now it follows from Lemma 4 that T has properties (A) and (C), and thus it is seen from Lemma 4.2 that

$$\text{(ii)} \qquad \sigma(E(\gamma)x) \subseteq \gamma\sigma(x), \qquad \sigma(E(\bar{\gamma}_n)x) \subseteq \bar{\gamma}_n\,\sigma(x).$$

These relations (i) and (ii) show that γ is in $\mathscr{S}_2(T)$. Since γ is an arbitrary closed subinterval of Γ_0 whose end points are regular relative to T, this proves that the closure $\bar{\gamma}_n$ is also in $\mathscr{S}_2(T)$. Thus it follows from (i) and

Definition 3.7 that γ is in $\mathscr{S}(T)$. It follows that every Borel subset of Γ_0 is measurable T. By Lemma 3.9 every subset of $\rho(T)$ is in $\mathscr{S}(T)$ and hence in a set measurable T. Thus, since $\sigma(T) \subseteqq \Gamma_0$, every Borel set in the plane is measurable T. Q.E.D.

In view of Lemma 10, it behooves us to seek conditions under which the points on Γ_0 which are regular relative to T are dense on Γ_0. Some results in this direction will be found in the next four lemmas.

11 LEMMA (G). *The union of all intervals of constancy relative to T is an open set dense in Γ_0.*

PROOF. It is clear that the union of intervals of constancy is open. To see that it is dense, let γ be a closed subarc of Γ_0 having positive length and let

$$\begin{aligned}\gamma_n &= \{\lambda_0 \,|\, \lambda_0 \in \gamma, |\lambda - \lambda_0|^n \,|R(\lambda; T)| \leqq 1, \lambda_0 \neq \lambda \in \Delta_{\lambda_0}\} \\ &= \{\lambda_0 \,|\, \lambda_0 \in \gamma, |\xi(\lambda_0, \delta) - \lambda_0|^n R(\xi(\lambda_0, \delta); T)| \leqq 1, 0 < |\delta| \leqq 1\}.\end{aligned}$$

It is clear from the second expression for γ_n that it is closed, and it follows from (G) that every point in γ is in one of the sets γ_n. Thus, by the Baire category theorem (cf. I.6.9), one of the sets γ_n contains a non-trivial subinterval of γ. Q.E.D.

12 LEMMA (G). *If the point spectrum of the adjoint T^* contains no non-trivial subarc of Γ_0, then the set of points regular relative to T is dense in Γ_0.*

PROOF. If λ is not in the point spectrum of the adjoint of T, then there is no functional $x^* \neq 0$ for which $x^*(\lambda I - T)\mathfrak{X} = 0$. In view of the Hahn-Banach theorem, this means that $(\lambda I - T)\mathfrak{X}$ is dense in $\mathfrak{X}$. Hence if λ is also in an interval of constancy for T, then λ is regular relative to T. The present lemma thus follows from Lemma 11. Q.E.D.

13 LEMMA (G). *If the space $\mathfrak{X}$ is reflexive and if the function identically one on Γ_0 is an index function for T, then every point of Γ_0 is regular relative to T.*

PROOF. It is clear that Γ_0 itself is an interval of constancy relative to T. Let x be an arbitrary vector in $\mathfrak{X}$, λ_0 an arbitrary point on Γ_0, and $\{\lambda_n\}$ a sequence of points in the transversal Δ_{λ_0} with $\lambda_n \neq \lambda_0$ and $\lambda_0 = \lim \lambda_n$. Since $\mathfrak{X}$ is reflexive, the bounded sequence $\{(\lambda_n - \lambda_0)R(\lambda_n; T)x\}$ contains a subsequence weakly convergent to an element y of $\mathfrak{X}$ (cf. Theorem II.3.28). By replacing the sequence $\{\lambda_n\}$ by a suitably chosen subsequence, it may therefore be assumed that, in the weak topology, we have

$$\lim_n (\lambda_n - \lambda_0) R(\lambda_n; T)x = y.$$

Then $(\lambda_0 I - T)y$ is the weak limit of

$$(\lambda_n - \lambda_0)(\lambda_0 I - T)R(\lambda_n; T)x = (\lambda_n - \lambda_0)x - (\lambda_0 - \lambda_n)^2 R(\lambda_n; T)x,$$

and this limit is clearly zero. Thus $(\lambda_0 I - T)y = 0$.

It will next be shown that the vector $x - y$ is in the closure of the manifold $(\lambda_0 I - T)\mathfrak{X}$. To see this it will, in view of Corollary II.3.13, suffice to show that $x^*(x - y) = 0$ for every linear functional x^* which vanishes on $(\lambda_0 I - T)\mathfrak{X}$. If x^* is such a functional, then $T^*x^* = \lambda_0 x^*$,

$$R(\lambda_n; T)^*x^* = \frac{x^*}{\lambda_n - \lambda_0},$$

and so

$$x^*y = \lim_{n\to\infty} x^*(\lambda_n - \lambda_0)R(\lambda_n; T)x = x^*x,$$

and $x^*(x - y) = 0$. It has been shown that an arbitrary vector x in $\mathfrak{X}$ is the sum of a vector y with $(\lambda_0 I - T)y = 0$ and a vector $x - y$ in the closure of $(\lambda_0 I - T)\mathfrak{X}$. Since λ_0 is interior to an interval of constancy relative to T, it is therefore a regular point relative to T. Q.E.D.

14 LEMMA (G). *If $\mathfrak{X}$ is reflexive and if the adjoint T^* satisfies the boundedness condition* (B), *then the regular points relative to T are dense in Γ_0 and, in particular, every interval of constancy relative to T consists entirely of regular points.*

PROOF. In view of Lemma 11 it suffices to show that a point λ_0 in an interval of constancy relative to T is regular. Since $\sigma(T^*) = \sigma(T)$ and $R(\lambda; T^*) = R(\lambda; T)^*$, it follows that T^* also satisfies the growth condition (G) and that every index function for T is also an index function for T^* and vice versa. By applying Corollary 9 to T^* and the interval consisting of the single point λ_0, it is seen that, for N sufficiently large, every element in the manifold $(\lambda_0 I^* - T^*)^N\mathfrak{X}^*$ may be approximated by elements z^* with $\lambda_0 \notin \sigma(z^*)$. To prove that λ_0 is regular relative to T it will be shown that, for such N, the manifold

$$\text{(i)} \qquad (\lambda_0 I - T)^N\mathfrak{X} + \{x \,|\, (\lambda_0 I - T)^N x = 0\}$$

is dense in $\mathfrak{X}$. To do this it will, in view of the Hahn-Banach theorem (cf. Corollary II.3.13), suffice to show that the functional $x^* = 0$ is the only functional which vanishes on the manifold (i). Let x^* be such a functional. Then

$$\text{(ii)} \qquad (\lambda_0 I^* - T^*)^N x^* = 0.$$

To see that $x^* = 0$ it will first be shown that

$$\text{(iii)} \qquad x^* \in \overline{(\lambda_0 I^* - T^*)^N \mathfrak{X}^*}.$$

If (iii) is not true, then, since $\mathfrak{X}$ is reflexive, it follows from the Hahn-Banach theorem (cf. Corollary II.3.13) that there is an x in $\mathfrak{X}$ with $x^*x \neq 0$ and $[(\lambda_0 I^* - T^*)^N \mathfrak{X}^*]x = 0$ which means that $(\lambda_0 I - T)^N x = 0$. Since x^* vanishes on the manifold (i), we have $x^*x = 0$, which contradicts the inequality $x^*x \neq 0$ and establishes (iii).

Now from (ii) and (iii) together it may be concluded that $x^* = 0$. To do this, note first that, according to Lemma 7, $\sigma(x^*) = \{\lambda_0\}$. By Corollary 9, applied to T^* and the interval consisting of the single point λ_0, there is a sequence $\{x_n^*\}$ converging to x^* with $\lambda_0 \notin \sigma(x_n^*)$. Since T^* satisfies condition (B), we have

$$|x^*| \leqq K\,|x^* - x_n^*| \to 0,$$

and so $x^* = 0$. This completes the proof that the manifold (i) is dense in $\mathfrak{X}$ and thus proves that λ_0 is regular relative to T. Q.E.D.

The preceding lemmas, when combined with the general criteria given in Theorems 4.5 and 4.7, allow us to summarize a set of conditions that are sufficient to guarantee that an operator is a spectral operator. This will be done in the following two results.

→ 15 THEOREM. *If a bounded linear operator in a weakly complete space satisfies the boundedness condition* (B) *and the growth condition* (G), *then it is a spectral operator provided that any one of the following conditions holds.*

(a) *The point spectrum of the adjoint contains no non-trivial subarc of* Γ_0.

(b) *The space is reflexive and the function*

$$\nu(\lambda) = 1, \qquad \lambda \in \Gamma_0,$$

is an index function for the operator.

(c) *The space is reflexive and the adjoint operator satisfies the boundedness condition* (B).

PROOF. If the bounded linear operator T in a weakly complete space has properties (B) and (G) then, by Lemma 4, it has properties (A) and (C). Thus, in view of Theorem 4.5, to prove the present theorem it suffices to show that T has property (D). According to Lemma 10 condition (D) will be satisfied if the points regular relative to T are dense on Γ_0. Thus Lemmas 12, 13, 14 give the desired conclusions. Q.E.D.

16 THEOREM. *Let T be a bounded linear operator in the B-space $\mathfrak{X}$ which satisfies conditions* (B) *and* (G) *and let $\mathscr{B}$ be the field of Borel sets in the plane. Then the adjoint T^* is a spectral operator of class $(\mathscr{B}, \mathfrak{X})$ provided that any one of the conditions* (a), (b), (c) *of the preceding theorem hold.*

PROOF. The proof is the same as that of the preceding theorem except that Theorem 4.7 is used instead of Theorem 4.5. Q.E.D.

It is to be expected from analogies with finite matrices, as well as by comparison with Theorem XV.6.7, that a spectral operator T which satisfies the growth condition (G) will be a spectral operator of type $m-1$ if and only if the constant function $\nu(\xi) \equiv m$ is an index function. For convenience of reference this finite rate of growth condition is stated formally as follows.

17 DEFINITION. The bounded linear operator T is said to satisfy the *growth condition* (G_m) if its spectrum lies in the curve Γ_0 and if, for some constant M,

$$|(\xi-\xi_\delta)^m R(\xi_\delta; T)| \leqq M, \qquad \xi \in \Gamma_0, \qquad 0 < |\delta| \leqq 1,$$

where ξ_δ is the intersection of the transversal Δ_ξ with the curve Γ_δ.

→ 18 THEOREM. *A bounded linear operator T in Hilbert space whose spectrum lies in the Jordan curve Γ_0 will be a spectral operator of type $m-1$ if and only if both T and its adjoint satisfy conditions* (B) *and* (G_m).

PROOF. If T is a spectral operator in the Hilbert space $\mathfrak{X}$ then, since $\mathfrak{X}$ is reflexive, the adjoint T^* is, by Lemma 4.6, a spectral operator. By Theorem 4.4 both T and T^* satisfy the boundedness condition (B). Now if T is of type $m-1$ with radical part N and resolution of the identity E, then

$$R(\xi; T) = \sum_{n=0}^{m-1} N^n \int_{\sigma(T)} \frac{E(d\lambda)}{(\xi-\lambda)^{n+1}}, \quad \xi \in \rho(T),$$

from which it is apparent that T, and consequently T^* also, satisfies the condition $(\mathrm{G_m})$.

Conversely, let T and its adjoint satisfy conditions (B) and $(\mathrm{G_m})$. It is clear from Theorem 15(c) that T is a spectral operator, and so it suffices to prove only that T is of type $m-1$.

The proof will involve Riemann integrals of the type

$$\text{(i)} \qquad \int_{\sigma(T)} F(\xi)E(d\xi),$$

where E is the resolution of the identity for T and F is an operator valued function defined on $\sigma(T)$ continuous in the uniform operator topology, and for which

$$\text{(ii)} \qquad F(\xi)T = TF(\xi), \qquad \xi \in \sigma(T).$$

It is seen from Corollary XV.3.7 that $F(\xi)$ also commutes with the projections in the range of E, that is,

$$\text{(iii)} \qquad F(\xi)E(\sigma) = E(\sigma)F(\xi), \qquad \xi \in \rho(T),$$

for every Borel set σ. If $\pi = \{\sigma_1, \ldots, \sigma_n\}$, $\pi' = \{\sigma_1', \ldots, \sigma'_{n'}\}$ are two partitions of $\sigma(T)$ and if $\xi_i \in \sigma_i$, $\xi_j' \in \sigma_j'$, $i = 1, \ldots, n$, $j = 1, \ldots, n'$, then using (iii) and Lemma XV.6.8, it is seen that, for some constant K,

$$\begin{aligned} &\left| \sum_{i=1}^{n} F(\xi_i)E(\sigma_i) - \sum_{j=1}^{n'} F(\xi_j')E(\sigma_j') \right| \\ &\qquad = \left| \sum_{i=1}^{n} \sum_{j=1}^{n'} \{F(\xi_i) - F(\xi_j')\}E(\sigma_i \sigma_j') \right| \\ &\qquad \leqq K \sup |F(\xi_i) - F(\xi_j')|, \end{aligned}$$

where the supremum is taken over those i and j for which $\sigma_i \sigma_j'$ is not void. If by the norm $|\pi|$ is understood the quantity

$$|\pi| = \max_{1 \leqq i \leqq n} \operatorname{diam} \sigma_i,$$

it is seen that the limit

$$\lim_{|\pi| \to 0} \sum_{i=1}^{n} F(\xi_i)E(\sigma_i)$$

exists in the uniform topology of operators for every function F on $\sigma(T)$ which is continuous in the uniform operator topology and satisfies (ii). This limit defines the Riemann integral (i). It is clear that the integral is linear in F and satisfies the inequality

$$\text{(iv)} \qquad \left|\int_{\sigma(T)} F(\xi)E(d\xi)\right| \leqq K \sup_{\xi\in\sigma(T)} |F(\xi)|.$$

If F and G are two continuous operator valued functions both satisfying (ii) then, since

$$\left(\sum_{i=1}^{n} F(\xi_i)E(\sigma_i)\right)\left(\sum_{j=1}^{n} G(\xi_j)E(\sigma_j)\right) = \sum_{i=1}^{n} F(\xi_i)G(\xi_i)E(\sigma_i),$$

it follows that

$$\text{(v)} \quad \left\{\int_{\sigma(T)} F(\xi)E(d\xi)\right\}\left\{\int_{\sigma(T)} G(\xi)E(d\xi)\right\} = \int_{\sigma(T)} F(\xi)G(\xi)E(d\xi).$$

It will next be shown that

$$\text{(vi)} \qquad \int_{\sigma(T)} (\xi I - T)^{2m}E(d\xi) = 0.$$

The proof of (vi) will use the integral

$$\text{(vii)} \qquad I(\mu, \gamma) = \frac{1}{2\pi i}\int_{C_\delta(\mu,\gamma)} (\mu - \xi)^m(\gamma - \xi)^m R(\xi; T)\, d\xi,$$

where μ, γ are points on Γ_0 and, for $0 < \delta \leqq 1$, the contour $C_\delta(\mu, \gamma)$ is the positively oriented contour through μ, γ which is defined as follows. For an arbitrary ξ in Γ_0 let ξ_δ be the intersection of the transversal through ξ with the curve Γ_δ. The curve $C_\delta(\mu, \gamma)$ is a positively oriented Jordan curve through the points μ_δ, γ_δ, $\gamma_{-\delta}$, $\mu_{-\delta}$ and consisting of subarcs of

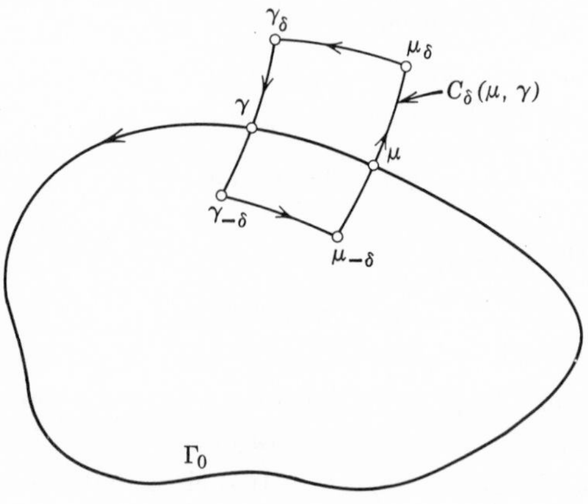

the curves $\Gamma_\delta, \Delta_\gamma, \Gamma_{-\delta}, \Delta_\mu$. The curve $C_\delta(\mu, \gamma)$ is defined so as to contain in its interior the interior of the directed segment (μ, γ) on the oriented contour Γ_0. The integrand $f(\xi)$ in $I(\mu, \gamma)$ is defined and continuous at every point of $C_\delta(\mu, \gamma)$ except at μ and γ. In view of the condition $(\mathrm{G_m})$, $f(\xi)$ is bounded on $C_\delta(\mu, \gamma)$. Hence $I(\mu, \gamma)$ exists and is clearly independent of δ since $f(\xi)$ has its only singularities on Γ_0. It will first be shown that $I(\mu, \gamma) \to 0$ as $|\mu-\gamma| \to 0$. Let $\varepsilon > 0$ be arbitrary and fix $\delta > 0$ so that the arcs $\mu_{-\delta}\mu_\delta$ and $\gamma_{-\delta}\gamma_\delta$ both have length less than ε. Fix K_ε so that

$$|R(\xi; T)| \leqq K_\varepsilon, \qquad \xi \in \Gamma_\delta \cup \Gamma_{-\delta},$$

and fix $\alpha_\varepsilon > 0$ so that the arcs $\gamma_{-\delta}\mu_{-\delta}$ and $\mu_\delta\gamma_\delta$ both have length less than $\varepsilon/K_\varepsilon$ provided that $|\mu-\gamma| < \alpha_\varepsilon$. Then, for $|\mu-\gamma| < \alpha_\varepsilon$, we have

$$|I(\mu, \gamma)| \leqq \frac{1}{2\pi}\left[2\varepsilon M + 2K_\varepsilon \frac{\varepsilon}{K_\varepsilon}\right],$$

which shows that $I(\mu, \gamma) \to 0$ as $|\mu-\gamma| \to 0$.

It will next be observed that the spectrum of any vector of the form $I(\mu, \gamma)x$ is contained in the directed closed subarc $\mu\gamma$. To see this, let λ lie in $\rho(T)$ and outside $C_\delta(\mu, \gamma)$. Then it is seen from the resolvent equation that

$$R(\lambda; T)I(\mu, \gamma)x = \frac{1}{2\pi i}\int_{C_\delta(\mu, \gamma)} \frac{(\mu-\xi)^m(\gamma-\xi)^m}{\lambda-\xi} R(\xi; T)x\, d\xi$$

and the integral on the right of this equation gives an analytic extension of $R(\lambda; T)I(\mu, \gamma)x$ to all points outside of $C_\delta(\mu, \gamma)$. Thus the spectrum $\sigma(I(\mu, \gamma)x)$ is contained in the closed arc $\mu\gamma$. Thus if the closed intervals $[\mu, \gamma]$ and $[\mu', \gamma']$ of Γ_0 are disjoint we have, from Corollary XV.3.3,

$$E([\mu, \gamma])I(\mu', \gamma') = 0.$$

Let $\mu_n, \mu, \gamma, \gamma_n$ be an ordered set on the oriented curve Γ_0 with $\mu_n \to \mu$ and $\gamma_n \to \gamma$. Then $E([\mu, \gamma])I(\gamma_n, \mu_n) = 0$. Since

$$(\mu I - T)^m(\gamma I - T)^m = I(\mu_n, \mu) + I(\mu, \gamma) + I(\gamma, \gamma_n) + I(\gamma_n, \mu_n),$$

and $I(\mu_n, \mu) \to 0$, $I(\gamma, \gamma_n) \to 0$, it follows that

$$E([\mu, \gamma])(\mu I - T)^m(\gamma I - T)^m = I(\mu, \gamma).$$

Now let Γ_0 be divided into n disjoint subintervals $\sigma_1, \ldots, \sigma_n$ each of length L/n where L is the length of Γ_0, and let $\xi_1, \xi_2, \ldots, \xi_n, \xi_{n+1} = \xi_1$

be their positively ordered sequence of end points. Then, according to the preceding equation,

$$\text{(viii)} \qquad E(\sigma_i)(\xi_j I - T)^m(\xi_{j+1}I - T)^m = I(\xi_j, \xi_{j+1}).$$

Now let $\delta = 1/n$ so that the length of $C_\delta(\xi_j, \xi_{j+1})$ is of the order of $1/n$. It follows from (vii) and (viii) that

$$|E(\sigma_j)(\xi_j I - T)^m(\xi_{j+1}I - T)^m| \leq \frac{C_1}{n^{m+1}},$$

and thus that

$$\left|\sum_{j=1}^{n} (\xi_j I - T)^m(\xi_{j+1}I - T)^m E(\sigma_j)\right| \leq \frac{C_2}{n^m}.$$

An elementary continuity argument shows that the sum, whose norm appears on the left of the preceding inequality, approaches the integral $\int_{\sigma(T)} (\xi I - T)^{2m} E(d\xi)$ as $n \to \infty$, and thus this inequality establishes the equation (vi).

It will next be proved that

$$\text{(ix)} \qquad \int_{\sigma(T)} (\xi I - T)^j E(d\xi) = 0, \qquad j \geq m.$$

In view of (vi) this equation may be proved by induction downward. Thus it will be shown that (ix) holds for the integer $j \geq m$ provided that it holds for the integer $j+1$. To do this, let ξ_δ be the point of intersection of the curve Γ_δ with the transversal Δ_ξ through the point ξ on Γ_0 and let $R(\xi_\delta) = R(\xi_\delta; T)$. Then, from (iv) and the hypothesis ($\mathrm{G_m}$), we have

$$\left|\int_{\sigma(T)} (\xi - \xi_\delta)^{j+1} R(\xi_\delta) E(d\xi)\right| \leq C_1 |\xi - \xi_\delta|,$$

which shows that

$$\lim_{\delta \to 0} \int_{\sigma(T)} (\xi - \xi_\delta)^{j+1} R(\xi_\delta) E(d\xi_\delta) = 0.$$

On the other hand, it is seen by writing $(\xi - \xi_\delta)I = (\xi I - T) - (\xi_\delta I - T)$ that

$$\int_{\sigma(T)} (\xi - \xi_\delta)^{j+1} R(\xi_\delta) E(d\xi) = \int_{\sigma(T)} (\xi_\delta I - T)^{j+1} R(\xi_\delta) E(d\xi)$$
$$+ \sum_{r=1}^{j+1} (-1)^r \binom{j+1}{r} \int_{\sigma(T)} (\xi I - T)^{j+1-r} (\xi_\delta I - T)^{r-1} E(d\xi).$$

But, in view of (v) and the induction hypothesis,

$$\int_{\sigma(T)} (\xi I - T)^{j+1} R(\xi_\delta) E(d\xi) = 0, \qquad 0 < \delta \leqq 1.$$

Thus, since $\lim_{\delta \to 0} (\xi_\delta I - T)^{r-1} = (\xi I - T)^{r-1}$ in the uniform operator topology, it follows from (iv) that

$$\lim_{\delta \to 0} \int_{\sigma(T)} (\xi - \xi_\delta)^{j+1} R(\xi_\delta) E(d\xi) = -\int_{\sigma(T)} (\xi I - T)^j E(d\xi),$$

which establishes the equation (ix) for every $j \geqq m$. Now if S is the scalar part of T, then

$$\begin{aligned} 0 = \int_{\sigma(T)} (\xi I - T)^m E(d\xi) &= \sum_{r=0}^{m} \binom{m}{r} \int_{\sigma(T)} \xi^{m-r} E(d\xi)(-T)^r \\ &= \sum_{r=0}^{m} \binom{m}{r} S^{m-r}(-T)^r \\ &= (S - T)^m = (-N)^m, \end{aligned}$$

which shows that $N^m = 0$. The fact that T is of type $m-1$ now follows from Theorem XV.5.4. Q.E.D.

The spectral theorem for a bounded self adjoint operator T in Hilbert space follows from Theorem 18. This will be shown in the next section.

Another special case is one whose unbounded analogue (XVIII.2.39) is useful in the discussion of the nonselfadjoint second order singular differential boundary value problems discussed by Naĭmark. Because of its special nature, however, an independent proof, shorter than that of Theorem 18 is possible and will be given here. In this theorem the operator T has its spectrum in the smooth curve Γ_0 as described in the discussion of the growth condition (G). It will not be necessary, however, to assume the growth condition as it will follow from the other assumptions.

→ 19 THEOREM. *Let the operator T in the reflexive space $\mathfrak{X}$ have its spectrum in the curve Γ_0. In addition, let $\mathfrak{X}_0$, $\mathfrak{X}_0^*$ be dense linear manifolds in the spaces $\mathfrak{X}$, $\mathfrak{X}^*$, respectively, with the following three properties.*

(i) *For x_0 in $\mathfrak{X}_0$ and x_0^* in $\mathfrak{X}_0^*$ there is a constant $K(x_0^*, x_0)$ with*

$$|x_0^* R(\xi(t, \delta); T)x_0| \leqq K(x_0^*, x_0), \qquad -1 \leqq t, \delta \leqq 1, \delta \neq 0.$$

(ii) *For each x_0 in $\mathfrak{X}_0$ and x_0^* in $\mathfrak{X}_0^*$ the limits*

$$R^+(\lambda, x_0^*, x_0) = \lim_{\delta \to 0+} x_0^* R(\xi(t, \delta); T)x_0$$

and

$$R^-(\lambda, x_0^*, x_0) = \lim_{\delta \to 0-} x_0^* R(\xi(t, \delta); T)x_0$$

exist for each point $\lambda = \xi(t, 0)$ *in* Γ_0.

(iii) *There is a constant* M *depending only on* T *such that*

$$\int_{\sigma(T)} |R^+(\lambda, x_0^*, x_0) - R^-(\lambda, x_0^*, x_0)|\, ds \leqq M|x_0^*|\,|x_0|, \qquad x_0 \in \mathfrak{X}, x^* \in \mathfrak{X}^*,$$

where s *is the arc length on* Γ_0.

Then T *is a scalar type spectral operator whose spectral resolution is given by the formula*

$$\text{(iv)} \qquad x_0^* E(\sigma)x_0 = \frac{1}{2\pi i}\int_\sigma \{R^+(\lambda, x_0^*, x_0) - R^-(\lambda, x_0^*, x_0)\}\, d\lambda.$$

PROOF. Let $\delta > 0$ and let f be single valued and analytic in a neighborhood of the annular region bounded by the curves Γ_δ and $\Gamma_{-\delta}$. Then (cf. Definition VI.3.9)

$$f(T) = \frac{1}{2\pi i}\left[\int_{\Gamma_\delta} f(\lambda)R(\lambda; T)\, d\lambda - \int_{\Gamma_{-\delta}} f(\lambda)R(\lambda; T)\, d\lambda\right],$$

where both integrals are taken in the positive sense. Then, in view of hypotheses (i) and (ii), this formula may be written as

$$x_0^* f(T)x_0 = \frac{1}{2\pi i}\int_{\Gamma_0} f(\lambda)\{R^+(\lambda, x_0^*, x_0) - R^-(\lambda, x_0^*, x_0)\}\, d\lambda,$$

for $x_0 \in \mathfrak{X}_0$, $x_0^* \in \mathfrak{X}_0^*$. Since the integrand vanishes if λ is in the resolvent set.

$$\text{(v)} \qquad x_0^* f(T)x_0 = \frac{1}{2\pi i}\int_{\sigma(T)} f(\lambda)\{R^+(\lambda, x_0^*, x_0) - R^-(\lambda, x_0^*, x_0)\}\, d\lambda.$$

In view of hypothesis (iii) the right side of this equation (v) defines, for each bounded Borel function f on $\sigma(T)$, a continuous bilinear form (f, x_0^*, x_0) which, since $\mathfrak{X}_0$ and $\mathfrak{X}_0^*$ are dense, has a unique extension to a bounded bilinear form (f, x^*, x) defined for all x in $\mathfrak{X}$ and x^* in $\mathfrak{X}^*$. Since $\mathfrak{X}$ is reflexive, it follows that there is a unique operator $f(T)$ in $\mathfrak{X}$ for which (v) holds. The mapping $f \to f(T)$ is a homomorphism on the algebra of analytic functions f, and since every continuous function on Γ_0 is the uniform limit of analytic functions, it follows that this map is also a homomorphism on the algebra of continuous functions. To see that it is a homomorphism on the algebra of bounded Borel functions, note that for a fixed continuous function g the set of all bounded Borel functions f for which

$$(fg)(T) = f(T)g(T), \tag{vi}$$

includes all continuous functions. Furthermore, if the equation (vi) holds for each function in a uniformly bounded pointwise convergent sequence $\{f_n\}$, then it follows from (v) that it holds for the limit function $f = \lim f_n$. This shows that (vi) holds for every bounded Borel function f and every continuous function g. A repetition of this argument shows that it also holds if f and g are both bounded Borel functions. Thus the operators $f(T)$ and $g(T)$ commute and also, as (v) shows, satisfy the inequality

$$|f(T)| \leqq M \sup_{\lambda \in \sigma(T)} |f(\lambda)|. \tag{vii}$$

These facts show that the operator $E(\sigma)$ defined by (iv) is a bounded spectral measure which, in view of the Orlicz-Pettis Theorem IV.10.1, is countably additive in the strong operator topology. To see that E is a spectral resolution for T it will therefore be sufficient to show that, for each Borel subset σ of $\sigma(T)$, the spectrum of the restriction $T \mid E(\sigma)\mathfrak{X}$ is contained in $\bar{\sigma}$. For every $\xi \notin \bar{\sigma}$, let the bounded Borel function r_ξ be defined by the equation $r_\xi(\lambda) = (\xi - \lambda)^{-1}\chi_\sigma(\lambda)$, where $\chi_{\bar{\sigma}}$ is the characteristic function of σ. Then $r_\xi(T)(\xi I - T) = E(\sigma)$, which shows that $(T \mid E(\sigma)\mathfrak{X}) \subseteqq \bar{\sigma}$ and proves that T is a spectral operator. It follows from (vii) that T is a scalar operator. Q.E.D.

6. Self Adjoint Operators in Hilbert Space

It is the purpose of this section to show how the theory of spectral operators may be applied to yield the classical spectral theorem in Hilbert space, that is, the theorem asserting that a bounded self adjoint operator in Hilbert space is a scalar type spectral operator. To this end the reader should recall that the bounded operator T in Hilbert space $\mathfrak{X}$ is self adjoint if it coincides with its Hilbert space adjoint, that is, if $T = T^*$. The scalar product of the vectors x and y in $\mathfrak{X}$ will be denoted, as usual, by (x, y). In order to apply the theory of spectral operators, the following two lemmas will be needed. These lemmas will show that the hypotheses of Theorem 5.18 are satisfied by a self adjoint operator in Hilbert space.

1 LEMMA. *If T is a bounded self adjoint operator in Hilbert space, its spectrum is real and for every non-real α we have*

$$|R(\alpha;\, T)| \leqq \frac{1}{|\mathscr{I}(\alpha)|}.$$

PROOF. If α is not real, an expansion of the scalar product $((\alpha I - T)x, (\alpha I - T)x)$ shows that

$$|(\alpha I - T)x|^2 = |\mathscr{I}(\alpha)x|^2 + |(\mathscr{R}(\alpha)I - T)x|^2 \geqq |\mathscr{I}(\alpha)|^2\,|x|^2,$$

so that

$$|x| \leqq \frac{|(\alpha I - T)x|}{|\mathscr{I}(\alpha)|}.$$

This shows that $(\alpha I - T)^{-1}$ exists as a bounded operator, from which it readily follows that $(\alpha I - T)\mathfrak{X}$ is closed. Hence it remains only to prove that $(\alpha I - T)\mathfrak{X}$ is dense. If y is orthogonal to this manifold, then

$$0 = ((\alpha I - T)x, y) = (x, (\bar{\alpha} I - T)y), \qquad x \in \mathfrak{X},$$

from which it follows that $(\bar{\alpha} I - T)y = 0$. Since $\bar{\alpha}$ is not real, this shows that $y = 0$ and proves (cf. Lemma IV.4.4) that $(\alpha I - T)\mathfrak{X} = \mathfrak{X}$. Q.E.D.

2 LEMMA. *A self adjoint operator in Hilbert space satisfies the boundedness condition* (B).

PROOF. Since orthogonal vectors x, y satisfy the relation $|x + y|^2 = |x|^2 + |y|^2$, it will suffice to show that x is orthogonal to y if their spectra $\sigma(x)$ and $\sigma(y)$, relative to the self adjoint operator T, are disjoint. In this case the function

$$(R(\lambda;\, T)x, y) = (x, R(\bar{\lambda};\, T)y) = \overline{(R(\bar{\lambda};\, T)y, x)}$$

is analytic if either $\lambda \notin \sigma(x)$ or $\bar{\lambda} \notin \sigma(y)$. By Lemma 1, $\sigma(y)$ is real, so that the function $(R(\lambda;\, T)x, y)$ is everywhere analytic. The expansion (cf. VII.3.4)

$$(R(\lambda;\, T)x, y) = \frac{(x, y)}{\lambda} + \frac{(Tx, y)}{\lambda^2} + \cdots$$

shows that $(R(\lambda;\, T)x, y)$ vanishes at infinity and hence is identically zero. Thus the coefficient $(x, y) = 0$. Q.E.D.

→ 3 THEOREM. *A bounded self adjoint operator in Hilbert space is a scalar type spectral operator.*

PROOF. In view of Lemmas 1 and 2, Theorem 5.18 shows that a bounded self adjoint operator in Hilbert space is a spectral operator of type 0, which means that it is of scalar type. Q.E.D.

7. Exercises

Some of the exercises will use the following notation. The symbol T is a bounded linear operator on a complex B-space $\mathfrak{X}$. For each x in $\mathfrak{X}$ the symbol $[x]$ will be used for the closed linear manifold determined by all the vectors $R(\xi; T)x$ with ξ in $\rho(T)$. If σ is a closed set of complex numbers, the symbol $\mathfrak{M}(\sigma)$ will denote the set of all vectors x whose spectrum is contained in σ, and, as usual, $\mathscr{F}(\sigma)$ will denote the class of all complex functions which are single valued and analytic on an open set containing σ.

1 For every x in $\mathfrak{X}$,

(a) $$x \in [x];$$
(b) $$f(T)[x] \subseteq [x], \qquad f \in \mathscr{F}(\sigma(T));$$
(c) $$x(\xi) \in [x], \qquad \xi \in \rho(x);$$
(d) $$[y] \subseteq [x], \qquad y \in [x].$$
(e) If σ is a connected component of $\rho(x)$ containing a point of $\rho(T)$, then $x(\lambda) \in [x]$ for $\lambda \in \sigma$. Show that this may fail if $\sigma \cap \rho(T) = \phi$.

2 If T satisfies condition (C), then for every closed set σ of complex numbers the set $\mathfrak{M}(\sigma)$ is a closed linear manifold with $T\mathfrak{M}(\sigma) \subseteq \mathfrak{M}(\sigma)$ and

$$\sigma(T \mid \mathfrak{M}(\sigma)) \subseteq \sigma.$$

3 If T has the property (C), then for every pair σ_1, σ_2 of disjoint closed sets of complex numbers there is a constant K depending upon σ_1 and σ_2 and such that

$$|x(\xi)| \leq K\,|x|, \qquad \xi \in \sigma_1,\ x \in \mathfrak{M}(\sigma_2).$$

4 If T has the property (C), then for every x in $\mathfrak{X}$ we have $T[x] \subseteq [x]$ and $\sigma(T \mid [x]) = \sigma(x)$.

5 If T has the property (C), then

$$\sigma(y) \subseteq \sigma(x), \qquad y \in [x].$$

6 DEFINITION. An operator $T \in B(\mathfrak{X})$ is said to be a spectral operator of class (Γ) in case Γ is a linear manifold in $\mathfrak{X}^*$ which is total, i.e., $\Gamma x = 0$ only when $x = 0$, and T has a resolution of the identity E defined on the Borel sets in the complex plane such that x^*E is countably additive for every x^* in Γ.

7 (Fixman) Let $\mathfrak{X} = l_\infty$ and let T be defined for $x = (\xi_1, \xi_2, \ldots)$ in l_∞ to be the element $y = (\eta_1, \eta_2, \ldots)$, where

$$\eta_1 = \xi_1; \qquad \eta_n = \left(1 - \frac{1}{n}\right)\xi_n, \qquad n \geqq 2.$$

Let Γ be the linear manifold in $(l_\infty)^*$ spanned by the coordinate functionals

$$\gamma_n(x) = \xi_n, \qquad n \geqq 1.$$

Let $E(\sigma)x = y$ where $\eta_1 = \xi_1$ if and only if $1 \in \sigma$ and $\eta_n = \xi_n$, $n \geqq 2$, if and only if $1 - 1/n \in \sigma$. Show that T is a spectral operator of class (Γ), but that it is not a spectral operator of class $(\mathfrak{X}^*)$.

8 (Fixman) Let T be as in the preceding exercise and let φ in $(l_\infty)^*$ be a Banach limit as in Exercise II.4.22. Let A be defined on l_∞ by

$$Ax = A(\xi_1, \xi_2, \xi_3, \ldots) = (\varphi(x), 0, 0, \ldots).$$

Show that $A^2 = 0$ and that $\varphi(Tx) = \varphi(x)$ and $AT = TA$. However, if $\sigma = \{1\}$, then $AE(\sigma) = 0$ and $E(\sigma)A = A$. Hence A is a nilpotent operator which commutes with T but which does not commute with the resolution of the identity E.

9 (Fixman) Let T, E, and A be as in the preceding exercise and let F be defined for each Borel set σ by

$$F(\sigma) = E(\sigma) + AE(\sigma) - E(\sigma)A.$$

Show that F is a bounded spectral measure and that x^*F is countably additive for every x^* in the manifold Γ_1 spanned by the linear functionals

$$\{\gamma_1 - \varphi, \gamma_2, \gamma_3, \ldots\}.$$

Show that F is a resolution of the identity for T and that T is a spectral operator of class (Γ_1) but the resolutions E, F are not the same and they do not commute. [Hint: $AE(\sigma)A = 0$ and $F(\sigma) = E(\bar{\sigma})F(\sigma)$ for all Borel sets σ.]

8. Notes and Remarks

Characterization of spectral operators. This chapter is concerned with determining conditions which are sufficient to guarantee that an operator in $B(\mathfrak{X})$ is spectral. The results presented here are due to Dunford (see especially [16, 17, 19]). Some of the results from Chapter XVII can also be used to obtain sufficient conditions for spectral operators. For example, if $\mathfrak{X}$ is a weakly complete B-space and $T \in B(\mathfrak{X})$ has real spectrum, then T is a scalar type operator if and only if there exists $M > 0$ such that for every polynomial p one has

$$|p(T)| \leqq M \sup_{\lambda \in \sigma(T)} |p(\lambda)| .$$

(See Theorem XVII.2.5.) Similarly, an operator $T \in B(\mathfrak{X})$ is scalar type if and only if the set

$$\{p(T)x \mid p \text{ a polynomial}, \sup_{\lambda \in \sigma(T)} |p(\lambda)| \leqq 1\}$$

is weakly sequentially compact for each $x \in \mathfrak{X}$. (See Kluvánek [1].) Other results of this type were given by Fixman [1].

Although it is not difficult to give examples of operators which are *not* spectral operators, it is instructive to mention certain "well-behaved" operators which are not spectral. An operator $U \in B(\mathfrak{X})$ which is isometric and maps $\mathfrak{X}$ onto $\mathfrak{X}$ is sometimes said to be a *unitary operator* on $\mathfrak{X}$. It is easy to see that if $U \in B(\mathfrak{X})$ is unitary, then (i) $|U^n| = 1$ for $n = 0, \pm 1, \pm 2, \ldots$, (ii) $\sigma(U) \subseteq \{\lambda \in C \mid |\lambda| = 1\}$, and (iii) $|R(\lambda; U)| \leqq |1 - |\lambda||^{-1}$ for $\lambda \in \rho(U)$. Moreover, if U is a unitary operator which is spectral, then it is scalar type.

Let K be a compact Hausdorff space, then the most general unitary operator in $C(K)$ has the form

$$(Uf)(x) = \mu(x) f(\varphi(x)), \qquad x \in K,$$

where φ is a homeomorphism of K, $\mu \in C(K)$ and $|\mu(x)| = 1$. It was proved by Fixman [1] that if φ is not periodic, then the operator U is not a spectral operator.

Let $l_p(-\infty, \infty)$, $1 \leqq p \leqq \infty$, denote the space of complex sequences $x = \{x_k \mid k = 0, \pm 1, \pm 2, \ldots\}$ with $|x| = \{\sum |x_k|^p\}^{1/p} < \infty$, and let U be the *shift operator* defined by

$$U(\{x_k\}) = \{x_{k+1}\};$$

then Fixman [1] showed that if $p \neq 2$, the unitary operator U is not spectral. See also Krabbe [3] where it is also noted that the Hilbert transform on $l_p(-\infty, \infty)$ generated by the matrix $(1/(n-m))$ is not a scalar operator; on the other hand, Krabbe [6] shows that although certain convolution operators are not spectral operators, they have the property that they can be represented in the weak operator topology by an integral of the form $\int \lambda E(d\lambda)$ where E is an additive operator-valued function defined on *rectangles*.

In a Hilbert space the condition that $|e^{itT}| \leqq M$ for all $t \in R$ implies that T is equivalent to a self adjoint operator and hence is a scalar type operator with real spectrum. (This follows from Lemma XV.6.1 which implies that the bounded group $G = \{e^{itT} \mid t \in R\}$ is equivalent to a group of unitary operators. By Stone's theorem, the latter group has an infinitesimal generator iA, where A is self adjoint. Thus T is equivalent with A.) However, if $\mathfrak{X}$ is a B-space and $T \in B(\mathfrak{X})$ the condition $|e^{itT}| \leqq M$ for $t \in R$ need not imply that T is spectral, even when $\mathfrak{X}$ is reflexive. To see this, note that if S and T are commuting scalar operators with real spectrum then $|e^{itS}| \leqq M_1$ and $|e^{itT}| \leqq M_2$ for all $t \in R$. Hence we have

$$\begin{aligned} |e^{it(S+T)}| &= |e^{itS}e^{itT}| \\ &\leqq |e^{itS}|\,|e^{itT}| \leqq M_1 M_2 \end{aligned}$$

for all $t \in R$. Hence if this boundedness implied that $S + T$ was a spectral operator, we would have a contradiction to McCarthy's [2, I] modification of Kakutani's example. In the positive direction, Berkson [2] proved that if $\mathfrak{X}$ is reflexive, then $T \in B(\mathfrak{X})$ is a scalar type operator if and only if there exist commuting operators A, B such that $T = A + iB$ and such that $\{e^{iP} \mid P \in R(A, B)\}$ is bounded in $B(\mathfrak{X})$, where $R(A, B)$ denotes the set of polynomials in A, B with real coefficients. Similarly, Kantorovitz [7; p. 545] showed that if $\mathfrak{X}$ is weakly complete and if for some $M > 0$ and every polynomial p with real coefficients we have $|e^{ip(T)}| \leqq M$, then T is a scalar type spectral operator with real spectrum.

It was noted by Kantorovitz [4] that if T is a scalar type operator with real spectrum, then $e^{-2\pi itT}$ is the Fourier-Stieltjes transform of its resolution E of the identity; hence E can be considered to be the *inverse* Fourier-Stieltjes transform of the group $\{e^{-2\pi itT} \mid t \in R\}$. Assuming that this group is bounded in $B(\mathfrak{X})$ where $\mathfrak{X}$ is reflexive, Kantorovitz obtained a characterization of scalar type operators with real spectrum by adding appropriate analytic conditions to insure that the inverse Fourier-Stieltjes transform exists. The following result was proved by Kantorovitz [4]:

THEOREM. *Let $\mathfrak{X}$ be reflexive and let $T \in B(\mathfrak{X})$. Then the following statements are equivalent:*

(1) *T is a scalar type spectral operator with real spectrum.*

(2) *There exists an $M > 0$ such that for every $f \in L_1(R)$*

$$\left| \int_R f(t) e^{-2\pi i t T} \, dt \right| \leqq M \sup_{s \in R} |\hat{f}(s)|,$$

where $\hat{f}$ denotes the Fourier transform of f.

(3) *There exists an $M > 0$ such that for every real vector $(t_1, \ldots, t_n)$ and every complex vector $(c_1, \ldots, c_n)$, $n = 1, 2, \ldots$, we have*

$$\left| \sum_{k=1}^{n} c_k e^{-2\pi i t_k T} \right| \leqq M \sup_{s \in R} \left| \sum_{k=1}^{n} c_k e^{-2\pi i t_k s} \right|.$$

(4) *There exists an $M > 0$ such that for every $x \in \mathfrak{X}$, $x^* \in \mathfrak{X}^*$ with unit norm, then*

$$\sup_{\varepsilon > 0} \int_R |x^*\{R(t - i\varepsilon; T) - R(t + i\varepsilon; T)\}x| \, dt \leqq M.$$

Similarly, if $\mathfrak{X}$ is an arbitrary B-space, then T is a scalar type operator with real spectrum if and only if for each $x \in \mathfrak{X}$, the set

$$\left\{ \int_R f(t) e^{-2\pi i t T} x \, dt : |\hat{f}|_\infty \leqq 1, f \in L_1(R) \right\}$$

is weakly sequentially compact (see Kluvánek [2]).

Operators whose spectra lie on the real line R or the unit circle and whose resolvents satisfy a growth condition are specific examples of generalized scalar operators in the sense of Colojoară and Foiaş [4; Chapters 4, 5]. The reason for this is that, as observed by Wolf [4] and Tillmann [2], they satisfy an operational calculus. The earliest work of this type stems from E. R. Lorch [7] who considered operators T which are *power-bounded* in the sense that $\{|T^n| \mid n = 0, \pm 1, \pm 2, \ldots\}$ is a bounded set in $B(\mathfrak{X})$. Such operators have their spectra in $\{\lambda \mid |\lambda| = 1\}$ and their resolvents satisfy a growth condition of the form

$$|R(\lambda; T)| \leqq \frac{K}{|1 - |\lambda||}$$

for $|\lambda| \neq 1$. If $\mathfrak{X}$ is a Hilbert space, then it follows from the theorem of Sz.-Nagy [7] mentioned earlier that such an operator is similar to a unitary operator; hence it is a scalar type operator with a resolution E of the identity which can be taken to be defined on the unit circle. In the case of a B-space the unitary operators show that power-bounded operators need not be spectral. However, when $\mathfrak{X}$ is a reflexive B-space, Lorch was able to construct for each $\theta \in [0, 2\pi]$ a pair $(\mathfrak{E}_\theta, \mathfrak{F}_\theta)$ of closed subspaces which are invariant under T and which resemble the manifolds corresponding to a resolution of the identity. Alternatively, there exists a family $\{P_\theta\}$ of closed (but possibly unbounded) projections playing the role of a resolution of the identity in many ways. The work of Lorch was simplified and extended by Leaf [1] to cover operators satisfying (a) $|T^n| = o(|n|)$ as $|n| \to \infty$, and (b) $\sigma(T) = \sigma_c(T)$, without the hypothesis that $\mathfrak{X}$ be reflexive; these operators have their spectrum on the unit circle and satisfy a second-order growth condition. See also Leaf [2] where the case $|T^n| = O(|n|^q)$ is treated.

Bartle [6, 7] and Kocan [1] considered operators with real spectrum and a growth condition $|R(\lambda; T)| \leqq K/|\mathscr{I}(\lambda)|^n$ for $|\mathscr{I}(\lambda)| > 0$, and such that $\sigma(T) = \sigma_c(T)$. It is proved that for each real number t there exists a closed (but possibly unbounded) densely-defined projection operator P_t which commutes with T and any operator commuting with T, and such that the family $\{P_t \mid t \in R\}$ "increases" from 0 to I as t varies over R, and that $t_0 \in \rho(T)$ if and only if P_t is constant on a neighborhood of t_0. If $n = 1$ and $\mathfrak{X}$ is reflexive, then the condition that $\sigma(T) = \sigma_c(T)$ can be dropped and analogous results obtained. Similar results were also obtained by Dollinger [1] and Sine[1] under the hypothesis that T satisfies certain types of operational calculi, although their assumptions were quite different in nature. Stampfli [12] has considered operators with spectra in a C^2 curve and whose resolvents have finite rates of growth. He has obtained generalizations of the results of this chapter, as well as analogous theorems.

CHAPTER XVII

Algebras of Spectral Operators

1. Introduction

The sum and product of two commuting bounded normal operators in Hilbert space is normal and hence spectral. In Corollary XV.6.5 it was seen that this principle could be extended to the sum and product of two commuting spectral operators in Hilbert space. The attentive reader may have noticed, however, that no attempt has been made to prove the corresponding theorem in arbitrary B-spaces. In fact, as an example of Kakutani [15] shows, such a generalization would be false. In this chapter a systematic survey of the algebraic and analytic properties of families of spectral operators will be undertaken, in order to find sufficient conditions that the sum and product of two commuting spectral operators be spectral. At the same time, various sufficient conditions that the uniform, weak, and strong limits of sequences of commuting and non-commuting spectral operators be spectral will be developed.

In brief, the organization of the present chapter is as follows. In Section 2 a representation for uniformly closed commutative algebras of spectral operators is given. In Section 3 weakly and strongly closed commutative algebras of spectral operators are discussed. Finally, in Section 4, will be found two results concerning the strong limit of a sequence of non-commuting spectral operators.

To be more specific we take, in Section 2, a family τ of commuting spectral operators, and consider not only the algebra but the *full algebra* generated by τ, which is defined as follows.

1 Definition. A *full algebra* of operators is a uniformly closed algebra of operators which contains the inverse of each of its non-singular elements. The *full algebra generated by a family of operators* is the intersection of all the full algebras containing the given family of operators.

The reader should note that a full algebra of operators is closed in the uniform operator topology, while an algebra containing all inverses as defined in Definition XV.13.4 is not required to be closed.

Taking, then, a family τ of commuting spectral operators, adjoining to it the family of all projections in the spectral resolutions of the operators in τ, and forming the full algebra $\mathfrak{A}$ generated by this extended family of operators, we begin Section 2 by showing that under suitable hypotheses $\mathfrak{A}$ is decomposable as a vector direct sum

$$\mathfrak{A} = \mathfrak{S} \oplus \mathfrak{R},$$

where $\mathfrak{R}$ is the radical in $\mathfrak{A}$, and $\mathfrak{S}$ is a semi-simple full subalgebra of $\mathfrak{A}$ algebraically and topologically equivalent to the set of all continuous functions on its own space of maximal ideals. This theorem generalizes the Gelfand-Naĭmark theorem (cf. IX.3.7) to algebras of spectral operators. It is, moreover, similar to the second principal Wedderburn theorem of abstract algebra. Section 2 continues with a discussion of algebras of operators which are isomorphic to algebras of continuous functions, and with general theorems on functional calculi which generalize Corollaries X.2.9 and X.2.10 from Hilbert space to arbitrary B-spaces.

In Section 3 we take a commuting family τ of *scalar type* spectral operators, and ask the two following questions.

(i) When are all the operators in the weakly closed algebra generated by τ scalar type spectral operators?

(ii) What are the operators in the weakly closed algebra generated by τ?

As will be seen in Section 3, it is possible to give satisfactory answers to these questions in a surprisingly wide variety of cases.

It should also be noted that in the course of Section 3 a number of results about Boolean algebras of projections are proved, some of which are needed to answer questions (i) and (ii), and some of which are appended for the sake of completeness and for their own intrinsic interest. Many of these deep results were developed by W. G. Bade.

Finally, in Section 4, it is shown that, under certain conditions, the strong limit T of an arbitrary sequence $\{T_n\}$ of scalar type spectral operators is a spectral operator, and that the resolution of the identity of T is, in a suitable sense, the limit of the resolutions of the identity for the operators T_n.

2. The Structure of Commutative B-Algebras of Spectral Operators

This section will be concerned with commutative algebras of spectral operators which are closed in the uniform operator topology. It will first be shown that such algebras $\mathfrak{A}$ may be decomposed into vector direct sums of the form $\mathfrak{A} = \mathfrak{S} \oplus \mathfrak{R}$, where $\mathfrak{R}$ is the radical in $\mathfrak{A}$ and $\mathfrak{S}$ is an algebra of scalar type operators which is equivalent to an algebra of continuous functions.

1 LEMMA. *The uniformly closed algebra of operators generated by a bounded Boolean algebra of projection operators is a full algebra equivalent to the algebra of continuous functions on its own space of maximal ideals.*

PROOF. Let B be a Boolean algebra of projections, and $\mathfrak{A}_0(B)$ the set of all operators U of the form

$$\text{(i)} \qquad U = \sum_{i=1}^{n} \alpha_i E_i,$$

where

$$\text{(ii)} \quad 0 \neq E_i \in B, \quad \sum_{i=1}^{n} E_i = I, \quad E_i E_j = 0, \qquad i \neq j, \quad i, j = 1, \ldots, n.$$

Then $\mathfrak{A}_0(B)$ is clearly an algebra of operators containing B. Hence, if $\mathfrak{A}(B) = \overline{\mathfrak{A}_0(B)}$, then $\mathfrak{A}(B)$, is the uniformly closed algebra of operators generated by B.

To show that $\mathfrak{A}(B)$ is a full algebra, let the operator A in $\mathfrak{A}(B)$ have an inverse A^{-1} in the ring of bounded operators. Let $\{A_n\}$ be a sequence of elements of $\mathfrak{A}_0(B)$ converging uniformly to A. Then, by Lemma VII.6.1, A_n has an inverse for sufficiently large n, and $A_n^{-1} \to A^{-1}$ uniformly. To show that A^{-1} is in $\mathfrak{A}(B)$ it is consequently sufficient to show that A_n^{-1} is in $\mathfrak{A}(B)$. Thus we may suppose without loss of generality that A is in $\mathfrak{A}_0(B)$, that is, that A is of the form specified by equations (i), (ii). Then, if $\alpha_i = 0$ for some i, we have $E_i \neq 0$, $AE_i = 0$, contradicting the assumption that A^{-1} exists. Thus $\alpha_i \neq 0$, $i = 1, \ldots, n$, and so the operator $A^{-1} = \sum_{i=1}^{n} \alpha_i^{-1} E_i$ is in $\mathfrak{A}(B)$. This shows that $\mathfrak{A}(B)$ is a full algebra.

Let κ be the canonical mapping of $\mathfrak{A}(B)$ into the ring of continuous functions on its own space $\mathscr{M}$ of maximal ideals (cf. IX.2.9). To show that κ establishes an equivalence of $\mathfrak{A}(B)$ with all of $C(\mathscr{M})$, it is sufficient to show that κ^{-1} is bounded, and that $\kappa\mathfrak{A}(B)$ is dense in $C(\mathscr{M})$. To show that κ^{-1} is bounded, the existence of a finite constant M such that

$$|U| \leqq 4M \sup_{m \in \mathscr{M}} |U(m)|$$

will be established. Since both sides of this inequality are continuous functions of U, it is sufficient to establish the inequality for U in $\mathfrak{A}_0(B)$. Thus let U have the form given in equation (i) with the auxiliary conditions (ii) satisfied. For $1 \leqq k \leqq n$ it is possible, since E_k is not a quasi-nilpotent, to find a maximal ideal m in $\mathscr{M}$ such that $E_k(m) \neq 0$. Since $E_k E_j = 0$ for $j \neq k$, we have $E_j(m) = 0$ for $j \neq k$, and since $E_k^2(m) = E_k(m)$, we have $E_k(m) = 1$. It follows that $U(m) = \alpha_k$. On the other hand, it is clear that every maximal ideal m in $\mathscr{M}$ has the property that $E_i(m) = 1$ for one and only one integer $i \leqq n$ and that for other integers $j \leqq n$ we have $E_j(m) = 0$. Thus

$$\sup_{m \in \mathscr{M}} |U(m)| = \sup_{1 \leqq i \leqq n} |\alpha_i| .$$

Let M be an upper bound for the norms $|E|$ of the projections E in B. It will be shown that

$$\text{(i)} \qquad \sum_{i=1}^{n} |x^* E_i x| \leqq 4M |x| |x^*|, \qquad x \in \mathfrak{X},\ x^* \in \mathfrak{X}^*.$$

To see this, note that

$$\begin{aligned} \sum |\mathscr{R}(x^* E_i x)| &= \sum{}^{+} \mathscr{R}(x^* E_i x) - \sum{}^{-} \mathscr{R}(x^* E_i x) \\ &= \mathscr{R}(x^*(\textstyle\sum^{+} E_i)x) - \mathscr{R}(x^*(\textstyle\sum^{-} E_i)x) \\ &\leqq 2M |x| |x^*|, \end{aligned}$$

where $\sum^{+}$ ($\sum^{-}$) represents the sum over those i for which $\mathscr{R}(x^* E_i x) \geqq 0$ ($\leqq 0$). By treating the imaginary part in the same way and adding, it is seen that

$$\sum |x^* E_i x| \leqq 4M |x| |x^*| ,$$

so that (i) is proved. It follows immediately that

$$\begin{aligned} |U| = \Big|\sum_{i=1}^{n} \alpha_i E_i\Big| &= \sup_{|x^*|, |x| \leqq 1} \Big|\sum_{i=1}^{n} \alpha_i x^* E_i x\Big| \\ &\leqq 4M \sup |\alpha_i| \leqq 4M \sup_{m \in \mathscr{M}} |U(m)| , \end{aligned}$$

proving that κ^{-1} is bounded.

To show that $\kappa\mathfrak{A}(B)$ is dense in $C(\mathscr{M})$, it suffices to show, in view of the Weierstrass theorem (cf. IV.6.17), that $\kappa\mathfrak{A}_0(B)$ distinguishes between points of $\mathscr{M}$ and contains the complex conjugate of each of its elements. Since $\kappa\mathfrak{A}_0(B)$ is dense in $\kappa\mathfrak{A}(B)$, and $\mathscr{M}$ is the space of maximal ideals of $\mathfrak{A}(B)$, it follows immediately that $\kappa\mathfrak{A}_0(B)$ distinguishes between points of $\mathscr{M}$. If m is in $\mathscr{M}$ and E is in B, then $E(m)^2 = E(m)$, so that $E(m)$ is 0 or 1, and hence real. Thus

$$\overline{\left(\sum_{i=1}^{n} \alpha_i E_i\right)(m)} = \left(\sum_{i=1}^{n} \bar{\alpha}_i E_i\right)(m),$$

proving that $\kappa\mathfrak{A}_0(B)$, and hence $\kappa\mathfrak{A}(B)$, contains the complex conjugate of each of its elements. Q.E.D.

Some of the inequalities established during the preceding proof will be of use later, and consequently these are collected and stated explicitly in the following lemma.

2 LEMMA. *Let M be a bound for the Boolean algebra B of projection operators and let $\mathscr{M}$ be the structure space for the uniformly closed algebra $\mathfrak{A}(B)$ generated by B. Then*

$$\sup_{m\in\mathscr{M}} |U(m)| \leqq |U| \leqq 4M \sup_{m\in\mathscr{M}} |U(m)|, \qquad U \in \mathfrak{A}(B),$$

and, in particular,

$$\sup_{1\leqq i\leqq n} |\alpha_i| \leqq \left| \sum_{i=1}^{n} \alpha_i E_i \right| \leqq 4M \sup_{1\leqq i\leqq n} |\alpha_i|,$$

whenever the projections $E_1, \ldots, E_n$ are disjoint.

The preceding lemmas prepare us for a basic result describing the structure of a uniformly closed full algebra $\mathfrak{A}(\tau)$ generated by a commuting family τ of spectral operators together with the Boolean algebra B determined by their resolutions of the identity. As will be seen, under certain conditions, such an algebra is a vector direct sum $\mathfrak{A}(\tau) = \mathfrak{A}(B) \oplus \mathfrak{R}$, where $\mathfrak{R}$ is the radical in $\mathfrak{A}(\tau)$ and where $\mathfrak{A}(B)$, the uniformly closed algebra generated by B, is equivalent to the algebra of continuous functions on the space $\mathscr{M}$ of maximal ideals in $\mathfrak{A}(\tau)$. It should be noted, and it follows from this decomposition, that the space $\mathscr{M}$ is homeomorphic to the space $\mathscr{M}_1$ of maximal ideals in $\mathfrak{A}(B)$. To see this it suffices to note that every complex homomorphism (complex multiplicative linear functional) on $\mathfrak{A}(\tau)$

vanishes on the radical and thus the equation $\mathfrak{A}(\tau) = \mathfrak{A}(B) \oplus \mathfrak{R}$ determines a natural one-to-one correspondence between the non-zero complex homomorphisms on $\mathfrak{A}(\tau)$ with those on $\mathfrak{A}(B)$. It follows (cf. IX.2.2 and IX.2.7) that $\mathscr{M}$ and $\mathscr{M}_1$ are homeomorphic. If there is no radical in $\mathfrak{A}(\tau)$, as is the case with commutative B^*-algebras of operators in Hilbert space, then $\mathfrak{A}(\tau)$ is equivalent to the algebra of continuous functions on its structure space. Thus the following theorem is an extension of the Gelfand-Naĭmark theorem (cf. IX.2.7).

→ 3 THEOREM. *Let $\mathfrak{A}(\tau)$ be the full algebra generated by a family τ of commuting spectral operators together with their resolutions of the identity. If the Boolean algebra B generated by the resolutions of the identity of the operators in τ is bounded, then $\mathfrak{A}(\tau)$ is a vector direct sum*

$$\mathfrak{A}(\tau) = \mathfrak{A}(B) \oplus \mathfrak{R},$$

where $\mathfrak{R}$ is the radical in $\mathfrak{A}(\tau)$ and where $\mathfrak{A}(B)$, the uniformly closed algebra generated by B, is equivalent to the algebra of continuous functions on the structure space of $\mathfrak{A}(\tau)$.

PROOF. Once the decomposition is established, the final assertion concerning the nature of $\mathfrak{A}(B)$ will follow from Lemma 1 together with the remarks in the paragraph preceding the statement of the theorem.

The first part of the theorem will be established by constructing a projection P in the space $\mathfrak{A}(\tau)$ such that $P\mathfrak{A}(\tau) = \mathfrak{A}(B), (I-P)\mathfrak{A}(\tau) = \mathfrak{R}$. This will be accomplished in two steps. First P will be defined on a dense subset $\mathfrak{B}$ of $\mathfrak{A}(\tau)$, and then it will be shown that P is bounded on $\mathfrak{B}$ so that it may be extended by continuity to a projection defined on all of $\mathfrak{A}(\tau)$.

Note to begin with that if $T, U \in \tau$ have resolutions of the identity $E(\cdot\,; T)$, $E(\cdot\,; U)$, respectively, then for every pair σ, μ of Borel sets in the plane the projections $E(\sigma; T)$, $E(\mu; U)$ commute. This follows from a double application of Corollary XV.3.7. Thus the various projections $E(\sigma; T)$ determined by Borel sets σ and operators T in τ determine a Boolean algebra B.

Let $\mathfrak{B}$ denote the family of all elements U in $\mathfrak{A}(\tau)$ which have the form

$$\text{(i)} \qquad U = S + N,$$

where N is in the radical $\mathfrak{R}$ of $\mathfrak{A}(\tau)$, and

$$\text{(ii)} \qquad S = \sum_{i=1}^{n} \alpha_i E_i,$$

where

$$\text{(iii)} \qquad 0 \neq E_i \in B; \qquad E_i E_j = 0,\ i \neq j; \qquad E_1 + \cdots + E_n = I.$$

Let $PU = S$. By the uniqueness assertion of Theorem XV.4.5, P is a well-defined linear map of $\mathfrak{B}$ into itself. Moreover, P is evidently a projection in $\mathfrak{B}$.

The set $\mathfrak{B}$ is clearly a subalgebra of $\mathfrak{A}(\tau)$. Let $\overline{\mathfrak{B}}$ denote its closure in the uniform topology of operators. By Theorem XV.4.5 and the fact that a scalar type operator is clearly in the uniformly closed algebra generated by the projections in its resolution of the identity (cf. Definition XV.4.1), $\overline{\mathfrak{B}} \supseteq \tau$. It will be shown below that $\overline{\mathfrak{B}}$ is a full algebra, from which it follows that $\overline{\mathfrak{B}} = \mathfrak{A}(\tau)$. It will also be shown that P is bounded on $\mathfrak{B}$, so that P can be extended by continuity to a projection (which will still be denoted by the symbol P) of $\mathfrak{A}(\tau)$ into itself. Since it is obvious from the definition of P that $P\mathfrak{B} \subseteq \mathfrak{A}(B)$, it will follow that $P\mathfrak{A}(\tau) \subseteq \mathfrak{A}(B)$. On the other hand, since $P\mathfrak{B}$ contains all the elements of $\mathfrak{A}(B)$ which have the form (ii), and since this set of elements is dense in $\mathfrak{A}(B)$, it will be seen that $P\mathfrak{A}(\tau) = \mathfrak{A}(B)$. Moreover, it is clear that $(I - P)\mathfrak{B} = \mathfrak{R}$, and since $\mathfrak{R}$ is closed, it will follow that $(I - P)\mathfrak{A}(\tau) = \mathfrak{R}$.

Thus, to show that $\mathfrak{A}(\tau) = \mathfrak{A}(B) \oplus \mathfrak{R}$, it is sufficient to show that P is bounded on $\mathfrak{B}$, and that $\mathfrak{A}(B) \oplus \mathfrak{R}$ is a full algebra. To show that P is bounded on $\mathfrak{B}$, consider an element of the form (i), where (ii) and (iii) are satisfied. If $\mathscr{M}$ is the space of maximal ideals in $\mathfrak{A}(\tau)$, then since E_i is not quasi-nilpotent there is an element m in $\mathscr{M}$ with $E_i(m) \neq 0$. Since $E_i^2 = E_i$, $E_i E_j = 0$ for $i \neq j$, it is seen that $E_i(m) = 1$, $E_j(m) = 0$ for $i \neq j$. Hence $U(m) = S(m) + R(m) = S(m) = \alpha_i$; thus $\sup|\alpha_i| \leqq \sup_{m \in \mathscr{M}} |U(m)| \leqq |U|$, and by the preceding corollary,

$$|PU| = |S| \leqq 4M \sup|\alpha_i| \leqq 4M|U|,$$

which proves that P is bounded on $\mathfrak{B}$.

To show that $\mathfrak{A}(B) \oplus \mathfrak{R}$ is a full algebra, we let T be in $\mathfrak{A}(B) \oplus \mathfrak{R}$, and suppose that T^{-1} exists as an element of $B(\mathfrak{X})$. We wish to show that T^{-1} is in $\mathfrak{A}(B) \oplus \mathfrak{R}$. Let $T = S + N$, where S is in $\mathfrak{A}(B)$, and N is in $\mathfrak{R}$. Since N is a topological nilpotent, the operator $S = T - N$ has the inverse

$\sum_0^\infty (T^{-1})^{n+1}N^n$ and this inverse is, by Lemma 1, in $\mathfrak{A}(B)$. An elementary multiplication shows that the equation $(S+N)(S^{-1}+C)=I$ is satisfied by the operator $C=-NS^{-1}T^{-1}$. Since N is in the radical $\mathfrak{R}$, so is C, which shows that $T^{-1}=S^{-1}+C$ is in the algebra $\mathfrak{A}(B)\oplus\mathfrak{R}$ and proves that this algebra is a full algebra. Q.E.D.

Having established the general structure described in the preceding theorem, we now study in more detail the algebras of the type $\mathfrak{A}(B)$. It will be seen in the following theorem that an arbitrary isomorphic homeomorphism of the algebra $\mathfrak{A}(B)$ onto the algebra of continuous functions on its structure space may be represented by an integral with respect to a uniquely determined spectral measure. This is the analogue of the general spectral theorem for algebras of normal operators in Hilbert space (cf. X.2.1).

4 THEOREM. *Let $\mathfrak{A}$ be an algebra of operators in the complex B-space $\mathfrak{X}$ which is the homomorphic image under a continuous homomorphism S of the algebra $C(\Lambda)$ of all complex continuous functions on a compact space Λ. Then there is a unique spectral measure A in $\mathfrak{X}^*$ defined on the Borel sets in Λ such that $xA(\cdot)x^*$ is regular and countably additive for x in $\mathfrak{X}$ and x^* in $\mathfrak{X}^*$, and for which*

$$S^*(f)=\int_\Lambda f(\lambda)A(d\lambda),\qquad f\in C(\Lambda).$$

PROOF. For each x in $\mathfrak{X}$ and x^* in $\mathfrak{X}^*$ the number $x^*S(f)x$ depends linearly and continuously upon the function f in $C(\Lambda)$ and hence (cf. IV.6.3) determines a unique regular measure $\mu(\cdot, x, x^*)$ on the Borel sets in Λ for which

$$\text{(i)}\qquad x^*S(f)x=\int_\Lambda f(\lambda)\mu(d\lambda, x, x^*),\qquad f\in C(\Lambda).$$

The left side of this equation is bilinear in x and x^*, and since the measure $\mu(\cdot, x, x^*)$ is uniquely determined by this equation, the number $\mu(\delta, x, x^*)$ is consequently bilinear in x and x^*. Furthermore, since

$$|\mu(\delta, x, x^*)|\leqq v(\Lambda, \mu(\cdot, x, x^*))=\sup_{|f|=1}|x^*S(f)x|\leqq K|x|\,|x^*|,$$

it is seen that $\mu(\delta, x, x^*)$ is continuous in x and x^*. Thus, for each Borel set δ in Λ and each x^* in $\mathfrak{X}^*$, there is a vector $A(\delta)x^*$ in $\mathfrak{X}^*$ with

$$\mu(\delta, x, x^*)=xA(\delta)x^*,\qquad x\in\mathfrak{X}$$

It follows from the bilinearity and boundedness of μ that $A(\delta)x^*$ is linear and continuous in x^*, that is, $A(\delta)$ exists as a bounded linear operator in $\mathfrak{X}^*$. Since every function f in $C(\Lambda)$ is bounded and Borel measurable, the integral $\int f(\lambda)A(d\lambda)$ exists and it is seen from (i) that

$$\text{(ii)} \qquad S^*(f) = \int_\Lambda f(\lambda)A(d\lambda), \qquad f \in C(\Lambda).$$

It will now be shown that A is a spectral measure in $\mathfrak{X}^*$. By placing $f = 1$ in (ii), it is seen that $I^* = A(\Lambda)$ and, since $A(\delta)$ is additive in δ, that

$$A(\delta)' = I^* - A(\delta) = A(\Lambda) - A(\delta) = A(\delta').$$

To prove that A is a spectral measure it will therefore suffice to show that $A(\delta \cap \sigma) = A(\delta)A(\sigma)$ for every pair δ, σ of Borel subsets of Λ. Now, for every pair f, g of functions in $C(\Lambda)$,

$$\begin{aligned}\int_\Lambda f(\lambda) \int_\Lambda g(\mu)A(d\mu \cap d\lambda) &= \int_\Lambda f(\lambda)g(\lambda)A(d\lambda) = S^*(fg)\\ &= S^*(f)S^*(g) = \int_\Lambda f(\lambda)S^*(g)A(d\lambda)\\ &= \int_\Lambda f(\lambda) \int_\Lambda g(\mu)A(d\mu)A(d\lambda).\end{aligned}$$

Thus for x in $\mathfrak{X}$ and x^* in $\mathfrak{X}^*$,

$$\text{(iii)} \quad \int_\Lambda f(\lambda) \int_\Lambda g(\mu)xA(d\mu \cap d\lambda)x^* = \int_\Lambda f(\lambda) \int_\Lambda g(\mu)xA(d\mu)A(d\lambda)x^*.$$

Since the measures $xA(\cdot)x^*$ are all regular and the integral (as a functional on $C(\Lambda)$) uniquely determines the regular measure, it follows from (iii) that

$$\int_\Lambda g(\mu)xA(d\mu \cap \delta)x^* = \int_\Lambda g(\mu)xA(d\mu)A(\delta)x^*, \qquad g \in C(\Lambda).$$

This uniqueness argument may be repeated to conclude that

$$A(\sigma \cap \delta) = A(\sigma)A(\delta),$$

for every pair σ, δ of Borel subsets of Λ, and this proves that A is a spectral measure. Q.E.D.

The following theorem gives more specific information in case the space $\mathfrak{X}$ is weakly complete.

5 THEOREM. *Let $\mathfrak{A}$ be an algebra of operators in the weakly complete complex B-space $\mathfrak{X}$ which is the image under a continuous homomorphism S of the algebra $C(\Lambda)$ of all complex continuous functions on a compact space Λ. Then there exists a uniquely determined spectral measure E in $\mathfrak{X}$ defined on the Borel sets in Λ which is countably additive in the strong operator topology and for which*

$$S(f) = \int_\Lambda f(\lambda)E(d\lambda), \qquad f \in C(\Lambda).$$

PROOF. For a fixed x in $\mathfrak{X}$, consider the map $f \to S(f)x$ of $C(\Lambda)$ into $\mathfrak{X}$. Since every bounded linear map from $C(\Lambda)$ into a weakly complete space is weakly compact (cf. VI.7.6), it is seen from Theorem VI.7.3 that the map $f \to S(f)x$ uniquely determines a regular $\mathfrak{X}$-valued measure $\nu(\cdot, x)$ such that

$$\text{(i)} \qquad S(f)x = \int_\Lambda f(\lambda)\nu(d\lambda, x), \qquad f \in C(\Lambda).$$

Comparing the above equation with equation (ii) of the proof of Theorem 4, and using the uniqueness assertion of the Riesz representation theorem, it is found that

$$\text{(ii)} \qquad xA(\delta)x^* = x^*\nu(\delta, x), \qquad x^* \in \mathfrak{X}^*,$$

from which it follows that $\nu(\delta, x)$ is linear and continuous in x. Thus for every Borel set δ in Λ there is a bounded linear operator $E(\delta)$ in $\mathfrak{X}$ with $\nu(\delta, x) = E(\delta)x$. Equation (ii) shows that $E^*(\delta) = A(\delta)$, so that E is a spectral measure. The desired integral representation follows immediately from equation (i). Q.E.D.

Next, algebras representable in terms of a spectral measure as in Theorem 5 will be analyzed.

Throughout the remainder of the present section the letter E will stand for a countably additive spectral measure in the complex B-space $\mathfrak{X}$. Sometimes E will be defined on an arbitrary σ-field Σ of subsets of an abstract set Λ, and at other times its domain will be the field of Borel sets in the complex plane. In either case it will be convenient to use the notion of an E-essentially bounded function described in the following definition.

6 DEFINITION. Let E be a countably additive spectral measure defined on the σ-field Σ of subsets of a set Λ. Then a function f on Λ is said

to be *E-essentially bounded on Λ* if

$$E\text{-ess}\sup_{\lambda\in\Lambda}|f(\lambda)| = \inf_{E(\delta)=I}\ \sup_{\lambda\in\delta}|f(\lambda)|$$

is finite.

Since E is countably additive on Σ, there is a set δ_0 in Σ with $E(\delta_0) = I$ and

$$E\text{-ess}\sup_{\lambda\in\Lambda}|f(\lambda)| = \sup_{\lambda\in\delta_0}|f(\lambda)|,$$

and hence there is a bounded function f_0 on Λ with $f(\lambda) = f_0(\lambda)$ except for λ in an E-null set. If f is Σ-measureable, so is f_0, and it is clear that the integral $\int f_0(\lambda)E(d\lambda)$ is independent of the particular bounded function f_0 which coincides E-almost everywhere with f. Thus the integral of an E-essentially bounded Σ-measurable function f may be defined as the integral of a bounded Σ-measurable function which coincides with f, E-almost everywhere on Λ. It should also be noted that the space of all E-essentially bounded Σ-measurable functions on Λ is a B-algebra under the natural operations

$$\text{(i)}\qquad \begin{aligned} (\alpha f+\beta g)(\lambda) &= \alpha f(\lambda)+\beta g(\lambda), && \lambda\in\Lambda;\\ (fg)(\lambda) &= f(\lambda)g(\lambda), && \lambda\in\Lambda;\\ e(\lambda) &= 1, && \lambda\in\Lambda;\\ |f|_E &= E\text{-ess}\sup_{\lambda\in\Lambda}|f(\lambda)|. \end{aligned}$$

As in the case of a self adjoint spectral measure in Hilbert space, it is possible to define an operational calculus which establishes an isomorphism between the algebra of E-essentially bounded Σ-measurable functions on Λ and a B-algebra of spectral operators.

7 DEFINITION. Let $\mathfrak{X}$ be a complex B-space and let E be a spectral measure in $\mathfrak{X}$ which is defined and countably additive on the σ-field Σ of subsets of a set Λ. Then the algebra of all complex Σ-measurable E-essentially bounded functions on S will be denoted by $EB(\Lambda, \Sigma)$. The operations in $EB(\Lambda, \Sigma)$ are defined by the equations (i) above. If Λ is a topological space and if Σ is the field of Borel sets in Λ, then the symbol $EB(\Lambda)$ will sometimes be used instead of $EB(\Lambda, \Sigma)$.

The following lemma gives a change of measure principle slightly more general than that which has been used before. It will be used frequently in what follows.

8 LEMMA. *Let E be a spectral measure in the complex B-space $\mathfrak{X}$ which is defined and countably additive on a σ-field Σ of subsets of a set Λ and let g be a bounded Borel measurable function defined on the complex plane. Then*

$$\int_{\Lambda} g(f(\lambda))E(d\lambda) = \int_{f(\Lambda)} g(\mu)E(f^{-1}(d\mu)), \qquad f \in EB(\Lambda, \Sigma).$$

PROOF. Let f be in $EB(\Lambda, \Sigma)$, and for every Borel set δ in the complex plane let $E_1(\delta) = E(f^{-1}(\delta))$. If g is the characteristic function of such a set δ, then since E_1 vanishes outside $f(\Lambda)$,

$$\int_{f(\Lambda)} g(\mu)E_1(d\mu) = E_1(\delta) = E(f^{-1}(\delta)) = \int_{\Lambda} g(f(\lambda))E(d\lambda).$$

Now the set of bounded Borel measurable functions g for which

$$\int_{f(\Lambda)} g(\mu)E_1(d\mu) = \int_{\Lambda} g(f(\lambda))E(d\lambda)$$

is clearly linear and closed in the set of all bounded Borel functions. Since this set contains every characteristic function of a Borel set, it contains every bounded Borel function. Q.E.D.

9 LEMMA. *Let E be a spectral measure in the complex B-space $\mathfrak{X}$ which is defined and countably additive on a σ-field Σ of subsets of a set Λ. Then the map*

$$S(f) = \int_{\Lambda} f(\lambda)E(d\lambda), \qquad f \in EB(\Lambda, \Sigma),$$

is a continuous homomorphism of $EB(\Lambda, \Sigma)$ onto an algebra of scalar type spectral operators. Moreover, the resolution of the identity for $S(f)$ is given by the equation

$$E(\delta; S(f)) = E(f^{-1}(\delta)),$$

where δ is an arbitrary Borel set in the complex plane.

PROOF. It is clear that $S(f)$ is linear in f and that

(i) $$|S(f)| \leq 4K\,|f|_E, \qquad f \in EB(\Lambda, \Sigma),$$

where K is a bound for the numbers $|E(\sigma)|$, $\sigma \in \Sigma$ (cf. Lemma 2). Since $E(\sigma\delta) = E(\sigma)E(\delta)$, it is also clear that

$$\text{(ii)} \qquad S(fg) = S(f)S(g)$$

if f and g are characteristic functions of sets in Σ. For a fixed characteristic function f the set of g in $EB(\Lambda, \Sigma)$ for which (ii) holds is linear; and, in view of (i), it is closed. Thus, since it contains all characteristic functions of sets in Σ, it must contain every function in $EB(\Lambda, \Sigma)$. Hence for an arbitrary g in $EB(\Lambda, \Sigma)$, equation (ii) holds for every characteristic function f of a set in Σ. But the set of f for which (ii) holds is linear and closed and therefore coincides with $EB(\Lambda, \Sigma)$. Thus the map $f \to S(f)$ is a continuous homomorphism of $EB(\Lambda, \Sigma)$ onto an algebra of operators $S(f)$.

To see that $S(f)$ is a scalar type operator with the stated resolution of the identity, let f be in $EB(\Lambda, \Sigma)$ and, for every Borel set δ in the plane, let $E_1(\delta) = E(f^{-1}(\delta))$. Then, by taking $g(\mu) = \mu$ in Lemma 8,

$$\text{(iii)} \qquad S(f) = \int_\Lambda f(\lambda)E(d\lambda) = \int_{f(\Lambda)} \mu E_1(d\mu) = \int \mu E_1(d\mu),$$

where the last of the above integrals is taken over the whole complex plane. Now E_1 is defined and countably additive on the field of Borel sets and it commutes with $S(f)$. Thus to see that E_1 is the resolution of the identity for $S(f)$ it will suffice to show that

$$\text{(iv)} \qquad \sigma(S(f) \mid E_1(\delta)\mathfrak{X}) \subseteq \bar{\delta}$$

for every Borel set δ of complex numbers. Let δ be such a Borel set and let $\mu_0 \notin \bar{\delta}$. Then

$$\left(\int_\delta \frac{E_1(d\mu)}{\mu_0 - \mu}\right)\left(\int_\delta (\mu_0 - \mu)E_1(d\mu)\right) = \int_\delta E_1(d\mu) = E_1(\delta).$$

This shows that $\mu_0 \in \rho(S(f) \mid E_1(\delta)\mathfrak{X})$ and proves (iv). Since E_1 is the resolution of the identity for $S(f)$, it follows from (iii) that $S(f)$ is a scalar type operator. Q.E.D.

10 THEOREM. *Let $\mathfrak{X}$ be a complex B-space and let E be a spectral measure which is defined and countably additive on a σ-field Σ of subsets of a set Λ. For each E-essentially bounded Σ-measurable function f in Λ, let the operator $S(f)$ be defined by the equation*

$$S(f) = \int_\Lambda f(\lambda)E(d\lambda), \qquad f \in EB(\Lambda, \Sigma).$$

Then the map $f \to S(f)$ is an isomorphism between the B-algebra $EB(\Lambda, \Sigma)$ and a full B-algebra of scalar type spectral operators. The resolution of the identity for $S(f)$ is given by the equation

$$E(\sigma; S(f)) = E(f^{-1}(\sigma)),$$

where σ is an arbitrary Borel subset of the complex plane. Moreover, there is a constant K such that

$$|f|_E \leqq |S(f)| \leqq K|f|_E, \qquad f \in EB(\Lambda, \Sigma).$$

PROOF. From the preceding lemma it is seen that the map $f \to S(f)$ is a continuous homomorphism of $EB(\Lambda, \Sigma)$ onto an algebra of scalar type spectral operators. Thus for some constant K we have $|S(f)| \leqq K|f|$ and to prove the final inequality of the present theorem it will suffice to prove that $|f|_E \leqq |S(f)|$. Since both terms in this inequality are continuous functions of f, it will suffice to prove it for every function f in a set dense in $EB(\Lambda, \Sigma)$. Thus, let $f = \sum_{i=1}^{n} \alpha_i \chi_{\sigma_i}$, where the sets $\sigma_1, \ldots, \sigma_n$ are disjoint sets in Σ whose union is Λ. There is an $i_0 \leqq n$ with $E(\sigma_{i_0}) \neq 0$ and $|f|_E = |\alpha_{i_0}|$. If $E(\sigma_{i_0})x_0 = x_0 \neq 0$, then $S(f)x_0 = a_{i_0}x_0$ and

$$|S(f)| \geqq |\alpha_{i_0}| = |f|_E.$$

This establishes the final inequality of the theorem. From this inequality it is evident that the homomorphism $f \to S(f)$ is an isomorphism and that the algebra $\{S(f) | f \in EB(\Lambda, S)\}$ is a B-algebra.

To complete the proof of the theorem it only remains to show that the algebra of all the operators $S(f)$ with f in $EB(\Lambda, S)$ is a full algebra. This means that if $S(f)^{-1}$ exists as a bounded everywhere defined operator, then $S(f)^{-1} = S(g)$ for some g in $EB(\Lambda, \Sigma)$. It will be shown that if $S(f)^{-1}$ exists, then f^{-1} is E-essentially bounded and $S(f)^{-1} = S(f^{-1})$. For each $m = 1, 2, \ldots$, let $\delta_1, \ldots, \delta_{n_m}$ be disjoint Borel sets of diameter less than $1/m$ whose union is the disk $\{\lambda | |\lambda| \leqq |f|_E\}$. Let $\sigma_i = f^{-1}(\delta_i)$ and let λ_i be a point in σ_i. Then the functions

$$f_m = \sum_{i=1}^{n_m} f(\lambda_i)\chi_{\sigma_i}$$

have the properties

$$|f - f_m|_E \leqq \frac{1}{m}, \qquad |S(f) - S(f_m)| \leqq \frac{K}{m}.$$

Since $S(f)^{-1}$ exists as a bounded everywhere defined operator, it follows from Lemma VII.6.1 that, for sufficiently large values of m, the operators

$$S(f_m) = \sum_{i=1}^{n_m} f_m(\lambda_i)E(\sigma_i)$$

have bounded everywhere defined inverses. This shows that, for sufficiently large m, $f_m(\lambda_i) \neq 0$ if $E(\sigma_i) \neq 0$, which proves that f_m^{-1} is in $EB(\Lambda, \Sigma)$ and that $S(f_m)^{-1} = S(f_m^{-1})$. Now since

$$|f_m^{-1} - f_n^{-1}| \leq |S(f_m^{-1}) - S(f_n^{-1})| = |S(f_m)^{-1} - S(f_n)^{-1}| \to 0,$$

the sequence $\{f_n^{-1}\}$ converges in $EB(\Lambda, \Sigma)$ and it follows that f^{-1} is E-essentially bounded and that $S(f)^{-1} = \lim_m S(f_m)^{-1} = \lim_m S(f_m^{-1}) = S(f^{-1})$. Q.E.D.

11 Corollary. *The isomorphism $f \to S(f)$ of Theorem* 10 *has the following additional properties.*

(i) *The operator $S(f)$ has a bounded everywhere defined inverse if and only if f^{-1} is E-essentially bounded on Λ. In this case $S(f)^{-1} = S(f^{-1})$.*

(ii) *The spectrum of $S(f)$ is given by the formula*

$$\sigma(S(f)) = \bigcap_{E(\delta)=I} \overline{f(\delta)}, \qquad f \in EB(\Lambda, \Sigma).$$

(iii) *If $\{f_n\}$ is a bounded sequence in $EB(\Lambda, \Sigma)$ and if $f(\lambda) = \lim_n f_n(\lambda)$ except for those λ in a set of E-measure zero, then $S(f_n)x \to S(f)x$ for every x in $\mathfrak{X}$.*

Proof. The first statement was established at the end of the proof of Theorem 10, and the third one follows from Theorem IV.10.10. To prove the second, let δ be a set in Σ with $E(\delta) = I$. If $\lambda_0 \notin \overline{f(\delta)}$ then the function h defined by the equations

$$h(\lambda) = \begin{cases} \dfrac{1}{\lambda_0 - f(\lambda)}, & \lambda \in \delta \\ 0, & \lambda \notin \delta \end{cases}$$

is E-essentially bounded on Λ and

$$S(h)(\lambda_0 I - S(f)) = I,$$

which shows that λ_0 is in the resolvent set $\rho(S(f))$. Thus $\overline{f(\delta)} \supseteq \sigma(S(f))$ if $E(\delta) = I$, and hence

$$\bigcap_{E(\delta)=I} \overline{f(\delta)} \supseteq \sigma(S(f)).$$

Conversely, if $\lambda_0 \in \rho(S(f))$, it follows from (i) that $(\lambda_0 - f(\lambda))^{-1}$ is E-essentially bounded on Λ. Hence there is a set σ in Σ with $E(\sigma) = I$ and

$$\left| \frac{1}{\lambda_0 - f(\lambda)} \right| \leqq M, \qquad \lambda \in \sigma.$$

Thus, λ_0 is not in $\overline{f(\sigma)}$. This shows that

$$\sigma(S(f)) \supseteqq \overline{f(\sigma)} \supseteqq \bigcap_{E(\delta)=I} \overline{f(\delta)}$$

and completes the proof of the lemma. Q.E.D.

12 COROLLARY. *Let $\mathfrak{A}$ be an algebra of operators in a weakly complete B-space $\mathfrak{X}$. Suppose that $\mathfrak{A}$ is topologically and algebraically isomorphic to some B-algebra of bounded continuous functions. Then every operator in $\mathfrak{A}$ is a scalar type spectral operator.*

PROOF. This follows from Theorem 5 and Theorem 10. Q.E.D.

13 COROLLARY. *Every operator in the uniformly closed algebra generated by a bounded Boolean algebra of projection operators in a weakly complete B-space is a scalar type spectral operator.*

PROOF. This follows from Corollary 12 and Lemma 1. Q.E.D.

14 COROLLARY. *Let $\mathfrak{X}$ be a weakly complete B-space. Let $\mathfrak{A}(\tau)$ be the full algebra generated by a family τ of commuting spectral operators. If the Boolean algebra B generated by the resolutions of the identity of the operators in τ is bounded, then every operator in $\mathfrak{A}(\tau)$ is a spectral operator.*

PROOF. Since every operator in the radical of $\mathfrak{A}(\tau)$ is quasi-nilpotent, this follows immediately from Theorem 3, Corollary 13, and Theorem XV.4.5. Q.E.D.

15 COROLLARY. *The full algebra $\mathfrak{A}$ generated by a finite number $T_1, \ldots, T_n$ of commuting spectral operators in Hilbert space and their resolutions of the identity has the form*

$$\mathfrak{A} = \mathfrak{B} \oplus \mathfrak{R},$$

where $\mathfrak{R}$ is the radical in $\mathfrak{A}$ and where $\mathfrak{B}$ is the algebra generated by the projections in the resolutions of the identity for $T_1, \ldots, T_n$. Furthermore, every operator in $\mathfrak{A}$ is a spectral operator.

PROOF. It follows from Corollary XV.6.3 that the projections generating $\mathfrak{B}$ are all contained in a bounded Boolean algebra B of projections, and thus the desired decomposition of $\mathfrak{A}$ follows from Theorem 3. The final assertion follows from Corollary 14. Q.E.D.

The preceding corollary is not true if the Hilbert space is replaced by an arbitrary B-space, for it has been shown by Kakutani [15] that the sum of two commuting scalar type spectral operators in a space of continuous functions need not be a spectral operator. However, if $n = 1$ in Corollary 15, then the Hilbert space may be replaced by an arbitrary complex B-space. That is, the full algebra $\mathfrak{A}$ generated by a spectral operator T and the family B of projections in its resolution of the identity is not only the vector direct sum $\mathfrak{B} \oplus \mathfrak{R}$, where $\mathfrak{B}$ is the operator algebra generated by B and $\mathfrak{R}$ is the radical in $\mathfrak{A}$, but every operator in $\mathfrak{A}$ is a spectral operator. In this situation we have two representations for $\mathfrak{B}$, for $\mathfrak{B}$ is equivalent to the algebra of all continuous functions on the space of maximal ideals in $\mathfrak{A}$, and if E is the resolution of the identity for T, $\mathfrak{B}$ is also equivalent to the algebra of all E-essentially bounded Borel measurable functions on the spectrum of T. This latter representation is sometimes more useful because it furnishes an operational calculus of functions defined on the spectrum of T rather than an operational calculus of functions defined on the compact Hausdorff space of maximal ideals in $\mathfrak{A}$. This situation will now be described.

→ 16 THEOREM. *Let T be a spectral operator, E its resolution of the identity, and S its scalar part. Then the full algebra generated by T and the family B of projections in the range of E is the vector direct sum*

$$\mathfrak{A} = \mathfrak{A}(B) \oplus \mathfrak{R},$$

where $\mathfrak{A}(B)$ is the algebra generated by B and $\mathfrak{R}$ is the radical in $\mathfrak{A}$. Furthermore, $\mathfrak{A}(B)$ is a full operator algebra of scalar type spectral operators equivalent to the algebra of all E-essentially bounded Borel measurable functions on the spectrum $\sigma(T) = \sigma(S)$ and consists of all operators of the form

$$f(S) = \int_{\sigma(S)} f(\lambda)E(d\lambda), \qquad f \in EB(\sigma(S)).$$

Moreover, every operator in $\mathfrak{A}$ is a spectral operator.

PROOF. The first statement follows from Theorem 3. Let $\mathfrak{A}_1$ be the algebra of all operators of the form $\int f(\lambda)E(d\lambda)$ where f is E-essentially bounded on $\sigma(S)$. It follows from Theorem 10 that $\mathfrak{A}_1$ is a full algebra of scalar type spectral operators which is equivalent to $EB(\Lambda, \Sigma)$. It follows from the definition of the integral that $\mathfrak{A}_1 \subseteq \mathfrak{A}(B)$. On the other hand, every projection $E(\delta)$ in B is in $\mathfrak{A}_1$. Thus $\mathfrak{A}(B) \subseteq \mathfrak{A}_1$ and $\mathfrak{A}(B) = \mathfrak{A}_1$. The final statement follows from Theorem XV.4.5. Q.E.D.

3. Strongly Closed Algebras and Complete Boolean Algebras

In this section an attempt will be made to characterize the strong closure of a commutative algebra of spectral operators. It has been observed (cf. VI.1.5) that a convex set in the space of all bounded linear maps between two B-spaces has the same closure in the weak as in the strong operator topology. Thus *the strong and weak operator closures of an algebra of operators are the same.* This section, like the last one, starts with a commuting family τ of spectral operators together with their resolutions of the identity and endeavors to determine all operators in the strongly (or, equivalently, weakly) closed operator algebra generated by τ. The following discussion will be restricted to the case where all the operators in the given commuting family are spectral operators of scalar type. Since every scalar type operator is in the strong (even in the uniform) closure of the algebra generated by the projections in its resolution of the identity, the problem of the present section amounts to characterizing the operators in the strong (or weak) closure of the operator algebra generated by a Boolean algebra of projection operators. The basic result in this direction is the following theorem of W. G. Bade.

THEOREM. *Let B be a bounded Boolean algebra of projection operators in a weakly complete B-space $\mathfrak{X}$. Then the weakly* (*or, equivalently, strongly*) *closed operator algebra generated by B consists of all operators in $\mathfrak{X}$ which leave invariant every closed linear manifold which is left invariant by every member of B.*

The proof of this theorem, which will follow the lines of Bade, is based upon a careful analysis and comparison of various notions of completeness of a Boolean algebra. In fact, as will be seen in Corollary 8 to follow, a bounded Boolean algebra of projections in a weakly complete space is strongly closed if and only if it is complete in the sense of the following definition.

1 DEFINITION. A Boolean algebra B of projections in a B-space $\mathfrak{X}$ is said to be *complete* (*σ-complete*) *as an abstract Boolean algebra* if each subset (sequence) of B has a greatest lower bound and a least upper bound in B. The Boolean algebra B is said to be *complete* (*σ-complete*) if it is complete (σ-complete) as an abstract Boolean algebra and if, for every set (sequence) B_0 in B,

$$\Big(\bigvee_{E\in B_0} E\Big)\mathfrak{X} = \overline{\text{sp}}\{E\mathfrak{X} \mid E \in B_0\}, \qquad \Big(\bigwedge_{E\in B_0} E\Big)\mathfrak{X} = \bigcap_{E\in B_0} E\mathfrak{X}.$$

The Boolean algebra B is said to be *bounded* if there is a constant K with

$$|E| \leqq K, \qquad E \in B.$$

The first two lemmas to follow serve merely to show that a Boolean algebra which has the weakest of the four completeness properties of the preceding definition (that is, the property of being σ-complete as an abstract Boolean algebra) is necessarily bounded. The arguments used in proving these lemmas depart somewhat from the main line of thought involved in the proof of Bade's theorem, and the reader who wishes to get more rapidly to the core of the problem may omit these lemmas by simply adding the redundant hypothesis of boundedness throughout the discussion beginning with Lemma 4.

2 LEMMA. *If a set* $\{b_\alpha\}$ *in an abstract Boolean algebra* B *has a least upper bound, then for every* b *in* B *the set* $\{b \wedge b_\alpha\}$ *of intersections has a least upper bound and*

$$b \wedge \Big(\bigvee_\alpha b_\alpha\Big) = \bigvee_\alpha (b \wedge b_\alpha), \qquad b \in B.$$

PROOF. It is clear that $b \wedge (\bigvee_\alpha b_\alpha) \geqq b \wedge b_\alpha$ for each α. Now let $a \geqq b \wedge b_\alpha$ for each α so that

$$a \vee \Big[\Big(\bigvee_\alpha b_\alpha\Big) \wedge b'\Big] \geqq (b \wedge b_\alpha) \vee (b_\alpha \wedge b') = b_\alpha$$

and therefore

$$a \vee \Big[\Big(\bigvee_\alpha b_\alpha\Big) \wedge b'\Big] \geqq \bigvee_\alpha b_\alpha .$$

Thus, by intersecting both sides of this inequality with b and using the distributive law $b \wedge (a \vee c) = (b \wedge a) \vee (b \wedge c)$, we obtain

$$(b \wedge a) \vee \left\{b \wedge \left[\bigvee_{\alpha} b_{\alpha}\right] \wedge b'\right\} = b \wedge a \geqq b \wedge \left(\bigvee_{\alpha} b_{\alpha}\right)$$

and so

$$a \geqq b \wedge a \geqq b \wedge \left(\bigvee_{\alpha} b_{\alpha}\right),$$

which shows that $b \wedge (\bigvee_{\alpha} b_{\alpha})$ is the least upper bound for the set of all the elements $b \wedge b_{\alpha}$. Q.E.D.

3 LEMMA. *If a Boolean algebra of projections is σ-complete as an abstract Boolean algebra, it is bounded.*

PROOF. Suppose that the Boolean algebra B of projections is not bounded. It will first be shown that B contains a monotone decreasing sequence $\{E_n\}$ such that

$$|E_n| \geqq n + |E_{n-1}|, \qquad n = 2, 3, \ldots.$$

A projection E is said to have property (α) if $\sup_{F \leqq E} |F| = \infty$. Clearly, for any E in B, either E or $I - E$ has property (α), and if E has the property (α) and $F \leqq E$, then either F or $E - F$ has the property (α). Let E_1 have the property (α). Then there is an $F_1 \leqq E_1$ such that $|F_1| \geqq 2 + 2|E_1|$. Let E_2 be a member of the pair F_1, $E_1 - F_1$ with property (α). The inequality $|E_1 - F_1| \geqq |F_1| - |E_1|$ shows $|E_2| \geqq 2 + |E_1|$. An F_2 is now selected in E_2 such that $|F_2| \geqq 3 + 2|E_2|$, and so on. The construction proceeds by induction.

Now for each n let $G_n = E_n - E_{n+1}$. The projections G_n are disjoint and $\lim_{n \to \infty} |G_n| = \infty$. By selecting subsequences from the sequence $\{G_n\}$, a collection of mutually disjoint sequences of projections $\{H_{jk}\}$, $j = 1, 2, \ldots$, $k = 1, 2, \ldots$, is obtained such that

$$\lim_{k \to \infty} |H_{jk}| = \infty, \qquad j = 1, 2, \ldots.$$

Define $P_j = \bigvee_{k=1}^{\infty} H_{jk}$. Then it follows readily from Lemma 2 that $P_n P_m = 0$ for $n \neq m$. The relation

$$\frac{|H_{mn} x|}{|x|} = \frac{|H_{mn} P_m x|}{|x|} \leqq \frac{|P_m| \, |H_{mn} P_m x|}{|P_m x|}, \qquad P_m x \neq 0,$$

shows that

$$|H_{mn}|_{P_m\mathfrak{X}} \geqq \frac{|H_{mn}|}{|P_m|},$$

where the left side is the norm of H_{mn} as an operator in $P_m\mathfrak{X}$. Consequently,

$$\lim_{n\to\infty} |H_{mn}|_{P_m\mathfrak{X}} = \infty, \qquad m = 1, 2, \ldots .$$

Select a subsequence $\{n_i\}$ and unit vectors x_i in $P_i\mathfrak{X}$ such that $|H_{in_i}x_i| > i$, $i = 1, 2, \ldots$. The projection $Q = \bigvee_{i=1}^{\infty} H_{in_i}$ cannot be bounded since

$$|Qx_i| = |QP_ix_i| = |H_{in_i}x_i| > i, \qquad i = 1, 2, \ldots,$$

and this contradiction completes the proof. Q.E.D.

4 Lemma. *Let B be a complete (σ-complete) Boolean algebra of projections in the B-space $\mathfrak{X}$ and let $\{E_\alpha\}$ be a monotone generalized sequence (a monotone sequence) in B. Then, if $\{E_\alpha\}$ is increasing,*

$$\lim_\alpha E_\alpha x = \Big(\bigvee_\alpha E_\alpha\Big)x, \qquad x \in \mathfrak{X},$$

while if $\{E_\alpha\}$ is decreasing, then

$$\lim_\alpha E_\alpha x = \Big(\bigwedge_\alpha E_\alpha\Big)x, \qquad x \in \mathfrak{X}.$$

Conversely, if every monotone increasing generalized sequence (monotone increasing sequence) of elements of a Boolean algebra B of projections converges strongly to an element of B, then B is complete (σ-complete).

Proof. Let $E_0 = \bigvee_\alpha E_\alpha$ be the union of the increasing generalized sequence $\{E_\alpha\}$ and let $\varepsilon > 0$ and $x \in \mathfrak{X}$ be arbitrary. Since $E_0\mathfrak{X} = \overline{\text{sp}}\, E_\alpha\mathfrak{X}$ there is a vector $y = \sum_{i=1}^n z_i$ and indices α_i such that $E_{\alpha_i}z_i = z_i$ and $|E_0x - y| \leqq \varepsilon$. If $\alpha \geqq \alpha_i$, $i = 1, \ldots, n$, then $E_\alpha y = y$. Thus, since $E_\alpha E_0 = E_\alpha$ it follows that, for $\alpha \geqq \alpha_i$,

$$\begin{aligned}|E_\alpha x - E_0x| &\leqq |E_\alpha x - y| + |y - E_0x| \\ &= |E_\alpha(E_0x - y)| + |y - E_0x| < (K+1)\varepsilon,\end{aligned}$$

where K is a bound for the norms of the projections in B (cf. Lemma 3). This proves that $\lim_\alpha E_\alpha x = E_0x$. The dual statement concerning decreasing sequences now follows from the formula

$$\bigwedge_\alpha E_\alpha = I - \bigvee_\alpha (I - E_\alpha).$$

The proof in the σ-complete case is quite similar.

To prove the converse, let $\{E\}$ be a subset of B and let $\{F_\alpha\}$ be the generalized sequence of finite unions of elements of $\{E\}$ ordered in the natural increasing order of projections. A projection F is an upper bound for $\{E\}$ if and only if it is an upper bound for $\{F_\alpha\}$, and since

$$(E_1 \vee \cdots \vee E_n)\mathfrak{X} = \overline{\text{sp}}\left(\bigcup_{i=1}^{n} E_i \mathfrak{X}\right)$$

for any finite set of projections, to construct a least upper bound for $\{E\}$ with the property required by Definition 1, it suffices to make the corresponding construction for $\{F_\alpha\}$. To do this we let $F_\infty = \lim_\alpha F_\alpha$ in the strong operator topology. Since $F_\alpha F_\beta = F_\alpha$, if $\beta \geqq \alpha$, we have

$$F_\alpha F_\infty = \lim_\beta F_\alpha F_\beta = F_\alpha,$$

which shows that $F_\infty \geqq F_\alpha$ and thus that F_∞ is an upper bound for $\{F_\alpha\}$. If F is another upper bound, then

$$FF_\infty = \lim_\alpha FF_\alpha = \lim_\alpha F_\alpha = F_\infty,$$

which proves that $F \geqq F_\infty$ and thus that F_∞ is the least upper bound $\bigvee_\alpha F_\alpha$ of $\{F_\alpha\}$. Since

$$F_\infty x = \lim_\alpha F_\alpha x \in \overline{\text{sp}}\left\{\bigcup_\alpha F_\alpha \mathfrak{X}\right\},$$

it follows that $F_\infty \mathfrak{X} \subseteqq \overline{\text{sp}}\,\{\bigcup_\alpha F_\alpha \mathfrak{X}\}$. On the other hand, $F_\infty \mathfrak{X} \supseteqq F_\alpha \mathfrak{X}$ for all α, which shows that

$$F_\infty \mathfrak{X} = \overline{\text{sp}}\left\{\bigcup_\alpha F_\alpha \mathfrak{X}\right\}.$$

If every increasing generalized sequence $\{F_\alpha\}$ of projections in B converges strongly, it is evident (by considering the sequence $\{I - F_\alpha\}$) that every monotone decreasing generalized sequence of projections in B converges strongly also. With this observation in mind, a greatest lower bound for $\{E\}$ which has the property of Definition 1 may be constructed in a fashion analogous to that used for constructing the least upper bound. The proof in the σ-complete case follows similar lines. Q.E.D.

5 LEMMA. *A strongly closed bounded Boolean algebra of projections in a weakly complete B-space is complete.*

PROOF. If the strongly closed bounded Boolean algebra B of projections is not complete, there is, by the preceding lemma, a vector x and an increasing generalized sequence $\{E_\alpha\}$ of projections in B such that the limit $\lim_\alpha E_\alpha x$ does not exist. It follows from Lemma I.7.5 that $\{E_\alpha x\}$ is not a generalized Cauchy sequence. Consequently there is an $\varepsilon > 0$ and, for each α, a $\beta(\alpha) \geqq \alpha$ such that

$$|E_{\beta(\alpha)}x - E_\alpha x| > \varepsilon.$$

Let α_1 be arbitrary and, for $n \geqq 1$, let $\alpha_{n+1} = \beta(\alpha_n)$ and $E_n = E_{\alpha_n}$. Then $\{E_n\}$ is an increasing sequence of elements in B for which the $\lim_n E_n x$ does not exist. A contradiction of this statement will now be obtained by considering the uniformly closed operator algebra $\mathfrak{A}(B)$ generated by the projections in B. By Lemma 2.1 the algebra $\mathfrak{A}(B)$ is equivalent, under an isomorphism $S(f) \leftrightarrow f$, to the algebra $C(\Lambda)$ of continuous functions on the space Λ of maximal ideals in $\mathfrak{A}(B)$. Moreover, by Theorem 2.5, this isomorphism $S(f)$ is given in terms of a countably additive spectral measure E defined on the family of Borel sets in Λ by the equation

$$S(f) = \int_\Lambda f(\lambda)E(d\lambda), \qquad f \in C(\Lambda).$$

The projection E_n is the image under S of a continuous function f_n, and since $E_n^2 = E_n$ we have $f_n^2 = f_n$, which shows that f_n assumes only the values zero and one and is therefore the characteristic function of a set e_n. Since f_n is continuous, the set e_n is open and closed. Furthermore, since $E_{n+1}E_n = E_n$ we have $f_{n+1}f_n = f_n$ and $e_{n+1}e_n = e_n$, which shows that the sequence $\{e_n\}$ is increasing. Thus, since

$$E_n = S(f_n) = \int_\Lambda f_n(\lambda)E(d\lambda) = E(e_n),$$

the limit

$$\lim_n E_n x = \lim_n E(e_n)x$$

exists because of the countable additivity of the spectral measure E. This contradiction completes the proof of the lemma. Q.E.D.

6 LEMMA. *A complete Boolean algebra of projections contains every projection in the weakly closed operator algebra it generates.*

PROOF. Let $\mathfrak{S}(B)$ and $\mathfrak{W}(B)$ be the strongly closed and the weakly closed operator algebras generated by the complete Boolean algebra B of projections in the B-space $\mathfrak{X}$. Since, by Corollary VI.1.5, $\mathfrak{S}(B) = \mathfrak{W}(B)$, to prove the theorem it will suffice to show that every projection F in $\mathfrak{S}(B)$ is in B. The proof that F is in B will be made by showing that to each pair (y, z) where y is in $\mathfrak{M} = F\mathfrak{X}$ and z is in $\mathfrak{N} = (I - F)\,\mathfrak{X}$, there can be associated a projection E_{yz} in B such that $E_{yz}y = y = Fy$, and $F_{yz}z = 0 = Fz$. For, if this is granted, the projection

$$E = \bigwedge_{z \in \mathfrak{N}} \bigvee_{y \in \mathfrak{M}} E_{yz}$$

is in B since B is complete. If $x_0 = y_0 + z_0$, where y_0 is in $\mathfrak{M}$ and z_0 is in $\mathfrak{N}$, then $(\bigvee_{y \in \mathfrak{M}} E_{yz})y_0 = y_0$ for each z in $\mathfrak{N}$, and $(\bigvee_{y \in \mathfrak{M}} E_{yz_0})z_0 = 0$. Thus $Ey_0 = y_0$, $Ez_0 = 0$, and $E = F$.

The projections E_{yz} will now be constructed. It should be remarked that the construction uses only the fact that B is σ-complete. Let y and z be fixed elements of $\mathfrak{M}$ and $\mathfrak{N}$, respectively, and let ε be a given positive number. Since F is in the strongly closed operator algebra generated by B, there is an operator A having the form $A = \sum_{i=1}^{n} \alpha_i E_i$, where

$$0 \neq E_i \in B, \quad \sum_{i=1}^{n} E_i = I, \quad E_i E_j = 0, \qquad 1 \leqq i \neq j \leqq n,$$

$$|y - Ay| < \varepsilon, \qquad |Az| < \varepsilon.$$

By Lemma 3 there is a bound $K > 1$ for the norms of the projections in B, and thus if

$$E = \sum_{|\alpha_i| > 1/2} E_i,$$

it follows from Lemma 2.2 that

$$|Ez| = \Big|\Big(\sum_{|\alpha_i| > 1/2} \alpha_i^{-1} E_i\Big) Az\Big| \leqq 8K\varepsilon.$$

In the same way it is seen that

$$\begin{aligned} |y - Ey| &= \Big|\sum_{|\alpha_i| \leqq 1/2} E_i y\Big| \\ &= \Big|\Big(\sum_{|\alpha_i| \leqq 1/2} (1 - \alpha_i)^{-1} E_i\Big)(y - Ay)\Big| \leqq 8K\varepsilon. \end{aligned}$$

By choosing ε so that $8K\varepsilon < 2^{-n}$, we see that there is a sequence $\{E_n\}$ in B with

$$\text{(i)} \qquad |y - E_n y| < \frac{1}{2^n}, \qquad |E_n z| < \frac{1}{2^n}.$$

Let the projection $E_{n,m}$ be defined by the equation

$$E_{n,m} = \bigvee_{k=n}^{n+m} E_k .$$

Then $E_{n,m} \geqq E_n$, $(I - E_{n,m}) \leqq I - E_n$, and thus

$$(I - E_{n,m})y = (I - E_{n,m})(I - E_n)y,$$

from which it follows that

$$\text{(ii)} \qquad |y - E_{n,m}y| \leqq \frac{K}{2^n}.$$

Since

$$E_{n,m+1} = E_{n,m} + (I - E_{n,m})E_{n+m+1},$$

it follows inductively from (i) that

$$\text{(iii)} \qquad |E_{n,m}z| \leqq \frac{1}{2^n} + \frac{K}{2^{n+1}} + \cdots + \frac{K}{2^{n+m}} < \frac{K}{2^{n+1}}.$$

Now the sequence $\{E_{n,m}\}$ is increasing in m and the sequence

$$\left\{\bigvee_{m=0}^{\infty} E_{n,m}\right\} = \left\{\bigvee_{k=n}^{\infty} E_k\right\}$$

is decreasing. Thus, by Lemma 4, the limit

$$E_{yz} = \bigwedge_{n=0}^{\infty} \bigvee_{m=0}^{\infty} E_{n,m} = \lim_{n\to\infty} \lim_{m\to\infty} E_{n,m}$$

exists in the strong operator topology. It follows immediately from (ii) and (iii) that this limit has the desired properties as expressed by the equations $E_{yz}y = y$, $E_{yz} = 0$. Q.E.D.

7 Corollary. *A complete Boolean algebra of projections is strongly closed.*

Proof. Since every operator in the strong closure of a bounded Boolean algebra of projections is itself a projection, this corollary follows immediately from Lemmas 3 and 6. Q.E.D.

8 Corollary. *A bounded Boolean algebra of projections in a weakly complete space is complete if and only if it is strongly closed.*

Proof. This follows from Lemma 5 and Corollary 7. Q.E.D.

9 LEMMA. *Let $\mathfrak{A}(B)$ be the uniformly closed operator algebra generated by the complete Boolean algebra B of projections in a B-space $\mathfrak{X}$. Then $\mathfrak{A}(B)$ is equivalent to the algebra of continuous functions on its own set Λ of maximal ideals, and any homeomorphic isomorphism T between these algebras uniquely determines a regular countably additive spectral measure E in $\mathfrak{X}$ defined on the family of Borel sets in Λ and such that*

$$T(f) = \int_\Lambda f(\lambda)E(d\lambda), \qquad f \in C(\Lambda).$$

Moreover the range of E is precisely the Boolean algebra B.

PROOF. By Lemma 3 the Boolean algebra B is bounded, and by Lemma 2.1 the algebra $\mathfrak{A}(B)$ is equivalent to the algebra of continuous functions on the structure space Λ of $\mathfrak{A}(B)$. It follows from Theorem 2.4 that there is a spectral measure A in $\mathfrak{X}^*$ defined on the Borel sets in Λ and such that

$$xT(f)x^* = \int_\Lambda f(\lambda)xA(d\lambda)x^*, \qquad x \in \mathfrak{X},\ x^* \in \mathfrak{X}^*.$$

This spectral measure A is uniquely determined and has the property that, for each x in $\mathfrak{X}$ and each x^* in $\mathfrak{X}^*$, the numerical measure $xA(\cdot)x^*$ is regular and countably additive. It will now be shown that, because of the completeness of B, the operator $A(e)$ is, for every Borel set e, the adjoint of a projection in B. To this end let Σ be the family of those Borel sets e in Λ for which $A(e) = E(e)^*$ for some projection $E(e)$ in B. Since B is a Boolean algebra, the family Σ is a field. To see that Σ is a σ-field, let $\{e_n\}$ be an increasing sequence of sets in Σ. Since

$$E(e_{n+1})^*E(e_n)^* = A(e_{n+1}e_n) = A(e_n) = E(e_n)^*,$$

the sequence $\{E(e_n)\}$ is increasing, and since B is complete it follows from Lemma 4 that the strong limit $E = \lim_n E(e_n)$ exists and is a projection in B. Thus

$$\begin{aligned} xA\left(\bigcup_{n=1}^{\infty} e_n\right)x^* &= \lim_{n\to\infty} xA(e_n)x^* \\ &= \lim_{n\to\infty} xE(e_n)^*x^* = xE^*x^*, \qquad x \in \mathfrak{X},\ x^* \in \mathfrak{X}^*, \end{aligned}$$

which shows that the union $\bigcup_{n=1}^{\infty} e_n$ is in Σ and proves that Σ is a σ-field. Hence to see that Σ contains all Borel sets it suffices to show that Σ

contains every open set. Let e be an open subset of Λ, let x and x^* be elements of $\mathfrak{X}$, $\mathfrak{X}^*$, respectively, and let $\varepsilon > 0$. Then, because of the regularity of $xA(\cdot)x^*$, there is a closed subset δ of e such that

$$\text{(i)} \qquad |xA(\delta_1)x^* - xA(e)x^*| < \varepsilon,$$

for every Borel set δ_1 with $\delta \subseteq \delta_1 \subseteq e$. To each E in B corresponds a continuous function f_E which is uniquely determined by the equation $E = T(f_E)$. Since $E^2 = E$ we have $f_E^2 = f_E$, that is, f_E is the characteristic function of a set $\sigma(E)$. Since f_E is continuous, the set $\sigma(E)$ is open and closed. The mapping $E \leftrightarrow \sigma(E)$ is clearly an isomorphism between B and the Boolean algebra of all open and closed subsets of Λ. These open and closed sets $\sigma(E)$ form a basis for the topology in Λ. To see this, note that sets of the form $\{\lambda \,|\, |(T^{-1}A)(\lambda)| < \alpha\}$, with A in $\mathfrak{A}(B)$ form, by definition, a subbasis for the topology of Λ, and since B generates $\mathfrak{A}(B)$, the sets $\sigma(E)$ form a subbasis for the topology of Λ. Since $\sigma(E)\sigma(F) = \sigma(EF)$, the sets $\sigma(E)$ actually form a basis for the topology in Λ (cf. Theorem IX.2.11 and its proof).

Consider now a closed set δ_1 with $\delta \subseteq \delta_1 \subseteq e$. Each point λ in δ_1 is interior to some set $\sigma(E) \subseteq e$. Since δ_1 is compact, a finite number of sets $\sigma(E_1), \ldots, \sigma(E_n)$ cover δ_1, and thus if E_0 is the union of the projections $E_1, \ldots, E_n$, then

$$\delta \subseteq \delta_1 \subseteq \bigcup_{i=1}^{n} \sigma(E_i) = \sigma(E_0) \subseteq e.$$

Thus if E is a projection in B with $\sigma(E_0) \subseteq \sigma(E) \subseteq e$, it follows from (i) that

$$\text{(ii)} \qquad |xA(\sigma(E))x^* - xA(e)x^*| < \varepsilon.$$

Since

$$\begin{aligned} A(\sigma(E)) &= \int_{\sigma(E)} A(d\lambda) = \int_{\Lambda} \chi_{\sigma(E)}(\lambda) A(d\lambda) \\ &= T(\chi_{\sigma(E)}) = E, \end{aligned}$$

it follows from (ii) that

$$\text{(iii)} \qquad |x^*Ex - xA(e)x^*| < \varepsilon$$

for every projection E in B with $\sigma(E_0) \subseteq \sigma(E) \subseteq e$. Thus if $\{E_\alpha\}$ is the generalized sequence of projections E in B with $\sigma(E) \subseteq e$, directed in the natural increasing order of projections, it is seen from (iii) that

$$xA(e)x^* = \lim_{\alpha} x^*E_\alpha x.$$

On the other hand, it follows from Lemma 4 that

$$E_\infty = \bigvee_{\alpha} E_\alpha = \lim_{\alpha} E_\alpha$$

in the strong operator topology. Thus $x^*E_\infty x = xA(e)x^*$, which proves that e is in Σ and shows that Σ consists of all Borel sets in Λ. This means that for every Borel set e in Λ there is a projection $E(e)$ in B with $A(e) = E(e)^*$, and thus the proof is complete. Q.E.D.

10 COROLLARY. *A Boolean algebra of projections in a B-space is σ-complete if and only if it is the range of a countably additive regular spectral measure defined on a σ-field of subsets of a compact space.*

PROOF. Let B be a σ-complete Boolean algebra of projections. Then the proof of the preceding lemma shows that it is the range of a countably additive regular spectral measure E defined on a σ-field Σ of sets in the compact space Λ. Conversely, suppose that B is the range of a spectral measure E which is defined and countably additive on a σ-field Σ of subsets of a set Λ. Let $\{E_n\}$ be an increasing sequence of elements of B. If $E_n = E(e_n)$ then

$$E(e_n - e_{n+1}) = E(e_n) - E(e_{n+1}e_n) = E_n - E_{n+1}E_n = E_n - E_n = 0,$$

and so, except for a set of E-measure zero, we have $e_n \subseteq e_{n+1}$. Since E is countably additive, we have the limit

$$\lim_{n} E_n = \lim_{n\to\infty} E(e_n) = E\left(\bigcup_{n=1}^{\infty} e_n\right)$$

existing in the strong operator topology, and furthermore it is in B. It follows from Lemma 4 that B is σ-complete. Q.E.D.

11 COROLLARY. *The restriction to an invariant subspace of a σ-complete Boolean algebra of projections is σ-complete.*

PROOF. This follows immediately from the criterion for σ-completeness as given in Corollary 10. Q.E.D.

A Boolean algebra B of self adjoint projections in a Hilbert space $\mathfrak{H}$ has the property that for every x_0 in $\mathfrak{H}$ the scalar product (Ex_0, x_0) is non-negative and vanishes only if $Ex_0 = 0$. In order to continue the analysis of Boolean algebras in arbitrary B-spaces, this useful property of the scalar product in Hilbert space may be replaced by the following ingenious result of Bade.

12 LEMMA. *Let B be a σ-complete Boolean algebra of projections in the B-space $\mathfrak{X}$. Then, for each x_0 in $\mathfrak{X}$ there is a linear functional x_0^* in $\mathfrak{X}^*$ with the properties*

$$\text{(i)} \qquad x_0^*Ex_0 \geqq 0, \qquad E \in B;$$

(ii) *if $x_0^*Ex_0 = 0$ for some E in B, then $Ex_0 = 0$.*

PROOF. By Corollary 10, B is the range of a countably additive spectral measure E defined on a σ-field Σ. In view of Corollary 11 and the Hahn-Banach theorem, it may be assumed that $\mathfrak{X} = \overline{\text{sp}}\ \{Ex_0 \mid E \in B\}$. For every linear functional y^* in $\mathfrak{X}^*$, let the measure μ_{y^*} be defined on Σ by the equation

$$\mu_{y^*}(e) = y^*E(e)x_0, \qquad e \in \Sigma,$$

and call a set δ in Σ a y^*-carrier if $\mu_{y^*}(\delta) \neq 0$ and if every measurable subset e of δ upon which the total variation $v(\mu_{y^*}, e) = 0$ has $E(e) = 0$. We note that with this definition a y^*-carrier may have proper subsets which are y^*-carriers. This fact will be unimportant in what follows. An application of Zorn's lemma yields a maximal family $\{\delta_\alpha\}$ of disjoint sets each of which is a y_α^*-carrier for some y_α^* in $\mathfrak{X}^*$. It will first be observed that $\{\delta_\alpha\}$ is at most denumerable. To see this, note that, since the spectral measure E is strongly countably additive on Σ, every series $\Sigma_\alpha E(\delta_\alpha)x_0$ of a countable number of terms converges and hence contains at most a finite number of terms whose norms exceed a given positive number. Since

$$0 \neq \mu_{y^*}(\delta_\alpha) = y^*E(\delta_\alpha)x_0$$

for all α, it follows that the sequence $\{\delta_\alpha\}$ is at most countable and it will therefore be written as $\{\delta_n\}$ in what follows. Let $\Delta = \bigcup_{n=1}^{\infty} \delta_n$, so that the complementary set Δ' contains no carrier. It will next be shown that $E(\Delta') = 0$. For if $E(\Delta') \neq 0$ it follows readily, from the fact that $\{E(e)x_0 \mid e \in \Sigma\}$ spans $\mathfrak{X}$, that $E(\Delta')x_0 \neq 0$ and hence, for some functional y_0^*, that $y_0^*E(\Delta')x_0 \neq 0$. Let $y^* = E(\Delta')^*y_0^*$ so that μ_{y^*} vanishes on subsets of Δ, and thus the total variation $v(\mu_{y^*}, \Delta) = 0$. Now, by Lemma IV.10.5, the vector measure $E(\cdot)x_0$ is continuous with respect to a finite positive measure ν on Σ. The measure ν cannot be μ_{y^*}-singular, for if it were there would be a set e in Σ with $v(\mu_{y*}, e) = 0$ and $\nu(e') = 0$, in which case $E(e')x_0 = 0$, $v(\mu_{y*}, e') = 0$. It would follow that $v(\mu_{y*}, \Lambda) = v(\mu_{y*}, e) + v(\mu_{y*}, e') = 0$, which contradicts the fact that $\mu_{y*} \neq 0$. It follows from the Lebesgue decomposition theorem (III.4.14) that there is a set e_1 with $\mu_{y*}(e_1) \neq 0$ and such that ν, and hence E, vanishes on any subset δ of e_1 upon which the variation $v(\mu_{y*}, \delta) = 0$. If $\delta = e_1\Delta'$ we have $\mu_{y*}(\delta) = \mu_{y*}(e_1) \neq 0$, and at the same time $\delta_1 \subseteq \delta$ and $v(\mu_{y*}, \delta_1) = 0$ imply $E(\delta_1) = 0$. Thus δ is a carrier contained in Δ' which is a contradiction of the fact that Δ' contains no carriers. This completes the proof that $E(\Delta') = 0$.

Let y_n^* be a functional such that δ_n is a y_n^*-carrier. Consider now any measure of the form

$$(\alpha) \qquad \mu(e) = \sum_{n=1}^{\infty} c_n v(\mu_{y_n^*}, e), \qquad e \in \Sigma,$$

where $c_n > 0$, $n = 1, 2, \ldots$. If $\mu(e) = 0$ then $v(\mu_{y_n^*}, e) = 0$ for all n and thus $E(e\delta_n) = 0 = E(e\Delta)$. Since $E(\Delta') = 0$ we have $E(e) = 0$. Consequently the proof of the lemma may be completed by showing the existence of an x_0^* such that $\mu_{x_0^*}$ has the form given in equation (α).

Before constructing such an x_0^*, we note that the Radon-Nikodým theorem (III.10.7) yields a function f_n for which

$$v(\mu_{y_n^*}, e) = \int_e f_n(\lambda)\mu_{y_n^*}(d\lambda), \qquad e \in \Sigma, \qquad n \geqq 1.$$

Since, by Theorem III.2.20(a),

$$v(\mu_{y_n^*}, e) = \int_e |f_n(\lambda)|\, v(\mu_{y_n^*}, d\lambda), \qquad e \in \Sigma,$$

it follows that $|f_n(\lambda)| = 1$ for $\mu_{y_n^*}$-almost all λ. It may thus be assumed that $|f_n(\lambda)| = 1$ for all λ in Λ. Let $z_n^* = T_n^* y_n^*$ where the operators T_n are defined by the formula

$$T_n = \int_\Lambda f_n(\lambda)E(d\lambda), \qquad n \geqq 1.$$

Since the functions f_n are bounded, so are the operators T_n. Then

$$\begin{aligned}\mu_{z_n^*}(e) &= (T_n^* y_n^*)E(e)x_0 = y_n^* E(e) T_n x_0 \\ &= y_n^* \int_e f_n(\lambda)E(d\lambda)x_0 = \int_e f_n(\lambda)\mu_{y_n^*}(d\lambda) \\ &= v(\mu_{y_n^*}, e).\end{aligned}$$

Thus the functional

$$x_0^* = \sum_{n=0}^{\infty} \frac{1}{2^n(1+|z_n^*|)} z_n^*$$

has

$$\mu_{x_0^*} = \sum_{n=0}^{\infty} \frac{1}{2^n(1+|z_n^*|)} \mu_{z_n^*},$$

which is the form expressed in equation (α) above. Q.E.D.

13 LEMMA. *Let B be a σ-complete Boolean algebra of projections in the B-space $\mathfrak{X}$. Assume that for some x_0 in $\mathfrak{X}$ the vectors Ex_0 with E in B form a fundamental set in $\mathfrak{X}$. Let x_0^* be a functional with the properties* (i) *and* (ii) *of Lemma* 12. *Then the $\mathfrak{X}$-closure of the linear manifold in $\mathfrak{X}^*$ determined by the vectors $E^*x_0^*$ with E in B is the whole space $\mathfrak{X}^*$.*

PROOF. By Corollary 10, B is the range of a countably additive regular spectral measure E on a σ-field Σ of subsets of a space Λ. Thus, in order to show that the set $\text{sp}\{E^*x_0^* \mid E^* \in B^*\}$ is dense in $\mathfrak{X}^*$ with its $\mathfrak{X}$-topology, it suffices to show that $x = 0$ is the only vector x in $\mathfrak{X}$ for which

$$E^*(\sigma)x_0^*x = x_0^*E(\sigma)x = 0, \qquad \sigma \in \Sigma.$$

Since vectors of the form $E(\sigma)x_0$ with σ in Σ form a fundamental set in $\mathfrak{X}$, there is a sequence $\{f_n\}$ of finite linear combinations of characteristic functions such that

$$x = \lim_{n\to\infty} \int_\Lambda f_n(\lambda)E(d\lambda)x_0.$$

Thus, since B is bounded (cf. Lemma 3),

$$\lim_{n\to\infty} \int_\sigma f_n(\lambda)x_0^*E(d\lambda)x_0 = x_0^*E(\sigma)x = 0,$$

uniformly for σ in Σ. Since

$$\int_\Lambda |f_n(\lambda)|\, x_0^*E(d\lambda)x_0 \leqq 4 \sup_{\sigma\in\Sigma} \left| \int_\sigma f_n(\lambda)x_0^*E(d\lambda)x_0 \right|,$$

it is seen that f_n approaches zero in the space $L_1(\Lambda, \Sigma, x_0^*E(\cdot)x_0)$. Thus a subsequence $\{g_n\}$ of $\{f_n\}$ converges to zero almost everywhere and almost uniformly. Let $\{\delta_m\}$ be a decreasing sequence of sets in Σ with $x_0^*E(\delta_m)x_0 \to 0$ and such that g_n converges to zero uniformly on the complement of each of the sets δ_m, $m \geqq 1$. Then

$$x = \lim_{n\to\infty} \int_{\delta_m} g_n(\lambda)E(d\lambda)x_0 + \lim_{n\to\infty} \int_{\delta_m'} g_n(\lambda)E(d\lambda)x_0$$

$$= \lim_{n\to\infty} E(\delta_m) \int_{\Lambda} g_n(\lambda)E(d\lambda)x_0 = E(\delta_m)x_0, \qquad m \geqq 1,$$

which shows that

$$x = \lim_{m\to\infty} E(\delta_m)x_0 = E\left(\bigcap_{m=1}^{\infty} \delta_m\right)x_0.$$

Since $x_0^*E(\bigcap_{m=1}^{\infty} \delta_m)x_0 = 0$ we have $E(\bigcap_{m=1}^{\infty} \delta_m)x_0 = 0$ and hence $x = 0$. Q.E.D.

The following lemma is a weakened form of the theorem of Bade stated in the introduction to this section.

14 LEMMA. *Let B be a σ-complete Boolean algebra of projections in a B-space $\mathfrak{X}$. Suppose that for some x in $\mathfrak{X}$ the set $\{Ex \mid E \in B\}$ is fundamental in $\mathfrak{X}$. Then the uniformly closed operator algebra generated by B consists precisely of those bounded linear operators in $\mathfrak{X}$ which commute with every element of B.*

PROOF. It is clear that every element in the uniformly closed algebra generated by B commutes with every element of B, and so to prove the lemma it will suffice to show that every operator A which commutes with every element of B is in the uniformly closed operator algebra generated by B. By Corollary 10, B is the range of a countably additive spectral measure E on a σ-field Σ of subsets of a set Λ. Let y^* be associated with x as in Lemma 12. Then, by the Radon-Nikodým theorem, there is a Σ-measurable function h with

$$y^*AE(e)x = \int_e h(\lambda)y^*E(d\lambda)x, \qquad e \in \Sigma.$$

Let

$$e_n = \{\lambda \mid |h(\lambda)| \leqq n\}, \qquad A_n = \int_{e_n} h(\lambda)E(d\lambda),$$

so that

$$y^*\{E(e)E(e_n)AE(\delta) - E(e)A_nE(\delta)\}x$$

$$= \int_{e\delta e_n} (h(\lambda) - h(\lambda))y^*E(d\lambda)x = 0, \qquad e, \delta \in \Sigma.$$

Thus, since the set $\{E(\delta)x \mid \delta \in \Sigma\}$ is fundamental in $\mathfrak{X}$, we have

$$E(e)^*y^*E(e_n)Az = E(e)^*y^*A_n z, \qquad z \in \mathfrak{X}.$$

By Lemma 13 the manifold in $\mathfrak{X}^*$ spanned by vectors of the form $E(e)^*y^*$ with e in Σ is $\mathfrak{X}$-dense in $\mathfrak{X}^*$. It follows then from the preceding equation that $E(e_n)A = A_n$ and thus, since A is a bounded operator, the sequence $\{|A_n|\}$ is bounded. By Theorem 2.10,

$$\sup_n E\text{-ess}\sup_{\lambda \in \Lambda} |h_n(\lambda)| < \infty,$$

and hence h is E-essentially bounded. Consequently it may be assumed that h is bounded and therefore that $e_n = \Lambda$ for all large n. This means that

$$A = A_n = \int_\Lambda h(\lambda)E(d\lambda)$$

for sufficiently large integers n, from which it clearly follows that A is in the uniformly closed algebra generated by B. Q.E.D.

15 Corollary. *The weakly closed operator algebra generated by a σ-complete Boolean algebra B which satisfies the condition of the preceding lemma is the same as the uniformly closed algebra generated by B.*

Proof. It is clear that every element in the weakly closed operator algebra generated by B commutes wtih every element of B, and so the corollary follows immediately from Lemma 14. Q.E.D.

→ 16 Theorem. *The uniformly closed operator algebra generated by a complete Boolean algebra B of projections in a B-space $\mathfrak{X}$ consists of all bounded linear operators in $\mathfrak{X}$ which leave invariant every manifold which is invariant under every member of B.*

Proof. It is clear that every operator in the uniformly closed algebra generated by B has the required invariance property.

Before proving the converse, a few general observations will be made. If Λ is the structure space of the uniformly closed operator algebra $\mathfrak{A}$ generated by B then, by Lemma 9, there is a homeomorphic isomorphism T of $C(\Lambda)$ with $\mathfrak{A}$. Thus each projection E in B determines a function f_E in $C(\Lambda)$ for which $E = T(f_E)$. Since $E^2 = E$ we have $f_E^2 = f_E$, and thus f_E is the characteristic function of a set $\sigma(E)$ in Λ. Since f_E is continuous, the set $\sigma(E)$ is both open and closed. It will be observed that

(i) the sets $\sigma(E)$ form a base for the topology in Λ.

To see this, let λ_0 be a point of the open set N. Since distinct points in Λ may be distinguished by functions in $C(\Lambda)$ and since B generates the algebra $\mathfrak{A}$ which is equivalent to $C(\Lambda)$, it is seen that distinct points in Λ may be distinguished by characteristic functions of the sets $\sigma(E)$ with E in B. Thus if λ_1 is in the complement N', there is a projection E in B with $\lambda_0 \in \sigma(E)$, $\lambda_1 \notin \sigma(E)$. This shows that the intersection of the closed sets $\sigma(E)N'$, where E varies over all projections in B with λ_0 in $\sigma(E)$, is void. Since Λ is compact, some finite intersection $\sigma(E_1) \cdots \sigma(E_n)N'$ is void. Thus, if $E = E_1E_2 \cdots E_n$, we have $\lambda_0 \in \sigma(E) \subseteq N$, which proves the statement (i).

Since B is complete, the projections

$$\text{(ii)} \qquad E_x = \bigwedge_{Ex=x} E, \qquad x \in \mathfrak{X},$$

exist and the corresponding open and closed set $\sigma(E_x)$ will be denoted by σ_x. It follows from Lemma 4 that

$$\text{(iii)} \qquad E_x x = x.$$

It will next be observed that

(iv) if $f(\lambda) = 0$ for $\lambda \notin \sigma_x$ and if $T(f)x = 0$, then $f = 0$.

To prove this, let us suppose that $f \neq 0$ so that, by (i), there is an $E \neq 0$ in B such that $f(\lambda) \neq 0$ for λ in $\sigma(E)$. Since $\sigma(E)$ is compact, the function g defined by the equations

$$\begin{aligned} g(\lambda) &= \frac{1}{f(\lambda)}, \qquad \lambda \in \sigma(E), \\ &= 0, \qquad \lambda \notin \sigma(E), \end{aligned}$$

is continuous and $T(g)T(f) = E$. Now $Ex = T(g)T(f)x = 0$ and using (iii) we have $(E_x - E)x = x$, from which it follows that $E_x - E \geqq E_x$. But, since $f(\lambda) = 0$ on σ_x', we have $\sigma(E) \subseteq \sigma_x$ and thus $E \leqq E_x$ and $E_x - E \leqq E_x$. Hence $E_x - E = E_x$ and $E = 0$, a contradiction which establishes (iv).

Now let A be a bounded linear operator in $\mathfrak{X}$ which leaves invariant every manifold which is invariant under every member of B. Then $AE\mathfrak{X} \subseteq E\mathfrak{X}$ and $A(I-E)\mathfrak{X} \subseteq (I-E)\mathfrak{X}$, from which it follows that

$$EAE = AE, \qquad EA(I-E) = 0,$$

and thus that

$$AE = EAE = EAE + EA(I-E) = EA.$$

Hence

$$EA = AE, \qquad E \in B.$$

By Lemma 9 the Boolean algebra B is the range of a countably additive spectral measure E on the family Σ of Borel subsets of Λ. For each x in $\mathfrak{X}$, let

$$\mathfrak{X}(x) = \overline{\text{sp}}\, \{Ex \,|\, E \in B\}.$$

Since $\mathfrak{X}(x)$ is invariant under every member of B, it is also invariant under A, and in view of (v), the argument used in the proof of Lemma 14 shows that the restriction of A to $\mathfrak{X}(x)$ is given by the formula

$$A \,|\, \mathfrak{X}(x) = \int h_x(\lambda)(E \,|\, \mathfrak{X}(x))(d\lambda),$$

where h_x is a bounded Borel measurable function on Λ with

$$\sup_{\lambda} |h_x(\lambda)| = |(A \,|\, \mathfrak{X}(x))| \leqq |A|.$$

If the operator A_x is defined by the formula

$$A_x = E_x \int_{\Lambda} h_x(\lambda) E(d\lambda),$$

then clearly

$$A_x \in \mathfrak{A}, \qquad A_x x = Ax, \qquad A_x E_x = A_x, \qquad |A_x| \leqq K|A|,$$

where K is a bound for B. Since A_x is in $\mathfrak{A}$, there is a uniquely defined continuous function f_x in $C(\Lambda)$ with $A_x = T(f_x)$. This function f_x has the properties

$$\text{(vi)} \qquad T(f_x)x = Ax, \qquad f_x(\lambda) = 0, \qquad \lambda \notin \sigma_x.$$

Moreover, if g also has properties (vi), it follows from (iv) that $f_x = g$, so that the properties (vi) characterize f_x.

It will next be observed that

$$\text{(vii)} \qquad \sigma_{Fx} = \sigma(F)\sigma_x, \qquad F \in B.$$

To prove this, note first that $(E_x F)Fx = FE_x x = Fx$, so that $E_x F \geqq E_{Fx}$. Conversely, if $E_x F > E_{Fx}$, there is a $G \neq 0$ with $G \leqq E_x F$ and $GE_{Fx} = 0$. For such a G we clearly have $GFx = GE_{Fx} Fx = 0$ and $G(I - F) = 0$. Thus

$$Gx = GFx + G(I - F)x = 0$$

and hence $(I-G)x = x$, $I - G \geqq E_x$ and so $G \leqq I - E_x$. This shows that $G \leqq (I-E_x)E_x = 0$, which is a contradiction proving that $E_x F = E_{Fx}$, from which statement (vii) follows immediately. We next establish the equation

$$\text{(viii)} \qquad f_{Fx} = f_x \chi_{\sigma(F)}, \qquad F \in B.$$

To prove this, let $g = f_x \chi_{\sigma(F)}$ so that, using (vii), we have $g(\lambda) = 0$ for $\lambda \notin \sigma_x \sigma(F) = \sigma_{Fx}$. Also

$$\begin{aligned} T(g)Fx &= T(f_x)T(\chi_{\sigma(F)})Fx \\ &= FT(f_x)x = FAx = AFx. \end{aligned}$$

Thus statement (vi), with f_x replaced by g and x replaced by Fx, shows that $g = f_{Fx}$, which establishes (viii). We now establish the relation

$$\text{(ix)} \qquad f_x(\lambda) = f_y(\lambda), \qquad \lambda \in \sigma_x \sigma_y.$$

To prove (ix) it may be assumed that $\sigma_x = \sigma_y$, for if $z = E_y x$ and $w = E_x y$ it follows from (vii) that $\sigma_z = \sigma_y \sigma_x = \sigma_w$ and from (viii) that $f_z = \chi_{\sigma_y} f_x$ and $f_w = \chi_{\sigma_x} f_y$. Thus if $f_z(\lambda) = f_w(\lambda)$ on $\sigma_z = \sigma_w$, equation (ix) will be established. Hence we may and shall assume, in the proof of (ix), that $\sigma_x = \sigma_y$.

Since $A(x-y) = Ax - Ay$ it follows that

$$T(f_{x-y})(x-y) = T(f_x)x - T(f_y)y,$$

and consequently that

$$T(f_{x-y} - f_x)x = T(f_{x-y} - f_y)y.$$

To complete the proof of (ix) an indirect proof will be used by supposing that this equation is false; that is, we shall assume that the functions f_x and f_y are not identically equal and one of them, say f_x, differs from both f_y and f_{x-y} at some point λ_0 in σ_x. Thus from (i) there is a projection E and an $\varepsilon > 0$ such that $\sigma(E) \subseteq \sigma_x = \sigma_y$ and

$$|f_{x-y}(\lambda) - f_x(\lambda)| \geqq \varepsilon, \qquad \lambda \in \sigma(E).$$

Since $\sigma(E)$ is open and closed, the function

$$\begin{aligned} g(\lambda) &= \frac{1}{f_{x-y}(\lambda) - f_x(\lambda)}, \qquad && \lambda \in \sigma(E), \\ &= 0, && \lambda \notin \sigma(E), \end{aligned}$$

is continuous and hence in $C(\Lambda)$. Let $h(\lambda) = (f_{x-y}(\lambda) - f_y(\lambda))g(\lambda)$. Then

$$T(h)y = T(g)T(f_{x-y} - f_y)y = T(g)T(f_{x-y} - f_x)x$$
$$= T(\chi_{\sigma(E)})x = Ex,$$

and so

$$T(f_x)Ex = AEx = AT(h)y = T(f_y)T(h)y = T(f_y)Ex.$$

Thus $T(f_y \chi_{\sigma(E)})Ex = AEx$ and $f_y(\lambda)\chi_{\sigma(E)}(\lambda) = 0$ for $\lambda \notin \sigma_y \cap \sigma(E) = \sigma_x \cap \sigma(E) = \sigma_{Ex}$ (using (vii)). Since equations (vi) determine f_x uniquely, it is seen that

$$f_y \chi_{\sigma(E)} = f_{Ex}.$$

On the other hand, it follows from (viii) that $f_{Ex} = f_x \chi_{\sigma(E)}$. Thus $f_y(\lambda) = f_x(\lambda)$ for λ in $\sigma(E)$, and this contradiction proves (ix).

In view of (ix), the function

$$\begin{aligned} \mathfrak{F}(\lambda) &= f_x(\lambda), & \lambda &\in \sigma_x, \\ &= 0, & \lambda &\notin \bigcup_x \sigma_x, \end{aligned}$$

is defined and continuous on the open set $\bigcup_x \sigma_x$. Since

$$|f_x| = |T^{-1}A_x| \leqq |T^{-1}|\,|A_x| \leqq K|T^{-1}|\,|A|,$$

the functions f_x are uniformly bounded and $\mathfrak{F}$ is therefore a bounded Borel function. Let

$$\text{(x)} \qquad A_0 = \int_\Lambda \mathfrak{F}(\lambda)E(d\lambda).$$

Then

$$A_0 x = A_0 E_x x = \int_{\sigma x} \mathfrak{F}(\lambda)E(d\lambda)x$$
$$= \int_{\sigma x} f_x(\lambda)E(d\lambda)x = A_x E_x x = Ax, \qquad x \in \mathfrak{X}.$$

Since $A_0 = A$, equation (x) shows that A is in the uniformly closed algebra generated by the projections $E(e)$ and thus in the uniformly closed algebra generated by B. Q.E.D.

17 Corollary. *The weakly closed operator algebra generated by a complete Boolean algebra B of projections is the same as the uniformly closed operator algebra generated by B.*

PROOF. By Corollary VI.1.5, the weakly closed operator algebra $\mathfrak{W}(B)$ generated by B is the same as the strongly closed algebra generated by B. Thus every A in $\mathfrak{W}(B)$ is the strong limit of finite linear combinations of elements of B. It follows that A leaves invariant every closed linear manifold which is invariant under every member of B. The theorem shows that A is in the uniformly closed algebra $\mathfrak{A}(B)$ generated by B. Thus $\mathfrak{W}(B) \subseteq \mathfrak{A}(B)$. On the other hand, it is clear that $\mathfrak{A}(B) \subseteq \mathfrak{W}(B)$. Q.E.D.

The preceding material of the present section was largely preliminary to the following basic result of Bade, which was described in the introduction.

18 THEOREM. *Let B be a bounded Boolean algebra of projections in a weakly complete space. Then an operator is in the weakly closed algebra generated by B if and only if it leaves invariant every closed linear manifold which is invariant under every member of B.*

PROOF. It was observed in the preceding proof that an operator in the weakly closed algebra generated by B has the required invariance property. To prove the converse, let A leave invariant every closed linear manifold which is invariant under every element of B and let B_1 be the strong closure of B. Then it is clear that A leaves invariant every closed linear manifold which is invariant under every member of B_1. By Corollary 8, B_1 is complete, and by Theorem 16, A is in the strongly (or weakly) closed algebra generated by B. Q.E.D.

The following is a closely related result also due to Bade.

19 THEOREM. *Let B be a bounded Boolean algebra of projections in a weakly complete space. Then every operator in the weakly closed algebra generated by B is a spectral operator of scalar type.*

PROOF. Let B_1 be the strong closure of B. By Corollary 8, B_1 is complete, and by Corollary 17 the weakly closed operator algebra generated by B is the same as the uniformly closed operator algebra generated by B_1. The desired result now follows from Lemma 2.1 and Corollary 2.12. Q.E.D.

The next result shows that the invariance condition of Theorem 18 may be replaced by a simpler one of commutivity in the special case where some set of the form $\{Ex \mid E \in B\}$ is fundamental in $\mathfrak{X}$.

20 THEOREM. *Let B be a bounded Boolean algebra of projections in a weakly complete space and suppose that for some vector x the set $\{Ex \mid E \in B\}$*

is fundamental in $\mathfrak{X}$. Then a bounded operator is in the weakly closed operator algebra generated by B if and only if it commutes with every element of B.

PROOF. It is clear that every element in the weakly closed operator algebra generated by B commutes with every element of B. To prove the converse, let A commute with every element of B, and let B_1 be the strong closure of B. Clearly A commutes with every element of B_1, and by Lemma 5, B_1 is complete. Thus the conclusion follows from Lemma 14 and Corollary 15. Q.E.D.

21 LEMMA. *Let B be a bounded σ-complete Boolean algebra of projections in a B-space $\mathfrak{X}$. Suppose that, for some sequence $\{x_i\}$ in $\mathfrak{X}$,*

$$\mathfrak{X} = \overline{\text{sp}}\,\{Ex_i \,|\, E \in B,\ i \geqq 1\}.$$

Then B is complete.

PROOF. It will be established that every set of disjoint projections in B is at most countable. It will then follow from Lemma IV.11.5 that every set in B has a least upper bound which is the least upper bound of a countable subset. Thus the completeness of B will follow from its σ-completeness.

Let $\{E_\alpha\}$ be a family of disjoint elements of B. It follows from Lemma 4 that every series $\sum_{i=1}^\infty E_{\alpha_i}$, with $\alpha_i \neq \alpha_j$ if $i \neq j$, converges strongly. Hence, for every integer $n \geqq 1$ and every $\varepsilon > 0$, only a finite number of the vectors $E_\alpha x_n$ have norms greater than ε. This shows that, for all but a countable number of α, $E_\alpha x_n = 0$, for every integer $n \geqq 1$. Thus, with the exception of a countable number of α, $E_\alpha E x_n = 0$ for every E in B and $n \geqq 1$. Since the set $\{Ex_n \,|\, E \in B,\ n \geqq 1\}$ is fundamental in $\mathfrak{X}$, it follows that $E_\alpha = 0$ for all but a countable number of α. Thus $\{E_\alpha\}$ is a countable set. Q.E.D.

It is now possible to give the following classical result of J. von Neumann.

22 THEOREM. *Let T be a bounded self adjoint operator in a separable Hilbert space. Then the following four algebras are the same:*

(i) *the algebra of all bounded Borel functions of T;*

(ii) *the weakly closed operator algebra generated by T;*

(iii) *the set of all bounded linear operators which commute with every operator commuting with T;*

(iv) *the uniformly closed operator algebra generated by the projections in the resolution of the identity for T.*

PROOF. Let $\mathfrak{A}_1, \ldots, \mathfrak{A}_4$ denote the algebras defined by the statements (i), ..., (iv), respectively. It follows from Theorem 2.10 that $\mathfrak{A}_1$ is uniformly closed and thus that $\mathfrak{A}_4 \subseteq \mathfrak{A}_1$. It is clear from the definition of the integral $\int f(\lambda)E(d\lambda)$ that $\mathfrak{A}_1 \subseteq \mathfrak{A}_4$. Thus $\mathfrak{A}_1 = \mathfrak{A}_4$. It is clear from their definitions that $\mathfrak{A}_2 \subseteq \mathfrak{A}_3$. To prove that $\mathfrak{A}_4 \subseteq \mathfrak{A}_2$ it will suffice to show that all projections $E(e)$ in the range of the resolution of the identity for T lie in $\mathfrak{A}_2$. Since $\mathfrak{A}_2$ is an algebra, the family Σ of Borel sets e for which $E(e) \in \mathfrak{A}_2$ is a field. By Corollary 2.11(iii) it is also a σ-field. Since the characteristic function of a finite interval is the limit of a bounded sequence of polynomials, it follows from Corollary 2.11(iii) that every finite real interval is in Σ. Thus, since the spectrum of T is real (cf. Theorem X.4.2), Σ consists of all Borel sets. It has now been shown that $\mathfrak{A}_1 = \mathfrak{A}_4 \subseteq \mathfrak{A}_2 \subseteq \mathfrak{A}_3$, and so to complete the proof it will suffice to show that $\mathfrak{A}_3 \subseteq \mathfrak{A}_4$.

To do this, let A be an operator in $\mathfrak{A}_3$ and let $\mathfrak{H}_0$ be a closed subspace of Hilbert space which is invariant under every element $E(e)$ of the resolution of the identity for T. Thus, if P is the orthogonal projection onto $\mathfrak{H}_0$ then, since $E(e)$ leaves $\mathfrak{H}_0$ invariant, $PE(e)P = E(e)P$. If y is orthogonal to $\mathfrak{H}_0$,

$$(E(e)y, \mathfrak{H}_0) = (y, E(e)\mathfrak{H}_0) = 0,$$

so that $E(e)y$ is orthogonal to $\mathfrak{H}_0$. Thus $E(e)$ leaves the orthogonal complement of $\mathfrak{H}_0$ invariant and so

$$(I - P)E(e)(I - P) = E(e)(I - P).$$

Hence

$$\begin{aligned} PE(e) &= PE(e)P + PE(e)(I - P) \\ &= E(e)P + P(I - P)E(e)(I - P) = E(e)P, \end{aligned}$$

from which it follows immediately that $PT = TP$. Since A is in $\mathfrak{A}_3$, we have $AP = PA$ so that A leaves $\mathfrak{H}_0$ invariant. It has been shown that every operator in $\mathfrak{A}_3$ leaves invariant every closed linear manifold which is invariant under every projection in the resolution of the identity for T. Since the Hilbert space is separable by assumption, these projections form (cf. Lemma 21) a complete Boolean algebra. Thus it follows from Theorem 16 that $\mathfrak{A}_3 \subseteq \mathfrak{A}_4$. Q.E.D.

23 LEMMA. *The strongly closed Boolean algebra of projections generated by a σ-complete Boolean algebra of projections in a B-space is complete.*

PROOF. Let B be a σ-complete Boolean algebra of projections in a B-space $\mathfrak{X}$, and let B_1 be its strong closure. By Lemma 3, B is bounded and thus B_1 is also a bounded Boolean algebra of projections in $\mathfrak{X}$. Suppose that B_1 is not complete. By Lemma 4 there is a monotone increasing generalized sequence $\{E_\alpha\}$ in B_1 and an x in $\mathfrak{X}$ such that the limit $\lim_\alpha E_\alpha x$ does not exist. Let $\mathfrak{X}(x)$, $\mathfrak{X}_1(x)$ be the closed linear manifolds generated by the sets $\{Ex \mid E \in B\}$, $\{Ex \mid E \in B_1\}$, respectively. Since B_1 is the strong closure of B, each element Ex with E in B_1 is contained in $\mathfrak{X}(x)$ and thus $\mathfrak{X}(x) \supseteq \mathfrak{X}_1(x)$. Evidently $\mathfrak{X}(x) \subseteq \mathfrak{X}_1(x)$ and so $\mathfrak{X}(x) = \mathfrak{X}_1(x)$, from which it follows that $E\mathfrak{X}(x) \subseteq \mathfrak{X}(x)$ for each E in B_1. For each E in B_1, let $\tilde{E}$ denote the restriction of E to $\mathfrak{X}(x)$. It is clear that the set $\{\tilde{E} \mid E \in B_1\}$ is contained in the strong closure of $B(x) = \{\tilde{E} \mid E \in B\}$. Since the limit $\lim_\alpha E_\alpha x$ does not exist, the limit $\lim_\alpha \tilde{E}_\alpha$ fails to exist in the strong topology. It follows from Lemma 4 that the strong closure of $B(x)$ is not complete. Thus, by Corollary 7, $B(x)$ is not complete. On the other hand, $B(x)$ is, according to Corollary 11, σ-complete. These two statements contradict Lemma 21 and prove the lemma. Q.E.D.

24 COROLLARY. *A bounded linear operator is in the weakly closed operator algebra generated by a σ-complete Boolean algebra B of projections in a B-space if and only if it leaves invariant every closed linear manifold which remains invariant under every element of B.*

PROOF. Let B_1 be the strong closure of B. By Lemma 23, B_1 is complete and it follows from Theorem 16 and Corollary 17 that the weakly closed operator algebra generated by B_1 (which is clearly the same as that generated by B) consists of those operators which leave invariant every closed linear manifold that is left invariant by every member of B. Since it is evident that a closed linear manifold is invariant under every member of B_1 if and only if it is invariant under every member of B, the proof is complete. Q.E.D.

25 COROLLARY. *Every operator in the weakly closed operator algebra generated by a σ-complete Boolean algebra of projections in a B-space is a spectral operator of scalar type.*

PROOF. Let B_1 be the strong closure of the σ-complete Boolean algebra B so that B_1 is, by Lemma 23, complete. By Corollary 17 the weakly closed operator algebra generated by B_1 (which is clearly the same as the weakly closed operator algebra generated by B) is the same as the uniformly closed operator algebra generated by B_1. Every operator in such a uniformly closed algebra is, by Lemma 9, given in terms of a countably additive spectral measure by an expression of the form $\int f(\lambda)E(d\lambda)$. Such operators are, by Lemma 2.9, spectral operators of scalar type. Q.E.D.

26 COROLLARY. *Every operator in the weakly closed operator algebra generated by a spectral operator of scalar type and the projections in its resolution of the identity is a spectral operator of scalar type.*

PROOF. Since a spectral operator of scalar type is clearly in the weakly closed algebra $\mathfrak{A}$ generated by the projections in its resolution of the identity, it suffices to show that every operator in $\mathfrak{A}$ is a spectral operator of scalar type. This follows immediately from Corollaries 10 and 25. Q.E.D.

The section will be concluded with two theorems of Bade which belong to that interesting collection of results in B-spaces concerning conditions when weak convergence implies strong convergence.

27 THEOREM. *If a generalized sequence of projections in a σ-complete Boolean algebra of projections in a B-space converges weakly to a projection, then it converges strongly.*

PROOF. In view of Lemma 23, the proof may be restricted to the case where the Boolean algebra B is complete. Let $\{E_\alpha\}$ be a weakly convergent generalized sequence in B and suppose that its limit E is a projection. It must be shown that $\{E_\alpha\}$ converges strongly to E. By Lemma 6, E is in B and so a consideration of the sequence $\{E_\alpha - E\}$ shows that it may be assumed that $E = 0$. Thus, to make an indirect proof, it is assumed that the sequence $\{E_\alpha\}$ is weakly convergent to zero and, for some vector x_0, the sequence $\{E_\alpha x_0\}$ is not convergent to zero. Hence, by Lemma 9, B is the range of a countably additive spectral measure E defined on a σ-field Σ. Hence $E_\alpha = E(e_\alpha)$, where e_α is in Σ. By Lemma 12 there is a linear functional y^* such that the measure $\mu = y^*Ex_0$ has the property that $\mu(e) = 0$ implies $E(e)x_0 = 0$. It follows from the Theorem IV.10.1 of Pettis that $\lim_{\mu(e)\to 0} E(e)x_0 = 0$. Since $\{E(e_\alpha)\}$ is weakly convergent to zero, it follows that $\mu(e_\alpha) \to 0$, and hence it follows from the result of Pettis that $E(e_\alpha)x_0 \to 0$, a contradiction which completes the proof. Q.E.D.

28 COROLLARY. *If a generalized sequence in a bounded Boolean algebra*

of projections in a weakly complete B-space converges weakly to a projection, then it converges strongly.

PROOF. This follows from Theorem 27 and Lemma 5. Q.E.D.

4. Strong Limits of Spectral Operators: Non-Commutative Case

In this section conditions will be given to insure that the strong limit $T = \lim_\alpha T_\alpha$ of a generalized sequence $\{T_\alpha\}$ of scalar type spectral operators is itself a scalar type spectral operator. These same conditions will also insure that $f(T_\alpha) \to f(T)$ strongly for any bounded Borel function f whose discontinuities have measure zero relative to the resolution of the identity for T. This is a perturbation theorem of the type considered by Rellich.

1 THEOREM. *Let $\{T_\alpha\}$, $\alpha \in A$, be a strongly convergent generalized sequence of spectral operators in a B-space $\mathfrak{X}$. Suppose that there is a compact set V containing all the spectra $\sigma(T_\alpha)$. It is assumed that every complex continuous function on V is the uniform limit of rational functions and that the resolutions of the identity E_α for T_α satisfy the inequality*

$$|E_\alpha(\delta)| \leqq M, \qquad \alpha \in A,$$

for every δ in the family $\mathscr{B}$ of all Borel sets of complex numbers. If $T = \lim_\alpha T_\alpha$ and $\sigma(T) \subseteq V$, then T^* is a spectral operator of class $(\mathscr{B}, \mathfrak{X})$. If $\mathfrak{X}$ is weakly complete, then T is itself a spectral operator.

REMARK. In order that V have the property that every continuous function on V is the uniform limit of rational functions, it is clearly necessary that V have no interior points. However, it is known that not every compact nowhere dense set V has the required property, and the complete characterization of such sets seems to be an unsolved problem in approximation theory. Lavrentieff [1] (see also Mergelyan [1]) showed that if V is nowhere dense and does not separate the plane, then V has the required property. Hartogs and Rosenthal [1] showed that if V has plane measure zero, then V has the desired property. Walsh [1] showed that if V is the union of a finite number of Jordan arcs, no two of which intersect in more than a finite number of points, then V has the required property.

PROOF. Since the operators T_α are scalar type spectral operators whose spectra lie in V, their resolvents are given by the formula

$$R(\lambda; T_\alpha) = \int_V \frac{E_\alpha(d\zeta)}{\zeta - \lambda}, \qquad \lambda \notin V,$$

and it follows from Theorem 2.10 that

$$|R(\lambda; T_\alpha)| \leqq 4M \sup_{\zeta \in V} \frac{1}{|\zeta - \lambda|}, \qquad \lambda \notin V.$$

Thus it follows from the identity

$$\begin{aligned} R(\lambda; T_\alpha) - R(\lambda; T) &= R(\lambda; T_\alpha)(\lambda I - T)R(\lambda; T) - R(\lambda; T_\alpha)(\lambda I - T_\alpha)R(\lambda; T) \\ &= R(\lambda; T_\alpha)(T_\alpha - T)R(\lambda; T) \end{aligned}$$

that, for $\lambda \notin V$, $R(\lambda; T_\alpha) \to R(\lambda; T)$ strongly. Now let r be an element of the class $R(V)$ of rational functions which are continuous on V. Then $r(T_\alpha)$ is a finite product of terms of the form $\lambda I - T_\alpha$ and $R(\lambda; T_\alpha)$. Thus, since the product of bounded, strongly convergent generalized sequences is itself a strongly convergent generalized sequence, we see that $r(T_\alpha) \to r(T)$ strongly. Now, from Theorem 2.10,

$$|r(T_\alpha)| = \left| \int_V r(\lambda) E_\alpha(d\lambda) \right| \leqq 4M \sup_{\zeta \in V} |r(\zeta)|,$$

and since $r(T_\alpha) \to r(T)$ strongly, we have

$$|r(T)| \leqq 4M \sup_{\zeta \in V} |r(\zeta)|.$$

Since $R(V)$ is dense in $C(V)$, this inequality shows that the homomorphism $r \to r(T)$ on $R(V)$ has a unique continuous extension to a homomorphism of $C(V)$ onto the uniformly closed operator algebra generated by the operators $r(T)$ with r in $R(V)$. The present theorem then follows immediately from Theorems 2.4, 2.5, and 2.10. Q.E.D.

2 COROLLARY. *Under the hypotheses of the preceding theorem.*

$$f(T) = \lim_\alpha f(T_\alpha), \qquad f \in C(V),$$

in the strong operator topology.

PROOF. Let $\{r_n\}$ be a sequence of rational functions converging uniformly to the continuous function f on V. It was shown in the preceding proof that, for each α,

$$r_n(T)x = \lim_{\alpha} r_n(T_\alpha)x, \qquad x \in \mathfrak{X}.$$

On the other hand,

$$|r_n(T_\alpha) - f(T_\alpha)| \leqq 4M \sup_{\lambda \in V} |r_n(\lambda) - f(\lambda)|$$

and so

$$\lim_{n \to \infty} r_n(T_\alpha)x = f(T_\alpha)x$$

uniformly for $|x| \leqq 1$. Thus, by the Moore-Smith convergence theorem (cf. I.7.6),

$$f(T)x = \lim_n \lim_\alpha r_n(T_\alpha)x = \lim_\alpha \lim_n r_n(T_\alpha)x = \lim_\alpha f(T_\alpha)x. \qquad \text{Q.E.D.}$$

Using Corollary 2, we may obtain the following result which is analogous to a perturbation theorem of Rellich (cf. Theorem X.7.2) for normal operators on Hilbert space.

3 THEOREM. *Suppose that the hypotheses of Theorem 1 are satisfied and that E is the resolution of the identity for T. Then*

$$f(T)x = \lim_{\alpha} f(T_\alpha)x, \qquad x \in \mathfrak{X},$$

for every bounded Borel function f on V whose set of discontinuities is contained in a closed set upon which E vanishes.

PROOF. Let K be the set of discontinuities of f and let g_n be a continuous function vanishing on the closure $\bar{K}$, having the value 1 for those λ whose distance from $\bar{K}$ is greater than $1/n$, and satisfying the inequality $|g_n(\lambda)| \leqq 1$ elsewhere (cf. Theorem I.5.2). Then fg_n is continuous so that, by Corollary 2,

$$\lim_{\alpha} f(T_\alpha)g_n(T_\alpha)x = f(T)g_n(T)x, \qquad x \in \mathfrak{X}.$$

Similarly $\lim_\alpha g_n(T_\alpha) = g_n(T)$ in the strong operator topology. Since

$$|f(T_\alpha)| \leqq 4M \sup_{\lambda \in V} |f(\lambda)| \leqq L,$$

the inequality

$$|f(T_\alpha)g_n(T)x - f(T_\alpha)g_n(T_\alpha)x| \leqq L\,|g_n(T)x - g_n(T_\alpha)x|$$

shows that

$$\lim_{\alpha} f(T_\alpha)g_n(T_\alpha)x = \lim_{\alpha} f(T_\alpha)g_n(T)x, \qquad x \in \mathfrak{X}.$$

Thus

$$\lim_{\alpha} f(T_\alpha)y = f(T)y$$

for every y of the form $y = g_n(T)x$. But, by hypothesis, $E(\bar{K}) = 0$ and so $g_n(T)x \to E(\bar{K}')x = x$, which shows that the vectors $y = g_n(T)x$ are dense in $\mathfrak{X}$. Since the operators $f(T_\alpha)$ are bounded uniformly in α, it follows that $f(T_\alpha)x \to f(T)x$ for every x in $\mathfrak{X}$. Q.E.D.

5. Exercises

1 DEFINITION. If K is a compact set in the complex plane, then let $R(K)$ be the set of rational functions which are analytic on K and let $CR(K)$ be the closure of $R(K)$ in $C(K)$. A set K is said to be an *R-set* if $CR(K) = C(K)$.

2 Let $\mathfrak{X}$ be a reflexive B-space and let $T \in B(\mathfrak{X})$ and suppose that $\sigma(T)$ is an R-set. Then T is a scalar type spectral operator if and only if there exists a constant A such that

$$|f(T)| \leqq A\,|f|, \qquad f \in R(\sigma(T)).$$

3 Let $\mathfrak{X}$ be a reflexive B-space and $T \in B(\mathfrak{X})$ and suppose that $\sigma(T)$ is an R-set. Then T is a scalar type spectral operator if and only if there exists a constant A such that

$$|f(T)|^2 \leqq A\,|f(T)^2|, \qquad f \in R(\sigma(T)).$$

4 Let U be a unitary operator in $B(\mathfrak{X})$. If U is spectral, then there exists a constant A such that if

$$P(\lambda) = \sum_{-N}^{N} c_k \lambda^k,$$

then

$$|P(U)| \leqq A \sup_{|\lambda|=1} |P(\lambda)|.$$

Conversely, if $\mathfrak{X}$ is reflexive and U is a unitary operator satisfying this condition, then U is spectral.

5 (Fixman, Krabbe) The right shift operator U defined in $\mathfrak{X} = l_1(-\infty, \infty)$ by

$$U(\ldots, \xi_1, \xi_0, \xi_1, \ldots) = (\ldots, \xi_{-2}, \xi_{-1}, \xi_0, \ldots)$$

is not spectral. [Hint: Either examine $\sigma_r(U)$ or use the fact that there exists a continuous function whose Fourier series does not converge absolutely.]

6 (Fixman, Krabbe) The right shift operator U, defined in $l_p(-\infty, \infty)$, $1 < p < \infty$, $p \neq 2$, by

$$U(\ldots, \xi_{-1}, \xi_0, \xi_{+1}, \ldots) = (\ldots, \xi_{-2}, \xi_{-1}, \xi_0, \ldots),$$

is not spectral. [Hint: If $1 < p < 2$, then it is known (cf. Zygmund [1; p. 190]) that there exists a continuous function whose sequence of Fourier coefficients does not belong to l_p.]

7 (Fixman) Let $\alpha > 0$ and let r_α be positive rotation of the unit circle $S = \{\lambda \mid |\lambda| = 1\}$ through α radians. Then r_α is a homeomorphism of S onto S and is periodic if and only if α is a rational multiple of π. Let U_α be defined on $C(S)$ by $(U_\alpha f)(s) = f(r_\alpha(s))$. Show that U_α is a spectral operator on $C(S)$ if and only if α is a rational multiple of π. [Hint: If α/π is irrational, for each n there exists a point s_0 such that $s_k = r_\alpha^k(s_0)$, $k = 0, \pm 1, \ldots, \pm n$, have disjoint neighborhoods.]

8 Let $\mathfrak{A}$ be a commutative subalgebra of $B(\mathfrak{X})$ which contains I and is *full* (in the sense that if $A \in \mathfrak{A}$ and $A^{-1} \in B(\mathfrak{X})$ then $A^{-1} \in \mathfrak{A}$). (i) Show that the spectrum $\sigma(A, \mathfrak{A})$ of $A \in \mathfrak{A}$ as an element of $\mathfrak{A}$ coincides with the spectrum $\sigma(A, \mathfrak{X})$ of A as an element of $B(\mathfrak{X})$. (ii) If E is a projection in $\mathfrak{A}$, and $\mathfrak{A}_E = \{AE \mid A \in \mathfrak{A}\}$, then $\sigma(A, \mathfrak{A}_E) = \sigma(A, E\mathfrak{X})$.

9 Let $\mathfrak{A}$ be a commutative full subalgebra of $B(\mathfrak{X})$ containing I. Let E be a projection in $\mathfrak{A}$ and let $\mathfrak{A}_E = \{AE \mid A \in \mathfrak{A}\}$, and let $\mathfrak{M}$ and $\mathfrak{M}_E$ denote the set of all multiplicative linear functionals on $\mathfrak{A}$ and $\mathfrak{A}_E$, respectively. (i) Show that there is a one-to-one correspondence between $\mathfrak{M}_E$ and $\{m \in \mathfrak{M} \mid m(E) = 1\}$. (ii) Show that if $A \in \mathfrak{A}_E$, then

$$\sigma(A, E\mathfrak{X}) = \{m(A) \mid m \in \mathfrak{M}, m(E) = 1\}.$$

10 Let S, T be commuting operators in $B(\mathfrak{X})$. (i) Show that

$$\sigma(S + T) \subseteq \sigma(S) + \sigma(T).$$

In particular, if N is quasi-nilpotent, then $\sigma(S + N) = \sigma(S)$. (ii) If E is a projection operator which commutes with S and N, then

$$\sigma(S+N, E\mathfrak{X}) = \sigma(S, E\mathfrak{X}).$$

11 Use the preceding exercises to give another proof of Theorem XV.4.5 giving the canonical decomposition of a spectral operator into its scalar and radical parts.

12 Let E be a countably additive spectral measure and let $\mathfrak{A}$ be a full commuting algebra containing E. If m is a multiplicative linear functional on $\mathfrak{A}$, then there exists a unique complex number λ_m such that the scalar valued measure $mE(\cdot)$ equals 1 on each neighborhood of λ_m and equals 0 on each closed set not containing λ_m.

13 (Saffern) Let T be an operator in $B(\mathfrak{X})$ and let $\mathfrak{A}(T)$ be the full closed subalgebra of $B(\mathfrak{X})$ generated by T. Show that $\mathfrak{A}(T)$ is the uniform closure of

$$R(T) = \{f(T) \,|\, f \text{ is rational with poles in } \rho(T)\}.$$

14 (Kantorovitz) Let A and B be elements of $B(\mathfrak{X})$ with resolvent operators $R(\lambda; A)$ and $R(\lambda; B)$. Suppose that μ is not in the set

$$\sigma(A) + \sigma(B) = \{a + b \,|\, a \in \sigma(A),\ b \in \sigma(B)\}$$

and let C_μ be a positively oriented rectifiable contour surrounding $\sigma(A)$ and which is, together with its interior, contained in $\mu - \rho(B)$. Show that the integral

$$J(\mu; A, B) = \frac{1}{2\pi i}\int_{C_\mu} R(\mu-\lambda; B)R(\lambda; A)\, d\lambda$$

exists in $B(\mathfrak{X})$ and is independent of the contour C_μ. Show that this integral equals

$$\frac{1}{2\pi i}\int_{D_\mu} R(\lambda; B)R(\mu-\lambda; A)\, d\lambda,$$

where D_μ is a contour surrounding $\sigma(B)$ and which is, together with its interior, contained in $\mu - \rho(A)$.

15 (Kantorovitz) With the notation of the preceding exercise, show that if $AB = BA$, then

$$R(\mu; A+B) = J(\mu; A, B).$$

Moreover, if the distance between μ and $\sigma(A)$ exceeds the spectral radius

$|\sigma(B)|$, then

$$R(\mu; A+B) = \sum_{n=0}^{\infty} R(\mu; A)^{n+1}B^n.$$

16 (Stone) A completely regular space is called *extremally disconnected* if the closure of every open set is itself open. Prove that every complete Boolean algebra of projections in a B-space is isomorphic (as a Boolean algebra) to the Boolean algebra of all open and closed sets of an extremally disconnected compact Hausdorff space. [Hint: Use Theorem I.12.1.]

6. Notes and Remarks

The results presented in this chapter are taken primarily from Dunford [18] and Bade [3, 4, 5]. In addition to these papers, the following deal primarily with algebras of spectral operators, Boolean algebras of projections, or multiplicity theory. Dieudonné [19, 20, 21], Dowson [2, 4, 6], Edwards and Ionescu Tulcea [1], Feldzaman [1, 2], Foguel [3, 5, 7, 8], Lumer [2], McCarthy [1], McCarthy and Schwartz [1], McCarthy and Tzafriri [1], Moyal [1], Simpson [3], Tzafriri [2, 3, 4, 5, 6], and Walsh [1, 2, 3]. (However, it should be borne in mind that other papers dealing with spectral operators or operator algebras sometimes establish results that are related to the material discussed here.)

Section 2, which deals with uniformly closed commutative algebras of spectral operators, is taken from Dunford [18]. Most of the results in Section 3 are due to Bade [3, 4] although special cases of some of these theorems were proved earlier. For example, the fact that if B is a bounded Boolean algebra of projections in a reflexive space then the fact that the strong closure of B is complete follows from a theorem of Day [10] and was stated explicitly by Dunford [2] (see also Bade [3; p. 404]). Extending a theorem of Lorch [2] for sequences, Barry [1] proved that a bounded monotone generalized sequence of projections in a reflexive space is strongly convergent to a projection. In the case of Hilbert spaces, Lemma 3.14 was proved by Wolf [2]. It was observed by F. Riesz [21] that some earlier results of von Neumann implied that if A is a self adjoint operator in a separable Hilbert space $\mathfrak{H}$ and if B is a bounded operator which commutes with every operator commuting with A, then $B = f(A)$ for some bounded Borel function f. Riesz gave another proof of this result and his method was extended to unbounded operators by Mimura [1]. For other proofs, see Nakano [8, 9], Riesz and Sz.-Nagy [1; Sec. 129], and Sz.-Nagy [3, pp. 63–65]. The theorem fails if $\mathfrak{H}$ is not separable (cf. Nakano [9], Sz.-Nagy [3; p. 65], Wecken [2]). However, in the bounded case Segal [5, II] has found an extension by considering a larger class of functions. Segal's result is extended to spectral operators by Bade [3; p. 410]. Theorems 3.16 and 3.18 may also be regarded as being generalizations of this von Neumann-Riesz theorem.

It was proved by Berkson [3] that if $\mathfrak{X}$ is a uniformly convex B-space and B is a bounded Boolean algebra of projections on $\mathfrak{X}$, then $\mathfrak{X}$ has an equivalent norm such that the functionals given by Lemma 3.12 can be expressed by means of a semi-inner product. On the other hand, Walsh [2; p. 315] has shown that in a complete metrizable locally convex space there may not be a *single* functional corresponding to x_0^*; however, some of the results presented here can still be generalized.

Walsh [1] showed that if $\mathfrak{A}$ is a complex B-algebra with the property that for every $a \in \mathfrak{A}$ there is a homomorphism h_a of $C(\sigma(a))$ into $\mathfrak{A}$ which sends the function $f(\lambda) \equiv \lambda$ into a, then $\mathfrak{A}$ is semi-simple, commutative, and isomorphic to $C(\mathscr{M})$, where $\mathscr{M}$ is the space of maximal ideals of $\mathfrak{A}$.

It was proved by Dowson [2] that if S is a scalar operator whose spectrum is nowhere dense and does not separate the plane, then the resolution of the identity of S is contained in the weakly sequentially closed algebra generated by S and I. However, this conclusion fails if $\sigma(S)$ has non-empty interior or separates the plane.

The results of Section 4 on strong limits of spectral operators are due to Bade [3; pp. 397–403] [4]. Theorem 4.3 generalizes theorems of Kaplansky [7] and Rellich [2, II]. Further results along this line have been obtained by Foguel [1, 4], Kantorovitz [4, 5], and Tzafriri [3, 4]. Analogous results were obtained in locally convex spaces by Simpson [1]. Theorem 4.1 is false without the hypothesis that $\sigma(T) \subseteqq V$. This has been shown by a counterexample constructed by C. Foiaş. Berkson [5] has shown that this hypothesis is satisfied if the complement of V is connected.

CHAPTER XVIII

Unbounded Spectral Operators

1. Introduction

It was shown in the course of Chapters XII, XIII, and XIV that in order to apply the spectral theory of Hermitian operators to ordinary and partial differential operators it is first necessary to extend the spectral theory of bounded operators to a theory of unbounded operators. The present chapter is an attempt to make the corresponding step in the theory of spectral operators.

We begin by defining a closed spectral operator and its resolution of the identity, and showing that the latter is unique. Then a functional calculus of spectral operators is developed, first for analytic functions of general spectral operators, and next for arbitrary unbounded Borel functions of scalar type spectral operators. In the latter case it is possible to extend the very general theory developed for unbounded Hermitian operators in Definition XII.2.5 through Theorem XII.2.9 to arbitrary B-spaces. Next we show by example that it is possible for an analytic function $f(T)$ of a *non-scalar type* spectral operator not to be spectral, and give sufficient conditions on f that $f(T)$ be spectral. The relation between a general spectral operator and its scalar part is shown to be less close for unbounded than for bounded operators. The present chapter ends with a theorem establishing a sufficient condition that an unbounded operator be spectral; this condition will be used in proving the principal result of the following chapter.

It should be remarked that no analysis of the extension theory of spectral operators paralleling Section XII.4 is given in the present chapter. Such a theory would doubtless be useful in the study of nonselfadjoint differential operators, and its omission must be regarded as a serious gap in the theory.

2. Unbounded Spectral Operators

1 DEFINITION. Let $\mathscr{B}$ denote the σ-field of Borel subsets of the complex plane. Let T be a linear operator whose domain and range are contained in a complex B-space $\mathfrak{X}$. Then T is said to be a *spectral operator* if it is closed and if there is a regular countably additive spectral measure E defined on $\mathscr{B}$, such that

(i) $$\mathfrak{D}(T) \supseteq E(\sigma)\mathfrak{X} \qquad \text{if } \sigma \text{ is bounded;}$$

(ii) $$E(\sigma)\mathfrak{D}(T) \subseteq \mathfrak{D}(T)$$

and

$$TE(\sigma)x = E(\sigma)Tx, \qquad x \in \mathfrak{D}(T),\ \sigma \in \mathscr{B};$$

(iii) the restriction $T \mid E(\sigma)\mathfrak{X}$ with domain $\mathfrak{D}(T) \cap E(\sigma)\mathfrak{X}$ has its spectrum

$$\sigma(T \mid E(\sigma)\mathfrak{X}) \subseteq \bar{\sigma}, \qquad \sigma \in \mathscr{B}.$$

The spectral measure E is said to be a *resolution of the identity for* T.

It is immediate that a spectral operator is closed and densely defined.

It will first be shown that the resolution of a spectral operator is unique. This is done in Theorem 5 to follow, and the preliminary lemmas follow the same line of argument as was used in the case of bounded spectral operators.

2 LEMMA. *If σ is a Borel set, and T is a spectral operator with resolution of the identity E, then the restriction $T \mid E(\sigma)\mathfrak{X}$ of T to $E(\sigma)\mathfrak{X}$ is a spectral operator whose resolution of the identity is the restriction of E to $E(\sigma)\mathfrak{X}$. If σ is bounded, $T \mid E(\sigma)\mathfrak{X}$ is bounded.*

PROOF. If σ is bounded, then by Definition 1(i), the restriction R of T to $E(\sigma)\mathfrak{X}$ is a closed everywhere defined operator, so that, by the closed graph theorem, R is bounded. In any case, R is closed. If F is the restriction of E to $E(\sigma)\mathfrak{X}$, it is clear from Definition 1(ii) that $E(e)\mathfrak{D}(R) \subseteq \mathfrak{D}(R)$, and that $F(e)R \subseteq RF(e)$ for each Borel set e. Using the notation of Definition 1(iii), it is clear that $R \mid E(e)E(\sigma)\mathfrak{X} = T \mid E(e\sigma)\mathfrak{X}$, so that, by 1(iii), $\sigma(R \mid E(e)E(\sigma)\mathfrak{X}) \subseteq \overline{e\sigma} \subseteq \bar{\sigma}$. Thus, R is spectral, and has the indicated resolution of the identity. Q.E.D.

The next lemma is the "unbounded" form of Theorem XV.3.4.

3 LEMMA. *Let T be a spectral operator with resolution of the identity E, and let σ be a compact subset of the plane. Then $E(\sigma)\mathfrak{X}$ is the family of all vectors x such that there exists an analytic vector function y_λ defined for $\lambda \notin \sigma$ such that*

$$(\lambda I - T)y_\lambda = x, \qquad \lambda \notin \sigma.$$

PROOF. Suppose that a y_λ with the indicated properties exists for $\lambda \notin \sigma$. Let e be an arbitrary bounded Borel set. Then by Definition 1(ii) we have

$$(\lambda I - T)E(e)y_\lambda = E(e)x, \qquad \lambda \in \sigma.$$

It follows (in the notation of Definition 1(iii) and Theorem XV.3.4) that the spectrum $\sigma(E(e)x)$ of the element $E(e)x$ of $E(e)\mathfrak{X}$ relative to the operator $T|E(e)\mathfrak{X}$ is contained in σ. By Lemma 2 and Theorem XV.3.4 we then have $E(\sigma)E(e)x = E(e)x$. Taking the limit of this equation as e runs through an increasing sequence of bounded Borel sets whose union is the entire complex plane, we find that $E(\sigma)x = x$, so that x is in $E(\sigma)\mathfrak{X}$.

Conversely, let x be in $E(\sigma)\mathfrak{X}$. Then, by Definition 1(iii), the resolvent $R(\lambda; T \,|\, E(\sigma)\mathfrak{X})$ is defined and analytic for λ not in σ. Putting $y_\lambda = R(\lambda; T \,|\, E(\sigma)\mathfrak{X})x$, we clearly have $(\lambda I - T)y_\lambda = x$. Q.E.D.

4 COROLLARY. *Let T be a closed spectral operator with resolution of the identity E, and let A be a bounded operator such that $A\mathfrak{D}(T) \subseteq \mathfrak{D}(T)$ and $ATx = TAx$ for x in $\mathfrak{D}(T)$. Then $AE(e) = E(e)A$ for each Borel set e.*

PROOF. Let σ be a compact set, and let x be a vector in $E(\sigma)\mathfrak{X}$. Then, by Lemma 3, there is an analytic vector function y_λ defined for $\lambda \notin \sigma$ such that

$$(\lambda I - T)y_\lambda = x, \qquad \lambda \notin \sigma.$$

It follows that

$$(\lambda I - T)Ay_\lambda = Ax, \qquad \lambda \notin \sigma.$$

Hence, by Lemma 3, Ax is in $E(\sigma)\mathfrak{X}$. This shows that $E(\sigma)AE(\sigma) = AE(\sigma)$ and that $(I - E(\sigma))AE(\sigma) = 0$ for each compact set σ. If we take the limit of this equation as σ runs through an increasing sequence of compact sets with union e, it follows that $E(e)AE(e) = AE(e)$ for each open set e. Thus

$$\begin{aligned} AE(\sigma) = E(\sigma)AE(\sigma) &= E(\sigma)AE(\sigma) + (I - E(\sigma'))AE(\sigma') \\ &= E(\sigma)A(E(\sigma) + E(\sigma')) = E(\sigma)A \end{aligned}$$

for each compact set σ. Since the family of sets e for which $E(e)A = AE(e)$ is evidently a σ-field, it follows that $E(e)A = AE(e)$ for each Borel set e. Q.E.D.

5 THEOREM. *The resolution of the identity of a closed spectral operator is unique.*

PROOF. Let E and E_1 be two resolutions of the identity for the closed spectral operator T. Then, by Corollary 4, $E(e)$ and $E_1(f)$ commute for each pair of Borel sets e and f. Let σ be a compact set. It follows that each projection $E(e)$ leaves $E_1(\sigma)\mathfrak{X}$ invariant. By Lemma 2 the restriction S of T to $E_1(\sigma)\mathfrak{X}$ is a bounded spectral operator, whose resolution of the identity is the restriction F_1 of E_1 to $E_1(\sigma)\mathfrak{X}$. We will show that the restriction F of E to $E_1(\sigma)\mathfrak{X}$ is also a resolution of the identity for S. Once this is done, it will follow from the uniqueness theorem given in Corollary XV.3.8 that $F_1 = F$, so that $E_1(e)E(\sigma) = E(e)E(\sigma)$ for every Borel set e and compact set σ. Since $E(\sigma)$ is countably additive in σ, it will follow that $E_1(e) = E(e)$, which expresses the desired uniqueness.

Thus it suffices to show that F is a resolution of the identity for S. It is clear that $F(e)S = SF(e)$ for every Borel set e, so that we have only to show that the spectrum of the restriction $S \mid E(e)E_1(\sigma)\mathfrak{X} = T \mid E(e)E_1(\sigma)\mathfrak{X}$ is entirely contained in $\bar{e}$.

It follows from Definition 1(iii) that for $\lambda \notin \bar{e}$ the operator $\lambda I - T$ is a one-to-one map of $E(\sigma)\mathfrak{X}$ into itself. Thus it suffices to show that, for $\lambda \notin \bar{e}$, the operator $\lambda I - T$ maps the space $E(e)E_1(\sigma)\mathfrak{X} = E_1(\sigma)E(e)\mathfrak{X}$ onto itself. From this it will follow that $(\lambda I - T)^{-1}$ is everywhere defined, closed, and hence bounded, which means that $\lambda \notin \sigma(T \mid E(e)E_1(\sigma)\mathfrak{X})$. But, by Definition 1, if $\lambda \notin \bar{e}$ then $(\lambda I - T)(\mathfrak{D}(T) \cap E(e)\mathfrak{X}) = E(e)\mathfrak{X}$. Thus

$$\begin{aligned} E_1(\sigma)E(e)\mathfrak{X} &= E_1(\sigma)(\lambda I - T)(\mathfrak{D}(T) \cap E(e)\mathfrak{X}) \\ &= (\lambda I - T)E_1(\sigma)(\mathfrak{D}(T) \cap E(e)\mathfrak{X}) \\ &\subseteq (\lambda I - T)E_1(\sigma)E(e)\mathfrak{X} \end{aligned}$$

by Definition 1. Q.E.D.

Turning to the theory of functions of unbounded spectral operators, we begin the analysis with a useful general lemma.

6 LEMMA. *Let E be a spectral measure in the B-space $\mathfrak{X}$ which is defined and countably additive on the σ-field Σ of subsets of a set S. Suppose that the subfamily Σ_0 of Σ contains the union of every finite collection of its*

elements, as well as every subset of any one of its elements provided that the subset belongs to Σ. Let Q_0 be a linear operator whose domain is the linear set $\bigcup_{e\in\Sigma_0} E(e)\mathfrak{X}$. It is assumed that, for each e in Σ_0, the operator $Q_0 E(e)$ is bounded and that

$$E(e)Q_0 x = Q_0 E(e)x, \qquad e \in \Sigma_0,\ x \in \mathfrak{D}(Q_0).$$

For a given increasing sequence $\{e_n\}$ in Σ_0 for which $E(\bigcup_{n=1}^{\infty} e_n) = I$, let the operator Q be defined by the equations

$$\mathfrak{D}(Q) = \{x \mid \lim_{n\to\infty} Q_0 E(e_n)x \text{ exists}\},$$

$$Qx = \lim_{n\to\infty} Q_0 E(e_n)x, \qquad x \in \mathfrak{D}(Q).$$

Then Q is a closed, densely defined operator which is independent of the particular increasing sequence $\{e_n\}$ in Σ_0 provided only that $E(\bigcup_{n=1}^{\infty} e_n) = I$.

PROOF. It is clear that Q is a linear operator. If $x = E(e_n)y$, where y is in $\mathfrak{X}$, then $Q_0 E(e_m)x = Q_0 x$ for $m \geqq n$, so that $\lim_{m\to\infty} Q_0 E(e_m)x = Q_0 x = Q_0 E(e_n)x$. This shows that x is in $\mathfrak{D}(Q)$. Consequently, $\mathfrak{D}(Q) \supseteqq E(e_n)\mathfrak{X}$ for each n, so that since $E(\bigcup_{n=1}^{\infty} e_n) = I$, $\mathfrak{D}(Q)$ is dense.

Next let z be an arbitrary vector in $\mathfrak{D}(Q)$. Then

$$\lim_{n\to\infty} Q_0 E(e_n)E(e)z = \lim_{n\to\infty} E(e)Q_0 E(e_n)z = E(e)Qz, \qquad e \in \Sigma_0,$$

which proves that $QE(e)z = E(e)Qz$ for every e in Σ_0 and z in $\mathfrak{D}(Q)$. It is clear, moreover, that $QE(e_n)z = Q_0 E(e_n)z$ for each $n \geqq 1$ and each z in $\mathfrak{X}$.

Let $x_m \in \mathfrak{D}(Q)$, $\lim_{m\to\infty} x_m = x$, and $\lim_{m\to\infty} Qx_m = y$. Then

$$\begin{aligned} y &= \lim_{m\to\infty} Qx_m = \lim_{m\to\infty} \lim_{n\to\infty} Q_0 E(e_n)x_m \\ &= \lim_{m\to\infty} \lim_{n\to\infty} E(e_n)Qx_m \end{aligned}$$

by the argument in the preceding paragraph. Since the norms $|E(e_n)|$ are bounded in n (by the uniform boundedness theorem and Corollary IV.10.2), it is seen that $\lim_{m\to\infty} E(e_n)Qx_m = E(e_n)y$ uniformly in n. It follows by the Moore-Smith convergence theorem (I.7.6) and the boundedness of $Q_0 E(e_n)$ that

$$\begin{aligned} y &= \lim_{n\to\infty} \lim_{m\to\infty} E(e_n)Qx_m \\ &= \lim_{n\to\infty} \lim_{m\to\infty} Q_0 E(e_n)x_m \\ &= \lim_{n\to\infty} Q_0 E(e_n)x. \end{aligned}$$

Thus x is in $\mathfrak{D}(Q)$ and $y = Qx$. This shows that Q is closed.

Let $\{\tilde{e}_n\}$ be a second increasing sequence of elements of Σ_0 such that $E(\bigcup_{n=1}^{\infty} \tilde{e}_n) = I$, and let $\tilde{Q}$ be defined by the equations

$$\mathfrak{D}(\tilde{Q}) = \{x \mid \lim_{n\to\infty} Q_0 E(\tilde{e}_n)x \text{ exists}\},$$

$$\tilde{Q}x = \lim_{n\to\infty} Q_0 E(\tilde{e}_n)x, \qquad x \in \mathfrak{D}(\tilde{Q}).$$

To complete the proof of the theorem it will be shown that $Q = \tilde{Q}$. If x is in $\mathfrak{D}(\tilde{Q})$, then $\lim_{n\to\infty} E(\tilde{e}_n)x = x$ and by the argument of the second paragraph of the present proof,

$$\tilde{Q}x = \lim_{n\to\infty} Q_0 E(\tilde{e}_n)x = \lim_{n\to\infty} QE(\tilde{e}_n)x.$$

Since Q is closed, it follows that x is in $\mathfrak{D}(Q)$ and that $Qx = \tilde{Q}x$. Thus $\tilde{Q} \subseteq Q$. It follows from symmetry that $\tilde{Q} = Q$. Q.E.D.

7 COROLLARY. *Under the hypotheses of the preceding theorem it follows that* $E(e)\mathfrak{D}(Q) \subseteq \mathfrak{D}(Q)$ *and* $QE(e)x = E(e)Qx$ *for every* x *in* $\mathfrak{D}(Q)$ *and every* e *in* Σ.

PROOF. Let $\{e_n\}$ be as in the preceding proof, and let x be in $\mathfrak{D}(Q)$. Then $\lim_{n\to\infty} E(e)E(e_n)x = E(e)x$, and

$$\begin{aligned}\lim_{n\to\infty} QE(e_n)E(e)x &= \lim_{n\to\infty} QE(e_n e)x \\ &= \lim_{n\to\infty} E(e_n e)Qx = E(e)Qx\end{aligned}$$

by the second paragraph of the preceding proof. Thus, since Q is closed, it follows that $E(e)x$ is in $\mathfrak{D}(Q)$ and that $E(e)Qx = QE(e)x$. Q.E.D.

Functions of an unbounded spectral operator can now be defined. Let T be a closed spectral operator with resolution of the identity E and let f be a function analytic on an open set U such that $E(\sigma(T) - U) = 0$. Let e be a bounded Borel set whose closure is contained in U. Then by Lemma 2, $T \mid E(e)\mathfrak{X}$ is a bounded linear operator whose spectrum is contained in U. According to the functional calculus developed in Chapter XV.5, or even according to that defined in Chapter VII.3, $f(T \mid E(e)\mathfrak{X})$ may be defined.

Next it is noted that if x is in $E(e)\mathfrak{X} \cap E(\tilde{e})\mathfrak{X}$, then $f(T \mid E(e)\mathfrak{X})x = f(T \mid E(\tilde{e})\mathfrak{X})x$. Since $e \supseteq e\tilde{e} \subseteq \tilde{e}$ it is sufficient to establish this in the special case where $\tilde{e} \subseteq e$. This case follows readily from the definition (cf. VII.3.9) of an analytic function of a bounded operator by observing that

$$R(\lambda; T \mid E(\tilde{e})\mathfrak{X})x = R(\lambda; T \mid E(e)\mathfrak{X})x, \qquad \lambda \in \bar{e}, \qquad x \in E(\tilde{e})\mathfrak{X}.$$

By virtue of the equation $f(T \mid E(e)\mathfrak{X})x = f(T \mid E(\tilde{e})\mathfrak{X})x$ which has been established for x in $E(e)\mathfrak{X} \cap E(\tilde{e})\mathfrak{X}$, we may define a single valued linear operator Q_0 on $\bigcup_e E(e)\mathfrak{X}$, where e varies over the family of bounded Borel sets whose closures are in U, by the equation

$$Q_0 x = f(T \mid E(e)\mathfrak{X})x, \qquad x \in E(e)\mathfrak{X}.$$

Now, using the machinery established in Lemma 6, $f(T)$ may be defined as follows.

8 Definition. Let T be a spectral operator with resolution of the identity E, and let f be a function analytic in an open set U such that $E(U) = I$. Let $\{e_n\}$ be an arbitrary increasing sequence of bounded Borel sets with closures contained in U, such that $E(\bigcup_{n=1}^{\infty} e_n) = I$. The operator $f(T)$ is defined by the equations

$$\mathfrak{D}(f(T)) = \{x \mid \lim_{n \to \infty} f(T \mid E(e_n)\mathfrak{X})E(e_n)x \text{ exists}\},$$

$$f(T)x = \lim_{n \to \infty} f(T \mid E(e_n)\mathfrak{X})E(e_n)x, \qquad x \in \mathfrak{D}(f(T)).$$

9 Theorem. *The operator $f(T)$ of Definition 8 is closed, linear, and independent of the particular sequence of Borel sets used to define it.*

(i) *For each Borel set e and each x in $\mathfrak{D}(f(T))$, $E(e)\mathfrak{D}(f(T)) \subseteq \mathfrak{D}(f(T))$ and $E(e)f(T)x = f(T)E(e)x$.*

(ii) *If e is any Borel set, $f(T \mid E(e)\mathfrak{X}) = f(T) \mid E(e)\mathfrak{X}$. If, in particular, e is a bounded Borel set with closure contained in U, $f(T) \mid E(e)\mathfrak{X}$ is bounded.*

(iii) *If the set Z of zeros of f has spectral measure $E(Z) = 0$, then $f(T)$ has an inverse and $f(T)^{-1} = (1/f)(T)$.*

(iv) *In case f is a polynomial, Definition 8 agrees with Definition* VII.9.6.

(v) *If T is bounded and f is analytic on $\sigma(T)$, $f(T)$ is bounded and Definition 8 agrees with Definition* VII.3.9. *If f is analytic on $\sigma(T)$ and at infinity, $f(T)$ is bounded.*

Moreover, if g is a second function analytic in an open set V such that $E(V) = I$, then

(vi) $\mathfrak{D}(g(T) + f(T)) = \mathfrak{D}((g+f)(T)) \cap \mathfrak{D}(g(T))$ *and*
$g(T)x + f(T)x = (g+f)(T)x$ *for x in $\mathfrak{D}(g(T) + f(T))$.*

(vii) $\mathfrak{D}(g(T)f(T)) = \mathfrak{D}((gf)(T)) \cap \mathfrak{D}(f(T))$ *and*
$g(T)f(T)x = (gf)(T)x$ *for* x *in* $\mathfrak{D}(g(T)f(T))$.

(viii) *If* $\alpha \neq 0$, *then* $\mathfrak{D}((\alpha f)(T)) = \mathfrak{D}(f(T))$ *and*
$(\alpha f)(T)x = \alpha f(T)x$ *for* x *in* $\mathfrak{D}(f(T))$.

PROOF. The first statement follows from Definition 8 and the three paragraphs of explanation which precede it, and from Lemma 6. Statement (i) follows from Corollary 7. If e is a bounded Borel set with closure contained in U, it may be supposed, since $f(T)$ is independent of $\{e_n\}$, that $e = e_1$. If x is in $E(e)\mathfrak{X}$, then, by the paragraph preceding Definition 8, $f(T \mid E(e_n)\mathfrak{X})x = f(T \mid E(e)\mathfrak{X})x$ for $n \geqq 1$ so that $f(T)x = \lim_{n\to\infty} f(T \mid E(e_n)\mathfrak{X})x = f(T \mid E(e)\mathfrak{X})x$. This proves statement (ii) in case e is bounded and has closure contained in U.

To prove (ii) in the general case, let e be a Borel set, and note that by Lemma 2, $T \mid E(e)\mathfrak{X}$ is a spectral operator whose resolution of the identity F is the restriction of E to $E(e)\mathfrak{X}$. Clearly $F(U) = I$ and so $f(T \mid E(e)\mathfrak{X})$ is defined. Let $\{e_n\}$ be an increasing sequence of bounded Borel sets with closures contained in U and such that $E(\bigcup_{n=1}^{\infty} e_n) = I$. Then $F(\bigcup_{n=1}^{\infty} e_n) = I$ also. Let x be in $E(e)\mathfrak{X}$ and let x be in $\mathfrak{D}(f(T \mid E(e)\mathfrak{X}))$. Then by Definition 8, since (ii) has already been established for bounded Borel sets with closures contained in U,

$$\begin{aligned} f(T \mid E(e)\mathfrak{X})x &= \lim_{n\to\infty} f(T \mid F(e_n)E(e)\mathfrak{X})F(e_n)x \\ &= \lim_{n\to\infty} f(T \mid E(ee_n)\mathfrak{X})E(e_n)x \\ &= \lim_{n\to\infty} f(T)E(e_n)x. \end{aligned}$$

Since $E(e_n)x \to x$, and since $f(T)$ is closed, it follows that x is in $\mathfrak{D}(f(T))$ and that $f(T)x = f(T \mid E(e)\mathfrak{X})x$. Conversely, let x be in $E(e)\mathfrak{X}$ and x be in $\mathfrak{D}(f(T))$. Then since $E(e)f(T)x = f(T)E(e)x = f(T)x$, it follows that $f(T)x$ is in $E(e)\mathfrak{X}$. Hence, using (i), and the fact that (ii) has been established for bounded Borel sets with closures contained in U, it is seen that

$$\begin{aligned} f(T)x &= \lim_{n\to\infty} f(T)E(ee_n)x \\ &= \lim_{n\to\infty} f(T \mid E(ee_n)\mathfrak{X})E(e_n)x \\ &= \lim_{n\to\infty} f(T \mid F(e_n)E(e)\mathfrak{X})F(e_n)x. \end{aligned}$$

It follows from Definition 8 that x is in $\mathfrak{D}(f(T \mid E(e)\mathfrak{X}))$ and that $f(T \mid E(e)\mathfrak{X})x = f(T)x$. This proves (ii) in the general case.

Statement (vii) will next be proved. Let x be in $\mathfrak{D}(g(T)f(T))$. Since $E(\sigma(T) - (U \cap V)) = 0$ it may be assumed without loss of generality that $U = V$. Let $\{e_n\}$ be an increasing sequence of bounded Borel sets with closures contained in U such that $E(\bigcup_{n=1}^{\infty} e_n) = I$. Then, since $T \mid E(e_n)\mathfrak{X}$ is a bounded operator, we may apply the functional calculus of bounded operators (cf. VII.3.10) and conclude, using (i) and (ii), that

$$\begin{aligned}\lim_{n\to\infty} (gf)(T \mid E(e_n)\mathfrak{X})E(e_n)x &= \lim_{n\to\infty} g(T \mid E(e_n)\mathfrak{X})f(T \mid E(e_n)\mathfrak{X})E(e_n)x \\ &= \lim_{n\to\infty} g(T \mid E(e_n)\mathfrak{X})E(e_n)f(T)x \\ &= \lim_{n\to\infty} E(e_n)g(T)f(T)x \\ &= g(T)f(T)x.\end{aligned}$$

Thus it follows from Definition 8 that x is in $\mathfrak{D}(gf(T))$ and that $(gf)(T)x = g(T)f(T)x$.

Next let x be in $\mathfrak{D}(f(T)) \cap \mathfrak{D}((gf)(T))$. Then, using (i) and (ii), and what has already been established, it is seen that

$$\begin{aligned}\lim_{n\to\infty} g(T \mid E(e_n)\mathfrak{X})E(e_n)f(T)x &= \lim_{n\to\infty} g(T)E(e_n)f(T)x \\ &= \lim_{n\to\infty} g(T)f(T)E(e_n)x \\ &= \lim_{n\to\infty} (gf)(T)E(e_n)x \\ &= \lim_{n\to\infty} E(e_n)(gf)(T)x = (gf)(T)x,\end{aligned}$$

and it follows that x is in $\mathfrak{D}(g(T)f(T))$ and that $g(T)f(T)x = (gf)(T)x$. This proves (vii).

To prove (iii) we may argue as follows. Let $g = 1/f$. Then, by (vii),

$$\begin{aligned}&\mathfrak{D}(f(T)g(T)) = \mathfrak{D}(g(T)); \qquad \mathfrak{D}(g(T)f(T)) = \mathfrak{D}(f(T)); \\ &f(T)g(T)x = x, \qquad x \in \mathfrak{D}(g(T)); \\ &g(T)f(T)x = x, \qquad x \in \mathfrak{D}(f(T)).\end{aligned}$$

These equations show that $f(T)$ and $g(T)$ are one-to-one operators, and that the range of $f(T)$ is a subset of the domain of $g(T)$ and vice versa. If x is in $\mathfrak{D}(f(T))$, then $x = g(T)f(T)x$, so that x is in the range $\mathfrak{R}(g(T))$ of

$g(T)$. Thus $\mathfrak{R}(g(T)) = \mathfrak{D}(f(T))$. The equation $\mathfrak{R}(f(T)) = \mathfrak{D}(g(T))$ follows in the same way. This proves (iii).

The proof of (viii) is evident.

To prove (vi), let x be in $\mathfrak{D}(f(T) + g(T))$ and let $\{e_n\}$ be as above. Then, since $T \mid E(e_n)\mathfrak{X}$ is bounded, statements (i) and (ii) and the functional calculus of bounded operators (cf. VII.3.10) may be applied to conclude that

$$\begin{aligned}\lim_{n\to\infty} (f+g)(T \mid E(e_n)\mathfrak{X})E(e_n)x &= \lim_{n\to\infty} \{f(T \mid E(e_n)\mathfrak{X})E(e_n)x + g(T \mid E(e_n)\mathfrak{X})E(e_n)x\} \\ &= \lim_{n\to\infty} E(e_n)\{f(T)x + g(T)x\} \\ &= f(T)x + g(T)x.\end{aligned}$$

Thus it follows from Definition 8 that x is in $\mathfrak{D}((f+g)(T))$ and that $(f+g)(T)x = f(T)x + g(T)x$. This shows $\mathfrak{D}((f+g)(T)) \supseteq \mathfrak{D}(f(T) + g(T))$, and since $\mathfrak{D}(f(T)) \supseteq \mathfrak{D}(f(T) + g(T))$, it is seen that $\mathfrak{D}(f(T) + g(T)) \subseteq \mathfrak{D}((f+g)(T)) \cap \mathfrak{D}(f(T))$. Using (viii), the same argument shows that

$$\mathfrak{D}(g(T)) \supseteq \mathfrak{D}((f+g)(T) - f(T)) = \mathfrak{D}((f+g)(T)) \cap \mathfrak{D}(f(T)),$$

and since $\mathfrak{D}(f(T) + g(T)) = \mathfrak{D}(f(T)) \cap \mathfrak{D}(g(T))$, it is apparent that $\mathfrak{D}(f(T) + g(T)) \supseteq \mathfrak{D}((f+g)(T)) \cap \mathfrak{D}(f(T))$. This completes the proof of (vi).

Statement (iv) follows readily from (vi), (vii), and Definition VII.9.6.

Only statement (v) remains to be proved. In the first case described in statement (v), $\sigma(T)$ is a compact subset of U, and by Corollary XV.3.5, $E(\sigma(T)) = I$ so that without loss of generality each of the sets e_n, $n \geqq 1$, may be taken to be $\sigma(T)$. Then, by Definition 8, $f(T)$ is the same as $f(T \mid E(\sigma(T))\mathfrak{X})E(\sigma(T))$, and since $E(\sigma(T)) = I$, this is the same as the operator $f(T)$ of Definition VII.3.9. This proves the first part of (v).

To prove the second part of (v) it is sufficient, since $f(T)$ is closed, to show that $f(T)$ is everywhere defined (cf. the closed graph theorem (II.2.4)). By (ii), $\mathfrak{D}(f(T)) \supseteq E(e)\mathfrak{X}$, where e is any bounded Borel set. Consequently, it is sufficient to show that $\mathfrak{D}(f(T)) \supseteq E(b_r)\mathfrak{X}$, where r is any positive number and $b_r = \{z \mid |z| \geqq r\}$. Choose r so large that b_{r-2} is entirely contained in the domain of analyticity of f. It follows from (ii) that it suffices to show that $\mathfrak{D}(f(T \mid E(b_r)\mathfrak{X})) \supseteq E(b_r)\mathfrak{X}$. Since, by Lemma 2,

$T \mid E(b_r)\mathfrak{X}$ is itself a spectral operator whose spectral resolution F satisfies the equation $F(b_r) = I$, it is clear and will henceforth be assumed that $T = T \mid E(b_r)\mathfrak{X}$, and that $E(b_r) = I$.

Let C denote the circle $|z| = r - 1$, let $R > r$, and let C_R denote the circle $|z| = R$. Suppose that C_R is oriented in the positive sense, and C in the negative sense. Then (by VII.3.9)

$$[*] \qquad f(S) = \frac{1}{2\pi i}\left\{\int_C - \int_{C_R}\right\} f(\lambda)(\lambda I - S)^{-1}\, d\lambda$$

for each bounded operator S with spectrum contained between C and C_R. Noting that

$$(\lambda I - S)^{-1} = \sum_{n=0}^{\infty} \lambda^{-n-1} S^n.$$

for sufficiently large λ (cf. VII.3.4), it is clear that we may let $R \to \infty$ in formula [*] and obtain

$$f(S) = -f(\infty)I + \frac{1}{2\pi i}\int_C f(\lambda)(\lambda I - S)^{-1}\, d\lambda.$$

Let $\{e_n\}$ be an increasing sequence of compact subsets of b_r whose union is b_r. Then, using the above formula and (ii), it is seen that

$$\begin{aligned} f(T \mid E(e_n)\mathfrak{X}) &= -f(\infty)I \mid E(e_n)\mathfrak{X} \\ &\quad + \frac{1}{2\pi i}\int_C f(\lambda)\{(\lambda I - T)^{-1} \mid E(e_n)\mathfrak{X}\}\, d\lambda \\ &= \{-f(\infty)I + \frac{1}{2\pi i}\int_C f(\lambda)(\lambda I - T)^{-1}\, d\lambda\} \mid E(e_n)\mathfrak{X}. \end{aligned}$$

Since

$$-f(\infty)I + \frac{1}{2\pi i}\int_C f(\lambda)(\lambda I - T)^{-1}\, d\lambda$$

is clearly a bounded operator, this formula makes it obvious that the limit $\lim_{n\to\infty} f(T \mid E(e_n)\mathfrak{X})E(e_n)x$ exists for all x in $\mathfrak{X}$, so that (v) is proved. Q.E.D.

In Theorem 9 we have developed an operational calculus for an arbitrary spectral operator. The functions f to which an operator $f(T)$ was assigned were analytic. We shall now show that when the spectral operator is of scalar type, a much larger class of functions may be admitted. We begin by extending the very general functional calculus of XII.2.5 through XII.2.9 to arbitrary B-spaces.

10 DEFINITION. Let S be a set, Σ a σ-field of subsets of S, and E a strongly countably additive spectral measure defined on Σ. Let f be a Σ-measurable function defined E-almost everywhere on S. Then the operator $T(f)$ is defined by the equations

$$\mathfrak{D}(T(f)) = \{x \,|\, \lim_n T(f_n)x \text{ exists}\},$$

where

$$f_n(\lambda) = f(\lambda), \qquad |f(\lambda)| \leqq n; \qquad f_n(\lambda) = 0, \qquad |f(\lambda)| > n,$$

and

$$T(f_n) = \int_S f_n(s)E(ds); \qquad T(f)x = \lim_{n\to\infty} T(f_n)x, \qquad x \in \mathfrak{D}(T(f)).$$

→ 11 THEOREM. *Let (S, Σ, E), f, and $T(f)$ be as in Definition 10. Then the operator $T(f)$ is a closed operator with dense domain. Moreover,*

(a) $\mathfrak{D}(T(f)) = \mathfrak{D}(T(|f(\cdot)|))$;

(b) $\mathfrak{D}(T(f)) \subseteq \mathfrak{D}(T(g))$ *if* $|f(s)| \geqq |g(s)|$ *E-almost everywhere*;

(c) $T(f)$ *is bounded if and only if f is E-essentially bounded, and*

$$E\text{-ess}\sup_{s\in S} |f(s)| \leqq |T(f)| \leqq 4M\ E\text{-ess}\sup_{s\in S} |f(s)|;$$

(d) $T(\alpha f) = \alpha T(f)$;

(e) $T(f+g) \supseteqq T(f) + T(g)$ *and*

$$\mathfrak{D}(T(f)+T(g)) = \mathfrak{D}(T(f+g)) \cap \mathfrak{D}(T(f));$$

(f) $T(fg) \supseteqq T(f)T(g)$; $\mathfrak{D}(T(f)T(g)) = \mathfrak{D}(T(fg)) \cap \mathfrak{D}(T(g))$;

(g) $T(f)E(e) \supseteqq E(e)T(f)$ *for each e in Σ*;

(h) $T(f)$ *has an inverse if and only if* $E(f^{-1}(0)) = 0$, *in which case,* $T(f)^{-1} = T(1/f)$;

(i) $$x^*T(f)x = \int_S f(s)x^*E(ds)x, \qquad x \in \mathfrak{D}(T(f)),\ x^* \in \mathfrak{X}^*.$$

PROOF. To see that $T(f)$ is closed and densely defined we use Lemma 6. The family Σ_0 of that lemma will be taken as the family of all sets e in Σ upon which f is bounded. The operator Q_0 of Lemma 6, whose domain is $\bigcup_{e\in\Sigma_0} E(e)\mathfrak{X}$, is defined by the equation

$$Q_0 x = T(f\chi_e)x, \qquad x \in E(e)\mathfrak{X}, \qquad e \in \Sigma_0.$$

Since $f\chi_e$ is a bounded function, the operator $T(f\chi_e)$ is a bounded operator. If x is in $E(\tilde{e})\mathfrak{X}$ as well as in $E(e)\mathfrak{X}$, it follows from the operational calculus for bounded functions (cf. XVII.2.10) that

$$\begin{aligned} T(f\chi_e)x &= T(f\chi_e)E(\tilde{e})x = T(f\chi_{e\tilde{e}})x \\ &= T(f\chi_{\tilde{e}})E(e)x = T(f\chi_{\tilde{e}})x, \end{aligned}$$

so that Q_0 is well defined on $\bigcup_{e\in\Sigma_0} E(e)\mathfrak{X}$. It thus follows from Lemma 6 that $T(f)$ is a closed, densely defined operator. Moreover, statement (g) follows from Corollary 7.

Statement (d) is obvious. Letting $e \in \Sigma_0$ and $x \in E(e)\mathfrak{X}$, we have

$$\begin{aligned} T(f\chi_e)x &= \lim_{n\to\infty} T(f\chi_e)E(e_n)x = \lim_{n\to\infty} T(f\chi_e\chi_{e_n})x \\ &= \lim_{n\to\infty} T(f\chi_{e_n})x = T(f)x \end{aligned}$$

by the operational calculus for bounded functions (cf. XVII.2). Hence, for $x \in E(e)\mathfrak{X}$, we have $T(f\chi_e)x = T(f)E(e)x$. On the other hand, if $x \in E(S-e)\mathfrak{X}$, $T(f)E(e)x = 0$ and

$$T(f\chi_e)x = T(f\chi_e^2)x = T(f\chi_e)E(e)x = 0$$

by the operational calculus for bounded functions, so that $T(f\chi_e)x = T(f)E(e)x$ for $x \in E(S-e)$ also. Thus $T(f\chi_e) = T(f)E(e)x$ in all cases. If $T(f)$ is bounded, it follows that the family of operators $T(f\chi_e)$ is uniformly bounded and hence, by Theorem XVII.2.10, that the family of functions $f\chi_e$, $e \in \Sigma$, is uniformly E-essentially bounded. Thus, if $T(f)$ is bounded, f if E-essentially bounded. This proves part of (c); the rest follows immediately from Theorem XVII.2.10.

Statement (a) clearly will follow from statement (b). Statement (b) follows from statement (f); indeed, let $|f(s)| \geqq |g(s)|$, and put $h(s) = g(s)/f(s)$ if $f(s) \neq 0$, and $h(s) = 0$ otherwise. Then $|h(s)| \leqq 1$, so that by (c) the operator $T(h)$ is bounded. We have $g = hf$; since $\mathfrak{D}(T(h))$ is the whole space, it follows from (f) that $\mathfrak{D}(T(f)) = \mathfrak{D}(T(h)T(f)) = \mathfrak{D}(T(f)) \cap \mathfrak{D}(T(g))$, so that $\mathfrak{D}(T(f)) \supseteq \mathfrak{D}(T(g))$ and (b) follows.

We next prove (f). Let $e_n = \{s \mid |f(s)| \leqq n \text{ and } |g(s)| \leqq n\}$. The first paragraph of the present proof and Lemma 6 show that

$$\mathfrak{D}(T(g)) = \{x \mid \lim_{n\to\infty} T(g\chi_{e_n})x \text{ exists}\},$$

$$T(g)x = \lim_{n\to\infty} T(g\chi_{e_n})x, \qquad x \in \mathfrak{D}(T(g)),$$

where $T(g\chi_{e_n})$ is a *bounded* operator defined by the functional calculus of bounded functions (cf. XVII.2.10). Exactly similar statements may be made about $T(gf)$ and $T(f)$. Consequently, if x is in $\mathfrak{D}(T(g)T(f))$, we have (by XVII.2.10)

$$\lim_{n\to\infty} T(gf\chi_{e_n})x = \lim_{n\to\infty} T(g\chi_{e_n})T(f\chi_{e_n})E(e_n)x.$$

Now, by (g), $T(f)E(e_n)z = \lim_{m\to\infty} T(f\chi_{e_m}\chi_{e_n})E(e_n)z = T(f\chi_{e_n})E(e_n)z$ for $z \in \mathfrak{D}(T(f))$. Similarly, $T(g)E(e_n)z = T(g\chi_{e_n})E(e_n)z$ for $z \in \mathfrak{D}(T(g))$. Thus we find by using (g) that

$$\begin{aligned}\lim_{n\to\infty} T(gf\chi_{e_n})x &= \lim_{n\to\infty} T(g\chi_{e_n})T(f)E(e_n)x\\ &= \lim_{n\to\infty} T(g)E(e_n)T(f)x\\ &= \lim_{n\to\infty} E(e_n)T(g)T(f)x = T(g)T(f)x.\end{aligned}$$

This shows that x is in $\mathfrak{D}(T(gf))$ and that $T(gf)x = T(g)T(f)x$. Conversely, let x be in $\mathfrak{D}(T(gf)) \cap \mathfrak{D}(T(f))$. Then, using (g), we have

$$\begin{aligned}\lim_{n\to\infty} T(g\chi_{e_n})T(f)x &= \lim_{n\to\infty} T(g\chi_{e_n})E(e_n)T(f)x\\ &= \lim_{n\to\infty} T(g\chi_{e_n})T(f)E(e_n)x\\ &= \lim_{n\to\infty} T(g\chi_{e_n})T(f\chi_{e_n})x\\ &= \lim_{n\to\infty} T(gf\chi_{e_n})x = T(gf)x.\end{aligned}$$

Thus $T(f)x$ is in $\mathfrak{D}(T(g))$, and $T(g)T(f)x = T(gf)x$. This proves (f).

Statement (e) may be proved in the same way, as follows. Let $\{e_n\}$ be as in the preceding paragraph. Let x be in $\mathfrak{D}(T(f) + T(g))$. Then, by XVII.2.10,

$$\begin{aligned}\lim_{n\to\infty} T((f+g)\chi_{e_n})x &= \lim_{n\to\infty} T(f\chi_{e_n})x + T(g\chi_{e_n})x\\ &= T(f)x + T(g)x,\end{aligned}$$

so that x is in $\mathfrak{D}(T(f+g))$ and $T(f+g)x = T(f)x + T(g)x$. Conversely, if x is in $\mathfrak{D}(T(f+g)) \cap \mathfrak{D}(T(g))$, then

$$\begin{aligned}\lim_{n\to\infty} T(g\chi_{e_n})x &= \lim_{n\to\infty} T((f+g)\chi_{e_n} - f\chi_{e_n})x\\ &= \lim_{n\to\infty} T((f+g)\chi_{e_n})x - T(f\chi_{e_n})x\\ &= T(f+g)x - T(f)x,\end{aligned}$$

so that x is in $\mathfrak{D}(T(g))$ and hence in $\mathfrak{D}(T(f)+T(g))=\mathfrak{D}(T(f))\cap\mathfrak{D}(T(g))$, and (e) is proved.

If $E(f^{-1}(0))\neq 0$, there is an $x\neq 0$ such that $E(f^{-1}(0))x=x$. It is then clear from Definition 10 that x is in $\mathfrak{D}(T(f))$ and that $T(f)x=0$; consequently $T(f)$ has no inverse. Conversely, if $E(f^{-1}(0))=0$, so that $1/f$ is defined E-almost everywhere and Σ-measurable, then by (f) we have

$$\mathfrak{D}(T(f)T(1/f))=\mathfrak{D}(T(1/f));\ \mathfrak{D}(T(1/f)T(f))=\mathfrak{D}(T(f));$$
$$T(f)T(1/f)x=x, \qquad x\in\mathfrak{D}(T(1/f));$$
$$T(1/f)T(f)x=x, \qquad x\in\mathfrak{D}(T(f)).$$

These equations show that $T(f)$ and $T(1/f)$ are one-to-one operators and that the range of $T(f)$ is a subset of the domain of $T(1/f)$ and vice versa. If x is in $\mathfrak{D}(T(f))$, then $x=T(1/f)T(f)x$, and so x is in the range $\mathfrak{R}(T(1/f))$ of $T(1/f)$. Thus $\mathfrak{R}(T(1/f))=\mathfrak{D}(T(f))$. The equation $\mathfrak{R}(T(f))=\mathfrak{D}(T(1/f))$ follows the same way. This proves (h).

To prove (i), let x be in $\mathfrak{D}(T(f))$, x^* in $\mathfrak{X}^*$, and let M be an upper bound for the norms $|E(e)|$ with e in Σ. Then, for each subset e of e_n,

$$\begin{aligned}\left|\int_e f(s)x^*E(ds)x\right| &= |x^*E(e)T(f_n)x|\\ &= |x^*E(e)T(f)x| \leqq M|x^*|\,|T(f)x|.\end{aligned}$$

Thus, by III.2.20(a) and III.1.5,

$$\int_{e_n}|f(s)|v(x^*E(\cdot)x, ds)\leqq 4M|x^*|\,|T(f)x|.$$

It follows by letting $n\to\infty$ that f is $x^*E(\cdot)x$-integrable. Then, by the Lebesgue dominated convergence theorem (III.6.16), it follows, since f_n is bounded, that

$$\begin{aligned}\int_S f(s)x^*E(ds)x &= \lim_{n\to\infty}\int_{e_n} f(s)x^*E(ds)x\\ &= \lim_{n\to\infty} x^*\left\{\int_{e_n} f(s)E(ds)\right\}x\\ &= \lim_{n\to\infty} x^*T(f_n)x = x^*T(f)x. \qquad \text{Q.E.D.}\end{aligned}$$

12 Definition. Let T be a bounded or unbounded linear operator in a B-space $\mathfrak{X}$. If there exists a countably additive resolution of the identity E defined on the field of Borel subsets of the plane such that

$T = T(f)$ in the sense of Definition 10, where $f(z) = z$, then T is said to be an unbounded *spectral operator of scalar type*. The projection valued measure E is said to be *the resolution of the identity for* T.

13 LEMMA. *An unbounded spectral operator of scalar type in the sense of Definition* 12 *is a spectral operator in the sense of Definition* 1. *Moreover, the resolution of the identity for T in the sense of Definition* 12 *is the same as the resolution of the identity for T in the sense of Definition* 1.

PROOF. Using the notations of Definitions 12 and 1, let T and E have all the properties of Definition 12. It is clear from Definition 10 that $\mathfrak{D}(T) \supseteq E(e)\mathfrak{X}$ for each bounded Borel set e. By Theorem 11(g), $E(e)\mathfrak{D}(T) \subseteq \mathfrak{D}(T)$ and $TE(e)x = E(e)Tx$ for each vector x in $\mathfrak{D}(T)$.

Let e be a given Borel set, and $\lambda \notin \bar{e}$. Let $f(z) = (\lambda - z)^{-1}$ for $z \in e$ and $f(z) = 0$ for $z \notin e$. Then, by Theorem 11, $T(f)$ is an everywhere defined bounded operator, $T(f)(\lambda I - T)x = E(e)x$ for $x \in \mathfrak{D}(T)$, and $(\lambda I - T)T(f)x = E(e)x$ for all x. Moreover, $E(e)T(f) = T(f)E(e)$. It follows that the restriction of $T(f)$ to the space $E(e)\mathfrak{X}$ is the inverse of the restriction $(\lambda I - T) \mid E(e)\mathfrak{X}$ of Definition 1. Thus T and E have all the properties of Definition 1. Q.E.D.

14 COROLLARY. *The resolution of the identity of a scalar type spectral operator is unique.*

PROOF. This follows from Lemma 13 and Theorem 5. Q.E.D.

15 DEFINITION. If f is a complex valued Borel measurable function of a complex variable, and T is a scalar type spectral operator, then by $f(T)$ we shall understand the operator $T(f)$ of Definition 10, the resolution of the identity in Definition 10 being taken to be the resolution of the identity for T.

Note that it follows from Corollary 14 that this definition is unique. To show that no contradiction arises between Definition 15 and Definition 8, we introduce the following lemma.

16 LEMMA. *Let T be a scalar type spectral operator, E its resolution of the identity, and f a function analytic in an open set U such that $E(U) = I$. Definitions* 15 *and* 8 *of $f(T)$ agree.*

PROOF. Let $f_1(T)$ and $f_2(T)$ be the values assigned to the symbol $f(T)$ by Definitions 8 and 15, respectively. Let $\{e_n\}$ be an increasing sequence of compact subsets of U such that $\bigcup_{n=1}^{\infty} e_n = U$. Since, by Theorem 11,

$$Tx = \int_{e_n} \lambda E(d\lambda)x, \qquad x \in E(e_n)\mathfrak{X},$$

it follows from Definition 12 that $T \mid E(e_n)\mathfrak{X}$ is a bounded scalar type spectral operator whose resolution of the identity F is defined by the formula $F(e) = E(ee_n) \mid E(e_n)\mathfrak{X}$. Since $\sigma(T \mid E(e_n)\mathfrak{X}) \subseteq e_n$, and f is analytic on e_n, it follows from Theorem XV.5.1 that $f(T \mid E(e_n)\mathfrak{X}) = \int_{e_n} f(\lambda)F(d\lambda)$, so that

$$f(T \mid E(e_n)\mathfrak{X})E(e_n)x = \int_{e_n} f(\lambda)E(d\lambda)x, \qquad x \in \mathfrak{X}.$$

It follows by Theorem 11 that

$$f(T \mid E(e_n)\mathfrak{X})E(e_n)x = f_2(T)E(e_n)x.$$

Now let x be in $\mathfrak{D}(f_1(T))$. Then it follows from Definition 8 and the above formula that

$$f_1(T)x = \lim_{n \to \infty} f_2(T)E(e_n)x.$$

Since $\lim_{n \to \infty} E(e_n)x = x$, and since, by Theorem 11, $f_2(T)$ is closed, it follows that x is in $\mathfrak{D}(f_2(T))$ and that $f_1(T)x = f_2(T)x$. This shows that $f_1(T) \subseteq f_2(T)$.

Let $\sigma_n \{\lambda \mid |f_n(\lambda)| \leq n\}$. It will be shown that $\mathfrak{D}(f_1(T)) \supseteq E(\sigma_m)\mathfrak{X}$. Indeed, if x is in $E(\sigma_m)\mathfrak{X}$, then, by Theorem 11 and what has just been proved,

$$\begin{aligned} f(T \mid E(e_n)\mathfrak{X})E(e_n)x &= \int_{e_n} f(\lambda)E(d\lambda)x \\ &= E(e_n) \int_{\sigma_m} f(\lambda)E(d\lambda)x \\ &= E(e_n) f_2(T)x. \end{aligned}$$

Consequently, the limit $f_1(T)x = \lim_{n \to \infty} f(T \mid E(e_n)\mathfrak{X})E(e_n)x$ exists and equals $f_2(T)x$. Thus, by Theorem 11, for each x in $\mathfrak{D}(f_2(T))$,

$$f_1(T)E(\sigma_m)x = f_2(T)E(\sigma_m)x = E(\sigma_m)f_2(T)x.$$

Consequently, $\lim_{m \to \infty} f_1(T)E(\sigma_m)x = f_2(T)x$. Since, by Theorem 9, $\lim_{m \to \infty} E(\sigma_m)x = x$, and $f_1(T)$ is closed, it follows that x is $\mathfrak{D}(f_1(T))$ and that $f_1(T)x = f_2(T)x$. This shows that $f_2(T) \subseteq f_1(T)$. Q.E.D.

17 THEOREM. *Let S be a set, Σ a σ-field of subsets of S, and E a strongly countably additive resolution of the identity defined on Σ. Let f be a Σ-measurable function defined on S. Then the operator $T(f)$ of Definition* 10 *is a scalar type spectral operator, whose resolution of the identity E_1 is given by the formula*

$$E_1(e) = E(f^{-1}(e)).$$

PROOF. Let $f_n(s) = f(s)$ if $|f(s)| \leqq n$ and $f_n(s) = 0$ if $|f(s)| > n$. Then

$$\mathfrak{D}(T(f)) = \{x \mid \lim_{n\to\infty} T(f_n)x \text{ exists}\},$$

$$T(f)x = \lim_{n\to\infty} T(f_n)x, \qquad x \in \mathfrak{D}(T).$$

According to Corollary 14, Lemma 13, and Definition 12, it will suffice to show that

$$\mathfrak{D}(T(f)) = \{x \mid \lim_{n\to\infty} \int_{|\lambda|\leqq n} \lambda E_1(d\lambda)x \text{ exists}\},$$

$$T(f)x = \lim_{n\to\infty} \int_{|\lambda|\leqq n} \lambda E_1(d\lambda)x, \qquad x \in \mathfrak{D}(T(f)).$$

To see this we note that, by Lemma XVII.2.9,

$$T(f_n) = \int_{|\lambda|\leqq n} \lambda E_1(d\lambda),$$

from which the desired conclusion follows. Q.E.D.

18 THEOREM. *Let E be a countably additive spectral measure on the sigma field Σ of subsets of a set S. Let $g, f_1, f_2, \ldots,$ be a pointwise convergent sequence of Σ-measurable functions with*

$$|f_n(s)| \leqq |g(s)|, \qquad s \in S,$$

$$f(s) = \lim_{n\to\infty} f_n(s), \qquad s \in S,$$

and let x be in $\mathfrak{D}(T(g))$. Then x is in each of the domains $\mathfrak{D}(T(f))$, $\mathfrak{D}(T(f_n))$, $n \geqq 1$, and

$$T(f)x = \lim_{n\to\infty} T(f_n)x.$$

PROOF. The first conclusion follows immediately from Theorem 11(b). Let $h_n(s) = f_n(s)/g(s)$ if $g(s) \neq 0$, $h_n(s) = 0$ if $g(s) = 0$; let $h(s) = f(s)/g(s)$ if $g(s) \neq 0$, $h(s) = 0$ if $g(s) = 0$. Then $|h_n(s)| \leqq 1$, $|h(s)| \leqq 1$, $h_n g = f_n$, and

$hg = f$. It follows by Theorem 11(f) that $T(f_n)x = T(h_n)T(g)x$, $T(f)x = T(h)T(g)x$. Hence it suffices to show that $T(h_n)$ converges to $T(h)$ in the strong operator topology. This follows immediately from Theorem IV.10.10. Q.E.D.

It would be valuable to extend Theorem 17 from scalar type spectral operators to arbitrary spectral operators. Unfortunately, no very literal extension can be given, as is shown by the following example of a spectral operator T such that $\sin \pi T$ is not a spectral operator. For $n \geqq 1$, let $\mathfrak{H}_{(n)}$ be an n-dimensional unitary space with orthonormal basis $(x_1^{(n)}, \ldots, x_n^{(n)})$. Let $A_{(n)}x = nx$ for $x \in \mathfrak{H}_n$, and let $B_{(n)}$ be defined by $B_{(n)}x_i^{(n)} = 2^{-1}nx_{i-1}^{(n)}$, $1 < i \leqq n$, $B_{(n)}x_1^{(n)} = 0$. Let $\mathfrak{H}$ be the direct sum $\sum_{n=1}^{\infty} \mathfrak{H}_{(n)}$. Let A and B be the operators in $\mathfrak{H}$ defined by

$$\mathfrak{D}(A) = \{x^{(1)} \oplus x^{(2)} \oplus \cdots \mid \textstyle\sum_{n=1}^{\infty} |A_{(n)}x^{(n)}|^2 < \infty\},$$
$$\mathfrak{D}(B) = \{x^{(1)} \oplus x^{(2)} \oplus \cdots \mid \textstyle\sum_{n=1}^{\infty} |B_{(n)}x^{(n)}|^2 < \infty\},$$
$$A(x) = A(x^{(1)} \oplus x^{(2)} \oplus \cdots) = A_{(1)}x^{(1)} \oplus A_{(2)}x^{(2)} \oplus \cdots, \qquad x \in \mathfrak{D}(A),$$
$$B(x) = B(x^{(1)} \oplus x^{(2)} \oplus \cdots) = B_{(1)}x^{(1)} \oplus B_{(2)}x^{(2)} \oplus \cdots, \qquad x \in \mathfrak{D}(B),$$

respectively. It is easily seen that A and B are closed operators, and it is clear that $\mathfrak{D}(A) = \mathfrak{D}(B)$. Thus, if $C = A + B$ then $\mathfrak{D}(C) = \mathfrak{D}(A)$. Since the operator C will give service in all of the counterexamples to be developed in the rest of this section, it will be examined rather carefully.

First it will be shown that $\sigma(C)$ is the set of positive integers. Suppose that λ is not a positive integer. Then, if $I_{(n)}$ is the identity mapping in $\mathfrak{H}_{(n)}$, it follows, since $B_{(n)}$ is nilpotent, that the operator

$$(\lambda I_{(n)} - (A_{(n)} + B_{(n)}))^{-1} = ((\lambda - n)I_{(n)} - B_{(n)})^{-1}$$

exists. Moreover,

$$|((\lambda - n)I_{(n)} - B_{(n)})^{-1}| = \frac{1}{n}\left|\left(\left(1 - \frac{\lambda}{n}\right)I_{(n)} + \frac{1}{n}B_n\right)^{-1}\right|,$$

and, since $|n^{-1}B_n| = \frac{1}{2}$,

$$\limsup_{n \to \infty} n\,|(\lambda I_{(n)} - (A_{(n)} + B_{(n)}))|^{-1} \leqq 2.$$

Thus, if we let $R(\lambda)$ be defined by the equation

$$R(\lambda)(x^{(1)} \oplus x^{(2)} \oplus \cdots)$$
$$= R^{(1)}(\lambda; A_{(1)} + B_{(1)})x^{(1)} \oplus R^{(2)}(\lambda; A_{(2)} + B_{(2)})x^{(2)} \oplus \cdots,$$

it follows that $R(\lambda)$ is a bounded operator whose range is contained in the domain of C. It is clear then that $(\lambda I - C)R(\lambda)x = x$ for x in $\mathfrak{H}$ and $R(\lambda)(\lambda I - C)x = x$ for x in $\mathfrak{D}(C)$, so that $R(\lambda) = R(\lambda; C)$ and $\lambda \notin \sigma(C)$. On the other hand, if $\lambda = n$ and x is in $\mathfrak{H}_{(n)}$, then $(\lambda I - C)^n x = B_{(n)}^n x = 0$, so that λ is in $\sigma(C)$. Thus $\sigma(C)$ is the set of all positive integers. It may be shown in precisely the same way that if J is a subset of the class of positive integers, and $\mathfrak{H}_J = \sum_{n \in J} \oplus \mathfrak{H}_{(n)}$, then the spectrum of the restriction $C \mid \mathfrak{H}_J$ is J. Since $(iI - C)^{-1}$ is bounded, $iI - C$ is closed. Thus C is closed.

Let $P_{(n)}$ denote the orthogonal projection of $\mathfrak{H}$ on $\mathfrak{H}_{(n)}$. For each Borel set e, let $E(e) = \sum_{n \in e} P_{(n)}$. Then E is a strongly countably additive resolution of the identity. If J is the set of integers in e, it is clear that $E(e)\mathfrak{H} = \mathfrak{H}_J$. Thus $\sigma(C \mid E(e)\mathfrak{H}) = J \subseteq \bar{e}$, so that Definition 1(iii) is verified. Since parts (i) and (ii) of Definition 1 are obvious, C is a spectral operator with spectral resolution E.

The function $f(z) = \sin \pi z$ is entire. It will be shown that $\sin \pi C$ is not a spectral operator. Suppose that it were, and let F be its resolution of the identity. It follows from Corollary 4 and Theorem 9(i) that F commutes with E. By Theorem 9(ii), the restriction $(\sin \pi C) \mid P_{(n)}\mathfrak{H}$ is the same as

$$\begin{aligned}\sin(\pi C \mid P_{(n)}\mathfrak{H}) &= \sin \pi(A_{(n)} + B_{(n)}) = \sin \pi(nI_{(n)} + B_{(n)}) \\ &= (-1)^n \sin \pi B_{(n)} = B_{(n)} h(B_{(n)}),\end{aligned}$$

where $h(z) = \pm z^{-1} \sin \pi z$. Since $B_{(n)}^n = 0$, it follows that $(\sin \pi C)^n P_{(n)}\mathfrak{H} = \{0\}$. Let e be a closed set in the complex plane not containing zero. Then, by Theorem 9,

$$\begin{aligned}((\sin \pi C)F(e))^n P_{(n)}\mathfrak{H} &= (\sin \pi C)^n F(e) P_{(n)}\mathfrak{H} \\ &= (\sin \pi C)^n P_{(n)} F(e)\mathfrak{H} = \{0\}.\end{aligned}$$

Since, by Definition 1, $(\sin \pi C) \mid F(e)\mathfrak{H}$ is a one-to-one map, it follows that $P_{(n)} F(e) = 0$. Summing over n, it follows that $F(e) = 0$. Consequently, $F(\{0\}) = I$. It follows from Definition 1 that $\sigma(\sin \pi C) = 0$. This, however, may be proved false as follows. It will be shown that

$$\lim_{n \to \infty} |(\lambda I_{(n)} - (-1)^n \sin \pi B_{(n)})^{-1}| \to \infty, \qquad \lambda \neq 0,$$

from which it will follow that there exists a sequence of elements $y_{(n)}$ in $\mathfrak{H}_{(n)}$ such that $|y_{(n)}| = 1$ and $(\lambda I - \sin \pi C)y_{(n)} \to 0$, so that $\sigma(\sin \pi C)$ covers the whole complex plane. To show this, note that since $(\sin \pi B_{(n)})x_1^{(n)} = 0$

and $(\sin \pi B_{(n)})x_2^{(n)} = (\pi/2n)x_1^{(n)}$, we have

$$(\lambda I - (-1)^n \sin \pi B_{(n)})^{-1}x_2^{(n)} = \lambda^{-1}(x_2^{(n)} + (-1)^n \frac{n\pi}{2\lambda} x_1^{(n)}),$$

from which the statement $|\lambda I - (-1)^n\pi \sin B_{(n)})^{-1}| \to \infty$ is obvious. This concludes the proof that $\sin \pi C$ is not a spectral operator.

We note for future use that it may be shown in exactly the same way that $\sigma(B)$ covers the whole complex plane. Moreover, BA^{-1} is evidently a bounded operator of norm $1/2$, whose nth power has norm 2^{-n}.

19 LEMMA. *Let T be a spectral operator, and let E be its resolution of the identity. Let e be an open set. Then*

$$\sigma(T) \cap e \subseteq \sigma(T \,|\, E(e)\mathfrak{X}) \subseteq \sigma(T) \cap \bar{e}.$$

PROOF. By Definition 1, $\sigma(T \,|\, E(e)\mathfrak{X}) \subseteq \bar{e}$. If $\lambda \notin \sigma(T)$, so that $\lambda I - T$ has the bounded inverse R, then, letting R_0 be the restriction of $E(e)R$ to $E(e)\mathfrak{X}$,

$$R_0(\lambda I - T)x = E(e)R(\lambda I - T)x = E(e)x = x, \qquad x \in \mathfrak{D}(T \,|\, E(e)\mathfrak{X}),$$

and

$$(\lambda I - T)R_0 x = E(e)(\lambda I - T)Rx = E(e)x = x, \qquad x \in E(e)\mathfrak{X},$$

by Definition 1. Thus $\lambda \notin \sigma(T \,|\, E(e)\mathfrak{X})$, which shows that $\sigma(T \,|\, E(e)\mathfrak{X}) \subseteq \sigma(T) \cap \bar{e}$.

Conversely, let $\lambda \in e$, and suppose that $\lambda \notin \sigma(T \,|\, E(e)\mathfrak{X})$; then there exists a bounded operator $R_0 = (\lambda I - T \,|\, E(e)\mathfrak{X})^{-1}$. Let e' be the complement of e. Since $\lambda \notin e'$, it follows from Definition 1 that there exists a bounded operator $R_1 = (\lambda I - T \,|\, E(e')\mathfrak{X})^{-1}$. If we let $Rx = R_0 E(e)x + R_1 E(e')x$ for x in $\mathfrak{X}$, it is evident that $R = R(\lambda; T)$, so that $\lambda \notin \sigma(T)$. Thus $e \cap \sigma(T) \subseteq \sigma(T \,|\, E(e)\mathfrak{X})$. Q.E.D.

20 COROLLARY. *Let T satisfy the hypotheses of Lemma 19, and let E be a projection such that $E\mathfrak{D}(T) \subseteq \mathfrak{D}(T)$ and $ETx = TEx$ for x in $\mathfrak{D}(T)$. Then $\sigma(T) \supseteq \sigma(T \,|\, E\mathfrak{X})$.*

PROOF. Only these two properties of $E(e)$ were used in the second paragraph of the preceding proof. Q.E.D.

The following theorem gives sufficient conditions in order that a function of a general spectral operator shall be a spectral operator.

→ 21 THEOREM. *Let T be a spectral operator, and E its resolution of identity. Let f be a function analytic in a domain U which, when taken together with a finite number of exceptional points p, includes a neighborhood of $\sigma(T)$ and a neighborhood of the point at infinity. Suppose that each exceptional point p satisfies $E(p)=0$, and that f has at most a pole at each of the exceptional points p and at infinity. Then $f(T)$ is a spectral operator whose resolution of the identity E_1 is given by the formula $E_1(e)=E(f^{-1}(e))$ and whose spectrum is determined by the formula $\sigma(f(T))=\overline{f(\sigma(T))}$.*

PROOF. Let E_1 be defined by the equation $E_1(e)=E(f^{-1}(e))$. It follows immediately from Theorem 9 that $E_1(e)\mathfrak{D}(f(T)) \subseteq \mathfrak{D}(f(T))$ and that $f(T)E_1(e)x = E_1(e)f(T)x$ for x in $\mathfrak{D}(f(T))$.

Let e be a bounded Borel set. We wish first to verify the condition (i) of Definition 1, that is, to show that $\mathfrak{D}(f(T)) \supseteq E_1(e)\mathfrak{X}$. If f is not analytic at infinity, then $e_1 = f^{-1}(e)$ is bounded, and it follows from Theorem 9(ii) that $\mathfrak{D}(f(T)) \supseteq E(e_1)\mathfrak{X} = E_1(e)\mathfrak{X}$. Suppose now that f is analytic at infinity. Then it is clear that f is analytic in a neighborhood of $\bar{e}_1 \cap \sigma(T)$. Since $\sigma(T \mid E_1(e)\mathfrak{X}) = \sigma(T \mid E(e_1)\mathfrak{X}) \subseteq \bar{e}_1 \cap \sigma(T)$ by Lemma 19, this means that f is analytic in a neighborhood of $\sigma(T \mid E_1(e)\mathfrak{X})$ and at infinity. Thus, by Theorem 9(v), $\mathfrak{D}(f(T \mid E_1(e)\mathfrak{X})) = \mathfrak{D}(f(T \mid E(e_1)\mathfrak{X})) \supseteq E(e_1)\mathfrak{X}$. By Theorem 9(ii) this means that $\mathfrak{D}(f(T)) \supseteq E_1(e)\mathfrak{X}$.

Consequently, in order to prove the first two assertions of the theorem it need only be shown that, for every Borel set e, $\sigma(f(T) \mid E_1(e)\mathfrak{X}) \subseteq \bar{e}$ or, in view of Theorem 9(ii), that $\sigma(f(T \mid E(e_1)\mathfrak{X})) \subseteq \overline{f(e_1)}$. Since $T \mid E(e_1)\mathfrak{X}$ is a spectral operator with spectrum contained in $\bar{e}_1$, it is enough to show that $\sigma(f(A)) \subseteq \overline{f(\sigma(A))}$ for each spectral operator A and for each f in the class of functions defined by the hypotheses of the theorem. Let $\lambda \notin \overline{f(\sigma(A))}$; it remains to show that $\lambda \notin \sigma(f(A))$. It may be assumed, without loss of generality, that $\lambda = 0$, in which case it is necessary to show that $f(A)$ has a bounded inverse.

By Lemma 3, $E(A, \sigma(A)) = I$. Hence, if $\sigma(A)$ is bounded, then it follows by Lemma 2 that A is bounded. Since $1/f$ is bounded on $\sigma(A)$ and is known to have only poles as singularities in $\sigma(A)$, it follows that $1/f$ is analytic on $\sigma(A)$. Thus, by the operational calculus for bounded operators (cf. VII.3.10(b)), $f(A)$ has the inverse $(1/f)(A)$.

Next consider the case in which $\sigma(A)$ is unbounded. By Theorem 9(v), $f(A)$ has the inverse $(1/f)(A)$. Since $0 \notin \overline{f(\sigma(A))}$ by hypothesis, $1/f$ is bounded on $\sigma(A)$. Thus, since $1/f$ has at most a pole at each point of $\sigma(A)$ and at infinity, and since $\sigma(A)$ is unbounded, $1/f$ is analytic on $\sigma(A)$ and at infinity, and it follows immediately from Theorem 9(v) that $(1/f)(A)$ is bounded. This proves the first two assertions of the present theorem, and also the inclusion $\sigma(f(T)) \subseteq \overline{f(\sigma(T))}$, which is part of the final assertion of the present theorem.

Since $\sigma(f(T))$ is closed, to prove the equality $\sigma(f(T)) = \overline{f(\sigma(T))}$, it is sufficient to prove that $\sigma(f(T)) \supseteq f(\sigma(T))$. Thus let λ be in $f(\sigma(T))$. We may and shall assume that $\lambda = 0$. Let Z be the set of points in $\sigma(T)$ where f assumes the value zero. Since f is analytic at every point of $\sigma(T)$ and at infinity except for possible poles, Z is a finite, and hence a bounded, set. Let e be a neighborhood of Z with compact closure contained in the domain of analyticity of f. Then, by Theorem 9(ii) and Lemma 19, $T \mid E(e)\mathfrak{X}$ is a bounded operator and $\sigma(T \mid E(e)\mathfrak{X}) \supseteq Z$. By the spectral mapping theorem for bounded operators (Theorem VII.3.11) and by Theorem 9(ii), Theorem 9(i), and Corollary 20, it is seen that

$$0 \in \sigma(f(T \mid E(e)\mathfrak{X})) = \sigma(f(T) \mid E(e)\mathfrak{X}) \subseteq \sigma(f(T)). \qquad \text{Q.E.D.}$$

22 Corollary. *A polynomial function of a closed spectral operator is a closed spectral operator.*

23 Corollary. *Let λ be in the resolvent set of the closed operator T. Then T is a spectral operator if and only if $(\lambda I - T)^{-1}$ is a spectral operator, whose spectral measure $E_1(\cdot)$ satisfies the equation $E_1(\{0\}) = 0$.*

Proof. Let T be a spectral operator. Then, by Theorem 9 and Theorem 21, $(\lambda I - T)^{-1}$ is a spectral operator whose spectral measure $E_1(\cdot)$ satisfies the equations

$$E_1(\{0\}) = E(\{z \mid (\lambda - z)^{-1} = 0\}) = 0.$$

Conversely, let $(\lambda I - T)^{-1}$ be a spectral operator whose spectral measure satisfies the equation $E_1(\{0\}) = 0$. Then, by Theorem 9 and Theorem 21, $(\lambda I - T) = ((\lambda I - T)^{-1})^{-1}$ is a spectral operator, and hence T is a spectral operator. Q.E.D.

→ 24 Theorem. *Let T and f satisfy the hypotheses of Theorem 21. Let E_1 be the spectral resolution of the spectral operator $f(T)$, and let g be a function analytic in an open set U such that $E_1(U) = I$. Then, if $g(f(z)) = h(z)$,*

$$g(f(T)) = h(T).$$

PROOF. Let E be the spectral resolution of T. Then h is analytic in $f^{-1}(U)$, and since $E(f^{-1}(U)) = E_1(U)$ by Theorem 21, $h(T)$ is a well-defined closed operator.

If f is analytic at infinity, let $q = \{f(\infty)\}$; if f is not analytic at infinity, let q be the null set. Let q' be the open set $U - q$. The conclusion of the theorem will be divided into two parts.

(a) $$g(f(T)) \mid E_1(q)\mathfrak{X} = h(T) \mid E_1(q)\mathfrak{X}.$$

(b) $$g(f(T)) \mid E_1(q')\mathfrak{X} = h(T) \mid E_1(q')\mathfrak{X}.$$

To prove (a), note that, by Theorem 9(ii) and by Theorem 21,

$$g(f(T)) \mid E_1(q)\mathfrak{X} = g(f(T \mid E(f^{-1}(q))\mathfrak{X})).$$

Since $f^{-1}(q)$ is a finite set of points, $T \mid E(f^{-1}(q))\mathfrak{X}$ is bounded by 9(ii), and hence it follows by Theorem VII.3.12 and another application of 9(ii) that

$$\begin{aligned} g(f(T \mid E(f^{-1}(q))\mathfrak{X})) &= h(T \mid E(f^{-1}(q))\mathfrak{X}) \\ &= h(T) \mid E(f^{-1}(q))\mathfrak{X}, \end{aligned}$$

proving (a).

To prove (b), note that, by 9(ii), (b) is equivalent to the assertion that

$$g(f(T \mid E(f^{-1}(q'))\mathfrak{X})) = h(T \mid E(f^{-1}(q'))\mathfrak{X}).$$

By Lemma 2 no generality is lost in passing from consideration of T to consideration of $T \mid E(f^{-1}(q'))\mathfrak{X}$, that is, in assuming that $E(f^{-1}(q')) = I$. This amounts to taking $U = q'$, that is, to assuming that, even if f is analytic at infinity, it does not assume its value at infinity anywhere in $f^{-1}(U)$.

Now let e_n be an increasing sequence of compact subsets of U with union U and let $\tilde{e}_n = f^{-1}(e_n)$. Since e_n does not contain $f(\infty)$, $\tilde{e}_n$ is compact. Let x be in $\mathfrak{D}(g(f(T)))$. Then, using Theorem 9(ii), Definition 8, and Theorem VII.3.12,

$$\begin{aligned} g(f(T))x &= \lim_{n\to\infty} g(f(T) \mid E_1(e_n))E_1(e_n)x \\ &= \lim_{n\to\infty} g(f(T \mid E(\tilde{e}_n)))E(\tilde{e}_n)x \\ &= \lim_{n\to\infty} h(T \mid E(\tilde{e}_n))E(\tilde{e}_n)x. \end{aligned}$$

Hence it follows from Definition 8 that x is in $\mathfrak{D}(h(T))$ and that $h(T)x = g(f(T))x$. On the other hand, if it is assumed that x is in $\mathfrak{D}(h(T))$, all these steps may be reversed, and it may be concluded that x is in $\mathfrak{D}(g(f(T)))$. Thus (b) is also proved. Q.E.D.

25 LEMMA. *Let T be a closed spectral operator and E its resolution of identity. Then*

$$\sigma(T) = \bigcap_{E(e)=I} \bar{e}.$$

PROOF. If $E(e) = I$, then, by Definition 1, $\sigma(T) = \sigma(T \mid E(e)\mathfrak{X}) \subseteq \bar{e}$. Hence, to obtain the desired result it is sufficient to show that $E(\sigma(T)) = I$. Let σ' be any compact subset of the complement of $\sigma(T)$. By Lemma 19, $\sigma(T \mid E(\sigma')\mathfrak{X}) \subseteq \sigma(T) \cap \sigma' = \phi$. From Theorem 9(ii) and Lemma VII.3.4 it follows that $E(\sigma') = 0$. Since E is countably additive, the complement of $\sigma(T)$ has itself E-measure zero. Thus $E(\sigma(T)) = I$. Q.E.D.

A slight gap left in the original statement of Theorem 9(v) may now be filled in.

26 COROLLARY. *Let T be a spectral operator, and let f be analytic on $\sigma(T)$ and in a neighborhood of infinity. Then Definition* 8 *of $f(T)$ agrees with Definition* VII.9.3.

PROOF. Using Theorem 9(v), and examining Definition VII.9.3, it is apparent that it suffices to show that $f(T)$, as given by Definition 8, satisfies the equation $f(T) = g((\lambda I - T)^{-1})$, where λ is in the resolvent set of T and where g is defined by the equation $f(z) = g((\lambda - z)^{-1})$. But this follows immediately from Theorem 24 and Lemma 25. Q.E.D.

A decomposition for unbounded spectral operators which gives a satisfactory extension of Theorem XV.4.5 does not seem to be known. The scalar part of a general spectral operator may easily be defined, but no satisfactory characterization of the difference between a spectral operator and its scalar part is known.

27 DEFINITION. Let T be a closed spectral operator, and E its resolution of the identity. The closed scalar type spectral operator S defined by the equations

$$\mathfrak{D}(S) = \{x \mid \lim_{n\to\infty} \int_{|\lambda| \leq n} \lambda E(d\lambda)x \text{ exists}\},$$

$$Sx = \lim_{n\to\infty} \int_{|\lambda| \leq n} \lambda E(d\lambda)x, \qquad x \in \mathfrak{D}(S),$$

is called *the scalar part of T*.

An attempt to follow the development in the bounded case by writing $N = T - S$ runs into difficulties. The operator N, although easily seen by Lemma 2 to have a quasi-nilpotent restriction to each space $E(\sigma)\mathfrak{X}$ with σ bounded, need not be quasi-nilpotent itself. It may be bounded and not quasi-nilpotent, and it may even be unbounded. It is even possible that NS^{-1} should fail to be quasi-nilpotent. If, for instance, we consider the spectral operator C introduced as an example in the paragraphs of discussion preceding Lemma 19, we see from the discussion given there that its scalar part is the operator A introduced in those paragraphs, and the closure of $C - A$ is the operator B. It was remarked just before the statement of Lemma 19 that $\sigma(B)$ is the whole complex plane, which shows how far B is from being a quasi-nilpotent operator. It was also remarked that $|(BA^{-1})^n| = 2^{-n}$ for $n \geqq 1$, so that $\lim_{n\to\infty} \sqrt[n]{|(BA^{-1})^n|} = \frac{1}{2}$, and BA^{-1} is not quasi-nilpotent. All this merely illustrates the variety of types of pathological behavior of the operator $N = T - S$.

No general structural characterization of the difference between a spectral operator and its scalar part seems to be known. However, the following theorem gives a partial generalization of Theorem XV.4.5.

28 THEOREM. *Let S be a scalar type spectral operator, and let E be its spectral resolution. Let N be a bounded operator which commutes with all the projections $E(e)$ and whose restriction to each of the spaces $E(e)\mathfrak{X}$ with e bounded, is quasi-nilpotent. Then $S + N$ is a spectral operator with resolution of the identity E.*

PROOF. Since N is bounded, it is clear that $T = S + N$ is closed, that $\mathfrak{D}(T) \supseteq E(\sigma)\mathfrak{X}$ if σ is bounded, and that $TE(\sigma) \supseteq E(\sigma)T$ for each Borel set σ. Hence, by Definition 1, to prove the theorem it will suffice to show that $\sigma(T \mid E(e)\mathfrak{X}) \subseteq \bar{e}$ for every Borel set e.

Suppose that $\lambda \notin \bar{e}$. Then the operator

$$M = N \int_e (\lambda - z)^{-1} E(dz)$$

is quasi-nilpotent. To see this, let $\varepsilon > 0$ be given and let $e_1 \subseteq e$ be a bounded Borel set so large that $|N||\lambda - z|^{-1} < \varepsilon$ for z in $e - e_1$. Then

$$M = (N \mid E(e_1)\mathfrak{X}) \int_{e_1} (\lambda - z)^{-1} E(d\lambda) + N \int_{e - e_1} (\lambda - z)^{-1} E(d\lambda).$$

Since N commutes with every projection in the range of E, it follows from Theorem 11 that

$$M^n = (N \mid E(e^1)\mathfrak{X})^n \int_{e_1} (\lambda - z)^{-n} E(d\lambda) + N^n \int_{e-e_1} (\lambda - z)^{-n} E(d\lambda), \qquad n \geqq 1.$$

Since $N \mid E(e_1)\mathfrak{X}$ is quasi-nilpotent,

$$\begin{aligned} \limsup_{n\to\infty} |M^n|^{1/n} &\leqq \limsup_{n\to\infty} |N| \left| \int_{e-e_1} (\lambda - z)^{-n} E(d\lambda) \right|^{1/n} \\ &\leqq |N| \max_{z \in e - e_1} |\lambda - z|^{-1} < \varepsilon, \end{aligned}$$

by Theorem 11(c). Thus, since ε is arbitrary, M is quasi-nilpotent. Let

$$R = \sum_{n=0}^{\infty} M^n \int_e (\lambda - z)^{-1} E(dz).$$

Then R is a bounded operator by Theorem 11(c). If x is in $\mathfrak{D}(T \mid E(e)\mathfrak{X})$ then, using Theorem 11 and the fact that N commutes with every projection in the range of E, we have

$$\begin{aligned} R(\lambda I - S - N)x &= \sum_{n=0}^{\infty} N^n \int_e (\lambda - z)^{-n} E(dz)x \\ &\quad - \sum_{n=0}^{\infty} N^{n+1} \int_e (\lambda - z)^{-n-1} E(dz)x = x. \end{aligned}$$

Next it should be observed that $SN \supseteq NS$. Indeed, if the limit $\lim_{n\to\infty} \int_{|z|\leqq n} zE(dz)x$ exists, then

$$\lim_{n\to\infty} \int_{|z|\leqq n} zE(dz)Nx = \lim_{n\to\infty} N \int_{|z|\leqq n} zE(dz)x$$

exists and equals $N \lim_{n\to\infty} \int_{|z|\leqq n} zE(dz)x$.

It follows that if x is in $\mathfrak{D}(T \mid E(e)\mathfrak{X})$, we have

$$\begin{aligned} &\lim_{k\to\infty} (\lambda I - S - N) \sum_{n=0}^{k} N^n \int_e (\lambda - z)^{-n-1} E(dz)x \\ &\quad = \lim_{k\to\infty} \sum_{n=0}^{k} N^n \int_e (\lambda - z)^{-n} E(dz)x - \sum_{n=0}^{k} N^{n+1} \int_e (\lambda - z)^{-n-1} E(dz)x \\ &\quad = \lim_{k\to\infty} (x - M^{k+1}x) = x. \end{aligned}$$

Since M is quasi-nilpotent and $\lambda I - T$ is closed, it follows that Rx is in

$\mathfrak{D}(T)$, and that $(\lambda I - S - N)Rx = x$. This shows that $R \,|\, E(e)\mathfrak{X} = (\lambda I - T \,|\, E(e)\mathfrak{X})^{-1}$, so that $\lambda \notin \sigma(T \,|\, E(e)\mathfrak{X})$. Thus $\sigma(T \,|\, E(e)\mathfrak{X}) \subseteq \bar{e}$. Q.E.D.

In the special case in which the operator N of Theorem 28 is restricted to be quasi-nilpotent, Theorem 9(v) can be significantly improved.

→ 29 THEOREM. *Let the operator T have the form $T = S + N$, where S is of scalar type and N is a bounded quasi-nilpotent operator which commutes with S. Let E be the resolution of the identity for T and let the function f be bounded and analytic on a domain including all points within some fixed positive distance from the spectrum of T. Then $f(T)$ is a bounded spectral operator given by the series*

$$f(T) = \sum_{n=0}^{\infty} \frac{N^n}{n!} \int_{\sigma(T)} f^{(n)}(\lambda) E(d\lambda),$$

whose terms are all bounded operators and which converges in the uniform operator topology.

PROOF. Let U be a domain upon which f is bounded and analytic and which contains all complex numbers whose distance from the spectrum $\sigma(T)$ is less than the positive number d. Let $|f(z)| \leqq M$ for z in U. If z is in $\sigma(T)$, then

$$f^{(n)}(z) = \frac{n!}{2\pi i} \int_C f(\zeta)(\zeta - z)^{-n-1}\, d\zeta,$$

where C is a circle of radius $d/2$ and center z, so that

$$(n!)^{-1} |f^{(n)}(z)| \leqq M 2^n d^{-n}.$$

Thus, by Theorem 11,

$$\left|(n!)^{-1} \int_{\sigma(T)} f^{(n)}(\lambda) E(d\lambda)\right| \leqq 4MK 2^n d^{-n},$$

where $K = \sup_e |E(e)|$. Since N is quasi-nilpotent, the series

$$\sum_{n=0}^{\infty} \frac{N^n}{n!} \int_{\sigma(T)} f^{(n)}(\lambda) E(d\lambda)$$

converges uniformly to a bounded operator R. Let $\{e_m\}$ be an increasing family of compact subsets of U with union U. Then $T \,|\, E(e_m)\mathfrak{X} = S \,|\, E(e_m)\mathfrak{X} + N \,|\, E(e_m)\mathfrak{X}$. It is clear that

$$S \,|\, E(e_m)\mathfrak{X} = \int_{e_m} \lambda F(d\lambda),$$

where $F(e) = E(e) \mid E(e_m)\mathfrak{X}$. Thus, by Theorem XV.5.1,

$$f(T \mid E(e_m)\mathfrak{X})x = \sum_{n=0}^{\infty} \frac{(N \mid E(e_m)\mathfrak{X})^n}{n!} \int_{e_m} f^{(n)}(\lambda)F(d\lambda)x$$
$$= \sum_{n=0}^{\infty} \frac{(N \mid E(e_m)\mathfrak{X})^n}{n!} \int_{e_m} f^{(n)}(\lambda)E(d\lambda)x$$

for x in $E(e_m)\mathfrak{X}$. Thus, since N commutes with $E(\sigma)$ by Corollary 4, we have

$$f(T \mid E(e_m)\mathfrak{X})E(e_m)x = E(e_m) \sum_{n=0}^{\infty} \frac{N^n}{n!} \int_{\sigma(T)} f^{(n)}(\lambda)E(d\lambda)x$$
$$= E(e_m)Rx, \qquad x \in \mathfrak{X},\ m \geqq 1.$$

By Definition 8, $f(T)$ is defined for all x in $\mathfrak{X}$ and $f(T) = R$. Q.E.D.

The present section will be concluded with the establishment of a criterion by which a closed operator with totally disconnected spectrum may be recognized to be spectral. This criterion will be of fundamental importance in the analysis of the next chapter.

30 Definition. Let T be a closed operator, and σ a compact subset of $\sigma(T)$ which is open and closed in the relative topology of $\sigma(T)$. Then σ is called a *compact spectral set* of T. Let f_σ be the function which is identically 1 in a neighborhood of σ and identically zero in a neighborhood of $\sigma(T) - \sigma$. Then f_σ is analytic in a neighborhood of $\sigma(T)$ and in a neighborhood of infinity, so that VII.9.3 $f_\sigma(T)$ is a well-defined bounded operator. We shall write $f_\sigma(T) = E(\sigma; T)$ or, if T is understood, write $f_\sigma(T) = E(\sigma)$.

31 Lemma. *The operator $E(\sigma)$ of Definition 30 is a projection. We have $E(\phi) = 0$, $E(\sigma_1)E(\sigma_2) = E(\sigma_1\sigma_2)$, and $E(\sigma_1 \cup \sigma_2) = E(\sigma_1) \vee E(\sigma_2)$. Finally,*

$$E(\sigma) = \frac{1}{2\pi i} \int_C (\lambda I - T)^{-1}\, d\lambda,$$

where C denotes the union of a finite collection of closed Jordan curves, oriented in the customary positive sense of complex variable theory, which together bound a finite domain containing every point of σ and no point of $\sigma(T) - \sigma$.

Proof. The first three assertions follow immediately from the functional calculus established in Theorem VII.9.5. Using the notations of Definition 30, it follows by Theorem VII.9.4 that

$$[*] \qquad f_\sigma(T) = f_\sigma(\infty)I + \frac{1}{2\pi i} \int_{C_1} f_\sigma(\zeta)(\zeta I - T)^{-1}\, d\zeta,$$

where C_1 is a finite collection of closed Jordan curves bounding a domain D_1 containing the union of $\sigma(T)$ and a neighborhood of infinity, C_1 being oriented in the customary positive sense of complex variable theory. The curves C of the present lemma bound a finite domain D containing every point of σ and no point of $\sigma(T)-\sigma$. It is clear that we can find a finite collection of curves C_2 which bound a domain D_2 containing the union of $\sigma(T)-\sigma$ and a neighborhood of infinity, and such that $C_2 \cup D_2$ is disjoint from $C \cup D$. Thus it may be assumed without loss of generality that $D_1 = D \cup D_2$, $C_1 = C \cup C_2$. Since $f_\sigma(\zeta) = 0$ for $\zeta \in C_2 \cup D_2$ and $f_\sigma(\zeta) = 1$ for $\zeta \in C \cup D$, it follows immediately from [*] that

$$f_\sigma(T) = \frac{1}{2\pi i} \int_C (\zeta I - T)^{-1}\, d\zeta. \qquad \text{Q.E.D.}$$

→ 32 THEOREM. *Let T be an operator in a weakly complete B-space $\mathfrak{X}$. Suppose that $\sigma(T)$ is totally disconnected. Then T is a spectral operator if and only if*

(a) *the family of projections $E(\sigma; T)$ corresponding to compact spectral sets of T is uniformly bounded, and*

(b) *no non-zero x in $\mathfrak{X}$ satisfies the equation $E(\sigma)x = 0$ for every compact spectral set σ of T.*

PROOF. Suppose first that (a) and (b) are satisfied. Since $\sigma(T)$ is totally disconnected, there exists a complex number $\lambda \notin \sigma(T)$. It is clear that without any loss of generality we may assume that $\lambda = 0$. Let $R = T^{-1}$. Then, by Theorem VII.9.5, $\sigma(R) = \{z \mid z^{-1} \in \sigma(T)\} \cup \{0\}$. Since $\sigma(T)$ is totally disconnected, each point λ in $\sigma(T)$ is contained in an arbitrarily small compact subset σ of $\sigma(T)$ which is open in the relative topology of $\sigma(T)$. It follows that the set $\tau(\sigma) = \{z \mid z^{-1} \in \sigma\}$ is a compact subset of $\sigma(R)$, open in the relative topology of $\sigma(R)$. Thus each point in $\sigma(R)$ different from zero is contained in an arbitrarily small compact set open in the relative topology of $\sigma(R)$.

Suppose now that U is any neighborhood of zero in $\sigma(R)$. Then, as observed in the preceding paragraph, every point p in $\sigma(R) - U$ is contained in an open and closed subset σ_p of $\sigma(R)$ not containing zero. A finite collection $\sigma_{p1}, \ldots, \sigma_{pn}$ of those sets cover $\sigma(R) - U$; if we let

$$\sigma = \sigma(R) - \sigma_{p1} - \sigma_{p2} - \cdots - \sigma_{pn},$$

then σ is an open and closed subset of zero contained in U. Thus, each point in $\sigma(R)$, including zero, is contained in an arbitrarily small compact set open in the relative topology of $\sigma(R)$. Hence, $\sigma(R)$ is totally disconnected.

It follows by Definition VII.9.3 that $E(\sigma; T) = E(\tau(\sigma); R)$ for each compact spectral set σ of T. Moreover, it is clear that as σ runs over the family K of all compact open subsets of $\sigma(T)$, $\tau(\sigma)$ runs over the family of all compact open subsets of $\sigma(R)$ which do not contain 0. Since by assumption $|E(\sigma; T)|$ is uniformly bounded for σ in K, since each compact open subset σ_1 of $\sigma(R)$ is either a compact open subset of $\sigma(R)$ not containing zero or the complement of such a set, and since $E(\sigma(R) - \sigma_1; R) = I - E(\sigma_1; R)$ by VII.3.10, it follows immediately that $|E(\sigma_1; R)|$ is uniformly bounded as σ_1 runs over the family of all compact open subsets of $\sigma(R)$. Thus, by Theorem XVI.5.2, R is spectral.

It is also evident that, assuming T to be a spectral operator and making use of Corollary 23, we can use all the steps of this argument in reverse and conclude that, conversely, (a) must be satisfied.

To continue with the main line of our argument, however: since R is a spectral operator, and since by (b) no non-zero x in $\mathfrak{X}$ satisfies the equations $E(\tau; R)x = E(\tau(\sigma); R)x = E(\sigma; T)x = 0$ for every compact open subset τ of $\sigma(R)$, it follows immediately from the countable additivity of the resolution of the identity for R, and the fact that $\sigma(R)$ is totally disconnected, that the resolution of the identity E_1 of R satisfies the equation $E_1(\{0\}) = 0$. Thus, by Corollary 23, T is a spectral operator.

Conversely, if T is a spectral operator, it follows from Corollary 23 that $E_1(\{0\}) = 0$. Clearly the steps in the argument of the preceding paragraph may be reversed, to prove the necessity of (b). Q.E.D.

→ 33 Corollary. *Let T be an operator in a weakly complete B-space $\mathfrak{X}$. Suppose that $\sigma(T)$ consists of a denumerable collection $\{\lambda_i\}$ of points such that $|\lambda_i| \to \infty$ as $i \to \infty$. Then T is spectral if and only if*

(a) *the family of sums of finite collections of projections $E(\lambda_i; T)$ corresponding to single points λ_i in $\sigma(T)$ is uniformly bounded, and*

(b) *no non-zero x in $\mathfrak{X}$ satisfies all of the equations $E(\lambda_i)x = 0$, $i \geqq 1$.*

Proof. Under the present assumption about $\sigma(T)$, it is evident that the most general compact spectral set of T is a union of finitely many points λ_i. The present corollary now follows immediately from Theorem 32 and Lemma 31. Q.E.D.

The following theorem is at once an "unbounded version" and a generalization of Theorem XVI.5.19. We shall find it useful in our study of applications of the general spectral theory in Chapter XX.

→ 34 THEOREM. *Let the closed unbounded operator S in the reflexive space $\mathfrak{X}$ have a spectrum $\sigma(S)$ which is the disjoint union of a finite set $\{\lambda_1, \ldots, \lambda_n\}$ of points and of a set which is contained in a simple Jordan curve C. Suppose that C is parametrized by a one-to-one function $z(t)$, defined for $-\infty < t < +\infty$, which is smooth, differentiable, and such that $z'(t)$ approaches a limit α as $t \to \pm\infty$. Then C divides the complex sphere into two regions R_1 and R_2, surrounding the points of R_1 (resp., R_2) in the positive (resp., the negative) sense of complex function theory; in what follows, we shall refer to the intersection of R_1 (resp., R_2) with a neighborhood of C as the left hand edge (resp., right hand edge) of C. Let ν_0 be an "exceptional point in σ, with the special significance appearing in* (i) *below. Suppose that there exist dense linear manifolds $\mathfrak{X}_0$, $\mathfrak{X}_0^*$ in the spaces $\mathfrak{X}$, $\mathfrak{X}^*$, respectively, with the following three properties.*

(i) *For x_0 in $\mathfrak{X}_0$ and x_0^* in $\mathfrak{X}_0^*$ there is a constant $K(x_0^*, x_0)$ which bounds the quantity $|\lambda - \nu_0|\,|x_0^* R(\lambda; S)x_0|$ for all λ in an open set including all points of a a neighborhood of the point $\lambda = \infty$ of the complex plane which do not lie in the curve C.*

(ii) *For each x_0 in $\mathfrak{X}_0$ and x_0^* in $\mathfrak{X}_0^*$ the function $x_0^* R(\lambda; S)x_0$ has limiting values $R^+(\hat{\lambda}, x_0^*, x_0)$ and $R^-(\hat{\lambda}, x_0^*, x_0)$ as λ approaches almost any point $\hat{\lambda} \in C$ from the left or right hand edge of the curve C, respectively, in a non-tangential direction.*

(iii) *There is a constant M depending only on S such that*

$$\int_\sigma |R^+(\lambda, x_0^*, x_0) - R^-(\lambda, x_0^*, x_0)|\, ds \leqq M\,|x_0^*|\,|x_0|, \qquad x_0 \in \mathfrak{X}_0,\ x_0^* \in \mathfrak{X}_0^*,$$

where s is the arc length along C.

Then S is a spectral operator. For each bounded set $e \subseteq \sigma$ not containing the exceptional point ν_0, the spectral projection $E(e)$ is given by the formula

$$x_0^* E(e) x_0 = \frac{1}{2\pi i} \int_e \{R^+(\lambda, x_0^*, x_0) - R^-(\lambda, x_0^*, x_0)\}\, d\lambda.$$

PROOF. Our general approach is as follows. First, we show that the presence of the finite point spectrum $\{\lambda_1, \ldots, \lambda_n\}$ lends no essential complication to the situation at hand, so that, without loss of generality, we may assume that $n = 0$, that is, that the whole of $\sigma(S)$ is contained in the curve C. Then, essentially by considering the inverse $T = S^{-1}$, we construct a bounded operator to which a slightly modified version of the proof of Theorem XVI.5.19 will apply, from which we can establish that T is a spectral operator. Finally, the asserted formula for the spectral resolution of S will be proved.

Now to details.

Passing from the consideration of S to that of $S + zI$ where z is suitably chosen, we may assume without loss of generality that neither $\sigma(S)$ nor C includes the point $\lambda = 0$, so that S has a bounded inverse $T = S^{-1}$. This will be assumed without additional comment in the remainder of the present proof.

Let f be the analytic function which is 1 on a neighborhood of $\{\lambda_1, \ldots, \lambda_n\}$, zero on σ, and zero in a neighborhood of the point at infinity. Making use of the functional calculus developed in section VII.9 (cf. especially Theorem VII.9.5), we put $E = f(S)$. By Theorem VII.9.5, E is a projection; by Theorem VII.9.8, E maps the domain $\mathfrak{X} = \mathfrak{D}(I)$ into $\mathfrak{D}(S)$ and $ESx = SEx$ for $x \in \mathfrak{X}$. Therefore, putting $\mathfrak{X}_1 = E\mathfrak{X}$, $\mathfrak{X}_2 = (I - E)\mathfrak{X}$, it follows easily that the restrictions $S_i = S \mid \mathfrak{X}_i \cap \mathfrak{D}(T)$, $i = 1, 2$, are closed operators, and that for their resolvents we have $R(\lambda; S_i) = R(\lambda; S) \mid \mathfrak{X}_i$, $i = 1, 2$. The operator S_1 is everywhere defined and hence bounded. The reader will readily verify, using Theorems VII.9.5 and VII.9.8, that $\sigma(S_1) = \{\lambda_1, \ldots, \lambda_n\}$ and $\sigma(S_2) = \sigma$. It is plain from Definition VII.3.17 (cf. also the paragraph which follows this definition), from Theorem VII.3.19, and from Definitions XV.2.2 and XV.2.5 that S_1 is a spectral operator, so that S_1^{-1} is a spectral operator. We shall show in what follows that S_2^{-1} is also a spectral operator. It will then follow immediately from Theorem XV.3.10 that S^{-1}, which is the direct sum of S_1^{-1} and S_2^{-1}, is also spectral, so that, by Corollary 2.23, it will follow that S is a spectral operator.

Write $T = S_2^{-1}$, so that T is a bounded operator in the space $\mathfrak{Y} = \mathfrak{X}_2$. Let $\mathfrak{Y}_0^*$ be the set of restrictions to $\mathfrak{Y}$ of the functionals in the space $\mathfrak{X}_0^*$, and let $\mathfrak{Y}_0 = (I - E)\mathfrak{X}_0$. Plainly, $\mathfrak{Y}_0^*$ is dense in $\mathfrak{Y}^*$ and $\mathfrak{Y}_0$ is dense in $\mathfrak{Y}$. Since $R(\lambda; S_2)$ is the restriction of $R(\lambda; S)$ to the space $\mathfrak{Y}$, the following statements result immediately from our hypotheses.

(i′) For $y_0 \in \mathfrak{Y}_0$ and $y_0^* \in \mathfrak{Y}_0^*$ there is a constant $K(y_0^*, y_0)$ which bounds the quantity $|\lambda - \nu_0| \, |y_0^* R(\lambda; S_2) \, y_0|$ for all λ in an open set including the spectrum $\sigma(S_2)$ and also including all points of a complete neighborhood of the point $\lambda = \infty$ in the complex plane which do not lie in the curve C.

(ii′) For each $y_0 \in \mathfrak{Y}_0$ and $y_0^* \in \mathfrak{Y}_0^*$ the function $y_0^* R(\lambda; S_2) y_0$ has limiting values $R^+(\hat{\lambda}, y_0^*, y_0)$ and $R^-(\hat{\lambda}, y_0^*, y_0)$ as λ approaches almost any point $\hat{\lambda} \in C$ from the left or right hand edge of the curve C, respectively, in a non-tangential direction.

(iii′) There is a constant M depending only on S such that

$$\int_{\sigma(S_2)} |R^+(\lambda, y_0^*, y_0) - R^-(\lambda, y_0^*, y_0)| \, ds \leqq M |y_0^*| \, |y_0|, \qquad y_0 \in \mathfrak{Y}_0, \; y_0^* \in \mathfrak{Y}_0^*,$$

where s is the arc length along C.

We now note (cf. Theorem VII.9.5 and formula [*] of the proof of Lemma VII.9.2) that $\sigma(T)$ is contained in the smooth Jordan curve $\Gamma_0 = \{\lambda^{-1} \mid \lambda \in C\} \cup \{0\}$, and that

$$(1) \qquad R(\lambda; T) = \lambda^{-2} R(\lambda^{-1}; S_2) + \lambda^{-1} I.$$

We may therefore restate the above properties (i′), (ii′), (iii′) as follows, in a form applicable directly to the operator T.

(i″) For $y_0 \in \mathfrak{Y}_0$ and $y_0^* \in \mathfrak{Y}_0^*$ there is a constant $K(y_0^*, y_0)$ which bounds the quantity $|\lambda - \nu_0^{-1}| \, |\lambda|^2 \, |y_0^* R(\lambda; T) y_0|$ for all λ in an open set including the spectrum $\sigma(T)$ of T.

(ii″) For each $y_0 \in \mathfrak{Y}_0$ and $y_0^* \in \mathfrak{Y}_0^*$ the function $y_0^* R(\lambda; T) y_0$ has limiting values $R_1^+(\hat{\lambda}, y_0^*, y_0)$ and $R_1^-(\hat{\lambda}, y_0^*, y_0)$ as λ approaches almost any point $\hat{\lambda} \in \Gamma_0$, from the left or right hand edge of the curve Γ_0, respectively, in a non-tangential direction.

(iii″) There is a constant M depending only on S such that

$$\int_{\sigma(T)} |R_1^+(\lambda, y_0^*, y_0) - R_1^-(\lambda, y_0^*, y_0)| \, ds \leqq M |y_0^*| \, |y_0|, \qquad y_0 \in \mathfrak{Y}_0, \; y_0^* \in \mathfrak{Y}_0^*,$$

where S is the arc length along Γ_0.

In regard to the proof of (iii″), we note that

$$(2) \quad R_1^+(\lambda, y_0^*, y_0) - R_1^-(\lambda, y_0^*, y_0) = \lambda^{-2}\{R^+(\lambda^{-1}, y_0^*, y_0) - R^-(\lambda^{-1}, y_0^*, y_0)\}$$

by the identity (1), while the arc length dt along C at a point μ and the arc length ds along Γ_0 at the corresponding point $\lambda = \mu^{-1}$ are related by $ds = |\lambda|^2 \, dt$, so that (iii″) follows very readily from (iii′).

If f is any function single valued and analytic in a neighborhood of Γ_0, we may write

$$f(T) = \frac{1}{2\pi i}\int_{\Gamma_\delta} f(\lambda)R(\lambda; T)\, d\lambda,$$

where Γ_δ is a closed Jordan curve, or pair of closed Jordan curves, surrounding Γ_0 at distance δ. If f vanishes at least to second order at $\lambda = 0$, and at least to first order at ν_0^{-1}, then, in view of (i″) and (ii″), we may let $\delta \to 0$ and obtain

$$y_0^* f(T)y_0$$
$$= \frac{1}{2\pi i}\int_{\Gamma_0} f(\lambda)\{R_1^+(\lambda, y_0^*, y_0) - R_1^-(\lambda, y_0^*, y_0)\}\, d\lambda, \qquad y_0 \in \mathfrak{Y}_0,\ y_0^* \in \mathfrak{Y}_0^*.$$

Since the integrand vanishes if λ is in the resolvent set, we may also write

(3) $$y_0^* f(T)y_0$$
$$= \frac{1}{2\pi i}\int_{\sigma(T)} f(\lambda)\{R^+(\lambda, y_0^*, y_0) - R^-(\lambda, y_0^*, y_0)\}\, d\lambda, \qquad y_0 \in \mathfrak{Y}_0,\ y_0^* \in \mathfrak{Y}_0^*$$

for f vanishing at least to second order at $\lambda = 0$ and at least to first order at $\lambda = 1$. In view of (iii″) the right side of (3) defines, for each bounded Borel function f on $\sigma(T)$, a continuous bilinear form (f, y_0^*, y_0), which, since $\mathfrak{Y}_0$ and $\mathfrak{Y}_0^*$ are dense, has a unique extension to a bounded bilinear form (f, y^*, y) defined for all $y \in \mathfrak{Y}$ and $y^* \in \mathfrak{Y}^*$. Since $\mathfrak{Y}$ is reflexive, it follows that there is a unique operator $T(f)$ for which

(4) $$y_0^* T(f)y_0$$
$$= \frac{1}{2\pi i}\int_{\sigma(T)} f(\lambda)\{R^+(\lambda, y_0^*, y_0) - R^-(\lambda, y_0^*, y_0)\}\, d\lambda, \qquad y_0 \in \mathfrak{Y}_0,\ y_0^* \in \mathfrak{Y}_0^*.$$

The mapping $f \to T(f)$ is a homomorphism on the algebra of analytic functions with a double zero at $\lambda = 0$ and a single zero at ν_0^{-1}, and since every continuous function on Γ_0 which vanishes at 0 and at ν_0^{-1} is the uniform limit of such functions, it follows that this map is also a homomorphism on the algebra of all such continuous functions. To see that it is a homomorphism on the algebra of all bounded Borel functions, note that for a fixed continuous function g vanishing at zero and at ν_0^{-1}, the set of all bounded Borel functions f for which

(5) $$T(fg) = T(f)T(g)$$

includes all continuous functions vanishing at $\lambda = 0$ and at $\lambda = \nu_0^{-1}$. Furthermore, if equation (5) holds for each function in a uniformly bounded pointwise convergent sequence $\{f_n\}$, then it follows from (4) that it holds for the limit function

$$f = \lim_{n \to \infty} f_n .$$

Hence (5) holds if f is any bounded Borel function vanishing at $\lambda = 0$ and at $\lambda = \nu_0^{-1}$. A repetition of this argument shows that (5) still holds if f and g are both bounded Borel functions vanishing at $\lambda = 0$ and at $\lambda_0 = \nu_0^{-1}$. Since it is obvious from (4) that $T(f) = 0$ if f is any Borel function vanishing except at $\lambda = 0$ and at $\lambda = \nu_0^{-1}$, it follows that (5) is valid for every pair of bounded Borel functions f and g.

The operator $E_0 = T(1)$ is therefore a projection for which $T^2(T - \nu_0^{-1}I)T(1) = T(g)T(1) = T(g) = T^2(T - \nu_0^{-1}I)$, where we have taken g to be defined by $g(\lambda) \equiv \lambda^2(\lambda - \nu_0^{-1})$. Thus, if $E = I - E_0$, we have $T^2(T - \nu_0^{-1}I)E = 0$, and since $T = S_2^{-1}$ is invertible, we have also $(T - \nu_0^{-1}I)E = 0$. For each bounded Borel function f, put $\bar{T}(f) = T(f) + f(\nu_0^{-1})E$. Then $T(f)E = T(f)(I - T(1)) = T(f) - T(f) = 0$, $f \to \bar{T}(f)$ is still a homomorphism on the algebra of all bounded Borel functions, and we have $\bar{T}(1) = I$.

If $(T - \nu_0^{-1})y = 0$, then $R(\lambda; T)y$ is analytic except at $\lambda = \nu_0^{-1}$, so that plainly, from (4), $T(f)y = 0$ for all bounded Borel functions f. Thus $E_0 y = 0$, $Ey = y$. Therefore, if we write $h(\lambda) \equiv \lambda$, and note that

$$T^2(T - \nu_0^{-1}I)T(h) = T(g)T(h) = T(gh) = T^3(T - \nu_0^{-1}I)$$

by (3) and (4), we find on using the fact that T is invertible that $(T - \nu_0^{-1})(T(h) - T) = 0$, that is, $E_0(T(h) - T) = 0$. We have also $E(T(h) + \nu_0^{-1}E - T) = 0$; thus

$$(T(h) + \nu_0^{-1}E - T) = E_0(T(h) - T) + E(T(h) + \nu_0^{-1} - T) = 0,$$

so that $\bar{T}(h) = T$. Using the above facts, it follows that if f is analytic on $\sigma(T)$ and has a double zero at $\lambda = 0$ and we let $f_1(\lambda) = (\lambda - \nu_0^{-1})f(\lambda)$, then

$$(T - \nu_0^{-1}I)f(T) = f_1(T) = T(f_1) = (T - \nu_0^{-1}I)T(f).$$

Thus $E_0(f(T) - T(f)) = 0$. Since $E(f(T) - f(\nu_0^{-1})E) = 0$, it follows that $f(T) = \bar{T}(f)$ if f is analytic on $\sigma(T)$ and has a double zero at $\lambda = 0$. Combining all the above facts, we find that $f(T) = \bar{T}(f) = T(f) + f(\nu_0^{-1})E$ for all functions f analytic on the spectrum of T. From this, and proceeding almost precisely as in the final paragraph of Theorem XVI.5.19, we conclude that T is a scalar operator whose spectral resolution is given by the formula

$$y_0^* E_1(e) y_0^* = E + \frac{1}{2\pi i} \int_e \{R_1^+(\lambda, y_0^*, y_0) - R_1^-(\lambda, y_0^*, y_0)\}\, d\lambda$$

if the Borel set e includes the point ν_0^{-1}, or, alternatively, if e omits this point, by the formula

$$y_0^* E_1(e) y_0 = \frac{1}{2\pi i} \int_e \{R_1^+(\lambda, y_0^*, y_0) - R_1^-(\lambda, y_0^*, y_0)\}\, d\lambda.$$

Changing the variable of integration in this last formula from λ to $\mu = \lambda^{-1}$, and using the identity (2), we may write this last formula as

$$y_0^* E_1(e^{-1}) y_0$$
$$= \frac{1}{2\pi i} \int_e \{R^+(\lambda, y_0^*, y_0) - R^-(\lambda, y_0^*, y_0)\}\, d\lambda, \qquad y_0 \in \mathfrak{Y}_0,\ y_0^* \in \mathfrak{Y}_0^*,$$

provided that $\nu_0^{-1} \notin e^{-1}$. It follows at once using Theorem 21 that the spectral resolution of $S_2 = T^{-1}$ satisfies the relationship

$$y_0^* E(e) y_0 = \frac{1}{2\pi i} \int_e \{R^+(\lambda, y_0^*, y_0) - R^-(\lambda, y_0^*, y_0)\}\, d\lambda, \qquad y_0 \in \mathfrak{Y}_0,\ y_0^* \in \mathfrak{Y}_0^*,$$

if $\nu_0 \notin e$, from which, by use of Theorem XV.3.10, we readily obtain the final assertion of the present theorem. Q.E.D.

→ 35 Corollary. *Under the hypotheses of the preceding theorem, every point $\lambda \in \sigma$, with the sole possible exception of the point ν_0, belongs to the continuous spectrum of the spectral operator S. The point ν_0 belongs to the point or to the continuous spectrum of the operator S depending on whether ν_0 is or is not an eigenvalue of S.*

Proof. It follows from Lemma 19 that if $\lambda \in \sigma$ and if δ is a closed neighborhood of λ while P denotes the whole complex plane, then $E(P - \delta)\mathfrak{X}$ is contained in the range of $S - \lambda I$. Thus, $E(P - \{\lambda\})\mathfrak{X} = (I - E(\{\lambda\}))\mathfrak{X}$ is contained in the closure of the range of $S - \lambda I$. Therefore, if $E(\{\lambda\}) = 0$, then the closure of the range of $S - \lambda I$ is dense. Moreover, since any vector x such that $(S - \lambda I)x = 0$ must satisfy $E(P - \delta)x = 0$ for all closed neighborhoods δ of λ, we must have $E(\{\lambda\})x = x$. Thus, if $E(\{\lambda\}) = 0$, then λ belongs neither to the point nor to the residual spectrum of S. Using the formula for the spectral resolution of S given by the preceding theorem, we find that $E(\{\lambda\}) = 0$ for $\lambda \neq \nu_0$, $\lambda \in \sigma$; thus, each such λ must belong to the continuous spectrum of S.

We saw in the course of the proof of the preceding theorem that S is the direct sum of two operators S_1 and S_2, of which $\sigma(S_1)$ is disjoint from ν_0 and S_2 is spectral. Then plainly ν_0 belongs to the point (respectively residual or continuous) spectrum of S if and only if it belongs to the point (respectively residual or continuous) spectrum of S_2. Since S_2 is a spectral operator of scalar type, we have $(S_2 - \nu_0 I)E_2(\{\nu_0\}) = 0$, $E_2(\cdot)$ being the spectral resolution of S_2. Thus, unless ν_0 belongs to the point spectrum of S_2, we have $E_2(\{\nu_0\}) = 0$, and it follows, as above, that ν_0 belongs to the continuous spectrum of S_2. This verifies the final assertion of the present corollary. Q.E.D.

3. Multiplicity Theory and Spectral Representation

The methods and results of this section are due to Bade and are intimately dependent upon the ideas introduced in Section XVII.3. A multiplicity theory for Boolean algebras of projections in a B-space $\mathfrak{X}$ will be developed and applied to show that, under certain restrictions, a scalar type operator S in $\mathfrak{X}$ is similar, $S = A^{-1}VA$, to a normal operator V in an appropriate Hilbert space $\mathfrak{H}$ where the operator $A : \mathfrak{X} \to \mathfrak{H}$ and its inverse are closed and densely defined.

We begin by defining the multiplicity function for a complete (see Definition XVII.3.1) Boolean algebra of projections in Banach space.

1 DEFINITION. Let B be a complete Boolean algebra of projections in the real or complex B-space $\mathfrak{X}$. A function m on B whose values are cardinals will be called a *multiplicity function for B* if

(i) $$m(0) = 0, \quad \text{and}$$

(ii) $$m(\bigvee_\alpha E_\alpha) = \bigvee_\alpha m(E_\alpha), \qquad \{E_\alpha\} \subseteq B.$$

The number $m(E)$ is called the *multiplicity* of E. We say $E \in B$ has *uniform multiplicity* n if $m(F) = n$ whenever $0 \neq F \leq E$.

A subset D of B is an *ideal* if E, $F \in D$ implies $E \vee F \in D$ and $G \leq E$, $E \in D$, implies $G \in D$. The ideal D is *dense* if every element of B is a union of elements of D. A *σ-ideal* is an ideal closed under countable unions.

2 LEMMA. *Let D be a dense ideal in B and m be a function on D to cardinals such that $m(0) = 0$ and $m(\bigvee_\alpha E_\alpha) = \bigvee_\alpha m(E_\alpha)$ for each family $\{E_\alpha\} \subseteq D$ for which $\bigvee_\alpha E_\alpha \in D$. Then there is a unique multiplicity function on B which is an extension of m on D.*

PROOF. If $E \in B$ define $m(E) = \bigvee_\alpha m(B_\alpha)$, where $\{B_\alpha\}$ is any family in D such that $E = \bigvee B_\alpha$. If also $E = \bigvee C_\beta$, $C_\beta \in D$, then for each α, β, $B_\alpha = \bigvee_\beta B_\alpha C_\beta$, and $C_\beta = \bigvee_\alpha B_\alpha C_\beta$. Thus

$$\begin{aligned}\bigvee_\alpha m(B_\alpha) &= \bigvee_\alpha m(\bigvee_\beta B_\alpha C_\beta) = \bigvee_{\alpha,\beta} m(B_\alpha C_\beta) \\ &= \bigvee_\beta m(\bigvee_\alpha B_\alpha C_\beta) = \bigvee_\beta m(C_\beta),\end{aligned}$$

showing the extension is well defined. To see the extension is a multiplicity function, let $E_0 \in B$, $E_0 = \bigvee_\gamma E_\gamma$, $E_\gamma \in B$. For each γ we have $E_\gamma = \bigvee_\gamma G_{\gamma\delta}$, where $G_{\gamma\delta} \in D$. Then

$$\begin{aligned}m(E_0) &= \bigvee_{\gamma,\delta} m(G_{\gamma\delta}) = \bigvee_\gamma \bigvee_\delta m(G_{\gamma\delta}) \\ &= \bigvee_\gamma m(E_\gamma).\end{aligned}$$

That the extension is unique follows easily from the distributivity. Q.E.D.

3 THEOREM. *Let m be a multiplicity function on a complete Boolean algebra B of projections in a B-space $\mathfrak{X}$. Then there is a unique family $\{E_n\}$ of disjoint elements in B, n running over the cardinals $\leqq m(I)$, such that*

(i) *$I = \bigvee E_n$.*

(ii) *If $E_n \neq 0$, E_n has uniform multiplicity n.*

PROOF. We show first that each nonzero $E \in B$ bounds a nonzero $G \in B$ of uniform multiplicity. Let $n_0 = \min\{m(F) \mid 0 \neq F \leqq E\}$. Since the cardinals are well ordered, there exists a projection $G \leqq E$ with $m(G) = n_0$. Clearly G has uniform multiplicity n_0. Now from Zorn's lemma we obtain a maximal family $\mathfrak{F}$ of disjoint elements of B each having uniform multiplicity. For each cardinal n let E_n be the supremum of those $F \in \mathfrak{F}$ with $m(F) = n$. If $0 \neq G \leqq E_n$, then G is the union of elements of B of multiplicity n. Hence E_n has uniform multiplicity n. Finally suppose also that $I = \bigvee F_n$ where the F_n are disjoint and satisfy (ii). Suppose $E_k \neq 0$. Since we may write $E_k = \bigvee E_k F_n$, each element $E_k F_n$ is zero or has uniform multiplicity k. Thus $E_k F_n = 0$ if $n \neq k$. Consequently $F_k \neq 0$ and $E_k \leqq F_k$. Correspondingly if $F_k \neq 0$, $E_k \neq 0$ and $F_k \leqq E_k$. Thus for each n, $E_n = F_n$, showing the uniqueness. Q.E.D.

In defining a natural multiplicity function for B it will be convenient to use a few concepts which we introduce formally in the following definition.

4 DEFINITION. For each $x \in \mathfrak{X}$ the projection $\bigwedge \{E \mid Ex = x\}$ will be called the *carrier projection* of x. (Note that if G is the carrier projection of x and $0 \neq F \leqq G$, then $Fx \neq 0$.) The *cyclic subspace* $\mathfrak{M}(x)$ spanned by a vector x is $\overline{\text{sp}}\{Ex \mid E \in B\}$. A projection $E \in B$ will be said to satisfy the *countable chain condition* if every family of disjoint projections in B bounded by E is at most countable. We shall denote by $\mathscr{C}$ the set of all $E \in B$ satisfying this condition.

It will be shown that $\mathscr{C}$ is a dense σ-ideal in B and thus, in defining the multiplicity on B, Lemma 2 permits us to restrict our attention to $\mathscr{C}$.

5 LEMMA. *The set $\mathscr{C}$ is a dense σ-ideal in B. A projection belongs to $\mathscr{C}$ if and only if it is the carrier projection of a vector in $\mathfrak{X}$.*

PROOF. We note first the purely algebraic fact that $\mathscr{C}$ is a σ-ideal. That it is an ideal is clear. Now suppose $F = \bigvee_{n=1}^{\infty} F_n$, where $F_n \in \mathscr{C}$, and let F be the union of a family $\{E_\gamma \mid \gamma \in \Gamma\}$ of disjoint elements of B. Then for each fixed n, at most countably many of the products $F_n E_\gamma$, $\gamma \in \Gamma$, are different from zero. It follows that $\{\gamma \mid E_\gamma F_n \neq 0 \text{ for some } n\}$ is at most countable. But this is the set of γ for which $E_\gamma \neq 0$, since for every γ, $E_\gamma = \bigvee_{n=1}^{\infty} E_\gamma F_n$.

Now let E be the carrier projection of a vector x and suppose E is the union of a family $\{E_\alpha\}$, $\alpha \in A$, of disjoint elements of B. By Lemma XVII.3.4 we have $\lim_\sigma \sum_{\alpha \in \sigma} E_\alpha x = x$, where σ runs over the finite subsets of A, ordered by inclusion. For each $\varepsilon > 0$ there exists a σ such that $|x - \sum_\sigma E_\alpha x| < \varepsilon M^{-1}$, so if $\beta \notin \sigma$, then $|E_\beta x| = |E_\beta(x - \sum_\sigma E_\alpha x)| < \varepsilon$. It follows that at most countably many of the vectors $E_\alpha x$, $\alpha \in A$, are not zero. As $E_\alpha \leqq E$, the equation $E_\alpha x = 0$ implies $E_\alpha = 0$ as noted in Definition 4. Thus the carrier of any vector belongs to $\mathscr{C}$.

It will next be shown that $\mathscr{C}$ is a dense ideal. If $E \in B$, then it bounds the carrier projection of every vector in its range. By Zorn's lemma we can find a maximal family of disjoint carrier projections E_γ bounded by E. If the projection $E - \bigvee_\gamma E_\gamma$ is not zero it bounds a carrier projection, so the maximality of the family implies $E = \bigvee E_\gamma$. Moreover, each E_γ is in $\mathscr{C}$ as we have just shown.

Finally it must be shown that each $E \in \mathscr{C}$ is the carrier projection of a vector. Since $E \in \mathscr{C}$ we may express E as the union of a *sequence* of disjoint projections E_n each of which is the carrier of a vector x_n with $|x_n| \leqq 1$. Define $x_0 = \sum_{n=1}^{\infty} 2^{-n} x_n$. Clearly $Ex_0 = x_0$. We assert E is the carrier of x_0. For suppose that $F \in B$ and $Fx_0 = x_0$. Then for each n,

$$2^{-n} x_n = E_n x_0 = E_n F x_0 = F E_n x_0 = 2^{-n} F x_n .$$

Thus $Fx_n = x_n$, $n = 1, 2, \ldots$, showing $F \geqq E_n$, $n = 1, 2, \ldots$. It follows that $F \geqq E$. Q.E.D.

6 Definition. For each E in $\mathscr{C}$ the *multiplicity* $m(E)$ of E is defined to be the smallest cardinal power of a set A of vectors such that

$$E\mathfrak{X} = \overline{\text{sp}}\{\mathfrak{M}(x) \mid x \in A\}.$$

7 Lemma. *If E, $F \in \mathscr{C}$ and $F \leqq E$, then $m(F) \leqq m(E)$. If $\{E_\alpha\} \subseteqq \mathscr{C}$ and $E_0 = \bigvee E_\alpha \in \mathscr{C}$, then $m(E_0) = \bigvee_\alpha m(E_\alpha)$.*

Proof. The first statement is clear since if $E\mathfrak{X} = \overline{\text{sp}}\{\mathfrak{M}(x) \mid x \in A\}$, then $F\mathfrak{X} = \overline{\text{sp}}\{\mathfrak{M}(Fx) \mid x \in A\}$. It follows that $\bigvee_\alpha m(E_\alpha) \leqq m(E_0)$. We now observe that there exists a sequence $\{G_k\}$ of disjoint elements in $\mathscr{C}$ such that each G_k is bounded by some E_{α_k} and $E_0 = \bigvee_{k=1}^{\infty} G_k$. For we may consider families of disjoint projections in B each member of which is bounded by an E_α, and order these families by inclusion. By Zorn's lemma there exists a maximal family which must be countable, since $E_0 \in \mathscr{C}$. Clearly the union of this maximal family is E_0. Now let $n_0 = \bigvee m(E_\alpha)$. There exist families of vectors x_n^k, $k = 1, 2, \ldots$, (some of whose members may be zero) such that $|x_n^k| \leqq 1$, $G_k x_n^k = x_n^k$, and

$$G_k \mathfrak{X} = \overline{\text{sp}}\{\mathfrak{M}(x_n^k) \mid 1 \leqq n \leqq n_0\}, \qquad k = 1, 2, \ldots .$$

For each cardinal $n \leqq n_0$ define $y_n = \sum_{k=1}^{\infty} 2^{-k} x_n^k$. Then $G_k y_n = 2^{-k} x_n^k$ and

$$\begin{aligned} E_0 \mathfrak{X} &= \overline{\text{sp}}\{G_k \mathfrak{X} \mid k = 1, 2, \ldots\} = \overline{\text{sp}}\{\mathfrak{M}(x_n^k) \mid 1 \leqq k < \infty, n \leqq n_0\} \\ &= \overline{\text{sp}}\{\mathfrak{M}(y_n) \mid n \leqq n_0\}. \end{aligned}$$

Thus $m(E_0) \leqq n_0$, proving that $m(E_0) = \bigvee m(E_\alpha)$. Q.E.D.

→ 8 Theorem. *Let B be a complete Boolean algebra of projections in $\mathfrak{X}$. There exists a unique multiplicity function m defined on B with the property that for each E in B satisfying the countable chain condition, $m(E)$ is the* *least cardinal power of a set of cyclic subspaces spanning the range of* E. *There is a unique decomposition of the identity* $I = \bigvee E_n$ *into disjoint projections such that if* $E_n \neq 0$, E_n *has uniform multiplicity* n.

PROOF. This theorem follows immediately from the preceding theorem, Lemma 2 and Theorem 3. Q.E.D.

We shall next examine the structure of the cyclic subspaces determined by the complete Boolean algebra B of projections on the B-space $\mathfrak{X}$. The results of this examination will then yield a spectral representation for certain classes of scalar type (bounded or unbounded) operators.

We begin by recalling that Lemma XVII.3.9 shows that B is the range of a spectral measure E defined on the Borel sets of the structure space Λ whose points are the maximal ideals in the uniformly closed operator algebra $\mathfrak{A}(B)$ generated by B. Actually we may if we wish take $\mathfrak{A}(B)$ to be the weakly closed operator algebra generated by B for, as was seen in Corollary XVII.3.17, the uniformly closed and the weakly closed operator algebras generated by a complete Boolean algebra of projections are the same. First let us suppose that $\mathfrak{X}$ is a Hilbert space and that B is a complete Boolean algebra of self adjoint projections. We recall that for a complex valued Borel measurable function f the operator

$$S(f) = \int_\Lambda f(\lambda)E(d\lambda)$$

has domain

$$\mathfrak{D}(S(f)) = \{y \,|\, \int_\Lambda |f(\lambda)|^2 \, (E(d\lambda)y, y) < \infty\}.$$

For a vector x in $\mathfrak{X}$ the cyclic subspace $\mathfrak{M}(x)$ is the same as the closed linear manifold spanned by all vectors of the form $S(f)x$ where f varies over all Borel measurable functions for which $\mathfrak{D}(S(f))$ contains x. In fact (Lemma XII.3.1) the correspondence $S(f)x \to f$ defines an isometric isomorphism of the cyclic subspace $\mathfrak{M}(x)$ onto all of $L_2(\Lambda, \mathscr{B}, \mu)$, where $\mathscr{B}$ denotes the family of Borel sets in Λ and $\mu = (E(\cdot)x, x)$. Furthermore the operators $S(g)$ on $\mathfrak{M}(x)$ act under this isomorphism as multiplication by g in $L_2(\Lambda, \mathscr{B}, \mu)$. If a projection in B satisfies the countable chain condition and has multiplicity n, we may span its range by n orthogonal cyclic subspaces and the isomorphism extends to an isometric mapping of $\sum_{i \leqq n} \mathfrak{M}(x_i)$ onto $\sum_{i \leqq n} L_2(\Lambda, \mathscr{B}, \mu_i)$, $\mu_i = (E(\cdot)x_i, x_i)$. Our aim here is to recapture as much as possible of this structure when $\mathfrak{X}$ is an arbitrary Banach space. The main obstacle is the lack of orthogonal complementation, and one must supply a substitute for the measures $(E(\cdot)x, x)$. A suitable representation for the range of E will be obtained only in the case $m(E) < \infty$. However, there the similarity with the Hilbert space case is striking.

Our first objective will be to obtain a characterization of cyclic subspaces similar to that in Hilbert space. Hereafter B will be a complete Boolean algebra of projections in a Banach space $\mathfrak{X}$. Regarding B as a spectral measure we recall (cf. the discussion on pp. 891-892) that for any bounded Borel function f, the integral $S(f) = \int_\Lambda f(\lambda)E(d\lambda)$ exists in the uniform operator topology and the correspondence $f \to S(f)$ determines a homomorphism of the algebra of bounded Borel functions on Λ into the algebra of bounded operators on $\mathfrak{X}$. When f is unbounded the operator $S(f)$ has been defined in XIII.2.10 as the operator with domain

$$\mathfrak{D}(S(f)) = \{y \mid \lim_{m\to\infty} \int_{e_m} f(\lambda)E(d\lambda)y \text{ exists}\},$$

where $e_m = \{\lambda \mid |f(\lambda)| \leqq m\}$ and

$$S(f)y = \lim_{m\to\infty} \int_{e_m} f(\lambda)E(d\lambda)y, \qquad y \in \mathfrak{D}(S(f)).$$

An operational calculus based upon this definition of the unbounded operators $S(f)$ has been given in Theorem 2.11 and we shall use these elementary results freely as needed without necessarily making explicit reference to this theorem.

If $x \in \mathfrak{X}$, it is clear from the definition of an integral that $\mathfrak{M}(x) \supseteqq \{S(f)x \mid x \in \mathfrak{D}(S(f))\}$. A certain amount of machinery will be required to prove that this inclusion is an equality. The fundamental tool is the separation theorem (Theorem 12) proved below. For its proof we shall need some properties of the family B^* of adjoints of elements of B. If $E \in B$, then E^* is a projection in $\mathfrak{X}^*$ whose range is the $\mathfrak{X}$-closed manifold $\{x^* \mid x^*(I-E)\mathfrak{X} = 0\}$. Since $E \vee F = E + F - EF$ and $E \wedge F = EF$ it follows that $(E \vee F)^* = E^* \vee F^*$ and $(E \wedge F)^* = E^* \wedge F^*$. Also $E^* \leqq F^*$ if and only if $E \leqq F$. It follows that B^* is a Boolean algebra of projections with the same bound as B. The sense in which the completeness of B implies that of B^* will be made clear in the following definition and lemma.

9 DEFINITION. Let D be a Boolean algebra of projections in $\mathfrak{X}^*$ and let $F\mathfrak{X}^*$ be $\mathfrak{X}$-closed for every $F \in D$. Then D is called *$\mathfrak{X}$-complete* (*$\mathfrak{X}$-σ-complete*) if for every subset (sequence) $\{F_\alpha\} \subseteq D$, the following two conditions hold.

(a) $\mathfrak{X}^*$ admits the direct sum decomposition $\mathfrak{X}^* = \mathfrak{M} \oplus \mathfrak{N}$ where

$$\mathfrak{M} = \mathfrak{X}\ \mathrm{sp}\{F_\alpha \mathfrak{X}^*\}, \qquad \mathfrak{N} = \bigcap_\alpha (I - F_\alpha)\mathfrak{X}^*,$$

where here and later we use the notation $\mathfrak{X}\ \mathrm{sp}\{A_\alpha\}$ for the least $\mathfrak{X}$-closed linear manifold in $\mathfrak{X}^*$ containing all the sets A_α.

(b) The projection F_0 with range $\mathfrak{M}$ and null manifold $\mathfrak{N}$ defined by this decomposition belongs to D.

It is easily verified that $\mathfrak{X}$-completeness implies completeness for D as an abstract Boolean algebra, and that F_0 in the definition above is the projection $\bigvee_\alpha F_\alpha$. Moreover

$$(\bigwedge_\alpha F_\alpha)\mathfrak{X}^* = \bigcap_\alpha F_\alpha \mathfrak{X}^*, \qquad (I - \bigwedge_\alpha F_\alpha)\mathfrak{X}^* = \mathfrak{X}\ \mathrm{sp}\{(I - F_\alpha)\mathfrak{X}^*\}.$$

10 LEMMA. *Let B be a complete (σ-complete) Boolean algebra of projections in $\mathfrak{X}$ and let B^* be the Boolean algebra of adjoints in $\mathfrak{X}^*$ of elements of B. Then B^* is $\mathfrak{X}$-complete ($\mathfrak{X}$-σ-complete) in $\mathfrak{X}^*$. In particular the manifolds $E^*\mathfrak{X}^*$, $E^* \in B^*$, are $\mathfrak{X}$-closed.*

PROOF. That $E^*\mathfrak{X}^* = \{x^* \mid (I - E)^*x^* = 0\}$ is $\mathfrak{X}$-closed is a standard elementary property of the null manifold of an adjoint operator. Let $\{E_\alpha\}$ be an arbitrary set (sequence) of elements of B and $E_0 = \bigvee E_\alpha$. Since we may replace $\{E_\alpha\}$ by the increasing generalized sequence of its finite unions, we may suppose $\{E_\alpha\}$ is an increasing generalized sequence. Then $E_0 x = \lim_\alpha E_\alpha x$, $x \in \mathfrak{X}$, by Lemma XVII.3.4. Consequently $E_0^* x^*(x) = x^* E_0 x = \lim x^* E_\alpha x = \lim E_\alpha^* x^* x$, $x \in \mathfrak{X}$. Thus $E_0^*\mathfrak{X}^* \subseteq \mathfrak{X}\ \mathrm{sp}\{E_\alpha^*\mathfrak{X}^*\}$. But $E_0^*\mathfrak{X}^*$ is $\mathfrak{X}$-closed. Thus $E_0^*\mathfrak{X}^* = \mathfrak{X}\ \mathrm{sp}\{E_\alpha^*\mathfrak{X}^*\}$. One shows similarly that $(I^* - E_0^*)\mathfrak{X}^* = \bigcap_\alpha (I^* - E_\alpha^*)\mathfrak{X}^*$. Q.E.D.

11 DEFINITION. A closed linear manifold $\mathfrak{M}$ in $\mathfrak{X}$ is called an *invariant subspace* if $S(f)x \in \mathfrak{M}$ whenever $x \in \mathfrak{M}$ and $x \in \mathfrak{D}(S(f))$.

→ 12 THEOREM. *Let $\mathfrak{M}$ be a closed invariant subspace in $\mathfrak{X}$ and $x_0 \in \mathfrak{X}$ with $\mathfrak{M}(x_0) \cap \mathfrak{M} = (0)$. There exists a functional $x_0^* \in \mathfrak{X}^*$ with the properties*

(1) $x_0^*(\mathfrak{M}) = (0)$,

(2) $x_0^* S(f)x_0 > 0$ if f is a non-negative bounded Borel function for which $S(f)x_0 \neq 0$.

PROOF. It is clear that without loss of generality we may suppose that the carrier projection of x_0 is the identity I. Thus if $0 \neq E \in B$, we have $Ex_0 \neq 0$, and I satisfies the countable chain condition. It will be convenient to isolate a portion of the argument.

13 PROPOSITION. *If $0 \neq F \in B$ there exists a projection G in B and a nonzero functional $y^* \in \mathfrak{X}^*$ such that*

(i) $G = \bigwedge \{E \mid y^*Ex = y^*x,\ x \in \mathfrak{M}(x_0)\}$,

(ii) $y^*(\mathfrak{M}) = (0)$, *and*

(iii) $0 \neq G \leqq F$.

PROOF OF THE PROPOSITION. Let $z^* \in \mathfrak{X}^*$ be chosen so that $z^*(\mathfrak{M}) = (0)$, while $z^*(Fx_0) \neq 0$. Let $\mathscr{G} = \{E \in B \mid z^*Ex = z^*Fx,\ x \in \mathfrak{M}(x_0)\}$. Now if $E_1, E_2 \in \mathscr{G}$, $z^*E_1E_2x = z^*FE_2x = z^*E_2Fx = z^*F^2x = z^*Fx$, $x \in \mathfrak{M}(x_0)$, so $\mathscr{G}$ is closed under finite intersections. Thus $\mathscr{G}$ is a decreasing generalized sequence when its members are directed by inclusion and therefore, by Lemma XVII.3.4, converges strongly to the projection

$$G = \bigwedge \{E \mid E \in \mathscr{G}\}.$$

Consequently

$$G^*z^*x = z^*Gx = \lim_{E \in \mathscr{G}} z^*Ex = z^*Fx, \qquad x \in \mathfrak{M}(x_0),$$

and

$$G^*z^*y = \lim_{E \in \mathscr{G}} z^*Ey = 0, \qquad y \in \mathfrak{M}.$$

If we set $y^* = G^*z^*$, then clearly G and y^* satisfy (ii) and (iii). Let $H = \bigwedge\{E \mid y^*Ex = y^*x, x \in \mathfrak{M}(x_0)\}$. Clearly $G \geqq H$. However, if $y^*Ex = y^*x$ for all $x \in \mathfrak{M}(x_0)$, then

$$\begin{aligned} z^*GEx = y^*Ex = y^*x &= F^*G^*z^*x \\ &= G^*z^*Fx = z^*Fx, \qquad x \in \mathfrak{M}(x_0), \end{aligned}$$

showing $GE \geqq G$. Thus $G \leqq E$, from which it follows $G \leqq H$. Hence $G = H$. Q.E.D.

Returning now to the proof of Theorem 12, let us call a projection $G^* \in B^*$ the x_0-*carrier* of the functional y^* if condition (i) of the Proposition is satisfied. The Proposition shows the existence of families of disjoint projections in B^*, each of whose members is the x_0-carrier of a nonzero functional which vanishes on $\mathfrak{M}$. By Zorn's lemma there exists a maximal family $\mathscr{F}$ of such disjoint projections. Since $I \in \mathscr{C}$, the family $\mathscr{F}$ is countable. Let $\mathscr{F} = \{G_n^*\}$ where G_n^* is the x_0-carrier of y_n^*. Clearly $\bigvee G_n^* = I^*$, since otherwise the Proposition, applied to $I - \bigvee G_n$, would contradict the maximality of $\mathscr{F}$. We may suppose $\sum_{n=1}^{\infty} y^*$ converges and set $y_0^* = \sum_{n=1}^{\infty} y_n^*$. We assert that I^* is the x_0-carrier of y_0^*. For if $y_0^* Ex = y_0^* x$, $x \in \mathfrak{M}(x_0)$, then for each n,

$$\begin{aligned} y_n^* Ex &= G_n^* y_0^* Ex = y_0^* E G_n x \\ &= y_0^* G_n x = y_n^* x, \qquad x \in \mathfrak{M}(x_0), \end{aligned}$$

so $E^* \geqq G_n^*$, $n = 1, 2, \ldots$.

Finally we obtain x_0^* from y_0^*. It is convenient again to regard B as a spectral measure. Let ν denote the total variation of the measure $y_0^* E(\cdot)x_0$. Then there exists a function g in $L_1(\Lambda, \mathscr{B}, \nu)$ such that

$$\nu(e) = \int_e g(\lambda) y_0^* E(d\lambda) x_0, \qquad e \in \mathscr{B}.$$

Thus

$$\nu(e) = \int_e |g(\lambda)| \, \nu(d\lambda), \qquad e \in \mathscr{B},$$

showing $|g(\lambda)| = 1$, except possibly on a ν-null set. Redefining g to be zero on this set we form $S(g) = \int_\Lambda g(\lambda) E(d\lambda)$ and define $x_0^* = S(g)^* y_0^*$. Thus $x_0^* E(\cdot)x_0 = \nu(\cdot)$, and $x_0^*(\mathfrak{M}) = y_0^*(S(g)\mathfrak{M}) \subseteq y_0^*(\mathfrak{M}) = (0)$. Suppose now that f is a bounded non-negative Borel function and that $x_0^* S(f) x_0 = \int_\Lambda f(\lambda) x_0^* E(d\lambda) x_0 = 0$. Clearly then

$$y_0^* E(e) x_0 = 0, \qquad e \in \mathscr{B},\ e \subseteq e_0,$$

where $e_0 = \{\lambda \mid f(\lambda) > 0\}$. Thus $y_0^*(\mathfrak{M}(E(e_0)x_0)) = (0)$, so $y_0^*(I - E(e_0))x = y_0^* x$, $x \in \mathfrak{M}(x_0)$. Since I^* is the x_0-carrier of y_0^* we must have $E(e_0) = 0$, showing that $S(f)x_0 = 0$. Q.E.D.

→ 14 COROLLARY. *Let B be a complete Boolean algebra of projections in $\mathfrak{X}$ and let $x_0 \in \mathfrak{X}$. There exists a functional x_0^* in $\mathfrak{X}^*$ such that the measure $x_0^* E(\cdot)x_0$ is positive and dominates the vector measure $E(\cdot)x_0$.*

→ 15 THEOREM. *If $x_0 \in \mathfrak{X}$, then $\mathfrak{M}(x_0) = \{S(f)x_0 \mid x_0 \in \mathfrak{D}(S(f))\}$.*

PROOF. Let x_0^* be chosen as in Corollary 14 and let $\mu_0 = x_0^* E(\cdot)x_0$. Now $\mu_0(e) = 0$ implies that $E(d)x_0 = 0$ for $d \subseteq e$, and thus that $\mathfrak{M}(E(e)x_0) = (0)$. Thus if $x \in \mathfrak{M}(x_0)$, $x_0^* E(e)x = 0$. By the Radon-Nikodým theorem there exists a function $f \in L_1(\Lambda, \mathscr{B}, \mu_0)$ such that

$$x_0^* E(e) x = \int_e f(\lambda) \mu_0(d\lambda), \qquad e \in \mathscr{B}.$$

The correspondence $T: x \to f$ defines a linear map of $\mathfrak{M}(x_0)$ into $L_1(\Lambda, \mathscr{B}, \mu_0)$. It is continuous since $\int_\Lambda |f(\lambda)| \mu_0(d\lambda) = \text{tot. var. } x_0^* E(\cdot)x \leqq 4M|x_0^*||x|$. For each m let $e_m = \{\lambda | |f(\lambda)| \leqq m\}$. We shall show that $E(e_m)x = S(f \cdot \chi_{e_m})x_0$, the symbol χ_e denoting the characteristic function of the set e. If $x^* \in \mathfrak{X}^*$, then $x^*E(\cdot)x_0$ is dominated by μ_0, so

$$x^*E(e)x_0 = \int_e p(\lambda)\mu_0(d\lambda), \qquad e \in \mathscr{B},$$

where $p \in L_1(\Lambda, \mathscr{B}, \mu_0)$. Let d be a subset of e_m on which p is bounded, and let $\{g_n\}$ be a sequence of bounded functions such that $x = \lim S(g_n)x_0$. Then

$$\begin{aligned} x^*S(g_n)E(d)x_0 &= \int_d g_n(\lambda)p(\lambda)\mu_0(d\lambda) \to \int_d f(\lambda)p(\lambda)\mu_0(d\lambda) \\ &= x^*S(f \cdot \chi_d)x_0, \end{aligned}$$

by continuity of T. However, $x^*S(g_n)E(d)x_0 \to x^*E(d)x$. Thus $x^*S(f \cdot \chi_d)x_0 = x^*E(d)x$. By a limiting argument $x^*S(f \cdot \chi_{e_m})x_0 = x^*E(e_m)x$. However, x^* is arbitrary so $E(e_m)x = S(f \cdot \chi_{e_m})x_0$. Since $x = \lim E(e_m)x$, $x_0 \in \mathfrak{D}(S(f))$ and $x = S(f)x_0$. Q.E.D.

16 Corollary. *If f is a Borel function and $x_0 \in \mathfrak{D}(S(f))$, then $f \in L_1(\Lambda, \mathscr{B}, x_0^*E(\cdot)x_0)$ for every x_0^* satisfying the conditions of Corollary* 14.

Next we shall examine the range of a projection E of finite uniform multiplicity. Since every projection in B is the union of disjoint projections in $\mathscr{C}$, it will be sufficient to suppose that E is in $\mathscr{C}$. To avoid notational difficulties we shall suppose the identity is in $\mathscr{C}$ and has finite uniform multiplicity n. Then there exist vectors $x_1, \ldots, x_n$ such that $\mathfrak{X} = \bigvee_{i=1}^n \mathfrak{M}(x_i)$ and no smaller number of cyclic subspaces will span $\mathfrak{X}$. Here we have denoted by $\bigvee_{i=1}^n \mathfrak{M}(x_i)$ the smallest closed linear manifold containing the manifolds $\mathfrak{M}(x_i)$, $i = 1, \ldots, n$. We shall see that the manifolds $\mathfrak{M}(x_i)$ for such a minimal spanning set are all disjoint and that we may embed $\mathfrak{X}$ in a direct sum of L_1 spaces $\sum_{i=1}^n L_1(\Lambda, \mathscr{B}, \mu_i)$ analogous to the embedding in a sum of L_2 spaces in the classical Hilbert space case. The measures μ_i have the form $x_i^*E(\cdot)x_i$; $x_i^* \in \mathfrak{X}^*$. This fact is the key to the later discussion of the duality between multiplicity in $\mathfrak{X}$ and in $\mathfrak{X}^*$.

The first lemma to follow is independent of the simplifying assumptions made in the preceding paragraph.

17 Lemma. *Let $\mathfrak{M}$ be a closed invariant subspace and let $y_0 \in \mathfrak{X}$. There exists a maximal projection $E_0 \in B$ such that $E_0 y_0 \in \mathfrak{M}$. Moreover,*

$$\mathfrak{M} \cap \mathfrak{M}((I - E_0)y_0) = (0).$$

PROOF. If $z \in \mathfrak{M}(y_0) \cap \mathfrak{M}$ then, by Theorem 15, $z = S(f)y_0$ for some Borel function f. If e is any set in $\mathscr{B}$ on which f satisfies an inequality $K^{-1} \leqq |f(\lambda)| \leqq K$, then $E(e)y_0 = S(g)z$, where $g = f^{-1}\chi_e$. Thus if $\mathfrak{M}(y_0) \cap \mathfrak{M} \neq (0)$, there exist projections E with $0 \neq Ey_0 \in \mathfrak{M}(y_0) \cap \mathfrak{M}$. Let E_0 be the supremum of such projections. If $\mathfrak{M}((I - E_0)y_0) \cap \mathfrak{M} \neq (0)$, the same argument may be used to contradict the maximality of E_0. Finally if $\mathfrak{M}(y_0) \cap \mathfrak{M} = (0)$, let E_0 be defined as the complement of the carrier projections of y_0. So $(I - E_0)y_0 = y_0$, $E_0 y_0 = 0$ and thus $E_0 y_0 \in \mathfrak{M}$ and

$$\mathfrak{M} \cap \mathfrak{M}((I - E_0)y_0) = 0.$$

To see that E_0 is maximal we note (cf. remark in Definition 4) that if $E_1 > E$ then $(I - E_1)y_0 \neq 0$, and since $\mathfrak{M}$ is an invariant subspace

$$0 \neq (I - E_0)y_0 \in \mathfrak{M} \cap \mathfrak{M}((I - E_0)y_0),$$

which proves the maximality of E_0 in this final case. Q.E.D.

18 LEMMA. *Let the identity satisfy the countable chain condition and have finite uniform multiplicity n. If $x_1, \ldots, x_n$ is any set of vectors such that $\mathfrak{X} = \bigvee_{i=1}^{n} \mathfrak{M}(x_i)$, then*

(1) *the carrier of each x_i is I,*

(2) $\mathfrak{M}(x_i) \cap \bigvee_{j \neq i} \mathfrak{M}(x_j) = (0), \qquad i = 1, \ldots, n,$

(3) *for each i there exists a linear functional x_i^* such that $x_i^*(\bigvee_{j \neq i} \mathfrak{M}(x_j)) = (0)$, while $x_i^* S(f)x_i > 0$ for each non-negative Borel function f such that $S(f)x_i \neq 0$.*

PROOF. To prove (2) suppose, for example, that there is a non-zero vector $z \in \mathfrak{M}(x_n) \cap \bigvee_{i=1}^{n-1} \mathfrak{M}(x_i)$. By Lemma 17 there is a projection E with $0 \neq Ex_n \in Ex_n \in \bigvee_{i=1}^{n-1} \mathfrak{M}(x_i)$. Hence $E\mathfrak{X} = \bigvee_{i=1}^{n-1} \mathfrak{M}(Ex_i)$ which shows that $m(E) \leqq n - 1$ and contradicts the assumption of uniform multiplicity. Statement (1) is proved in a similar way, for if E is the carrier of x_i then $(I - E)\mathfrak{X} \subseteq \bigvee_{j \neq i} \mathfrak{M}((I - E)x_j)$. Statement (3) follows from Theorem 12. Q.E.D.

→ 19 THEOREM. *Let the hypotheses of Lemma 18 be satisfied. There exists a linear continuous one-to-one map T of $\mathfrak{X}$ onto a dense subspace of the direct sum $\sum_{i=1}^{n} L_1(\Lambda, \mathscr{B}, \mu_i)$, $\mu_i = x_i^* E(\cdot)x_i$; such that if $Tx = [f_1, \ldots, f_n]$ then*

(a) $x_i^* E(e)x = \int_e f_i(\lambda)\mu_i(d\lambda), \qquad e \in \mathscr{B},\ i = 1, \ldots, n.$

(b) $x = \lim_{m \to \infty} \sum_{i=1}^{n} S(f_i \cdot \chi_{e_m})x_i$

where

$$e_m = \{\lambda \,|\, |f_i(\lambda)| \leqq m,\ i = 1, \ldots, n\}.$$

PROOF. The proof proceeds very much like that of Theorem 15. Given $x \in \mathfrak{X}$, there exist bounded functions g_i^m such that

$$x = \lim_{m\to\infty} \sum_{i=1}^{n} S(g_i^m)x_i.$$

For each $e \in \mathscr{B}$

$$x_i^*E(e)x = \lim_{m\to\infty} x_i^*S(g_i^m \cdot \chi_e)x_i.$$

Thus $\mu_i(e) = 0$ implies that $E(d)x_i = 0$, if $d \subseteq e$, and so $x_i^*E(e)x = 0$. Consequently, for each i there is a function $f_i \in L_1(\Lambda, \mathscr{B}, \mu_i)$ such that

$$x_i^*E(e)x = \int_e f_i(\lambda)\mu_i(d\lambda).$$

Since

$$\sup_i \int_\Lambda |f_i(\lambda)|\,\mu_i(d\lambda) = \sup_i \text{tot. var.}\{x_i^*E(\cdot)x\} \leqq 4M\,|x| \sup_i |x_i^*|,$$

the linear map $T: x \to [f_1, \ldots, f_n]$ is a continuous map of $\mathfrak{X}$ into $\sum_{i=1}^{n} L_1(\Lambda, \mathscr{B}, \mu_i)$. Now let $e_m = \{\lambda \,|\, |f_i(\lambda)| \leqq m,\ i = 1, \ldots, n\}$. We show next that

$$(*) \qquad E(e_m)x = \sum_{i=1}^{n} S(f_i \cdot \chi_{e_m})x_i, \qquad m = 1, 2, \ldots .$$

Fix m and let $x^* \in \mathfrak{X}^*$. We may find functions $p_i \in L_1(\Lambda, \mathscr{B}, \mu_i)$ such that

$$x^*E(e)x_i = \int_e p_i(\lambda)\mu_i(d\lambda), \qquad e \in \mathscr{B}.$$

If d is an arbitrary subset of e_m on which $p_1, \ldots, p_n$ are all bounded

$$\begin{aligned}\sum_{i=1}^{n} x^*S(g_i^k)E(d)x_i &= \sum_{i=1}^{n} \int_d g_i^k(\lambda)p_i(\lambda)\mu_i(d\lambda)\\ &\to \sum_{i=1}^{n} \int_d f_i(\lambda)p_i(\lambda)\mu_i(d\lambda)\\ &= \sum_{i=1}^{n} x^*S(f_i \cdot \chi_d)x_i.\end{aligned}$$

However also

$$\lim_{k\to\infty} \sum_{i=1}^{n} x^*S(g_i^k)E(d)x_i = x^*E(d)x.$$

A limiting argument shows

$$x^*E(e_m)x = \sum_{i=1}^{n} x^*S(f_i \cdot \chi_{e_m})x_i.$$

Since x^* is arbitrary we have proved (*). Statement (b) is obtained by letting $m \to \infty$. Also (b) shows that if $Tx = 0$, then $x = 0$. The range of T is dense, as it contains all n-tuples of bounded functions. Q.E.D.

20 COROLLARY. *Under the hypotheses of the theorem* $[\bigvee_{i \in \sigma} \mathfrak{M}(x_i)] \cap [\bigvee_{j \in \sigma'} \mathfrak{M}(x_j)] = (0)$ *for any partition* $[\sigma, \sigma']$ *of the set* $[1, \ldots, n]$.

PROOF. Suppose for example that $0 \neq z \in \bigvee_{i=1}^{p} \mathfrak{M}(x_i) \cap \bigvee_{i=p+1}^{n} \mathfrak{M}(x_i)$. Then there exists a set $e \in \mathscr{B}$ and bounded functions g_i such that

$$E(e)z = \sum_{i=1}^{n} S(g_i)E(e)x_i, \qquad E(e)z \neq 0.$$

Suppose without loss of generality that $S(g_n)E(e)x_n \neq 0$. Then it belongs to $\bigvee_{i=1}^{n-1} \mathfrak{M}(x_i)$, contradicting the fact $\mathfrak{M}(x_n) \cap \bigvee_{i=1}^{n-1} \mathfrak{M}(x_i) = (0)$. Q.E.D.

We now turn our attention to defining a multiplicity function on the complete Boolean algebra B^* in $\mathfrak{X}^*$ composed of all E^* for which E is in the complete Boolean algebra B. It has already been observed in the remarks preceding Definition 11 that B^* is a Boolean algebra of projections in $\mathfrak{X}^*$ isomorphic with B, that is, $(E \vee F)^* = E^* \vee F^*$, $(E \wedge F)^* = E^* \wedge F^*$ (from which it follows immediately that for an arbitrary set $\{E_\alpha\} \subset B$, $(\bigvee E_\alpha)^* = \bigvee E_\alpha^*$, $(\bigwedge E_\alpha)^* = \bigwedge E^*$). It was also observed that B and B^* have the same bound, that $E \leqq F$ if and only if $E^* \leqq F^*$ and the range of a projection E^* in B^* is the $\mathfrak{X}$-closed linear manifold $\{x^* \mid x^*(I - E)\mathfrak{X} = 0\}$. In Lemma 10 the completeness properties of B^* were given and it was observed following Definition 9 that if $\{E_\alpha\} \subset B$, then

$$(\bigvee_\alpha E_\alpha^*)\mathfrak{X}^* = \mathfrak{X}\ \mathrm{sp}\{E_\alpha^*\mathfrak{X}^*\}$$

$$(\bigwedge_\alpha E_\alpha^*)\mathfrak{X}^* = \bigcap_\alpha E_\alpha^*\mathfrak{X}^*$$

$$(I - \bigwedge_\alpha E_\alpha^*)\mathfrak{X}^* = \mathfrak{X}\ \mathrm{sp}\{(I - E_\alpha^*)\mathfrak{X}^*\}.$$

From these properties and from Lemma XVII.3.4 it follows that if $\{E_\alpha^*\}$ is an increasing generalized sequence in B^*, then

$$\lim_\alpha (E_\alpha^* x^*)x = ((\bigvee_\alpha E_\alpha^*)x^*)x, \qquad x \in \mathfrak{X},\ x^* \in \mathfrak{X}^*,$$

whereas if $\{E_\alpha\}$ is decreasing

$$\lim_\alpha (E_\alpha^* x^*)x = ((\bigwedge_\alpha E_\alpha^*)x^*)x, \qquad x \in \mathfrak{X},\ x^* \in \mathfrak{X}^*.$$

To verify these convergence properties we have by Lemma XVII.3.4 (for an increasing generalized sequence) strong convergence of $\{E_\alpha x\}$ and, *a fortiori*,

$$\lim_\alpha x^* E_\alpha x = x^*(\bigvee_\alpha E_\alpha)x, \qquad x \in \mathfrak{X},\ x^* \in \mathfrak{X}^*,$$

and since $(\bigvee E_\alpha)^* = \bigvee E_\alpha^*$,

$$\lim_\alpha (E_\alpha^* x^*)x = ((\bigvee_\alpha E_\alpha)^* x^*)x = ((\bigvee_\alpha E_\alpha^*)x^*)x.$$

A similar argument holds if $\{E_\alpha\}$ is decreasing.

For a given $x^* \in \mathfrak{X}^*$ we define $\mathfrak{N}(x^*)$ to be the span in the $\mathfrak{X}$-topology of the set of vectors $\{E^*x^* \mid E^* \in B^*\}$, and call it the *cyclic subspace* generated by x^*. These subspaces will be the building blocks for the multiplicity theory in $\mathfrak{X}^*$. The symbol $\mathscr{C}^*$ will denote the family of projections in B^* which satisfy the countable chain condition. Clearly $E^* \in \mathscr{C}^*$ if and only if $E \in \mathscr{C}$. The projection $\bigwedge \{E^* \mid E^*x^* = x^*\}$ is called the *carrier projection* of x^*.

21 Lemma. *The family $\mathscr{C}^*$ is a dense σ-ideal in B^*. Each $E^* \in \mathscr{C}^*$ is the carrier projection of a vector in $\mathfrak{X}^*$.*

Proof. The first statement follows from Lemma 5. To prove the second let $E^* \in \mathscr{C}^*$ and recall that E is the carrier of a vector $x_0 \in \mathfrak{X}$, by Lemma 5. Let x_0^* be chosen as in Corollary 14. Replacing x_0^* if necessary by $E^*x_0^*$, we may suppose $E^*x_0^* = x_0^*$. Suppose that $F^* \in B^*$ and $F^*x_0^* = x_0^*$. Then $(I^* - F^*)x_0^* = 0$. Thus $x_0^*(I - F)x_0 = 0$, from which it follows $Fx_0 = x_0$. Thus $F \geqq E$ so $F^* \geqq E^*$, which shows that E^* is the carrier of x_0^*. Q.E.D.

Not every carrier projection in B^* belongs to $\mathscr{C}^*$ as one sees from the following example. Let R be the real line, Γ be the σ-field of all subsets of R and γ be the measure which assigns to each point in R the mass one. Taking $\mathfrak{X} = L_1(R, \Gamma, \gamma)$, $\mathfrak{X}^* = M(R)$, the Banach space of all bounded functions on R. For $e \in \Gamma$ define $(E(e)f)(r) = \delta_r^e f(r)$ where $\delta_r^e = 0$ if $r \notin e$, $\delta_r^e = 1$ if $r \in e$. The Boolean algebra of such projections is complete and $\mathscr{C}$ corresponds to the countable sets. If x^* is the function identically one in $M(R)$, its carrier is I^* which is not in $\mathscr{C}^*$.

22 DEFINITION. If $E^* \in \mathscr{C}^*$ we define $m(E^*)$, the *multiplicity* of E^*, to be the smallest cardinal power of a set A of vectors such that $E^*\mathfrak{X}^*$ is spanned in the $\mathfrak{X}$-topology by the manifolds $\{\mathfrak{M}(x^*) \mid x^* \in A\}$.

The next lemma establishes the required continuity properties.

23 LEMMA. *If E^*, $F^* \in \mathscr{C}^*$ and $F^* \leqq E^*$, then $m(F^*) \leqq m(E^*)$. If $\{E_\alpha^*\} \subseteqq \mathscr{C}^*$ and $E_0^* = \bigvee E_\alpha^* \in \mathscr{C}^*$, then $m(E_0^*) = \bigvee m(E_\alpha^*)$.*

PROOF. The proof is almost identical to that of Lemma 7. For example, in the proof of the second statement one shows, using Zorn's lemma as before, that $E_0^* = \bigvee_{n=1}^{\infty} G_n^*$, where each G_n^* belongs to $\mathscr{C}^*$ and dominates some $E_{\alpha_n}^*$. There exist families $\{x_k^{*k}\}$ such that $G_k^*\mathfrak{X}^*$ is spanned in the $\mathfrak{X}$-topology by the manifolds $\{\mathfrak{M}(x_n^{*k}) \mid 1 \leqq n \leqq n_0\}$, $n_0 = \bigvee m(E_\alpha^*)$. The functionals $y_n^* = \sum_{k=1}^{\infty} 2^{-k} x_n^{*k}$ span $E_0^*\mathfrak{X}$. Q.E.D.

Now Theorem 3 yields

24 THEOREM. *There exists a unique multiplicity function m defined on B^* with the property that for each $F^* \in \mathscr{C}^*$, $m(F^*)$ is the least cardinal power of a set of cyclic subspaces spanning $F^*\mathfrak{X}^*$ in the $\mathfrak{X}$-topology. There is a unique decomposition of the identity $I^* = \bigvee F_n^*$ into disjoint projections such that if $F_n^* \neq 0$, F_n^* has uniform multiplicity n.*

We may also consider B^* as a projection valued measure on the Borel sets $\mathscr{B}$ of Λ. Now the measures $E^*(\cdot)x^*x$ are countably additive for each $x \in \mathfrak{X}$, $x^* \in \mathfrak{X}^*$. However, the vector measures $E^*(\cdot)x^*$ may not be countably additive unless $\mathfrak{X}$ is reflexive. If f is a bounded Borel function the integral $S^*(f) = \int_\Lambda f(\lambda)E^*(d\lambda)$ exists in the uniform operator topology and clearly the operator $S^*(f)$ is just the adjoint $S(f)^*$ of $S(f) = \int_\Lambda f(\lambda)E(d\lambda)$. Thus if f is an unbounded function, we are led to define $S^*(f)$ to be the adjoint of the closed, densely defined operator $S(f)$. That is, $\mathfrak{D}(S^*(f))$ is the set of all x^* such that $z^*(x) = x^*S(f)x$ is continuous for x in $\mathfrak{D}(S(f))$ and $S^*(f)x^* = z^*$ for $x^* \in \mathfrak{D}(S^*(f))$. The proof of the following lemma is straightforward and will be omitted.

25 LEMMA. (a) *The operator $S^*(f)$ is closed and $\mathfrak{D}(S^*(f))$ is $\mathfrak{X}$-dense in $\mathfrak{X}^*$.*

(b) *Let $e_m = \{\lambda \mid |f(\lambda)| \leqq m\}$. Then $x^* \in \mathfrak{D}(S^*(f))$ if and only if*

$$\lim_{n \to \infty} S^*(f \cdot \chi_{e_m})x^*x \text{ exists for each } x \in \mathfrak{X}.$$

If this limit exists it equals $S^*(f)x^*x$.

(c) *If* g *is a bounded function and* $x^* \in \mathfrak{D}(S^*(f))$, *then* $S^*(g)x^* \in \mathfrak{D}(S^*(f))$ *and* $S^*(g)S^*(f)x^* = S^*(f)S^*(g)x^*$.

The next task is to prove that $\mathfrak{N}(x^*) = \{S^*(f)x^* \mid x^* \in \mathfrak{D}(S^*(f))\}$. For this and other purposes we shall need a separation theorem analogous to Theorem 12.

26 THEOREM. *Let* $\mathfrak{N}$ *be an* $\mathfrak{X}$*-closed invariant subspace in* $\mathfrak{X}^*$ *and let* x_0^* *be a functional whose carrier is in* $\mathscr{C}^*$. *If* $\mathfrak{N} \cap \mathfrak{N}(x_0^*) = (0)$, *then there exists a vector* $x_0 \in \mathfrak{X}$ *such that*

(1) $y^*(x_0) = 0$ *for all* $y^* \in \mathfrak{N}$,

(2) $S^*(f)x_0^*x_0 > 0$ *for each bounded non-negative Borel function* f *such that* $S^*(f)x_0^* \neq 0$.

PROOF. We shall merely outline the proof, as it is similar to that of Theorem 12. As before, it is sufficient to suppose the carrier of x_0^* is the identity I^*. It is well known (cf. Theorem V.3.9) that the $\mathfrak{X}$-continuous linear functionals on $\mathfrak{X}^*$ are precisely those induced by elements of $\mathfrak{X}$. Also it was shown in the remarks following Corollary 20 that if $\{G_\alpha^*\}$ is a decreasing generalized sequence in B^*, $\lim G_\alpha^* x^*(x) = (\bigwedge G_\alpha^*)x^*(x)$ for $x \in \mathfrak{X}$, $x^* \in \mathfrak{X}^*$. Using these two facts it follows much as before that if $0 \neq F^* \in B^*$, there exists a projection G^* in B^* and a nonzero vector y in $\mathfrak{X}$ such that

(i) $G^* = \bigwedge \{E^* \mid E^*x^*y = x^*y,\ x^* \in \mathfrak{N}(x_0^*)\}$,

(ii) $y^*y = 0$ for $y^* \in \mathfrak{N}$,

(iii) $0 \neq G^* \leqq F^*$.

Now using Zorn's lemma, express I^* as the union of a sequence of disjoint projections G_n^*, each of which satisfies (i) for an appropriate vector y_n of norm one. The vector x_0 is obtained from $y_0 = \sum_{n=1}^{\infty} 2^{-n} y_n$ by the formula $x_0 = S(f)y_0$, where f is the Radon-Nikodým derivative of total variation $x_0^*E(\cdot)y_0$ with respect to $x_0^*E(\cdot)y_0$. Q.E.D.

→ 27 THEOREM. *If* $x_0^* \in \mathfrak{X}^*$, *then*

$$\mathfrak{N}(x_0^*) = \{S^*(f)x_0^* \mid x_0^* \in \mathfrak{D}(S^*(f))\}.$$

PROOF. It is clear from Lemma 25 that each functional $S^*(f)x_0^*$ is in $\mathfrak{N}(x_0^*)$. To prove the converse it will be sufficient to suppose the carrier of

x_0^* is in $\mathscr{C}^*$, since in any case it is the union of disjoint projections in $\mathscr{C}^*$. If $z^* \in \mathfrak{N}(x_0^*)$, there exists a generalized sequence $\{g_\alpha\}$ of finite linear combinations of characteristic functions such that

$$S^*(g_\alpha)x_0^*(x) \to z^*(x), \qquad x \in \mathfrak{X}.$$

By Theorem 26 there exists a vector x_0 such that

$$\nu_0 = E^*(\cdot)x_0^*x_0 = x_0^*E(\cdot)x_0 \geqq 0,$$

while if $E^*(e_0)x_0^*x_0 = 0$ then $E^*(e_0)x_0^* = 0$. Clearly, $z^*E(\cdot)x_0$ is dominated by ν_0 so there exists an $f \in L_1(\Lambda, \mathscr{B}, \nu_0)$ such that

$$z^*E(e)x_0 = \int_e f(\lambda)\nu_0(d\lambda), \qquad e \in \mathscr{B}.$$

If h is any bounded Borel function

$$\begin{aligned}\int_\Lambda f(\lambda)h(\lambda)\nu_0(d\lambda) &= z^*S(h)x_0\\ &= \lim_\alpha S^*(g_\alpha)x_0^*S(h)x_0\\ &= \lim_\alpha \int_\Lambda g_\alpha(\lambda)h(\lambda)\nu_0(d\lambda),\end{aligned}$$

showing $g_\alpha \to f$ weakly in $L_1(\Lambda, \mathscr{B}, \nu_0)$.

It remains to show that $z^* = S^*(f)x_0^*$. If $x \in \mathfrak{X}$, then ν_0 dominates $x_0^*E(\cdot)x = E^*(\cdot)x_0^*x$, since $\nu_0(e) = 0$ implies $E^*(e)x_0^* = 0$, as remarked above. Let $p \in L_1(\Lambda, \mathscr{B}, \nu_0)$ be chosen so that

$$x_0^*E(e)x = \int_e p(\lambda)\nu_0(d\lambda), \qquad e \in \mathscr{B},$$

and set $e_m = \{\lambda \mid |f(\lambda)| \leqq m\}$, $m = 1, 2, \ldots$. If d is any subset of e_m on which p is bounded,

$$\begin{aligned}S^*(g_\alpha)x_0^*E(d)x &= \int_d g_\alpha(\lambda)p(\lambda)\nu_0(d\lambda)\\ &\to \int_d f(\lambda)\nu_0(d\lambda),\end{aligned}$$

as $g_\alpha \to f$ weakly. Thus $z^*E(d)x = S^*(f \cdot \chi_d)x_0^*x$. By a limiting argument $z^*E(e_m)x = S^*(f \cdot \chi_{e_m})x_0^*x$, and since x is arbitrary, $E^*(e_m)z^* = S^*(f \cdot \chi_{e_m})x_0^*$. It now follows easily that $x_0^* \in \mathfrak{D}(S^*(f))$ and $z^* = S^*(f)x_0^*$. Q.E.D.

Analogous to Lemma 17 we have

28 LEMMA. *Let $\mathfrak{N}$ be an $\mathfrak{X}$-closed invariant subspace in $\mathfrak{X}^*$ and let $y_0^* \in \mathfrak{X}^*$. There exists a maximal projection $E_0^* \in B^*$ such that $E_0^* y_0^* \in \mathfrak{N}$. Moreover $\mathfrak{N} \cap \mathfrak{N}((I^* - E_0^*)x_0^*) = (0)$.*

The proof is similar to that of Lemma 17 and will be omitted.

We turn now to the representation of $E^*\mathfrak{X}^*$ when $E^* \in \mathscr{C}^*$ and has finite uniform multiplicity n. The results are parallel to those in $\mathfrak{X}$. The weaker form of completeness enjoyed by B^* leads to a weaker form of the representation theorem, as the relevant topology in $\mathfrak{X}^*$ is now the $\mathfrak{X}$-topology. As before we shall suppose for convenience that E^* is the identity I^*.

29 LEMMA. *Let the identity I^* in B^* satisfy the countable chain condition and have finite uniform multiplicity n. If $x_1^*, \ldots, x_n^*$ is any set of functionals such that $\mathfrak{X}^* = \bigvee_{i=1}^n \mathfrak{N}(x_i^*)$ (here we are using the symbol $\bigvee_{i=1}^n \mathfrak{N}(x_i^*)$ in place of $\mathfrak{X}$ sp$\{\mathfrak{N}(x_i^*), i = 1, \ldots, n\}$), then*

(1) *the carrier of each x_i^* is I^*;*

(2) $\mathfrak{N}(x_i^*) \cap \bigvee_{j \neq i} \mathfrak{N}(x_j^*) = 0, \qquad i = 1, \ldots, n;$

(3) *for each i there exists a vector $x_i \in \mathfrak{X}$ such that $y^*(x_i) = 0$ for $y^* \in \bigvee_{j \neq i} \mathfrak{N}(x_j^*)$, while $S^*(f)x_i^* x_i > 0$ for each non-negative bounded Borel function f such that $S^*(f)x_i^* \neq 0$.*

PROOF. This lemma is proved from Lemma 28 and Theorem 26 exactly as Lemma 18 was proved from Lemma 17 and Theorem 12. Q.E.D.

→ 30 THEOREM. *Let the hypotheses of Lemma 29 be satisfied. There exists a linear one-to-one map U of $\mathfrak{X}^*$ onto a (norm) dense subspace of the direct sum $\sum_{i=1}^n L_1(\Lambda, \mathscr{B}, \nu_i)$, $\nu_i = E^*(\cdot)x_i^* x_i$, such that if $Ux^* = [f_1, \ldots, f_n]$ then*

$$\text{(a)} \quad E^*(e)x^*x_i = \int_e f_i(\lambda)\nu_i(d\lambda), \qquad e \in \mathscr{B}, i = 1, \ldots, n.$$

$$\text{(b)} \quad x^*(x) = \lim_{m\to\infty} \sum_{i=1}^n S^*(f_i \cdot \chi_{e_m})x_i^*(x), \qquad x \in \mathfrak{X},$$

where

$$e_m = \{\lambda \mid |f_i(\lambda)| \leqq m, i = 1, \ldots, n\}.$$

The map U is continuous when $\mathfrak{X}^*$ and $\sum_{i=1}^n L_1(\Lambda, \mathscr{B}, \nu_i)$ have their norm topologies and also when $\mathfrak{X}^*$ has its $\mathfrak{X}$-topology and $\sum_{i=1}^n L_1(\Lambda, \mathscr{B}, \nu_i)$ has its weak topology.

PROOF. Most of the proof of this theorem is a straightforward modification of the proof of Theorem 19. Given $x^* \in \mathfrak{X}^*$ there exist generalized sequences g_i^α of bounded functions such that

$$x^*(x) = \lim \sum_{i=1}^{n} S^*(g_i^\alpha)x^*(x), \qquad x \in \mathfrak{X}.$$

For each $e \in \mathscr{B}$

$$x^*E(e)x_i = \lim_a x^*S(g_i^\alpha \cdot \chi_e)x_i,$$

and the Radon-Nikodým theorem yields functions $[f_1, \ldots, f_n]$ such that

$$x^*E(e)x_i = \int_e f_i(\lambda)\nu_i(d\lambda), \qquad e \in \mathscr{B},$$

yielding (a). The argument proceeds much as before to yield (b) and the continuity of U for the norm topologies.

Suppose now that $z_\alpha^*(x) \to z_0^*(x)$ for each $x \in \mathfrak{X}$. Let $[h_1, \ldots, h_n]$ be any n-tuple of bounded Borel functions. Then for each $i = 1, \ldots, n$

$$z_a^*S(h_i)x_i \to z_0^*S(h_i)x_i.$$

Let $Uz_\alpha^* = [f_1^\alpha, \ldots, f_n^\alpha]$ and $Uz_0^* = [f_1^0, \ldots, f_n^0]$. Then using Lemma 29, we have

$$\begin{aligned} z_\alpha^*S(h_i)x_i &= \lim_{m\to\infty} \sum_{j=1}^{n} S^*(f_j^\alpha \cdot \chi_{e_m})x_j^*S(h_i)x_i \\ &= \lim_{m\to\infty} S^*(f_i^\alpha \cdot \chi_{e_m})x_i^*S(h_i)x_i \\ &= \lim_{m\to\infty} \int_{e_m} f_i^\alpha(\lambda)h_i(\lambda)\nu_i(d\lambda) \\ &= \int_\Lambda f_i^\alpha(\lambda)h_i(\lambda)\nu_i(d\lambda). \end{aligned}$$

Similarly

$$z_0^*S(h_i)x_i = \int_\Lambda f_i^0(\lambda)h_i(\lambda)\nu_i(d\lambda).$$

Consequently,

$$\lim_\alpha \int_\Lambda f_i^\alpha(\lambda)h_i(\lambda)\nu_i(d\lambda) = \int_\Lambda f_i^0(\lambda)h_i(\lambda)\nu_i(d\lambda), \qquad i = 1, \ldots, n$$

showing the continuity of U when $\mathfrak{X}^*$ has its $\mathfrak{X}$-topology and $\sum_{i=1}^n L_1(\Lambda, \mathscr{B}, \nu_i)$ has its weak topology. Q.E.D.

31 COROLLARY. *Under the hypotheses of Theorem* 30, $[\bigvee_{i\in\sigma} \mathfrak{N}(x_i^*)] \cap [\bigvee_{j\in\sigma'} \mathfrak{N}(x_j^*)] = (0)$ *for any partition* $[\sigma, \sigma']$ *of the set* $[1, \ldots, n]$.

We are now able to prove the following important theorem.

→ 32 THEOREM. *Let B be a complete Boolean algebra of projections in a Banach space $\mathfrak{X}$ and let B^* be the Boolean algebra of adjoints in B^*. Then a projection E in B has finite uniform multiplicity n if and only if its adjoint E^* in B^* has finite uniform multiplicity n.*

PROOF. It is sufficient to suppose E and E^* satisfy the countable chain condition. Also since each projection is the union of projections of uniform multiplicity, it is enough to show (a) if E has finite uniform multiplicity n, then $m(E^*) \leqq n$; and (b) if E^* has finite uniform multiplicity n, then $m(E) \leqq n$. To prove (a) suppose $E\mathfrak{X} = \bigvee_{i=1}^{n} \mathfrak{M}(x_i)$ and choose x_i^* as in Lemma 18. If $E^*\mathfrak{X}^* \neq \bigvee_{i=1}^{n} \mathfrak{N}(x_i^*)$, then by the Hahn-Banach theorem for the $\mathfrak{X}$-topology of $\mathfrak{X}^*$ (cf. Corollary V.3.12) there exists a nonzero vector $x \in E\mathfrak{X}$ such that $y^*x = 0$ for $y^* \in \bigvee_{i=1}^{n} \mathfrak{N}(x_i^*)$. Then $x_i^*E(e)x = 0$, $e \in \mathscr{B}$, $i = 1, \ldots, n$. Thus $x = 0$ by the one-to-one character of the map T of Theorem 19. The proof of (b) follows in the same way from the fact U is one-to-one (Theorem 30). Q.E.D.

33 COROLLARY. *If n is an integer and $E \in B$, then $m(E) = n$ if and only if $m(E^*) = n$.*

34 COROLLARY. *If $\mathfrak{X}$ is separable, then $m(E) = m(E^*)$ for every $E \in B$.*

PROOF. For each *integer* n let E_n and F_n^* be respectively the projections of uniform multiplicity n of Theorems 8 and 24. Then, in view of Theorem 32, $F_n^* = E_n^*$, the adjoint of E_n. Since $\mathfrak{X}$ is separable, $E_0 = I - \bigvee_{n=1}^{\infty} E_n$ is zero or has uniform multiplicity $\aleph_0$. To complete the proof it is sufficient to prove that if $E_0 \neq 0$, then $m(E_0^*) = \aleph_0$. Now Theorem V.4.2 shows that the unit sphere in $\mathfrak{X}^*$ is compact, and since $\mathfrak{X}$ is separable, Theorem V.5.1 shows that it is a metric space and hence separable in the $\mathfrak{X}$-topology. Thus $m(E_0^*) \leqq \aleph_0$ and from the definition of the projections F_n^* it is seen that E_0^* can have no part of finite nonzero multiplicity. Q.E.D.

We shall now discuss a problem of similarity equivalence for certain spectral operators of scalar type. Let Q be a scalar type operator with resolution of the identity $E(\cdot)$. Since $E(\cdot)$ is countably additive in the strong operator topology, the Boolean algebra B consisting of the range of $E(\cdot)$ is σ-complete. Thus its closure $\bar{B}^s$ in the strong operator topology is complete (cf. Lemma XVII.3.23) and the preceding multiplicity theory is applicable to $\bar{B}^s$. We shall assume throughout what follows that $\bar{B}^s$ contains no projections of infinite uniform multiplicity. Our objective will be to show that, with this restriction, the operator Q is, in a suitable sense, similarity equivalent to a normal operator on a Hilbert space. However, rather than seek the maximum generality, it will be convenient to suppose that *B is itself complete and satisfies the countable chain condition.* Both these properties hold for B if $\mathfrak{X}$ is separable, so this will be assumed for the rest of this section.

By Theorem 8 there exist disjoint projections E_n, $n = 1, 2, \ldots$, such that $I = \bigvee_{n=1}^{\infty} E_n$ and each E_n is either zero or has uniform multiplicity n. (By our assumption above $E_{\aleph_0} = 0$.) Corresponding to each n there is a Borel set e_n in the plane such that $E_n = E(e_n)$. We may suppose the sets e_n are disjoint and $\bigcup_{n=1}^{\infty} e_n = \mathfrak{P}$, the whole plane.

Now consider n fixed and suppose $E_n \neq 0$. It follows from Lemma 18 and Theorem 19 that there exist vectors $x_1, \ldots, x_n$ and functions $x_1^*, \ldots, x_n^*$ such that $E_n \mathfrak{X} = \bigvee_{i=1}^{n} \mathfrak{M}(x_i)$, $x_i^*(\bigvee_{j \neq i} \mathfrak{M}(x_j)) = 0$, and the measures $\mu_i = x_i^* E(\cdot) x_i$, now defined on the Borel sets of the plane $\mathfrak{P}$ are positive and vanish outside e_n. Moreover, there is a natural continuous linear map T_n of $E_n \mathfrak{X}$ into $\sum_{i=1}^{n} L_1(\mathfrak{P}, \mathscr{B}, \mu_i)$ with densely defined inverse. Let W_n denote the identity map of $\mathfrak{H}_n = \sum_{i=1}^{n} L_2(\mathfrak{P}, \mathscr{B}, \mu_i)$ into $\sum_{i=1}^{n} L_1(\mathfrak{P}, \mathscr{B}, \mu_i)$. As the measures μ_i are finite, W_n is defined and continuous, and the map $A_n = W_n^{-1} T_n$ is a densely defined closed map of $E_n \mathfrak{X}$ into $\mathfrak{H}_n$ with densely defined inverse. We suppose the norm of $[h_1, \ldots, h_n]$ in $\mathfrak{H}_n$ is defined to be

$$\left[\sum_{i=1}^{n} \int_{\mathfrak{P}} |h_i(\lambda)|^2 \, \mu_i(d\lambda)\right]^{1/2}$$

so that $\mathfrak{H}_n$ is a Hilbert space. It is easily seen that

$$\mathfrak{D}(A_n) = \{x \in E_n \mathfrak{X} \,|\, (T_n x)_i(\cdot) \in L_2(\mathfrak{P}, \mathscr{B}, \mu_i),\ i = 1, \ldots, n\},$$

$$\mathfrak{D}(A_n^{-1}) = \{[h_1, \ldots, h_n] \in \mathfrak{H}_n \,|\, \lim_{m \to \infty} \sum_{i=1}^{n} S(h_i \cdot \chi_{e_m}) x_i \text{ exists}\},$$

where $e_m = \{\lambda \,|\, |h_i(\lambda)| \leqq m,\ i = 1, \ldots, n\}$. Moreover, if g is a bounded Borel function on $\mathfrak{P}$, the closure in $\mathfrak{H}_n$ of the densely defined operator $A_n S(g) A_n^{-1}$ sends $[h_1, \ldots, h_n]$ into $[gh_1, \ldots, gh_n]$, that is, corresponds to multiplication by g. Taking for g the characteristic function χ_e of a Borel set, there results an isomorphic correspondence $E(e) \to \chi_e$ between subprojections of E_n in B and a Boolean algebra $\tilde{B}_n$ of self adjoint projections in $\mathfrak{H}_n$. Moreover, $\tilde{B}_n$ is the resolution of the identity of the bounded normal operator $\tilde{Q}_n = \int_{e_n} \lambda \tilde{E}(d\lambda)$ which is the closure of $A_n E(e_n) Q A_n^{-1}$.

Now let the Hilbert space $\mathfrak{H} = \sum_{n=1}^{\infty} \mathfrak{H}_n$ be the direct sum of the Hilbert spaces $\mathfrak{H}_n$. Elements of $\mathfrak{H}$ will be denoted by the single letter h. We shall define a map A of $\mathfrak{X}$ into $\mathfrak{H}$. Let

$$\mathfrak{D}(A) = \{x \mid E_n x \in \mathfrak{D}(A_n) \text{ for all } n \text{ and } \sum_{n=1}^{\infty} A_n E_n x \text{ converges in } \mathfrak{H}\}.$$

Define

$$Ax = \sum_{n=1}^{\infty} A_n E_n x, \qquad x \in \mathfrak{D}(A).$$

The properties of A are collected in the next lemma.

35 LEMMA. *The operator A is a densely defined closed linear map of $\mathfrak{X}$ into $\mathfrak{H}$ with densely defined inverse. Moreover if P_n denotes the perpendicular projection of $\mathfrak{H}$ onto $\mathfrak{H}_n$, then*

$$\mathfrak{D}(A^{-1}) = \{h \mid P_n h \in \mathfrak{D}(A_n^{-1}) \textit{ for all } n \textit{ and } \sum_{n=1}^{\infty} A_n^{-1} P_n h \textit{ converges in } \mathfrak{X}\}$$

and

$$A^{-1}h = \sum_{n=1}^{\infty} A_n^{-1} P_n h, \qquad h \in \mathfrak{D}(A^{-1}).$$

PROOF. Let $y_m \in \mathfrak{D}(A)$, $y_m \to y_0$ and $Ay_m \to h_0$. Then $E_n y_m \to E_n y_0$, and $P_n A y_m = A_n E_n y_m \to P_n h_0$, $n = 1, 2, \ldots$. Since A_n is closed, $E_n y_0 \in \mathfrak{D}(A_n)$ and $A_n E_n y_0 = P_n h_0$. Now

$$\sum_{n=1}^{\infty} A_n E_n y_0 = \sum_{n=1}^{\infty} P_n h_0 = h_0.$$

Thus $y_0 \in \mathfrak{D}(A)$ and $Ay_0 = h_0$, showing that A is closed. If $Ax = 0$, then $A_n E_n x = E_n A x = 0$. Since A_n^{-1} exists, $E_n x = 0$, so $x = \sum_{n=1}^{\infty} E_n x = 0$ proving that A^{-1} exists. Since each A_n^{-1} has dense domain, the same is true of A^{-1}. The formula for $\mathfrak{D}(A^{-1})$ is easily verified. Q.E.D.

The next theorem is a straightforward consequence of the preceding discussion.

→ 36 THEOREM. *Let $\mathfrak{X}$ be a separable complex Banach space and Q be a bounded scalar type spectral operator in $\mathfrak{X}$. Let B denote the range of the* *resolution of the identity* $E(\cdot)$ *of* Q *and suppose that* B *contains no projection of infinite uniform multiplicity. If* A *is the closed densely defined linear map of* $\mathfrak{X}$ *into the Hilbert space* $\mathfrak{H}$ *of Lemma* 35, *then the closure* $\tilde{Q}$ *of* AQA^{-1} *is a bounded normal operator. For each bounded Borel function* g *on the plane let* $\tilde{S}(g)$ *denote the closure of* $AS(g)A^{-1}$. *The correspondence* $\tau : S(g) \to \tilde{S}(g)$ *preserves the operational calculus and*

$$\tilde{Q} = \int_{\sigma(Q)} \lambda \tilde{E}(d\lambda),$$

where $\tau E(e) = \tilde{E}(e)$, $e \in \mathscr{B}$.

Our final objective is to show that the operator A induces a one-to-one correspondence between the subspaces of $\mathfrak{X}$ invariant under B and those of $\mathfrak{H}$ invariant under $\tilde{B}$, the resolution of the identity of Q.

37 DEFINITION. If $\mathfrak{M}$ is a closed invariant subspace in $\mathfrak{X}$, we denote by $\Phi(\mathfrak{M})$ the closure in $\mathfrak{H}$ of the linear set $A(\mathfrak{M} \cap \mathfrak{D}(A))$. Similarly if $\mathfrak{K}$ is a closed invariant subspace in $\mathfrak{H}$, $\Psi(\mathfrak{K})$ denotes the closure in $\mathfrak{X}$ of $A^{-1}(\mathfrak{K} \cap \mathfrak{D}(A^{-1}))$.

It follows from Theorem 19(b) and Lemma 35 that $\mathfrak{M} \cap \mathfrak{D}(A)$ is dense in $\mathfrak{M}$. Similarly $\mathfrak{K} \cap \mathfrak{D}(A^{-1})$ is dense in $\mathfrak{K}$. It is conceivable that $\Phi(\mathfrak{M})$ contains elements of $\mathfrak{D}(A^{-1})$ not belonging to $A(\mathfrak{M} \cap \mathfrak{D}(A))$, in which case $\Psi(\Phi(\mathfrak{M}))$ contains $\mathfrak{M}$ properly. Our objective is to show that this does not happen. In fact $\Psi(\Phi(\mathfrak{M})) = \mathfrak{M}$, $\Phi(\Psi(\mathfrak{K})) = \mathfrak{K}$, for all $\mathfrak{M}$ and $\mathfrak{K}$. This property of the invariant subspace is a consequence of the spectral structure of $E_n \mathfrak{X}$, $n = 1, 2, \ldots$, and it will be convenient to prove a lemma first about the maps A_n.

38 LEMMA. (a) *If* $\mathfrak{M}$ *is an invariant subspace of* $E_n \mathfrak{X}$ *then*

$$\Phi(\mathfrak{M}) \cap \mathfrak{D}(A_n^{-1}) \subseteq A_n[\mathfrak{M} \cap \mathfrak{D}(A_n)].$$

(b) *If* $\mathfrak{K}$ *is an invariant subspace of* $\mathfrak{H}_n$ *then*

$$\Psi(\mathfrak{K}) \cap \mathfrak{D}(A_n) \subseteq A_n^{-1}[\mathfrak{K} \cap \mathfrak{D}(A_n^{-1})].$$

PROOF. To prove (a) suppose $h^0 = [h_1^0, \ldots, h_n^0]$ is an element of $\Phi(\mathfrak{M}) \cap \mathfrak{D}(A_n^{-1}) \subseteq \mathfrak{H}_n$, and that $h^0 = A_n y_0$. We must show $y_0 \in \mathfrak{M}$. There exist vectors y_m in $\mathfrak{M} \cap \mathfrak{D}(A_n)$ such that $A_n y_m = [h_1^m, \ldots, h_n^m] \to [h_1^0, \ldots, h_n^0]$ in $\mathfrak{H}_n = \sum_{i=1}^n L_2(\mathfrak{P}, \mathscr{B}, \mu_i)$. By extracting subsequences we may suppose $h_i^m \to h_i^0$ almost uniformly with respect to μ_i, $i = 1, \ldots, n$. Given $\varepsilon > 0$, there exists a Borel set e_ε such that $\mu_i(e_\varepsilon') < \varepsilon$, $i = 1, \ldots, n$, and $h_i^m \to h_i^0$ uniformly on e_ε. The functions h_i^m are bounded on e_ε and

$$E(e_\varepsilon)y_m = \sum_{i=1}^{n} S(h_i^m \cdot \chi_{e_\varepsilon})x_i$$

$$\to \sum_{i=1}^{n} S(h_i^0 \cdot \chi_e) = E(e_\varepsilon)y_0,$$

so $E(e_\varepsilon)y_0 \in \mathfrak{M}$. An obvious limiting argument shows that $y_0 \in \mathfrak{M}$, establishing (a).

If $y_0 \in \Psi(\mathfrak{K}) \cap \mathfrak{D}(A_n)$ and $y_0 = A_n^{-1}h_0$, there is a sequence $\{h^m\}$ of n-tuples in $\mathfrak{K} \cap \mathfrak{D}(A^{-1})$ such that $y_m = A_n^{-1}h^m \to y_0$. Then $T_n y_m = [h_1^m, \ldots, h_n^m] \to [h_1^0, \ldots, h_n^0]$ in $\sum_{i=1}^n L_1(\mathfrak{P}, \mathscr{B}, \mu_i)$. Consequently, given $\varepsilon > 0$, $\tilde{E}(e_\varepsilon)h^m \to \tilde{E}(e_\varepsilon)h^0$ in subsequence, where $\mu_i(e_\varepsilon') < \varepsilon$, $i = 1, \ldots, n$. Thus $h_0 \in \mathfrak{K}$, so $y_0 \in A_n^{-1}[\mathfrak{K} \cap \mathfrak{D}(A_n^{-1})]$. Q.E.D.

39 THEOREM. *Let the hypotheses of Theorem 36 be satisfied and let the maps Φ and Ψ be defined as in Definition 37. Then there exists a one-to-one correspondence between the invariant subspaces $\mathfrak{M}$ of $\mathfrak{X}$ and the invariant subspaces $\mathfrak{K}$ of $\mathfrak{H}$ given by the relations*

$$\Psi(\Phi(\mathfrak{M})) = \mathfrak{M}, \qquad \Phi(\Psi(\mathfrak{K})) = \mathfrak{K}, \qquad \mathfrak{M} \subseteq \mathfrak{X}, \qquad \mathfrak{K} \subseteq \mathfrak{H}.$$

PROOF. The desired relations follow immediately from the two equations

$$\text{(a)} \qquad \Phi(\mathfrak{M}) \cap \mathfrak{D}(A^{-1}) = A(\mathfrak{M} \cap \mathfrak{D}(A))$$

and

$$\text{(b)} \qquad \Psi(\mathfrak{K}) \cap \mathfrak{D}(A) = A^{-1}(\mathfrak{K} \cap \mathfrak{D}(A^{-1})).$$

We shall prove (a), the proof of (b) being similar. Clearly the right side of (a) is contained in the left. It follows from Lemma 38 that

$$(\#) \qquad \Phi(E_n\mathfrak{M}) \cap \mathfrak{D}(A^{-1}) \subseteq A[(E_n\mathfrak{M}) \cap \mathfrak{D}(A)], \qquad n = 1, 2, \ldots.$$

Now

$$(E_n\mathfrak{M}) \cap \mathfrak{D}(A) = E_n(\mathfrak{M} \cap \mathfrak{D}(A))$$

and

$$P_n A(\mathfrak{M} \cap \mathfrak{D}(A)) = AE_n(\mathfrak{M} \cap \mathfrak{D}(A)).$$

Taking closures of both sides of the last equation shows that $P_n\Phi(\mathfrak{M}) = \Phi(E_n\mathfrak{M})$, $n = 1, 2, \ldots$. Now let $h \in \Phi(\mathfrak{M}) \cap \mathfrak{D}(A^{-1})$ and $h = Ax$. For each n

$$P_n h \in P_n\Phi(\mathfrak{M}) = \Phi(E_n\mathfrak{M}).$$

Thus by (#)

$$A^{-1}P_n h = E_n x \in E_n \mathfrak{M},$$

so

$$x = \sum_{n=1}^{\infty} E_n x \in \mathfrak{M} \cap \mathfrak{D}(A^{-1}). \qquad \text{Q.E.D.}$$

4. Notes and Remarks

The work presented in this chapter is an elaboration of the theory presented in Bade [2]. As was mentioned earlier, Ionescu Tulcea [3] and Schaefer [10] discussed spectral operators in certain locally convex spaces without always assuming that the domain was the entire space; thus much of their work also applies to unbounded operators.

Nel [1] improved the canonical decomposition of an unbounded spectral operator by giving the following characterization:

THEOREM. *A closed operator T is spectral if and only if it is the minimal closed extension of an operator of the form $S + N$ where S is a scalar type operator with resolution of the identity E and N satisfies the conditions:*

(a) *If b is a bounded Borel set, then $E(b)\mathfrak{X} \subseteq D(N)$, $E(b)NE(b) = NE(b)$, and $N \mid E(b)\mathfrak{X}$ is a quasi-nilpotent operator.*

(b) *If c is a Borel set, $\beta \notin \bar{c}$, and $b_n = \{\lambda \mid |\lambda| \leqq n\}$, then the sequence*

$$(\beta I - T)^{-1}E(b_n \cap c)x = \sum_{k=0}^{+\infty} N^k \int_{b_n \cap c} (\beta - \gamma)^{-k-1} E(d\gamma)x$$

is bounded, for each $x \in \mathfrak{X}$.

The multiplicity theory as presented in Section 3 is due to W. G. Bade [5]. J. Dieudonné [20] had previously obtained a multiplicity theory in the case where the adjoint $\mathfrak{X}^*$ of the underlying Banach space $\mathfrak{X}$ is separable (which implies the separability of $\mathfrak{X}$). In connection with Lemma 3.18 we note that in general $\mathfrak{X}$ is not the algebraic direct sum of the manifolds $\mathfrak{M}(x_i)$. In fact J. Dieudonné [21] has constructed an ingenious example of a space $\mathfrak{X}$ and Boolean algebra of projections such that every nonzero $E \in B$ has multiplicity two. However, for no choice of x_1 and x_2 or $E \in B$ is $E\mathfrak{X}$ the algebraic sum of $\mathfrak{M}(Ex_1)$ and $\mathfrak{M}(Ex_2)$.

A fundamental question remaining unanswered concerns the relation between the multiplicity functions in B and B^* on projections of infinite multiplicity. It was shown in Corollary 3.33 that $m(E) = m(E^*)$ for all projections of finite multiplicity. We have no information on this question. It is undecided even whether E^* has infinite uniform multiplicity when E has infinite uniform multiplicity. A related question concerns invariant subspaces. If $\mathfrak{M}$ is an invariant subspace we may define the multiplicity of B in $\mathfrak{M}$. If $\mathfrak{M}_1 \subseteqq \mathfrak{M}_2$ it is to be expected that $m_1(E) \leqq m_2(E)$, $E \in B$. This is true if B contains no projections of infinite uniform multiplicity, but the question is open in general.

CHAPTER XIX

Perturbations of Spectral Operators with Discrete Spectra

1. Introduction

In this chapter we discuss the problem raised by Theorem XVI.5.2; that is, we establish certain sufficient conditions for the Boolean algebra of projections $E(\sigma; T)$ associated with the components of the spectrum of an operator having totally disconnected spectrum to be uniformly bounded. This will give analytic conditions on an operator which are sufficient to insure that it is a spectral operator. The basic idea of the method is the following: if T is spectral, and if P is, in some suitable sense, sufficiently small relative to T, then $T+P$ will also be spectral. It turns out that in order to make P sufficiently small relative to T, it is convenient to let T be unbounded and thus most of the present chapter deals with perturbations of an unbounded spectral operator T, and is directed toward the perturbation of differential operators.

The main theoretical work of the chapter is done in Section 2, and especially in the proof of Theorem 2.7. In this theorem the perturbation method is used to show that if T is spectral, if the spectrum of T is a discrete sequence with a certain regular distribution, if all but a finite number of the points in $\sigma(T)$ correspond to spectral projections with a one-dimensional range, and if P is small relative to T in a suitable sense, then $T+P$ is spectral. The next two sections deal with applications of this main result to non-symmetric ordinary differential operators. We let the unperturbed operator T be a simple nth order differential operator $i^{-n}(d/dt)^n$, restricted by a set of boundary conditions subject to certain broad conditions of regularity. It then follows that if P is any arbitrary differential operator of lower order, $T+P$ is spectral. In this way it is shown that a large class of nonselfadjoint differential operators are spectral. The general proof of this fact is given in Section 4. Unfortunately, the computational details of this proof are extremely lengthy. For this reason, and also because of the special importance of second order operators in applications, a separate proof of the spectral character of $-(d/dx)^2$ (subject to a set of boundary conditions which make it nonselfadjoint) is given in Section 3. Even though the main result of Section 3 is greatly generalized in Section 4, we feel that it is worth inserting the simpler case of Section 3 as a kind of preparation for the more complicated case.

The question of deciding when the generalized eigenfunctions of an operator span the whole space in which it acts is discussed in Section 5. It is seen that this is the case for a much larger class of operators than those covered by Theorem 2.7.

The case in which one has conditionally (as contrasted to unconditionally) convergent eigenvalue expansions is a case whose generality falls between that of Theorem 2.7 and the general results of Section 5. A number of results concerning the conditional convergence of eigenvalue expansions are mentioned in Section 6 below.

2. The Principal Abstract Perturbation Theorem

In this section we shall study perturbations of an operator whose resolvent is compact. The principal result gives conditions under which such an operator, if spectral, remains spectral after a perturbation. The results are directed toward the applications to nonselfadjoint differential operators which follow in the next two sections. Since the notion of an operator with compact resolvent occurs so frequently in this section, it will be convenient to introduce, in the following definition, a special term for such operators.

→ 1 Definition. An operator T is *discrete* if there is a number λ in its resolvent set for which the resolvent $R(\lambda; T) = (\lambda I - T)^{-1}$ is compact.

Remark. Except in the trivial case where $\mathfrak{X}$ is finite dimensional, a discrete operator T cannot be bounded. For, otherwise, by Theorem VI.5.4, the identity operator

$$I = (\lambda I - T)R(\lambda; T)$$

is compact and thus, by Theorem IV.3.5, $\mathfrak{X}$ is finite dimensional.

The first two lemmas that follow are reminiscent of analogous results established in Chapter VII for bounded operators.

2 LEMMA. *If T is discrete, then*

(a) *its spectrum is a denumerable set of points with no finite limit point;*

(b) *the resolvent $R(\lambda; T)$ is compact for every $\lambda \notin \sigma(T)$;*

(c) *every λ_0 in $\sigma(T)$ is a pole of finite order $\nu(\lambda_0)$ of the resolvent and if, for some positive integer k, f satisfies the equation*

$$(T - \lambda_0 I)^k f = 0,$$

then f satisfies the equation

$$(T - \lambda_0 I)^{\nu(\lambda_0)} f = 0.$$

The set of all vectors f satisfying the equation $(T - \lambda_0 I)^{\nu(\lambda_0)} f = 0$ is a finite dimensional linear space, called the space of generalized eigenvectors of T corresponding to the eigenvalue λ_0. If $E(\lambda_0; T) = E(\lambda_0)$ is the idempotent function of T corresponding to the analytic function which is one near λ_0 and zero elsewhere near the spectrum of T and near infinity, then $E(\lambda_0)$ projects $\mathfrak{X}$ onto the space of generalized eigenvectors corresponding to λ_0.

PROOF. We may suppose without loss of generality that $0 \notin \sigma(T)$, and that T^{-1} is compact. Then, by Theorem VII.4.5, $R(\mu) = (\mu I - T^{-1})^{-1}$ exists for every complex number μ except zero and at most a denumerable sequence of points $\mu_n \neq 0$ tending to zero. For each of these points μ_n, there exists a non-zero vector x_n such that $T^{-1} x_n = \mu_n x_n$. It follows that $T x_n = \mu_n^{-1} x_n$, so that $\sigma(T) \supseteqq \{\mu_n^{-1}\}$. On the other hand, if $\lambda \neq 0$, $\lambda \neq \mu_n^{-1}$, $n \geqq 0$, we have (putting $\lambda = \mu^{-1}$)

$$-(\mu^{-1} I - T)\mu T^{-1} R(\mu) = I$$

and

$$\begin{aligned} -\mu T^{-1} R(\mu)(\mu^{-1} I - T)x &= -R(\mu)\mu T^{-1}(\mu^{-1} I - T)x \\ &= -R(\mu)(T^{-1} - \mu I)x \\ &= x, \qquad x \in \mathfrak{D}(T). \end{aligned}$$

Consequently, $\lambda = \mu^{-1} \in \rho(T)$, and

[*] $$(\mu^{-1} I - T)^{-1} = -\mu T^{-1} R(\mu).$$

This shows that $\sigma(T) = \{\mu_n^{-1}\}$ and establishes (a). Since T^{-1} is compact, it follows from [*] that $(\mu^{-1} I - T)^{-1}$ is compact for $\mu^{-1} \notin \sigma(T)$. This establishes (b).

Since by Theorem VII.4.5 each point μ_n is a pole of some finite order $\nu(\mu_n)$ of $R(\mu)$, it follows from [*] that each point μ_n^{-1} is a pole of finite order $\nu(\mu_n)$ of $(\mu^{-1}I - T)^{-1}$. This proves the first assertion of (c). If a non-zero vector f satisfies the equation $(T - \lambda_0 I)^k f = 0$, then $(\lambda_0^{-1}I - T^{-1})^k f = 0$, so that, by Theorem VII.1.7, f satisfies the equation $(\lambda_0^{-1}I - T^{-1})^{\nu(\lambda_0^{-1})} f = 0$. Now we may easily show inductively that for $\mu \neq 0$ and $k \geqq 0$, $T^k(\mu I - T^{-1})^k = (\mu T - I)^k$ in the strict sense that the operators on both sides of the equation have the same domain. Indeed, this is clearly true for $k = 1$, and since $P(T)Q(T) = Q(T)P(T)$ for any two polynomials in T, we have

$$\begin{aligned} T^k(\mu I - T^{-1})^k &= T(\mu T - I)^{k-1}(\mu I - T^{-1}) \\ &= (\mu T - I)^{k-1} T(\mu I - T^{-1}) \\ &= (\mu T - I)^k. \end{aligned}$$

Thus since $(\lambda_0^{-1}I - T^{-1})^{\nu(\lambda_0^{-1})} f = 0$, it follows that $(T - \lambda_0 I)^{\nu(\lambda_0)} f = 0$, proving the second assertion of (c). Conversely, if $(T - \lambda_0 I)^k f = 0$, we have, upon multiplying by $(T^{-1})^k$, $(\lambda_0^{-1}I - T^{-1})^k f = 0$. It follows then, by Theorem VII.4.5, that

$$\{f \mid (T - \lambda_0 I)^{\nu(\lambda_0)} f = 0\} = E(\lambda_0^{-1}; T^{-1})\mathfrak{X},$$

and, in particular, that $\{f \mid (T - \lambda_0 I)^{\nu(\lambda_0)} f = 0\}$ is finite dimensional. This proves the third assertion of (c).

To prove the last part of (c) we may argue as follows. If C is a small closed curve surrounding the point λ_0 and traversed once in the positive sense, then, by Lemma XVIII.2.31 and by [*],

$$\begin{aligned} E(\lambda_0; T) = E(\lambda_0) &= \frac{1}{2\pi i} \int_C (\lambda I - T)^{-1}\, d\lambda \\ &= \frac{1}{2\pi i} \int_C T^{-1}(T^{-1} - \lambda^{-1} I)^{-1} \lambda^{-1}\, d\lambda \\ &= \frac{1}{2\pi i} \int_{C'} \mu^{-1} T^{-1} (\mu I - T^{-1})^{-1}\, d\mu, \end{aligned}$$

where C' is a small curve surrounding λ_0^{-1} and traversed in the positive sense. By the functional calculus of bounded operators (cf. Theorem VII.3.10) this last expression is

$$\begin{aligned} T^{-1} \frac{1}{2\pi i} \int_{C'} \mu^{-1}(\mu I - T^{-1})^{-1}\, d\mu &= \frac{1}{2\pi i} \int_{C'} (\mu I - T^{-1})^{-1}\, d\mu \\ &= E(\lambda_0^{-1}; T^{-1}). \qquad \text{Q.E.D.} \end{aligned}$$

REMARK. It is to be noted that we have actually proved a little more than is stated in Lemma 2. It has, in fact, been proved that the points of $\sigma(T)$ and the non-zero points of $\sigma(T^{-1})$ are in one-to-one correspondence through the map $\mu \to \mu^{-1}$, that $(\mu I - T)^{-1} = -\mu^{-1}T^{-1}(\mu^{-1}I - T^{-1})^{-1}$ for $\mu \notin \sigma(T)$, and that the corresponding projections are related by the equation $E(\mu; T) = E(\mu^{-1}; T^{-1})$. This is, of course, merely a generalization to unbounded discrete operators of the spectral mapping Theorem VII.3.19. Compare also Theorem VII.3.11, which gives a much more general result than the result $\sigma(T^{-1}) = \overline{\sigma(T)^{-1}}$ which we have proved.

3 LEMMA. *Let T be an unbounded discrete operator. Let σ be a compact open subset of $\sigma(T)$. Then $\mathfrak{D}(T) \supseteq E(\sigma; T)\mathfrak{X}$. The space $E(\sigma; T)\mathfrak{X}$ is invariant under T, and $\sigma(T \mid E(\sigma; T)\mathfrak{X}) = \sigma$. If σ is non-void, $E(\sigma; T) \neq 0$. If $E(\sigma; T)$ has a k-dimensional range, σ consists of at most k points; in particular, if $E(\sigma; T)$ has a one-dimensional range, σ consists of exactly one point.*

PROOF. Let C be a closed rectifiable curve lying in the resolvent set of T and bounding a domain whose intersection with $\sigma(T)$ is precisely σ. Then, by Lemma XVIII.2.31,

$$E(\sigma; T)x = \frac{1}{2\pi i}\int_C (\lambda I - T)^{-1}x\, d\lambda.$$

Since the integral

$$\frac{1}{2\pi i}\int_C T(\lambda I - T)^{-1}x\, d\lambda = \frac{1}{2\pi i}\int_C \lambda(\lambda I - T)^{-1}x\, d\lambda$$

exists, it follows from Theorem III.6.20 that $E(\sigma; T)\mathfrak{X} \subseteq \mathfrak{D}(T)$, and that

$$[*] \qquad TE(\sigma; T) = \frac{1}{2\pi i}\int_C \lambda(\lambda I - T)^{-1}\, d\lambda.$$

It now follows from Theorem VII.9.8 that $E(\sigma; T)Tx = Tx$ for $x \in E(\sigma; T)\mathfrak{X}$, so that $TE(\sigma; T)\mathfrak{X} \subseteq E(\sigma; T)\mathfrak{X}$. By formula [*],

$$[**] \qquad Tx = \frac{1}{2\pi i}\int_C \lambda(\lambda I - T)^{-1}x\, d\lambda, \qquad x \in E(\sigma; T)\mathfrak{X}.$$

For $\mu \notin \sigma$, let $f_\mu(z) = (\mu - z)^{-1}$ for z in a neighborhood of σ, $f_\mu(z) = 0$ for z in a neighborhood of infinity, and of $\sigma(T) - \sigma$. From Theorem VII.9.5 it follows that $f_\mu(T)E(\sigma; T)\mathfrak{X} \subseteq E(\sigma; T)\mathfrak{X}$ and that

$$f_\mu(T)(\mu I - T)x = (\mu I - T)f_\mu(T)x = x, \qquad x \in E(\sigma; T)\mathfrak{X}.$$

Hence $f_\mu(T) \mid E(\sigma; T)\mathfrak{X}$ is the inverse of $(\mu I - T) \mid E(\sigma; T)\mathfrak{X}$, which shows that $\sigma(T \mid E(\sigma; T)\mathfrak{X}) \subseteq \sigma$.

Suppose next that $E(\sigma; T) = 0$, but that σ is not void. Then $E(\lambda; T) = E(\lambda \cap \sigma; T) = E(\lambda; T)E(\sigma; T) = 0$ for λ in σ. It follows from the remark following Lemma 2 that $E(\lambda^{-1}; T^{-1}) = 0$. It then follows from the first assertion of Theorem VII.3.20 that λ^{-1} is in $\rho(T^{-1})$. Consequently, according to the remark following Lemma 2, $\lambda \notin \sigma(T)$, which is contrary to the assumption that $\sigma \subseteq \sigma(T)$. This contradiction proves that $E(\sigma; T) \neq 0$ if σ is not void.

Let λ be in σ. Then $(\lambda I - T)^{\nu(\lambda)}E(\lambda; T)\mathfrak{X} = 0$ by Lemma 2. Thus, since $E(\lambda; T)\mathfrak{X} \neq 0$, λ is in $\sigma(T \mid E(\sigma; T)\mathfrak{X})$. Finally, let σ consist of the k points $\lambda_1, \ldots, \lambda_k$. By Lemma XVIII.2.31, the non-zero subspaces $E(\lambda_i; T)\mathfrak{X}$ of $E(\sigma; T)\mathfrak{X}$ are linearly independent. Hence $E(\sigma; T)\mathfrak{X}$ is at least k-dimensional. This proves the remainder of the present lemma. Q.E.D.

4 Definition. Let T be an unbounded discrete operator in the B-space $\mathfrak{X}$, with spectrum $\{\lambda_i\}$. If $E(\lambda_i; T) = E(\lambda_i)$ for $\lambda_i \in \sigma(T)$, then the linear manifold $\mathfrak{S}_\infty(T)$ is defined by the equation

$$\mathfrak{S}_\infty(T) = \{f \mid E(\lambda_i)f = 0, \ 1 \leqq i < \infty\}.$$

5 Lemma. *The space $\mathfrak{S}_\infty(T)$ either is infinite dimensional or consists only of zero.*

Proof. We may suppose without loss of generality that $0 \notin \sigma(T)$. If $U = T^{-1}$, it then follows from the remark following Definition 1 that

$$\sigma(U) = \bigcup_{i=1}^{\infty} \{\lambda_i^{-1}\} \cup \{0\},$$

and that the projection $\hat{E}(\lambda_i^{-1}) = E(\lambda_i^{-1}; U)$ is defined by the equation

$$\hat{E}(\lambda_i^{-1}) = E(\lambda_i).$$

Hence, if $f \in \mathfrak{S}_\infty = \mathfrak{S}_\infty(T)$, we have

$$\hat{E}(\lambda_i^{-1})Uf = U\hat{E}(\lambda_i^{-1})f = 0,$$

so that $U\mathfrak{S}_\infty \subseteq \mathfrak{S}_\infty$. Moreover, by Theorem VII.3.20, $(U - \lambda I)^{-1}f$ is analytic at every point λ_i^{-1} if $f \in \mathfrak{S}_\infty$. Thus, if $f \in \mathfrak{S}_\infty$, $(U - \lambda I)^{-1}f$ has no singularity other than the origin. Hence the spectrum of the restriction of U to $\mathfrak{S}_\infty$ consists of zero alone. If the space $\mathfrak{S}_\infty$ contains a non-zero element, and is also finite dimensional, then, since $\sigma(U \mid \mathfrak{S}_\infty) = \{0\}$, there must exist an $x \neq 0$ in $\mathfrak{S}_\infty$ such that $Ux = 0$. Multiplying by T, however, gives $TUx = x = 0$. This contradiction proves the lemma. Q.E.D.

6 LEMMA. *Let T be a discrete operator. The space $\mathfrak{S}_\infty(T)$ is the set of all f in $\mathfrak{X}$ for which $(T-\lambda I)^{-1}f$ is an entire function of λ.*

PROOF. If $(T-\lambda I)^{-1}f$ is entire, then by letting C be a small circle around $\lambda_i \in \sigma(T)$ we find that

$$0 = \frac{1}{2\pi i}\int_C (T-\lambda I)^{-1}\,d\lambda = -E(\lambda_i)f.$$

Conversely, let $E(\lambda_i)f = 0$. Then assuming without loss of generality that T^{-1} exists, it follows according to the remark following Lemma 2 that

$$(\mu I - T)^{-1}f = -\mu^{-1}T^{-1}(\mu^{-1}I - T^{-1})^{-1}f, \qquad \mu \notin \sigma(T).$$

Since, by this same remark, $E(\lambda_i^{-1};\, T^{-1})f = 0$, it follows by Theorem VII.3.20 that $(\mu^{-1}I - T^{-1})^{-1}f$ is analytic whenever $\mu \neq 0$. Thus the left-hand side of the equation displayed above is analytic for $\mu \neq 0$, and, since the point $\mu = 0$ is not in $\sigma(T)$, $(\mu I - T)^{-1}f$ is entire. Q.E.D.

Now we come to the fundamental theorem of the present section.

→ 7 THEOREM. *Let T be a discrete spectral operator in a weakly complete space $\mathfrak{X}$, and let E be its resolution of the identity. Suppose that for all but a finite number of spectral points λ the projection $E(\lambda)$ has a one-dimensional range. Let $\lambda_0 \in \rho(T)$, let $0 \leqq \nu < 1$, and let P be an operator such that $\mathfrak{D}(P) \supseteq \mathfrak{D}((T-\lambda_0 I)^\nu)$ and such that $P(T-\lambda_0 I)^{-\nu}$ is bounded. Let $\{\lambda_n\}$ be an enumeration of $\sigma(T)$, and let d_n be the distance from λ_n to $\sigma(T)-\{\lambda_n\}$. Then, if*

$$\sum_{n=1}^{\infty} d_n^{-1}(|\lambda_n| + d_n)^\nu < \infty,$$

the operator $T+P$ is a discrete spectral operator. If

$$\sum_{n=1}^{\infty} d_n^{-2}(|\lambda_n| + d_n)^{2\nu} < \infty$$

and if $\mathfrak{X}$ is Hilbert space, then $T+P$ is a discrete spectral operator.

REMARK. Since $\sigma(T)$ is a countable set of points which is discrete in any bounded portion of the complex plane, a straight half-line "branch cut" γ may be made from λ_0 to the point at infinity of the complex plane in a direction chosen so that γ does not intersect $\sigma(T)$. In the complement of γ, an everywhere analytic branch of $(z-\lambda_0)^\nu$ exists. Any two such branches differ by a constant factor $e^{2\pi i\nu}$. Thus $(T-\lambda_0 I)^\nu$ is defined by Definition XVIII.2.8, as soon as a branch cut γ and a branch of $(z-\lambda_0)^\nu$ in γ' are chosen. None of the following argumentation will depend upon the particular choices made, so that, without specifying any particular choice, we shall simply suppose in what follows that a choice has been made once and for all.

PROOF. Let E be the spectral resolution of T. By Lemma XVIII.2.25, $E(e)=0$ if e is a Borel set disjoint from $\sigma(T)$. Thus, for each Borel e, $E(e)=\sum_{\lambda\in e\cap\sigma(T)} E(\lambda)$, the series converging in the strong operator topology. Let S be the scalar part of T, and $N=T-S$. By Theorem XVIII.2.28 and Lemma XVIII.2.25, S and T have the same resolution of the identity and the same spectrum. By Definition XVIII.2.1,

$$\sigma(T\,|\,E(\lambda)\mathfrak{X})=\sigma(S\,|\,E(\lambda)\mathfrak{X})=\{\lambda\}$$

for $\lambda\in\sigma(T)$. Hence, if $E(\lambda)\mathfrak{X}$ is one-dimensional, we have $Tx=Sx=\lambda x$ for x in $E(\lambda)\mathfrak{X}$. Let σ_0 be the finite collection of points in $\sigma(T)$ for which $E(\lambda)$ fails to have a one-dimensional range, and $\sigma_0'=\sigma(T)-\sigma_0$. Then, by Definition XVIII.2.8, x is in $\mathfrak{D}(f(T))$ if and only if x is in $\mathfrak{D}(f(S))$. Moreover, $f(T)x=f(S)x$ for each analytic function f defined on $\sigma(T)$ and for each x in $E(\sigma_0')\mathfrak{X}$. In particular, x is in $\mathfrak{D}(T)$ if and only if x is in $\mathfrak{D}(S)$, and $Tx=Sx$ for each x in $E(\sigma_0')\mathfrak{X}$. By Theorem XVIII.2.9(ii), $E(\sigma_0)\mathfrak{X}$ is contained both in $\mathfrak{D}(f(T))$ and in $\mathfrak{D}(f(S))$, and by this same theorem this space is invariant both under $f(T)$ and under $f(S)$, which are both bounded in $E(\sigma_0)\mathfrak{X}$. This shows that $\mathfrak{D}(f(T))=\mathfrak{D}(f(S))$ and that, if we put

$$N_f x=f(T)x-f(S)x, \qquad x\in E(\sigma_0)\mathfrak{X},$$
$$N_f x=0, \qquad x\in E(\sigma_0')\mathfrak{X},$$

then N_f is a bounded operator, and $f(T)=f(S)+N_f$. In particular, we have $T=S+N$, where N is a bounded operator, and S is discrete. Moreover, if we let

$$Lx=x, \qquad x\in E(\sigma_0')\mathfrak{X},$$
$$Lx=(T-\lambda I)^\nu(S-\lambda I)^{-\nu}x, \qquad x\in E(\sigma_0)\mathfrak{X},$$

then it is clear that L is a bounded operator, and that $(S-\lambda I)^{-\nu}=(T-\lambda I)^{-\nu}L$. Hence

$$\begin{aligned}(P+N)(S-\lambda I)^{-\nu}&=P(S-\lambda I)^{-\nu}+N(S-\lambda I)^{-\nu}\\&=P(T-\lambda I)^{-\nu}L+N(S-\lambda I)^{-\nu}\end{aligned}$$

is a bounded operator. Since $T + P = S + (N + P)$, $\sigma(T) = \sigma(S)$, and N is bounded, it is then clear that we may assume without loss of generality that $T = S$, that is, that T is an operator of scalar type. In the same way, since $T + P = (T - \lambda_0 I) + (P + \lambda_0 I)$, and since $\sigma(T - \lambda_0 I) = \{z - \lambda_0 \mid z \in \sigma(T)\}$, it is clear that we may take $\lambda_0 = 0$.

Since $\mathfrak{D}(T^\nu) \supseteq \mathfrak{D}(T)$ by Theorem XVIII.2.11, $\mathfrak{D}(P) \supseteq \mathfrak{D}(T)$. Thus $\mathfrak{D}(T + P) = \mathfrak{D}(T)$. By hypothesis, $PT^{-\nu}$ is bounded; henceforth we shall denote the bounded operator $PT^{-\nu}$ by the symbol A.

Next consider the series

$$B(\mu) = R(\mu; T) \sum_{n=0}^{\infty} \{AT^\nu R(\mu; T)\}^n,$$

where $R(\mu; T) = (\mu I - T)^{-1}$. This series is well defined for $\mu \notin \sigma(T)$ and converges uniformly whenever $|T^\nu R(\mu; T)| < |A|^{-1}$. For values of μ for which the series $B(\mu)$ converges, we have

$$(\mu I - T - P)B(\mu) = \sum_{n=0}^{\infty} (AT^\nu R(\mu;T))^n - AT^\nu R(\mu;T) \sum_{n=0}^{\infty} (AT^\nu R(\mu;T))^n = I$$

and

$$\begin{aligned} B(\mu)(\mu I - T - P)x &= x + R(\mu; T)\left\{\sum_{n=0}^{\infty} (AT^\nu R(\mu; T))^{n-1}\right\} AT^\nu x \\ &\quad - R(\mu; T)\left\{\sum_{n=0}^{\infty} (AT^\nu R(\mu; T))^n\right\} AT^\nu x \\ &= x, \qquad x \in \mathfrak{D}(T) = \mathfrak{D}(T + P). \end{aligned}$$

These two equations show that whenever $|T^\nu R(\mu; T)| < |A|^{-1}$, the operator $(\mu I - T - P)^{-1}$ exists and equals $B(\mu)$.

Let C_n denote the circle of radius $d_n/2$ and center λ_n. Let $M/4$ be an upper bound for the norms of the projections in the range of E. For μ in C_n, it follows from Theorem XVIII.2.11 that

$$|T^\nu R(\mu; T)| \leqq \max_{1 \leqq k < \infty} M |\lambda_k|^\nu |\lambda_k - \mu|^{-1}.$$

Now $|\lambda_k - \mu| \geqq |\lambda_k - \lambda_n| - |\lambda_n - \mu| = |\lambda_k - \lambda_n| - d_n/2$. Moreover, $|\lambda_k| \leqq |\lambda_n| + |\lambda_k - \lambda_n|$. Since the function $(a + x)^\nu (x - b)^{-1}$ is evidently decreasing for $x > b$, $a > 0$, $b > 0$, we have $(a + x)^\nu (x - b)^{-1} < (a + 2b)^\nu b^{-1}$ for $x > 2b$, $a > 0$, $b > 0$. In particular,

$$\begin{aligned} \max_{1 \leqq k < \infty} |\lambda_k|^\nu |\lambda_k - \mu|^{-1} &\leqq \max_{1 \leqq k < \infty} (|\lambda_n| + |\lambda_k - \lambda_n|)^\nu \left(|\lambda_k - \lambda_n| - \frac{d_n}{2}\right)^{-1} \\ &\leqq 2(|\lambda_n| + d_n)^\nu d_n^{-1}. \end{aligned}$$

Thus

$$|T^{\nu}R(\mu; T)| < 2M(|\lambda_n| + d_n)^{\nu}\, d_n^{-1}, \qquad \mu \in C_n.$$

By hypothesis, the term on the right goes to zero as $n \to \infty$. Hence we find that for n sufficiently large, each μ in C_n is in $\rho(T+P)$ and that $R(\mu; T+P) = B(\mu)$. Since $B(\mu)$ is clearly the product of the compact operator $R(\mu; T)$ and a bounded operator, it follows that $T+P$ is a discrete operator. Moreover, since $(T+P-\mu I)^{-1}$ is bounded and hence closed, it follows immediately that $T+P$ is closed.

It follows in the same way that for n sufficiently large and for μ in C_n, we have

$$\begin{aligned}|R(\mu; T+P) - R(\mu; T)| &= |B(\mu) - R(\mu; T)| \\ &\leqq 2M\, d_n^{-1} \sum_{m=1}^{\infty} (|\lambda_n| + d_n)^{m\nu}\, d_n^{-m}\, (2M|A|)^m \\ &\leqq 8M^2 |A|\, d_n^{-2}(|\lambda_n| + d_n)^{\nu},\end{aligned}$$

since for sufficiently large n,

$$\sum_{m=0}^{\infty} (|\lambda_n| + d_n)^{m\nu}\, d_n^{-m}\, (2M|A|)^m = [1 - 2M|A|(|\lambda_n| + d_n)^{\nu}\, d_n^{-1}]^{-1}$$

will be less than 2. It follows that for sufficiently large n

$$\left| \frac{1}{2\pi i} \int_{C_n} R(\mu; T+P)\, d\mu - \frac{1}{2\pi i} \int_{C_n} R(\mu; T)\, d\mu \right| \leqq 4M^2 |A|\, d_n^{-1}(|\lambda_n| + d_n)^{\nu}.$$

It is clear from Lemma XVIII.2.31 that the first integral in the above formula is the projection $E(\sigma_n; T+P)$, where σ_n is that part of $\sigma(T+P)$ contained inside C_n. Thus

[*] $$|E(\sigma_n; T+P) - E(\lambda_n; T)| < 4M^2 |A|\, d_n^{-1}(|\lambda_n| + d_n)^{\nu}.$$

Since the expression on the right of this inequality approaches zero as $n \to \infty$, it follows by Lemma VII.6.7 that $E(\sigma_n; T+P)$ has a one-dimensional range for sufficiently large n. Thus, for sufficiently large n, say for $n \geqq K$, it follows from Lemma 3 that σ_n consists of a single point: $\sigma_n = \{\mu_n\}$.

Assume now that

$$\sum_{n=1}^{\infty} (|\lambda_n| + d_n)^{\nu}\, d_n^{-1} < \infty.$$

Then, since the collection of finite sums of projections $E(\lambda_n\,;T)$ is uniformly bounded, it is clear from [*] that the collection of finite sums of projections $E(\mu_n\,;T+P)$, $n \geqq K$, is uniformly bounded. Moreover, $\sum_{n=p}^{\infty}(E(\lambda_n\,;T) - E(\mu_n\,;T+P))$ clearly converges uniformly for $p \geqq K$ and approaches zero in norm as $p \to \infty$. Since $\sum_{n=p}^{\infty} E(\lambda_n\,;T)$ converges strongly (T being spectral), it follows that the series $E_p = \sum_{n=p}^{\infty} E(\mu_n\,;T+P)$ converges strongly, and that, if $\hat{E}_p$ is defined by the equation $\hat{E}_p = \sum_{n=p}^{\infty} E(\lambda_n\,;T)$, then we have $\lim_{p\to\infty} |\hat{E}_p - E_p| = 0$. Since $\hat{E}_p + \sum_{n=1}^{p-1} E(\lambda_n\,;T) = I$, $I - \hat{E}_p$ has a finite dimensional range for all p. Thus, by Lemma VII.6.7, $I - E_p$ has finite dimensional range for all sufficiently large p. Since E is a countably additive spectral resolution, we have $E(\mu;\,T+P)(I-E_p)=0$ if μ is not one of the points μ_n, $n \geqq K$. It follows from Lemma 3 that $\sigma(T+P)$ consists of the union of the points μ_n, $n \geqq K$, and a finite set of points. Consequently, the set of all finite sums of projections $E(\lambda;\,T+P)$ with λ in $\sigma(T+P)$ is uniformly bounded.

Since $I - E_p$ has a finite dimensional range for p sufficiently large, it follows that

$$\mathfrak{X}_0 = \{x \,|\, E(\lambda;\,T+P)x = 0,\ \lambda \in \sigma(T+P)\},$$

is finite dimensional. Then, by Lemma 6, $\mathfrak{X}_0 = \{0\}$. Hence, by Corollary XVIII.2.33, $T+P$ is a spectral operator.

In case $\mathfrak{X}$ is a Hilbert space and $\sum_{n=1}^{\infty} (|\lambda_n| + d_n)^{2\nu} d_n^{-2} < \infty$, the proof can be concluded in the same way once we establish, in the notations introduced above, that the collection of finite sums of projections $E(\mu_i;\,T+P)$ is uniformly bounded and that

$$\lim_{p\to\infty} \Big|\sum_{n=p}^{\infty} E(\lambda_n\,;T) - E(\mu_n\,;T+P)\Big| = 0.$$

This may be established as follows. First of all, by Lemma XV.6.2 there exists an automorphism of Hilbert space which makes every projection $E(\lambda_n\,;T)$ orthogonal. Thus it may be assumed without loss of generality that these projections are orthogonal projections to begin with. We have shown above that for n sufficiently large and $\mu \in C_n$, we have $R(\mu;\,T+P) = B(\mu)$. Thus

$$\begin{aligned}
&|R(\mu;\,T+P) - R(\mu;\,T) - R(\mu;\,T)AT^{\nu}R(\mu;\,T)| \\
&\qquad \leqq 2Md_n^{-1} \sum_{m=2}^{\infty} (2M)^m (|\lambda_n| + d_n)^{m\nu} d_n^{-m} |A|^m \\
&\qquad \leqq 16M^3 |A|^2 d_n^{-3} (|\lambda_n| + d_n)^{2\nu}
\end{aligned}$$

for sufficiently large n. Thus, for sufficiently large n,

$$[{*\atop *}]\quad \left|E(\lambda_n\,;\,T)-E(\mu_n\,;\,T+P)-\frac{1}{2\pi i}\int_{C_n}R(\mu;\,T)AT^\nu R(\mu;\,T)\,d\mu\right|$$
$$\leqq 8M^3\,|A|^2(|\lambda_n|+d_n)^{2\nu}\,d_n^{-2}.$$

Now, by Theorem XVIII.2.11,

$$R(\mu;\,T)=(\mu-\lambda_n)^{-1}E(\lambda_n\,;\,T)+\sum_{i\neq n}(\mu-\lambda_i)^{-1}E(\lambda_i\,;\,T).$$

The second term on the right, which we shall denote by the symbol $R_{(n)}(\mu;\,T)$, is analytic for μ in $\rho(T)$ and also for $\mu=\lambda_n$. Similarly,

$$T^\nu R(\mu;\,T)=\lambda_n^\nu(\mu-\lambda_n)^{-1}E(\lambda_n\,;\,T)+T^\nu R_{(n)}(\mu;\,T),$$

and $T^\nu R_{(n)}(T;\,\mu)$ is analytic for μ in $\rho(T)$ and also for $\mu=\lambda_n$. It follows that

$$\frac{1}{2\pi i}\int_{C_n}R(\mu;\,T)AT^\nu R(\mu;\,T)\,d\mu$$
$$=E(\lambda_n\,;\,T)AT^\nu R_{(n)}(\lambda_n\,;\,T)+\lambda_n^\nu\,R(\lambda_n\,;\,T)AE(\lambda_n\,;\,T).$$

Since $\sum_{n=1}^\infty(|\lambda_n|+d_n)^{2\nu}\,d_n^{-2}<\infty$ by hypothesis, it follows from [${*\atop *}$] that to prove the theorem it suffices to show that the collection of all finite sums of terms $E(\lambda_n\,;\,T)AT^\nu R_{(n)}(\lambda_n\,;\,T)$ and of terms $\lambda_n^\nu\,R_{(n)}(\lambda_n\,;\,T)$ $\times\,AE(\lambda_n\,;\,T)$ is uniformly bounded, and that

$$\lim_{p\to\infty}\left|\sum_{n=p}^{\infty}E(\lambda_n\,;\,T)AT^\nu R_{(n)}(\lambda_n\,;\,T)+\lambda_n^\nu\,R_{(n)}(\lambda_n\,;\,T)AE(\lambda_n\,;\,T)\right|=0.$$

Now, since $E(\lambda_n\,;\,T)x$ is orthogonal to $E(\lambda_m\,;\,T)x$ for $n\neq m$, we have, if $|x|\leqq 1$,

$$\left|\sum_{n\in J}E(\lambda_n\,;\,T)AT^\nu R_{(n)}(\lambda_n\,;\,T)x\right|^2$$
$$=\sum_{n\in J}|AT^\nu R_{(n)}(\lambda_n\,;\,T)x|^2\leqq(2M\,|A|)^2\sum_{n\in J}(|\lambda_n|+d_n)^{2\nu}\,d_n^{-2}$$

for each finite set J of integers. This last inequality follows, since just as for $|T^\nu R(\lambda_n\,;\,T)|$, we have the estimate

$$|T^\nu R_{(n)}(\lambda_n\,;\,T)|\leqq 2M(|\lambda_n|+d_n)^\nu\,d_n^{-1}.$$

By hypothesis, the collection of finite sums appearing on the right side of this inequality is uniformly bounded. Now it is clear from this inequality that the series

$$C_p = \sum_{n=p}^{\infty} E(\lambda_n; T) A T^{\nu} R_{(n)}(\lambda_n; T)$$

converges uniformly and that

$$|C_p|^2 \leqq (2M|A|)^2 \sum_{n=p}^{\infty} (|\lambda_n| + d_n)^{2\nu} d_n^{-2}.$$

Thus it follows from our hypothesis that $|C_p| \to 0$ as $p \to \infty$.

In the same way, since $|U| = |U^*|$,

$$\left| \sum_{n \in J} \lambda_n^{\nu} R_{(n)}(\lambda_n; T) A E(\lambda_n; T) \right|^2 = \left| \sum_{n \in J} \bar{\lambda}_n^{\nu} E(\lambda_n; T) A^* R_{(n)}(\lambda; T)^* \right|^2$$

$$\leqq (2M|A|)^2 \sum_{n \in J} (|\lambda_n| + d_n)^{2\nu} d_n^{-2},$$

for every finite set J of integers. Thus, as above, it is seen that the collection of finite sums

$$\sum_{n \in J} \lambda_n^{\nu} R_{(n)}(\lambda_n; T) A E(\lambda_n; T)$$

is uniformly bounded, and that

$$\lim_{p \to \infty} \left| \sum_{n=p}^{\infty} \lambda_n^{\nu} R_{(n)}(\lambda_n; T) A E(\lambda_n; T) \right| = 0.$$

Thus, Theorem 7 is fully proved. Q.E.D.

8 COROLLARY. *Let the hypotheses of Theorem 7 be satisfied. Then the spectrum $\sigma(T + P)$ consists of a sequence $\{\mu_n\}$ satisfying the inequality*

$$|\lambda_n - \mu_n| \leqq K_2 |\lambda_n|^{\nu}, \qquad n \geqq K_1,$$

where K_1 and K_2 are suitably chosen constants with $K_1 \geqq 1$. Furthermore, the range of each of the projections $E(\mu_n, T + P)$, $n \geqq K_1$, is one-dimensional.

PROOF. It was shown in the course of the proof of Theorem 7 that (with the notations of its proof) for n sufficiently large, there is just one point μ_n of $\sigma(T + P)$ within the circle C_n, and that $E(\mu_n; T + P)\mathfrak{X}$ is one-dimensional. It was also shown that if $|T^{\nu} R(\mu; T)| < |A|^{-1}$ then μ is in $\rho(T + P)$. Thus, only the inequality asserted by the corollary need be proved. Let C be a constant to be chosen later, and let $C|\lambda_n|^{\nu} \leqq |\mu - \lambda_n| \leqq d_n/2$. Then, since $(a + x)^{\nu}(x - b)^{-1}$ is decreasing for $x > b$, $a > 0$,

$b > 0$,

$$\begin{aligned}\max_{1 \leqq k < \infty} |\lambda_k|^\nu |\lambda_k - \mu|^{-1} &= \max_{1 \leqq k < \infty} (|\lambda_n| + |\lambda_n - \lambda_k|)^\nu (|\lambda_n - \lambda_k| - |\mu - \lambda_n|)^{-1} \\ &\leqq (|\lambda_n| + 2|\mu - \lambda_n|)^\nu |\mu - \lambda_n|^{-1} \\ &\leqq (|\lambda_n| + d_n)^\nu C^{-1} |\lambda_n|^{-\nu} \\ &\leqq C^{-1}\left(1 + \frac{d_n}{|\lambda_n|}\right)^\nu .\end{aligned}$$

Since $d_n < |\lambda_n - \lambda_0|$, $\lim_{n \to \infty} d_n/|\lambda_n| \leqq 1$. Thus, for n sufficiently large, we may write the displayed inequality as

$$\max_{1 \leqq k < \infty} |\lambda_k|^\nu |\lambda_k - \mu|^{-1} \leqq 2C^{-1}, \qquad C|\lambda_n|^\nu \leqq |\mu - \lambda_n| \leqq \frac{d_n}{2}.$$

It follows, by Theorem XVIII.2.11(c), that $|T^\nu R(\mu; T)| \leqq 2MC^{-1}$ for all μ in the annulus $C|\lambda_n|^\nu \leqq |\mu - \lambda_n| \leqq d_n/2$. Now choose C so that $2MC^{-1} < |A|^{-1}$; it follows from the above argument that no μ in the annulus belongs to $\sigma(T+P)$. Since μ_n is in $\sigma(T+P)$ and $|\mu_n - \lambda_n| \leqq d_n/2$, we have $|\mu_n - \lambda_n| \leqq C|\lambda_n|^\nu$, proving the present corollary. Q.E.D.

9 Corollary. *Let T be a discrete spectral operator in a weakly complete space. Suppose that E is its resolution of the identity, and suppose that $\{\lambda_n\}$ is an enumeration of its spectrum. Let d_n denote the distance from λ_n to $\sigma(T) - \{\lambda_n\}$. Suppose that for all but a finite number of n, $E(\lambda_n)$ has a one-dimensional range. Let B be a bounded operator. Then*

(a) *if $\sum_{n=1}^\infty d_n^{-1} < \infty$, then $T + B$ is a spectral operator;*
(b) *if T acts in Hilbert space and $\sum_{n=1}^\infty d_n^{-2} < \infty$, then $T + B$ is a spectral operator.*

Proof. This corollary follows by placing $\nu = 0$ in Theorem 7. Q.E.D.

To illustrate the preceding results, we shall give a number of examples. First, suppose that the B-space $\mathfrak{X}$ is the sequence space l_p, $p \geqq 1$, whose elements are sequences $\xi = \{\xi_n\}$, $n \geqq 1$. Let T denote the closed operator defined by

$$\mathfrak{D}(T) = \{\{\xi_n\} \mid \{n\xi_n\} \in l_p\},$$
$$T\{\xi_n\} = \{n\xi_n\}.$$

Then it is clear that T is a spectral operator of scalar type, that $\sigma(T) = \{n, n \geqq 1\}$, and that $E(n; T)$ has a one-dimensional range for each $n \geqq 1$. Let $\alpha > 0$, and let $\{p_{mn}\}$ be an infinite matrix subject to the conditions

$$\sum_{n=1}^{\infty} |p_{mn}| n^{-\alpha} < K, \qquad m \geqq 1,$$

$$\sum_{m=1}^{\infty} |p_{mn}| n^{-\alpha} < K, \qquad n \geqq 1,$$

K being some positive constant. Define an operator P in terms of the matrix $\{p_{mn}\}$ as follows:

$$\mathfrak{D}(P) = \mathfrak{D}(T^\alpha),$$

$$P\{\xi_m\} = \left\{\sum_{n=1}^{\infty} p_{mn} \xi_n\right\}, \qquad \{\xi_m\} \in \mathfrak{D}(P).$$

It is clear from this definition that P may be written as $P = QT^\alpha$, where Q is the bounded operator such that

$$Q\{\xi_m\} = \left\{\sum_{n=1}^{\infty} p_{mn} n^{-\alpha} \xi_n\right\}.$$

Indeed, the boundedness of Q follows from Exercise VI.9.54. For the sake of completeness, we give an independent proof here. If $\xi = \{\xi_i\}$ is bounded, $\sup_{1 \leqq i < \infty} |\xi_i| = |\xi|_\infty$, we have

$$\begin{aligned} |Q\xi|_\infty &= \sup_{1 \leqq m < \infty} \left| \sum_{n=1}^{\infty} p_{mn} n^{-\alpha} \xi_n \right| \\ &\leqq \sup_{1 \leqq m < \infty} |\xi|_\infty \sum_{n=1}^{\infty} |p_{mn}| n^{-\alpha} \leqq K |\xi|_\infty . \end{aligned}$$

In the same way, if $\xi = \{\xi_i\} \in l_1$, $\sum_{i=1}^{\infty} |\xi_i| = |\xi|_1$, we have

$$\begin{aligned} |Q\xi|_1 &= \sum_{m=1}^{\infty} \left| \sum_{n=1}^{\infty} p_{mn} n^{-\alpha} \xi_n \right| \\ &\leqq \sum_{n=1}^{\infty} \sum_{m=1}^{\infty} |p_{mn}| n^{-\alpha} |\xi_n| \\ &\leqq K \sum_{n=1}^{\infty} |\xi_n| = K |\xi|_1. \end{aligned}$$

The boundedness of Q as an operator in l_p then follows from the Riesz convexity theorem (VI.10.11). It follows from Theorem 7 that if $\beta > \alpha + 2$, $T^\beta + P$ is a discrete spectral operator, while if $p = 2$, so that $\mathfrak{X} = l_2$ is Hilbert space, and if $\beta > \alpha + \frac{3}{2}$, $T^\beta + P$ is a discrete operator.

A more special but less artificial example can be given as follows. Let us consider the formal differential operator

$$-\frac{d}{dt}(1-t^2)\frac{d}{dt}+\frac{2\alpha^2}{1+t}+\frac{2\beta^2}{1-t}$$

of the first part of Section XIII.8, and again make the assumption made there that $\alpha \geqq \frac{1}{2}$, $\beta \geqq \frac{1}{2}$. Then by the theory of Chapter XIII (compare the discussion in Section XIII.8), our formal differential operator defines a unique self adjoint operator L in the Hilbert space $L_2(-1, +1)$. According to the analysis of Section XIII.8, $\sigma(L)$ is the set of numbers $\lambda_n = (n + \alpha + \beta + 1)(n + \alpha + \beta)$, and each eigenspace corresponding to these eigenvalues is one-dimensional. It follows immediately from Corollary 9 that $L + B$ is a spectral operator for each bounded operator B. In particular, we may state the following conclusion. Let $q(z)$ be a function defined and analytic in a complex neighborhood of the segment $[-1, +1]$, except for the points $-1, +1$, where we suppose q to have first order poles. Suppose that the residue of q at $z = 1$ (at $z = -1$) is real and not less than $\frac{1}{2}$ (and not greater than $-\frac{1}{2}$). Let M be the operator whose domain is the set of all $f \in L_2(-1, +1)$ which have absolutely continuous first derivatives and which are such that

$$-((1-t^2)f'(t))' + q(t)f(t) \in L_2(+1, -1)$$

and put

$$(Mf)(t) = -((1-t^2)f'(t))' + q(t)f(t), \qquad f \in \mathfrak{D}(M).$$

Then M is a discrete spectral operator.

If q has real residues at ± 1 not lying in the indicated intervals, then a similar result may readily be stated; but in this case boundary conditions must be imposed at the end points.

As a final example, we let T be the self adjoint operator in Hilbert space $L_2(0, 1)$ defined (in the sense of Chapter XIII) by the formal differential operator id/dt, and by the "periodic" boundary condition $f(0) = f(1)$. The eigenvalues of this operator are $\lambda_n = 2\pi n$, $n = 0, \pm 1, \pm 2, \ldots$, and the corresponding eigenfunctions (the corresponding projections having one-dimensional ranges) are $e^{2\pi i n x}$, $n = 0, \pm 1, \pm 2, \ldots$. It is then clear that for odd $k = 2m + 1$, each projection $E(\lambda; T^k)$ corresponding to a point $\lambda \in \sigma(T^k) = \{\sigma(T)\}^k$ has a one-dimensional range. Theorem 7 shows that if P is a closed operator such that $\mathfrak{D}(P) \supseteq \mathfrak{D}(T^{k\nu})$, where

$$\sum_{n=-\infty}^{+\infty} |n|^{-4m} |n|^{2\nu(2m+1)} < \infty,$$

then $T+P$ is a spectral operator. This condition is equivalent to $2\nu(2m+1)-4m<-1$, that is, to $k\nu<k-3/2$. Let us consider in particular those cases in which $\mathfrak{D}(P) \supseteq \mathfrak{D}(T^{k-2})$. It is readily seen from the definition of $\mathfrak{D}(T)$ that $\mathfrak{D}(T^j)$ is the set of all functions $f \in L_2(0, 1)$ which have $j-1$ absolutely continuous derivatives, which satisfy the equations

$$f(0)=f(1), \ldots, f^{(j-1)}(0)=f^{(j-1)}(1),$$

and which are such that $f^{(j)}$ is in $L_2(0, 1)$. Hence we may take P to be any operator of the form

$$B_{k-2}\left(\frac{d}{dt}\right)^{k-2}+\cdots+B_0,$$

the coefficients B_j being arbitrary bounded operators in Hilbert space. A number of bizarre and amusing possibilities are worth mentioning at this point, simply to emphasize the variety of operators, in addition to formal differential operators, which are covered by Theorem 7. We can, for instance, take $(B_0 f)(t)=f(t/2)$; or letting $\varphi(t)$ be the fractional part of $t+\alpha$, put $(B_0 f)(t)=f(\varphi(t))$. In fact, if $t \to \varphi(t)$ is any measure preserving transformation of (0, 1) into itself, we may take $(B_0 f)(t)=f(\varphi(t))$. Since every function in $\mathfrak{D}(T^{k-2})$ has $k-3$ continuous derivatives, it is evident that we can take

$$(B_{k-2}f^{(k-2)})(t)=\int_0^1 f^{(k-3)}(s)\mu(ds), \qquad 0 \leqq t \leqq 1,$$

μ being any Borel measure on [0, 1]. Thus, for example, the operator which is formally defined by

$$\begin{aligned}(Tf)(t)=\left(\frac{d}{dt}\right)^{2m+1} f(t) &+ \sum_{k=0}^{2m-1} a_k(t)\left(\frac{d}{dt}\right)^k f(t)\\ &+\sum_{k=0}^{2m-2} c_k f^{(k)}(t\rho)\\ &+\sum_{j=0}^{2m-1}\int_0^1 K_j(t, s) f^{(j)}(s)\, ds\\ &+\sum_{i=0}^{2m-1} f^{(i)}(\varphi_i(t)),\end{aligned}$$

$f(0)=f(1), \ldots, f^{(2m)}(0)=f^{(2m)}(1)$, is a discrete spectral operator provided that $0 \leqq \rho \leqq 1$, the kernels K_j are such that

$$\int_0^1 K_j(t, s)\, ds + \int_0^1 K_j(t, s)\, dt$$

is a bounded function, and the φ_i are measure preserving transformations of (0, 1) into itself.

The last example above illustrates a problem which will be of great concern to us in the remainder of this chapter: the problem of finding which formal differential operators and sets of boundary conditions lead to spectral operators. As our treatment of this example makes clear, it suffices to study differential operators of the simple form $(d/dt)^k$ and then to use Theorem 7 to introduce quite general lower order terms as "perturbations." In Section 4 below, we shall carry out this idea in detail. However, since the technical difficulties in the cases $k > 2$ are so annoying, we shall, in the next section, treat the simple case $k = 2$ separately, so as to illustrate in a more easily comprehended way the (fundamentally simple) analytic method to be employed in Section 4.

3. Separated Boundary Conditions for the Second Order Operator

In this section Corollary 2.9 will be applied to show that operators of the form $T + B$, where B is an arbitrary bounded operator and where T is the unbounded operator in Hilbert space defined by the formal differential operator $-(d/dt)^2$ and with arbitrary separated boundary conditions, are spectral operators. We begin with a general lemma relating the spectrum and the point spectrum of a differential operator. The reader to whom the notation and terminology of the next few lemmas appear strange should reexamine the definitions given in Chapter XIII, especially those in Sections 2–5.

1 Lemma. *Let τ be a formally symmetric formal differential operator of order n defined on an interval I. Let the deficiency indices of τ both be equal to an integer m. Let A_i, $i = 1, \ldots, m$, be a set of m linearly independent boundary values for τ. Let S be the unbounded operator in $L_2(I)$ derived from τ by imposition of the boundary conditions $A_i(f) = 0$, $i = 1, \ldots, m$. Then every point λ in the finite complex plane which is in the spectrum $\sigma(S)$ of S but not in $\sigma_e(\tau)$ is in the point spectrum $\sigma_p(S)$ of S.*

PROOF. Suppose that $\lambda \notin \sigma_p(S)$. By Corollary XIII.6.8, at least m linearly independent solutions of the equation $\tau\sigma = \lambda\sigma$ lie in $L_2(I)$. Let the space spanned by these solutions be called Σ. Since $\lambda \notin \sigma_p(S)$, the mapping $\sigma \to [A_1(\sigma), \ldots, A_m(\sigma)]$ is a one-to-one mapping of Σ. Since it maps Σ into unitary m-space, and since the space Σ is at least m-dimensional, it must map Σ onto all of unitary m-space. Suppose that we can show that the operator $T_1(\tau) - \lambda I$ maps $\mathfrak{D}(T_1(\tau))$ onto $L_2(I)$. Then, given a g in $L_2(I)$, there exists a φ in $\mathfrak{D}(T_1(\tau))$ such that $\tau\varphi - \lambda\varphi = g$. Let $A_i(\varphi) = \gamma_i$, $i = 1, \ldots, n$. Then by the above there exists a solution σ of $\tau\sigma - \lambda\sigma = 0$ such that $A_i(\sigma) = \gamma_i = A_i(\varphi)$, $i = 1, \ldots, m$. Let $\psi = \varphi - \sigma$, so $\psi \in \mathfrak{D}(S)$ and $(S - \lambda I)\psi = g$. Thus $S - \lambda I$ is a one-to-one mapping of $\mathfrak{D}(S)$ onto $L_2(I)$. Since $T_1(\tau)$ is closed by Theorem XIII.2.10 and by Lemma XII.1.6(a), it follows from Definition XIII.2.17 and the remark preceding Definition XIII.2.29, that S is closed. From this, it follows immediately that $S - \lambda I$ is closed, and hence from Lemma XII.1.5, that $(S - \lambda I)^{-1}$ is closed. Since this latter operator is everywhere defined, it follows by the closed graph theorem (II.2.4) that it is bounded. Consequently, λ is in the resolvent set of S.

Hence we have only to show that $T_1(\tau) - \lambda I$ maps $\mathfrak{D}(T_1(\tau))$ onto $L_2(I)$. Since $\lambda \notin \sigma_e(\tau)$, it follows that $(T_1(\tau) - \lambda I)\mathfrak{D}(\tau)$ is closed. Hence, in view of Lemma XII.1.6(d), it suffices to show that $\{(T_1(\tau) - \lambda I)\mathfrak{D}(T_1(\tau))\}^{\perp} = \{f \mid T_1(\tau)^* f = \bar{\lambda} f\}$ contains only the zero vector. Since, by Theorem XIII.2.10, $T_1(\tau) = T_0(\tau)^*$, it follows, by Lemma XII.4.8 and Lemma XII.7.1, that $T_1(\tau)^* = T_0(\tau)^{**} = \overline{T_0(\tau)}$.

Consider any f which satisfies the equation $\overline{T_0(\tau)} f = \bar{\lambda} f$. If λ is real, then, since, by Definition XIII.2.17 and the remark preceding Definition XIII.2.29, we have $S \supseteq \overline{T_0(\tau)}$, it follows that $Sf = \lambda f$, which, since $\lambda \in \rho(S)$, means that $f = 0$. On the other hand, if λ is complex then, since $(\overline{T_0(\tau)})$ is symmetric by Lemma XII.4.8, it follows from Lemma XII.2.1 that $f = 0$. This completes the proof of the lemma. Q.E.D.

2 COROLLARY. *Let τ be a formally symmetric formal differential operator defined on a bounded closed interval I. Let S be an unbounded operator in $L_2(I)$ derived from τ by the imposition of a finite set of boundary conditions. Then every λ in the finite complex plane in the spectrum $\sigma(S)$ of S is an eigenvalue of S.*

Proof. It is seen from Theorems XIII.4.1, XIII.4.2, XIII.6.5, and Corollary XIII.6.4 that $\sigma_c(\tau)$ is void, and the present lemma follows immediately from Lemma 1. Q.E.D.

3 Lemma. *Let τ be a formal differential operator on a finite interval I, and let S be the operator derived from τ by the imposition of a certain set of boundary conditions. Then, if the spectrum $\sigma(S)$ is not the whole plane, S is a discrete operator.*

Proof. We may pass without any essential change in the situation from consideration of τ to consideration of $\tau - \lambda$, so that without loss of generality we may assume that the point $\lambda = 0$ is in the resolvent set $\rho(S)$. Then, if U denotes the unit sphere of $L_2(I)$, and $\{f_n\}$ a sequence of elements of $S^{-1}U$, then f_n may be written as $f_n = S^{-1}g_n$ with g_n in U. The space $L_2(I)$ being reflexive, the sequence $\{g_n\}$ has a weakly convergent subsequence with weak limit g. Hence, by Lemma XIII.2.16, the sequence $\{f_n\}$ has a subsequence $\{f_{n_i}\}$ which converges weakly in the topology of $C(I)$ to some element f in $C(I)$. Thus, $\{f_{n_i}\}$ is uniformly bounded and converges to f at each point x in I, so that $|f_{n_i} - f|^2 = \int_I |f_{n_i}(x) - f(x)|^2\, dx \to 0$ by Lebesgue's bounded convergence theorem. That is, $\{f_n\}$ has a strongly convergent subsequence. Thus $S^{-1}U$ is compact, proving that S^{-1} is a compact operator. Q.E.D.

The analytic side of our method involves the notion of an asymptotic series. Our next step is consequently to define such series and develop some of their basic properties.

4 Definition. (a) Let R be an unbounded subset of the complex plane. Let R_1 be a subset of the complex plane, and f a function defined in $R \times R_1$. Let $\{g_n\}$ be a sequence of functions defined in the set R_1. Suppose that, for each N,

$$\lim_{\substack{|z| \to \infty \\ z \in R}} |z|^N \left| f(z, w) - \sum_{n=0}^{N} g_n(w) z^{-n} \right| = 0$$

uniformly for $w \in R_1$. Then we say that f has the *uniform asymptotic expansion* $\sum_{n=0}^{\infty} g_n(w) z^{-k}$ at $z = \infty$ in $R \times R_1$, and write

$$f(z, w) \sim \sum_{n=0}^{\infty} g_n(w) z^{-n}$$

uniformly in $R \times R_1$.

(b) Let R_1 be a subset of the complex plane, and let $\{f_m\}$ be a sequence of functions all defined in R_1. Let $\{g_n\}$ be a sequence of functions defined in R_1, and suppose that, for each N,

$$\lim_{m\to\infty} m^N \left| f_m(w) - \sum_{n=0}^{N} g_n(w)m^{-n} \right| = 0$$

uniformly for w in R_1. Then we say that the sequence $\{f_m\}$ has the *uniform asymptotic representation* $\sum_{n=0}^{\infty} g_n(w)m^{-n}$ in R_1, and write

$$f_m(w) \sim \sum_{n=0}^{\infty} g_n(w)m^{-n}$$

uniformly in R_1.

If the set R_1 consists of a single point, these notions specialize to those of the asymptotic representation at infinity of a function in an unbounded set R and the asymptotic representation of a sequence, respectively.

Thus we say that a function $f(z)$ defined in an unbounded set R has the *asymptotic series* $\sum_{n=0}^{\infty} g_n z^{-n}$ if

$$\lim_{\substack{|z|\to\infty \\ z\in R}} |z|^N |f(z) - \sum_{n=0}^{N} g_n z^{-n}| = 0, \qquad N \geqq 1,$$

and that a sequence $\{f_m\}$ has the *asymptotic series* $\sum_{n=0}^{\infty} g_n m^{-n}$ if

$$\lim_{m\to\infty} m^N \left| f_m - \sum_{n=0}^{N} g_n m^{-n} \right| = 0, \qquad N \geqq 1.$$

5 LEMMA. *Let R_1 be the closure of a bounded domain in the complex plane, and let $\{f_m\}$ be a sequence of continuous functions defined in R_1 and analytic in its interior. Let*

$$f_m(z) \sim \sum_{n=0}^{\infty} g_n(z)m^{-n}$$

uniformly in R_1. Let g_0 have a single simple zero ζ_0 in R_1 and let ζ_0 be interior to R_1. Then, for all sufficiently large m, f_m has a single simple zero ξ_m in R_1, and the sequence ξ_m has an asymptotic series representation

$$\xi_m \sim \sum_{n=0}^{\infty} \zeta_n m^{-n},$$

where $\zeta_1, \zeta_2, \ldots$ are certain coefficients.

PROOF. Note first that it follows immediately from Definition 4 that the functions g_m are continuous in R_1 and analytic in its interior. Thus we can meaningfully assert that ζ_0 is a simple zero of g_0. Let U denote a small circular disk with ζ_0 as center, and C denote the periphery. Let $U \cup C$ be contained in R_1. Since g_0 is not zero in $R_1 - U$ and is continuous there, there exists a positive $\varepsilon > 0$ such that $|g_0(z)| \geqq \varepsilon$ for z in $R_1 - U$. Since $\lim_{m\to\infty} |f_m(z) - g_0(z)| = 0$ uniformly in R_1, it follows that for m sufficiently large $f_m(z)$ has all its zeros (if any) in U. By Cauchy's integral formula, the number of zeros of f_m in U (counted according to multiplicity) is $(2\pi i)^{-1} \int_C \{f_m'(z)/f_m(z)\}\, dz$. Since this integral is integer valued and tends toward $(2\pi i)^{-1} \int_C \{g_0'(z)/g_0(z)\}\, dz = 1$ as $m \to \infty$, it follows immediately that for sufficiently large m, f_m has one simple zero ξ_m in R_1, which lies in U. By Cauchy's integral formula we have

$$\xi_m = \frac{1}{2\pi i} \int_C \frac{z f_m'(z)}{f_m(z)}\, dz.$$

If, in this integral, we make use of the asymptotic relation

$$\left| f_m(z) - \sum_{n=0}^{N} g_n(z) m^{-n} \right| = o(m^{-N}),$$

which clearly implies that $|f_m'(z) - \sum_{n=0}^{N} g_n'(z) m^{-n}| = o(m^{-N})$, it follows immediately that $\{z f_m'(z)/f_m(z)\}$ may be represented on C by an asymptotic series

$$\frac{z f_m'(z)}{f_m(z)} \sim \sum_{n=0}^{\infty} h_n(z) m^{-n}$$

uniformly for z in C, where $h_0(z) = z g_0'(z)/g_0(z)$. But then, by placing $\zeta_n = (2\pi i)^{-1} \int_C h_n(z)\, dz$, it is seen that $\xi_m \sim \sum_{n=0}^{\infty} \zeta_n m^{-n}$. Q.E.D.

In what follows, we shall have use for the general notion of the adjoint of an unbounded operator.

6 Definition. Let T be a linear operator defined on a dense subset $\mathfrak{D}(T)$ of a B-space $\mathfrak{X}$. The *domain* $\mathfrak{D}(T^*)$ of the adjoint T^* consists of those linear functionals y^* in $\mathfrak{X}^*$ for which y^*Tx is continuous on $\mathfrak{D}(T)$. Since $\mathfrak{D}(T)$ is dense in $\mathfrak{X}$, there is a uniquely determined functional z^* in $\mathfrak{X}^*$ for which $y^*Tx = z^*x$ for x in $\mathfrak{D}(T)$. The *adjoint* T^* of T is defined by the equation

$$T^*y^* = z^*, \qquad y^* \in \mathfrak{D}(T^*).$$

7 LEMMA. *The closure of the range of a densely defined linear operator* T *is the set of all* x *such that* $y^*x = 0$ *whenever* $T^*y^* = 0$.

PROOF. If $T^*y^* = 0$, then $y^*y = y^*Tz = (T^*y^*)z = 0$ for all $y = Tz$ in the range of T, and hence for all y in the closure of the range of T. On the other hand, if y is not in the closure of the range of T, then by the Hahn-Banach theorem (II.3.13) there exists a y^* in $\mathfrak{Y}^*$ such that $y^*(y) \neq 0$, $y^*Tz = 0$ for z in $\mathfrak{D}(T)$. Then it follows from Definition 6 that $T^*y^* = 0$ while $y^*y \neq 0$. Q.E.D.

8 LEMMA. *Let* λ_0 *be a spectral point of the discrete operator* T *in a B-space* $\mathfrak{X}$. *Let* $f_1^*, f_2^*, \ldots, f_n^*$ *be a basis for the solutions of the equation* $(T^* - \lambda_0)f^* = 0$ *and let* Σ *be the space of solutions of the equation* $(T - \lambda_0)\sigma = 0$. *Then* λ_0 *is a multiple pole of the resolvent* $R(\lambda; T)$ *if and only if some non-zero* σ *in* Σ *satisfies the equations* $f_i^*(\sigma) = 0$, $i = 1, 2, \ldots, n$.

PROOF. We can readily see from Lemma 1(c), that λ_0 is a multiple pole of the resolvent if and only if there exists a solution g of the equation $(T - \lambda_0)^2 g = 0$ which is not a solution of the equation $(T - \lambda_0)g = 0$, that is, if and only if some non-zero σ in Σ is in the range of $(T - \lambda_0 I)$. The lemma will consequently follow from Lemma 7 once it is seen that the range $\mathfrak{R}(T)$ of $(T - \lambda_0 I)$ is closed. To see this, we may proceed as follows. Since, by Theorem VII.9.8, $E(\lambda_0; T)\mathfrak{R}(T) \subseteq \mathfrak{R}(T)$, it follows that $\mathfrak{R}(T) = E(\lambda_0; T)\mathfrak{R}(T) \oplus (I - E(\lambda_0; T))\mathfrak{R}(T)$. The first of these direct summands is finite dimensional by Lemma 2.2 and hence closed by Corollary IV.3.2. Hence it suffices to show that the second of the direct summands is closed. By VII.9.8(b),

$$\begin{aligned}(I - E(\lambda_0; T))(T - \lambda_0 I)\mathfrak{D}(T) &\supseteq (I - E(\lambda_0; T))(T - \lambda_0 I)h(T)\mathfrak{X} \\ &= (I - E(\lambda_0; T))\mathfrak{X},\end{aligned}$$

where $h(\lambda) = 0$ for λ in a neighborhood of λ_0, and $h(\lambda) = (\lambda - \lambda_0)^{-1}$ elsewhere. Since $(I - E(\lambda_0; T))(T - \lambda_0 I)\mathfrak{X} \subseteq (I - E(\lambda_0; T))\mathfrak{X}$ is clear, it follows that

$$(I - E(\lambda_0; T))\mathfrak{R}(T) = (I - E(\lambda_0; T))\mathfrak{X},$$

proving by the above that $\mathfrak{R}(T)$ is closed. Q.E.D.

We shall return later in the present chapter to the general theory of unbounded adjoint operators in a Banach space. It is well to remark, however, that in the special case in which the Banach space is Hilbert space, the usual distinction between the definitions of adjoint operators arises. This is because in Hilbert space we use the *Hermitian* rather than the pure Banach space adjoint, contrary to our practice in other Banach spaces. This has the effect of introducing complex conjugates in many of the Hilbert-space formulas where the corresponding Banach-space formulas do not have complex conjugates. None of this should cause the reader any essential difficulty.

9 Lemma. *Let E be a projection of a B-space $\mathfrak{X}$ onto a finite dimensional range and let $E^*: \mathfrak{X}^* \to \mathfrak{X}^*$ be its adjoint. Then, if $\varphi_1, \varphi_2, \ldots, \varphi_n$ is a basis for $E\mathfrak{X}$, we can find a unique basis $\psi_1^*, \psi_2^*, \ldots, \psi_n^*$ of $E^*\mathfrak{X}^*$ such that $\psi_i^*(\varphi_j) = \delta_{ij}$, and then*

$$Ef = \sum_{i=1}^{n} \varphi_i \psi_i^*(f), \qquad f \in \mathfrak{X}.$$

Proof. Any element Ef can be written uniquely as

$$Ef = \sum_{i=1}^{n} \varphi_i \alpha_i(f),$$

where the $\alpha_i(f)$ are linear functionals. If $f_m \to f$ and $\alpha_i(f_m) \to \alpha_i$, it is clear that $\alpha_i = \alpha_i(f)$. Hence, by the closed graph theorem, the uniquely determined linear functionals α_i are continuous. Hence $\alpha_i(f) = \psi_i^*(f)$ for some $\psi_i^* \in \mathfrak{X}^*$.

From the equation

$$Ef = \sum_{i=1}^{n} \varphi_i \psi_i^*(f)$$

it follows readily that

$$E^*\psi^* = \sum_{i=1}^{n} \psi_i^* \psi^*(\varphi_i),$$

so that $\psi_1^*, \psi_2^*, \ldots, \psi_n^*$ span $E^*\mathfrak{X}^*$. To see that $\psi_1^*, \psi_2^*, \ldots, \psi_n^*$ are linearly independent, let $\sum_{i=1}^{n} \beta_i \psi_i^* = 0$; then

$$\beta_j = \left(\sum_{i=1}^{n} \beta_i \psi_i^* \right) \varphi_j = 0,$$

so that Lemma 9 is completely proved. Q.E.D.

We now turn to the details of the analysis of second order differential operators arising out of the formal differential operator

$$\tau = -\left(\frac{d}{dt}\right)^2 + q(t).$$

The perturbation result given in Corollary 2.9 reduces the study of this operator to the study of the much simpler operator $-(d/dt)^2$ on $L_2(0, 1)$. The information needed about the latter operator is summarized, however, in the following lemma.

10 LEMMA. *Let k_0, k_1 be arbitrary constants and let T be the unbounded operator in $L_2(0, 1)$ defined by the formal differential operator $\tau = -(d/dt)^2$ and the boundary conditions*

$$f(0) - k_0 f'(0) = 0, \qquad f(1) - k_1 f'(1) = 0.$$

Then T is a spectral operator satisfying all the hypotheses of case (b) *of Corollary* 2.9.

REMARK. We may also admit the boundary conditions determined by $k_0 = \infty$ and/or $k_1 = \infty$, that is, the conditions $f'(0) = 0$ and $f'(1) = 0$, respectively.

PROOF. Since it is easy to treat all special cases in which one of k_0 or k_1 is zero or infinity by a separate argument much like the argument given below, we shall assume for simplicity that we have none of these special cases to deal with. If we put $\lambda = s^2$, the general solution of the equation

$$-f''(t) - \lambda f(t) = 0$$

is $\sin s(t + \alpha)$, where α is an arbitrary constant. This satisfies the boundary condition at $t = 0$ if

$$\tan s\alpha = k_0 s,$$

and satisfies the boundary conditions at $t = 1$ if

$$\tan s(1 + \alpha) = k_1 s.$$

Thus, using the addition formula for the tangent function, $T - \lambda I$ can only fail to have an inverse if $\lambda = s^2$, where s is a root of the equation

$$\text{(i)} \qquad \tan s = \frac{(k_1 - k_0)s}{1 + k_0 k_1 s^2} = \frac{cs}{1 + ds^2}, \qquad d \neq 0.$$

By Corollary 2, λ is in the spectrum $\sigma(T)$ if and only if s satisfies the equation (i). Since not every s satisfies (i), it follows immediately from Lemma 3 that T is discrete. Since the function $\tan s$ is periodic of period π, the roots of (i) in the period strip $(2n-1/2)\pi \leqq \mathscr{R}s \leqq (2n+3/2)\pi$ are the sums of $2n\pi$ and the zeros of

$$\text{(ii)} \qquad \tan s - \frac{c(s+2n\pi)}{1+d(s+2n\pi)^2}$$

in the strip $-(1/2)\pi \leqq \mathscr{R}s \leqq (3/2)\pi$. Since $|\tan s| \to 1$ as $|s| \to \infty$ in this strip, and since

$$\lim_{n\to\infty} \left| \frac{c(s+2n\pi)}{1+d(s+2n\pi)^2} \right| = 0,$$

it follows that there exists a finite K such that for n sufficiently large all the zeros of (ii) lie in the portion $|\mathscr{I}s| \leqq K$ of our strip. The function $\tan s$ having the two simple zeros $s=0$ and $s=\pi$ in the strip $-(1/2)\pi \leqq \mathscr{R}s \leqq (3/2)\pi$, it follows from Lemma 5 that, for large n, (ii) has exactly two zeros in $-(1/2)\pi \leqq \mathscr{R}s \leqq (3/2)\pi$, and that these zeros s_n and s'_n have asymptotic series representations

$$s_n \sim \sum_{k=1}^{\infty} a_k n^{-k}, \qquad s'_n \sim \sum_{k=1}^{\infty} a'_k n^{-k}$$

in terms of coefficients a_k and a'_k. Substituting this series in (ii), we find that $a'_1 = a_1 = c(d\pi)^{-1}$. Thus the roots s_n of (ii) may be enumerated, after omission of a finite number, in such a way that $s_n = n\pi + c(d\pi n)^{-1} + O(n^{-2})$. We thus obtain an enumeration λ_n ($n = k, k+1, \ldots$; note that k need not be 1) of the eigenvalues of T such that

$$\lambda_n = (n\pi)^2 + 2cd^{-1} + O(n^{-1}).$$

Hence, if d_n is the distance from λ_n to the remainder of the spectrum,

$$d_n = \pi^2(2n-1) + O(n^{-1}),$$

so that

$$\sum_{n=k}^{\infty} d_n^{-2} < \infty.$$

It is evident from the form of the boundary conditions defining our operator that each λ_n can correspond to at most one function φ_n (up to a scalar multiple), which satisfies the equation

$$(T - \lambda_n I)\varphi_n = 0.$$

Thus, if $E(\lambda_n)$ is to be anything but a projection onto a one-dimensional range, it follows from Lemma 2.2 that λ_n must be a multiple pole of the resolvent. By Lemma 8, the condition for this is $(\varphi_n, \psi_n) = 0$, where ψ_n is a solution of

$$(T^* - \bar{\lambda}_n I^*)\psi_n = 0.$$

Since by Theorems XIII.2.10 and XII.4.28, T^* is defined by the formal differential operator $-(d/dt)^2$ and the complex-conjugate boundary conditions of those that define T, it is clear that there are constants $\gamma_n \neq 0$ such that

$$\gamma_n \psi_n(t) = \overline{\varphi_n(t)}.$$

Hence, λ_n can only be a multiple pole of the resolvent of T if

$$\int_0^1 (\varphi_n(t))^2 \, dt = 0.$$

Now, we have

$$\varphi_n(t) = \sin s_n(t + \alpha_n) = \sin(s_n t + \beta_n),$$

where β_n must be determined so as to satisfy

$$k_0^{-1} s_n^{-1} \sin \beta_n = \cos \beta_n .$$

It follows readily that

$$\beta_n = \frac{\pi}{2} - (n\pi k_0)^{-1} + O(n^{-2}),$$

so that

$$\varphi_n(t) = \cos(s_n t + \delta_n), \qquad \delta_n = (n\pi k_0)^{-1} + O(n^{-2}).$$

It follows that

$$\int_0^1 (\varphi_n(t))^2 \, dt \sim \int_0^1 \cos^2 n\pi t \, dt = \tfrac{1}{2},$$

so that only a finite set of λ_n can be multiple poles of the resolvent of T. For those λ_n which are simple poles of the resolvent of T, the projection $E(\lambda_n)$ is, by Lemma 9, the operator determined by the integral kernel

$$\hat{\varphi}_n(t)\hat{\varphi}_n(u) = E_n(t, u),$$

where $\hat{\varphi}_n$ is a scalar multiple of φ_n, the scalar being so chosen as to make

$$\int_0^1 |\hat{\varphi}_n(t)|^2 \, dt = 1.$$

We have $\hat{\varphi}_n = c_n \varphi_n$, and a simple computation reveals that

$$c_n = 2^{-1/2} + O(n^{-2}).$$

Hence it follows that

$$E_n(x, y) = \frac{1}{2} \cos n\pi x \cos n\pi y - \frac{(cd^{-1}x + k_0^{-1})}{2n\pi} \sin n\pi x \cos n\pi y$$
$$- \frac{(cd^{-1}y + k_0^{-1})}{2n\pi} \sin n\pi y \cos n\pi x + O(n^{-2}),$$

which gives a decomposition of E_n into four terms

$$\text{(iii)} \qquad E_n = \hat{E}_n + A_n + B_n + \Delta_n .$$

It is now easy to find a uniform bound for

$$\left| \sum_{n \in J} E_n \right|,$$

where J is an arbitrary finite set of integers, by making use of the decomposition (iii). We have

$$\left| \sum_{n \in J} \hat{E}_n \right| \leqq 1,$$

since the $\hat{E}_n$ are a family of orthogonal projections. We have

$$\left| \sum_{n \in J} \Delta_n \right| \leqq M,$$

since

$$|\Delta_n| = O(n^{-2}) \quad \text{and} \quad \sum_{n=1}^{\infty} n^{-2} < \infty.$$

The operators A_n and B_n have the form

$$A_n = \hat{E}_n \hat{A}_n \quad \text{and} \quad B_n = \hat{B}_n \hat{E}_n ,$$

where

$$|\hat{A}_n| = O(n^{-1}) \quad \text{and} \quad |\hat{B}_n| = O(n^{-1}),$$

a situation studied above in the last paragraph of the proof of Theorem 2.7, where the argument given proves, with suitable slight modifications, not only the uniform boundedness of $\sum_{n \in J} A_n$, but also that

$$\text{(iv)} \qquad \lim_{n \to \infty} \left| \sum_{m=n}^{\infty} A_m \right| = 0.$$

All that remains to complete the proof of our lemma is a proof that

$$\sum_{i=k}^{\infty} E(\lambda_i) = I.$$

By Lemma 2.5, the projection

$$E_{\infty} = I - \sum_{i=k}^{\infty} E(\lambda_i)$$

either projects onto an infinite dimensional space or is zero. But, by (iv),

$$\lim_{m\to\infty} \left| \left(I - \sum_{n=m}^{\infty} E(\lambda_n)\right) - \left(I - \sum_{n=m}^{\infty} \hat{E}_n\right) \right| = 0.$$

Hence, by Lemma VII.6.7,

$$I - \sum_{n=m}^{\infty} E(\lambda_n)$$

has a finite dimensional range for all sufficiently large m, and hence, *a fortiori*, E_{∞} has a finite dimensional range. Q.E.D.

→ 11 THEOREM. *Let T be the unbounded operator in Hilbert space defined by the formal differential operator $\tau = -(d/dx)^2$ and the boundary conditions*

[*] $$f(0) - k_0 f'(0) = 0, \qquad f(1) - k_1 f'(1) = 0,$$

where k_0 and k_1 are arbitrary, possibly infinite, complex numbers. Then if B is an arbitrary bounded operator, $T + B$ is a spectral operator.

PROOF. This follows from Lemma 10 and Corollary 2.9. Q.E.D.

12 COROLLARY. *Let q be a bounded measurable function and T be the unbounded differential operator defined by a formal differential operator*

$$\tau = -\left(\frac{d}{dx}\right)^2 + q(x)$$

and by the boundary conditions [*]. *Then T is a spectral operator.*

PROOF. This follows immediately from Theorem 11. Q.E.D.

As the final lemma of this section, we state a useful elementary principle in the theory of spectral differential operators.

13 LEMMA. *Let τ be a formal differential operator of order n defined on an interval I. Let T be a discrete spectral operator obtained from τ by*

imposition of a finite set of boundary conditions. Let $\{\lambda_i\}$ *be an enumeration of* $\sigma(T)$, *and* J *a compact subinterval of* I. *Then, if* f *is in the domain* $\mathfrak{D}(T)$, *the series expansion*

$$\sum_{i=1}^{\infty} E(\lambda_i\,;\,T)f$$

converges to f *unconditionally in the topology of* $A^{(n)}(J)$.

PROOF. The series $\sum_{i=1}^{\infty} E(\lambda_i\,;\,T)f$ certainly converges unconditionally in the topology of $L_2(J)$. On the other hand, so does the series

$$T\left(\sum_{i=1}^{\infty} E(\lambda_i\,;\,T)f\right) = \sum_{i=1}^{\infty} E(\lambda_i\,;\,T)(Tf)$$

(cf. VII.9.8). Hence, by Lemma XIII.2.16, the original series converges unconditionally in the topology of $A^{(n)}(J)$. Q.E.D.

14 COROLLARY. *Let* T *be as in Corollary* 12 *and let* f *be in the domain* $\mathfrak{D}(T)$. *Then if* $\{\lambda_i\}$ *is an enumeration of* $\sigma(T)$, *the series*

$$\sum_{i=1}^{\infty} E(\lambda_i)f$$

converges unconditionally in the topology of $A^{(2)}(J)$.

PROOF. This follows immediately from Corollary 12 and Lemma 13. Q.E.D.

4. Spectral Properties of $\left(\dfrac{1}{i}\dfrac{d}{dx}\right)^n$

In this section it will be shown that an nth order differential operator on a finite interval will give rise to a spectral operator under each of a large class of boundary conditions. The method used for establishing such results is that of applying the general perturbation theory of Section 2 by considering an arbitrary nth order differential operator (whose leading term is $(-id/dt)^n$) on $[0, 1]$ as arising by perturbing the fundamental operator

$$\tau_n = i^{-n}\frac{d^n}{dt^n}$$

on the interval $[0, 1]$. When the integer n is understood, the symbol τ will sometimes be used in place of τ_n.

Throughout this section, B_i, $i = 1, \ldots, n$, will denote a set of n linearly independent boundary values for τ. Thus, by Corollary XIII.2.23, there exist two $n \times n$ matrices α_{ij} and β_{ij} such that

$$B_i(f) = \sum_{j=0}^{n-1} \alpha_{ij} f^{(j)}(0) + \sum_{j=0}^{n-1} \beta_{ij} f^{(j)}(1), \qquad i = 1, \ldots, n. \tag{1}$$

The symbol T will denote the operator whose domain is

$$\mathfrak{D}(T) = \{f | f \in H^{(n)}(I),\ B_i(f) = 0,\ i = 1, \ldots, n\}, \tag{2}$$

and which is defined by

$$Tf = \tau_n f, \qquad f \in \mathfrak{D}(T). \tag{3}$$

We wish to study the resolvent of T; to do this we shall begin by studying the set of eigenvalues of T.

It is convenient to make a certain normalization of the set $\{B_i\}$ of boundary conditions at the outset. Suppose that if B is a boundary value, we agree to call the order of the highest derivative at 0 which appears with a non-zero coefficient in the unique expression of B in the form $Bf = \sum_{j=0}^{n-1} \gamma_j f^{(j)}(0) + \sum_{j=0}^{n-1} \tilde{\gamma}_j f^{(j)}(1)$, the *order of the boundary value B at* 0. *The order of the boundary value B at* 1 is defined similarly. The maximum of these two orders will be called the *order of the boundary value B*. If we are given two boundary conditions $B(f) = 0$, $C(f) = 0$, the boundary values B and C having the same order m at 0, then it is evident that by subtracting a suitable multiple of B from C, we can pass to an equivalent set of boundary conditions $B(f) = 0$, $(C - \rho B)(f) = 0$, where the boundary value $C - \rho B$ has order less than m at 0. The same observation holds relative to orders at 1. It then follows that if any set of three boundary conditions $B(f) = 0$, $C(f) = 0$, and $D(f) = 0$ is given, and if B, C, and D all have order not greater than m, we can find an equivalent set $\tilde{B}(f) = 0$, $\tilde{C}(f) = 0$, $\tilde{D}(f) = 0$, $\tilde{B}$ and $\tilde{C}$ having order at most equal to m, and $\tilde{D}$ having order at most equal to $m - 1$. If $\tilde{B}$ and $\tilde{C}$ do actually both have order m, then, subtracting a suitable multiple of $\tilde{B}$ from $\tilde{C}$, and a suitable multiple of $\tilde{C}$ from $\tilde{B}$, we find an equivalent set of three boundary conditions in which at most one element has order m at 0, and at most one element has order m at 1. Applying these observations inductively to the set B_i of boundary conditions, we see that we can suppose without loss of generality that *the orders m_i of B_i form a non-increasing sequence of integers, not containing any three successive equal terms, and that if B_i and B_{i+1} have equal orders, B_i has a higher order at* 0 *than B_{i+1}, and B_{i+1} has a higher order at 1 than B_i.*

We denote the sum $\sum_{i=1}^{n} m_i$ *by the symbol* p.

It is easily seen in this case that the pair of terms of order m_i in the boundary value B_i is uniquely defined up to a constant multiple, independent of the particular way in which we normalize the original set of boundary conditions. We now specify this constant factor by specifying that if B_i contains $f^{(m_i)}(0)$ with a non-zero coefficient, this coefficient is to be 1, whereas if B_i does not contain $f^{(m_i)}(0)$ with a non-zero coefficient, the coefficient of $f^{(m_i)}(1)$ in B_i is to be 1.

The details of the analysis of eigenvalues are slightly different when n is odd from these details when n is even. For this reason, we consider these two cases separately.

Case 1. *n even.*

Let $n = 2\nu$. Let w_j, $j = 0, \ldots, n-1$, denote the nth roots of unity, enumerated in such a way that $w_0 = 1$, $w_\nu = -1$, the imaginary part of w_i is positive for $0 < i < \nu$, and the imaginary part of w_i is negative for $\nu < i < 2\nu$.

Let λ be an arbitrary complex number. Let $\mu = \mu(\lambda)$ denote that unique nth root of λ which lies in the sector $\pi/n \geqq \arg \mu > (-\pi)/n$ of the complex plane. Put

$$\begin{aligned} \sigma_k(t, \mu) &= e^{i\mu w_k t}, & 0 \leqq k \leqq \nu, \\ \sigma_k(t, \mu) &= e^{i\mu w_k (t-1)}, & \nu < k < 2\nu. \end{aligned} \tag{4}$$

Then $\sigma_0(t, \mu(\lambda)), \ldots, \sigma_{n-1}(t, \mu(\lambda))$ clearly form a basis for the set of solutions of the equation $\tau_n f = \lambda f$.

Put

$$B_i(\sigma_k(\mu)) = M_{ik}(\mu). \tag{5}$$

The eigenvalues of T are clearly the set of values λ for which there exists a set of constants $c_0, \ldots, c_{n-1}$, not all zero, for which

$$B_i\left(\sum_{k=0}^{n-1} c_k \sigma_k(\mu(\lambda))\right) = \sum_{k=0}^{n-1} c_k M_{ik}(\mu(\lambda)) = 0, \qquad i = 1, \ldots, n.$$

Thus if we put

$$M(\mu) = \det(M_{ij}(\mu)), \tag{6}$$

the eigenvalues of T are the zeros of $M(\mu(\lambda)) = 0$.

We shall study the roots of the equation $M(\mu) = 0$, and see that, under a suitable assumption on the set $\{B_i(f) = 0\}$ of boundary conditions defining T, they have a simple asymptotic representation.

It is evident from the form (4) of $\sigma_k(t, \mu)$ and the form (1) of B_i that $B_i(\sigma_k(t, \mu)) = M_{ik}(\mu)$ has the form

$$
\begin{aligned}
M_{ik}(\mu) &= P_{ik}(\mu) + Q_{ik}(\mu)e^{iw_k\mu}, \qquad 0 \leqq k \leqq \nu, \\
M_{ik}(\mu) &= P_{ik}(\mu) + Q_{ik}(\mu)e^{-iw_k\mu}, \qquad \nu < k < 2\nu,
\end{aligned} \tag{7}
$$

where P_{ik} and Q_{ik} are polynomials in μ of order at most m_i for all $1 \leqq i \leqq n$, $0 \leqq k \leqq n-1$.

Let A denote the angular sector

$$A = \left\{ z \,\middle|\, |\arg z| \leqq \frac{\pi}{n} \right\} \tag{8}$$

in the complex plane. We shall investigate the zeros of $M(\mu)$ in the sector A. Since $\mu(\lambda)$ lies in the sector A for all λ, this will give us complete information on the zeros of $M(\mu(\lambda))$. Since for $0 < k < \nu$, w_k is an nth root of unity with positive imaginary part, it follows that for $0 < k < \nu$, $iw_k\mu$ ranges over an angle entirely contained in the left half-plane as μ ranges over A. Similarly, for $\nu < k < 2\nu$, $iw_k\mu$ ranges over an angle entirely contained in the right half-plane as μ ranges over A. Thus, if we put

$$
\begin{aligned}
N_{ik}(\mu) &= P_{ik}(\mu), \qquad 0 < k < \nu \quad \text{or} \quad \nu < k < 2\nu, \\
N_{i0}(\mu) &= M_{i0}(\mu) = P_{i0}(\mu) + Q_{i0}(\mu)e^{i\mu}, \\
N_{i\nu}(\mu) &= M_{i\nu}(\mu) = P_{i\nu}(\mu) + Q_{i\nu}(\mu)e^{-i\mu},
\end{aligned} \tag{9}
$$

it follows that there exists a number $a > 0$ such that

$$|N_{ik}(\mu) - M_{ik}(\mu)| = O(e^{-a|\mu|}), \tag{10}$$

as $|\mu| \to \infty$, μ remaining in the angular sector A. Let $N(\mu) = \det(N_{ik}(\mu))$. From the form (9) of the matrix $N_{ik}(\mu)$ it follows that $N(\mu)$ has the form

$$N(\mu) = \pi_1(\mu)e^{i\mu} + \pi_2(\mu)e^{-i\mu} + \pi_3(\mu), \tag{11}$$

where π_1, π_2, and π_3 are polynomials in μ of order at most $p = \sum_{i=1}^{n} m_i$. From relation (10) it follows that for any positive number $b < a$ we have

$$|N(\mu) - M(\mu)| = O(e^{-b|\mu|}e^{|\mathscr{I}\mu|}) \tag{12}$$

as $|\mu| \to \infty$, μ remaining in the angular sector A.

In order to bring our analysis to a successful conclusion, we must now make the following hypothesis.

1 REGULARITY HYPOTHESIS FOR EVEN ORDER CASE. *The polynomials* π_1 *and* π_2 *are of order* p.

If Regularity Hypothesis 1 is satisfied, we may write

$$\begin{aligned} \pi_1(\mu) &= a_p \mu^p + a_{p-1}\mu^{p-1} + \cdots + a_0, \qquad a_p \neq 0, \\ \pi_2(\mu) &= b_p \mu^p + b_{p-1}\mu^{p-1} + \cdots + b_0, \qquad b_p \neq 0, \qquad (13) \\ \pi_3(\mu) &= c_p \mu^p + c_{p-1}\mu^{p-1} + \cdots + c_0. \end{aligned}$$

The labor of verifying Hypothesis 1 may be simplified by using the following elementary observations. By equations (1), (4), and our preliminary normalization of boundary conditions, the terms of order p in the determinant $N(\mu)$ are evidently the same as the terms of order p in the determinant $\hat{N}(\mu)$ of the matrix $(\hat{N}_{ik}(\mu))$ defined by the equations

$$\begin{aligned} \hat{N}_{ik}(\mu) &= \alpha_{i,m_i}(i\mu w_k)^{m_i}, \qquad 0 < k < \nu, \\ \hat{N}_{ik}(\mu) &= \beta_{i,m_i}(i\mu w_k)^{m_i}, \qquad \nu < k < 2\nu, \\ \hat{N}_{i0}(\mu) &= \alpha_{i,m_i}(i\mu)^{m_i} + \beta_{i,m_i}(i\mu)^{m_i}e^{i\mu}, \\ \hat{N}_{i\nu}(\mu) &= \alpha_{i,m_i}(-i\mu)^{m_i} + \beta_{i,m_i}(-i\mu)^{m_i}e^{-i\mu}. \end{aligned}$$

Consequently, by factoring out a factor $(i\mu)^p$, the coefficients a_p, b_p, c_p of the terms of order p in the determinant $N(\mu)$ become the same as the coefficients a_p, b_p, c_p in the determinant $\hat{\hat{N}}(\mu) = a_p e^{i\mu} + b_p e^{-i\mu} + c_p$ of the matrix $(\hat{\hat{N}}_{ik})$ defined by the equations

$$\begin{aligned} \hat{\hat{N}}_{ik} &= \alpha_{i,m_i} w_k^{m_i}, \qquad 0 < k < \nu, \\ &= \beta_{i,m_i} w_k^{m_i}, \qquad \nu < k < 2\nu, \\ \hat{\hat{N}}_{i0} &= \alpha_{i,m_i} + \beta_{i,m_i} e^{i\mu}, \\ \hat{\hat{N}}_{i\nu} &= (\alpha_{i,m_i} + \beta_{i,m_i} e^{-i\mu})(-1)^{m_i}. \end{aligned}$$

Whether or not Hypothesis 1 is satisfied for a normalized set of boundary conditions consequently depends only on the "leading" coefficients in the boundary conditions, that is, on the coefficients α_{i,m_i} and β_{i,m_i} of the highest derivative occurring with a non-zero coefficient.

At any rate, if Regularity Hypothesis 1 is satisfied, it follows from (11), from (12), and from (13) that

$$\begin{aligned} \mu^{-p}M(\mu) &= a_p e^{i\mu} + b_p e^{-i\mu} + c_p + p_1(\mu)e^{i\mu} \qquad (14) \\ &\quad + p_2(\mu)e^{-i\mu} + p_3(\mu) + O(e^{-b|\mu| + |\mathscr{I}\mu|}) \end{aligned}$$

uniformly as $|\mu| \to \infty$, μ remaining in A, where p_1, p_2, and p_3 are polynomials in μ^{-1} without constant terms. Let t and s be so large that

$$|p_1(\mu)| + |p_2(\mu)| + |p_3(\mu)| < \tfrac{1}{3}|a_p|, \quad |\mu| > t,$$
$$|b_p e^{-i\mu}| + |c_p| < \tfrac{1}{3}|a_p e^{i\mu}|, \qquad -\mathscr{I}(\mu) > s,$$

and such that the final term on the right of (14) is less than $\frac{1}{3}|a_p e^{i\mu}|$ for $|\mu| > t$, $-\mathscr{I}\mu > s$. Then it is clear from (14) that $M(\mu)$ is non-zero in the subregion $\{\mu \in A \mid -\mathscr{I}\mu > s, |\mu| > t\}$ of A. In the same way, it follows that we may choose t and s so large that $M(\mu)$ is non-zero in the subregion $\{\mu \in A \mid \mathscr{I}\mu > s, |\mu| > t\}$ of A. This proves the following lemma.

2 LEMMA. *Let the set* (1) *of boundary conditions satisfy Regularity Hypothesis* 1. *Then there exist real numbers t and s so large that every zero z in the angle A (defined by formula* (8)) *of the determinant $M(\mu)$ (defined by formula* (6)) *satisfies either $|z| < t$ or $|\mathscr{I}z| < s$.*

Let α be chosen so that

$$a_p e^{i\alpha} = -b_p e^{-i\alpha} = k. \tag{15}$$

Then we may write

$$\begin{aligned} a_p e^{i\mu} + b_p e^{-i\mu} + c_p &= k(e^{i(\mu-\alpha)} - e^{-i(\mu-\alpha)}) + c_p \\ &= 2ik\{\sin(\mu - \alpha) - \beta\}, \end{aligned} \tag{16}$$

where

$$\beta = \frac{c_p}{2ik}. \tag{17}$$

We shall show below that the zeros of $M(\mu)$ are much like those of $\sin(\mu - \alpha) - \beta$. If $\beta \neq \pm 1$, all the zeros of this latter function are simple. If $\beta = \pm 1$, the zeros of $\sin(\mu - \alpha) \mp 1 = 0$ are $\mu = \alpha \pm \pi/2 + 2n\pi$, and are all double. Consequently, we shall divide our analysis into two cases, depending on whether $\beta = \pm 1$ or $\beta \neq \pm 1$.

Case 1A: *The constant β of formula* (16) *is different from* ± 1.

Then the period-strip $0 \leq \mathscr{R}z < 2\pi$ contains exactly two roots z_1 and z_2 of the equation $\sin z = \beta$. To see this, note that the mapping $w \to h(w) = w - (1/w)$ maps the w-plane with $w = 0$ removed onto the whole z-plane, and that each point z except $z = \pm 2i$ is the image of exactly two points w. The mapping $z \to e^{iz}$ is a one-to-one mapping of the period strip

$0 \leqq \Re z < 2\pi$ onto the w-plane with zero removed. Since $\sin z = (1/2i)h(e^{iz})$, our original observation is evident. The roots of $\sin(z - \alpha) - \beta = 0$ are consequently simple and fall into two arithmetic sequences, one of the form $z = 2\pi n + \alpha + z_1$, the other of the form $z = 2\pi n + \alpha + z_2$, where $2\pi > \mathscr{R}z_1 \geqq \mathscr{R}z_2 \geqq 0$, $z_1 \neq z_2$. In order to obtain information on the zeros of $M(\mu)$ from this, we now use Lemma 3.

We know by Lemma 2 that all but a finite number of roots of $M(\mu)$ lie in a strip $|\mathscr{I}z| < s$. Let α_1 be chosen in such a way that both zeros $z_1 + \alpha$ and $z_2 + \alpha$ of $\sin(z - \alpha) = \beta$ lie interior to the period strip $2\pi(n+1) \geqq \mathscr{R}(z - \alpha_1) \geqq 2\pi n$, and choose y so large that $y > |\mathscr{I}(z_1 + \alpha)| + |\mathscr{I}(z_2 + \alpha)|$. Divide the rectangle $2\pi \geqq \mathscr{R}(z - \alpha_1) \geqq 0$ into two rectangles $R^{(1)}$ and $R^{(2)}$, each of which contains exactly one zero of $\sin(z - \alpha) = \beta$. Put $R_m^{(1)} = R^{(1)} + 2m\pi$, $R_m^{(2)} = R^{(2)} + 2m\pi$. Taking our cue from the known form $2\pi m + \alpha + z_1$, $2\pi m + \alpha + z_2$, $z_1 \neq z_2$, $2\pi > \mathscr{R}z_1 \geqq \mathscr{R}z_2 \geqq 0$ of the zeros of $\sin(z - \alpha) - \beta$, we wish to show that for m sufficiently large each rectangle $R_m^{(1)}$ and each rectangle $R_m^{(2)}$ contains precisely one zero of $M(\mu)$. This, however, follows from Lemma 3.5 and from formulas (16) and (14). It also follows, from Lemma 3.5, formula (16), and formula (14), that the zero ξ_n of $M(\mu)$ in $R_m^{(1)}$ has the asymptotic representation

$$\xi_n \sim 2\pi n + \alpha + z_2 + \sum_{m=1}^{\infty} \zeta_m n^{-m},$$

and that the zero $\tilde{\xi}_n$ of $M(\mu)$ in $R_n^{(2)}$ has the asymptotic representation

$$\tilde{\xi}_n \sim 2\pi n + \alpha + z_2 + \sum_{m=1}^{\infty} \tilde{\zeta}_m n^{-m},$$

where the ζ_m and $\tilde{\zeta}_m$ are certain coefficients. Since the strip $|\mathscr{I}z| < s$, $\mathscr{R}z \geqq (z_1 + z_2)/2 - \pi$ is the union of the rectangles $R_n^{(1)}$, $n = 1, 2, \ldots$ and the rectangles $R_n^{(2)}$, $n = 1, 2, \ldots$, this establishes the following result.

3 Lemma. *Let the set* (1) *of boundary conditions satisfy Regularity Hypothesis* 1. *Let the constant β of formula* (16) *be different from* ± 1. *Then, if we neglect a finite number of roots, the remaining roots in the angle A (defined by formula* (8)) *of the determinant $M(\mu)$ (defined by formula* (6)) *are all simple and may be enumerated in two sequences $\{\xi_m\}$, $\{\tilde{\xi}_m\}$, in such a way that we have asymptotic expressions*

$$(18) \qquad \xi_m \sim 2\pi m\left(1 + \sum_{n=1}^{\infty} c_n m^{-n}\right), \qquad \tilde{\xi}_m \sim 2\pi m\left(1 + \sum_{n=1}^{\infty} \tilde{c}_n m^{-n}\right),$$

where

$$\tilde{c}_1 = c_1 + z, \qquad z \neq 0,\ 1 > \mathscr{R}z \geqq 0.$$

If we note that $\lambda = (\mu(\lambda))^n$, then the roots of $M(\mu(\lambda)) = 0$ are simply the nth powers of the roots of $M(\mu) = 0$. Hence, by Corollary 3.2, the spectrum of T is the set of roots of $M(\mu(\lambda)) = 0$. We obtain the following lemma directly from Lemma 3 and Corollary 3.2.

4 LEMMA. *Let the set* (1) *of boundary conditions satisfy Regularity Hypothesis* 1. *Let the constant* β *of formula* (16) *be different from* ± 1. *Then the spectrum of the operator* T *defined by formulas* (2) *and* (3) *consists entirely of isolated points, all of which lie in the point spectrum of* T. *If we neglect a finite number of these points, the remaining ones may be enumerated in two sequences* $\{\lambda_m\}$, $\{\tilde{\lambda}_m\}$ *in such a way that we have asymptotic expressions*

$$(19) \qquad \lambda_m \sim (2\pi m)^n\left(1 + \sum_{k=1}^{\infty} d_k m^{-k}\right), \qquad \tilde{\lambda}_m \sim (2\pi m)^n\left(1 + \sum_{k=1}^{\infty} \tilde{d}_k m^{-k}\right),$$

where

$$d_1 = \tilde{d}_1 + z, \qquad z \neq 0, \qquad n > \mathscr{R}z \geqq 0.$$

5 COROLLARY. *With the hypotheses and in the notation of Lemma* 4, *the operator* T *is discrete.*

PROOF. This is an immediate consequence of Lemmas 4 and 3. Q.E.D.

In order to apply the perturbation theory of Section 2 to T, we must show that for all but a finite number of points λ_0 in $\sigma(T)$ the projection $E(\lambda_0)$ has a one-dimensional range. This is accomplished by means of the following lemma.

6 LEMMA (*G. D. Birkhoff*). *A simple zero* λ_0 *of* $M(\mu(\lambda))$ *is a simple pole of the resolvent of* T. *Moreover, the projection* $E(\lambda_0) = E(\lambda_0; T)$ *associated with the spectral set* λ_0 *has a one-dimensional range.*

PROOF. Let I denote the interval $[0, 1]$. Let f be in $L_2(I)$ and let $G(\lambda)f$ denote that unique solution g of the equation $(\lambda - \tau_n)g = f$ which satisfies the initial conditions $g(0) = g'(0) = \cdots = g^{(n-1)}(0) = 0$. By Corollary XIII.1.5, $G(\lambda)f$ depends analytically on λ. Since $G(\lambda)$ is clearly a closed mapping of $L_2(I)$ into $H^{(n)}(I)$, it follows by the closed graph theorem (II.2.4) that $G(\lambda)$ is a bounded mapping of $L_2(I)$ into $H^{(n)}(I)$. Thus, since the topology of $H^{(n)}(I)$ is stronger than the relative topology of $H^{(n)}(I)$ as a subspace of $L_2(I)$, $G(\lambda)$ may be regarded as a bounded mapping either of $L_2(I)$ into $H^{(n)}(I)$ or of $L_2(I)$ into itself which depends analytically on the parameter λ. Let $\tilde{M}_{ik}(\mu)$ be the cofactor of the element $M_{ik}(\mu)$ in the matrix whose elements are given by the equation (5), so that the elements $M(\mu)^{-1}\tilde{M}_{ik}(\mu)$ are those of the inverse matrix of the matrix defined by the equation (5). Then $\tilde{M}_{ik}(\mu)$ depends analytically on μ, $1 \leq i, k \leq n$. Consider the element

$$H(\lambda)f = G(\lambda)f - M(\mu(\lambda))^{-1} \sum_{i,k=1}^{n} \tilde{M}_{ik}(\mu(\lambda))(B_i G(\lambda)f)\sigma_k(\mu(\lambda))$$

of $H^{(n)}(I)$. Since $(\tau - \lambda)\sigma_k(\mu(\lambda)) = 0$, we have $(\lambda - \tau)H(\lambda)f = (\lambda - \tau)G(\lambda)f = f$. Moreover, by (5),

$$\begin{aligned} B_j H(\lambda)f &= B_j G(\lambda)f - M(\mu(\lambda))^{-1} \sum_{i,k=1}^{n} \tilde{M}_{ik}(\mu(\lambda))M_{kj}(\mu(\lambda))B_i G(\lambda)f \\ &= B_j G(\lambda)f - B_j G(\lambda)f = 0, \qquad j = 1, \ldots, n. \end{aligned}$$

Thus $H(\lambda)f$ is in $\mathfrak{D}(T)$, so that $H(\lambda)$ is simply the resolvent $R(\lambda; T)$. This establishes the formula

$$\text{(20)} \quad R(\lambda; T)f = G(\lambda)f - M(\mu(\lambda))^{-1} \sum_{i,k=1}^{n} \tilde{M}_{ik}(\mu(\lambda))(B_i G(\lambda)f)\sigma_k(\mu(\lambda))$$

for every λ such that $M(\mu(\lambda)) \neq 0$. From this formula, it is evident that a simple zero of $M(\mu(\lambda))$ is a simple pole of the resolvent $R(\lambda; T)$, so that the first part of the lemma is proved.

To prove the second part, let λ_0 be a simple zero of $M(\mu(\lambda))$ and hence a simple pole of $R(\lambda; T)$, and suppose that $E(\lambda_0)$ has a two-dimensional range. Then, since by Lemma 2 and Corollary 5, $(T - \lambda_0 I)f = 0$ for each f in the range of $E(\lambda_0)$, it follows that there exist at least two linearly independent solutions φ_1, φ_2 of the equation $\tau\varphi = \lambda_0 \varphi$ such that $B_i(\varphi_1) = B_i(\varphi_2) = 0$, $i = 1, \ldots, k$. Then φ_1 may be represented uniquely as $\varphi_1 = \sum_{j=1}^{n} c_{1j}\sigma_j(\mu(\lambda))$ and, similarly, $\varphi_2 = \sum_{j=1}^{n} c_{2j}\sigma_j(\mu(\lambda))$. The vectors $[c_{11}, \ldots, c_{1n}]$ and $[c_{21}, \ldots, c_{2n}]$ are clearly linearly independent. Let $[c_{31}, \ldots, c_{3n}], \ldots, [c_{n1}, \ldots, c_{nn}]$ be chosen so as to form a basis for Euclidean n-space when taken together with these two vectors. Then the matrix c_{ij} is non-singular. Put $\sigma_i(\lambda) = \sum_{j=1}^{n} c_{ij}\sigma_j(\mu(\lambda))$, $\Phi_{ij}(\lambda) = B_i(\varphi_j(\lambda))$, and $\Phi(\lambda) = \det\{\Phi_{ij}(\lambda)\}$. Then, if C denotes the non-zero determinant of the matrix c_{ij}, we have $\Phi(\lambda) = CM(\mu(\lambda))$ by (5) and (6). Now, the first two rows of the matrix $\Phi_{ij}(\lambda)$ vanish when $\lambda = \lambda_0$. Since the determinant Φ of Φ_{ij} is a linear combination of products each of which contains one factor from each of the rows of Φ_{ij}, it follows that Φ has a double zero at $\lambda = \lambda_0$. This is contrary to assumption, and thus we have a contradiction which proves the lemma. Q.E.D.

We shall now establish an asymptotic form for the projections associated with the various eigenvalues of T.

Put

$$(21)\qquad g(\mu; t, s) = n^{-1}\mu^{1-n}\sum_{j=0}^{\nu}(iw_j)e^{iw_j\mu(t-s)}, \qquad t > s,$$
$$= -n^{-1}\mu^{1-n}\sum_{j=\nu+1}^{2\nu-1}(iw_j)e^{iw_j\mu(t-s)}, \qquad t < s.$$

Then it is clear that g is an infinitely differentiable function of t and s for $t \neq s$. We have

$$(\tau_n - \mu^n)g(\mu; t, s_0) = 0, \qquad t \neq s_0.$$

Moreover,

$$g^{(k)}(\mu; s_+, s) - g^{(k)}(\mu; s_-, s) = n^{-1}\mu^{1-n}\sum_{j=0}^{2\nu-1}(iw_j)^{k+1}.$$

Since the sum of kth powers of the nth roots of unity is 0 for $k \neq n$ and n for $k = n$, this means that

$$(22)\qquad g^{(k)}(\mu; s_+, s) - g^{(k)}(\mu; s_-, s) = 0, \qquad 0 \leqq k < n-1,$$
$$g^{(n-1)}(\mu; s_+, s) - g^{(n-1)}(\mu; s_-, s) = i^n.$$

Thus $g(\mu; s_0)$ is in $C^{(n-2)}[0, 1]$, but is not in $C^{(n-1)}[0, 1]$. Put

$$(G_\mu f)(t) = \int_0^1 g(\mu; t, s)f(s)\,ds.$$

Then

$$(G_\mu f)^{(k)}(t) = \int_0^1 g^{(k)}(\mu; t, s)f(s)\,ds, \qquad 0 \leqq k < n,$$

and, differentiating once more, we have

$$(G_\mu f)^{(n)}(t) = \int_0^1 g^{(n)}(\mu; t, s)f(s)\,ds$$
$$+ (g^{(n-1)}(\mu; t, t_-) - g^{(n-1)}(\mu; t, t_+))f(t)$$
$$= \int_0^1 g^{(n)}(\mu; t, s)f(s)\,ds + i^n f(t).$$

Thus

$$(\tau_n - \mu^n)G_\mu f = f.$$

It follows, precisely as in the derivation of formula (20) of the proof of Lemma 6, that

$$(23)\qquad R(\mu^n; T)f = G_\mu f - M(\mu)^{-1}\sum_{i,k=1}^{n} \tilde{M}_{ik}(\mu)(B_i G_\mu f)\sigma_k(\mu),$$

where $\tilde{M}_{ik}(\mu)$ is the cofactor of the element $M_{ik}(\mu)$ in the matrix whose elements are given by equation (5).

By (9) and (10), the cofactor determinant $\tilde{M}_{jk}(\mu)$ has the asymptotic form

$$(24)\qquad \tilde{M}_{jk}(\mu) \sim \mu^{p-m_j}(\tilde{\pi}_{jk}(\mu)e^{i\mu} + \tilde{\pi}'_{jk}(\mu)e^{-i\mu} + \tilde{\pi}''_{jk}(\mu)),$$

$\tilde{\pi}_{jk}$, $\tilde{\pi}'_{jk}$, and $\tilde{\pi}''_{jk}$ being finite polynomials in $1/\mu$. It follows immediately from formulas (21) and (4) that $(B_i G_\mu f)$ can be written in the form

$$(25)\qquad (B_i G_\mu f) = \mu^{1-n}\sum_{k=1}^{n} T_{ik}(\mu)\int_0^1 \sigma_k(1-s, \mu)f(s)\,ds,$$

where T_{ik} is a polynomial in μ of order at most m_i. Thus, using formulas (11), (23), (24), and (25), we see that the resolvent $R(\mu^n; T)$ is represented asymptotically in the following way:

$$(26)\quad R(\mu^n; T)f(x) \sim G_\mu f(x)$$

$$-\mu^{1-n}\frac{\sum_{k,j=1}^{n}(A_{kj}(\mu)e^{i\mu} + A'_{kj}(\mu)e^{-i\mu} + A''_{kj}(\mu))\sigma_j(x,\mu)\int_0^1 \sigma_k(1-s,\mu)f(s)\,ds}{A(\mu)e^{i\mu} + A'(\mu)e^{-i\mu} + A''(\mu)},$$

A, A', A'' and A_{kj}, A'_{kj} and A''_{kj} being polynomials in $1/\mu$. Moreover, formula (26) may clearly be differentiated any number of times with respect to x, and still remain a valid asymptotic expression. The coefficients a, a', a'' in the constant terms of $A(\mu)$, $A'(\mu)$, and $A''(\mu)$ are of course the same constants as a_p, b_p, c_p of formulas (13) and (16). The projections $E_m = E(\lambda_m; T)$ and $\tilde{E}_m = E(\tilde{\lambda}_m; T)$ are the residues at the points ξ_m and $\tilde{\xi}_m$, respectively, of the contour integral

$$\int R(\lambda; T)\,d\lambda = \int n\mu^{n-1}R(\mu^n; T)\,d\mu.$$

Since G_μ is analytic by formula (21), we have only to take the residues at ξ_m and $\tilde{\xi}_m$ of the contour integral

$$\int (n\mu^{n-1}R(\mu^m; T) - n\mu^{n-1}G_\mu)\, d\mu. \tag{27}$$

By Lemma 3, if we take any sufficiently small $\varepsilon > 0$, and let C_m denote the circle with radius ε and center $2\pi m + 2\pi c_1$, C_m contains only the single singularity ξ_m of $R(\mu^n; T)$. Consequently, the contour integral (27), extended over the contour C_m, gives the residue $E_m = E(\lambda_m; T)$. It follows from formula (26) that E_m is given asymptotically by the formula

$$(28)\quad (E_m f)(t) \sim$$

$$\frac{n}{2\pi i}\int_{C_0}\int_0^1 \frac{\left\{\sum_{k,j=1}^{n} A_{kj}(\mu+2\pi n)e^{i\mu} + A'_{kj}(\mu+2\pi n)e^{-i\mu} + A''_{kj}(\mu+2\pi n)\right\}}{A(\mu+2\pi n)e^{i\mu} + A'(\mu+2\pi n)e^{-i\mu} + A''(\mu+2\pi n)}$$

$$\times\, \sigma_j(t, \mu+2\pi n)\sigma_k(1-s, \mu+2\pi n) f(s)\, ds\, d\mu.$$

Moreover, both sides of this formula may be differentiated any number of times with respect to x and the asymptotic relation will still remain valid. It is clear that since C_0 contains no zeros of $ae^{i\mu} + a'e^{-i\mu} + a'' \equiv a_p e^{i\mu} + b_p e^{-i\mu} + c_p$, we have

$$|(A(\mu+2\pi n)e^{i\mu} + A'(\mu+2\pi n)e^{-i\mu} + A''(\mu+2\pi n))^{-1}$$

$$-(ae^{-i\mu} + a'e^{-i\mu} + a'')^{-1}| = O(n^{-1}),$$

uniformly on C_0. Thus, formula (28) may be rewritten as follows:

$$(29)\quad (E_m f)(\cdot) \sim \sum_{k,j=0}^{n-1}\int_{C_0}\int_0^1 \frac{(a_{kj}e^{-i\mu} + a'_{kj}e^{-i\mu} + a''_{kj})}{ae^{i\mu} + a'e^{i\mu} + a''}$$

$$\times\, \sigma_j(\cdot, \mu+2\pi m)\sigma_k(1-y, \mu+2\pi m) f(y)\, dy\, d\mu$$

$$+ \sum_{k,j=0}^{n-1}\int_{C_0}\int_0^1 F_{kj}^{(m)}(\mu)\sigma_j(\cdot, \mu+2\pi m)\sigma_k(1-s, \mu+2\pi m) f(s)\, ds\, d\mu,$$

where a_{kj}, a'_{kj}, a''_{kj} are certain constants, and $F_{kj}^{(m)}$, $0 \leqq k, j < n$, $m \geqq 1$, is a sequence of functions of μ which is uniformly $O(m^{-1})$ on C_0 as $m \to \infty$. We shall show, using formula (29), that the family of all sums

$$|\sum_{m\in J} E_m|,$$

J ranging over all finite sets of integers, is uniformly bounded. To do this, it is evidently sufficient, in view of formula (29) and the fact that

$\sigma_j(\cdot\,, \mu + 2\pi m)$ is uniformly bounded for μ in C_0 and $m \geqq 1$ (by formula (4)) and that $\sigma_k(1-y,\ \mu+2\pi m) = \sigma_k(1-y,\ \mu)\sigma_n(1-y,\ 2\pi m)$ (by this same formula), to establish the following two separate assertions.

(a) For $0 \leqq k, j < n$, $m \geqq 1$, and $\mu \in C_0$, let $A_{kj}^{(m)}(\mu)$ be the operator defined by

$$A_{kj}^{(m)}(\mu)f = \sigma_k(\mu + 2\pi m)\int_0^1 \sigma_j(1-s,\ \mu + 2\pi m)f(s)\,ds. \tag{30}$$

Then, as μ ranges over C_0, k and j over the set of all integers between 0 and $n-1$, and J over the family of all finite sets of integers, the family of sums

$$\sum_{m \in J} A_{kj}^{(m)}(\mu)$$

remains uniformly bounded.

(b) For $0 \leqq k < n$, let $b_m^{(k)}(f)$ be defined by

$$b_m^{(k)}(f) = \int_0^1 \sigma_j(1-s,\ 2\pi m)f(s)\,ds.$$

Then there exists a constant M such that

$$\sum_{m=1}^{\infty} \frac{|b_m^{(k)}(f)|}{m} \leqq M|f|, \qquad 0 \leqq k < n, f \in L_2(0, 1).$$

Now (b) will follow from the uniform boundedness theorem (Corollary II.3.21) and Lemma IV.4.1 once we show that

$$\sum_{m=1}^{\infty} |b_m^{(k)}(f)|^2 < \infty, \qquad 0 \leqq k < n, f \in L_2(0, 1). \tag{31}$$

We propose to do this in the lemma immediately following the next paragraph.

If $U_\mu^{(k)}$ and $V_\mu^{(k)}$ denote the operations of multiplication by $\sigma_k(t, \mu)$ and $\sigma_k(1-t, \mu)$, respectively, it is obvious from formula (30) and formula (4) that

$$A_{kj}^{(m)}(\mu) = U_\mu^{(k)} A_{kj}^{(m)}(0) V_\mu^{(k)}; \tag{32}$$

since the operators $U_\mu^{(k)}$ and $V_\mu^{(k)}$ clearly remain uniformly bounded as μ ranges over the circle C_0, (a) will be proved if we show that as J ranges over the family of finite collections of integers, the family of sums

$$\sum_{m \in J} A_{kj}^{(m)}(0), \qquad 0 \leqq k, j < n,$$

remains uniformly bounded. In view of the uniform boundedness principle (Corollary II.3.21), and of formulas (31) and (30), it is sufficient for us to establish that as J ranges over the family of all finite sets of integers, the family of sums

$$\sum_{m \in J} b_m \sigma_j(t, 2\pi m)$$

remains uniformly bounded in $L_2(0, 1)$; here $0 \leqq j < n$, and $\{b_m\}$ is some given sequence in l_2.

Thus both (a) and (b) are consequences of the following lemma.

7 LEMMA. *Let* $\alpha \neq 0$, $\mathscr{R}\alpha \leqq 0$. *Then*

(a) *for each* f *in* $L_2(0, 1)$, *the sequence* $\{a_m\}$ *of coefficients*

$$a_m(f) = \int_0^1 e^{2\pi m\alpha t} f(t)\, dt$$

is in l_2;

(b) *for each sequence* $\{b_m\}$ *in* l_2 *the family of all finite sums of terms in the sequence* $\{b_m \exp(2\pi m\alpha x)\}$ *is bounded in* $L_2(0, 1)$.

PROOF. First we shall prove (a). If $\alpha = i\beta$ is purely imaginary, then we decompose f into a finite sum of square integrable functions, each vanishing outside an interval of length at most $1/\beta$, and immediately apply the ordinary theory of orthogonal expansions in Hilbert space (cf. Theorem IV.4.13) to obtain the desired result. If $\beta = -\mathscr{R}(\alpha) > 0$, then we note that

$$|a_m(f)| \geqq \left| \int_0^1 e^{-2\pi\beta t} f(t)\, dt \right|.$$

Hence, putting $f(t) = 0$ for $t > 1$, and making the change of variable $s = -2\pi\beta t$, it is clearly sufficient for us to show that for each f in $L_2(0, \infty)$, the sequence $b_m(f)$ defined for $m \geqq 1$ by the formula

$$b_m(f) = \int_0^\infty e^{-mt} f(t)\, dt$$

is in l_2. We may clearly suppose without loss of generality that f is positive, in which case $b_m(f)$ is clearly a decreasing function of m. Consequently, $|b_m(f)|^2 \leqq \int_{m-1}^m |b_t(f)|^2\, dt$, so that we have only to show that

$$\int_0^\infty b_m(f)^2\, dm < \infty.$$

Hence, in view of the Fubini theorem, the proof of (a) will be complete as soon as the quantity

$$\int_0^\infty \int_0^\infty \int_0^\infty e^{-mt}e^{-ms}f(t)f(s)\,dt\,ds\,dm = \int_0^\infty \int_0^\infty \frac{f(t)f(s)}{t+s}\,dt\,ds$$

is proved to be finite.

Since f is in $L_2(0, \infty)$, it is sufficient by Schwarz's inequality to show that

$$g(t) = \int_0^\infty \frac{f(s)}{s+t}\,ds = \int_0^\infty \frac{f(ut)}{1+u}\,du$$

belongs to $L_2(0, \infty)$. Putting $f_t(x) = f(tx)$, we may use Theorem III.11.17 to obtain the following formula, in which we write g as the integral of a vector valued function:

$$g = \int_0^\infty \frac{f_u}{1+u}\,du.$$

Now $|f_u| = (\int_0^\infty f(tu)^2\,dt)^{1/2} = u^{-1/2}|f|$. Thus

$$|g| \leqq \int_0^\infty \frac{u^{-1/2}|f|}{1+u}\,du,$$

and since

$$\int_0^\infty \frac{u^{-1/2}}{1+u}\,du < \infty,$$

it follows that $|g| < \infty$, proving that g is in $L_2(0, \infty)$, which completes the proof of (a).

Statement (b) can be derived from statement (a) as follows. By the uniform boundedness theorem (II.3.21) and by Theorem IV.8.1, (b) will follow if only we establish that for each f in $L_2(0, 1)$, the set of all finite sums of terms $b_m \int_0^1 \exp(2\pi m\alpha t)f(t)\,dt = b_m a_m(f)$ is uniformly bounded. Since $\{b_m\} \in l_2$ by hypothesis and $\{a_m(f)\}$ is in l_2 by (a), this is evident. Q.E.D.

As remarked above, it follows immediately from Lemma 7 that the collection of all finite sums of projections E_m is bounded. In the same way it follows that the collection of all finite sums of projections $\tilde{E}_m$ is bounded. Thus we see:

The collection of all finite sums of projections $E(\lambda; T)$, $\lambda \in \sigma(T)$, is uniformly bounded.

(Here we have used the uniform boundedness theorem (Corollary II.3.21). In all this, and in what follows, the reader should compare the final paragraph of the proof of Lemma 3.10.)

We shall now improve this assertion by showing that

$$\sum_{\lambda \in \sigma(T)} E(\lambda; T) = I, \tag{33}$$

so that in Case 1A, T is a spectral operator. This we do as follows. By Lemmas XVII.3.5 and XVII.3.4, the series $\sum_{\lambda \in \sigma(T)} E(\lambda; T)f$ converges strongly and unconditionally for each f in $L_2(0, 1)$. Let

$$g = f - \sum_{\lambda \in \sigma(T)} E(\lambda; T)f;$$

we wish to show that $g = 0$. It is clear that $E(\lambda; T)g = 0$ for λ in $\sigma(T)$. By Definition 2.4 and Lemma 2.6, $R(\lambda; T)g$ is an entire function of λ. On the other hand, it follows from the asymptotic formula (26) and from the formula (21) giving the form of the kernel defining the operator G_μ that if we excise from the angle A of the μ-plane (defined by formula (8)) a circle of radius ε about each of the roots $2\pi m + 2\pi c_1$, $2\pi m + 2\pi \tilde{c}_1$ of the function $ae^{i\mu} + a'e^{-i\mu} + a'' \equiv a_p e^{i\mu} + b_p e^{-i\mu} + c_p$, then $R(\mu^n; T)g$ is uniformly bounded in the resulting domain and approaches zero at least as fast as $|\mu|^{1-n}$. Thus, by the maximum modulus theorem and by Liouville's theorem, the entire function $R(\lambda; T)g$ is constant. Differentiating with respect to λ, we find that $R(\lambda; T)^2 g = 0$; since $R(\lambda; T)$ is one-to-one for λ in $\rho(T)$, $g = 0$. This proves that, in Case 1A, T is spectral.

We sum up the preceding discussion in the following theorem.

8 THEOREM. *Let n be even. Let the set* (1) *of boundary conditions satisfy Hypothesis* 1. *Let the constant β of formula* (16) *be different from* ± 1. *Then the operator T defined by formulas* (2) *and* (3) *is a discrete spectral operator, all but a finite number of whose eigenvalues λ correspond to one-dimensional projections $E(\lambda; T)$. The eigenvalues of T have the asymptotic distribution as stated in Lemma* 4.

We now turn to consider the case in which n is odd.

Case 2: n odd.

Let $n = 2\nu + 1$. Let w_j, $j = 0, \ldots, n-1$ denote the nth roots of unity, enumerated in such a way that $w_0 = 1$, the imaginary part of w_i is positive for $0 < i \leqq \nu$, and the imaginary part of w_i is negative for $\nu < i \leqq 2\nu$.

Let λ be an arbitrary complex number. If λ is in the right half-plane, let $\mu = \mu(\lambda)$ denote that unique nth root of λ which lies in the angular sector $\{\mu \mid \pi/2n \geqq \arg \mu > -\pi/2n\}$. If λ is in the left half-plane, let $\mu(\lambda)$ denote that unique nth root of λ which lies in the angular sector $\{\mu \mid \pi/2n \geqq (\arg \mu) - \pi > -\pi/2n\}$ of the complex plane. Put

$$\begin{aligned} \sigma_k(t, \mu) &= e^{i\mu w_k t}, & 0 \leqq k \leqq \nu, \\ &= e^{i\mu w_k (t-1)}, & \nu < k \leqq 2\nu. \end{aligned} \tag{34}$$

Then $\sigma_0(t, \mu(\lambda)), \ldots, \sigma_{n-1}(t, \mu(\lambda))$ is clearly a basis for the set of solutions of the equation $\tau f = f$.

Put

$$B_i(\sigma_k(\mu)) = M_{ik}(\mu),\ M(\mu) = \det(M_{ij}(\mu)). \tag{35}$$

(See (1) for the definition of the boundary conditions B_i.) Then, just as in the even order case studied above, the eigenvalues of T are the zeros of $M(\mu(\lambda)) = 0$. We shall study the zeros of this equation, considering the right and the left half-planes separately.

It is evident from the form (34) of $\sigma_k(t, \mu)$ and the form (1) of B_i that $M_{ik}(\mu)$ has the form

$$\begin{aligned} M_{ik}(\mu) &= P_{ik}(\mu) + Q_{ik}(\mu) e^{i w_k \mu}, & 0 \leqq k \leqq \nu, \\ M_{ik}(\mu) &= P_{ik}(\mu) + Q_{ik}(\mu) e^{-i w_k \mu}, & \nu < k \leqq 2\nu, \end{aligned} \tag{36}$$

where P_{ik} and Q_{ik} are polynomials in μ of order at most m_i for all $1 \leqq i \leqq n$, $0 \leqq k \leqq n-1$.

Let A_1 and A_2 be the angular sectors in the complex plane which are defined as follows:

$$\begin{aligned} A_1 &= \left\{ z \,\middle|\, |\arg z| \leqq \frac{\pi}{2n} \right\}; \\ A_2 &= \left\{ z \,\middle|\, |(\arg z) - \pi| \leqq \frac{\pi}{2n} \right\}. \end{aligned} \tag{37}$$

We shall first investigate the zeros of $M(\mu)$ in A_1, and then those in A_2. Since $\mu(\lambda)$ lies in $A_1 \cup A_2$ for all λ, this will give us complete information on the zeros of $M(\mu(\lambda))$. Since for $0 < k \leqq \nu$, w_k is an nth root of unity with positive imaginary part, it follows that for $0 < k \leqq \nu$, $iw_k\mu$ ranges over an angle entirely contained in the left half-plane as μ ranges over A_1. Similarly, for $\nu < k \leqq 2\nu$, $iw_k\mu$ ranges over an angle entirely contained in the right half-plane as μ ranges over A_2. Thus if we put

$$N_{i0}(\mu) = M_{i0}(\mu) = P_{i0}(\mu) + Q_{i0}(\mu)e^{i\mu} \tag{38}$$

$$N_{ik}(\mu) = P_{i0}(\mu), \qquad 0 < k \leqq 2v,$$

it follows that there exists a number $a > 0$ such that

$$|N_{ik}(\mu) - M_{ik}(\mu)| = O(e^{-a|\mu|}), \tag{39}$$

as $|\mu| \to \infty$, μ remaining in the angular sector A_1. Let $N(\mu) = \det(N_{ik}(\mu))$. From form (38) of the matrix $N_{ik}(\mu)$ it follows that $N(\mu)$ has the form

$$N(\mu) = \pi_1(\mu)e^{i\mu} + \pi_2(\mu), \tag{40}$$

where π_1 and π_2 are polynomials in μ of order at most p. From relation (39) it follows that for any positive number $b < a$ we have

$$|N(\mu) - M(\mu)| = O(e^{-b|\mu|}e^{|\mathscr{I}\mu|}), \tag{41}$$

as $|\mu| \to \infty$, μ remaining in the angular sector A_1. In order to continue with the analysis, we must now make the following assumption.

9 FIRST REGULARITY HYPOTHESIS FOR ODD ORDER CASE. *The polynomials π_1 and π_2 have the precise order p.*

Thus we may write

$$\begin{aligned} \pi_1(\mu) &= a_p\mu^p + \cdots + a_0, \qquad a_p \neq 0, \\ \pi_2(\mu) &= b_p\mu^p + \cdots + b_0, \qquad b_p \neq 0. \end{aligned} \tag{42}$$

It then follows from (42), from (40), and from (41) that

$$\mu^{-p}M(\mu) = a_p e^{i\mu} + b_p + P_1(\mu)e^{i\mu} + P_2(\mu) + O(e^{-b|\mu| + |\mathscr{I}\mu|}) \tag{43}$$

uniformly as $|\mu| \to \infty$, μ remaining in A_1, where $P_1(\mu)$ and $P_2(\mu)$ are polynomials in μ^{-1} without constant terms. Let t and s be so large that

$$|P_1(\mu)| + |P_2(\mu)| < \tfrac{1}{3}|a_p|, \qquad |\mu| > t,$$

$$|b_p| < \tfrac{1}{9}|a_p e^{i\mu}|, \qquad -\mathscr{I}\mu > s,$$

and such that the final term on the right of (43) is less than $\frac{1}{3}|a_p e^{i\mu}|$ for $|\mu| > t$ and $-\mathscr{I}\mu > s$. Then it is clear from (43) that $M(\mu)$ is non-zero in the subregion $\{\mu \in A_1 \mid -\mathscr{I}\mu > s,\ |\mu| > t\}$ of A_1. In the same way, it follows that we may choose t and s so large that $M(\mu)$ is non-zero in the subregion $\{\mu \in A_1 \mid \mathscr{I}\mu > s,\ |\mu| > t\}$ of A_1.

It is quite easy to make the corresponding analysis for the angular sector A_2. Put

$$\tilde{N}_{i0}(\mu) = M_{i0} = P_{i0}(\mu) + Q_{i0}(\mu)e^{i\mu},$$
$$\tilde{N}_{ik}(\mu) = Q_{ik}(\mu), \qquad 0 < k \leqq 2\nu. \tag{44}$$

Then there exists a number $a > 0$ such that

$$|N_{ik}(\mu) - e^{-iw_k\mu}M_{ik}(\mu)| = O(e^{-a|\mu|}), \qquad 0 < k \leqq \nu, \tag{45}$$

$$|N_{ik}(\mu) - e^{iw_k\mu}M_{ik}(\mu)| = O(e^{-a|\mu|}), \qquad \nu < k \leqq 2\nu, \tag{46}$$

as $|\mu| \to \infty$, μ remaining in the angular sector A_2. Let $\tilde{N}(\mu) = \det(\tilde{N}_{ik}(\mu))$. Then from (46), if b is any positive number less than a, we have

$$|\tilde{N}(\mu) - e^{i\eta\mu}M(\mu)| = O(e^{-b|\mu|}e^{|\mathscr{I}\mu|})$$

as $|\mu| \to \infty$, μ remaining in the angular sector A_1; here we have put

$$\eta = -i\sum_{k=1}^{\nu} w_k + i\sum_{k=\nu+1}^{2\nu} w_k.$$

From form (44) of the matrix $\tilde{N}_{ik}(\mu)$ it follows that $\tilde{N}(\mu)$ has the form

$$\tilde{N}(\mu) = \tilde{\pi}_1(\mu)e^{i\mu} + \tilde{\pi}_2(\mu), \tag{47}$$

where $\tilde{\pi}_1$ and $\tilde{\pi}_2$ are polynomials in μ of order at most p. Now we make another regularity hypothesis.

10 Second Regularity Hypothesis for Odd Order Case. *The polynomials $\tilde{\pi}_1$ and $\tilde{\pi}_2$ have the precise order p:*

$$\tilde{\pi}_1(\mu) = \tilde{a}_p\mu^p + \cdots + \tilde{a}_0, \qquad \tilde{a}_p \neq 0,$$
$$\tilde{\pi}_2(\mu) = \tilde{b}_p\mu^p + \cdots + \tilde{b}_0, \qquad \tilde{b}_p \neq 0. \tag{48}$$

Just as in Case 1, we may simplify the labor of verifying our Regularity Hypotheses 9 and 10 by using the following observations. If $\hat{N}(\mu) = ae^{i\mu} + b$ is the determinant of the matrix $\hat{N}_{ik}(\mu)$ defined by the equations

$$\hat{N}_{ik}(\mu) = \alpha_{i,m_i}(w_k)^{m_i}, \qquad 0 < k \leqq \nu,$$
$$\hat{N}_{ik}(\mu) = \beta_{i,m_i}(w_k)^{m_i}, \qquad \nu < k \leqq 2\nu,$$
$$\hat{N}_{ik}(\mu) = \alpha_{i,m_i} + \beta_{i,m_i}e^{i\mu},$$

then, just as in Case 1, the coefficients a and b are identical (up to a common factor of modulus 1) with the coefficients a_p and b_p of formula (42). Similarly, if $\hat{\tilde{N}}(\mu) = \tilde{a}e^{i\mu} + \tilde{b}$ is the determinant of the matrix $\hat{\tilde{N}}_{ik}(\mu)$ defined by the equations

$$\hat{\tilde{N}}_{ik}(\mu) = \beta_{i,m_i}(w_k)^{m_i}, \qquad 0 < k \leqq \nu,$$

$$\hat{\tilde{N}}_{ik}(\mu) = \alpha_{i,m_i}(w_k)^{m_i}, \qquad \nu < k \leqq 2\nu,$$

$$\hat{\tilde{N}}_{i0}(\mu) = \alpha_{i,m_i} + \beta_{i,m_i}e^{i\mu},$$

then the coefficients $\tilde{a}$ and $\tilde{b}$ are identical up to a common factor of modulus 1 with the coefficients $\tilde{a}_p$ and $\tilde{b}_p$ of equation (48). Thus, for a normalized set of boundary conditions, Hypotheses 9 and 10 depend only on the "leading" coefficients in the boundary conditions, that is, on the coefficients α_{i,m_i} and β_{i,m_i} of the highest derivative occurring with a non-zero coefficient. If Regularity Hypotheses 9 and 10 are satisfied, it follows just as in the above analysis of A_1 that numbers t and s may be chosen so large that $M(\mu)$ is non-zero in the subregion $\{\mu \in A_2 \mid |\mathscr{I}\mu| > s, |\mu| > t\}$ of A_1. Summing up, it is seen that we have proved the following lemma.

11 LEMMA. *Let the set* (1) *of boundary conditions satisfy Regularity Hypotheses* 9 *and* 10. *Then there exist real numbers t and s so large that every zero z in the angle $A = A_1 \cup A_2$ (defined by formula* (37)) *of the determinant $M(\mu)$ (defined by formula* (35) *ff.) satisfies either $|z| < t$ or $|\mathscr{I}z| < s$.*

Note that the roots of the equation $a_p e^{i\mu} + b_p = 0$ are all simple and form an arithmetic sequence of the form $2\pi n i + z_0$. Using this observation, Lemma 3.5, and the preceding lemma, we find, just as in Case 1, that all but a finite number of the roots of $M(\mu) = 0$ which lie in A_1 are simple, and that they may be enumerated in a sequence ξ_m in such a way that we have an asymptotic expression

$$\xi_m \sim 2\pi m\left(1 + \sum_{n=1}^{\infty} c_n m^{-n}\right). \tag{49}$$

The analysis of the angle A_2 is exactly similar, and leads in exactly the same way to the conclusion that all but a finite number of the roots of $M(\mu) = 0$ which lie in A_2 are simple, and that they may be enumerated in such a way that we have an asymptotic expression

$$\tilde{\xi}_m \sim -2\pi m\left(1 + \sum_{n=1}^{\infty} \tilde{c}_n m^{-n}\right). \tag{50}$$

If we then pass from the μ-plane to the λ-plane, where $\lambda = \mu^n$, and make use of Lemmas 3.5 and 4.6, we obtain the following statement, which is the analogue for the odd order case of a combination of Lemma 4 and Corollary 5.

12 LEMMA. *Let the set* (1) *of boundary conditions satisfy Regularity Hypotheses* 9 *and* 10. *Then the operator* T *defined by formulas* (2) *and* (3) *is discrete. The spectrum of* T *consists entirely of isolated points. If we neglect a finite number of these points, the remaining ones are all simple poles of the resolvent of* T, *and all have associated projections with one-dimensional ranges. Moreover, the points of* $\sigma(T)$ *may be enumerated in two sequences* $\{\lambda_m\}$, $\{\tilde{\lambda}_m\}$, *in such a way that they have asymptotic expressions*

$$\lambda_m \sim (2\pi m)^n\left(1 + \sum_{k=1}^{\infty} d_k m^{-k}\right), \tag{51}$$

$$\tilde{\lambda}_m \sim -(2\pi m)^n\left(1 + \sum_{k=1}^{\infty} \tilde{d}_k m^{-k}\right).$$

To show that T is a spectral operator we may proceed just as in Case 1A. Put

$$\begin{aligned} g(\mu; t, s) &= n^{-1}\mu^{1-n} \sum_{j=0}^{\nu} (iw_j)e^{iw_j\mu(t-s)}, \qquad t > s, \\ &= -n^{-1}\mu^{1-n} \sum_{j=\nu+1}^{2\nu} (iw_j)e^{iw_j\mu(t-s)}, \qquad t < s. \end{aligned} \tag{52}$$

Then g is an infinitely differentiable function of t and s for $t \neq s$. It follows just as in Case 1A above that if we put

$$(G_\mu f)(t) = \int_0^1 g(\mu; t, s)f(s)\, ds,$$

then

$$(\tau_n - \mu^n)G_u f = f,$$

so that

$$R(\mu^n; T)f = G_\mu f - M(\mu)^{-1} \sum_{i,k=1}^{n} \tilde{M}_{ik}(\mu)(B_i G_\mu f)\sigma_k(\mu), \tag{53}$$

where $\tilde{M}_{ik}(\mu)$ is the cofactor of the element $M_{ik}(\mu)$ in the matrix whose elements are given by equation (5).

By (38), the cofactor determinant $\tilde{M}_{jk}(\mu)$ has the asymptotic form

$$\tilde{M}_{jk}(\mu) \sim \mu^{p-m_j}(\tilde{\pi}_{jk}(\mu)e^{i\mu} + \tilde{\pi}'_{jk}(\mu)), \tag{54}$$

where $\tilde{\pi}_{jk}$ and $\tilde{\pi}'_{jk}$ are finite polynomials in $1/\mu$. It follows immediately from formulas (52) and (34) that $(B_i G_\mu f)$ can be written in the form

$$(B_i G_\mu f) = \mu^{1-n} \sum_{k=1}^{n} T_{ik}(\mu) \int_0^1 \sigma_k(1-s, \mu) f(s)\, ds, \tag{55}$$

where T_{ik} is a polynomial in μ of order at most m_i. Thus, using formulas (40), (41), (53), (54), and (55), we see that the resolvent $R(\mu^n; T)$ is represented asymptotically in the following way in the sector A_1:

$$R(\mu^n; T) f(x) \sim G_\mu f(x) - \mu^{1-n} \frac{\sum_{k,j=1}^{n} (A_{kj}(\mu) e^{i\mu} + A'_{kj}(\mu)) \sigma_j(x, \mu) \int_0^1 \sigma_k(1-s, \mu) f(s)\, ds}{A(\mu) e^{i\mu} + A'(\mu)}, \tag{56}$$

A, A', A_{kj}, and A'_{kj} being polynomials in $1/\mu$. Moreover, formula (56) may clearly be differentiated any number of times with respect to x, and still remain a valid asymptotic expression. The coefficients a, a' in the constant terms of $A(\mu)$ and $A'(\mu)$ are of course the same constants at a_p and b_p of formula (42). It now follows just as in Case 1A that the projection $E_m = E(\lambda_m; T)$ is given asymptotically by the formula

$$(E_m f)(x) \sim \frac{n}{2\pi i} \int_{C_0} \int_0^1 \frac{\left\{ \sum_{k,j=1}^{n} A_{kj}(\mu + 2\pi m) e^{i\mu} + A'_{kj}(\mu + 2\pi m) \right\}}{A(\mu + 2\pi m) e^{i\mu} + A'(\mu + 2\pi m)} \times \sigma_j(x, \mu + 2\pi m) \sigma_k(1-s, \mu + 2\pi m) f(s)\, ds\, d\mu, \tag{57}$$

C_0 denoting a sufficiently small circle about the point $2\pi m c_1$, and c_1 being as in formula (49). Moreover, both sides of this formula may be differentiated any number of times with respect to x and the asymptotic relation will still remain valid. It is then clear, since C_0 contains no zeros of $ae^{i\mu} + a' = a_p e^{i\mu} + b_p$, that we have $|(A(\mu + 2\pi m) e^{i\mu} + A(\mu + 2\pi m))^{-1} - (ae^{i\mu} + a')^{-1}| = O(m^{-1})$ uniformly on C_0. Thus, formula (57) may be rewritten as follows:

(58)

$$(E_m f) \sim \sum_{k,j=0}^{n-1} \int_{C_0} \int_0^1 \frac{(a_{kj} e^{i\mu} + a'_{kj})}{ae^{i\mu} + a} \sigma_j(\cdot, \mu + 2\pi m) \sigma_k(1-s, \mu + 2\pi m) f(s)\, ds\, d\mu$$
$$+ \sum_{k,j=1}^{n-1} \int_{C_0} \int_0^1 F_{kj}^{(m)}(\mu) \sigma_j(\cdot, \mu + 2\pi m) \sigma_k(1-s, \mu + 2\pi m) f(s)\, ds\, d\mu,$$

where a_{kj} and a'_{kj} are certain constants, and $F_{kj}^{(m)}$, $0 \leqq k, j < n$, $m \geqq 1$, is a sequence of functions which is uniformly $O(m^{-1})$ on C_0 as $m \to \infty$. We wish to show, using formula (58), that the family of all sums

$$\left|\sum_{m \in J} E_m\right|,$$

J ranging over all finite sets of integers, is uniformly bounded. This follows from (58) by an argument using Lemma 7, which is similar to the corresponding argument used in the discussion of Case 1A. It follows in the same way that the collection of all finite sums of projections $E(\tilde{\lambda}_m; T)$ is uniformly bounded. Hence it is seen that:

the collection of all finite sums of projections $E(\lambda; T)$, *with* λ *in* $\sigma(T)$, *is bounded.*

We now wish to improve this assertion by showing that

$$\sum_{\lambda \in \sigma(T)} E(\lambda; T) = I,$$

so that in Case 2, T is a spectral operator. Just as in the corresponding proof in Case 1A we see that it suffices to show that if $R(\lambda; T)g$ is entire, then $g = 0$. It follows from the asymptotic formula (56) and from formula (52) giving the form of the kernel G_μ that if we excise from the angle A_1 of the μ-plane (defined by the formula immediately following (36)) a circle about each of the roots $2\pi m + 2\pi c_1$ of the function $ae^{i\mu} + a' = a_p e^{i\mu} + b_p$, then $R(\mu^n; T)g$ is bounded in the resulting domain. By the maximum modulus theorem, $R(\mu^n; T)g$ is bounded in the whole angle A_1. We can show by an exactly corresponding argument that $R(\mu^n; T)g$ is bounded in the angle A_2. Thus $R(\mu^n; T)g$ is uniformly bounded in the whole complex plane, and it follows just as in Case 1 that $g = 0$ and that T is a spectral operator.

We sum up the preceding discussion in the following theorem.

13 THEOREM. *Let* n *be odd. Let the set* (1) *of boundary conditions satisfy Hypotheses* 9 *and* 10. *Then the operator* T *defined by formulas* (2) *and* (3) *is a discrete spectral operator, all but a finite number of whose eigenvalues* λ *correspond to one-dimensional projections* $E(\lambda; T)$. *The eigenvalues of* T *have the asymptotic distribution described in Lemma* 12.

An interesting special case in which the Regularity Hypothesis 1 is automatically satisfied was discovered by H. P. Kramer. Let $n = 2\nu$, and suppose that the set $\{B_i\}$ of boundary values falls into two parts, $B_1, \ldots, B_\nu$

being boundary values for τ at 0, $B_{\nu+1}, \ldots, B_{2\nu}$ being boundary values for τ at 1. By the remark following formula (3), it is no loss of generality to assume that the order of B_i is m_i, and that $m_1 > m_2 > \cdots > m_\nu$, $m_{\nu+1} > m_{\nu+2} > \cdots > m_{2\nu}$. Moreover, we may clearly assume that the coefficient of $f^{(m_i)}$ in the unique expression $B_i(f) = \sum_{j=0}^{n-1} \gamma_j f^{(j)}(0) + \sum_{j=0}^{n-1} \tilde{\gamma}_j f^{(j)}(1)$ is 1. Then, if $B_1, \ldots, B_\nu$ and $B_{\nu+1}, \ldots, B_{2\nu}$ are arranged together in "size place" according to their orders, we have a set of boundary values normalized according to the specifications of the paragraph following formula (3). Thus, all the above analysis applies without any other preliminary normalization to the set $B_1, \ldots, B_n$ of boundary values. By the remark following Regularity Hypothesis 1, we may then assume without loss of generality that $B_i f = f^{(m_i)}(0)$ for $1 \leqq i \leqq \nu$ and $B_i f = f^{(m_i)}(1)$ for $\nu < i \leqq 2\nu$. The matrix $\hat{\hat{N}}_{ik}(\mu)$ of the remark following Regularity Hypothesis 1 is consequently determined by the equations

$$
\begin{aligned}
\hat{\hat{N}}_{ik} &= 0 && \text{for} \quad 0 < k < \nu \quad \text{and} \quad \nu < i \leqq 2\nu, \\
&= 0 && \text{for} \quad \nu < k < 2\nu \quad \text{and} \quad 1 \leqq i \leqq \nu, \\
\hat{\hat{N}}_{ik} &= (i w_k)^{m_i} && \text{otherwise, if} \quad k \neq 0,\ k \neq \nu; \\
\hat{\hat{N}}_{i0} &= i^{m_i}, && 1 \leqq i \leqq \nu, \\
&= i^{m_i} e^{i\mu}, && \nu < i \leqq 2\nu, \\
\hat{\hat{N}}_{i\nu} &= (-i)^{m_i}, && 1 \leqq i \leqq \nu, \\
&= (-i)^{m_i} e^{-i\mu}, && \nu < i \leqq 2\nu.
\end{aligned}
$$

The determinant $\hat{\hat{N}}(\mu)$ of the remark following Regularity Hypothesis 1 is consequently of the following form.

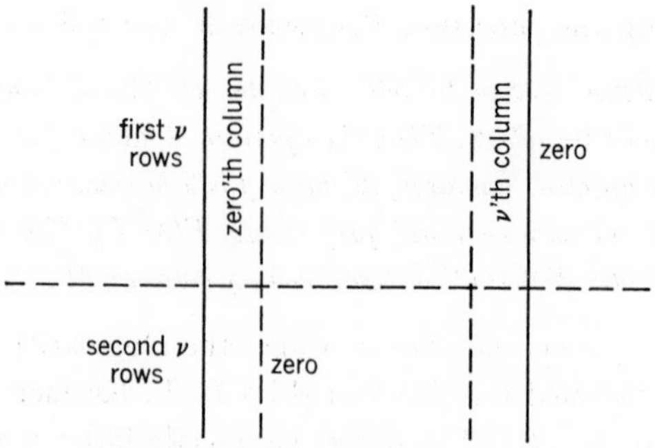

Consequently, if we use Lagrange's rule to expand this $2\nu \times 2\nu$ determinant by minors of order ν, we find that the expansion contains only two non-vanishing terms. Thus our $2\nu \times 2\nu$ determinant may be expressed as $P_1 P_2 \pm Q_1 Q_2$, where P_1 and P_2 are the $\nu \times \nu$ determinants of the first ν rows and columns and the second ν rows and columns, respectively, and where Q_1 is the $\nu \times \nu$ determinant of the first ν rows and the second through $(\nu+1)$th column, and Q_2 is the $\nu \times \nu$ determinant of the second ν rows and the first and $(\nu+2)$th through 2νth column. Consequently, $P_1 P_2$ is equal to the product

$$\begin{vmatrix} (iw_0)^{m_1} & (iw_1)^{m_1} & \cdots & (iw_{\nu-1})^{m_1} \\ (iw_0)^{m_2} & \cdots & & \cdots \\ \cdots & & & \cdots \\ (iw_0)^{m_\nu} & \cdots & & (iw_{\nu-1})^{m_\nu} \end{vmatrix}$$

$$\times \begin{vmatrix} e^{-i\mu}(iw_\nu)^{m_{\nu+1}} & (iw_{\nu+1})^{m_{\nu+1}} & \cdots & (iw_{2\nu-1})^{m_{\nu+1}} \\ e^{-i\mu}(iw_\nu)^{m_{\nu+2}} & (iw_{\nu+1})^{m_{\nu+2}} & & \cdots \\ \cdots & & & \cdots \\ e^{-i\mu}(iw_\nu)^{m_{2\nu}} & (iw_{\nu+1})^{m_{2\nu}} & \cdots & (iw_{2\nu-1})^{m_{\nu 2}} \end{vmatrix} = c_1 e^{-i\mu},$$

where c_1, as the product of two non-vanishing Vandermonde type determinants, is non-vanishing. Moreover, $Q_1 Q_2$ is equal to the product

$$\begin{vmatrix} e^{i\mu}(iw_0)^{m_{\nu+1}} & (iw_{\nu+1})^{m_{\nu+1}} & \cdots & (iw_{2\nu-1})^{m_{\nu+1}} \\ e^{i\mu}(iw_0)^{m_{\nu+2}} & \cdots & \cdots & \\ \cdots & & \cdots & \\ e^{i\mu}(iw_0)^{m_{2\nu}} & \cdots & \cdots & (iw_{2\nu-1})^{m_{2\nu}} \end{vmatrix} \begin{vmatrix} (iw_1)^{m_1} & (iw_2)^{m_1} & \cdots & (iw_\nu)^{m_1} \\ (iw_1)^{m_2} & \cdots & \cdots & \\ \cdots & & & \cdots \\ (iw_1)^{m_\nu} & & \cdots & (iw_\nu)^{m_\nu} \end{vmatrix}$$

$$= c_2 e^{+i\mu},$$

where c_2 fails to vanish for the same reason. We then learn according to the remark following Regularity Hypothesis 1 that the constants a_p and b_p of formula (13) are non-zero, whereas the constant c_p of formula (13) is zero. Consequently, Hypothesis 1 is satisfied and the constant β of formula (16) is zero.

The conclusion of Theorem 8 is consequently valid in all cases in which $n = 2\nu$ is even and the boundary values (1) may be separated into a set of ν boundary values at zero and a set of ν boundary values at 1.

We state all this as a formal theorem.

14 THEOREM (*H. P. Kramer*). *Let n be even and let T be the closed operator defined by the formal differential operator $\tau_n = i^{-n}(d/dt)^n$ on the*

interval $[0, 1]$ *and by a set of* n *linearly independent boundary conditions, of which* $n/2$ *are boundary conditions at zero and* $n/2$ *are boundary conditions at* 1. *Then all the conclusions of Theorem* 8 *are valid.*

In this connection, it is also worth making a simple remark on the notion of regular boundary conditions in the case of a second order differential operator. If the boundary conditions are written in the normalized way described in the paragraph following formula (3), then we have three possibilities.

(a) $m_1 = 1$, $m_2 = 1$. In this case the boundary conditions in normalized form are

$$[*] \qquad u'(0) + \cdots = 0, \qquad u'(1) + \cdots = 0,$$

(neglecting boundary terms of order less than 1). By the remarks following formula (13) and Regularity Hypothesis 1, it follows, since the set of boundary conditions $u'(0) = 0$, $u'(1) = 0$ is evidently regular, that the boundary conditions [*] are regular.

(b) $m_1 = 0$, $m_2 = 0$. In this case, the boundary conditions in normalized form are

$$u(0) = 0, \qquad u(1) = 0,$$

and are evidently regular.

(c) $m_1 = 1$, $m_2 = 0$. Then, assuming for the sake of definiteness that $u'(0)$ occurs with a non-zero coefficient in the first boundary value, our boundary conditions may be written as

$$u'(0) + ku'(1) + \cdots = 0,$$
$$au(0) + bu(1) = 0.$$

If $a = 0$, these boundary conditions can be written as

$$u'(0) + ku'(1) + \cdots = 0,$$
$$u(1) = 0,$$

and are readily seen to be regular. If $a \neq 0$, the boundary conditions may be written as

$$u'(0) + ku'(1) + \cdots = 0,$$
$$u(0) + k'u(1) = 0,$$

and a simple computation based on the condition given in Regularity Hypothesis 1 shows that such a set of boundary conditions is regular if and only if $k + k' \neq 0$. Thus a set of two linearly independent boundary conditions for the second order operator is irregular if and only if it has one of the forms

$$u'(0) + ku'(1) + \cdots = 0,$$
$$u(0) - ku(1) = 0$$

or

$$u'(1) + ku'(0) + \cdots = 0,$$
$$u(1) - ku(0) = 0.$$

Before proving the main result stated as Theorem 16 below, we must establish the following lemma.

15 LEMMA. *Let T be a closed operator in the Hilbert space $L_2(0, 1)$ derived from the formal differential operator $((1/i)(d/dt))^n$ by the imposition of a set of n boundary conditions. If n is even, let these boundary conditions satisfy Regularity Hypothesis* 1 *and let the constant β of formulas* (16) *and* (17) *be different from* 1. *If n is odd, let these boundary conditions satisfy Regularity Hypotheses* 9 *and* 10. *Let $\lambda \in \sigma(T)$, and let k be an integer. Then $(T - \lambda I)^{-k/n}$ is a continuous linear mapping of $L_2(0, 1)$ into $H^{(k)}(0, 1)$.*

PROOF. First note that, by the first paragraph of the proof of Theorem 2.7 (which is valid word for word for any discrete operator T all but a finite number of the points of whose spectrum correspond to projections with one-dimensional ranges), $(T - \lambda I)^{-k/n}$ is a bounded operator. Let σ_0 be a finite collection of points in $\sigma(T)$, including all those points for which $E(\lambda; T)$ fails to have a one-dimensional range, and also including zero, if $0 \in \sigma(T)$. Let $\sigma_0' = \sigma(T) - \sigma_0$. Then

$$(T - \lambda I)^{-k/n} = E(\sigma_0)(T - \lambda I)^{-k/n}E(\sigma_0) + (T - \lambda I)^{-k/n}E(\sigma_0')$$

by Theorem XVIII.2.9. We shall show separately that the operators $E(\sigma_0)(T - \lambda I)^{-k/n}E(\sigma_0)$ and $(T - \lambda I)^{-k/n}E(\sigma_0')$ are continuous mappings of $L_2(0, 1)$ into $H^{(k)}(0, 1)$, and in this way prove the lemma.

By Lemmas 4 and 12, we may suppose without loss of generality that $\sigma_0' = \{\lambda_K, \tilde{\lambda}_K, \lambda_{K+1}, \tilde{\lambda}_{K+1}, \ldots\}$, K being greater than 1, and the sequences

$\{\lambda_j\}$ and $\{\tilde{\lambda}_j\}$ having the asymptotic forms given in those lemmas. By the first paragraph of the proof of Theorem 2.7,

$$\text{(i)} \qquad (T-\lambda I)^{-k/n}E(\sigma_0')f = \lim_{p\to\infty}\left\{\sum_{i=K}^{p}(\lambda_i-\lambda)^{-k/n}E(\lambda_i)f + \sum_{i=K}^{p}(\tilde{\lambda}_i-\lambda)^{-k/n}E(\tilde{\lambda}_i)f\right\}.$$

By Lemma 3.9, the one-dimensional projections $E(\lambda_i)$ and $E(\tilde{\lambda}_i)$ are all bounded mappings of $L_2(0, 1)$ into $H^{(k)}(0, 1)$. The limit on the right-hand side of (i) clearly exists in the topology of $H^{(k)}(0, 1)$ for each f in $E(\sigma_0)L_2(0, 1)$ and for each f of the form $f=\{E(\lambda_i)+E(\tilde{\lambda}_i)\}$ and, consequently, for each f in a dense subset of $L_2(0, 1)$. By Theorems 8, 13, and II.3.6, the continuity of $(T-\lambda I)^{-k/n}$ as a map of $L_2(0, 1)$ into $H^{(k)}(0, 1)$ will be established once it is shown that the set of mappings

$$\text{(ii)} \qquad \sum_{i=K}^{p}(\lambda_i-\lambda)^{-k/n}E(\lambda_i)+\sum_{i=K}^{p}(\tilde{\lambda}_i-\lambda)^{-k/n}E(\tilde{\lambda}_i), \qquad p \geqq K,$$

of $L_2(0, 1)$ into $H^{(k)}(0, 1)$ is uniformly bounded. We shall first establish that the set of mappings

$$\sum_{i=K}^{p}(\lambda_i-\lambda)^{-k/n}E(\lambda_i), \qquad p \geqq K,$$

of $L_2(0, 1)$ into $H^{(k)}(0, 1)$ is uniformly bounded. The corresponding result for the mappings

$$\sum_{i=K}^{p}(\tilde{\lambda}_i-\lambda)^{-k/n}E(\tilde{\lambda}_i), \qquad p \geqq K,$$

will follow similarly. From these two results, (ii) evidently follows.

By Lemmas 4 and 12, we have the asymptotic relation $(\lambda_m-\lambda)^{-k/n} \sim$ const. m^{-k}. Thus (by Theorem XVIII.2.11) the map

$$A=\sum_{i=K}^{\infty}(\lambda_i-\lambda)^{-k/n}m^kE(\lambda_i)$$

of $L_2(0, 1)$ into itself is bounded and has a bounded inverse. Since, by Theorem XVIII.2.11, we have evidently

$$\left(\sum_{i=k}^{p}(\lambda_i-\lambda)^{-k/n}E(\lambda_i)\right)=\left(\sum_{m=K}^{p}m^{-k}E(\lambda_i)\right)A,$$

it suffices to show that the family

$$\text{(iii)} \qquad \sum_{m=K}^{p} m^{-k}E(\lambda_i), \qquad p \geqq K,$$

of maps of $L_2(0, 1)$ into $H^{(k)}(0, 1)$ is uniformly bounded. An equivalent norm in $H^{(k)}(0, 1)$ is

$$\text{(iv)} \qquad |f| = \left\{\int_0^1 |f(t)|^2\, dt\right\}^{1/2} + \left\{\int_0^1 |f^{(k)}(t)|^2\, dt\right\}^{1/2}$$

The family (iii) of maps is evidently uniformly bounded as a set of maps of $L_2(0, 1)$ into itself. Consequently, it suffices to show (using the uniform boundedness theorem, Corollary II.3.21) that for each f in L_2, the set of functions

$$\text{(v)} \qquad \sum_{m=K}^{p} m^{-k}\left(\frac{d}{dx}\right)^k (E(\lambda_m)f)(t), \qquad p \geqq K,$$

is uniformly bounded in $L_2(0, 1)$. It follows from formula (28) (in Case 1A) and from the corresponding formula (57) (in Case 2) that

$$m^{-k}\left(\frac{d}{dx}\right)^k (E(\lambda_m)f)(t)$$

is represented asymptotically by a formula of exactly the same sort, (28) or (57), as $(E(\lambda_m)f)(t)$. Thus, we may show that the set of functions (v) is uniformly bounded in $L_2(0, 1)$ in the same way that we showed that the set of functions

$$\sum_{m=K}^{p} E(\lambda_m)f, \qquad p \geqq K,$$

is bounded in the course of proving Theorem 8 and Theorem 13. This observation completes the proof of the present lemma. Q.E.D.

→ 16 THEOREM. *Let T be the closed operator in the Hilbert space $L_2(0, 1)$ derived from the formal differential operator $(-id/dt)^n$ by the imposition of a set of n boundary conditions. If n is even, let these boundary conditions satisfy Regularity Hypothesis* 1 *and let the constant β of formulas* (16) *and* (17) *be different from* 1. *If n is odd, let these boundary conditions satisfy Regularity Hypotheses* 9 *and* 10. *Let P be an operator whose domain is the subset $H^{(n-1)}[(0, 1)]$ of $L_2(0, 1)$ and which is defined by the formula*

$$(Pf)(t) = a_{n-1}(x)f^{(n-1)}(t) + \sum_{j=0}^{n-2} (B_j f^{(j)})(t),$$

where $a_{n-1}(x)$ is a function in $C^\infty[(0, 1)]$ and $B_0, \ldots, B_{n-2}$ are arbitrary bounded operators in $L_2(0, 1)$. Then $T + P$ is a discrete spectral operator, all but a finite number of whose spectral points are simple poles of the resolvent giving rise to projections $E(\lambda; T + P)$ with one-dimensional range.

PROOF. First suppose that $a_{n-1}(t) \equiv 0$. Then the present result would follow from Theorem 2.7, Corollary 2.8, Theorem 8, Theorem 13, and Lemma 15 once we showed that (using the notations of the hypothesis of Theorem 2.7) $\sum_{m=1}^{\infty} d_m^{-2}(|\lambda_m| + d_m)^{2(n-2)/n} < \infty$. Using the asymptotic form for λ_n given in Lemmas 4 and 12, we see that we have only to show that $d_m \geqq Km^{n-1}$ for sufficiently large m and some constant $K > 0$ in order to obtain the desired result. Consider, for example, the case in which n is even. Then, by Lemma 4, upon neglecting a finite number of eigenvalues, the remaining ones fall into two sequences λ_m, $\tilde{\lambda}_m$ such that λ_m differs by a bounded quantity from

$$(*) \qquad \mu_m = (2\pi m)^n\left(1 + \sum_{k=1}^{n-1} c_k m^{-k}\right)$$

and $\tilde{\lambda}_m$ differs by a bounded quantity from

$$({*\atop *}) \qquad \tilde{\mu}_m = (2\pi m)^n\left(1 + \sum_{k=1}^{n-1} \tilde{c}_k m^{-k}\right).$$

Moreover,

$$({*\atop{*\atop *}}) \qquad \tilde{c}_1 = c_1 + nz, \qquad z \neq 0, \qquad n > \mathscr{R}z \geqq 0.$$

We shall show that for large m the distance between μ_m and the closest of the $\mu_{m'}$, $m' \neq m$, and of the $\tilde{\mu}_{m'}$, is bounded below by a function of the form $Km^{n'-1}$, and shall establish a similar result for the $\tilde{\mu}_m$. This will clearly give the desired result.

Note first that if we multiply μ_n and $\tilde{\mu}_n$ by a constant α of modulus 1, we may assume without loss of generality that $\mathscr{R}c_1 \neq \mathscr{R}\tilde{c}_1$, $\mathscr{R}c_1 = \mathscr{R}c_2 + z'$, $n > \mathscr{R}z' > 0$. Since $|\mu - \lambda| \geqq |\mathscr{R}\mu - \mathscr{R}\lambda|$, it suffices to establish our assertion for the sequences

$$\mathscr{R}\mu_m = (2\pi m)^n\left(1 + \sum_{k=1}^{n-1} (\mathscr{R}c_k)m^{-k}\right)$$

and $\mathscr{R}\tilde{\mu}_m$; thus we may assume without loss of generality that c_k and $\tilde{c}_k$, and hence μ_m and $\tilde{\mu}_m$, are real. It is then evident that for m sufficiently large, μ_m and $\tilde{\mu}_m$ form increasing sequences. Hence, for m sufficiently large,

$$\min_{k \neq m} |\mu_m - \mu_k| \geqq \max(|\mu_m - \mu_{m-1}|, |\mu_m - \mu_{m+1}|).$$

Since, by $(*)$,

$$|\mu_m - \mu_{m+1}| \sim 2\pi n (2\pi m)^{n-1},$$

it is clear that $\min_{k \neq m} |\mu_n - \mu_k|$ is bounded below by a function of the form Km^{n-1} for sufficiently large m. Since $c_1 \neq \tilde{c}_1$, we have either $c_1 > \tilde{c}_1$ or $\tilde{c}_1 > c_1$. Suppose for the sake of definiteness that $c_1 > \tilde{c}_1$. Then it is clear that $\mu_m > \tilde{\mu}_m$ for sufficiently large m. On the other hand, since

$$\begin{aligned}\tilde{\mu}_m - \mu_{m-1} &= 2\pi(\tilde{c}_1 - c_1)(2\pi m)^{n-1} + 2\pi n(2\pi m)^{n-1} + O(m^{n-2}) \\ &= 2\pi(n + \tilde{c}_1 - c_1)(2\pi m)^{n-1} + O(m^{n-2}),\end{aligned}$$

and since we have seen above that $n + \tilde{c}_1 - c_1$ is positive, it follows that $\mu_m > \tilde{\mu}_m > \mu_{m-1}$ for sufficiently large m. Consequently,

$$\cdots > \tilde{\mu}_{m+1} > \mu_m > \tilde{\mu}_m > \mu_{m-1} > \tilde{\mu}_{m-1} > \cdots.$$

Thus

$$\min_{k} |\tilde{\mu}_m - \mu_k| = \max(|\mu_m - \tilde{\mu}_m|, \quad |\mu_m - \tilde{\mu}_{m+1}|).$$

It now follows from $(_*)$, $(^*_*)$, and $(^{*}_{**})$ that $\min_k |\mu_m - \tilde{\mu}_k|$ is bounded below by a function of the form Km^{n-1}, which proves our theorem in case $a_{n-1}(t) \equiv 0$ and n is even.

If n is odd, a similar calculation based on Lemma 12 establishes the corresponding results in the same way, the calculations in this case being even easier.

If $a_{n-1}(t)$ is not identically zero, we may argue as follows. Let $b(t) = n^{-1} \int_0^t a_{n-1}(s)\, ds$, and $h(t) = \exp(b(t))$. Then $h(t) \neq 0$, so that the map U defined by $(Uf)(t) = h(t)f(t)$ is an automorphism of Hilbert space $L_2(0, 1)$ into itself. We have, formally,

$$\left(U^{-1}\left(\frac{d}{dt}\right) Uf\right)(t) = \left(\left(\frac{d}{dt}\right) + b'(t)\right) f(t).$$

It is easy to see from this and from Definition XIII.2.17 that if S is the operator in Hilbert space defined by the formal differential operator $(-id/dt)^n$ and a certain set $C_i(f)$, $i = 1 \ldots n$, of boundary conditions, then

$U^{-1}TU$ is the same as the operator in Hilbert space defined by the formal differential operator $(-id/dt + b'(t))^n$ and the set $C_i(Uf) = 0$, $i = 1 \ldots n$, of boundary conditions. Since

$$\left((-i)^n\left(\frac{d}{dt}\right) + b(t)\right)^n = (-i)^n\left(\frac{d}{dt}\right)^n + a_{n-1}(t)\left(\frac{d}{dt}\right)^{n-1} + \cdots,$$

the operator

$$T + P = T + a_{n-1}(t)\left(\frac{d}{dt}\right)^{n-1} + \sum_{j=0}^{n-2} B_j\left(\frac{d}{dt}\right)^j$$

of the hypothesis of the theorem may be written as

$$S + \sum_{j=0}^{n-2} \tilde{B}_j\left(\frac{d}{dt}\right)^j$$

$\tilde{B}_j$, $j = 0, \ldots, n-2$ being bounded operators. Hence, $U(T+P)U^{-1}$ has the form $T + P'$, P' being an operator of the form $\sum_{j=0}^{n-2} \tilde{\tilde{B}}_j(d/dt)^j$, $\tilde{\tilde{B}}_j$, $0 \leqq j \leqq n-2$, being bounded operators. Since $T + P$ has the same spectrum, and, in fact, identical properties as an abstract operator as $U(T+P)U^{-1}$, it follows that it is no loss of generality (provided that the set of boundary conditions $C_i(Uf) = 0$ is regular) to assume that $a_{n-1}(t) = 0$. But, under the condition that the set of boundary conditions $C_i(Uf) = 0$ is regular, this has been established above.

Thus, to complete the proof, it suffices to show that the set C of boundary conditions $C_i(Uf) = 0$, $i = 1, \ldots, n$, is regular. Let $C_i(f)$ be given by the formula

$$C_i(f) = \sum_{j=0}^{m_i} \alpha_{ij} f^{(j)}(0) + \sum_{j=0}^{m_i} \beta_{ij} f^{(j)}(1).$$

Then, since $U^{-1}(d/dt)Uf = ((d/dt) + b'(t))f(t)$, it follows that

$$C_i(Uf) = \sum_{j=0}^{m_i} \hat{\alpha}_{ij} f^{(j)}(0) + \sum_{j=0}^{m_i} \hat{\beta}_{ij} f^{(j)}(1),$$

where $\hat{\alpha}_{im_i} = \alpha_{im_i}$, $\hat{\beta}_{im_i} = \beta_{im_i}$. The regularity of the set C of boundary conditions now follows immediately from the regularity of our original set of boundary conditions and the remarks following Regularity Hypotheses 1, 9 and 10. Q.E.D.

5. Approximation by Generalized Eigenfunctions

In many cases in which the spectrum of a discrete spectral operator T is distributed in the complex plane in too chaotic a way for the very restrictive hypothesis of Theorem 2.7 to apply, a variant of the proof of Theorem 2.7 enables us to show that the set of solutions of $(T+P-\lambda)^k f = 0$, $k \geqq 1$, $\lambda \in \sigma(T+P)$, is fundamental in the B-space $\mathfrak{X}$. The present section is devoted to stating and proving results of this generalized sort. We begin with a basic definition and with a preliminary investigation of the concept of adjoints for closed operators in a general B-space.

1 Definition. Let $\mathfrak{X}$ be a B-space, and T a discrete operator in $\mathfrak{X}$. Then by $\overline{\text{sp}}(T)$, the *spectral span* of T, we denote the smallest closed manifold containing all the manifolds $E(\lambda; T)\mathfrak{X}$ with λ in $\sigma(T)$.

Remark. It follows, by Lemma 2.2, that $\overline{\text{sp}}(T)$ is the smallest closed manifold containing the solutions f of all of the equations $(T-\lambda I)^k f = 0$, $k \geqq 1$, $\lambda \in \sigma(T)$, and is also the smallest closed manifold containing all the solutions f of all of the equations $(T-\lambda)^{\nu(\lambda)} f = 0$ with λ in $\sigma(T)$.

The adjoint T^* of a densely defined unbounded linear operator was given in Definition 3.6. Now we shall develop more of the theory of adjoint operators than was done at that point, paralleling the theory in Hilbert space as given in Section XII.1.

Let $\mathfrak{X}$ be a B-space, and $\mathfrak{X} \oplus \mathfrak{X}$ the vector direct sum of $\mathfrak{X}$ with itself. In $\mathfrak{X} \oplus \mathfrak{X}$, let the norm be defined by

$$|[x, y]| = |x| + |y| \,.$$

Then $\mathfrak{X} \oplus \mathfrak{X}$ is evidently a B-space. The space $\mathfrak{X} \oplus \mathfrak{X}$ admits the evident automorphisms

$$A_1 \colon [x, y] \to [y, x],$$

$$A_2 \colon [x, y] \to [-y, x].$$

We have

$$A_1^2 = -A_1^2 = I, \qquad A_1 A_2 = -A_2 A_1.$$

If M is a subset of a Banach space $\mathfrak{Y}$, its annihilator $M^\perp$ is the closed subspace of $\mathfrak{Y}^*$ defined by

$$M^\perp = \{y^* \in \mathfrak{Y}^* \,|\, y^*(M) = 0\}.$$

The graph of a transformation T in the B-space $\mathfrak{X}$ is, as before, the set

$$\Gamma(T) = \{[x, Tx] \,|\, x \in \mathfrak{D}(T)\}.$$

Let φ be a continuous linear functional on $\mathfrak{X} \oplus \mathfrak{X}$. Then the equations $\varphi_1(x) = \varphi([x, 0])$ and $\varphi_2(y) = \varphi([0, y])$ clearly define continuous linear functionals on $\mathfrak{X}$, and it is readily seen that the map $\varphi \to [\varphi_1, \varphi_2]$ is an isomorphism of $(\mathfrak{X} \oplus \mathfrak{X})^*$ onto all of $\mathfrak{X}^* \oplus \mathfrak{X}^*$. Consequently we may regard $\mathfrak{X}^* \oplus \mathfrak{X}^*$ as being equivalent to the conjugate space of $\mathfrak{X} \oplus \mathfrak{X}$, in which case we have

$$[x^*, y^*][x, y] = x^*x + y^*y.$$

It is also evident from this consideration that *if $\mathfrak{X}$ is reflexive, then $\mathfrak{X} \oplus \mathfrak{X}$ is reflexive.*

By Definition 3.6 we see that

$$\begin{aligned}\Gamma(T^*) &= \{[x^*, y^*] \,|\, x^*(Tx) = y^*(x),\ x \in \mathfrak{D}(T)\},\\ &= \{[x^*, y^*] \,|\, x^*y - y^*x = 0,\ [x, y] \in \Gamma(T)\}.\end{aligned}$$

Thus

$$\Gamma(T^*) = [A_2 \Gamma(T)]^{\perp}.$$

This shows in particular that T^* is a closed operator.

2 LEMMA. *Let T be a densely defined operator. Then*

(a) *T and T^* both have bounded everywhere defined inverses if either does, and $(T^{-1})^* = (T^*)^{-1}$;*

(b) *if B is a bounded operator, $(T + B)^* = T^* + B^*$.*

PROOF. Suppose that for the purpose of the present proof we consider multivalued linear operators T, which we suppose to be defined when their graphs $\Gamma(T)$ are given as arbitrary linear subspaces of $\mathfrak{X} \oplus \mathfrak{X}$, and take T^{-1} and T^* to be (possibly multivalued) linear operators *defined* by the formulas

$$\Gamma(T^{-1}) = A_1\Gamma(T), \qquad \Gamma(T^*) = [A_2 \Gamma(T)]^{\perp}.$$

Then to prove (a) we have only to observe that

$$\Gamma((T^*)^{-1}) = A_1\Gamma(T^*) = A_1(A_2\Gamma(T))^{\perp} = (A_2(A_1\Gamma(T)))^{\perp} = \Gamma((T^{-1})^*).$$

Thus $(T^*)^{-1} = (T^{-1})^*$ even if either or both of the transformations are unbounded, multivalued, or not everywhere defined, so that (a) follows as a special case.

To prove (b), we first suppose that x^* is in $\mathfrak{D}(T^* + B^*)$. Then, since B^* is bounded, x^* is in $\mathfrak{D}(T^*)$ and

$$\begin{aligned}\{(T^* + B^*)x^*\}(x) &= (T^*x^*)(x) + (B^*x^*)x \\ &= x^*\{(T + B)x\}, \qquad x \in \mathfrak{D}(T).\end{aligned}$$

Thus x^* is in $\mathfrak{D}((T + B)^*)$ and $(T + B)^*x^* = T^*x^* + B^*x^*$. On the other hand, if x^* is in $\mathfrak{D}((T + B)^*)$, so that

$$x^*((T + B)y) = (T + B)^*x^*(y), \qquad y \in \mathfrak{D}(T),$$

then

$$x^*(Ty) = \{(T + B)^*x^* - B^*x^*\}(y), \qquad y \in \mathfrak{D}(T).$$

Thus x^* is in $\mathfrak{D}(T)^*$, and

$$T^*x^* + B^*x^* = (T + B)^*x^*. \qquad \text{Q.E.D.}$$

3 Lemma. *Let $\mathfrak{X}$ be a reflexive space, and T a closed densely defined linear operator in $\mathfrak{X}$. Let T^* be the adjoint of T. Then*

(a) *$\mathfrak{D}(T^*)$ is dense;*

(b) *the second adjoint T^{**} of T is the same as T.*

Proof. If $\mathfrak{D}(T^*)$ is not dense, then, by the Hahn-Banach theorem, Corollary II.3.14, and the reflexivity of $\mathfrak{X}$, there is an x in $\mathfrak{X}$ such that $x\mathfrak{D}(T^*) = 0$, while $x \neq 0$. Then

$$A_2[0, x] = [-x, 0] \in \Gamma(T^*)^{\perp}.$$

Thus

$$-[0, x] = A_2^2[0, x] \in (A_2\Gamma(T^*))^{\perp} = \Gamma(T^{**}).$$

Thus, if (b) is established, it follows that $[0, x] \in \Gamma(T)$, and since T is single valued, that $x = 0$. This shows that (a) follows from (b).

To prove (b) we shall first show that for closed linear submanifolds $\mathfrak{M}$ of a reflexive B-space, we have $(\mathfrak{M}^{\perp})^{\perp} = \mathfrak{M}$. Indeed, it is clear that $\mathfrak{M} \subseteq (\mathfrak{M}^{\perp})^{\perp}$. On the other hand, let $x \notin \mathfrak{M}$. Then by the Hahn-Banach theorem there exists a y^* in $\mathfrak{M}^{\perp}$ such that $y^*(x) \neq 0$, proving that $x \notin (\mathfrak{M}^{\perp})^{\perp}$.

Thus

$$\Gamma(T^{**}) = (A_2\Gamma(T^*))^{\perp} = [A_2(A_2\Gamma(T))^{\perp}]^{\perp} = A_2^2(\Gamma(T)^{\perp})^{\perp}$$

$$= -\Gamma(T) = \Gamma(T). \qquad \text{Q.E.D.}$$

4 LEMMA. *Let T be a densely defined operator in a B-space $\mathfrak{X}$. Then*
(a) *if one of T and T^* is discrete, both are;*
(b) *we have $\sigma(T) = \sigma(T^*)$;*
(c) *if T and T^* are discrete, then $E(\lambda;\, T)^* = E(\lambda; T^*)$ for each λ in $\sigma(T) = \sigma(T^*)$;*
(d) *if $\mathfrak{X}$ is reflexive, if T and T^* are discrete, and one is spectral, then so is the other.*

PROOF. By Lemma 2, we have

$$((T-\lambda I)^{-1})^* = ((T-\lambda I)^*)^{-1} = (T^*-\lambda I^*)^{-1}$$

with the operators on both sides of this equation existing as bounded operators for exactly the same λ. This proves (b) and (a), since by Theorem VI.5.2 an operator and its adjoint are either both compact or both not compact.

To prove (c), we note that $E(\lambda;\, T)$ may be characterized as

$$E(\lambda;\, T) = -\frac{1}{2\pi i}\int_C (T-\mu I)^{-1}\, d\mu,$$

where C is a sufficiently small circle about λ. But then

$$E(\lambda;\, T^*) = -\frac{1}{2\pi i}\int_C (T^*-\mu I)^{-1}\, d\mu$$

$$= -\frac{1}{2\pi i}\int_C ((T-\mu I)^{-1})^*\, d\mu$$

$$= -\frac{1}{2\pi i}\left(\int_C (T-\mu I)^{-1}\, d\mu\right)^*$$

$$= E(\lambda;\, T)^*,$$

by (b) and Lemma 2. This proves (c).

To prove (d), note that it follows from the preceding lemma that we have only to show that if T is spectral, then T^* is spectral. It follows from (c) that if T is spectral, T^* satisfies condition (a) in Corollary XVIII.2.33, so that by that theorem it suffices to show that if $E(\lambda; T^*)x^* = 0$ for each $\lambda \in \sigma(T)$, then $x^* = 0$. To do this, note that in this case, since T is spectral, it follows from (c) that

$$x^*(x) = x^*\Big(\sum_{\lambda\in\sigma(T)} E(\lambda; T)x\Big) = \sum_{\lambda\in\sigma(T)} (E(\lambda; T)^*x^*)(x)$$
$$= \sum_{\lambda\in\sigma(T)} 0 = 0, \qquad x \in \mathfrak{X},$$

so that $x^* = 0$. Q.E.D.

We are now able to relate the spectral span $\overline{\text{sp}}(T)$ of an operator T to the manifold $\mathfrak{S}_\infty(T^*)$ given by Definition 2.4.

5 LEMMA. *If T is a discrete operator in the reflexive Banach space $\mathfrak{X}$, then $\overline{\text{sp}}(T) = \mathfrak{S}_\infty(T^*)^\perp$.*

PROOF. It is clear that if λ is in $\sigma(T)$ and we have $E(\lambda)f = f$, while $E(\mu)^*g^* = 0$ for every μ in $\sigma(T) = \sigma(T^*)$, then

$$g^*(f) = g^*(E(\lambda)f) = E(\lambda)^*g^*(f) = 0.$$

Thus it is clear that $\overline{\text{sp}}(T) \subseteq \mathfrak{S}_\infty(T^*)^\perp$. Conversely, if $f \notin \overline{\text{sp}}(T)$, there exists a functional g^* in $\mathfrak{X}^*$ such that

$$g^*(f) = 1, \qquad g^*(\overline{\text{sp}}(T)) = 0.$$

Since $g^*(E(\lambda)f') = 0$ for any f' in $\mathfrak{X}$ and any λ in $\sigma(T)$, it follows that

$$E(\lambda)^*g^* = 0, \qquad \lambda \in \sigma(T) = \sigma(T^*).$$

Thus, by (c) of the preceding lemma, g^* is in $\mathfrak{S}_\infty(T^*)$; and since $g^*(f) = 1$, it follows that $f \notin \mathfrak{S}_\infty(T^*)^\perp$. Q.E.D.

REMARK. It should be emphasized that the condition for $\overline{\text{sp}}(T) = \mathfrak{X}$ given in the preceding lemma is $\mathfrak{S}_\infty(T^*) = 0$ and not $\mathfrak{S}_\infty(T) = 0$. Indeed, H. Hamburger [1] has constructed an example of a compact operator U in Hilbert space $\mathfrak{X}$ whose generalized eigenvectors span $\mathfrak{X}$, and which is such that an infinite dimensional closed subspace $\mathfrak{X}_0$ of $\mathfrak{X}$ exists such that $U\mathfrak{X}_0 \subseteq \mathfrak{X}_0$, and U is quasi-nilpotent in $\mathfrak{X}_0$. If we put $T = (U^*)^{-1}$, we have $\text{sp}(T) \neq \mathfrak{X}$, while $\mathfrak{S}_\infty(T) = 0$.

The next theorem gives the central result of the present section.

→ 6 THEOREM. *Let T be a discrete spectral operator in the B-space $\mathfrak{X}$. Suppose that all but a finite number of points in $\sigma(T)$ are simple poles of the resolvent function $R(\lambda; T)$. Let U_i be a sequence of bounded domains covering* *the whole complex plane, which is such that* $\min_{z \in U_i} |z| \to \infty$ *as* $i \to \infty$. *Let* V_i *be the boundary of* U_i. *Let* $0 \leqq \nu < 1$, *and put*

$$\mu_i = \max_{\lambda \in V_i, \mu \in \sigma(T)} |\mu|^{\nu} |\lambda - \mu|^{-1}.$$

Let $\lambda_0 \notin \sigma(T)$, *and let* P *be an operator such that* $P(T - \lambda_0 I)^{-\nu}$ *is bounded. Then*

(a) *if* $\mu_i \to 0$, $T + P$ *is discrete and* $\mathfrak{S}_\infty(T + P) = 0$;

(b) *if* $\limsup_{i\to\infty} \mu_i \leqq K \leqq \infty$, *there exists a* $\delta = \delta(K, T) > 0$ *such that if* $|P(T - \lambda_0 I)^{-\nu}| < \delta$, *then* $T + P$ *is discrete and* $\mathfrak{S}_\infty(T + P) = 0$;

(c) *if* $\limsup_{i\to\infty} \mu_i < \infty$, *and* $P(T - \lambda_0 I)^{-\nu}$ *is compact, then* $T + P$ *is discrete and* $\mathfrak{S}_\infty(T + P) = 0$.

PROOF. Let S be the scalar part of T, and $N = T - S$. Let E be the resolution of the identity for T. Then by Definition XVIII.2.27, Lemma XVIII.2.13, and Lemma XVIII.2.25, S and T have the same spectrum. By Lemma 2.2, $T|E(\lambda)\mathfrak{X} = \lambda I|E(\lambda)\mathfrak{X}$ if λ is a simple pole of the resolvent of $R(\lambda; T)$. Thus, by Definition XVIII.2.27, $T|E(\lambda)\mathfrak{X} = S|E(\lambda)\mathfrak{X}$ for each simple pole λ of $R(\lambda; T)$. Let σ_0 be the finite collection of points in $\sigma(T)$ for which λ is not a simple pole of $R(\lambda; T)$, and $\sigma_0' = \sigma(T) - \sigma_0$. Then by Definition XVIII.2.8, a vector x in $E(\sigma_0')\mathfrak{X}$ is in $\mathfrak{D}(f(T))$ if and only if it is in $\mathfrak{D}(f(S))$, and, moreover $f(T)x = f(S)x$, for each analytic function f defined on $\sigma(T)$ and for $x \in E(\sigma_0)\mathfrak{X}$. In particular, x is in $\mathfrak{D}(T)$ if and only if x is in $\mathfrak{D}(S)$; and $Tx = Sx$ for each x in $E(\sigma_0')\mathfrak{X}$. By Theorem XVIII.2.9, $E(\sigma_0)\mathfrak{X}$ is contained both in $\mathfrak{D}(f(T))$ and in $\mathfrak{D}(f(S))$, and by this same theorem this space is invariant both under $f(T)$ and under $f(S)$, which are both bounded in $E(\sigma_0)\mathfrak{X}$. This shows that $\mathfrak{D}(f(T)) = \mathfrak{D}(f(S))$, and that if we put

$$N_f x = f(T)x - f(S)x, \qquad x \in E(\sigma)_0\mathfrak{X},$$
$$N_f x = 0, \qquad x \in E(\sigma_0')\mathfrak{X},$$

N_f is a bounded operator, and $f(T) = f(S) + N_f$. Since N_f has a finite dimensional range, it is compact. Thus, for each analytic function f defined on $\sigma(T)$, the operator $f(T)$ is the sum of $f(S)$ and a compact operator. In particular, $T = S + N$, N being compact. Consequently, S is discrete if and only if T is discrete. Moreover, if we let

$$Lx = x, \qquad x \in E(\sigma_0')\mathfrak{X},$$
$$Lx = (T - \lambda I)^{\nu}(S - \lambda I)^{-\nu}x, \qquad x \in E(\sigma_0)\mathfrak{X},$$

then it is clear that L is a bounded operator and that $(S-\lambda I)^{-\nu} = (T-\lambda I)^{-\nu}L$. Hence

$$\begin{aligned}(P+N)(S-\lambda I)^{-\nu} &= P(S-\lambda I)^{-\nu} + N(S-\lambda I)^{-\nu} \\ &= P(T-\lambda I)^{-\nu}L + N(S-\lambda I)^{-\nu}\end{aligned}$$

is a bounded operator which is compact if $P(T-\lambda I)^{-\nu}$ is compact (cf. VI.5.4). In all cases (a), (b), (c) of the theorem, we may consequently pass from consideration of the operators T and $T+P$ to consideration of the operators S and $S+(N+P)$. Which is to say, we may and shall suppose, without loss of generality, that $T=S$ is of scalar type.

Now, under this assumption, $(T-\lambda_0 I)^{-\nu} = \Sigma_{\lambda\in\sigma(T)}(\lambda-\lambda_0)^{-\nu}E(\lambda; T)$ and by Theorem XVIII.2.11, the series is the uniform limit of the finite sums

$$\sum_{\substack{\lambda\in\sigma(T)\\ |\lambda-\lambda_0|\leqq n}} (\lambda-\lambda_0)^{-\nu}E(\lambda; T),$$

for each $\nu>0$. Since, by Lemma 2.2, each of these finite sums has a finite dimensional range and is hence compact, it follows from Lemma VI.5.3 that for $\nu>0$ the operator $(T-\lambda_0 I)^{-\nu}$ is compact. Thus, if $\nu>0$, then since $P+T=(P+\lambda_0 I)+(T-\lambda_0 I)$, and since $(P+\lambda_0 I)(T-\lambda_0 I)^{-\nu}$ is clearly compact if and only if $P(T-\lambda_0 I)^{-\nu}$ is compact, we may pass from consideration of the pair of operators T, $T+P$ to consideration of the pair of operators $T-\lambda_0 I$, $(T-\lambda_0 I)+(P+\lambda_0 I)$; that is, we may and shall assume, without loss of generality, that $\lambda_0=0$. If $\nu=0$, this is no longer strictly valid; nevertheless, to avoid notational complications, we shall assume that $\lambda_0=0$ in this case also, and leave to the reader the task of making the simple elementary modifications necessary in the more general case of arbitrary λ_0.

It will be shown below that $T+P$ is discrete. Suppose for the moment that this is established. Let f be in $\mathfrak{S}_\infty(T+P)$, so that, by Lemma 2.6,

$$F(\lambda) = (T+P-\lambda I)^{-1}f$$

is an entire function. We shall show that $F(\lambda)$ is uniformly bounded; it will follow from Liouville's theorem that $F(\lambda)\equiv g$ is constant and thus that $f=(T+P-\lambda I)g$ for all λ. From this it will follow that $0=(\lambda_1-\lambda_2)g$ for all λ_1 and λ_2, and hence that $g=0$, so $f=0$. Hence, the theorem will be proved.

To show that $F(\lambda)$ is uniformly bounded, we proceed as follows. Since $\mathfrak{D}(T^\nu) \supseteq \mathfrak{D}(T)$ by Theorem XVIII.2.9(vii), we have $\mathfrak{D}(P) \supseteq \mathfrak{D}(T)$. Thus $\mathfrak{D}(T+P) = \mathfrak{D}(T)$. By hypothesis, the operator $PT^{-\nu}$ is bounded and it will henceforth be denoted by the symbol A.

Next we observe that it follows exactly as in the third paragraph of the proof of Theorem 2.7 that for any μ such that $|T^\nu R(\mu; T)| < |A|^{-1}$ the operator $(\mu I - T - P)^{-1}$ exists as an everywhere defined operator and is given by the series

$$\text{(i)} \qquad B(\mu) = R(\mu; T) \sum_{n=0}^{\infty} \{AT^\nu R(\mu; T)\}^n.$$

By Theorem XVIII.2.11, there exists a constant $M = M(T)$ such that

$$|T^\nu R(\mu; T)| \leqq M \max_{\lambda \in \sigma(T)} |\lambda|^\nu |\mu - \lambda|^{-1}.$$

Thus it is clear that if, in case (a), we take i sufficiently large, no point μ in V_i belongs to $\sigma(T+P)$, and that $B(\mu) = (\mu I - T - P)^{-1}$ for μ in V_i. The same reasoning is valid in case (b), provided only that we consider an integer $i \geqq 1$ for which $M\mu_i |A| \leqq 1$; that is, provided that we take $\delta \leqq (MK)^{-1}$ in the hypothesis of (b).

From (i) and Theorem XVIII.2.11,

$$\begin{aligned} |B(\mu)| &\leqq \sup_{\lambda \in \sigma(T)} M |\mu - \lambda|^{-1} \sum_{n=0}^{\infty} (M |A| \mu_i)^n \\ &= M(1 - M|A| \mu_i)^{-1} \sup_{\lambda \in \sigma(T)} |\mu - \lambda|^{-1}, \qquad \mu \in V_i. \end{aligned}$$

Since 0 is not in $\sigma(T)$ by assumption, there exists a constant K_1 such that

$$\sup_{\lambda \in \sigma(T)} |\mu - \lambda|^{-1} \leqq K_1 \sup_{\lambda \in \sigma(T)} |\lambda|^\nu |\mu - \lambda|^{-1}.$$

Thus we have

$$\text{(ii)} \qquad |B(\mu)| \leqq MK_1 \mu_i (1 - M|A| \mu_i)^{-1}, \qquad \mu \in V_i.$$

Since we have assumed in both cases (a) and (b) that $M|A| \mu_i < 1$ for all but a finite number of indices i, the inequality (ii) gives us an immediate bound for the function

$$f(\mu) = R(\mu; T+P)f = B(\mu)f$$

in all but a finite number of the sets V_i. By the maximum modulus principle this entire function has the same uniform upper bound in all but a finite number of the sets U_i. Since each of the finite number of exceptional sets U_i is bounded, and since $\bigcup_{i=1}^{\infty} U_i$ is the whole complex plane, it follows immediately that $f(\mu)$ is uniformly bounded. Moreover, equation (i) shows that the operator $T + P$ is discrete. This establishes the desired result in cases (a) and (b).

If hypothesis (c) holds, we argue along essentially similar lines, as follows. Let $\{\lambda_n\}$ be an enumeration of $\sigma(T)$, let $\sigma_n = \bigcup_{i=1}^{n} \lambda_i$, and let $\sigma_n' = \sigma(T) - \sigma_n$. It will first be shown that $\lim_{n\to\infty} |E(\sigma_n')A| = 0$ where E is the resolution of the identity for T. If this is false, there exists an $\varepsilon > 0$, an increasing sequence $\{n_i\}$ of integers, and a sequence x_i of vectors of norm 1 such that $|E(\sigma_{n_i}')Ax_i| \geqq \varepsilon$, $i = 1, 2, \ldots$. Since A is compact, we may suppose, without loss of generality, that the limit $\lim_{i\to\infty} Ax_i = y$ exists. Now, if M is a uniform bound for $|E(\sigma_{n_i}')|$, we have

$$\begin{aligned} \varepsilon \leqq \limsup_{i\to\infty} |E(\sigma_{n_i}')Ax_i| &\leqq \limsup_{i\to\infty} |E(\sigma_{n_i}')(Ax_i - y)| \\ &\quad + \limsup_{i\to\infty} |E(\sigma_{n_i}')y| \\ &\leqq M \limsup_{i\to\infty} |Ax_i - y| + 0 \\ &= 0 + 0 = 0. \end{aligned}$$

This contradiction proves our assertion.

Next observe that the series $B(\mu)$ of equation (i) may be written as

$$\begin{aligned} \text{(iii)}\quad B(\mu) &= T^{-\nu}(T^{\nu}R(\mu; T) + T^{\nu}R(\mu; T)AT^{\nu}R(\mu; T) \\ &\quad + T^{\nu}R(\mu; T)AT^{\nu}R(\mu; T)AT^{\nu}R(\mu; T) + \cdots) \\ &= T^{-\nu}\left\{\sum_{n=0}^{\infty} (T^{\nu}R(\mu; T)A)^n\right\}T^{\nu}R(\mu; T). \end{aligned}$$

Thus the series $B(\mu)$ converges for all μ for which $|T^{\nu}R(\mu; T)A| < 1$, and if $K_1 = |T^{-\nu}|$, we have

$$|B(\mu)| \leqq K_1 |T^{\nu}R(\mu; T)| (1 - |T^{\nu}R(\mu; T)A|)^{-1}.$$

We have already noted that $|T^{\nu}R(\mu; T)| < M\mu_i$ for μ in V_i. Thus, for μ in V_i and $|T^{\nu}R(\mu; T)A| < 1$, it follows that $B(\mu) = (\mu I - T - P)^{-1}$ exists, and that

$$|B(\mu)| \leqq K_1 M\mu_i(1 - |T^{\nu}R(\mu; T)A|)^{-1}.$$

It will be shown below that $|T^{\nu}R(\mu;\, T)A| \leqq \frac{1}{2}$ for μ in V_i and i sufficiently large. From this it will then follow as above that the function $f(\mu) = R(\mu;\, T+P)f$ is uniformly bounded. It will also be shown that $T^{-\nu}$ is compact. From this, (iii), and Theorem VI.5.4, it will follow that $B(\mu) = R(\mu;\, T+P)$ is compact for μ in V_i and i sufficiently large, so that the theorem will be proved.

Let μ be in V_i. To show that $|T^{\nu}R(\mu;\, T)A| \leqq \frac{1}{2}$, for i sufficiently large, we argue as follows. By the above, $|T^{\nu}R(\mu;\, T)| \leqq M\mu_i$; thus $|T^{\nu}R(\mu;\, T)| \leqq MM'$, where $M' = \sup_{1 \leqq i < \infty} \mu_i$. We have shown above that $\lim_{n\to\infty} |E(\sigma_n')A| = 0$. Thus there exists an N such that $|E(\sigma_N')A| \leqq (4MM')^{-1}$. Then

$$\begin{aligned} |T^{\nu}R(\mu;\, T)A| &\leqq |T^{\nu}R(\mu;\, T)E(\sigma_N)A| + |T^{\nu}R(\mu;\, T)E(\sigma_N')A| \\ &\leqq |T^{\nu}R(\mu;\, T)E(\sigma_N)A| + \tfrac{1}{4}. \end{aligned}$$

It follows from Theorem XVIII.2.11 that

$$|T^{\nu}R(\mu;\, T)E(\sigma_N)A| \leqq M \sup_{\lambda \in \sigma_N} |\lambda|^{\nu}|\lambda - \mu|^{-1}|A|.$$

Since $\inf_{z \in U_i} |z| \to \infty$ as $i \to \infty$, it follows that

$$|T^{\nu}R(\mu;\, T)E(\sigma_N)A| \leqq \tfrac{1}{4}$$

for μ in V_i and i sufficiently large. Thus $|T^{\nu}R(\mu;\, T)A| \leqq \frac{1}{2}$ for μ in V_i and i sufficiently large, as asserted.

To see that $T^{-\nu}$ is compact if $\nu > 0$, note that, by Theorem XVIII.2.11, $\lim_{n\to\infty} |T^{-\nu} - \sum_{\lambda \in \sigma_N} \lambda^{-\nu}E(\lambda)| = 0$, so that $T^{-\nu}$ is the limit in the uniform operator topology of a sequence of operators with finite dimensional ranges. Since an operator with a finite dimensional range is clearly compact, $T^{-\nu}$ is compact by Lemma VI.5.3. This completes our proof if $\nu > 0$. If $\nu = 0$, we argue as follows. By (iii),

$$B(\mu) = \left\{\sum_{n=0}^{\infty} (R(\mu;\, T)A)^n\right\} R(\mu;\, T);$$

since $R(\mu;\, T)$ is compact, $B(\mu) = (\mu I - T - P)^{-1}$ is compact in this case also, so that $T + P$ is discrete even if $\nu = 0$, and the theorem still holds when $\nu = 0$. Q.E.D.

From Theorem 6, we obtain the following theorem as a corollary.

7 THEOREM. *Let T be a discrete spectral operator in the reflexive B-space $\mathfrak{X}$. Suppose that all but a finite number of points in the spectrum $\sigma(T)$*

are simple poles of the resolvent function $R(\mu; T)$. *Let* U_i *be a sequence of bounded domains covering the whole complex plane, and suppose that* $\lim_{i\to\infty} \min_{z\in U_i} |z| = \infty$. *Let* V_i *be the boundary of* U_i. *Let* $0 \leqq \nu < 1$, *and put*

$$\mu_i = \max_{\lambda\in V_i,\ \mu\in\sigma(T)} |\mu|^\nu |\lambda - \mu|^{-1}.$$

Let $\lambda_0 \notin \sigma(T)$, *and let* P *be an operator such that* $P(T-\lambda_0 I)^{-\nu}$ *is bounded. Then*

(a) *if* $\mu_i \to 0$, $T+P$ *is discrete and* $\overline{\mathrm{sp}}(T+P) = \mathfrak{X}$;

(b) *if* $\limsup_{i\to\infty} \mu_i \leqq K < \infty$, *there exists a* $\delta = \delta(K, T) > 0$ *such that if* $|P(T-\lambda_0 I)^{-\nu}| < \delta$, *then* $T+P$ *is discrete and* $\overline{\mathrm{sp}}(T+P) = \mathfrak{X}$;

(c) *if* $\limsup_{i\to\infty} \mu_i < \infty$, *and* $P(T-\lambda_0 I)^{-\nu}$ *is compact, then* $T+P$ *is discrete and* $\overline{\mathrm{sp}}(T+P) = \mathfrak{X}$.

Proof. By Theorem 6 and Lemma 5, it suffices to show that in each of the cases (a), (b), (c) the operator $(T+P)^*$ is discrete and $\mathfrak{S}_\infty((T+P)^*) = 0$. By Lemma 4 and Theorem 6, $T+P$ is discrete. Thus, as in the third paragraph of the proof of Theorem 6, it is sufficient to show that

$$F^*(\lambda) = ((T+P)^* - \lambda I)^{-1} f^*$$

is uniformly bounded for each $f \in \mathfrak{S}_\infty((T+P)^*)$. However, in the course of the proof of Theorem 6 it was established that the function $|((T+P)-\lambda I)^{-1}|$ is uniformly bounded for λ in $\bigcup_{i=N}^\infty V_i$, provided that N is chosen sufficiently large. By Lemma 3,

$$|((T+P)^* - \lambda I)^{-1}| = |(T+P-\lambda I)^{-1}|$$

for such λ. Thus, $F^*(\lambda)$ is uniformly bounded for λ in $\bigcup_{i=N}^\infty V_i$, and hence, by the maximum modulus theorem, is uniformly bounded for λ in $\bigcup_{i=N}^\infty U_i$. Since the set $\{U_i\}$, $i \geqq N$, covers the whole plane, with the possible exception of a bounded set, $|F^*(\lambda)|$ is uniformly bounded. Q.E.D.

8 Corollary *Let* T *be a discrete spectral operator in the Banach space* $\mathfrak{X}$. *Suppose that all but a finite number of points in the spectrum* $\sigma(T)$ *are simple poles of the resolvent function* $R(\lambda; T)$. *Let* $\{U_i\}$ *be a sequence of bounded domains whose union is the entire plane and which is such that* $\lim_{i\to\infty} \min_{z\in U_i} |z| = \infty$. *It is assumed that the boundary* V_i *of* U_i *is disjoint from the spectrum. Let* d_i *be the distance from* V_i *to the spectrum* $\sigma(T)$ *and let* B *be a bounded operator. Then*

(a) *if* $d_i \to \infty$, *then* $T + B$ *is discrete and* $\mathfrak{S}_\infty(T + B) = 0$;

(b) *if* $\liminf_{i\to\infty} d_i \geqq K > 0$, *then there is a number* $\varepsilon = \varepsilon(K, T) > 0$ *such that* $T + B$ *is discrete and* $\mathfrak{S}_\infty(T + B) = 0$ *whenever* $|B| \leqq \varepsilon$;

(c) *if* $\liminf_{i\to\infty} d_i > 0$, *and* B *is compact, then* $T + B$ *is discrete and* $\mathfrak{S}_\infty(T + B) = 0$.

PROOF. This follows from Theorem 6 by placing $\nu = 0$. Q.E.D.

9 COROLLARY. *Let* T *be a discrete spectral operator in the reflexive B-space* $\mathfrak{X}$. *Suppose that all but a finite number of the points in* $\sigma(T)$ *are simple poles of the resolvent function* $R(\lambda; T)$; *let* $\{U_i\}$ *be a sequence of bounded domains covering the entire plane, such that* $\lim_{i\to\infty} \min_{z\in U_i} |z| = \infty$. *It is assumed that the boundary* V_i *of* U_i *is disjoint from the spectrum. Let* d_i *be the distance from* V_i *to the spectrum* $\sigma(T)$ *and let* B *be a bounded operator. Then*

(a) *if* $d_i \to \infty$ *then* $T + B$ *is discrete and* $\overline{\mathrm{sp}}(T + B) = \mathfrak{X}$;

(b) *if* $\liminf_{i\to\infty} d_i \geqq K > 0$, *there exists an* $\varepsilon = \varepsilon(K, T) > 0$ *such that* $T + B$ *is discrete and* $\overline{\mathrm{sp}}(T + B) = \mathfrak{X}$ *whenever* $B \leqq \varepsilon$;

(c) *if* $\liminf_{i\to\infty} d_i > 0$ *and* B *is compact, then* $T + B$ *is discrete and* $\overline{\mathrm{sp}}(T + B) = \mathfrak{X}$.

Theorem 7 gives us a fairly general insight into a range of situations in which a "spectral density" property $\overline{\mathrm{sp}}(T) = \mathfrak{X}$ is to be expected of an operator T. However, in applying these results it is convenient to be able to deal, wherever possible, with solutions of the equation $(T - \lambda I)f = 0$, rather than with solutions of the equation $(T - \lambda I)^k f = 0$. The next lemma describes a simple case in which this is possible.

10 LEMMA. *Let* T *be a discrete spectral operator in the Banach space* $\mathfrak{X}$. *Suppose that all but a finite number of the countable set* $\{\lambda_n\}$ *of points in* $\sigma(T)$ *are simple poles of the resolvent function and correspond to projections with one-dimensional ranges. Let* U_i, $i \geqq 1$, *be a sequence of bounded open domains covering the whole complex plane which is such that as* $i \to \infty$, $\min_{z\in\bar{U}_i} |z| \to \infty$, *and suppose that all but a finite number of the domains* U_i *contain at most one point of* $\sigma(T)$. *Let* V_i *be the boundary of* U_i. *Let* $0 \leqq \nu < 1$, *and put*

$$\mu_i = \max_{\lambda\in V_i,\ \mu\in\sigma(T)} |\mu|^\nu |\lambda - \mu|^{-1}.$$

Let $\lambda_0 \in \rho(T)$, *and let* P *be an operator such that* $P(T - \lambda_0 I)^{-\nu}$ *is bounded.*

Then all but a finite number of points in $\sigma(T+P)$ *are simple poles of the resolvent* $R(\lambda;\ T+P)$ *corresponding to one-dimensional eigenspaces if any one of the following conditions holds:*

(a) μ_i *approaches zero;*

(b) $\limsup_{i\to\infty} \mu_i \leqq K < \infty$, *and* $|P(T-\lambda_0 I)^{-\nu}| < \delta = \delta(K, T)$;

(c) $\mathfrak{X}$ *is reflexive,* $\limsup_{i\to\infty} \mu_i < \infty$, *and* $P(T-\lambda_0 I)^{-\nu}$ *is compact.*

PROOF. By Theorem 6, $T+P$ is discrete. It may be shown exactly as in the proof of Theorem 6 that, under any of the hypotheses (a), (b), or (c),

$$\limsup_{i\to\infty} \sup_{\lambda \in V_i} |PR(\lambda;\ T)| = 1-\varepsilon < 1.$$

Thus, putting $\delta = \varepsilon/2$, there exist only a finite number of indices i for which the relation

$$\sup_{\lambda\in V_i} |PR(\lambda;\ T)| \leqq 1-\delta$$

fails. Let W_0 be the union of all the corresponding domains U_i, of all domains U_j containing more than one point of $\sigma(T)$, and of all domains U_k containing a point of $\sigma(T)$ which is not a simple pole of the resolvent function or which corresponds to projections with ranges of more than one dimension. Let W_1 be the union of $\bar{W}_0$ and of all domains U_j which intersect $\bar{W}_0$. If μ is in the boundary $\hat{W}_1$ of W_1, then it is evidently in the boundary of one of these domains U_j. Hence we have

$$\sup_{\lambda\in\hat{W}_1} |PR(\lambda;\ T)| \leqq 1-\delta.$$

If we now enumerate the sets U_n not contained in W_1 as $W_2, W_3, \ldots$, we see that in our original hypothesis we may assume without loss of generality that

(i) $\sup_{\lambda\in V_i} |PR(\lambda;\ T)| \leqq 1-\delta$ for all $i \geqq 1$;

(ii) if $i \geqq 2$, U_i contains at most one point of $\sigma(T)$, which is a simple pole of the resolvent function and corresponds to a projection with a one-dimensional range.

Let $\delta_1 = \frac{1}{2}\{(1-\delta)^{-1}-1\}$. Then $\sup_{\lambda\in V_i} |\eta PR(\lambda;\ T)| < 1$ for $|\eta| < 1+\delta_1$, and all $i \geqq 1$. Let λ_0 be some conveniently chosen point in some V_i. It follows as in the proof of Theorem 6 that for $|\eta| < 1+\delta_1$,

$$K(\eta) = (\lambda_0 I - T - \eta P)^{-1}$$

exists and is given by the formula

$$K(\eta) = R(\lambda_0\,;\,T)(I - \eta PR(\lambda_0\,;\,T))^{-1}.$$

Since $R(\lambda_0\,;\,T)$ is compact by Lemma 2.2, it follows from Lemma VII.6.6 and Theorem VI.5.4 that $K(\eta)$ is a compact operator which depends analytically on η for $|\eta| < 1 + \delta_1$. It follows from Lemma VII.6.6 that if O is any bounded open set whose boundary does not contain any point of $\sigma(K(\eta_0))$, and $|\eta_0| < 1 + \delta_1$, then $E(O; K(\eta))$ depends analytically on η for η sufficiently near to η_0. From this and from Theorem VII.9.5 it follows immediately that $E(U_i\,;\,T + \eta P)$ depends analytically on η for $|\eta| < 1 + \delta_1$, and for $i \geqq 1$. Since, by (ii), $E(U_i\,;\,T)$ has a one-dimensional range for $i \geqq 2$, it follows from Lemma VII.6.7 that $E(U_i\,;\,T + P)$ has a one-dimensional range for $i \geqq 2$. It follows from Theorem VII.9.5 and Theorem VII.3.18 that, for $i \geqq 2$, U_i contains precisely one point in $\sigma(T + P)$, and that this point is a simple pole of the resolvent function $R(\lambda;\, T + P)$. Since U_i is bounded and $T + P$ is regular, it follows from Lemma 2.2 that U_i contains only a finite number of points of $\sigma(T + P)$. Thus the present lemma is proved. Q.E.D.

The next result is of a different type, depending as it does on the Carleman inequality of Section XI.6.

11 THEOREM. *Let T be a discrete unbounded self adjoint operator in Hilbert space $\mathfrak{H}$. Let $\{\lambda_n\}$ be the set of eigenvalues of T, each eigenvalue being repeated in this enumeration a number of times equal to the dimension of $E(\lambda_n\,;\,T)\mathfrak{H}$. Suppose that k is an integer such that*

$$\sum_{n=1}^{\infty} |\lambda_n|^{-2k} < \infty.$$

Let P be an operator (possibly unbounded) such that every product A of $l \leqq k$ factors P and T containing at least one factor P satisfies the conditions $\mathfrak{D}(A) \supseteq \mathfrak{D}(T^l)$ and

$$\lim_{\mu \to +\infty} |A\{R(\mu i;\, T)\}^l| = 0.$$

Then $T + P$ is discrete and $\overline{\text{sp}}(T + P) = \mathfrak{H}$.

PROOF. Let μ be so large that $|PR(\mu i;\, T)| < \frac{1}{2}$. Then, by Lemma VII.3.4, $(I - PR(\mu i;\, T))^{-1} = B$ exists and is bounded. Since T is discrete,

$R(\mu i; T)$ is compact by Lemma 2.2. Thus, by Theorem VI.5.4,

$$R(\mu i\,; T)B = C$$

is compact. We have

$$\begin{aligned}(\mu i I - T - P)Cx &= (B - PR(\mu i; T)B)x \\ &= (I - PR(\mu i; T))Bx = x, \qquad x \in \mathfrak{H},\end{aligned}$$

and

$$\begin{aligned}C(\mu I - T - P)x &= C(I - PR(\mu i; T))(\mu i I - T)x \\ &= R(\mu i; T)(\mu i\, I - T)x = x, \qquad x \in \mathfrak{D}(T) = \mathfrak{D}(T + P).\end{aligned}$$

Thus $R(\mu i; T + P)$ exists and equals C, proving by Definition 2.1 that $T + P$ is discrete.

Since by hypothesis $\mathfrak{D}(T^{l-1}P) \supseteq \mathfrak{D}(T^l)$ for $l \leq k$, it follows that $P\mathfrak{D}(T^l) \subseteq \mathfrak{D}(T^{l-1})$. Thus $\mathfrak{D}((T + P)^k) \supseteq \mathfrak{D}(T^k)$, and for each μ and $x \in \mathfrak{D}(T^k)$, $(\mu I - T - P)^k x$ may be expanded as $(\mu I - T - P)^k x = (\mu I - T)^k x + \sum_{l=0}^{k} \mu_l^{k-l} A_l x$, each operator A_l being a sum of products of l factors $\pm T$, $\pm P$, each product containing at least one factor P. Consequently, $|A_l\{R(\mu i;\ T)\}^l| \to 0$ as $\mu \to +\infty$ by assumption. Since $|(\mu i)^{k-l}(R(\mu i;\ T))^{k-l}|$ is bounded as $\mu \to +\infty$ by Theorem XII.2.6, it follows that

$$\lim_{\mu\to+\infty} \left| \left(\sum_{l=0}^{k} (\mu i)^{k-l} A_l \right) (R(\mu i;\ T))^k \right| = 0.$$

It now follows by an argument like that given in the first paragraph of the present proof that if μ is sufficiently large so that the operator $O_\mu = (\sum_{l=0}^{k} (\mu i)^{k-l} A_l)(R(\mu i; T))^k$ has norm less than 1, then the operator $((\mu i I - T - P)^k)^{-1}$ exists and is equal to $R(\mu i;\ T)^k (I + O_\mu)^{-1}$. This shows, in particular, that $\mathfrak{D}((T + P)^k) \subseteq \mathfrak{D}(T^k)$, so that $\mathfrak{D}((T + P)^k) = \mathfrak{D}(T^k)$.

Next we shall show that if $|\mu|$ is sufficiently large and μ lies in the sector $|\arg \mu - \pi/2| \leqq \pi/2 - \varepsilon$, $\varepsilon > 0$ of the upper half plane, then $(\mu I - (T + P)^k)^{-1}$ exists and satisfies the rate of growth condition

$$\text{(i)} \qquad |(\mu I - (T + P)^k)^{-1}| \leqq O(|\mathscr{I}\mu|^{-1}).$$

By what has been shown above, we have $\mathfrak{D}((T + P)^k) = \mathfrak{D}(T^k)$, and $(T + P)^k = T^k + A_0$, where $\mathfrak{D}(A_0) \supseteq \mathfrak{D}(T^k)$, and

$$\text{(ii)} \qquad \lim_{\mu\to+\infty} |A_0(R(\mu i;\ T))^k| = 0.$$

We have

(iii) $$A_0 R(\mu; T^k) = A_0(R(|\mu|^{1/k}i; T))^k(|\mu|^{1/k}iI - T)^k R(\mu; T^k).$$

By Theorem XII.2.6,

$$|(|\mu|^{1/k}iI - T)^k R(\mu; T^k)| \leqq \sup_{-\infty < \lambda < +\infty} \left| \frac{(\lambda - |\mu|^{1/k}i)^k}{\lambda^k - \mu} \right|$$
$$= \sup_{-\infty < \lambda < +\infty} \left| \frac{(\lambda - i)^k}{\lambda^k - \mu/|\mu|} \right|.$$

It is clear that $\lim_{|\lambda| \to \infty} |(\lambda - i)^k/(\lambda^k - \alpha)| = 1$ uniformly for $|\alpha| = 1$. Thus there exists a number $c > 2$ sufficiently large so that

$$\sup_{|\lambda| \geqq c} \left| \frac{(\lambda - i)^k}{\lambda^k - \mu/|\mu|} \right| \leqq 2.$$

On the other hand, $|(\lambda - i)^k|$ is bounded in the region $|\lambda| \leqq c$, and $(\lambda^k - \mu/|\mu|)$ is bounded away from zero as λ varies in the interval $[-c, c]$ of the real axis, and $\mu \neq 0$ varies in the angle $|\arg \mu - \pi/2| \leqq \pi/2 - \varepsilon$ of the upper half-plane. Thus

$$|(T - |\mu|^{1/k}iI)^k R(\mu; T^k)|$$

is uniformly bounded as $\mu \neq 0$ varies over the angular sector $|\arg \mu - \pi/2| \leqq (\pi/2) - \varepsilon$ of the upper half-plane. It follows from (ii) and (iii) that

(iv) $$\lim_{|\mu| \to \infty, \left|\arg \mu - \frac{\pi}{2}\right| \leqq \frac{\pi}{2} - \varepsilon} |A_0 R(\mu; T^k)| = 0.$$

The same argument can be applied to show that we also have

(v) $$\lim_{|\mu| \to \infty, \left|\arg \mu + \frac{\pi}{2}\right| \leqq \frac{\pi}{2} - \varepsilon} |A_0 R(\mu; T^k)| = 0.$$

It now follows by an argument like that given in the first paragraph of the present proof that for $|\arg \mu - \pi/2| \leqq \pi/2 - \varepsilon$, and for $|\mu|$ sufficiently large such that $|A_0 R(\mu; T)^k| < 1$, the resolvent $R(\mu; (T + P)^k) = R(\mu; T^k + A_0)$ exists and is given by the formula

$$R(\mu; (T + P)^k) = R(\mu; T^k)(I - A_0 R(\mu; T^k))^{-1}.$$

As $|\mu| \to \infty$, μ remaining in the indicated sectors of the upper and lower half-planes, it follows from Lemma VII.6.1 that $|(I - A_0 R(\mu; T^k))^{-1}| \to 1$.

Since $|R(\mu;\ T^k)| = O(|\mathscr{I}\mu|^{-1})$, it follows from Lemma XII.2.2 and the formula displayed immediately above that

$$\text{(vi)} \qquad |R(\mu;\ (T+P)^k)| = O(|\mathscr{I}\mu|^{-1}),$$

as $|\mu| \to \infty$, μ remaining in a sector $|\arg \mu \pm \pi/2| \leqq \pi/2 - \varepsilon$ of the upper or the lower half-plane.

Next, choose some μ_0 for which $R_0 = R(\mu_0\,;\ (T+P)^k)$ exists and is given by the formula

$$R(\mu_0\,;\ (T+P)^k) = R(\mu_0\,;\ T^k)(I - A_0\, R(\mu_0\,;\ T^k))^{-1}.$$

Let $\{\varphi_n\}$ be a complete orthonormal basis for Hilbert space $\mathfrak{H}$ consisting of eigenvalues of T (cf. the proof of Theorem XIII.4.2) and let $T\varphi_n = \lambda_n \varphi_n$. Then it is seen from Theorem XII.2.6 that $E(\lambda_n)\varphi_n = \varphi_n$, and

$$R(\bar{\mu}_0\,;\ T^k)\varphi_n = (\bar{\mu}_0 - \lambda_n^k)^{-1}\varphi_n\,.$$

Hence, since $\sum_{n=1}^{\infty} |\lambda_n|^{-2k}$ converges by assumption, the series

$$\sum_{n=1}^{\infty} |R(\bar{\mu}_0\,;\ T^k)\varphi_n|^2 = \sum_{n=1}^{\infty} |\bar{\mu}_0 - \lambda_n^k|^{-2}$$

converges also. Thus, by Definition XI.6.1, $R(\bar{\mu}_0\,;\ T^k)$ is of Hilbert-Schmidt type, that is, $R(\bar{\mu}_0\,;\ T^k) \in HS$. Moreover,

$$R_0^* = \{(I - A_0\, R(\bar{\mu}_0\,;\ T^k))^{-1}\}^* R(\bar{\mu}_0\,;\ T^k)$$

so

$$\sum_{n=1}^{\infty} |R_0^*\varphi_n|^2 \leqq |(I - A_0\, R(\mu_0\,;\ T^k))^{-1}|^2 \sum_{n=0}^{\infty} |R(\bar{\mu}_0\,;\ T^k)\varphi_n|^2 < \infty.$$

This proves that $R_0^* \in HS$, so that, by Lemma XI.6.2, $R_0 \in HS$.

Since $R_0 = (\mu_0 I - (T+P)^k)^{-1}$, it follows from Lemma VII.9.2. that $\sigma(R_0) = \{(\mu_0 - \xi)^{-1} \mid \xi \in \sigma((T+P)^k)\} \cup \{0\}$, and that

$$(\mu I - R_0)^{-1} = \mu^{-1} I - \mu^{-2} R(\mu_0 + \mu^{-1};\ (T+P)^k).$$

It follows immediately from this formula and (vi) that for $\mu \neq 0$, μ in the sector $|\arg\ \mu - \pi/2| \leqq \pi/2 - \varepsilon$ of the upper half-plane or in the sector $|\arg\ \mu + \pi/2| \leqq \pi/2 - \varepsilon$ of the lower half-plane, and $|\mu|$ sufficiently large, the operator $(\mu I - R_0)^{-1}$ exists, and has its norm

$$|(\mu I - R_0)^{-1}| = O(|\mu|^{-1}).$$

It follows immediately from Theorem XI.6.29 that the set of vectors

$$\{f \mid (R_0 - \lambda I)^k f = 0,\ \lambda \in \sigma(R_0),\ k \geqq 1\},$$

is fundamental in $\mathfrak{H}$. Since $R_0 = (\mu_0 I - (T+P)^k)^{-1}$, R_0 is one-to-one, and hence $R_0^k f = 0$ implies $f = 0$. Thus

$$\{f \mid (R_0 - \lambda I)^k f = 0,\ \lambda \in \sigma(R_0),\ \lambda \neq 0,\ k \geqq 1\},$$

is fundamental in $\mathfrak{H}$. By Definition VII.9.3 this means that

$$\{f \in E(\lambda; R_0)\mathfrak{H} \mid \lambda \in \sigma(R_0),\ \lambda \neq 0\}$$

is fundamental in $\mathfrak{H}$. By Theorem VII.9.5,

$$\begin{aligned}\{f \in E(\lambda; R_0)\mathfrak{H} \mid \lambda \in \sigma(R_0),\ \lambda \neq 0\}&\\ &= \{f \in E(\mu; (T+P)^k)\mathfrak{H} \mid \mu \in \sigma((T+P)^k)\}\\ &= \{f \in E(\hat{\mu}; T+P)\mathfrak{H} \mid \hat{\mu} \in \sigma(T+P)\}.\end{aligned}$$

Thus, by Definition 1, $\overline{\text{sp}}(T+P) = \mathfrak{H}$. Q.E.D.

We conclude the present section with the following improvement of Theorem 4.16.

12 THEOREM. *Let T be the closed operator in the Hilbert space $L_2(0, 1)$ derived from the formal operator $(-id/dx)^n$ by the imposition of a set of n boundary conditions. If n is even, let these boundary values satisfy Regularity Hypothesis* 4.1 *and let the constant β of formulas* (4.16) *and* (4.17) *be different from* 1. *If n is odd, let these boundary conditions satisfy Regularity Hypotheses* (4.9) *and* (4.10). *Let P be an operator whose domain is the subset $H^{(n)}(0, 1)$ of $L_2(0, 1)$, and which is defined by a formula*

$$Pf = Af^{(n)} + \sum_{j=0}^{n-1} B_j f^{(j)}, \qquad f \in H^{(n)},$$

where A is *compact, and B_j is bounded, $j = 0, \ldots, n-1$. Then $T + P$ is discrete and $\overline{\text{sp}}(T+P) = \mathfrak{H}$.*

PROOF. Let us first note that Pf may be written in the form

$$Pf = A_1 f^{(n)} + B_\infty f, \qquad f \in H^{(n)},$$

where A_1 and B_∞ are compact. To show this we see from Theorem VI.5.4 that it is sufficient to prove that, for each j,

$$f^{(j-1)} = (Cf^{(j)} + Bf), \qquad f \in H^{(j)},$$

where C and B are compact. Now, $f^{(j-1)} = \text{Int}(f^{(j)}) + f^{(j-1)}(0)$, where Int is the operation of integration defined by

$$(\text{Int} f)(t) = \int_0^t f(s)\, ds, \qquad f \in L_2[0, 1].$$

Since, by the Schwarz inequality $|(\text{Int } f)(t) - (\text{Int } f)(s)| \leqq |f|\, |s-t|^{1/2}$, the compactness of Int follows from Theorem IV.6.5. It is consequently sufficient to show that $f^{(j-1)}(0)$ may be written as

$$[\dagger] \qquad f^{(j-1)}(0) = \hat{C} f^{(j)} + \hat{B} f, \qquad f \in H^{(j)},$$

where $\hat{C}$ and $\hat{B}$ are continuous, and hence compact, linear functionals on Hilbert space. Now, if φ is any function in $C^\infty[0, 1]$ which vanishes in the neighborhood of 1 and has the value 1 at $t = 0$, and all of whose derivatives vanish at $t = 0$, we have

$$f^{(j)}(0) = \int_0^1 \varphi(s) f^{(j)}(s)\, ds + (-1)^{j-1} \int_0^1 \varphi^{(j)}(s) f(s)\, ds$$

by Green's formula, Theorem XIII.2.4, which proves [†].

Let λ_n be the eigenvalues of T, and $E_n = E(\lambda_n; T)$ the corresponding projections. Let $\mu_n = \mathscr{R}\lambda_n$. By Theorems VII.9.5, VII.4.5, 4.8, 4.13, and Lemma XVII.2.2, $\hat{B} = \sum_{n=1}^\infty \mu_n E_n - T$ is the sum of a finite dimensional and a bounded operator, and hence is bounded. Let $\hat{T} = \sum_{n=1}^\infty \mu_n E_n$ in the sense of the functional calculus of Theorem XVIII.2.11. By Lemma XV.6.2 we can introduce an equivalent inner product into Hilbert space, in terms of which all the projections E_n are orthogonal. Hence, without loss of generality, we may take $\hat{T}$ to be self adjoint. Since $\mu_n \to \infty$ (and even $\sum_{n=1}^\infty |\mu_n|^{-2} < \infty$ by Theorems 4.8 and 4.13), it follows that $\hat{T}$ is discrete. Since, by what has been proved above

$$(T + P) f = \hat{T} f + A_1 f^{(n)} + \tilde{B} f,$$

where A_1 is compact and $\tilde{B}$ is bounded, and since $|\tilde{B} R(\mu i;\ \hat{T})| \to 0$ as $\mu \to +\infty$ by Theorem XII.2.6, the present theorem will follow immediately from Theorem 11 once it is shown that

$$\left| A_1 \left(\frac{d}{dt}\right)^n R(\mu i;\ \hat{T}) \right| \to 0 \qquad \text{as } \mu \to +\infty.$$

To prove this assertion, we argue as follows. Since $\hat{T} = T + \hat{B}$, where $\hat{B}$ is bounded, it follows (cf. the proof of Theorem 6) that if μ is so large that

$|\hat{B}R(\mu i;\, \hat{T})| < \frac{1}{2}$, then $\mu \notin \sigma(T)$ and

$$R(\mu i;\, T) = R(\mu i;\, \hat{T})(I - \hat{B}R(\mu i;\, \hat{T}))^{-1}.$$

This equation shows that for μ sufficiently large so that $|\hat{B}R(\mu i;\, \hat{T})| < \frac{1}{2}$, we have $|R(\mu i;\, T)| < 2\mu^{-1}$. Thus, arguing once more as in the proof of Theorem 6, we have also

$$R(\mu i;\, \hat{T}) = R(\mu i;\, T)(I + \hat{B}R(\mu i;\, T))$$

for sufficiently large μ. What must then be shown to conclude the present proof is that

$$\lim_{\mu \to \infty} |A_1 T R(\mu i;\, T)| = 0,$$

that is, that

$$\lim_{\mu \to \infty} |A_1(\mu i R(\mu i;\, T) - I)| = 0.$$

Now,

$$\begin{aligned} |\mu i(R(\mu i;\, T) - R(\mu i;\, \hat{T}))| &= |\mu i R(\mu i;\, T)\hat{B}R(\mu i;\, \hat{T})| \\ &\leqq 2|\hat{B}|\,\mu^{-1}. \end{aligned}$$

Hence, it is sufficient, in order to complete the proof, to show that

$$\lim_{\mu \to \infty} |A_1(\mu i R(\mu i;\, \hat{T}) - \mu i I)| = 0.$$

Suppose that this is false. Then there exists a sequence $\{\mu_j\}$ of real numbers approaching $+\infty$, a sequence $\{x_n\}$ of vectors of unit length, and an $\varepsilon > 0$ such that

$$|A_1(\mu_j i R(\mu_j i;\, \hat{T}) - \mu_j i I)x_j| \geqq \varepsilon.$$

Since, by Theorem XII.2.6, $\{z_n\} = \{(\mu_j i R(\mu_j i;\, \hat{T}) - \mu_j i I)x_j\}$ is a bounded sequence and A_1 is a compact operator, we can assume without loss of generality that $\{z_n\}$ is a convergent sequence with limit z. Clearly, $|z| \geqq \varepsilon$, so that $z \neq 0$. On the other hand, it follows by Theorem XII.2.6 and the Lebesgue dominated convergence theorem that

$$(z_n, y) = \int_{-\infty}^{+\infty} \frac{\lambda}{\mu_n i - \lambda} (E(d\lambda)x_n, A_1^* y)$$

converges to zero as $n \to \infty$, for any $y \in \mathfrak{H}$. Hence $(z, y) = 0$ for all y in $\mathfrak{H}$, so that $z = 0$. This contradiction proves the theorem. Q.E.D.

6. Notes and Remarks

The theory developed in the present chapter was initiated by G. D. Birkhoff in 1908 (cf. Birkhoff [3, 6, 7]), and continued by Tamarkin [2] in 1912; there is, however, related earlier work (1896) by H. Poincaré [2]. Investigations along the lines initiated by Birkhoff and Tamarkin were continued by Tamarkin [3], who studied a considerably generalized problem for nth order differential operators, by Birkhoff and Langer [1], who treated the case of a first order system of differential equations, and by C. E. Wilder [1, 2], who studied the case in which linear conditions are imposed at interior points of the interval of definition of a formal differential operator.

The abstract operator-theoretic approach via perturbation theorems used in Section 2 was introduced by J. Schwartz [2], and extended by H. P. Kramer [2] to the general case. Some results similar to those presented here are given in Turner [2] and Clark [1]. The main results of Schwartz were generalized by Maeda [1] to locally convex spaces.

Birkhoff and Tamarkin study not the problem of unconditional convergence in the mean, on which prime emphasis has been placed in the present chapter, but the problem of conditional convergence at given points. Birkhoff [3] showed that if the set of boundary conditions is subject to the regularity hypotheses of Section 4, the eigenvalue expansion of a function f of bounded variation converges to $\frac{1}{2}\{f(t+0)+f(t-0)\}$ at an interior point t of the interval $[0, 1]$ on which the formal differential operator is defined, and to $af(0+)+bf(1-)$ at an end point of $[0, 1]$; here the constants a and b are determined by the particular boundary conditions imposed. Tamarkin [3; Theorem 12] established a generalization of the following equiconvergence theorem, which had been demonstrated previously in the second order case by W. Steckloff [1] and by A. Haar [3].

THEOREM. *Let T be a linear operator determined by the formal differential operator*

$$\tau = \left(\frac{d}{dt}\right)^n + a_{n-2}(t)\left(\frac{d}{dt}\right)^{n-2} + \cdots + a_0(t)$$

on the interval $[0, 1]$, and by a set of n boundary conditions satisfying the regularity hypotheses of Section 4. Let T_0 be the linear operator defined *by these same boundary conditions and by the formal differential operator.*

$$\tau_0 = \left(\frac{d}{dt}\right)^n.$$

If $f \in L_1(0, 1)$, let $\sigma_n(f, t)$ denote the nth partial sum of the expansion of f in eigenfunctions and generalized eigenfunctions of T, and let $\hat{\sigma}_n(f, t)$ denote the nth partial sum of the expansion of f in eigenfunctions and generalized eigenfunctions of T_0; in each case, we suppose terms to be summed in the order of increasing moduli for the corresponding eigenvalues. Then

$$\lim_{n\to\infty} \max_{0\leq t\leq 1} |\sigma_n(f, t) - \hat{\sigma}_n(f, t)| = 0, \qquad f \in L_1(0, 1).$$

This most interesting theorem allows the detailed study of the convergence of eigenfunction expansions to be reduced to the theory of Fourier series in many cases; in the theory of Fourier series, of course, a colossal literature is available. For proofs of versions of the equiconvergence theorem in the second order case, see Titchmarsh [16; Chapter I] and Coddington-Levinson [1; Chapter 12, Theorem 3.2]. A resumé of the Birkhoff-Tamarkin theory is also given in the recent book of Naĭmark [5]. Related but weaker theorems on the convergence of the eigenvalue series are given by W. E. Milne [2].

An approach to the problem of convergence and summability of eigenfunction expansions in a number of singular second order nonselfadjoint cases in which the operators under investigation have purely discrete spectrum is given by Joanne Elliott [1, 2]. The theory developed by Elliott is based on the Feller-Phillips-Hille theory of semi-groups in L_1 and C defined by parabolic partial differential equations.

The most general set of two linearly independent boundary conditions for

$$\tau = \left(\frac{d}{dt}\right)^2$$

on the interval [0, 1] falls into one of four cases.

(a) *Regular* boundary conditions, satisfying the regularity hypotheses of Section 4.

(b) *Totally degenerate* boundary conditions like

$$f(0) = 0; \qquad f'(0) = 0,$$

which exclude the whole complex plane from the spectrum of the operator they determine.

(c) *Totally degenerate* boundary conditions like

$$f(0) = f(1); \qquad f'(0) = -f'(1),$$

which determine operators including the whole complex plane in their point spectrum.

(d) *Intermediate* boundary conditions like

$$f(0) = 0; \qquad f'(0) = f(1).$$

These boundary conditions determine discrete differential operators having eigenvalues $\lambda = s^2$ determined by equations of the form

$$\sin s = ks.$$

In this case the eigenvalues are located asymptotically at the points $an^2 + ibn \ln n + \ldots$, the ratio of a and b being real. The eigenfunction expansion theory of these intermediate cases is the subject of Steven P. Hoffman, Jr. [1].

B. Friedman and L. I. Mishoe [1] studied expansions in terms of the eigenfunctions of the equation

$$u'' + q(t)u + \lambda(p(t)u - u') = 0,$$

subject to the boundary conditions

$$u(0) = 0, \qquad u(1) = 0.$$

By the change of variable $v(t) = u'(t) - p(t)u(t)$, this may be reduced to a more conventional form of eigenvalue problem. Friedman and Mishoe deduce the corresponding expansion theorems. Mishoe and Ford [1] give a number of related results.

The final Theorem 5.12 of Section 5 is related to the completeness theorem proved by Browder [6], who used the fundamental inequality of Carleman in a manner similar to the manner in which it is used in the proof of Theorem 5.12. Cf. Theorem XIV.6.28.

Keldyš [1] announced a generalization of the following theorem, which is related to completeness Theorems XI.6.29 and XI.9.29, and applied it to the study of the eigenfunctions of systems of ordinary and partial differential equations.

THEOREM. *Let T be an unbounded discrete self adjoint operator in Hilbert space $\mathfrak{H}$, and let A be compact. Suppose that T^{-1} exists and is bounded, and that for some sufficiently large positive integer m, T^{-m} is of Hilbert-Schmidt type. Then $(I-A)T$ is a discrete operator, and* $\overline{\text{sp}}((I-A)T) = \mathfrak{H}$.

DIFFERENTIAL OPERATORS. For further references on differential operators and boundary value problems, the reader is referred to Askerov, Kreĭn, and Laptev [1], Balslev [1], Birman [5, 6], Butler [2, 3], Coddington and Gilbert [1], Ercolano and Schechter [1], R. S. Freeman [1], Gehtman and Stankevič [1], Gilbert and Kramer [1], Glazman [5], Goldberg [2], Greiner [1], Hellwig [1], Hörmander [2], Huige [1], I. S. Kac [1], Keldyš and Lidskiĭ [1], Kemp [1], Kesel'man [2], Ljance [4, 6], McGarvey [1], Marčenko [3], Marčenko and Rofe-Beketov [1], Martirosjan [4], Pavlov [1], Štraus [6], and Weidmann [2].

EIGENFUNCTION EXPANSIONS. The question of the expansion of a given function in terms of eigenfunctions (or generalized eigenfunctions) of an operator has been discussed by many authors. In addition, it is desirable to know when the (generalized) eigenvectors of an operator are fundamental in a given space. In the self adjoint case, much information can be found in the treatise of Berezanskiĭ [5]; for other references, the reader may consult Aleksandrjan [1], Allahverdiev [1 through 5], Berezanskiĭ [3], Foiaş (4 through 6, 18], Gerlach [1], Giertz [1], Gohberg and Kreĭn [6], Greiner [1], Harazov [5], Hirschfeld [1], I. S. Kac [1], Kacnel'son [1], Kreĭn [22], Kuroda [8], Lidskiĭ [1, 2], Ljance [6], Macaev [2, 3], Marčenko [3], Marčenko and Rofe-Beketov [1], Markus [1 through 4], Maurin [1 through 7], Nelson [1], Novosel'skiĭ [1], Palant [1], Pincus [2, 3], Pustyl'nik [1, 2], Rofe-Beketov [1], Shizuta [1], Smart [3], Turner [2], and Viziteĭ [1 through 3].

CHAPTER XX

Spectral Operators with Continuous Spectra: Applications of the General Theory

The present chapter is devoted to the examination of a number of cases in which it may be shown that a particular operator with a continuous spectrum is a spectral operator, or has one of a variety of closely related weaker or stronger properties. Here we are very much at the boundary of the known theory. The material in the present chapter is accordingly somewhat fragmentary, and is presented in the hope that it may stimulate research on the topics treated.

In the first section, we show by the use of Theorem XVIII.2.34, following an idea first suggested by Naĭmark, that under suitable hypotheses an operator T determined by a singular second order formal differential operator

$$-\left(\frac{d}{dt}\right)^2 + q(t), \qquad 0 \leqq t < \infty,$$

is a spectral operator. In this section, we use the following idea: The steps for calculating the spectral resolution of T, which we know by the Weyl-Kodaira theorem (XIII.5.13) to be valid if T is self adjoint, can in any case be carried out formally. If this is done, we arrive at formal expressions for a family of operators $E(e)$ which are logical candidates for the spectral resolution of T, if T has in fact any spectral resolution. If these operators $E(e)$ can be shown to form a uniformly bounded family, then we can in fact conclude that T is a spectral operator; this is the content of Theorem XVIII.2.34. To show that the operators $E(e)$ form a uniformly bounded family, we have only to compare them to the corresponding operators calculated for the "unperturbed" formally self adjoint operator

$$-\left(\frac{d}{dt}\right)^2, \qquad 0 \leqq t < \infty.$$

If the coefficient q is sufficiently small, it is reasonable to expect that this procedure will show T to be spectral. In the first section we shall show that this program can be carried through successfully if q is of a degree of smallness conforming with the requirement

$$\int_0^\infty (1+t^2)|q(t)|\,dt < \infty.$$

The idea involved is of quite general applicability; a number of other cases in which this general procedure can be carried out are stated in the exercises at the end of the chapter. Here the following open problem seems of interest. For what a, b, c, and q can the operator

$$-\left(\frac{d}{dt}\right)^2 + a(1+t)^{-1} + b(1+t)^{-2} + c(1+t)^{-3} + q(t), \qquad 0 \leqq t < \infty,$$

$$q(t) = O(t^{-4}) \text{ as } t \to \infty,$$

be treated by the method of Section 1? For what a, b, c is this operator spectral?

In Section 2 we study an interesting idea due to Friedrichs. If an "unperturbed" operator T and a "perturbation" K are given, one attempts to find an operator U such that $T+K = U^{-1}TU$, in this way reducing the spectral theory of $T+K$ to that of T. We shall see that this idea can be applied with success to a wide variety of operators and, in particular, to certain partial differential operators of the form $-\nabla + V(x)$. Applications of this same method to other operators are given as exercises in Section 5.

In Section 3 we generalize the Friedrichs technique to operators with discrete spectra, along lines first developed by Turner. Here one begins with an unperturbed operator T in an explicitly "diagonal" form and attempts to solve a perturbation equation of the form $T+K = U^{-1}(T+D)U$, where D is a "purely diagonal" operator. The general methods developed in Section 3 are applied to various particular operators with discrete spectrum; additional applications are given as exercises in Section 5.

A quite different idea is developed in Section 4. Given a self adjoint operator T and a perturbed self adjoint operator $T+K$, the "wave operator" $U = \lim_{s\to\infty} e^{isT}e^{-is(T+K)}$, if it exists and is unitary, may easily be seen to satisfy the operator equation $T+K = U^{-1}TU$. In Section 4 we develop this "wave operator method," deriving general theorems of Kato and Kuroda concerning the existence and properties of the wave operator U, and also deriving associated relations between the spectra of T and $T+K$. Section 4 contains general theorems rather than concrete applications. However, a number of applications of the results of Section 4 are given as exercises in Section 5.

It is worth noting that the material covered in the present chapter is closely related to certain profound investigations in quantum mechanics, quantum field theory, and various branches of the classical theory of wave phenomena, especially electromagnetic theory; these studies have played a central role in twentieth century physics. In quantum theory, the eigenfunctions and the eigenvalues of certain self adjoint operators determine the energy levels of physical systems. In the classical linear wave theory, these same functions are the key to a successful analysis of wave propagation and dispersion. In cases where closed-form solutions are not available, physicists have been accustomed to discussing the eigenfunction theory of the partial differential operators which arise in these subject areas using perturbation-theoretic methods. Quite typically, an operator of the form $-\nabla + V(x)$ is treated as a perturbation of the operators $-\nabla$; an operator of the form

$$-\left(\frac{\partial}{\partial x_1}\right)^2 - \cdots - \left(\frac{\partial}{\partial x_6}\right)^2 + V(x_1, x_2, x_3) + V(x_4, x_5, x_6) + V(x_1 - x_4, x_2 - x_5, x_3 - x_6)$$

as a perturbation of

$$(\dagger) \qquad -\left(\frac{\partial}{\partial x_1}\right)^2 - \cdots - \left(\frac{\partial}{\partial x_6}\right)^2;$$

and even more general multi- and infinite-dimensional cases are treated using the same idea. As the unperturbed operators $-\nabla$ and (†) have continuous spectra covering a half-axis, the proper mathematical treatment of the continuous-spectrum perturbation problems which arise here are by no means easy. The various mathematical investigations on which we report represent a few of salient cases in which it is possible to develop rigorous versions of informal reasoning based upon empirical principles. Additional references to related physical and mathematical-physics literature are found in the section of Notes and Remarks at the end of this last chapter. Friedrichs [17], Faddeev [2 through 4], and Lax-Phillips [2] will be particularly interesting to the reader wishing to follow the most recent developments in this area.

1. Spectral Differential Operators of Second Order

In the present section we shall apply Theorem XVIII.2.34 to show the spectral character of a class of operators in Hilbert space determined by formal differential operators τ of second order,

$$\tau = -\left(\frac{d}{dt}\right)^2 + q(t), \qquad 0 \leqq t < \infty, \tag{1}$$

having coefficients which are small at infinity in a sufficiently strong sense. The main part of our analytic work will consist in making sufficiently fine asymptotic estimates of the solutions of the equation $\tau\sigma = \lambda\sigma$ to be able to verify the hypotheses of Theorem XVIII.2.34 by direct computation. This task, while somewhat tedious, is not of any great difficulty in principle.

Suppose then that there is given a formal differential operator τ of the form of equation (1) and that $q \in C^\infty[0, \infty)$. Let

$$A(f) = 0 \tag{2}$$

be a nontrivial boundary condition for τ at zero. By Corollary XIII.2.23 the boundary condition (2) can be written in one of the two forms

$$f(0) = 0, \tag{2a}$$

$$f'(0) + kf(0) = 0. \tag{2b}$$

Let T be the closed operator in $L_2(0, \infty)$ determined by τ and by the boundary condition (2). (Cf. Definition XIII.2.17.)

The first condition which we shall impose on the coefficient q of (1) appears in the following lemma.

1 LEMMA. *Let the coefficient q in* (1) *satisfy the inequality*

$$\int_0^\infty |q(t)|\, dt < \infty.$$

Let $P^+ = \{\mu \mid \mu \geqq 0\}$; $P_\varepsilon^+ = \{\mu \in P^+ \mid |\mu| > \varepsilon\}$. Then the equation $\tau\sigma = \mu^2\sigma$ has a solution $\sigma_1(t, \mu)$ defined for $(t, \mu) \in [0, \infty) \times P_0^+$ with the following properties:

(i) $\sigma_1(t, \mu)$ *is* C^∞ *in* t, $\qquad \mu \in P_0^+$;

(ii) $\sigma_1(t, \mu)$ *and* $\sigma_1'(t, \mu)$ *are analytic in* μ *for* μ *interior to* P_0^+ *and continuous in* μ *for* $\mu \in P_0^+$, $0 \leqq t < \infty$. *Moreover,* $\sigma_1(t, \mu)$ *satisfies the following asymptotic relationships;*

$$\left.\begin{aligned} \sigma_1(t, \mu) &\sim e^{it\mu} \\ \sigma_1'(t, \mu) &\sim i\mu e^{it\mu} \end{aligned}\right\} \begin{array}{l} \textit{as } t \to \infty, \textit{ uniformly for } \mu \in P_\varepsilon^+,\ \varepsilon > 0; \textit{ and also as} \\ |\mu| \to \infty,\ \mu \textit{ remaining in } P^+, \textit{ uniformly in } 0 \leqq t < \infty. \end{array}$$

PROOF. Let $a \geqq 0$ and let L_μ be the operator in the B-space $C[a, \infty)$ of bounded continuous functions on $[a, \infty)$ determined by the formula

$$(3) \qquad (L_\mu h)(t) = \frac{1}{2i\mu} \int_0^\infty [e^{2i\mu(s-t)} - 1] q(s) h(s)\, ds, \qquad \mu \in P^+.$$

Then it is clear that $|L_\mu| \leqq |\mu|^{-1} \int_a^\infty |q(s)|\, ds$, $\mu \in P^+$. Moreover,

$$\begin{aligned} |(L_\mu h)(t) - (L_\nu h)(t)| &\leqq \frac{1}{|\mu|} |\mu(L_\mu h)(t) - \nu(L_\nu h)(t)| + \left|\frac{1}{\mu} - \frac{1}{\nu}\right| |\nu(L_\nu h)(t)| \\ &\leqq \frac{|h|}{2|\mu|} \int_t^\infty |e^{2i\mu(s-t)} - e^{2i\nu(s-t)}|\, |q(s)|\, ds \\ &\quad + \left|\frac{1}{\mu} - \frac{1}{\nu}\right| |h| \int_a^\infty |q(s)|\, ds. \end{aligned}$$

Thus, if b is chosen so that $a < b < \infty$, it follows that

$$\begin{aligned} |L_\mu - L_\nu| \leqq \left|\frac{1}{\mu} - \frac{1}{\nu}\right| \int_a^\infty |q(s)|\, ds + \frac{1}{2|\mu|} \sup_{0 \leqq r \leqq b-a} |e^{2i\mu r} - e^{2i\nu r}| \int_a^b |q(s)|\, ds \\ + \frac{1}{|\mu|} \int_b^\infty |q(s)|\, ds. \end{aligned}$$

Let ε be an arbitrary positive number. If $\mu \neq 0$ is in P^+, and we choose b sufficiently large so that $|\mu|^{-1} \int_b^\infty |q(s)|\, ds < \varepsilon$, it follows from the above that

$$\limsup_{\substack{\nu \to \mu \\ \nu \in P^+}} |L_\nu - L_\mu| \leqq \varepsilon.$$

Since ε is arbitrary, it follows that L_μ depends continuously on μ in the uniform topology for $\mu \in P^+$, $\mu \neq 0$. It is easy to see by a similar argument that L_μ is analytic in μ for μ interior to P^+.

We have

$$(4)\qquad (L_\mu h)'(t) = -\int_t^\infty e^{2i\mu(s-t)}\, q(s)h(s)\, ds, \qquad h \in C[a, \infty),$$

$$(L_\mu h)''(t) = q(t)h(t) + 2i\mu \int_t^\infty e^{2i\mu(s-t)} q(s)h(s)\, ds, \qquad h \in C[a, \infty),$$

so that

$$(5)\qquad (L_\mu h)''(t) + 2i\mu(L_\mu h)'(t) = q(t)h(t), \qquad h \in C[a, \infty).$$

Let 1 denote the function which is identically equal to the constant one and put

$$(6)\qquad h_\mu = (I - L_\mu)^{-1}1$$

(assuming for the moment that the indicated inverse exists). Then

$$(7)\qquad h_\mu = L_\mu h_\mu + 1,$$

so that, by (5), h_μ will also satisfy the differential equation

$$(8)\qquad -h_\mu''(t) - 2i\mu h_\mu'(t) + q(t)h_\mu(t) = 0,$$

and the function

$$(9)\qquad g_\mu(t) = e^{i\mu t} h_\mu(t)$$

will satisfy the differential equation $\tau g_\mu = \mu^2 g_\mu$.

We now construct the solution $\sigma_1(t, \mu)$ required by our lemma. First let ε be a fixed positive number and let a_ε be the least non-negative value of t such that

$$(10)\qquad \varepsilon^{-1} \int_t^\infty |q(s)|\, ds \leqq \tfrac{1}{2}.$$

Then by the inequality stated in the sentence following (3) above, we have

$$|L_\mu| \leqq \frac{1}{|\mu|} \int_{a_\varepsilon}^\infty |q(s)|\, ds \leqq \frac{1}{2}, \qquad \mu \in P_\varepsilon^+$$

in the B-space $C[a_\varepsilon, \infty)$. Consequently, by Lemmas VII.6.1, VII.6.3, and VII.6.4, the inverse $(I - L_\mu)^{-1}$ exists, is bounded in norm by 2, is continuously dependent on μ for $\mu \in P^+$, and is analytically dependent on μ for μ interior to P_ε^+. Thus h_μ and g_μ exist, are bounded in norm by 2, and depend in the same way on μ. Observe also that the solution $h_\mu \in C[a, \infty)$

of (7) is necessarily unique for $\mu \in P^+$. Now define $\sigma_1(t, \mu)$ to be the unique solution of $(\tau - \mu^2)\sigma = 0$, $(t, \mu) \in [0, \infty) \times P_\varepsilon^+$, which coincides with $g_\mu(t)$ for $t \geqq a_\varepsilon$ (cf. Corollary XIII.1.5).

If δ is some other positive constant and $\delta < \varepsilon$, then $a_\delta \geqq a_\varepsilon$. Let $\hat{h}_\mu(t)$, $(t, \mu) \in [a_\delta, \infty) \times P_\delta^+$, be the solution of equation (7) which exists in $C[a_\delta, \infty)$, by the above. Then, by the uniqueness of the solution of (7), we have $\hat{h}_\mu(t) \equiv h_\mu(t)$, $(t, \mu) \in [a_\varepsilon, \infty) \times P_\varepsilon^+$, and thus $e^{i\mu t}\hat{h}_\mu(t) \equiv \sigma_1(t, \mu)$ if $(t, \mu) \in [0, \infty) \times P_\varepsilon^+$. This makes it clear that by letting ε approach zero we can extend the definition of $\sigma(t, \mu)$ to the whole set $[0, \infty) \times P_0^+$ and obtain a function which is continuous in μ for $\mu \in P_0^+$ and analytic in μ for μ interior to P_0^+.

It remains to verify the asymptotic properties asserted for the solution $\sigma_1(t, \mu)$ which we have constructed and for its derivative $\sigma_1'(t, \mu)$. Given $\varepsilon > 0$, we note, using (7), that

$$\sigma(t, \mu)e^{-i\mu t} - 1 = h_\mu(t) - 1 = (L_\mu h_\mu)(t), \qquad (t, \mu) \in [a, \infty) \times P_\varepsilon^+. \tag{11}$$

Moreover, by (3),

$$|(L_\mu h_\mu)(t)| \leqq \frac{2}{|\mu|} \int_0^\infty |q(s)|\, ds \to 0 \tag{12}$$

as t approaches infinity, uniformly for $\mu \in P_\varepsilon^+$. Moreover,

$$|(L_\mu h_\mu)(t)| \leqq \frac{2}{|\mu|} \int_0^\infty |q(s)|\, ds \to 0 \tag{13}$$

as $|\mu| \to \infty$, $\mu \in P^+$, uniformly for $0 \leqq t < \infty$. This establishes the asymptotic properties of $\sigma_1(t, \mu)$ asserted in the statement of the present lemma.

To prove the asserted asymptotic properties of the derivative σ_1', we argue as follows. Using formulas (4) and (7), we have

$$\begin{aligned} |h_\mu'(t)| = |(L_\mu h_\mu)'(t)| &= \left| \int_t^\infty e^{2i\mu(s-t)} q(s)[(L_\mu h_\mu)(s) + 1]\, ds \right| \\ &\leqq |L_\mu h_\mu| \int_t^\infty |q(s)|\, ds + \left| \int_t^\infty e^{2i\mu(s-t)} q(s)\, ds \right|. \end{aligned}$$

Thus

$$|h_\mu'(t)| \leqq \frac{2Q}{\mu} \int_t^\infty |q(s)|\, ds + \left| \int_t^\infty e^{2i\mu(s-t)} q(s)\, ds \right|, \tag{14}$$

where $Q = \int_0^\infty |q(s)|\, ds$; this follows since we have shown above that $|L_\mu h_\mu| \leqq 2Q|\mu|^{-1}$.

We shall now show that the second expression on the right of (14) converges to zero as $|\mu| \to \infty$, μ remaining in P^+, uniformly for $0 \leqq t < \infty$. Let $\{q_n\}$ be a sequence of functions in $C^\infty[0, \infty)$, each vanishing outside a bounded subset of $[0, \infty)$, such that

$$\int_0^\infty |q_n(t) - q(t)|\, dt \to 0 \quad \text{as} \quad n \to \infty.$$

Then clearly

$$\int_t^\infty e^{2i\mu(s-t)} q_n(s)\, ds \to \int_t^\infty e^{2i\mu(s-t)} q(s)\, ds$$

uniformly for $\mu \in P^+$ and $0 \leqq t < \infty$. On the other hand, integrating by parts, we see that for each n

$$\int_t^\infty e^{2i\mu(s-t)} q_n(s)\, ds = -\frac{1}{2i\mu} q_n(t) - \frac{1}{2i\mu} \int_t^\infty e^{2i\mu(s-t)} q_n'(s)\, dt \to 0$$

as $|\mu| \to \infty$, $\mu \in P^+$, uniformly in $0 \leqq t < \infty$. Thus our assertion follows from the E. H. Moore theorem, Lemma I.7.6.

Formula (14) now implies that $|h_\mu'(t)| \to 0$ as $|\mu| \to \infty$, μ remaining in P^+, uniformly in $0 \leqq t < \infty$. It is also clear from (14) that $|h_\mu'(t)| \to 0$ as $t \to \infty$, uniformly for $\mu \in P_\varepsilon^+$. Since $\sigma_1'(t, \mu) = g_\mu'(t) = i\mu e^{i\mu t} h_\mu(t) + e^{i\mu t} h_\mu'(t)$ by (9), we conclude that $\sigma_1'(t, \mu) \cong g_\mu'(t) \sim i\mu e^{i\mu t}$ as $t \to \infty$, uniformly for $\mu \in P_\varepsilon^+$, and also as $|\mu| \to \infty$, μ remaining in P^+, uniformly in $0 \leqq t < \infty$. Q.E.D.

Imposing a stronger restriction on the coefficient q, we can make a number of crucial improvements in Lemma 1.

2 COROLLARY. *Let the coefficient q in* (2) *satisfy*

$$\int_0^\infty (1 + t)|q(t)| dt < \infty. \tag{15}$$

Then the solution σ_1 of Lemma 1 may be chosen so as to be defined for $\mu = 0$ in such a way as to satisfy conditions (i) *and* (ii) *of Lemma* 1 *for $\mu \in P^+$, and so that $\sigma_1(t, 0) \sim 1$ as $t \to \infty$, and also so as to satisfy the inequality*

$$|\sigma_1(t, \mu) - e^{i\mu t}| \leqq K(1 + |\mu|)^{-1} \int_t^\infty (1 + s)|q(s)|\, ds, \qquad 0 \leqq t < \infty, \tag{16}$$

for $\mu \in P^+$, where K is a constant depending only on q.

PROOF. Let $\varphi(a)$ denote the function $(2i\alpha)^{-1}(e^{2i\alpha}-1)$. Then it is clear that $\varphi(a)$ is entire, and that its modulus is bounded by 1 in the upper half-plane P^+. We may write equation (3), which defines the operator L_μ, in the variant form

$$(L_\mu h)(t) = \int_t^\infty \varphi(\mu(s-t))(s-t)q(s)h(s)\, ds. \tag{17}$$

It is clear from (17) that if $a \geqq 0$ the mapping $L_\mu\colon C[a, \infty) \to C[a, \infty)$ has norm bounded by $\int_a^\infty (1+s)|q(s)|\, ds$. Moreover, if $a < b < \infty$,

$$|L_\mu - L_\nu| \leqq \sup_{0 \leqq t \leqq b-a} |\varphi(\mu t) - \varphi(\nu t)| \int_a^b (1+s)|q(s)| ds + \int_b^\infty (1+s)|q(s)|\, ds,$$

from which it follows readily, as in the proof of Lemma 1, that L_μ is continuous for $\mu \in P^+$. The first assertion of our lemma may now be deduced by exactly those arguments used to establish the corresponding assertion of Lemma 1; details are left to the reader. We shall only note that $\sigma_1(t, \mu) = e^{i\mu t}h_\mu(t)$ for any t such that

$$\int_t^\infty (1+s)|q(s)|\, ds < \tfrac{1}{2}, \tag{18}$$

h_μ being (cf. (7)) the unique solution of the integral equation

$$h_\mu(t) = 1 + \int_t^\infty \varphi(\mu(s-t))(s-t)q(s)h_\mu(s)\, ds.$$

Since the norm of the mapping $L_\mu\colon C[t, \infty) \to C[t, \infty)$ is less than $\frac{1}{2}$ under condition (18), it follows (cf. VII.6.1) that $|h_\mu(t)| \leqq 2$. Thus

$$|h_\mu(t) - 1| \leqq 2 \int_t^\infty (1+s)|q(s)|\, ds \tag{19}$$

under condition (18). We find in the same way from (17), and using an argument parallel to that of Lemma 1 (cf. (11) and (12)), that

$$|h_\mu(t) - 1| \leqq 2|\mu|^{-1} \int_t^\infty |q(s)|\, ds \leqq 2|\mu|^{-1} \int_t^\infty (1+s)|q(s)|\, ds \tag{20}$$

under the condition (18). It is evident from (19) and (20) that there exist constants A and K_1 so large that

$$|h_\mu(t) - 1| \leqq K_1(1+|\mu|)^{-1} \int_t^\infty (1+s)|q(s)|\, ds, \qquad 0 \leqq t < \infty,\ \mu \in P^+,$$

unless $|\mu| \leqq A$ and $0 \leqq t \leqq A$. However, since the extension $\sigma_1(t, \mu)$ of $g_\mu(t)$ to $[0, \infty) \times P^+$ is continuous in μ and t, it follows that there exists a constant K so large that

$$|\sigma_1(t, \mu)e^{-i\mu t} - 1| \leqq K(1 + |\mu|)^{-1} \int_t^\infty (1 + s)|q(s)| \, ds, \qquad 0 \leqq t < \infty, \mu \in P^+$$

and this proves the second assertion of our lemma. The fact that $\sigma_1(t, 0) \sim 1$ as $t \to \infty$ follows readily from this inequality. Q.E.D.

For the spectral analysis of the operator T, we shall also need asymptotic information on the "second solution" of the differential equation $\tau\sigma = \mu^2\sigma$, that is, the solution asymptotic to $e^{-i\mu t}$ as $t \to \infty$. Since, in contrast to σ_1, such a solution is not uniquely determined by its asymptotic form, we meet a number of additional, slight, technical complications in the asymptotic analysis of such solutions. For this reason, we shall develop the asymptotic analysis of the second solution only in a somewhat fragmentary way, establishing its properties one by one as they become necessary for the spectral analysis of the operator T.

3 LEMMA. *Let the hypotheses of Lemma 1 be satisfied. Let $\varepsilon > 0$ and $P_\varepsilon^+ = \{\mu \mid \mathscr{I}\mu \geqq 0, |\mu| > \varepsilon\}$. The equation $\tau\sigma = \mu^2\sigma$ has a solution $\sigma_2(t, \mu)$ defined for $(t, \mu) \in [0, \infty) \times P_\varepsilon^+$ and having the following properties:*

(i) *$\sigma_2(t, \mu)$ is C^∞ in t,* $\qquad 0 \leqq t < \infty, \mu \in P^+$.

(ii) *$\sigma_2(t, \mu)$ is analytic in μ for μ interior to P_ε^+, and $\sigma_2(t, \mu)$ and $\sigma_2'(t, \mu)$ are continuous in t and μ,* $\qquad \mu \in P^+, t \in [0, \infty)$.

Moreover, $\sigma_2(t, \mu)$ satisfies the following asymptotic relationships:

(iii) *$\sigma_2(t, \mu) \sim e^{-it\mu}$; $\sigma_2'(t, \mu) \sim -ie^{-it\mu}$ as $t \to \infty$, uniformly for μ in any given bounded subset of the interior of P_ε^+.*

(iv) *$e^{-it\mu}\sigma_2(t, \mu)$ is bounded for $0 \leqq t < \infty$, uniformly for μ in any given bounded subset of P_ε^+.*

(v) *There exists a continuous function c defined for μ real, $\mu \in P_\varepsilon^+$, such that* $\lim_{t\to\infty} |\sigma_2(t, \mu) - e^{-it\mu} - c(\mu)e^{it\mu}| = 0,$ $\qquad$ *μ real, $\mu \in P^+$.*

PROOF. We proceed very much as in the proof of Lemma 1. Let a be chosen so large that

$$\varepsilon^{-1} \int_a^\infty |q(s)| \, ds < \tfrac{1}{2}.$$

Let K_μ be the linear operator in $C[a, \infty)$ defined by the formula

$$(21)\qquad (K_\mu f)(t) = \frac{1}{2i\mu}\left[\int_a^t e^{-2i\mu(s-t)} q(s) f(s)\, ds + \int_t^\infty q(s) f(s)\, ds\right].$$

It follows, as in the proof of the corresponding facts in Lemma 1, that $|K_\mu| < \frac{1}{2}$, $\mu \in P_\varepsilon^+$, and that K_μ is analytic for μ interior to P_ε^+. Formula (21) shows that $(K_\mu f)(t)$ is continuous in μ, uniformly for $\mu \in P_\varepsilon^+$ and t in any finite interval. By Lemma VII.6.1, $(I - K_\mu)^{-1}$ exists, and by Lemmas VII.6.3 and VII.6.4, the vector $f_\mu \in C[a, \infty)$ defined by

$$f_\mu = (I - K_\mu)^{-1} 1$$

satisfies $|f_\mu| \leqq 2$, $\mu \in P^+$, and is continuous in μ for $\mu \in P_\varepsilon^+$ and analytic in μ for μ interior to P_ε^+. We have

$$(K_\mu f)'(t) = \int_a^t e^{-2i\mu(s-t)} q(s) f(s)\, ds, \qquad f \in C[a, \infty),$$

$$(22)\qquad (K_\mu f)''(t) = q(t) f(t) + 2i\mu \int_a^t e^{-2i\mu(s-t)} q(s) f(s)\, ds, \qquad f \in C[a, \infty),$$

so that

$$(K_\mu f)''(t) - 2i\mu (K_\mu f)'(t) = q(t) f(t).$$

Consequently, since f_μ satisfies

$$(23)\qquad f_\mu = K_\mu f_\mu + 1,$$

f_μ also satisfies

$$-f_\mu''(t) + 2i\mu f_\mu'(t) + q(t) f_\mu(t) = 0,$$

and thus the function

$$(24)\qquad \hat{g}_\mu(t) = e^{-i\mu t} f_\mu(t)$$

will satisfy the differential equation $\tau \hat{g}_\mu = \tau^2 \hat{g}_\mu$. By (9) and Lemma VII.6.1, $|f_\mu(t)| \leqq 2$, $a \leqq t < \infty$. Thus, by (21),

$$(25)\qquad |(K_\mu f_\mu)(t)|$$

$$\leqq |\mu|^{-1}\left[\int_\alpha^t e^{2\delta(s-t)} |q(s)|\, ds + \int_t^\infty |q(s)|\, ds\right], \qquad a \leqq t < \infty,\ \mathscr{I}\mu \geqq \delta.$$

Therefore, by (23) and (24),

$$(26)\quad \begin{cases} f_\mu(t) \sim 1 \text{ as } t \to \infty, & \text{uniformly for } \mu \in P_\varepsilon^+,\ \mathscr{I}\mu \geqq \delta > 0, \\ \hat{g}_\mu(t) \sim e^{-i\mu t} \text{ as } t \to \infty, & \text{uniformly for } \mu \in P_\varepsilon^+,\ \mathscr{I}\mu \geqq \delta > 0. \end{cases}$$

By (23), (22), and the Lebesgue dominated convergence theorem,

$$(27)\qquad \limsup_{t\to\infty} |f_\mu'(t)| \leq \limsup_{t\to\infty} \int_a^t e^{2\delta(s-t)}|f_\mu(s)|\,|q(s)|\,ds = 0,$$

uniformly for $\mu \in P_\varepsilon^+$, $\mathscr{I}\mu \geqq \delta > 0$. Since, by (24),

$$\hat{g}_\mu(t) = -i\mu e^{-i\mu t} f_\mu(t) + e^{-i\mu t} f_\mu'(t),$$

it follows from (26) and (27), that

$$(28)\quad \hat{g}_\mu'(t) \sim -i\mu e^{-i\mu t} \text{ as } t \to \infty, \quad \text{uniformly for } \mu \in P_\varepsilon^+,\ \mathscr{I}\mu \geqq \delta > 0.$$

Since $|f_\mu(t)| \leqq 2$, $a \leqq t < \infty$, we have $|e^{it\mu}\hat{g}_\mu(t)| \leqq 2$, $\mu \in P_\varepsilon^+$, $a \leqq t < \infty$. By (23), (21), and the Lebesgue dominated convergence theorem,

$$(29)\qquad \lim_{t\to\infty} \left| f_\mu(t) - 1 - (2i\mu)^{-1} e^{2i\mu t} \int_a^\infty e^{-2i\mu s} q(s) f_\mu(s)\,ds \right|$$

$$= \lim_{t\to\infty} \left| \frac{1}{2i\mu} \int_t^\infty [1 - e^{-2i\mu(s-t)}] q(s) f_\mu(s)\,ds \right| = 0$$

for each $\mu \in P_\varepsilon^+$. Now define

$$(30)\qquad c(\mu) = (2i\mu)^{-1} \int_a^\infty e^{-2i\mu s} q(s) f_\mu(s)\,ds, \qquad \mu \text{ real}, \ \mu \in P_\varepsilon^+.$$

Since f_μ is continuous in μ for $\mu \in P_\varepsilon^+$, $c(\mu)$ is continuous in μ for real $\mu \in P_\varepsilon^+$ by the Lebesgue dominated convergence theorem. By (29) and (24),

$$(31)\qquad \lim_{t\to\infty} |\hat{g}_\mu(t) - e^{-it\mu} + c(\mu) e^{it\mu}| = 0$$

for real $\mu \in P_\varepsilon^+$.

Letting $\sigma_2(t, \mu)$ be the unique solution of $\tau\sigma = \mu^2\sigma$ on $[0, \infty)$ such that $\sigma_2(t, \mu) = \hat{g}_\mu(t)$, $a \leqq t < \infty$, we now see that all the statements of the present lemma follow, using Corollary XIII.1.5, from what has been proved above (cf. the corresponding portion of the proof of Lemma 1). Q.E.D.

Lemmas 1 and 3 enable us to make some preliminary spectral-theoretic assertions concerning T.

4 LEMMA. *Let the hypotheses of Lemma 1 be satisfied. Let $A(f)=0$ be the boundary condition (2), and σ_1 be the function described in Lemma 1. For each complex λ, let $\mu=\mu(\lambda)$ be the unique square root of λ lying in $P^+=\{\mu \mid \mathscr{I}\mu \geqq 0\}$ but not lying on the negative real axis. Let $A(\lambda)=A(\sigma_1(\cdot, \mu(\lambda)))$. Then*

(i) *$A(\lambda)$ is analytic for non-zero λ in the complement of the positive real axis $R=\{\lambda \mid 0<\lambda<\infty\}$, and approaches continuous limits $A^+(\lambda)$ and $A^-(\lambda)$ as λ approaches R from above and from below;*

(ii) *$A(\lambda) \sim 1$ or $A(\lambda) \sim i\mu(\lambda)$ as $|\lambda| \to \infty$, the first (respectively second) asymptotic formula being valid if formula (2a) (respectively formula (2b)) above is valid;*

(iii) *if $\lambda_0 \neq 0$ and $\lambda_0 \notin R$, then $\lambda_0 \in \sigma(T)$ if and only if $A(\lambda_0)=0$, in which case λ_0 is an isolated point of $\sigma(T)$ which belongs to the point spectrum of T, and is a pole of the resolvent of T.*

PROOF. Statements (i) and (ii) follow from Lemma 1 and formulas (2a) and (2b). Let $\lambda_0 \notin R \cup \{0\}$. If $A(\lambda_0)=0$, then $\sigma_1(\cdot, \mu(\lambda_0))$ is an eigenvector of T belonging to the eigenvalue λ_0, so λ_0 belongs to the point spectrum of T. Conversely, if $v(\cdot, \mu(\lambda_0))$ is an eigenvector of T corresponding to an eigenvalue $\lambda_0 \notin R \cup \{0\}$, then v must be a scalar multiple of $\sigma_1(\cdot, \mu(\lambda_0))$, since the equation $(\tau-\lambda_0)\sigma=0$ has, by Lemma 3, a second solution which is exponentially large as $t \to \infty$. Thus $A(\lambda_0)=0$. We conclude that λ_0 belongs to the point spectrum of T if and only if $A(\lambda_0)=0$. It remains to prove that the part of $\sigma(T)$ belonging to the complement of $R \cup \{0\}$ is a set of isolated points, each of which is a zero of $A(\lambda)$ and is a pole of the resolvent.

Choose an arbitrary $\varepsilon \geqq 0$, let $|\lambda|>\varepsilon^2$, and suppose $\lambda \notin R$, $A(\lambda) \neq 0$. For convenience write μ in place of $\mu(\lambda)$. Consider the integral operator $R(\lambda)$ defined by the equation

$$R(\lambda)(f, s)=\int_0^\infty R(s, t; \lambda) f(t)\, dt, \qquad f \in L_2(0, \infty),$$

where

$$\begin{aligned}(32) \quad R(s, t; \lambda) &= (2i\mu A(\lambda))^{-1}\{A(\lambda)\sigma_2(s, \mu)-B(\lambda)\sigma_1(s, \mu)\}\sigma_1(t, \mu), \qquad s<t, \\ &= (2i\mu A(\lambda))^{-1}\{A(\lambda)\sigma_2(t, \mu)-B(\lambda)\sigma_1(t, \mu)\}\sigma_1(s, \mu), \qquad s \geqq t.\end{aligned}$$

Here $B(\lambda) = A(\sigma_2(\cdot, \mu))$, $\sigma_2(\cdot, \mu)$ being the function of Lemma 3. We note that $R(s, t; \mu) = R(t, s; \mu)$. Suppose for the moment that we know the quantity $\sup_{0 \leqq s < \infty} \int_0^\infty |R(s, t; \lambda)|\, dt$ to be finite. Then, for $f \in L_2(0, \infty)$, we have

$$\begin{aligned}
|R(\lambda)f|_2^2 &\leqq \int_0^\infty \left\{ \int_0^\infty |R(s, t; \lambda)|\, |f(t)|\, dt \right\}^2 ds \\
&\leqq \int_0^\infty \left(\left\{ \int_0^\infty |R(s, t; \lambda)|\, dt \right\} \left\{ \int_0^\infty |R(s, t; \lambda)|\, |f(t)|^2\, dt \right\} \right) ds \\
&\leqq \left\{ \sup_{0 \leqq s < \infty} \int_0^\infty |R(s, t; \lambda)|\, dt \right\} \int_0^\infty \left\{ \int_0^\infty |R(s, t; \lambda)|\, ds \right\} |f(t)|^2\, dt \\
&\leqq \sup_{0 \leqq s < \infty} \left\{ \int_0^\infty |R(s, t; \lambda)|\, dt \right\}^2 |f|_2^2
\end{aligned}$$

by Hölder's inequality, Fubini's theorem, and the symmetry of $R(s, t; \lambda)$. From this it also follows by Fubini's theorem that

$$\int_0^\infty R(s, t; \lambda) f(t)\, dt$$

exists for almost all s and belongs to $L_2(0, \infty)$. Thus the integral operator $R(\lambda)$ with kernel defined by (32) maps $L_2(0, \infty)$ into itself, and its norm may be bounded above as will be seen from the inequality

$$|R(\lambda)| \leqq \sup_{0 \leqq s < \infty} \int_0^\infty |R(s, t; \lambda)|\, dt. \tag{33}$$

To show that this supremum is finite we argue as follows. Let Λ be a bounded subset of the set $\{\lambda \mid |\lambda| > \varepsilon^2\}$. By Lemmas 1 and 3 there exists a finite constant K such that

$$|\sigma_1(t, \mu)| \leqq K e^{-t\mathscr{I}(\mu)}, \quad |\sigma_2(t, \mu)| \leqq K e^{t\mathscr{I}(\mu)}, \qquad 0 \leqq t < \infty, \lambda \in \Lambda. \tag{34}$$

Hence

$$|\sigma_1(t, \mu)| \int_0^t |\sigma_2(s, \mu)|\, ds \leqq K^2 e^{-t\mathscr{I}(\mu)} \int_0^t e^{s\mathscr{I}(\mu)}\, ds \leqq \frac{K^2}{\mathscr{I}(\mu)}, \tag{35}$$

and similarly

$$|\sigma_2(t, \mu)| \int_t^\infty |\sigma_1(s, \mu)|\, ds \leqq \frac{K^2}{\mathscr{I}(\mu)}. \tag{36}$$

Put

$$C_1(\lambda) = \max\{|2\mu A(\lambda)|^{-1}\,|B(\lambda)|,\ |2\mu|^{-1}\} \tag{37}$$

and let $C_2(\lambda)$ be the common least upper bound for

$$|\sigma_1(t, \mu)| \int_0^t |\sigma_2(s, \mu)|\, ds \quad \text{and} \quad |\sigma_2(t, \mu)| \int_0^\infty |\sigma_1(s, \mu)|\, ds, \qquad 0 \leqq t < \infty.$$

It follows from (35) and (36) that $|\mathscr{I}(\mu)|\, C_2(\lambda)$ is uniformly bounded in each compact set of P_ε^+ and, in particular, that $C_2(\lambda)$ is uniformly bounded in each compact subset of the interior of P_ε^+. From this and from (32) and (33) it follows readily that $R(\lambda)$ is an operator bounded by $4C_1(\lambda)C_2(\lambda)$; using Lemmas 1 and 3, it follows that $R(\lambda)$ is defined and analytically dependent on λ for each λ such that $\lambda \notin R$, $|\lambda| > \varepsilon^2$, and $A(\lambda) \neq 0$. Moreover, it follows similarly that at a point $\lambda_0 \notin R$ for which $|\lambda_0| > \varepsilon^2$ but for which $A(\lambda_0) = 0$, $R(\lambda)$ has a pole of order at most equal to the order of the zero of $A(\lambda)$ at λ_0. Finally, if $0 < \lambda_1 < \infty$ and $A^+(\lambda_1) \neq 0$, $A^-(\lambda_1) \neq 0$, then $|\mathscr{I}(\lambda)|\,|R(\lambda)|$ is bounded on a short transversal drawn to the real axis through λ_1. The simple details of the verification of these assertions on the basis of (32), (33), (34), and Lemmas 1 and 3, are left to the reader.

We shall now show that if λ satisfies the conditions $\lambda \notin R$, $|\lambda| > \varepsilon^2$ and $A(\lambda) \neq 0$, then the resolvent $R(\lambda; T)$ of T exists and is equal to $R(\lambda)$. Taken together with what has been established above, this will prove (iii) and complete the proof of the present lemma.

Suppose that f is continuous on $[0, \infty)$ and vanishes outside a bounded set. Since $R(s, t; \lambda)$ is continuous at $s = t$,

$$(R(\lambda)f)'(s) = \int_0^\infty \frac{\partial}{\partial s} R(s, t; \lambda) f(t)\, dt.$$

However, since $\partial/\partial s[R(s, t; \lambda)]$ has a jump discontinuity at $s = t$,

$$(R(\lambda)f)''(s) = \int_0^\infty \left(\frac{\partial}{\partial s}\right)^2 R(s, t; \lambda) f(t)\, dt - \frac{1}{2i\mu} W(\mu, s) f(s),$$

where $-(1/2i\mu)W(\mu, s)$ is the discontinuity

$$(\partial R/\partial s)(s, s-0) - (\partial R/\partial s)(s, s+0)$$

in the first partial derivative of R. Thus, in virtue of (32), we see that $W(\mu, s)$ is the Wronskian determinant

$$W(\mu, s) = \sigma_1(s, \mu)\sigma_2'(s, \mu) - \sigma_2(s, \mu)\sigma_1'(s, \mu).$$

The Wronskian determinant of two solutions of the equation $\tau\sigma = \mu^2\sigma$ is independent of s, and by Lemmas 1 and 3,

$$\sigma_1(s, \mu) \sim e^{i\mu s}, \qquad \sigma_1'(s, \mu) \sim i\mu e^{i\mu s},$$

$$\sigma_2(s, \mu) \sim e^{-i\mu s}, \qquad \sigma_2'(s, \mu) \sim -i\mu e^{-i\mu s}.$$

Consequently $W(\mu) = -2i\mu$. Thus

$$(R(\lambda)f)''(s) = \int_0^\infty \left(\frac{\partial}{\partial s}\right)^2 R(s, t; \lambda) f(t)\, dt + f(s), \tag{38}$$

showing that $R(\lambda)f$ is a function in $L_2(0, \infty)$ with continuous first and second derivatives. Using (38) and the definition of $R(\lambda)$, it follows that

$$(\lambda - \tau)(R(\lambda)f)(s) = \int_0^\infty \left(\lambda - \frac{\partial^2}{\partial s^2} - q(s)\right) R(s, t; \lambda) f(t)\, dt + f(s).$$

Since by (32) we have

$$\left(\lambda - \frac{\partial^2}{\partial s^2} - q(s)\right) R(s, t; \lambda) = 0,$$

it follows that $(\lambda - \tau)(R(\lambda)f)(s) = f(s)$. Thus $R(\lambda)f \in \mathfrak{D}(\lambda I - T_1(\tau)) = \mathfrak{D}(T_1(\tau))$. Moreover, using (32), we see that

$$(R(\lambda)f)(0) = \frac{1}{2i\mu A(\lambda)} \{A(\lambda)\sigma_2(0, \mu) - B(\lambda)\sigma_1(0, \mu)\} \int_0^\infty f(t)\sigma_1(t, \mu)\, dt$$

and

$$(R(\lambda)f)'(0) = \frac{1}{2i\mu A(\lambda)} \{A(\lambda)\sigma_2'(0, \mu) - B(\lambda)\sigma_1'(0, \mu)\} \int_0^\infty f(t)\sigma_1(t, \mu)\, dt,$$

so that

$$\begin{aligned} A(R(\lambda)f) &= \frac{1}{2i\mu A(\lambda)} \{A(\lambda)A(\sigma_2(\cdot, \mu)) - B(\lambda)A(\sigma_1(\cdot, \mu))\} \int_0^\infty f(t)\sigma_1(t, \mu)\, dt \\ &= \frac{1}{2i\mu A(\lambda)} \{A(\lambda)B(\lambda) - B(\lambda)A(\lambda)\} \int_0^\infty f(t)\sigma_1(t, \mu)\, dt = 0. \end{aligned}$$

Consequently, $R(\lambda)f \in \mathfrak{D}(T)$ for all continuous f vanishing outside a bounded subset of $[0, \infty)$. Since T is closed, it follows immediately that $R(\lambda)f \in \mathfrak{D}(T)$ and $(\lambda I - T)R(\lambda)f = f$ for all $f \in L_2(0, \infty)$.

Next let $f \in \mathfrak{D}(T)$, let $A(\lambda) \neq 0$, $\lambda \notin R$, and consider $g = R(\lambda)(\lambda I - T)f$. Then $g \in \mathfrak{D}(T)$, and

$$(\lambda I - T)g = (\lambda I - T)R(\lambda)(\lambda I - T)f = (\lambda I - T)f.$$

Since $A(\lambda) \neq 0$, δ is not in the point spectrum of T. Consequently $f = g$. This shows that $f = R(\lambda)(\lambda I - T)f$ for each $f \in \mathfrak{D}(T)$. Thus, if $\lambda \in R$, $|\lambda| > \varepsilon^2$, and $A(\lambda) \neq 0$, then $(\lambda I - T)^{-1} = R(\lambda; T)$ exists and equals $R(\lambda)$, completing the proof of the present lemma. Q.E.D.

5 Corollary. *Let the hypotheses of Lemma 4 be satisfied and let $0 < \lambda_1 < \infty$. Suppose in the notation of Lemma 4 that $A^+(\lambda_1) \neq 0$, $A^-(\lambda_1) \neq 0$. Then for $\lambda = \lambda_1$ lying on any sufficiently short transversal to the real axis through λ_1, $R(\lambda; T) = (\lambda I - T)^{-1}$ exists and $|\mathscr{I}(\lambda)|\,|R(\lambda; T)|$ is bounded.*

For use in what is to follow we record the formula for $R(\lambda; T)$ obtained in the proof of Lemma 4 and also observe that in constructing the kernel $R(s, t; \lambda)$ in Lemma 4 we may replace the particular "second solution" $\sigma_2(t, \lambda)$ constructed in Lemma 3 by any solution asymptotic to $e^{-i\lambda t}$ as $t \to \infty$. Appropriate choice of this second solution will enable us to calculate finer properties of the resolvent as needed below. The following corollary summarizes the necessary facts in a form convenient for later use.

6 Corollary. *Let $A(\lambda)$ and $\mu(\lambda)$ be as in Lemma 4. Let $\lambda \notin R$, $\lambda \neq 0$, and $A(\lambda) \neq 0$ so that $\lambda \notin \sigma(T)$. If σ is any solution of $\tau\sigma = \lambda\sigma$ which is asymptotic to $e^{-i\mu(\lambda)t}$ as $t \to \infty$, and $B(\lambda) = A(\sigma(\cdot, \mu(\lambda)))$, then $R(\lambda; T)$ is an integral operator with the kernel*

$$\begin{aligned} &R(s, t; \lambda) \\ &\quad = (2i\mu(\lambda)A(\lambda))^{-1}\{A(\lambda)\sigma(s, \mu(\lambda)) - B(\lambda)\sigma_1(s, \mu(\lambda))\}\sigma_1(t, \mu(\lambda)), \quad s < t \\ &\quad = (2i\mu(\lambda)A(\lambda))^{-1}\{A(\lambda)\sigma(t, \mu(\lambda)) - B(\lambda)\sigma_1(t, \mu(\lambda))\}\sigma_1(s, \mu(\lambda)), \quad s \geqq t. \end{aligned}$$

7 Lemma. *Let the hypotheses of Corollary 2 be satisfied, and let $\sigma_1(t, \lambda)$ and $A^+(\lambda)$ and $A^-(\lambda)$ be the functions of Lemma 1, Corollary 2, and Lemma 4. Then:*

(i) *$A^+(\lambda)$ and $A^-(\lambda)$ are continuous for all λ such that $0 \leqq \lambda < \infty$.*

(ii) *$A^+(\lambda) \sim \pm A^-(\lambda)$ as $\lambda \to \infty$, the plus sign being valid if formula* (2a) *describes the boundary condition defining T, and the minus sign being valid if formula* (2b) *describes the boundary condition defining T.*

If in addition $A^+(\lambda) \neq 0$ *and* $A^-(\lambda) \neq 0$ *for all* $0 \leq \lambda < \infty$, *then:*

(iii) $\sigma(T)$ *is the union of* $\{\lambda \mid 0 \leq \lambda < \infty\}$ *and a finite set of points* λ_0 *not lying in this set, each of which is a pole of the resolvent* $R(\lambda; T)$ *and corresponds to a finite dimensional projection* $E(\lambda_0; T)$.

(iv) *For each pair of functions* $f, g \in C[0, \infty)$, *which vanish outside a bounded set, and for each real* $\lambda > 0$, *the limits*

$$B^+(f, g, \lambda) = \lim_{\delta \to 0+} (R(\lambda + i\delta; T)f, g),$$

$$B^-(f, g, \delta) = \lim_{\delta \to 0-} (R(\lambda - i\delta; T)f, g)$$

exist and are expressible by the formulas

$$\begin{aligned} B^+(f, g, \lambda) = (2i\mu(\lambda)A^+(\lambda))^{-1} \iint_{0<s<t<\infty} &\{A^+(\lambda)\sigma_1(s, -\mu(\lambda)) \\ &- A^-(\lambda)\sigma_1(s, \mu(\lambda))\}\sigma_1(t, \mu(\lambda))f(t)g(s)\, dt\, ds \\ + (2i\mu(\lambda)A^+(\lambda))^{-1} \iint_{0<t<s<\infty} &\{A^+(\lambda)\sigma_1(t, -\mu(\lambda)) \\ &- A^-(\lambda)\sigma_1(t, \mu(\lambda))\}\sigma_1(s, \mu(\lambda))f(t)\overline{g(s)}\, dt\, ds; \end{aligned}$$

$$\begin{aligned} B^-(f, g, \lambda) = -(2i\mu(\lambda)A^-(\lambda))^{-1} \iint_{0<s<t<\infty} &\{A^-(\lambda)\sigma_1(s, \mu(\lambda)) \\ &- A^+(\lambda)\sigma_1(s, -\mu(\lambda))\}\sigma_1(t, -\mu(\lambda))f(t)\overline{g(s)}\, dt\, ds \\ - (2i\mu(\lambda)A^-(\lambda))^{-1} \iint_{0<t<s<\infty} &\{A^-(\lambda)\sigma_1(t, \mu(\lambda)) \\ &- A^+(\lambda)\sigma_1(t, -\mu(\lambda))\}\sigma_1(s, -\mu(\lambda))f(t)\overline{g(s)}\, dt\, ds. \end{aligned}$$

PROOF. It is clear from Lemma 4 and Corollary 2 that $A^+(\lambda) = A(\sigma_1(\cdot, \mu(\lambda))$ and $A^-(\lambda) = A(\sigma_1(\cdot, -\mu(\lambda)))$ for $0 \leq \lambda < \infty$. Statements (i) and (ii) now follow immediately from Lemma 1 (ii) and Lemma 4 (ii) (cf. formulas (2a) and (2b)).

Now make the additional hypothesis that $A^+(\lambda) \neq 0$, $A^-(\lambda) \neq 0$, for all $0 \leq \lambda < \infty$. By Lemmas 1 and 3, the equation $\tau\sigma = \lambda\sigma$ has no solution in $L_2(0, \infty)$ for $0 < \lambda < \infty$. It follows from Lemma XIII.3.1 that for $0 < \lambda < \infty$, λ belongs to the spectrum $\sigma(T)$. Consequently, by Lemma 4(iii), $\sigma(T)$ consists of the union of the interval $0 \leq \lambda < \infty$ and a set Z of isolated points of the plane slit by the removal of the positive real axis; moreover, each point in Z is a pole of the resolvent of T and a zero of the function $A(\lambda)$. Since $A(\lambda) \sim 1$ or $A(\lambda) \sim \mu(\lambda)$ as $|\lambda| \to \infty$, Z is a bounded set. Moreover, since $A^+(\lambda) \neq 0$, $A^-(\lambda) \neq 0$, $0 \leqq \lambda < \infty$, Z has no limit points on the real axis. Consequently, Z is a finite set. Now let $\lambda_0 \in Z$ and let ν_0 be the order of λ_0 as a pole of the resolvent. By Theorems VII.3.18 and VII.3.24, each $f \in E(\lambda_0; T)L_2(0, \infty)$ satisfies $(T - \lambda_0 I)^{\nu_0} f = 0$. It then follows from Theorem XIII.1.3 and Corollary XIII.1.4 that the range of $E(\lambda; T)$ has dimension at most equal to $2\nu_0$. (Remark: an easy additional argument would show that the dimension of the range of $E(\lambda_0; T)$ is exactly ν_0.)

It only remains to prove (iv). Let f and g be functions in $C[0, \infty)$ vanishing outside a bounded set. If $|\mathscr{I}\eta| > 0$, define $B(\eta) = A(\sigma_2(\cdot, \mu(\eta)))$, where σ_2 is the function of Lemma 3. If $\lambda > 0$ and $\delta > 0$, then by Corollary 6 we have

$$
\begin{aligned}
&(R(\lambda + i\delta)f, g)\\
&\quad = (2i\mu(\lambda + i\delta)A(\lambda + i\delta))^{-1} \iint_{0<s<t<\infty} \{A(\lambda + i\delta)\sigma_2(s, \mu(\lambda + i\delta))\\
&\qquad -B(\lambda + i\delta)\sigma_1(s, \mu(\lambda + i\delta))\}\sigma_1(t, \mu(\lambda + i\delta))f(t)\overline{g(s)}\, dt\, ds\\
&\qquad + (2i\mu(\lambda + i\delta)\, A(\lambda+i\delta)^{-1} \iint_{0<t<s<\infty} \{A(\lambda + i\delta)\sigma_2(t, \mu(\lambda + i\delta))\\
&\qquad -B(\lambda + i\delta)\sigma_1(t, \mu(\lambda + i\delta))\}\sigma_1(s, \mu(\lambda + i\delta))f(t)\overline{g(s)}\, dt\, ds.
\end{aligned}
$$

By Lemma 3(ii), $B(\lambda + i\delta)$ has a limit $B^+(\lambda)$ as $\delta \to 0$. It is therefore evident that the limit $B^+(f, g, \lambda)$ in statement (iv) of the present lemma exists and is given by the expression

$$
\begin{aligned}
B^+(f, g, \lambda) &= (2i\mu(\lambda)A^+(\lambda))^{-1} \iint_{0<s<t<\infty} \{A^+(\lambda)\sigma_2(s, \mu(\lambda))\\
&\quad -B^+(\lambda)\sigma_1(s, \mu(\lambda))\}\sigma_1(t, \mu(\delta))f(t)\overline{g(s)}\, dt\, ds\\
&\quad + (2i\mu(\lambda)A^+(\lambda))^{-1} \iint_{0<t<s<\infty} \{A^+(\lambda)\sigma_2(t, \mu(\lambda))\\
&\quad - B^+(\lambda)\sigma_1(t, \mu(\lambda))\}\sigma_1(s, \mu(\lambda))f(t)\overline{g(s)}\, dt\, ds.
\end{aligned} \tag{39}
$$

By Lemma 3, $\sigma_2 = \sigma_2(t, \mu(\lambda))$ has, for real λ, the same asymptotic form as the linear combination $\sigma_1(t, -\mu(\lambda)) + c(\mu(\lambda))\sigma_1(t, \mu(\lambda))$ of the solutions $\hat{\sigma}_1 = \sigma_1(t, -\mu(\lambda))$ and $\sigma_1 = \sigma_1(t, \mu(\lambda))$ of the equation $\tau\sigma = \lambda\sigma$. On the other hand, since these two solutions are linearly independent, we must have a linear relation $\sigma_2 = a\hat{\sigma}_1 + b\sigma_1$. It is clear from Lemma 1 that such a linear combination can only have the indicated asymptotic form if $a = 1$, $b = c(\mu(\lambda))$. Hence

$$\sigma_2(t, \mu(\lambda)) = \sigma_1(t, -\mu(\lambda)) + c(\mu(\lambda))\sigma_1(t, \mu(\lambda)).$$

Therefore $B^+(\lambda) = A^-(\lambda) + c(\mu(\lambda))A^+(\lambda)$, and

$$\begin{aligned} A^+(\lambda)\sigma_2(s, \mu(\lambda)) - B^+(\lambda)\sigma_1(s, \mu(\lambda)) \\ = A^+(\lambda)\sigma_1(s, -\mu(\lambda)) - A^-(\lambda)\sigma_1(s, \mu(\lambda)). \end{aligned}$$

From this, the conclusions of (iv) relative to $B^+(f, g, \lambda)$ are evident. The conclusions of (iv) relative to $B^-(f, g, \lambda)$ may be established in the same way. Q.E.D.

8 LEMMA. *Let the hypotheses of Corollary 2 be satisfied. Then there exists a solution $\sigma_3(t, \mu)$ of the equation $\tau\sigma = \mu^2\sigma$, defined for $0 \leqq t < \infty$ and for all sufficiently small $\mu \in P^+$, such that σ_3 and σ_3' are continuous in t and μ for $0 \leqq t < \infty$ and μ sufficiently small, and such that*

$$\sigma_3(t, \mu) \sim e^{-it\mu}; \qquad \sigma_3'(t, \mu) \sim -i\mu e^{-it\mu};$$

as $t \to \infty$, for all $\mu \in P^+$ such that $\mathcal{I}(\mu) \neq 0$, $|\mu|$ sufficiently small.

PROOF. Let σ_1 be the function of Corollary 2. Then $\sigma_1(t, 0) \sim 1$ as $t \to \infty$ by Corollary 2. Choose some t_0 such that $\sigma_1(t_0, 0) \neq 0$. Since $\sigma_1(t, \mu)$ is continuous in t and μ by Corollary 2, $\sigma_1(t_0, \mu) \neq 0$ for sufficiently small $\mu \in P^+$. Let $\sigma_3(t, \mu)$ be the unique solution of $\tau\sigma = \mu^2\sigma$ which satisfies the conditions $\hat{\sigma}_3(t_0, \mu) = 0$, $\hat{\sigma}_3'(t_0, \mu) = \sigma_1(t_0, \mu)^{-1}$. The Wronskian of $\hat{\sigma}_3$ and σ_1 is plainly equal to 1. By Corollary XIII.1.5, $\hat{\sigma}_3(t, \mu)$ is continuous in t and μ for $0 \leqq t < \infty$ and all sufficiently small $\mu \in P^+$. The functions $\hat{\sigma}_3$ and σ_1 are clearly linearly independent. If $\mu > 0$ then by Lemma 3 $\tau\sigma = \mu^2\sigma$ has a solution $\sigma_2(t, \mu)$ which is asymptotic to $e^{-it\mu}$ as $t \to \infty$ and whose derivative is asymptotic to $-i\mu e^{it\mu}$ as $t \to \infty$. Then $\hat{\sigma}_3(t, \mu) = a(\mu)\sigma_1(t, \mu) + b(\mu)\sigma_2(t, \mu)$ for suitable functions $a(\mu)$ and $b(\mu)$. It is evident from the asymptotic form of σ_2 and the fact that $b(\mu) \neq 0$ ($\hat{\sigma}_3$ and σ_1 being linearly independent), that $\hat{\sigma}_3(t, \mu) \sim b(\mu)e^{-it\mu}$ and $\hat{\sigma}_3'(t, \mu) \sim -i\mu b(\mu)e^{-it\mu}$ as $t \to \infty$ for $\mathcal{I}(\mu) > 0$. Since the Wronskian

$$W(t, \mu) = \hat{\sigma}_3'(t, \mu)\sigma_1(t, \mu) - \hat{\sigma}_3(t, \mu)\sigma_1'(t, \mu)$$

is independent of t, and since $W(t, \mu) \sim -2i\mu b(\mu)$, we have

$$1 = W(t_0, \mu) = -2i\mu b(\mu),$$

proving that $b(\mu) = -(2i\mu)^{-1}$. The function $\sigma_3(t, \mu(\lambda)) = -2i\mu\hat{\sigma}_3(t, \mu(\lambda))$ therefore satisfies the conditions of the present lemma. Q.E.D.

9 Corollary. *Let the hypotheses of Lemma 7 be satisfied, and in particular let $A^+(\lambda)$ and $A^-(\lambda)$ be non-vanishing for $0 \leqq \lambda < \infty$. Let f and g be a pair of functions in $C[0, \infty)$ which vanish outside a bounded set. Then $(R(\lambda; T)f, g)$ has a continuous extension to a neighborhood of the point $\lambda = 0$ with the point $\lambda = 0$ itself deleted, and this extension satisfies the estimate*

$$|(R(\lambda; T)f, g)| = O(|\lambda|^{-1})$$

as $|\lambda| \to 0$.

Proof. Let σ_3 be as in the preceding lemma. Let $A(\lambda) = A(\sigma_1(\cdot, \mu(\lambda)))$, $B(\lambda) = A(\sigma_3(\cdot, \mu(\lambda)))$ and represent $R(\lambda; T)$ as the integral operator whose kernel is given by Corollary 6, using σ_3 for the function σ of that lemma. Then $A(\lambda)$ and $B(\lambda)$ are bounded for non-real λ near $\lambda = 0$ and the asserted estimate follows from the formula for $(R(\lambda; T)f, g)$. Q.E.D.

10 Lemma. *Let the hypotheses of Lemma 3 be satisfied. Then there exists a solution $\sigma_4(t, \mu)$ of the equation $\tau\sigma = \mu^2\sigma$, defined for $0 \leqq t < \infty$ and for all $\mu \in P^+$ such that $|\mu|$ is sufficiently large, such that*

$$\sigma_4(t, \mu) \sim e^{-it\mu}; \qquad \sigma_4'(t, \mu) \sim -i\mu e^{-it\mu};$$

as $t \to \infty$, $\mu > 0$, and also as $|\mu| \to \infty$, μ remaining in P^+, uniformly for $0 \leqq t < \infty$.

Proof. Choose ε so large that

$$\varepsilon^{-1} \int_0^\infty |q(s)|\, ds < \tfrac{1}{2}.$$

Then, for $|\mu| > \varepsilon$, $\mu \geqq 0$, the functions f_μ and $\hat{g}_\mu$ of the proof of Lemma 3 are all defined for $0 \leqq t < \infty$. We put $\sigma_4(t, \mu) = \hat{g}_\mu(t)$, $0 \leqq t < \infty$, $\mu \in P_\varepsilon^+$. It follows as in the proof of Lemma 3 that $|\sigma_4(t, \mu)| \leqq 2$, $0 \leqq t < \infty$, $\mu \in P_\varepsilon^+$. By equations (21) and (23) of the proof of Lemma 3, we find that

$$|f_\mu(t) - 1| \leqq |\mu|^{-1} \int_0^\infty |q(s)|\, ds.$$

Then, by equations (22) and (23) of that same proof,

$$|f'_\mu(t)| \leqq |\mu|^{-1}\left\{\int_0^\infty |q(s)|\, ds\right\}^2 + \left|\int_0^t e^{-2i\mu(s-t)}q(s)\, ds\right|.$$

It follows from this formula just as in the proof of Lemma 1 (cf. the paragraph following formula (14)) that

$$\lim_{\substack{|\mu|\to\infty \\ \mu\in P+}} |f'_\mu(t)| = 0, \quad \text{uniformly for} \quad 0 \leqq t < \infty.$$

Hence, by formula (24) of the proof of Lemma 3,

$$\hat{g}_\mu(t) \sim e^{-it\mu}; \qquad \hat{g}'_\mu(t) = -i\mu e^{-it\mu} f_\mu(t) + e^{it\mu} f'_\mu(t)$$
$$\sim -i\mu e^{-it\mu},$$

as $\mu \to \infty$, μ remaining in P^+, uniformly for $0 \leqq t < \infty$. The remaining assertions of the present lemma follow immediately from Lemma 3. Q.E.D.

11 COROLLARY. *Let τ and T be as above, and let the hypotheses of Lemma 7 be satisfied. Let f and g be a pair of functions in $C[0, \infty)$ which vanish outside a bounded set. Then $(R(\lambda; T)f, g)$ is bounded for $|\lambda|$ sufficiently large, λ non-real.*

PROOF. Let σ_4 be as in the preceding lemma, put $A(\lambda) = A(\sigma_1(\cdot, \mu(\lambda)))$, and let $B(\lambda) = A(\sigma_4(\cdot, \mu(\lambda)))$ (cf. Lemma 4 for the definition of $\mu(\lambda)$). Then, by the preceding lemma, by Lemma 1, and by formulas (2a) and (2b), $|B(\lambda)| \sim |A(\lambda)|$ as $|\lambda| \to \infty$. The present corollary now follows from the preceding lemma and from Corollary 6 along the same lines of proof as Corollary 9. Q.E.D.

Having accomplished all the necessary asymptotic preliminaries, we are now able to state and prove our main theorem.

→ 12 THEOREM. *Let τ denote the formal differential operator*

$$-\left(\frac{d}{dt}\right)^2 + q(t), \qquad 0 \leqq t < \infty.$$

Suppose that $q \in C^\infty[0, \infty)$, and that

$$\int_0^\infty (1 + t^2)|q(t)|\, dt < \infty.$$

Let T be the operator in $L_2(0, \infty)$ defined by the restriction of $T_1(\tau)$ to the subspace of functions in $\mathfrak{D}(T_1(\tau))$ satisfying a non-trivial boundary condition $A(f) = 0$ at zero.

Let $\sigma_1(t, \mu)$ be the function of Corollary 2. For each λ, let $\mu(\lambda)$ be that square root of λ not lying on the negative real axis. Put $A^+(\lambda) = A(\sigma_1(\cdot, \mu(\lambda)))$ and $A^-(\lambda) = A(\sigma_1(\cdot, -\mu(\lambda)))$ for $0 \leqq \lambda < \infty$. Suppose that $A^+(\lambda) \neq 0$ and $A^-(\lambda) \neq 0$ for all $0 \leqq \lambda < \infty$.

Then T is a spectral operator, whose spectrum is the union of the set $\{\lambda \mid 0 \leqq \lambda < \infty\}$ and of a finite number of additional points not lying on this set; each of these points is a pole of the resolvent of T and corresponds to a finite dimensional projection $E(\{\lambda\}; T)$. The restriction of T to $E((0, \infty); T)$ is a spectral operator of scalar type.

PROOF. The theorem will follow as soon as it is shown that the hypotheses of Theorem XVIII.2.34 are satisfied. In the notation of that theorem $\sigma(T)$ is the union of the sets consisting of the arc $[0, \infty)$ and a finite set of isolated points. The exceptional point ν_0 is $\nu_0 = 0$. The dense subspaces $\mathfrak{X}_0$ and $\mathfrak{X}_0^*$ both may be taken to consist of those functions $f \in C(0, \infty)$ which vanish outside a bounded set. Hypothesis (i) of Theorem XVIII.2.34 is satisfied by virtue of Corollaries 9 and 11. Hypothesis (ii) has been established and is given by Lemma 7 (iv). It therefore only remains to establish hypothesis (iii) of Theorem XVIII.2.34. Let $C_0[0, \infty)$ denote the set of functions in $C[0, \infty)$ vanishing outside a bounded set. For convenience, and using (iv) of Lemma 7, write

$$\begin{aligned} C(f, g, \lambda) &= B^+(f, g, \lambda) - B^-(f, g, \lambda) \\ &= -(2i\mu(\lambda)A^+(\lambda)A^-(\lambda))^{-1} \\ &\qquad \int_0^\infty \int_0^\infty \{A^-(\lambda)\sigma_1(s, \mu(\lambda)) - A^+(\lambda)\sigma_1(s, -\mu(\lambda))\} \\ &\qquad \{A^-(\lambda)\sigma_1(t, \mu(\lambda)) - A^+(\lambda)\sigma_1(t, -\mu(\lambda))\} f(t)\overline{g(s)}\, dt\, ds, \\ &\qquad\qquad f, g \in C_0[0, \infty). \end{aligned} \tag{40}$$

We have only to show that there exists a $K < \infty$ such that

$$\int_0^\infty |C(f, g, \lambda)|\, d\lambda \leqq K |f|_2 |g|_2, \qquad f, g \in C_0[0, \infty). \tag{41}$$

Put

$$(\Phi^+ f)(\mu) = \int_0^\infty \sigma_1(s, \mu) f(s)\, ds,$$

(42)

$$(\Phi^- f)(\mu) = \int_0^\infty \sigma_1(s, -\mu) f(s)\, ds.$$

By (40), Lemma 7 (ii), and Hölder's inequality, (41) will follow if we can find a finite constant K_1 such that

(42a) $$\left\{\int_0^\infty |(\Phi^+ f)(\mu)|^2\, d\mu\right\}^{1/2} \leqq K_1 \left\{\int_0^\infty |f(s)|^2\, ds\right\}^{1/2}, \qquad f \in C_0[0, \infty)$$

and

(42b) $$\left\{\int_0^\infty |(\Phi^- f)(\mu)|^2\, d\mu\right\}^{1/2} \leqq K_1 \left\{\int_0^\infty |f(s)|^2\, ds\right\}^{1/2}, \qquad f \in C_0[0, \infty).$$

We shall prove the first of these inequalities in detail; the second may be treated in the same way. Let

$$(\Phi f)(\mu) = \int_0^\infty e^{i\mu s} f(s)\, ds.$$

Then, by Plancherel's theorem (XV.11.3),

(43) $$\left\{\int_0^\infty |\Phi f(\mu)|^2\, d\mu\right\}^{1/2} \leqq (2\pi)^{1/2} |f|_2 .$$

By Corollary 2 there exists a finite constant K_2 such that

(44) $$|(\Phi f)(\mu) - (\Phi^+ f)(\mu)| \leqq K_2 (1 + |\mu|)^{-1} \int_0^\infty \left\{\int_t^\infty (1+s)|q(s)|\, ds\right\} f(t)\, dt.$$

Now

$$\begin{aligned} L &= \int_0^\infty \left\{\int_t^\infty (1+s)|q(s)|\, ds\right\}^2 dt \\ &\leqq Q \int_0^\infty \int_t^\infty (1+s)|q(s)|\, ds\, dt \\ &= Q \int_0^\infty s(1+s)|q(s)|\, ds \leqq Q \int_0^\infty (1+s)^2 |q(s)|\, ds < \infty, \end{aligned}$$

where

$$Q = \int_0^\infty (1+t)|q(t)|\, dt.$$

By (44) and Hölder's inequality we have

$$(45)\quad \left\{\int_0^\infty |(\Phi f)(\mu) - (\Phi^+ f)(\mu)|^2\, d\mu\right\}^{1/2} \leqq K_2 \left\{\int_0^\infty \frac{d\mu}{(1+\mu)^2}\right\} L^{1/2} |f|_2 .$$

The inequality (42a) follows immediately from (43) and (45). Since the inequality (42b) may be deduced in a precisely similar way, the proof of our theorem is now complete. Q.E.D.

We conclude the present section with an additional remark concerning the spectrum of the operator T.

13 Lemma. *Let the hypotheses of Corollary 2 be satisfied. Then the equation $\tau\sigma = 0$ has two solutions σ_1 and σ_2 satisfying the asymptotic relationships*

$$\sigma_1(t) \sim 1, \qquad \sigma_2(t) \sim t \qquad \text{as } t \to \infty.$$

Proof. We saw in Corollary 2 that $\sigma_1(t) = \sigma_1(t, 0)$ satisfies the first of these asymptotic relationships. Let a be so large that $\sigma_1(t) \neq 0$ for $a \leqq t < \infty$. Then put

$$\hat{\sigma}_2(t) = \sigma_1(t) \int_a^t (\sigma_1(s))^{-2}\, ds.$$

It is readily verified that $\hat{\sigma}_2(t)$ satisfies $\tau\sigma_2 = 0$, and also satisfies the second of the above asymptotic relationships. Hence, if we let $\sigma_2(t)$ be the unique solution $\tau\sigma_2 = 0$ such that $\sigma_2(t) = \hat{\sigma}_2(t)$ for $a \leqq t < \infty$, our lemma is proved. Q.E.D.

14 Corollary. *With the hypotheses and in the notation of Theorem 12 every point in $\{0 \leqq \lambda < \infty\}$ belongs to the continuous spectrum of T, and satisfies $E(\{\lambda\}; T) = 0$.*

Proof. For $\lambda \neq 0$, this follows from Corollary XVIII.2.35. Since by the previous lemma, $\lambda = 0$ is not in the point spectrum of T, Corollary XVIII.2.35 also covers the case $\lambda = 0$. Q.E.D.

2. Friedrichs' Method of Similar Operators

In the present section we shall begin our discussion of an elegant method, due to K. O. Friedrichs, which makes it possible to show, in a variety of cases, that an operator is spectral, and even much more. The basic idea of Friedrichs' method may be expressed heuristically as follows. Let $\mathfrak{X}$ be a B-space, and let T be a linear operator in $\mathfrak{X}$; let K be a second linear operator in $\mathfrak{X}$ which is, in a sense to be made precise below, very small relative to T. Following Friedrichs, we may then surmise that $T+K$ and T are similar operators, that is, that there exists a bounded linear operator U with bounded inverse, such that $T+K = U^{-1}TU$. To verify this conjecture in any particular case, we must solve the *linear* operator equation $U(T+K) = TU$ for the "unknown" operator U, and then show that U has a bounded inverse. A precise, though abstract, description of a general situation in which this idea may be carried through is given in the following theorem.

→ 1 THEOREM. *Let $\mathfrak{X}$ be a B-space, and let $T \in B(\mathfrak{X})$. Let $\mathfrak{A}$ be an "auxiliary" B-space, with the norm $\|A\|$, $A \in \mathfrak{A}$. Let M_1 and M_2 be real numbers greater than zero. Suppose that:*

(a) *a continuous linear mapping $\varphi: \mathfrak{A} \to B(\mathfrak{X})$ of norm at most M_1 is given;*

(b) *a continuous linear mapping $\Gamma: \mathfrak{A} \to B(\mathfrak{X})$, of norm at most M_1, such that*

$$T\Gamma(A) - \Gamma(A)T = \varphi(A), \qquad A \in \mathfrak{A},$$

is defined;

(c) *a continuous bilinear mapping $\psi(A, A_1)$ of $\mathfrak{A} \times \mathfrak{A}$ into $\mathfrak{A}$ is defined and satisfies*

(i) $$\varphi(\psi(A, A_1)) = \Gamma(A)\varphi(A_1), \qquad A, A_1 \in \mathfrak{A};$$

(ii) $$\|\psi(A, A_1)\| \leq M_2 \|A\| \|A_1\|, \qquad A, A_1 \in \mathfrak{A}.$$

Then for each $A_1 \in \mathfrak{A}$ such that $\|A_1\| < (M_1 + M_2)^{-1}$ the operators $T + \varphi(A_1)$ and T are similar, that is, there exists an element $U \in B(\mathfrak{X})$ with a bounded inverse such that $T + \varphi(A_1) = U^{-1}TU$.

PROOF. We shall search for an operator U satisfying the equation

$$U(T + \varphi(A_1)) = TU, \tag{1}$$

assuming that U has the form $U = I + \Gamma(B)$, with $B \in \mathfrak{A}$. Taking U to be of this form, we see that equation (1) is equivalent to the equation

$$(I + \Gamma(B))(T + \varphi(A_1)) = T(I + \Gamma(B)), \tag{2}$$

that is, to

$$\Gamma(B)T - T\Gamma(B) = -\Gamma(B)\varphi(A_1) - \varphi(A_1), \tag{3}$$

or, using hypothesis (b), to

$$\varphi(B) - \Gamma(B)\varphi(A_1) = \varphi(A_1). \tag{4}$$

Using hypothesis (c), we may write this last equation as

$$\varphi(B - \psi(B, A_1)) = \varphi(A_1). \tag{5}$$

Now, by hypothesis, the map $B \to \psi(B, A_1)$ of $\mathfrak{A} \to \mathfrak{A}$ has norm at most $M_2(M_1 + M_2)^{-1}$. Thus, by Lemma VII.3.4, the equation

$$B - \psi(B, A_1) = A_1 \tag{6}$$

has a solution $B \in \mathfrak{A}$, whose norm satisfies

$$\|B\| < (M_1 + M_2)^{-1}(1 - M_2(M_1 + M_2)^{-1})^{-1} = M_1^{-1}.$$

Thus $|\Gamma(B)| < M_1^{-1}M_1 = 1$, so that, using Lemma VII.3.4 again, it follows that $U = I + \Gamma(B)$ has a bounded inverse. As we have seen, equation (6) implies equation (1), so that $T + \varphi(A_1) = U^{-1}TU$, and our theorem is proved. Q.E.D.

Since the notion of similarity of operators arises naturally in connection with Theorem 1, we define this notion formally.

2 DEFINITION. Two operators S and T in a B-space $\mathfrak{Y}$ are called *similar* if there exists a bounded operator U in $\mathfrak{Y}$ with a bounded inverse such that $S = U^{-1}TU$.

Note that Definition 2 applies to unbounded operators as well as to bounded operators.

3 COROLLARY. *Suppose, in addition to the hypotheses* (a), (b), *and* (c) *of Theorem* 1, *that*

(d) *the operator $\Gamma(A)$ is quasi-nilpotent for each $A \in \mathfrak{A}$;*

(e) *$A_1 \in \mathfrak{A}$, and the transformation $A \to \psi(A, A_1)$ of $\mathfrak{A}$ into itself is quasi-nilpotent.*

Then the operators T and $T + \varphi(A_1)$ are similar.

PROOF. Using hypothesis (e) and Lemma VII.3.4, we find that equation (6) above has a solution B. Similarly, using hypothesis (d), we find that $U = I + \Gamma(B)$ has a bounded inverse. Since equation (6) implies equation (1), we have $T + \varphi(A) = U^{-1}TU$ just as in the proof of Theorem 1, and our corollary is proved. Q.E.D.

It is sometimes convenient to use the following slight variant of Theorem 1 and Corollary 3.

4 COROLLARY. *Let $\mathfrak{X}$ be a B-space, and let $T \in B(\mathfrak{X})$. Let $\mathfrak{A}$ be an "auxiliary" B-space, with the norm $\|A\|$, $A \in \mathfrak{A}$. Let M_1 and M_2 be positive real numbers. Suppose that*

(a) *a continuous linear mapping $\varphi: \mathfrak{A} \to B(\mathfrak{X})$ of norm at most M_1 is given;*

(b) *a continuous linear mapping $\Gamma: \mathfrak{A} \to B(\mathfrak{X})$, of norm at most M_1, such that*

$$T\Gamma(A) - \Gamma(A)T = \varphi(A), \qquad A \in \mathfrak{A}$$

is defined;

(c) *a continuous linear mapping $\psi(A, A_1)$ of $\mathfrak{A} \times \mathfrak{A}$ into $\mathfrak{A}$ is defined, which satisfies*

(i) $$\varphi(\psi(A, A_1)) = \varphi(A)\Gamma(A_1), \qquad A, A_1 \in \mathfrak{A};$$

(ii) $$\|\psi(A, A_1)\| \leqq M_2 \|A\| \|A_1\|, \quad A, A_1 \in \mathfrak{A}.$$

Then for each $A \in \mathfrak{A}$ such that $\|A\| < (M_1 + M_2)^{-1}$, the operators $T + \varphi(A)$ and T are similar.

If in addition to (a), (b), *and* (c) *we suppose that*

(d) *the operator $\Gamma(A)$ is quasi-nilpotent for each $A \in \mathfrak{A}$;*

(e) *$A \in \mathfrak{A}$, and the transformation $A_1 \to \psi(A, A_1)$ of $\mathfrak{A}$ into itself is quasi-nilpotent,*

then we may conclude that T and $T + \varphi(A)$ are similar.

The proof of Corollary 4 is entirely parallel to the proofs of Theorem 1 and Corollary 3, and is left to the reader.

The remainder of the present section is devoted to the application of the preceding theorems and corollaries (and of generalizations of them to unbounded operators; cf. Theorem 8 and Corollary 9 below) in a variety of analytic contexts. Our main task in the development of these applications will be to establish the specific inequalities which correspond, in each concrete situation to be studied, to hypotheses (a) and (c) of Theorem 1. Our overall plan will be as follows. First we shall prove an inequality for integral operators (Lemma 5 below) which is elementary in the sense that it relates only to the norms of the integral kernels involved. We then use this inequality to apply Theorem 1 in an illustrative but somewhat artificial setting, essentially to multiplication operators whose spectral measures are absolutely continuous with respect to two-dimensional Lebesgue measure. A first application (Theorem 6) is made in this way, and immediately following upon this we develop a similar but considerably generalized result (Theorem 7) whose hypotheses accord, in a general way, with the facts described by the spectral representation theorem (XII.3.16) (see also Definition XII.3.15). In working out this application, we note the occurrence of a type of singular integral, which, however, in the two-dimensional case is only "mildly" singular and may be treated by the elementary means which Lemma 5 provides. Next we generalize Theorem 1 and its concrete application Theorem 7 to unbounded operators, obtaining Theorems 8 and 10. Theorem 10 completes the elementary, illustrative part of the present section.

We then begin the study, more significant for ultimate application, of the implications of Theorems 1 and 8 for self adjoint operators whose spectral measures are absolutely continuous with respect to one-dimensional Lebesgue measure. Here again singular integrals play a role. As compared to the singular integrals treated in the first set of applications, these singular integrals are more seriously divergent. Their treatment consequently requires the derivation of certain non-elementary inequalities concerning Hölder-continuous functions. Lemma 12 begins the technical preparation necessary for the desired application, which is finally attained, following a series of preliminary lemmas, in Theorem 21.

The following inequality will provide a technical basis for a first group of applications of Theorem 1. Aside from the slight complications occasioned by a number of measure theoretic technicalities, its proof amounts merely to an application of the Hölder inequality and the Riesz convexity theorem.

5 LEMMA. *Let $\mathfrak{X}$ be a B-space, and let (S, Σ, μ) be a σ-finite measure space. Let $1 \leqq p \leqq q \leqq r \leqq \infty$. Let $A(\cdot, \cdot)$ be a strongly $\mu \times \mu$-measurable function with values in the space $B(\mathfrak{X})$ of all bounded operators in $\mathfrak{X}$. Suppose*

that $|A(\cdot, \cdot)|$ is a $\mu \times \mu$-measurable function. For each number ρ such that $1 \leqq \rho \leqq \infty$, define ρ' by $(\rho')^{-1} + \rho^{-1} = 1$. Put

$$(7) \qquad \{A\}_\rho = \left\{ \int_S \left\{ \int_S |A(s, t)|^{\rho'} \mu(dt) \right\}^{\rho/\rho'} \mu(ds) \right\}^{1/\rho}, \qquad 2 \leqq \rho < \infty,$$

and

$$(8) \qquad \{A\}_\rho = \left\{ \int_S \left\{ \int_S |A(s, t)|^{\rho} \mu(ds) \right\}^{\rho'/\rho} \right\}^{1/\rho'}, \qquad 1 < \rho \leqq 2.$$

Moreover, in the extreme cases $\rho = 1$ and $\rho = \infty$, put

$$(9) \qquad \{A\}_1 = \mu\text{-ess}\sup_t \int_S |A(s, t)|\, \mu(ds)$$

and

$$(10) \qquad \{A\}_\infty = \mu\text{-ess}\sup_s \int_S |A(s, t)|\, \mu(dt).$$

Suppose that $\{A\}_p < \infty$ and $\{A\}_r < \infty$. Then, for each $f \in L_q(S, \Sigma, \mu, \mathfrak{X})$, the integral

$$(11) \qquad g(s) = \int_S A(s, t) f(t) \mu(dt)$$

exists for μ-almost all s, and defines a function g belonging to $L_q(S, \Sigma, \mu, \mathfrak{X})$. Moreover, the mapping

$$(12) \qquad f(\cdot) \rightarrow \int_S A(\cdot, t) f(t) \mu(dt)$$

is a bounded linear mapping in $L_q(S, \Sigma\, \mu, \mathfrak{X})$, having norm at most $\max[\{A\}_p, \{A\}_r]$.

PROOF. We first observe that if f is μ-measurable, the function $A(s, t)f(t)$ is a $\mu \times \mu$-measurable function with values in $\mathfrak{X}$. Indeed, by Corollary III.3.8, Theorem III.3.6, Corollary III.6.3, and Theorem III.6.12, and using the σ-finiteness of S, we see that there exists a sequence f_n of μ-integrable μ-simple functions f_n with values in $\mathfrak{X}$, such that $f_n(s) \rightarrow f(s)$ as $n \rightarrow \infty$ for μ-almost all s. Since it follows readily from our hypotheses that $A(s, t)f_n(t)$ is $\mu \times \mu$-measurable, we conclude that $A(s, t)f(t)$ is $\mu \times \mu$-measurable.

Next, let h be a non-negative μ-measurable function defined on S. Then, by Hölder's inequality we have

$$(13)\quad \int_S \left|\int_S |A(s,t)|h(t)\mu(dt)\right|^r \mu(ds)$$

$$\leqq \int_S \left\{\int_S |A(s,t)|^{r'}\mu(dt)\right\}^{r/r'} \left\{\int_S |h(t)|^r \mu(dt)\right\}\mu(ds) = \{A\}_r^r \int_S |h(t)|^r \mu(dr).$$

If $h \in L_r(S, \Sigma, \mu)$ and $2 \leqq r < \infty$, it follows that the integral

$$(14)\qquad (\tilde{A}h)(s) = \int_S |A(s,t)|\, h(t)\mu(dt)$$

exists for μ-almost all s, and that, writing $|f|_p$ for the norm of an element f of $L_p(S, \Sigma, \mu)$, we have $|\tilde{A}h|_r \leqq \{A\}_r |h|_r$. Thus, using Theorem III.2.22(a), we see that if $2 \leqq r < \infty$, and if g is any complex valued function in $L_r(S, \Sigma, \mu)$, then the integral

$$(15)\qquad (\tilde{A}g)(s) = \int_S |A(s,t)|\, g(t)\mu(dt)$$

exists for μ-almost all s. Since $|(\tilde{A}g)(s)| \leqq \int_S |A(s,t)|\, |g(t)|\, \mu(dt)$, we have $|\tilde{A}g|_r \leqq \{A\}_r |g|_r$. The same conclusion is readily obtained in the extreme case $r = \infty$; we leave it to the reader to give the necessary details.

Next let $1 \leqq r \leqq 2$. Let h_1 be a second non-negative, μ-measurable function defined on S. Then we conclude as above that the integral

$$(16)\qquad (\hat{A}h_1)(t) = \int_S |A(s,t)|\, h_1(s)\mu(ds)$$

exists for μ-almost all $t \in S$, and that, if $h_1 \in L_{r'}(S, \Sigma, \mu)$, we have $\hat{A}h_1 \in L_{r'}(S, \Sigma, \mu)$, and $|\hat{A}h_1|_{r'} \leqq \{A\}_r |h_1|_{r'}$. Therefore, if h is a non-negative μ-measurable function in $L_r(S, \Sigma, \mu)$, it follows by Tonelli's theorem (III.11.14) and Hölder's inequality that the integral (14) exists for μ-almost all s in the set $\{s \in S \mid h_1(s) > 0\}$, and that

$$(17)\qquad \int_S (\tilde{A}h)(s)h_1(s)\mu(ds) \leqq \{A\}_r |h|_r |h_1|_{r'}.$$

Since we may find a non-negative $h_1 \in L_r(S, \Sigma, \mu)$ which is positive everywhere, it follows that the integral (14) exists for μ-almost all $s \in S$.

If $r = 1$, we may take $h_1(s) \equiv 1$ to find that $|\tilde{A}h|_1 \leqq \{A\}_1 |h|_1$. The corresponding conclusion in the case $1 < r \leqq 2$ may be obtained as follows. If in (17) we take h_1 to be a function which is identically 1 on a certain set of finite μ-measure, and identically zero outside this set, we see that the function $\tilde{A}h$ is μ-integrable over each set $e \in \Sigma$ such that $\mu(e) < \infty$. By (17) and Theorem IV.8.1, there exists a function $\varphi \in L_r(S, \Sigma, \mu)$, of norm at most $\{A\}_r |h|_r$, such that

$$(18) \qquad \int_S (\tilde{A}h)(s)h_1(s)\mu(ds) = \int_S \varphi(s)h_1(s)\mu(ds), \qquad h_1 \in L_{r'}(S, \Sigma, \mu),\ h_1 \geqq 0.$$

But then

$$(19) \qquad \int_e ((\tilde{A}h)(s) - \varphi(s))\mu(ds) = 0$$

for each set $e \in \Sigma$ of finite μ-measure, so that by Theorem III.2.20 and the σ-finiteness of S, we find that $(\tilde{A}h)(s) = \varphi(s)$, μ-almost everywhere. This shows that $\tilde{A}h \in L_r(S, \Sigma, \mu)$, and that $|\tilde{A}h|_r \leqq \{A\}_r |h|_r$. If we now use Theorem III.2.22 and Theorem III.2.20(a) (cf. Lemma III.2.15), we see that, for all $1 \leqq r \leqq \infty$, and for every complex valued function $g \in L_r(S, \Sigma, \mu)$, the integral (15) exists for μ-almost all s, and $|\tilde{A}g|_r \leqq \{A\}_r |g|_r$.

A corresponding conclusion holds for $g \in L_p(S, \Sigma, \mu)$. Hence, using the Riesz convexity theorem (VI.10.11) and putting $M = \max[\{A\}_p, \{A\}_r]$, we find that there exists a bounded linear mapping $\tilde{A}$ of $L_q(S, \Sigma, \mu)$ into itself, of norm at most M, such that formula (15) holds for every μ-simple, μ-integrable function $\hat{g}$. Let $g \in L_q(S, \Sigma, \mu)$, and suppose that g vanishes outside a set e of finite μ-measure and is bounded. Then, by Corollaries III.3.8 and III.6.3, and by Theorems III.3.6 and III.6.12, g is the limit almost everywhere of a sequence $\{g_n\}$ of μ-integrable, μ-simple functions. Making an elementary modification of the functions of the sequence $\{g_n\}$, we may suppose that all of these functions vanish outside e, and that the sequence $\{g_n\}$ is uniformly bounded. Then, by the Lebesgue dominated theorem we have $\lim_{n\to\infty} |g_n - g|_q = 0$, so that $\lim_{n\to\infty} |\tilde{A}g_n - \tilde{A}g|_q = 0$. By what we have proved already, $\int_e |A(s, t)|\, \mu(dt) < \infty$ for μ-almost all s. Hence, using the Lebesgue dominated convergence theorem once more, we obtain

$$(20) \qquad \lim_{n\to\infty} \int_S |A(s, t)|\, g_n(t)\mu(dt) = \int_S |A(s, t)|\, g(t)\mu(dt)$$

for μ-almost all s. It therefore follows, using Corollaries III.3.8 and III.6.3, and Theorems III.3.6 and III.6.12, that

$$(\tilde{A}g)(s) = \int_S |A(s, t)|\, g(t)\mu(dt)$$

for μ-almost all s, so that (15) holds for each $g \in L_q(S, \Sigma, \mu)$ which is bounded and vanishes outside a set of finite μ-measure. Next let $h \in L_q(S, \Sigma, \mu)$ be non-negative. Then h is the limit almost everywhere of a monotone increasing sequence of non-negative functions h_n, each of which is bounded and vanishes outside a set of finite μ-measure. By the Lebesgue dominated convergence theorem we have $\lim_{n\to\infty} |h_n - h|_q = 0$, so that $\lim_{n\to\infty} |\tilde{A}h_n - \tilde{A}h|_q = 0$. By Fatou's lemma (III.6.17)

$$\lim_{n\to\infty} \int_S |A(s, t)|\, h_n(t)\mu(dt) = \int_S |A(s, t)|\, h(t)\mu(dt) \tag{21}$$

for each $s \in S$, where the integral on the right of (14) exists μ-almost everywhere. Arguing as above, we conclude that (14) holds for each non-negative $h \in L_q(S, \Sigma, \mu)$. Moreover, as we have already noted, $|\tilde{A}h|_q \leqq M\,|h|_q$.

Therefore, using Theorem III.2.22 and Theorem III.2.22(a), (cf. Lemma III.2.15), we see that the integral (11) exists μ-almost everywhere for each $f \in L_q(S, \Sigma, \mu, \mathfrak{X})$, and that the mapping defined by formula (12) has norm at most M. This proves the present lemma. Q.E.D.

To illustrate the use of Theorem 1 in a particular case, we may proceed as follows. Let D be a bounded domain in the complex plane, and let $L_2(D)$ denote the Hilbert space of all complex valued Lebesgue measurable functions f defined in D and satisfying

$$\int_D |f(x, y)|^2\, dx\, dy = \int_D |f(z)|^2\, dx\, dy = |f|^2 < \infty. \tag{22}$$

Take the operator T of Theorem 1 to be defined by the formula

$$(Tf)(x, y) = (x + iy)f(x, y), \quad \text{that is,} \quad (Tf)(z) = zf(z). \tag{23}$$

The reader will verify without difficulty that the spectrum of T is the closure of D. Take the space $\mathfrak{A}$ of Theorem 1 to be the space of all bounded Lebesgue measurable functions on $D \times D$, that is, the space of all Lebesgue measurable functions $A(x, y; x', y') = A(z, z')$ with

$$\|A\| = \sup_{z,\, z' \in D} |A(z, z')|. \tag{24}$$

Define the transformation φ of Theorem 1 by the formula

$$[\varphi(A)f](z) = \int_D A(z, z')f(z')\,dx'\,dy', \qquad f \in L_2(D). \tag{25}$$

Define the transformation Γ of Theorem 1 by the formula

$$[\Gamma(A)f](z) = \int_D \frac{A(z, z')}{z - z'} f(z')\,dx'\,dy', \qquad f \in L_2(D). \tag{26}$$

Even though the kernel of the integral operator $\Gamma(A)$ defined by (26) is unbounded, the operator $\Gamma(A)$ is a bounded operator in $L_2(D)$. This assertion follows at once from Lemma 5 and from the following two inequalities:

$$\int_D \frac{|A(z, z')|}{|z - z'|}\,dx'\,dy' \leqq \|A\| \int_D \frac{1}{|z - z'|}\,dx'\,dy' \leqq 2\pi \|A\| \operatorname{diam}(D) \tag{27}$$

$$\int_D \frac{|A(z, z')|}{|z - z'|}\,dx\,dy \leqq 2\pi \|A\| \operatorname{diam}(D). \tag{28}$$

Note that by (26) we have

$$\begin{aligned} \{T(\Gamma A)f\}(z) - \{(\Gamma A)Tf\}(z) &= \int_D \frac{(zA(z, z') - A(z, z')z')}{z - z'} f(z')\,dx'\,dy' \\ &= \int_D A(z, z')f(z')\,dx'\,dy' \\ &= [\varphi(A)f](z), \qquad f \in L_2(D), \end{aligned} \tag{29}$$

so that hypotheses (a) and (b) of Theorem 1 are satisfied.

To verify hypothesis (c) of Theorem 1, we put

$$[\psi(A, A_1)](z, z') = \int_D \frac{A(z, z_1)A_1(z_1, z')}{z_1 - z'}\,dx_1\,dy_1, \tag{30}$$

and note that since, by Fubini's theorem,

$$\begin{aligned} [\varphi(\psi(A, A_1))f](z) &= \int_D\int_D \frac{A(z, z_1)A_1(z_1, z')}{z - z_1} f(z')\,dx_1\,dy_1\,dx'\,dy' \\ &= \int_D \frac{A(z, z_1)}{z - z_1} \left\{\int_D A(z_1, z')f(z')\,dx'\,dy'\right\} dx_1\,dy_1 \\ &= [\Gamma(A)\varphi(A)f](z), \end{aligned} \tag{31}$$

we have $\varphi(\psi(A, A_1)) = \Gamma(A)\varphi(A_1)$ for $A, A_1 \in \mathfrak{A}$. Moreover, we have

$$\left|\int_D \frac{A(z, z_1)A_1(z_1, z')}{z - z_1}\, dx_1\, dy_1\right| \leqq \|A\|\,\|A_1\| \int_D \frac{1}{|z - z_1|}\, dx_1\, dy_1 \tag{32}$$

$$\leqq 2\pi\, \mathrm{diam}(D)\, \|A\|\,\|A_1\|,$$

so that hypothesis (c) is verified. We may therefore apply Theorem 1, to find that if $A \in \mathfrak{A}$, and if $\|A\|$ is sufficiently small, then the operator T and the operator $T + \varphi(A)$ are similar. Since T is obviously a spectral operator of scalar type, $T + \varphi(A)$ must also be a spectral operator of scalar type.

We now wish to state a theorem recording the result of the preceding analysis for future reference. Before doing so, however, let us note that our argument may be generalized in two obvious ways.

(a) Instead of considering, as above, the space $L_2(D)$ of complex valued Lebesgue measurable functions defined in D and satisfying (7), we may let $\mathfrak{X}$ be an arbitrary complex B-space, and can consider the space $L_2(D, \mathfrak{X})$ of $\mathfrak{X}$-valued Borel-Lebesgue measurable functions defined in D.

(b) We may replace the subscript 2 by any p such that $1 \leqq p < \infty$.

That is, instead of considering the space $L_2(D)$ as above, we may consider the space $L_p(D, \mathfrak{X})$ of all $\mathfrak{X}$-valued Borel-Lebesgue measurable functions defined in D and satisfying

$$\left\{\int_D |f(x, y)|^p\, dx\, dy\right\}^{1/p} = \left\{\int_D |f(z)|^p\, dx\, dy\right\}^{1/p} = |f| < \infty. \tag{33}$$

Since Lemma 5 applies not only to spaces of complex valued functions but also to spaces of $\mathfrak{X}$-valued functions, all the above arguments remain valid with only trivial notational changes. We may therefore state the following theorem.

6 THEOREM. *Let $\mathfrak{X}$ be a complex B-space. Let D be a bounded domain in the complex plane, let $1 \leqq p < \infty$, and let $L_p(D, \mathfrak{X})$ be the space of all $\mathfrak{X}$-valued Lebesgue measurable functions defined in D and satisfying formula* (33). *Let T be the operator in $L_p(D, \mathfrak{X})$ defined by the formula*

$$(Tf)(x, y) = (x + iy)f(x, y), \qquad f \in L_p(D, \mathfrak{X}). \tag{34}$$

Let $A(z, z')$ be a Lebesgue measurable function defined in $D \times D$, with values in the space $B(\mathfrak{X})$ of all bounded operators in $\mathfrak{X}$. Suppose that

$$\|A\| = \sup_{z, z' \in D} |A(z, z')| < \infty, \tag{35}$$

and let $\varphi(A)$ be the integral operator defined by the equation

$$(\varphi(A)f)(z) = \int_D A(z, z')f(z')\, dx'\, dy', \qquad f \in L_p(D, \mathfrak{X}). \tag{36}$$

Then there exists a positive constant $\varepsilon = \varepsilon(p, D)$ depending only on p and D such that if $\|A\| < \varepsilon(p, D)$ it follows that the operator $T + \varphi(A)$ is similar to the operator T and is therefore a spectral operator of scalar type.

The following theorem generalizes Theorem 6 in various useful directions. The form in which its hypotheses are stated is suggested by the spectral representation theorem (XII.3.16) (see also Definition XII.3.15).

→ 7 THEOREM. *Let $\mathfrak{X}$ be a complex B-space. Let D be a bounded domain in the complex plane, let $1 \leqq p < \infty$, and let $L_p(D, \mathfrak{X})$ be the space of all $\mathfrak{X}$-valued Lebesgue measurable functions defined in D and satisfying* (33). *Let $1/p + 1/p' = 1$. Let $e_1, e_2, \ldots,$ be a family of disjoint Borel sets with union D, and let $\mathfrak{X}_1, \mathfrak{X}_2, \ldots,$ be a family of closed subspaces of $\mathfrak{X}$. Let $\hat{L}_p(D, \mathfrak{X})$ be the subspace of $L_p(D, \mathfrak{X})$ consisting of all functions f such that $f(z) \in \mathfrak{X}_j$ for all $z \in e_j$, $1 \leqq j < \infty$. Let T be the operator in $\hat{L}_p(D, \mathfrak{X})$ defined by the formula*

$$(Tf)(x, y) = (x + iy)f(x, y), \qquad f \in \hat{L}_p(D, \mathfrak{X}). \tag{37}$$

Let $A(z, z')$ be a Lebesgue measurable function defined in $D \times D$, with values in the space $B(\mathfrak{X})$ of all bounded operators in $\mathfrak{X}$. Suppose that $A(z, z')x \in \mathfrak{X}_j$ for all $x \in \mathfrak{X}$ and $z \in e_j$, $1 \leqq j < \infty$. Let $c > 4$ be a constant such that $c \geqq p$ and $c \geqq p'$. Suppose that $|A(\cdot, \cdot)|$ is a Lebesgue measurable function, and that

$$\|A\| = \left\{\int_D \int_D |A(z, z')|^c\, dx\, dy\, dx'\, dy'\right\}^{1/c} < \infty. \tag{38}$$

Let $\hat{\varphi}(A)$ be the integral operator $\hat{L}_p(D, \mathfrak{X})$ defined by

$$(\hat{\varphi}(A)f)(z) = \int_D A(z, z')f(z')\, dx'\, dy', \qquad f \in \hat{L}_p(D, \mathfrak{X}). \tag{39}$$

Then there exists a positive constant $\varepsilon = \varepsilon(p, D, c)$ *depending only on* p, D, *and* c, *such that the inequality* $\|A\| < \varepsilon$ *implies that the operator* $T + \hat{\varphi}(A)$ *is similar to the operator* T *and is therefore a spectral operator of scalar type.*

REMARK. It follows from Theorem XII.3.16 that if T_0 is any bounded normal operator in Hilbert space whose spectral resolution $E(e)$ satisfies $E(e_0) = 0$ for each Borel subset e_0 of the plane such that $\mu(e_0) = 0$, μ denoting the Borel-Lebesgue measure in the plane, then T_0 is similar to an operator T in a space $\hat{L}_2(D, \mathfrak{X})$ of the sort described by Theorem 7. Thus, Theorem 7 can be applied to a fairly general class of bounded normal operators T_0, provided only that the perturbing operator $\varphi(A)$ is sufficiently "smooth" and "small" relative to the normal operator T_0.

PROOF OF THEOREM 7. *Let* $\mathfrak{A}$ be the space of all Lebesgue measurable functions A defined in $D \times D$, with values in the space $B(\mathfrak{X})$ of all bounded operators in $\mathfrak{X}$, such that (38) holds. Let $(c')^{-1} + (c)^{-1} = 1$. Since $c \geqq p$ and $c \geqq p'$, we have $c \geqq 2$ so that $c \geqq c'$, and thus the boundedness of the domain D implies via Hölder's inequality that there exists a finite constant M such that

$$\left\{\int_D \left\{\int_D |A(z, z')|^{c'} \, dx \, dy\right\}^{c/c'} dx' \, dy'\right\}^{1/c} < M \, \|A\| \tag{40}$$

and

$$\left\{\int_D \left\{\int_D |A(z, z')|^{c'} \, dx' \, dy'\right\}^{c/c'} dx \, dy\right\}^{1/c} < M \, \|A\|. \tag{41}$$

Thus, by Lemma 5, the integral operator $\hat{\varphi}(A)$ defined by (39) maps $L_p(D, \mathfrak{X})$ into itself, and has norm at most $M \, \|A\|$. Since $A(z, z')x \in \mathfrak{X}_i$, for all $x \in \mathfrak{X}_i$ and $z \in e_i$, it follows that $\hat{\varphi}(A)$ maps the subspace $\hat{L}_p(D, \mathfrak{X})$ of $L_p(D, \mathfrak{X})$ into itself.

If $A \in \mathfrak{A}$, define the transformation $\Gamma(A)$ by

$$(\Gamma(A)f)(z) = \int_D \frac{A(z, z')}{z - z'} f(z') \, dx' \, dy', \qquad f \in L_p(D, \mathfrak{X}). \tag{42}$$

By Hölder's inequality we have

$$\int_D \left|\frac{A(z, z')}{z - z'}\right|^{c'} dx' \, dy' \leqq \left\{\int_D |A(z, z')|^{c'c/c'} \, dx' \, dy'\right\}^{c'/c} \times \left\{\int_D \left|\frac{1}{z - z'}\right|^{c'c/(c - c')} dx' \, dy'\right\}^{(c - c')/c}. \tag{43}$$

Since $c > 4$, it follows that the quantity $e = c'c/(c - c') = ((c')^{-1} - c^{-1})^{-1}$ is less than $(\frac{3}{4} - \frac{1}{4})^{-1} = 2$. Thus the quantity

$$N = \sup_{z \in D} \left\{ \int_D \frac{1}{|z - z')|^e} \, dx' \, dy' \right\}^{(c-c')/c} \tag{44}$$

is finite. From (43) we have

$$\left\{ \int_D \left\{ \int_D \left| \frac{A(z, z')}{z - z'} \right|^{c'} dx' \, dy' \right\}^{c/c'} dx \, dy \right\}^{1/c} \leqq N^{1/c'} \|A\|. \tag{45}$$

We may prove in just the same way that

$$\left\{ \int_D \left\{ \int_D \left| \frac{A(z, z')}{z - z'} \right|^{c'} dx \, dy \right\}^{c/c'} dx' \, dy' \right\}^{1/c'} \leqq N^{1/c'} \|A\|. \tag{46}$$

Thus, by Lemma 5, the integral operator $\Gamma(A)$ defined by (42) maps $L_p(D, \mathfrak{X})$ into itself, and has norm at most $N^{1/c'} \|A\|$. Since $A(z, z')x \in \mathfrak{X}$ for all $x \in \mathfrak{X}$ and all $z \in e_i$, it follows that $\hat{\varphi}(A)$ maps the subspace $\hat{L}_p(D, \mathfrak{X})$ of $L_p(D, \mathfrak{X})$ into itself. Let $\hat{\Gamma}(A)$ denote the restriction of $\Gamma(A)$ to the subspace $\hat{L}_p(D, \mathfrak{X})$ of $L_p(D, \mathfrak{X})$.

By (42) we have

$$\begin{aligned} (T\hat{\Gamma}(A)f - \hat{\Gamma}(A)Tf)(z) &= \int_D \frac{(zA(z, z') - A(z, z')z')}{z - z'} f(z') \, dx' \, dy' \\ &= \int_D A(z, z') f(z') \, dx' \, dy' \\ &= [\hat{\varphi}(A)f](z), \qquad f \in \hat{L}_p(D, \mathfrak{X}). \end{aligned} \tag{47}$$

Next, put

$$[\psi(A, A_1)](z, z') = \int_D \frac{A(z, z_1)A_1(z_1, z')}{z_1 - z'} \, dx_1 \, dy_1, \qquad A, A_1 \in \mathfrak{A}. \tag{48}$$

Since $A(z, z_1)x \in \mathfrak{X}_i$ for each $x \in \mathfrak{X}$ and each $z \in \mathfrak{X}_i$, it is plain that $[(\psi(A, A_1))(z, z')]x \in \mathfrak{X}_i$ for each $x \in \mathfrak{X}$ and each $z \in \mathfrak{X}_i$. By Hölder's inequality and the definition (44) of the constant N we have (cf. (43))

$$\int_D \left| \frac{A_1(z_1, z')}{z_1 - z'} \right|^{c'} dx_1 \, dy_1 \leqq N \left\{ \int_D |A_1(z_1, z')|^c \, dx_1 \, dy_1 \right\}^{c'/c}. \tag{49}$$

Thus, by (48), and using Hölder's inequality once more, we have

$$(50)\qquad |\psi(A, A_1)(z, z')| \leqq N^{1/c'}\left\{\int_D |A(z, z_1)|^c\, dx_1\, dy_1\right\}^{1/c} \times \left\{\int_D |A_1(z_1, z')|^c\, dx_1\, dy_1\right\}^{1/c},$$

and therefore

$$(51)\qquad \|\psi(A, A_1)\| \leqq N^{1/c'}\, \|A\|\, \|A_1\|, \qquad A, A_1 \in \mathfrak{A}.$$

We may now apply Theorem 1 to conclude that there exists a positive constant ε depending only on p, D, and c, such that the operators T and $T + \hat{\varphi}(A)$ are similar whenever $\|A\| < \varepsilon$. Q.E.D.

Theorem 1 and its corollaries are readily generalized to apply to unbounded operators.

8 Theorem. *Let $\mathfrak{X}$ be a B-space, and let T be a closed, densely defined operator in $\mathfrak{X}$, with domain $\mathfrak{D}(T)$. Let $\mathfrak{A}$ be an "auxiliary" B-space, with the norm $\|A\|$, $A \in \mathfrak{A}$. Let M_1 and M_2 be finite positive real numbers. Suppose that*

(a) *for each $A \in \mathfrak{A}$, there is defined a linear mapping $\varphi(A)$ such that $\mathfrak{D}(\varphi(A)) \supseteq \mathfrak{D}(T)$;*

(b) *a continuous linear mapping $\Gamma: \mathfrak{A} \to B(\mathfrak{X})$ of norm at most M_1 such that $\Gamma(A)\mathfrak{D}(T) \subseteq \mathfrak{D}(T)$ and*

$$(T\Gamma(A) - \Gamma(A)T)x = \varphi(A)x, \qquad A \in \mathfrak{A},\ x \in \mathfrak{D}(T)$$

is given;

(c) *a continuous bilinear mapping $\psi(A, A_1)$ of $\mathfrak{A} \times \mathfrak{A}$ into $\mathfrak{A}$ is defined, which satisfies*

$$\text{(i)}\qquad \varphi(\psi(A, A_1)) = \varphi(A_1)\Gamma(A), \qquad A, A_1 \in \mathfrak{A},$$

$$\text{(ii)}\qquad \|\psi(A, A_1)\| \leqq M_2\, \|A\|\, \|A_1\|.$$

Then, for each $A_1 \in \mathfrak{A}$ such that $\|A_1\| < (M_1 + M_2)^{-1}$, the operator $T + \varphi(A_1)$ has a closed restriction which is similar to T.

Proof. By Lemma VII.3.4, the equation

$$(52)\qquad B - \psi(B, A_1) = A_1$$

has a solution $B \in \mathfrak{A}$, whose norm, by this same lemma, satisfies

$$\|B\| < (M_1 + M_2)^{-1}(1 - M_2(M_1 + M_2)^{-1})^{-1} = M_1^{-1}.$$

Thus $\Gamma(B) < M_1^{-1}M_1 = 1$ so that using Lemma VII.3.4 again, it follows that $U = I + \Gamma(B)$ has a bounded inverse.

Let $x \in \mathfrak{D}(T)$. Then, using (52) and hypothesis (c), we have

$$\varphi(B)x - \varphi(A_1)\Gamma(B)x, \tag{53}$$

and therefore, using hypothesis (b), we have

$$T\Gamma(B)x - \Gamma(B)Tx - \varphi(A_1)\Gamma(B)x = 0. \tag{54}$$

This last equation may be written as

$$(T + \varphi(A_1))(I + \Gamma(B))x = (I + \Gamma(B))Tx, \tag{55}$$

or

$$(T + \varphi(A_1))x = (I + \Gamma(B))T(I + \Gamma(B))^{-1}x, \qquad x \in (I+\Gamma(B))\, \mathfrak{D}(T). \tag{56}$$

Put $U = I + \Gamma(B)$. The operator UTU^{-1} then gives the closed restriction of $T + \varphi(A_1)$ whose existence is the conclusion of our theorem. Q.E.D.

9 COROLLARY. *Suppose, in addition to the hypotheses of Theorem 8, that there exists a complex number λ_0 such that $(\lambda_0 I - T)^{-1}$ exists and is everywhere defined and bounded, such that $\varphi(A_1)(\lambda_0 I - T)^{-1}$ is bounded, and such that $(I - \varphi(A_1)(\lambda_0 I - T)^{-1})^{-1}$ exists and is everywhere defined and bounded for some $A_1 \in \mathfrak{A}$ with $\|A_1\| \leqq (M_1 + M_2)^{-1}$. Then the operator $T + \varphi(A_1)$ is closed and is similar to T.*

PROOF. Let

$$B = (\lambda_0 I - T)^{-1}(I - \varphi(A_1)(\lambda_0 I - T)^{-1})^{-1}. \tag{57}$$

Then

$$\begin{aligned}(\lambda_0 I - T - \varphi(A_1))B \\ &= (I - \varphi(A_1)(\lambda_0 I - T)^{-1})(I - \varphi(A_1)(\lambda_0 I - T)^{-1})^{-1} = I.\end{aligned} \tag{58}$$

Moreover, if $x \in \mathfrak{D}(T) = \mathfrak{D}(T + \varphi(A_1))$ we have

$$\begin{aligned} B(\lambda_0 I - T - \varphi(A_1))x &= B(\lambda_0 I - T - \varphi(A_1))(\lambda_0 I - T)^{-1}(\lambda_0 I - T)x \\ &= B(I - \varphi(A_1)(\lambda_0 I - T)^{-1})(\lambda_0 I - T)x \\ &= (\lambda_0 I - T)^{-1}(\lambda_0 I - T)x = x.\end{aligned} \tag{59}$$

Thus $(\lambda_0 I - T - \varphi(A_1))^{-1} = B$ exists, and is bounded and everywhere defined. If T_1 is the closed restriction of $T + \varphi(A_1)$ given by Theorem 8, then since T_1 and T are similar, $(\lambda_0 I - T_1)^{-1}$ exists and is everywhere defined and bounded. Suppose that T_1 is a proper restriction of $T + \varphi(A_1)$, so that there exists an $x \in \mathfrak{D}(T + \varphi(A_1))$ such that $x \notin \mathfrak{D}(T_1)$. Since $\lambda_0 I - T_1$ maps its domain onto all of Hilbert space, there exists some $y \in \mathfrak{D}(T_1)$ such that $(\lambda_0 I - T_1)y = (\lambda_0 I - T - \varphi(A_1))x$. But then $(\lambda_0 I - T - \varphi(A_1))(x - y) = 0$, contradicting the existence of $(\lambda_0 I - T - \varphi(A_1))^{-1}$. This shows that $T + \varphi(A_1) = T_1$, and the proof of the corollary is complete. Q.E.D.

The following theorem, which is closely related to Theorem 7, is an application of Theorem 8.

10 THEOREM. *Let $\mathfrak{Y}$ be a complex B-space. Let D be a subdomain of the complex plane which is not dense in the complex plane. Let $1 \leqq p < \infty$, and let $L_p(D, \mathfrak{Y})$ be the space of all $\mathfrak{Y}$-valued, Borel-Lebesgue measurable functions defined in D and satisfying the condition*

$$(60) \qquad \left\{\int_D |f(x, y)|\, dx\, dy\right\}^{1/p} = \left\{\int_D |f(z)|^p dx\, dy\right\}^{1/p} = |f| < \infty.$$

Let $e_1, e_2, \ldots,$ be a family of disjoint Borel sets with union D, and let $\mathfrak{Y}_1, \mathfrak{Y}_2, \ldots,$ be a family of closed subspaces of $\mathfrak{Y}$. Let $\hat{L}_p(D, \mathfrak{Y})$ be the subspace of $L_p(D, \mathfrak{Y})$ consisting of all functions f such that $f(z) \in \mathfrak{Y}_j$ for all $z \in e_j$, $1 \leqq j < \infty$. Let T be the operator in $\hat{L}_p(D, \mathfrak{Y})$ whose domain $\mathfrak{D}(T)$ is defined by the formula

$$(61) \qquad \mathfrak{D}(T) = \{f \in \hat{L}_p(D, \mathfrak{Y}) \,\Big|\, \left(\int_D |zf(z)|^p\, dx\, dy\right)^{1/p} < \infty\}$$

and which is such that

$$(62) \qquad (Tf)(z) = (x + iy)f(z), \qquad f \in \hat{L}_p(D, \mathfrak{Y}).$$

Let $A(z, z')$ be a Lebesgue measurable function defined in $D \times D$, with values in the space $B(\mathfrak{Y})$ of all bounded operators in $\mathfrak{Y}$. Suppose that $A(z, z')x \in \mathfrak{Y}_j$ for all $x \in \mathfrak{Y}$ and $z \in e_j$, $1 \leqq j < \infty$. Suppose that $|A(\cdot, \cdot)|$ is a Lebesgue measurable function, and that

$$(63) \quad \|A\| = \sup_{z, z' \in D} |A(z, z')| + \sup_{z \in D} \int_D |A(z, z')|\, dx'\, dy' + \sup_{z' \in D} \int_D |A(z, z')|\, dx\, dy$$

is finite. Let $\varphi(A)$ *be the integral operator in* $\hat{L}_p(D, \mathfrak{Y})$ *defined by*

$$(64) \qquad (\varphi(A)f)(z) = \int_D A(z, z')f(z')\, dx'\, dy', \qquad f \in \hat{L}_p(D, \mathfrak{Y}).$$

Then there exists a positive constant $\varepsilon = \varepsilon(p, D, c)$ *depending only on* p, D, *and* c, *such that if* $\|A\| < \varepsilon$ *it follows that the operator* $T + \varphi(A)$ *is similar to the operator* T *and is therefore a spectral operator of scalar type.*

PROOF. We apply Theorem 8 and Corollary 9. Let the B-space $\mathfrak{X}$ of Theorem 8 be $\hat{L}_p(D, \mathfrak{Y})$. Let $\mathfrak{A}$ be the set of all bounded, measurable functions $A(\cdot, \cdot)$ defined in $D \times D$ with values in $B(\mathfrak{Y})$, such that (63) is finite, and such that $A(z, z')v \in \mathfrak{Y}_j$ for each $v \in \mathfrak{Y}$ and $z \in e_j$. Let $\varphi(A)$ be defined by (64) for each $A \in \mathfrak{A}$; note that since $A(z, z')f(z') \in \mathfrak{Y}_i$ for all $z \in e_i$, $\varphi(A)f \in \hat{L}_p(D, \mathfrak{Y})$. By Lemma 5, $\varphi(A)$ is a bounded linear transformation of $\hat{L}_p(D, \mathfrak{Y})$ into itself, of norm at most $\|A\|$; thus hypothesis (a) of Theorem 8 is satisfied.

The reader will verify without difficulty that $\mathfrak{A}$ is a B-space with the norm defined by (63). Put

$$(65) \quad (\Gamma(A)f)(z) = \int_D \frac{A(z, z')}{z - z'} f(z')\, dx'\, dy', \qquad A \in \mathfrak{A}, f \in \hat{L}_p(D, \mathfrak{Y}).$$

Since

$$(66) \quad \int_D \left| \frac{A(z, z')}{z - z'} \right| dx'\, dy' \leqq \int_{\substack{z \in D \\ |z - z'| > 1}} |A(z, z')|\, dx'\, dy' + \|A\| \int_{\substack{z \in D \\ |z - z'| \leqq 1}} \frac{1}{|z - z'|}\, dx'\, dy' \leqq (2\pi + 1)\, \|A\|,$$

and

$$(67) \qquad \int_D \left| \frac{A(z, z')}{z - z'} \right| dx\, dy \leqq (2\pi + 1)\, \|A\|,$$

similarly, it follows, by Lemma 5, that the integral (65) exists for almost all $z \in D$. Moreover, using Lemma 5 once more, we see that $\Gamma(A)$ is a mapping of $\hat{L}_p(D, \mathfrak{Y})$ into $L_p(D, \mathfrak{Y})$, of norm at most $(2\pi + 1)\|A\|$. Since $A(z, z')f(z') \in \mathfrak{Y}_j$ for each $z \in e_j$, it is clear that $\Gamma(A)f \in \hat{L}_p(D, \mathfrak{Y})$ for each

$f \in \hat{L}_p(D, \mathfrak{Y})$. Finally, it is plain from (64) and (65) that $T\Gamma(A) - \Gamma(A)T = \varphi(A)$, $A \in \mathfrak{A}$. Thus hypothesis (b) of Theorem 8 is satisfied.

If $A_1, A_2 \in \mathfrak{A}$, put

$$(\psi(A_1, A_2))(z_1, z_2) = \int_D \frac{A_1(z_1, z')A_2(z', z_2)}{z' - z_2} dx'\, dy'. \tag{68}$$

Since

$$\int_D \left| \frac{A_1(z_1, z')A_2(z', z_2)}{z' - z_2} \right| dx'\, dy' \leqq \|A_1\| \int_{\substack{z' \in D \\ |z' - z_2| > 1}} |A_2(z', z_2)|\, dx'\, dy' + \|A_1\|\,\|A_2\| \int_{\substack{z' \in D \\ |z' - z_2| \leqq 1}} \frac{1}{|z' - z_2|} dx'\, dy', \tag{69}$$

the integral (68) is defined, and

$$|\psi(A_1, A_2)(z_1, z_2)| \leqq (2\pi + 1)\|A_1\|\,\|A_2\|. \tag{70}$$

By Tonelli's theorem (III.11.14), we have

$$\begin{aligned} \int_D |\psi(A_1, A_2)(z_1, z_2)|\, dx_1\, dy_1 &\leqq \int_{\substack{z' \in D \\ |z' - z_2| > 1}} \left\{ \int_D |A(z_1, z')|\, dz_1 \right\} |A_2(z', z_2)|\, dx'\, dy' \\ &\quad + \|A_2\| \int_{\substack{z' \in D \\ |z' - z_2| \leqq 1}} \frac{1}{|z' - z_2|} \left\{ \int_D |A(z_1, z')|\, dx_1\, dy_1 \right\} dx'\, dy' \\ &\leqq (2\pi + 1)\|A_1\|\,\|A_2\| \end{aligned} \tag{71}$$

and similarly

$$\int_D |\psi(A_1, A_2)(z_1, z_2)|\, dx_2\, dy_2 \leqq (2\pi + 1)\|A_1\|\,\|A_2\|. \tag{72}$$

Since $A_1(z_1, z')A_2(z', z_2)v \in \mathfrak{Y}_j$ for each $v \in \mathfrak{Y}$ and $z_1 \in e_j$, it follows from (68) that $[\psi(A_1, A_2)(z_1, z_2)]v \in \mathfrak{Y}_j$ for each $v \in \mathfrak{Y}$ and $z_1 \in e_j$. Thus, using (70), (71), and (72), we see that $\psi(A_1, A_2) \in \mathfrak{A}$, and that $|\psi(A_1, A_2)| \leqq 3(2\pi + 1)\|A_1\|\,\|A_2\|$. By Lemma 5, the integral

$$\int_D \left| \frac{A_2(z, z')}{z - z'} \right| |f(z')|\, dx'\, dy' \tag{73}$$

exists almost everywhere for each $f \in \hat{L}_p(D, \mathfrak{Y})$, and, using Lemma 5 once more, the integral

$$\int_D |A_1(z'', z)| \left\{ \int_D \left| \frac{A_2(z, z')}{z - z'} \right| |f(z')| \, dx' \, dy' \right\} dx \, dy \tag{74}$$

exists for almost all $z'' \in D$. Therefore, by Tonelli's theorem (III.11.14), the integral

$$\int_D \int_D |A_1(z'', z)| \frac{|A_2(z, z')|}{|z - z'|} |f(z')| \, dx' \, dy' \, dx \, dy \tag{75}$$

exists for almost all $z'' \in D$. It follows from (65) and (68), and from Fubini's theorem (III.11.9), that $\varphi(A_1)\Gamma(A_2) = \varphi(\psi(A_1, A_2))$. Therefore hypothesis (c) of Theorem 8 is verified.

Since by hypothesis the subdomain D of the complex plane is not dense in the complex plane, there exists a complex number λ_0 whose distance from the domain D exceeds a certain positive real number K. It is plain that $\lambda_0 I - T$ has the everywhere defined bounded inverse $f(s) \to (\lambda_0 - s)^{-1} f(s)$, and that $|(\lambda_0 I - T)^{-1}| \leqq K^{-1}$. We have $|\varphi(A)(\lambda_0 I - T)^{-1}| \leqq K^{-1} \|A\|$; thus, if $K^{-1}\|A\| < 1$, it follows by Lemma VII.3.4 that $(I - \varphi(A)(\lambda_0 I - T)^{-1})^{-1}$ exists and is everywhere defined and bounded. Thus, if $K^{-1}\|A\|$ is sufficiently small, the hypothesis of Corollary 9 is satisfied.

We may now apply Theorem 8 and Corollary 9, and the conclusion of the present theorem follows immediately. Q.E.D.

As the reader must surely suspect, Theorems 1 and 8 can also be applied if we let T be the operator $f(s) \to sf(s)$ in the Hilbert space $L_2(-\infty, +\infty)$ formed with respect to the Borel-Lebesgue measure on the real axis. This, of course, is the case of Theorems 1 and 8 which would be needed to treat self adjoint operators with continuous spectrum, or operators which arise by perturbation from such self adjoint operators. It is therefore this case of Theorems 1 and 8, rather than the case of Theorems 1 and 8 given by Theorem 7 or Theorem 10, which is of the greatest interest. However, since, in contradistinction to the corresponding two-dimensional integral, the one-dimensional integral

$$\int_0^1 \frac{1}{|s - \sigma|} \, d\sigma \tag{1}$$

diverges whenever $0 \leqq s \leqq 1$, application of Theorems 1 and 8 in such cases will necessitate consideration of *singular* rather regular integrals. This circumstance introduces a number of annoying even if non-essential complications into our analysis; in particular, instead of proceeding directly to the desired application of Theorem 1, we must first prepare the ground with a number of definitions and lemmas taken from the theory of singular integrals.

11 Definition. Let γ and β be two real numbers, with $\gamma \geqq 0$ and $1 > \beta > 0$. Let $\mathfrak{X}$ be a B-space. The symbol A is used for an $\mathfrak{X}$-valued function defined on the real axis and we put

$$(2) \qquad \|A\|_{\gamma,\beta} = \sup_{-\infty < s < +\infty} (1+|s|)^{\gamma}|A(s)| + \sup_{\substack{-\infty < s < +\infty \\ h>0}} h^{-\beta}|A(s+h) - A(s)|.$$

12 Lemma. *Let γ and β be real numbers, with $\gamma > 0$ and $1 > \beta > 0$. Let $\mathfrak{X}$ be a B-space. Let A be an $\mathfrak{X}$-valued function defined on the real axis, and suppose that $\|A\|_{\gamma,\beta} < \infty$. Then:*

(a) *The improper integral*

$$(3) \qquad (TA)(s) = \mathscr{P}\int_{-\infty}^{+\infty} \frac{A(\sigma)}{s-\sigma}\, d\sigma = \lim_{\varepsilon \to 0+} \left\{\int_{-\infty}^{s-\varepsilon} + \int_{s+\varepsilon}^{\infty}\right\} \frac{A(\sigma)}{s-\sigma}\, d\sigma$$

exists.

(b) *There exists a finite constant $M_1(\gamma, \beta)$ depending only on γ and β such that*

$$(4) \qquad \|TA\|_{0,\beta} \leqq M_1(\gamma, \beta)\|A\|_{\gamma,\beta}.$$

Proof. Choose $\bar{\gamma}$ so that $0 < \bar{\gamma} \leqq \gamma$ and $\bar{\gamma} < 2$; let $(\hat{\gamma})^{-1} + \bar{\gamma}/2 = 1$. Since $(1+|\sigma|)^{\gamma}|A(\sigma)| \leqq \|A\|_{\gamma,\beta}$, we have

$$(5) \qquad \int_{|s-\sigma|>1} \left|\frac{A(\sigma)}{s-\sigma}\right| d\sigma \leqq \|A\|_{\gamma,\beta}\left(\int_{|s-\sigma|>1} (1+|\sigma|^{-\bar{\gamma}})^{2/\bar{\gamma}}\, d\sigma\right)^{\bar{\gamma}/2} \times \left(\int_{|s-\sigma|>1} |s-\sigma|^{-\hat{\gamma}}\, d\sigma\right)^{1/\hat{\gamma}} < \infty$$

by Hölder's inequality. Therefore $\int_{|s-\sigma|>1} (A(\sigma))/(s-\sigma)\, d\sigma$ exists. Since $|A(\sigma) - A(s)| \leqq \|A\|_{\gamma,\beta}\, |\sigma - s|^{\beta}$, and since

$$(6) \qquad \mathscr{P}\int_{-1}^{+1} \frac{d\sigma}{\sigma} = 0,$$

we have

$$(7)\qquad \left\{\int_{s-1}^{s-\varepsilon} + \int_{s+\varepsilon}^{s+1}\right\} \frac{A(\sigma)}{s-\sigma}\, d\sigma = \left\{\int_{s-1}^{s-\varepsilon} + \int_{s+\varepsilon}^{s+1}\right\} \frac{A(\sigma)-A(s)}{s-\sigma}\, d\sigma,$$

and

$$(8)\qquad \int_{s-1}^{s+1} \left|\frac{A(\sigma)-A(s)}{s-\sigma}\right| d\sigma \leqq \|A\|_{\gamma,\beta} \int_{s-1}^{s+1} |s-\sigma|^{\beta-1}\, d\sigma < \infty.$$

Therefore

$$(9)\qquad \lim_{\varepsilon\to 0+} \left\{\int_{s-1}^{s-\varepsilon} + \int_{s-\varepsilon}^{s+1}\right\} \frac{A(\sigma)}{s-\sigma}\, d\sigma = \int_{s-1}^{s+1} \frac{A(\sigma)-A(s)}{s-\sigma}\, d\sigma$$

exists. It follows that the limit (3) exists. Moreover, using (5) and (8), we find that there exists a finite constant $M_0(\gamma, \beta)$ depending only on γ and β such that

$$(10)\qquad \left|\mathscr{P}\int_{-\infty}^{+\infty} \frac{A(\sigma)}{s-\sigma}\, d\sigma\right| \leqq M_0(\gamma,\beta)\|A\|_{\gamma,\beta}, \qquad -\infty < s < +\infty.$$

We have

$$(TA)(s-h) - (TA)(s+h) = \mathscr{P}\int_{-\infty}^{+\infty} \frac{A(\sigma)}{s-h-\sigma}\, d\sigma - \mathscr{P}\int_{-\infty}^{+\infty} \frac{A(\sigma)}{s+h-\sigma}\, d\sigma$$

$$(11)\qquad = 2h\mathscr{P}\int_{-\infty}^{+\infty} \frac{A(\sigma)}{(s+h-\sigma)(s-h-\sigma)}\, d\sigma$$

$$= 2h\mathscr{P}\int_{-\infty}^{+\infty} \frac{A(\sigma+s)}{(\sigma+h)(\sigma-h)}\, d\sigma.$$

Thus, since $1/(\sigma+h)(\sigma-h)$ is an even function of σ, we find

$$(12)\qquad \mathscr{P}\int_{0}^{\infty} \frac{1}{(\sigma+h)(\sigma-h)}\, d\sigma = \mathscr{P}\int_{-\infty}^{0} \frac{1}{(\sigma+h)(\sigma-h)}\, d\sigma,$$

so that

$$(13)\qquad \mathscr{P}\int_{0}^{\infty} \frac{1}{(\sigma+h)(\sigma-h)}\, d\sigma = \frac{1}{2}\mathscr{P}\int_{-\infty}^{+\infty} \frac{1}{(\sigma+h)(\sigma-h)}\, d\sigma$$

$$= \frac{1}{4h}\mathscr{P}\int_{-\infty}^{+\infty} \left(\frac{1}{\sigma-h} - \frac{1}{\sigma+h}\right) d\sigma = 0.$$

Therefore, using (11), we find that for $h > 0$ we have

(14) $|(TA)(s+h) - TA(s-h)|$

$$\leqq 2h \left| \mathscr{P} \int_0^\infty \frac{A(\sigma+s)}{(\sigma+h)(\sigma-h)}\, d\sigma - \mathscr{P} \int_0^\infty \frac{A(h+s)}{(\sigma+h)(\sigma-h)}\, d\sigma \right|$$

$$+ 2h \left| \mathscr{P} \int_{-\infty}^0 \frac{A(\sigma+s)}{(\sigma+h)(\sigma-h)}\, d\sigma - \mathscr{P} \int_{-\infty}^0 \frac{A(-h+s)}{(\sigma+h)(\sigma-h)}\, d\sigma \right|$$

$$\leqq 2h \int_0^\infty \left| \frac{A(\sigma+s) - A(h+s)}{\sigma^2 - h^2} \right| d\sigma + 2h$$

$$\times \int_{-\infty}^0 \left| \frac{A(\sigma+s) - A(-h+s)}{\sigma^2 - h^2} \right| d\sigma$$

$$\leqq 4h\, \|A\|_{\gamma,\beta} \int_0^\infty \frac{|\sigma-h|^\beta}{|\sigma^2-h^2|}\, d\sigma$$

$$\leqq 4h\, \|A\|_{\gamma,\beta} \int_0^\infty \frac{|\sigma-h|^\beta}{|\sigma^2-h^2|}\, d\sigma.$$

It is clear from (14) that there exists a constant $M_\infty(\gamma, \beta)$ depending only on γ and β, such that $|(TA)(s+h) - (TA)(s)| \leqq M_\infty(\gamma, \beta)h^\beta$ for $-\infty < s < +\infty$. Using this statement, and using (10), part (b) of the present lemma follows at once. Q.E.D.

The following definition specifies a class of functions of two real variables closely related to the class of functions of one real variable singled out by Definition 11.

13 Definition. Let γ and β be two real numbers, with $\gamma \geqq 0$ and $1 > \beta > 0$. Let $\mathfrak{X}$ be a B-space. Let $A(\cdot, \cdot)$ be an $\mathfrak{X}$-valued function of two real variables. Put

(15) $$(E_h^{(1)}A)(s_1, s_2) = A(s_1 + h, s_2),$$

$$(E_h^{(2)}A)(s_1, s_2) = A(s_1, s_2 + h), \qquad -\infty < h < \infty;$$

(16) $$O_h^{(1)}A = h^{-\beta}(E_h^{(1)}A - A),\ O_h^{(2)}A = h^{-\beta}(E_h^{(2)}A - A), \qquad \infty > h > 0;$$

(17) $$(O_\infty^{(1)}A)(s_1, s_2) = (1 + |s_1|^\gamma)A(s_1, s_2);$$

$$(O_\infty^{(2)}A)(s_1, s_2) = (1 + |s_2|^\gamma)A(s_1, s_2).$$

Put

$$(18)\qquad \|A\|_{\gamma,\beta} = \sup_{\infty \geqq h_1, h_2 > 0} \ \sup_{\infty > s_1, s_2 > -\infty} |(O_{h_1}^{(1)} O_{h_2}^{(2)} A)(s_1, s_2)|,$$

and let $\mathfrak{A}_{\gamma,\beta}(\mathfrak{X})$ be the set of all $\mathfrak{X}$-valued functions A of two variables such that $\|A\|_{\gamma,\beta} < \infty$.

14 LEMMA. *The space $\mathfrak{A}_{\gamma,\beta}(\mathfrak{X})$ is a B-space with the norm $\|A\|_{\gamma,\beta}$.*

PROOF. We leave the proof that $\|A\|_{\gamma,\beta}$ is a norm in $\mathfrak{A}_{\gamma,\beta}(\mathfrak{X})$ to the reader, and only prove the completeness of $\mathfrak{A}_{\gamma,\beta}(\mathfrak{X})$. Let $\{A_m\}$ be a Cauchy sequence in $\mathfrak{A}_{\gamma,\beta}(\mathfrak{X})$. It is evident from (15)–(18) that, for each real pair $[s_1, s_2]$, $\{A_m(s_1, s_2)\}$ is a Cauchy sequence in $\mathfrak{X}$. Thus there exists a function A of two real variables such that $\lim_{m\to\infty} A_m(s_1, s_2) = A(s_1, s_2)$ for all $-\infty < s_1, s_2 < \infty$. By (15)–(17) we have

$$(19)\qquad \lim_{m\to\infty} (O_{h_1}^{(1)} O_{h_2}^{(1)} A_m)(s_1, s_2)$$
$$= (O_{h_1}^{(1)} O_{h_2}^{(1)} A)(s_1, s_2), \qquad -\infty < s_1, s_2 < \infty, \qquad \infty \geqq h_1, h_2 > 0.$$

Thus

$$(20)\qquad \sup_{\infty \geqq h_1, h_2 > 0} \ \sup_{\infty > s_1, s_2 > -\infty} |\{O_{h_1}^{(1)} O_{h_2}^{(2)} (A - A_l)\}(s_1, s_2)|$$
$$\leqq \limsup_{m\to\infty} \|A_m - A_l\|_{\gamma,\beta}, \qquad l \geqq 1.$$

It is plain from (20) that $A \in \mathfrak{A}_{\gamma,\beta}(\mathfrak{X})$ and that $\lim_{1\to\infty} \|A - A_l\|_{\gamma,\beta} = 0$, so that $\mathfrak{A}_{\gamma,\beta}(\mathfrak{X})$ is complete and our lemma is proved. Q.E.D.

The following lemma is a kind of "two variable case" of Lemma 12; it will also serve in what follows to verify hypothesis c(ii) of Theorem 8 in a concrete case which we aim to study.

15 LEMMA. *Let γ and β be two real numbers, with $\gamma > 0$ and $\beta > 0$. Let $A, B \in \mathfrak{A}_{\gamma,\beta}(\mathfrak{X})$, and let $\mathfrak{X}$ be a B-algebra. Then the improper integral*

$$(21)\qquad (\psi(A, B))(s_1, s_2) = \mathscr{P}\int_{-\infty}^{+\infty} \frac{A(s_1, \sigma)B(\sigma, s_2)}{\sigma - s_2}\, d\sigma$$

exists. Moreover, $\psi(A, B)$ belongs to $\mathfrak{A}_{\gamma,\beta}(\mathfrak{X})$, and there is a finite constant $M_2(\gamma, \beta)$ such that $\|\psi(A, B)\|_{\gamma,\beta} \leqq M_2(\gamma, \beta)\|A\|_{\gamma,\beta}\|B\|_{\gamma,\beta}$.

PROOF. Let $-\infty < s_1, s_2 < \infty$, $\infty \geqq h_1, h_2 > 0$. Plainly

$$|(O_{h_1}^{(1)} O_h^{(2)} A)(s_1, \sigma)| \leqq \|A\|_{\gamma,\beta}$$

for $-\infty < \sigma < \infty$ and $\infty \geqq h > 0$, while $|(O_h^{(1)} O_{h_2}^{(2)} B)(\sigma, s_2)| \leqq \|B\|_{\gamma,\beta}$ for $-\infty < \sigma < \infty$ and $\infty \geqq h > 0$. Thus, using (15)–(17) we find that

$$|(1 + |\sigma|^{\gamma})(O_{h_1}^{(1)} A)(s_1, \sigma)(O_{h_2}^{(2)} B)(\sigma, s_2)| \leqq \|A\|_{\gamma,\beta} \|B\|_{\gamma,\beta}. \tag{22}$$

Moreover, we have

$$\begin{aligned} h^{-\beta}|(O_{h_1}^{(1)} A)(s_1, \sigma + h)(O_{h_2}^{(2)} B)(\sigma + h, s_2) - (O_{h_1}^{(1)} A)(s_1, \sigma)(O_{h_2}^{(2)} B)(\sigma, s_2)| \\ \leqq h^{-\beta}|(O_{h_1}^{(1)} A)(s_1, \sigma + h) - (O_{h_1}^{(1)} A)(s_1, \sigma)|\,|(O_{h_2}^{(2)} B)(\sigma + h, s_2)| \\ + h^{-\beta}|(O_{h_1}^{(1)} A)(s_1, \sigma)|\,|(O_{h_2}^{(2)} B)(\sigma + h, s_2) - (O_{h_2}^{(2)} B)(\sigma, s_2)| \\ \leqq |(O_{h_1}^{(1)} O_h^{(2)} A)(s_1, \sigma)|\, \|B\|_{\gamma,\beta} + \|A\|_{\gamma,\beta}\, |(O_h^{(1)} O_{h_2}^{(2)} B)(\sigma + h, s_2)| \\ \leqq 2\, \|A\|_{\gamma,\beta}\, \|B\|_{\gamma,\beta}. \end{aligned} \tag{23}$$

Thus, if $M_1(\gamma, \beta)$ is the finite constant of Lemma 12, and if we put

$$(C)_s(s_1, s_2) = \mathscr{P} \int_{-\infty}^{+\infty} \frac{A(s_1, \sigma) B(\sigma, s_2)}{\sigma - s}\, d\sigma, \tag{24}$$

we have $(\psi(A, B))(s_1, s_2) = C_{s_2}(s_1, s_2)$, and have also

$$\begin{aligned} |O_{h_1}^{(1)} O_{h_2}^{(2)} C_s(s_1, s_2)| + h^{-\beta} |O_{h_1}^{(1)} O_{h_2}^{(2)} C_{s+h}(s_1, s_2) - O_{h_1}^{(1)} O_{h_2}^{(2)} C_s(s_1, s_2)| \\ \leqq 2M_1(\gamma, \beta) \|A\|_{\gamma,\beta}\, \|B\|_{\gamma,\beta}, \quad -\infty < s < \infty, \qquad 0 < h < \infty, \end{aligned} \tag{25}$$

by Lemma 12. Therefore

$$\begin{aligned} |(1 + |s_2|^{\gamma}) O_{h_1}^{(1)} C_{s_2}(s_1, s_2)| = |O_{h_1}^{(1)} O_{\infty}^{(2)} C_s(s_1, s_2)|_{s = s_2} \\ \leqq 2M_1(\gamma, \beta) \|A\|_{\gamma,\beta} \|B\|_{\gamma,\beta}. \end{aligned} \tag{26}$$

Moreover

$$\begin{aligned} h^{-\beta} |(O_{h_1}^{(1)} C_{s_2 + h})(s_1, s_2 + h) - (O_{h_1}^{(1)} C_{s_2})(s_1, s_2)| \\ \leqq h^{-\beta} |O_{h_1}^{(1)} C_{s_2 + h}(s_1, s_2 + h) - O_{h_1}^{(1)} C_{s_2}(s_1, s_2 + h)| + |O_{h_1}^{(1)} O_h^{(2)} C_{s_2}(s_1, s_2)| \\ \leqq 4M_1(\gamma, \beta)\, \|A\|_{\gamma,\beta}\, \|B\|_{\gamma,\beta}. \end{aligned} \tag{27}$$

By (18) we now have $\|\psi(A, B)\|_{\gamma,\beta} \leqq 4M_2(\gamma, \beta) \|A\|_{\gamma,\beta}\, \|B\|_{\gamma,\beta}$, and our proof is complete. Q.E.D.

The following lemma is, in effect, a generalization of the Hilbert inequality XI.7.8. It will serve in what follows to verify the first part of hypothesis (b) of Theorem 8.

16 LEMMA. *Let $\mathfrak{H}$ be a Hilbert space, and let $B(\mathfrak{H})$ be the B-space of all bounded operators in $\mathfrak{H}$. Let γ and β be two real numbers, with $\gamma > 0$ and $1 > \beta > 0$. Let $A \in \mathfrak{A}_{\gamma,\beta}(B(\mathfrak{H}))$, and let f be an infinitely differentiable $\mathfrak{H}$-valued function of a real variable which vanishes outside a bounded interval of the real axis. Then the improper integral*

$$(\Gamma(A)f)(s) = \mathscr{P}\int_{-\infty}^{+\infty} \frac{A(s, \sigma)f(\sigma)}{s-\sigma}\, d\sigma \tag{28}$$

exists. Moreover, there is a finite constant $M_3(\gamma, \beta)$, depending only on γ and β, such that

$$\left\{\int_{-\infty}^{+\infty} |(\Gamma(A)f)(s)|^2\, ds\right\}^{1/2} \leqq M_3(\gamma, \beta)\, \|A\|_{\gamma,\beta} \left\{\int_{-\infty}^{+\infty} |f(s)|^2\, ds\right\}^{1/2}. \tag{29}$$

PROOF. It is easily seen, using the differentiability of f, that for each s there exists a finite constant $K(s)$ such that $|A(s, \sigma+h)f(\sigma+h) - A(s, \sigma)f(\sigma)| \leqq K(s)h^{\beta}$ for all sufficiently small β. Since $A(s, \sigma)f(\sigma)$ vanishes if σ does not belong to a certain bounded subinterval of the real axis, the existence of the improper integral (28) follows from Lemma 12.

Let φ be the characteristic function of the interval $[-1, +1]$, and put

$$B(s, \sigma) = A(s, \sigma) - A(s, s)\varphi(s-\sigma), \qquad -\infty < s, \sigma < \infty. \tag{30}$$

Then plainly

$$\left|\frac{B(s, \sigma)}{s-\sigma}\right| = \left|\frac{A(s, \sigma) - A(s, s)}{s-\sigma}\right| \leqq \|A\|_{\gamma,\beta}\, |s-\sigma|^{\beta}, \qquad |s-\sigma| \leqq 1. \tag{31}$$

Next, let $\bar{\gamma}$ be chosen so that $0 < \bar{\gamma} \leqq \gamma$ and $\bar{\gamma} < 2$, and let $(\hat{\gamma})^{-1} + \bar{\gamma}/2 = 1$. Then, by Hölder's inequality, we have

$$\begin{aligned}
&\int_{|s-\sigma|\geqq 1} \left|\frac{B(s, \sigma)}{s-\sigma}\right| d\sigma + \int_{|s-\sigma|\geqq 1} \left|\frac{B(\sigma, s)}{\sigma - s}\right| d\sigma \\
&= \int_{|s-\sigma|\geqq 1} \left|\frac{A(s, \sigma)}{\sigma - s}\right| d\sigma + \int_{|s-\sigma|\geqq 1} \left|\frac{A(\sigma, s)}{s-\sigma}\right| d\sigma \\
&\leqq 2\, \|A\|_{\gamma,\beta} \int_{|s-\sigma|\geqq 1} \frac{(1+|\sigma|^{\gamma})^{-1}}{|s-\sigma|}\, d\sigma \\
&\leqq 2\, \|A\|_{\gamma,\beta} \left(\int_{|s-\sigma|\geqq 1} ((1+|\sigma|)^{-\gamma})^{2/\bar{\gamma}} d\sigma\right)^{\bar{\gamma}/2} \left(\int_{|s-\sigma|>1} |s-\sigma|^{-\hat{\gamma}} d\sigma\right)^{1/\hat{\gamma}}.
\end{aligned} \tag{32}$$

If we put

$$(\Gamma(B)f)(s) = \int_{-\infty}^{+\infty} \frac{B(s, \sigma)}{s - \sigma} f(\sigma)\, d\sigma, \tag{33}$$

it follows from Lemma 5 and from (31) and (32) that there exists a finite constant $N(\gamma, \beta)$ depending only on γ and β such that

$$\left\{\int_{-\infty}^{+\infty} |(\Gamma(B)f)(s)|^2\, ds\right\}^{1/2} \leqq N(\gamma, \beta)\, \|A\|_{\gamma,\beta} \left\{\int_{-\infty}^{+\infty} |f(s)|^2\, ds\right\}^{1/2}. \tag{34}$$

On the other hand, it follows from (30) that

$$(\Gamma(A)f)(s) = (\Gamma(B)f)(s) + A(s, s)\mathscr{P}\int_{-\infty}^{+\infty} \frac{\varphi(s - \sigma)}{s - \sigma} f(\sigma)\, d\sigma. \tag{35}$$

Put

$$\mathscr{P}\int_{-\infty}^{+\infty} \frac{\varphi(s - \sigma)}{s - \sigma} f(\sigma)\, d\sigma = g(s). \tag{36}$$

We shall prove that there exists a finite constant N' independent of γ, β and f such that

$$\left\{\int_{-\infty}^{+\infty} |g(s)|^2\, ds\right\}^{1/2} \leqq N' \left\{\int_{-\infty}^{+\infty} |f(s)|^2\, ds\right\}^{1/2}. \tag{37}$$

This we do as follows. Let $\{x_\alpha, \alpha \in A\}$ be a complete orthonormal basis for $\mathfrak{H}$. Then

$$\begin{aligned}
(g(s), x_\alpha) &= \left(\lim_{\varepsilon \to 0+} \left\{\int_{-\infty}^{s-\varepsilon} + \int_{s+\varepsilon}^{\infty}\right\} \varphi(s - \sigma) \frac{f(\sigma)}{s - \sigma}\, d\sigma,\, x_\alpha\right) \\
&= \lim_{\varepsilon \to 0} \left\{\int_{-\infty}^{s-\varepsilon} + \int_{s+\varepsilon}^{\infty}\right\} \varphi(s - \sigma) \frac{(f(\sigma), x_\alpha)}{s - \sigma}\, d\sigma \\
&= \mathscr{P}\int_{-\infty}^{+\infty} \frac{\varphi(s - \sigma)}{s - \sigma} (f(\sigma), x_\alpha)\, d\sigma \\
&= \mathscr{P}\int_{-\infty}^{+\infty} \frac{(f(\sigma), x_\alpha)}{s - \sigma}\, d\sigma - \int_{-\infty}^{+\infty} \frac{(1 - \varphi(s - \sigma))}{s - \sigma} (f(\sigma), x_\alpha)\, d\sigma.
\end{aligned} \tag{38}$$

Using Theorem XI.7.8 and Lemma XI.7.9, we find that there exists a finite constant Λ independent of f and α such that

$$\int_{-\infty}^{+\infty} |(g(s), x_\alpha)|^2\, ds \leqq \Lambda \int_{-\infty}^{+\infty} |(f(s), x_\alpha)|^2\, ds. \tag{39}$$

Put $N' = \Lambda^{1/2}$. Summing the inequalities (39) over all $\alpha \in A$ and using Theorem IV.4.13, we obtain the inequality (37). Since $|A(s, s)| \leqq \|A\|_{\gamma,\beta}$, it follows from (37) that

$$(40) \qquad \left\{\int_{-\infty}^{+\infty} |A(s, s)g(s)|^2\, ds\right\}^{1/2} \leqq N' \|A\|_{\gamma,\beta}\left\{\int_{-\infty}^{+\infty} |f(s)|^2\, ds\right\}^{1/2}.$$

Using (35), (36), and (40) we obtain (29). Q.E.D.

Definition 17 below and the two lemmas which follow it prepare slightly modified versions of Lemmas 15 and 16 in the precise form required for the culminating application of Theorem 8 given in Theorem 21.

17 DEFINITION. Let γ, β, and α be three real numbers, and let $\gamma \geqq 0$, $1 > \beta > 0$. Let $\mathfrak{X}$ be a B-space. If A is an $\mathfrak{X}$-valued function of two real variables, let

$$(41) \qquad (V(\alpha)A)(s_1, s_2) = (1 + |s_1|)^{\alpha} A(s_1, s_2)(1 + |s_2|)^{-\alpha}.$$

Put

$$(42) \qquad \|A\|_{\gamma,\beta,\alpha} = \|A\|_{\gamma,\beta} + \|V(\alpha)A\|_{\gamma,\beta}.$$

Let $\mathfrak{A}_{\gamma,\beta,\alpha}(\mathfrak{X})$ be the set of all $\mathfrak{X}$-valued functions A of two variables such that $\|A\|_{\gamma,\beta,\alpha} < \infty$.

18 LEMMA. *Let $\mathfrak{X}$ be a B-space. Let γ, β, and α be three real numbers, and let $\gamma > 0$ and $1 > \beta > 0$.*

(a) *We have $\mathfrak{A}_{\gamma,\beta,\alpha}(\mathfrak{X}) \subseteq \mathfrak{A}_{\gamma,\beta}(\mathfrak{X})$, and $\|A\|_{\gamma,\beta,\alpha} \geqq \|A\|_{\gamma,\beta}$ for $A \in \mathfrak{A}_{\gamma,\beta,\alpha}(\mathfrak{X})$.*

(b) *The space $\mathfrak{A}_{\gamma,\beta,\alpha}(\mathfrak{X})$ is a B-space with the norm $\|A\|_{\gamma,\beta,\alpha}$.*

(c) *Suppose that $\mathfrak{X}$ is a B-algebra. Let $A, B \in \mathfrak{A}_{\gamma,\beta,\alpha}(\mathfrak{X})$. Then the improper integral*

$$(43) \qquad (\psi(A, B))(s_1, s_2) = \mathscr{P}\int_{-\infty}^{+\infty} \frac{A(s_1, \sigma)B(\sigma, s_2)}{\sigma - s_2}\, d\sigma$$

exists. Moreover, $\psi(A, B)$ belongs to $\mathfrak{A}_{\gamma,\beta,\alpha}(\mathfrak{X})$, and there is a finite constant $M_2(\gamma, \beta)$ depending only on γ and β such that $\|\psi(A, B)\|_{\gamma,\beta,\alpha} \leqq 2M_2(\gamma, \beta)\|A\|_{\gamma,\beta,\alpha} \|B\|_{\gamma,\beta,\alpha}$.

PROOF. Statement (a) follows immediately from Definition 17. Let $\{A_m\}$ be a Cauchy sequence in $\mathfrak{A}_{\gamma,\beta,\alpha}(\mathfrak{X})$. Then, by (42), $\{A_m\}$ and $\{V(\alpha)A_m\}$ are Cauchy sequences in $\mathfrak{A}_{\gamma,\beta}(\mathfrak{X})$. By Lemma 14, there exist elements

A and $\hat{A}$ in $\mathfrak{A}_{\gamma,\beta}(\mathfrak{X})$, such that $\lim_{m\to\infty}\|A_m - A\|_{\gamma,\beta} = 0$ and $\lim_{m\to\infty}\|V(\alpha)A_m - \hat{A}\|_{\gamma,\beta} = 0$. It is plain from (15)–(18) of Definition 13 that $A(s_1, s_2) = \lim_{m\to\infty} A_m(s_1, s_2)$ and

$$\hat{A}(s_1, s_2) = \lim_{m\to\infty} (1 + |s_1|)^{\alpha} A_m(s_1, s_2)(1 + |s_2|)^{-\alpha}$$

for $-\infty < s_1, s_2 < \infty$. Thus $\hat{A} = V(\alpha)A$, proving that $A \in \mathfrak{A}_{\gamma,\beta,\alpha}(\mathfrak{X})$ and that $\lim_{m\to\infty}\|A_m - A\|_{\gamma,\beta,\alpha} = 0$. This proves (b).

It is plain from (41) and (43) that $V(\alpha)\psi(A, B) = \psi(V(\alpha)A, V(\alpha)B)$. Thus, by Lemma 15, $\psi(A, B) \in \mathfrak{A}_{\gamma,\beta,\alpha}(\mathfrak{X})$ and

$$\|\psi(A, B)\|_{\gamma,\beta} + \|V(\alpha)\psi(A, B)\|_{\gamma,\beta} \leqq 2M_2(\gamma, \beta)\|A\|_{\gamma,\beta,\alpha}\|B\|_{\gamma,\beta,\alpha},$$

proving (c). Q.E.D.

19 Lemma. *Let $\mathfrak{X}$ be a B-space, and let $B(\mathfrak{X})$ be the B-space of all bounded operators in $\mathfrak{X}$. Let γ, β, and α, be three real numbers, and let $\gamma > 0$, $1 > \beta > 0$, and $\frac{1}{2} + \gamma > \alpha \geqq \frac{1}{2}$. Let $A_1 \in \mathfrak{A}_{\gamma,\beta,\alpha}(\mathfrak{X})$. Let $f \in L_2((-\infty + \infty), \mathfrak{X})$, and suppose that*

$$\int_{-\infty}^{+\infty} |\sigma f(\sigma)|^2 \, d\sigma < \infty. \tag{44}$$

Then the integral

$$(\varphi_0(A_1)f)(s) = \int_{-\infty}^{+\infty} A_1(s, \sigma)f(\sigma)\, d\sigma \tag{45}$$

exists for almost all s. Moreover, there exists a finite constant $M_3(\gamma, \beta, \alpha)$, depending only on γ, β, and α such that

$$\int_{-\infty}^{+\infty} |\varphi_0(A_1)f(s)|^2 \, ds \leqq M_3(\gamma, \beta, \alpha)\int_{-\infty}^{+\infty} |\sigma f(\sigma)|^2 \, d\sigma. \tag{46}$$

Proof. By Definition 17 we have

$$|A_1(s, \sigma)| \leqq \|A\|_{\gamma,\beta,\alpha}(1 + |s|)^{-\alpha-\gamma}(1 + |\sigma|)^{\alpha-\gamma}; \tag{47}$$

thus

$$\int_{-\infty}^{+\infty}\int_{-\infty}^{+\infty} \frac{|A_1(s, \sigma)|^2}{(1 + |\sigma|)^2}\, ds d\sigma \tag{48}$$

$$\leqq \|A_1\|^2_{\gamma,\beta,\alpha}\int_{-\infty}^{+\infty}\int_{-\infty}^{+\infty} \frac{ds\, d\sigma}{(1 + |s|)^{2(\alpha+\gamma)}(1 + |\sigma|)^{2(\gamma-\alpha)}(1 + |\sigma|)^2} < \infty.$$

The existence almost everywhere of the integral (45), and also the inequality (46), now follow immediately from Lemma 5. Q.E.D.

20 COROLLARY. *With the hypotheses of the preceding lemma let* T_1 *be the unbounded linear operator in* $L_2((-\infty+\infty), \mathfrak{H})$ *whose domain is defined by the formula*

$$\mathfrak{D}(T_1) = \left\{ f \in L_2((-\infty, +\infty), \mathfrak{H}) \,\middle|\, \int_{-\infty}^{+\infty} |\sigma f(\sigma)|^2 \, d\sigma < \infty \right\} \tag{49}$$

and which satisfies

$$(T_1 f)(s) = s f(s), \qquad f \in \mathfrak{D}(T). \tag{50}$$

Let $\varphi_1(A_1)$ *be the integral operator whose domain* $\mathfrak{D}(\varphi(A_1))$ *is the same as* $\mathfrak{D}(T_1)$, *and which satisfies* $\varphi_1(A_1)f = \varphi_0(A_1)f$, $f \in \mathfrak{D}(T)$. *Then, if* K *is real and non-zero, the operator* $\varphi_1(A_1)(iK - T_1)^{-1}$ *is everywhere defined and bounded. Moreover,* $\lim_{|K| \to \infty} |\varphi_1(A_1)(iK - T_1)^{-1}| = 0$.

PROOF. That $\varphi_1(A_1)(iK - T_1)^{-1}$ is defined everywhere is obvious; that it is bounded follows from the inequality (46). Since $((iK - T_1)^{-1} f)(s) = (iK - s)^{-1} f(s)$ for each $f \in L_2((-\infty, +\infty), \mathfrak{H})$, we have

$$(\varphi_1(A_1)(iK - T_1)^{-1} f)(s) = \int_{-\infty}^{+\infty} \frac{A_1(s, \sigma)}{iK - \sigma} f(\sigma) \, d\sigma. \tag{51}$$

Thus, by Lemma 5 and by (47) we have

$$|\varphi_1(A_1)(iK - T_1)^{-1}|^2 \leqq \|A_1\|^2_{\gamma,\beta,\alpha} \int_{-\infty}^{+\infty} \int_{-\infty}^{+\infty} \times \frac{1}{(1+|s|)^{2(\alpha+\gamma)}(1+|\sigma|)^{2+2\gamma-2\alpha}} \frac{(1+|\sigma|)^2}{K^2+|\sigma|^2} \, ds \, d\sigma. \tag{52}$$

Since $\lim_{|K| \to \infty} (1+|\sigma|^2)(K^2+|\sigma|^2)^{-1} = 0$ for each σ, and since $(1+|\sigma|^2) \times (K^2+|\sigma|^2)^{-1}$ is bounded in the range $|K| \geqq 1$, $-\infty < \sigma < +\infty$, it follows by the Lebesgue dominated convergence theorem that the integral on the right of (52) converges to zero. It follows that $\lim_{|K| \to \infty} |\varphi_1(A_1)(iK - T_1)^{-1}| = 0$. Q.E.D.

It is now easy for us to apply Theorem 8.

21 THEOREM. *Let* $\mathfrak{H}$ *be a Hilbert space, and let* $B(\mathfrak{H})$ *be the B-space of all bounded linear operators in* $\mathfrak{H}$. *Let* D *be an interval of the real axis, and*

let $L_2(D, \mathfrak{H})$ *be the space of all* $\mathfrak{H}$*-valued, Borel-Lebesgue measurable functions defined in* D *and satisfying*

$$\left\{\int_D |f(s)|^2\, ds\right\}^{1/2} = |f| < \infty. \tag{53}$$

Let $e_1, e_2, \ldots,$ *be a family of disjoint Borel sets with union* D, *and let* $\mathfrak{H}_1, \mathfrak{H}_2, \ldots,$ *be a family of closed subspaces of* $\mathfrak{H}$. *Let* $\hat{L}_2(D, \mathfrak{H})$ *be the subspace of* $L_2(D, \mathfrak{H})$ *consisting of all functions* $f \in L_2(D, \mathfrak{H})$ *such that* $f(s) \in \mathfrak{H}_i$ *for each* $s \in e_i$. *Let* T *be the operator in* $\hat{L}_2(D, \mathfrak{H})$ *whose domain* $\mathfrak{D}(T)$ *is defined by the formula*

$$\mathfrak{D}(T) = \left\{ f \in \hat{L}_2(D, \mathfrak{H}) \,\middle|\, \int_D |sf(s)|^2\, ds < \infty \right\} \tag{54}$$

and which satisfies

$$(Tf)(s) = sf(s), \qquad f \in \mathfrak{D}(T). \tag{55}$$

Let $\gamma > 0$, $1 > \beta > 0$, *and* $\frac{1}{2} + \gamma > \alpha \geqq \frac{1}{2}$. *Let* $A \in \mathfrak{A}_{\gamma,\beta,\alpha}(B(\mathfrak{H}))$, *and suppose that* $A(s, s')v \in \mathfrak{X}_i$ *for all* $v \in \mathfrak{X}$ *and all* $s \in e_i$, $1 \leqq i < \infty$, *while* $A(s, s') = 0$ *if* $s \notin D$. *Let* $\varphi(A)$ *be the integral operator in* $\hat{L}_2(D, \mathfrak{H})$ *with domain* $\mathfrak{D}(A) = \mathfrak{D}(T)$, *and such that*

$$(\varphi(A)f)(s) = \int_D A(s, \sigma)f(\sigma)\, d\sigma, \qquad f \in \mathfrak{D}(T). \tag{56}$$

Then there exists a finite positive constant $\varepsilon = \varepsilon(\gamma, \beta, \alpha)$, *depending only on* γ, β, *and* α, *such that if* $\|A\|_{\gamma,\beta,\alpha} < \varepsilon$, *then* $T + \varphi(A)$ *is similar to* T, *and is thus a (possibly unbounded) operator of scalar type.*

PROOF. The space $\hat{L}_2(D, \mathfrak{H})$ is a Hilbert space with the inner product

$$(f, g) = \int_D (f(\sigma), g(\sigma))\, d\sigma, \qquad f, g \in \hat{L}_2(D, \mathfrak{H}). \tag{57}$$

It is plain from (54), (55), and (57) that the operator T is symmetric. Since

$$((\lambda I - T)^{-1}f)(s) = (\lambda - s)^{-1}f(s), \qquad f \in \hat{L}_2(D, \mathfrak{H}), \qquad \mathscr{I}\lambda \neq 0, \tag{58}$$

we conclude, using Lemma XII.1.2 and Corollary XII.4.13(b), that T is self adjoint. In particular, T is a spectral operator of scalar type.

We now wish to apply Theorem 8. Take the space $\mathfrak{X}$ of Theorem 8 to be $\hat{L}_2(D, \mathfrak{H})$, and the space $\mathfrak{A}$ of Theorem 8 to be the subspace $\hat{\mathfrak{A}}_{\gamma,\beta,\alpha}(B(\mathfrak{H}))$ of $\mathfrak{A}(_{\gamma,\beta,\alpha}B(\mathfrak{H}))$ consisting of all $B \in \mathfrak{A}_{\gamma,\beta,\alpha}(B(\mathfrak{H}))$ such that $B(s, \sigma)v \in \mathfrak{H}_i$ for $v \in \mathfrak{H}$ and $s \in e_i$, and such that $B(s, \sigma) = 0$ for all real $s \notin D$. Let $A_1 \in \mathfrak{A}$. Since $A_1(s, \sigma)f(\sigma)\mathfrak{H}_i \subseteq \mathfrak{H}_i$ for all $s \in e_i$, it is plain that the expression $(\varphi_0(A_1)f)(s)$ of (45) of Lemma 19 belongs to $\mathfrak{H}_i$ for almost all $s \in e_i$, so that $\varphi_0(A_1)f \in \hat{L}_2(D, \mathfrak{H}) \subseteqq L_2(D, \mathfrak{H})$ for all $f \in \mathfrak{D}(T)$. Let the operator $\varphi(A_1)$ be the restriction of $\varphi_0(A_1)$ to $\mathfrak{D}(T)$, so that by Lemma 19 and Corollary 20, $\varphi(A_1)$ is a linear mapping with $\mathfrak{D}(\varphi(A_1)) \supseteqq \mathfrak{D}(T)$, verifying hypothesis (a) of Theorem 8.

We also note for use somewhat below that, by Corollary 20,

$$\lim_{|K| \to \infty} |\varphi(A_1)(iK - T)^{-1}| = 0. \tag{59}$$

Next, in order to verify hypothesis (b) of Theorem 8, we combine the preceding lemmas with a straightforward density argument. The closure, in the topology of $L_2((-\infty, +\infty), \mathfrak{H})$, of the set $C_0^\infty(\mathfrak{H})$ of all infinitely differentiable functions f vanishing outside a bounded interval plainly includes every function which is constant on a certain bounded interval $J \subseteqq (-\infty, +\infty)$ and vanishes outside J. Thus, by Lemma III.8.3, $C_0^\infty(\mathfrak{H})$ is a dense subset of $L_2((-\infty, +\infty), \mathfrak{H})$. Therefore, by Lemma 18(a) and Lemma 16, there exists a bounded linear mapping Γ_1 of $\hat{\mathfrak{A}}_{\gamma,\beta,\alpha}(B(\mathfrak{H}))$ into $B(L_2(-\infty, +\infty), \mathfrak{H})$ such that

$$(\Gamma_1(A_1)f)(s) = \mathscr{P}\int_{-\infty}^{+\infty} \frac{A_1(s, \sigma)f(\sigma)}{s - \sigma}\, d\sigma, \qquad B \in \mathfrak{A}, f \in C_0^\infty(\mathfrak{H}). \tag{60}$$

Since $A_1(s, \sigma) = 0$ for $s \notin D$, and since $A_1(s, \sigma)v \in \mathfrak{H}_i$ for $s \in e_i$ and $v \in \mathfrak{H}$, it is apparent from (60) that $\Gamma_1(A_1)$ maps $C_0^\infty(\mathfrak{H})$ into $\hat{L}_2(D, \mathfrak{H})$; thus $\Gamma_1(A_1)$ maps $\hat{L}_2(D, \mathfrak{H})$ into itself. For each $A_1 \in \mathfrak{A}$, let $\Gamma(A_1)$ be the restriction $\Gamma_1(A_1) \mid \hat{L}_2(D, \mathfrak{H})$ of $\Gamma_1(A_1)$ to $\hat{L}_2(D, \mathfrak{H})$.

Let T_1 be the operator in $L_2((-\infty, +\infty), \mathfrak{H})$ defined by (49) and (50), so that T is the restriction to $\hat{L}_2(D, \mathfrak{H}) \supset \mathfrak{D}(T)$ of T_1. It is plain that $(iI - T_1)^{-1}$ is the bounded operator defined by $((iI - T_1)^{-1}f)(s) = (i - s)^{-1}f(s)$. Thus, by Lemma XII.1.2, T_1 is a closed operator. It is plain from (60) that

$$((T_1\Gamma_1(A_1) - \Gamma_1(A_1)T_1)f)(s) = \int_{-\infty}^{+\infty} A_1(s, \sigma)f(\sigma), \qquad f \in C_0^\infty(\mathfrak{H}). \tag{61}$$

Let μ be the Borel measure on the real axis defined by $\mu(e) = \int_e (1 + \sigma^2)\, d\sigma$, so that by Theorem III.10.4 we have

$$\int_{-\infty}^{+\infty} g_0(s)\mu(ds) = \int_{-\infty}^{+\infty} g_0(\sigma)(1 + \sigma^2)\, d\sigma \tag{62}$$

for each non-negative Borel measurable function g_0.

The closure, in the topology of $L_2(\mu, \mathfrak{H})$, of the set $C_0^\infty(\mathfrak{H})$ of all infinitely differentiable functions vanishing outside a bounded interval plainly includes every function which is constant on a certain bounded interval $J \subseteq (-\infty, +\infty)$ and vanishes outside J. Thus, by Lemma III.8.3, $C_0^\infty(\mathfrak{H})$ is a dense subset of $L_2(\mu, \mathfrak{H})$. If $g \in \mathfrak{D}(T) = L_2(\mu, \mathfrak{H})$, it follows that there exists a sequence $\{g_n\}$ of elements of $C_0^\infty(\mathfrak{H})$ such that

$$\lim_{n\to\infty} \int_{-\infty}^{+\infty} (1+\sigma^2)\,|g_n(\sigma) - g(\sigma)|^2\, d\sigma = 0.$$

Then $g_n \to g$ and $Tg_n \to Tg$ in the topology of $L_2((-\infty, +\infty), \mathfrak{H})$ as $n \to \infty$. Since equation (61) holds for $f \in C_0^\infty(\mathfrak{H})$, since $\Gamma_1(A_1)$ is bounded, since T_1 is closed and since $\varphi_0(A_1)(iK - T)^{-1}$ is bounded for real non-zero K by Corollary 20, it follows that

$$((T_1\Gamma_1(A_1) - \Gamma_1(A_1)T_1)g)(s) = \int_{-\infty}^{+\infty} A_1(s, \sigma)g(\sigma)\, d\sigma, \qquad g \in \mathfrak{D}(T_1). \tag{63}$$

It follows *a fortiori*, on restricting all the operators appearing in (63) to the subspace $\mathfrak{D}(T)$ of $\mathfrak{D}(T_1)$, that

$$(T\Gamma(A_1) - \Gamma(A_1)T)f = \varphi(A_1)f, \qquad A_1 \in \mathfrak{A}, f \in \mathfrak{D}(T). \tag{64}$$

This verifies hypothesis (b) of Theorem 8.

Next, let $A_1, A_2 \in \mathfrak{A}$, and let $f \in C_0^\infty(\mathfrak{H})$. Put

$$(\psi(A_1, A_2))(s_1, s_2) = \mathscr{P}\int_{-\infty}^{+\infty} \frac{A_1(s_1, \sigma)A_2(\sigma, s_2)}{\sigma - s_2}\, d\sigma; \tag{65}$$

recall that by Lemma 18 the improper integral on the right of (65) exists. Since $A_1(s_1, \sigma)A_2(\sigma, s_2)v \in \mathfrak{H}$ for $v \in \mathfrak{H}$ and $s_1 \in e_i$, while $A_1(s_1, \sigma)A_2(\sigma, s_2) = 0$ for $s_1 \notin D$, it is clear that $\psi(A_1, A_2) \in \mathfrak{A}$.

Our next aim is to show that $\varphi(\psi(A_1, A_2)) = \varphi(A_1)\Gamma(A_2)$, thereby verifying hypothesis (c)(i) of Theorem 8. Formally, we need only apply the integral kernel (65) to a function $f \in \hat{L}_2$ and then interchange the order of integration in the resulting double integral to prove this identity. However, the occurrence in (65) of a singular proper value integral complicates this basically straightforward argument, forcing us to argue more delicately and at some length. The necessary detailed argument is as follows. Let $\chi(\cdot)$ be the characteristic function of the interval $[-1, +1]$.

For $1 > \varepsilon > 0$ it follows from Definitions 13 and 17 that

$$(66)\quad \left| \left\{ \int_{-\infty}^{s-\varepsilon} \int_{s+\varepsilon}^{\infty} \right\} \frac{A_2(s, \sigma)}{s - \sigma} f(\sigma)\, d\sigma \right|$$

$$\leqq \int_{-\infty}^{+\infty} \frac{|A_2(s, \sigma) - A_2(s, s)\chi(s - \sigma)|}{|s - \sigma|} |f(\sigma)|\, d\sigma$$

$$\leqq \|A_2\|_{\gamma,\beta} \int_{|s-\sigma|<1} |s - \sigma|^{\beta - 1} |f(\sigma)|\, d\sigma$$

$$+ \|A_2\|_{\gamma,\beta,\alpha} (1 + |s|)^{-\alpha-\gamma} \int_{|s-\sigma|>1} \frac{|f(\sigma)|}{|s - \sigma|}\, d\sigma.$$

If we denote the sum of the two terms on the right of the inequality (66) by $F(s)$, it is plain, since f is bounded and vanishes outside a finite interval, that $(1 + s^2)|F(s)|^2$ is bounded by a multiple of $(1 + |s|)^{-2(\alpha+\gamma)}$; thus $\int_{-\infty}^{+\infty} (1 + s^2)|F(s)|^2\, ds < \infty$. Therefore, if we put

$$(67)\qquad f_\varepsilon(s) = \left\{ \int_{-\infty}^{s-\varepsilon} + \int_{s+\varepsilon}^{\infty} \right\} \frac{A^2(s, \sigma)}{s - \sigma} f(\sigma)\, d\sigma,$$

it follows from (66) and from the Lebesgue dominated convergence theorem that

$$(68)\qquad \lim_{\varepsilon\to\infty} \int_{-\infty}^{+\infty} (1 + \sigma^2) |f_\varepsilon(\sigma) - (\Gamma(A_2)f)(\sigma)|^2\, d\sigma = 0.$$

Hence we have $\lim_{\varepsilon\to 0} f_\varepsilon = \Gamma(A_2)f$ and $\lim_{\varepsilon\to 0} T_1 f_\varepsilon = T_1\Gamma(A_2)f$ in the topology of $L_2(D, \mathfrak{H})$. Since $\varphi_0(A_1)(iI - T_1)^{-1}$ is bounded, it follows that

$$(69)\quad (\varphi_0(A_1)\Gamma(A_2)f)(s_1) = \underset{\varepsilon\to 0}{\text{l.i.m.}} (\varphi_0(A_1)f_\varepsilon)(s_1)$$

$$= \underset{\varepsilon\to 0}{\text{l.i.m.}} \int_{-\infty}^{+\infty} \left\{ \int_{|s-\sigma|>\varepsilon} \frac{A_1(s_1, \sigma)A_2(\sigma, s)}{\sigma - s} f(s)\, ds \right\} d\sigma.$$

Since for each $\varepsilon > 0$ and for all s in any finite interval there exists a finite constant $K(\varepsilon)$ such that

$$(70)\qquad \left| \frac{A_1(s_1, \sigma)A_2(\sigma, s)}{\sigma - s} \right| < K(\varepsilon)(1 + |\sigma|)^{-\gamma-1}, \qquad |s - \sigma| > \varepsilon,$$

we may apply Fubini's theorem to conclude from (69) that

$$(71)\quad (\varphi_0(A_1)\Gamma(A_2)f)(s) = \underset{\varepsilon\to 0}{\text{l.i.m.}} \int_{-\infty}^{+\infty} \left\{ \int_{|s-\sigma|>\varepsilon} \frac{A_1(s_1, \sigma)A_2(\sigma, s)}{\sigma - s}\, d\sigma \right\} f(s)\, ds.$$

On the other hand, there exists a finite constant K such that for $1 > \varepsilon > 0$,

$$\left|\int_{|s-\sigma|>\varepsilon} \frac{A_1(s_1, \sigma)A_2(\sigma, s)}{\sigma - s} d\sigma\right| \tag{72}$$

$$\leqq \int_{|s-\sigma|>\varepsilon} \frac{|A_1(s_1, \sigma)A_2(\sigma, s) - A_1(s_1, s)A_2(s, s)\chi(\sigma - s)|}{|\sigma - s|} d\sigma$$

$$\leqq K \int_{|s-\sigma|>1} \frac{1}{|\sigma - s|} \frac{1}{(1 + |\sigma|^\gamma)} d\sigma$$

$$+ K \int_{|s-\sigma|<1} |\sigma - s|^{\beta - 1} d\sigma;$$

thus the function

$$\int_{|s-\sigma|>\varepsilon} \frac{A_1(s_1\ \sigma)A_2(\sigma, s)}{s - \sigma} d\sigma \tag{73}$$

is uniformly bounded. It follows by the Lebesgue dominated convergence theorem that for each $s_1 \in (-\infty, +\infty)$ we have

$$\lim_{\varepsilon\to 0} \int_{-\infty}^{+\infty} \left\{\int_{|s-\sigma|>\varepsilon} \frac{A_1(s_1, \sigma)A_2(\sigma, s)}{\sigma - s} d\sigma\right\} f(\sigma)\, d\sigma \tag{74}$$

$$= \int_{-\infty}^{+\infty} \left\{\mathscr{P} \int_{-\infty}^{+\infty} \frac{A_1(s_1, \sigma)A_2(\sigma, s)}{\sigma - s} d\sigma\right\} f(s)\, ds.$$

Since the "mean" limit

$$\underset{\varepsilon\to 0}{\text{l.i.m.}} \int_{-\infty}^{+\infty} \left\{\int_{|s-\sigma|>\varepsilon} \frac{A_1(s_1, \sigma)A_2(\sigma, s)}{\sigma - s} d\sigma\right\} f(s)\, ds \tag{75}$$

and "pointwise" limit

$$\lim_{\varepsilon\to 0} \int_{-\infty}^{+\infty} \left\{\int_{|s-\sigma|>\varepsilon} \frac{A_1(s_1, \sigma)A_2(\sigma, s)}{\sigma - s} d\sigma\right\} f(s), ds, \qquad -\infty < s_1 < +\infty \tag{76}$$

both exist, it follows by Corollaries III.3.8 and III.6.3 and Theorems III.3.6 and III.6.12 that these limits must be equal. Therefore, comparing (69), (74), and (65), we see that

$$\varphi_0(\psi(A_1, A_2))f = \varphi_0(A_1)\Gamma_1(A_2)f, \qquad f \in C_0^\infty(\mathfrak{H}). \tag{77}$$

Let $g \in \mathfrak{D}(T_1)$. We have seen above that there exists a sequence $\{g_n\}$ of elements of $C_0^\infty(\mathfrak{H})$ such that $g_n \to g$ and $T_1 g_n \to T_1 g$ in the topology of

$L_2((-\infty, +\infty), \mathfrak{H})$ as $n \to \infty$. Using the equation

$$T_1\Gamma_1(A_2) = \Gamma_1(A_2)T_1 + \varphi_0(A_2), \tag{78}$$

which has already been established, and using the fact that $\Gamma_1(A_2)$ is bounded and that $\varphi_0(A_2)(iI - T_1)^{-1}$ is bounded, we conclude that $\Gamma_1(A_2)g_n \to \Gamma_1(A_2)g$ as $n \to \infty$ and that $T_1\Gamma_1(A_2)g_n \to T_1\Gamma_1(A_2)g$. Thus, using (77), we conclude that $\varphi_0(\psi(A_1, A_2))g = \varphi_0(A_1)\Gamma_1(A_2)g$, $g \in \mathfrak{D}(T_1)$; therefore, restricting all operators to the subspace $\mathfrak{D}(T)$ of $\mathfrak{D}(T_1)$, it follows that $\varphi(\psi(A_1, A_2))f = \varphi(A_1)\Gamma(A_2)f$, $f \in \mathfrak{D}(T)$. This verifies hypothesis (c) of Theorem 8.

By Lemma VII.3.4, and by (59), $(I - \varphi(A_1)(iK - T)^{-1})^{-1}$ exists and is bounded and everywhere defined for K real and sufficiently large. The hypotheses of Corollary 9 are therefore satisfied. Applying Theorem 8 and Corollary 9, the present theorem follows at once. Q.E.D.

By using appropriate "diagonalizing" transformations (cf. Theorem XII.3.16), we may apply Theorem 21 to analyze a variety of operators. The following theorem exemplifies this remark, showing how Theorem 21 can be applied in the study of an interesting class of partial differential operators.

22 THEOREM. *Let n be an integer ≥ 3, and let E^n be the real n-dimensional Euclidean space.*

(a) *Let ∇ be the formal partial differential operator*

$$\nabla = \frac{\partial^2}{\partial x_1^2} + \cdots + \frac{\partial^2}{\partial x_n^2} \tag{1}$$

in E^n, and let T_0 be the unbounded operator in the Hilbert space $L_2(E^n)$ whose domain $\mathfrak{D}(T_0)$ consists of all twice continuously differentiable functions f such that

$$|f(x)| + \sum_{i=1}^{n} \left| \frac{\partial f}{\partial x_i}(x) \right| + \sum_{i,j=1}^{n} \left| \frac{\partial^2 f}{\partial x_i \partial x_j}(x) \right| = O(|x|^{-n-1}) \tag{2}$$

as $|x| \to \infty$, and which satisfies

$$T_0 f = -\nabla f, \qquad f \in \mathfrak{D}(T_0). \tag{3}$$

Let T be the closure of the operator T_0. Then T is a self adjoint operator whose spectrum is the positive real axis.

(b) *There exists a positive constant* $\varepsilon = \varepsilon(n)$ *depending only on* n *such that if* $V(\cdot)$ *is a complex valued function in* $C^n(E^n)$ *satisfying*

$$\sum_{|J| \leqq n+1} \int_{E^n} (1 + |x|^2)|\partial^J V(x)| \, dx < \varepsilon, \tag{4}$$

and if V *denotes the bounded operator in* $L_2(E^n)$ *defined by*

$$(Vf)(x) = V(x)f(x), \qquad f \in L_2(E^n), \tag{5}$$

then $T + V$ *is similar to* T, *and is in particular an unbounded spectral operator in* $L_2(E^n)$ *of scalar type.*

PROOF. As the proof is somewhat lengthy, we shall outline its main features before beginning any of its details. Our first step is to represent the unperturbed operator T_0 as the operator of multiplication by x in an L_2-space of vector valued functions $f(x)$. This is accomplished by using two successive transformations, of which the first is the Fourier transform, which represents T_0 as a multiplication operator, and of which the second is an elementary change of variables, which reduces this multiplication operator to the standardized operator $f(x) \to xf(x)$. The necessary arguments, up to this point, are elementary and measure theoretic. Next, calculating the effect of the two elementary transformations on the perturbing V, we prove that V appears, in the T_0-diagonalizing representation, as an integral operator. The conditions assumed for V are then shown to imply the hypothesis of Theorem 21, and this brings us to the desired goal.

Having set forth this brief outline, we are ready to being our detailed proof.

Let $\hat{f}$ denote the Fourier transform of the function $f \in L_2(E^n)$, so that

$$\hat{f}(k) = (2\pi)^{-n/2} \operatorname*{l.i.m.}_{N \to \infty} \int_{|x| \leqq N} f(x) e^{ik \cdot x} \, dx, \tag{6}$$

and so that the map $W : f \to \hat{f}$ is a unitary mapping of $L_2(E^n)$ onto itself. If $f \in L_2(E^n) \cap L_1(E^n)$, then by Theorem XV.11.3 we have

$$\hat{f}(k) = (2\pi)^{-n/2} \int_{E^n} f(x) e^{ik \cdot x} \, dx. \tag{7}$$

If $f \in \mathfrak{D}(T_0)$, it follows on integrating twice by parts (a procedure justified in virtue of (2)) that

$$(8) \qquad \widehat{T_0 f}(k) = -\int_{E^n} (\nabla f)(x) e^{ik\cdot x}\, dx$$
$$= |k|^2 \int_{E^n} f(x) e^{ik\cdot x}\, dx.$$

Let S be the operator in $L_2(E^n)$ whose domain $\mathfrak{D}(S)$ is defined by

$$(9) \qquad \mathfrak{D}(S) = \left\{ f \in L_2(E^n) \,\middle|\, \int_{E^n} |k|^4 \, |f(k)|^2 \, dk < \infty \right\}$$

and which satisfies

$$(10) \qquad (Sf)(k) = |k|^2 f(k), \qquad k \in E^n.$$

It follows from (8) that $WT_0W^{-1} \subseteq S$.

Next, let $C_0^\infty(E^n)$ denote the set of all infinitely often differentiable complex valued functions defined in E^n and vanishing outside a bounded set, let $g \in C_0^\infty(E^n)$, and let $g = \hat{f}$, with $f \in L_2(E^n)$, so that, by the Plancherel theorem (XV.11.3), we have

$$(11) \qquad f(x) = (2\pi)^{-n/2} \int_{E^n} g(k) e^{-ik\cdot x}\, dk.$$

Integrating (11) repeatedly by parts, we find that

$$(12) \qquad (-1)^r |x|^{2r} f(x) = (2\pi)^{-n/2} \int_{E^n} (\nabla^r g)(k) e^{-ik\cdot x}\, dk,$$

so that $|f(x)| = O(|x|^{-2r})$ for each integer r. Since formula (11) may be differentiated arbitrarily often with respect to any of the coordinates $x_1, \ldots, x_n$ of x, it follows in the same way that every partial derivative $\partial^J f$ of f satisfies $|\partial^J f(x)| = O(|x|^{-2r})$ for each integer r. Thus $f \in \mathfrak{D}(T_0)$. This shows that if S_0 denotes the restriction of S to the subspace $C_0^\infty(E^n)$ of $\mathfrak{D}(S)$, then $S_0 \subseteq WT_0W^{-1} \subseteq S$.

Let ν denote the measure on E^n defined by $\nu(e) = \int_e (1 + |x|^4)\, dx$, so that by Theorem III.10.4 we have

$$(13) \qquad \int_{E^n} g_0(x)\nu(dx) = \int_{E^n} (1 + |x|^4) g_0(x)\, dx$$

for each non-negative Borel measurable function g_0. The closure, in the topology of $L_2(E^n, \nu)$, of the set $C_0^\infty(E^n)$ plainly includes every function which is constant on any bounded "rectangle"

$$(14) \qquad J = [a_1, b_1] \times [a_2, b_2] \times \cdots \times [a_n, b_n]$$

in E^n and vanishes outside J. Thus, by Lemma III.8.3, $C_0^\infty(E^n)$ is a dense subset of $L_2(E^n, \mu)$. If $g \in \mathfrak{D}(S) = L_2(E^n, \nu)$, it follows that there exists a sequence $\{g_n\}$ of elements of $C_0^\infty(E^n)$ such that

$$\lim_{n\to\infty} \int_{E^n} (1+|x|^4)|g_n(x)-g(x)|^2\, dx = 0. \tag{15}$$

Then $g_n \to g$ and $Sg_n \to Sg$ in the topology of $L_2(E^n)$ as $n \to \infty$. It follows (cf. Lemma XII.4.8(a)) that the closure S_0 of the operator $\bar{S}_0$ is an extension of the operator S. On the other hand, it is plain that if k is a complex number which is not real and non-negative, then the operator $(kI - S)$ admits the bounded everywhere defined inverse defined by

$$((kI-S)^{-1}f)(x) = (k-|x|^2)^{-1}f(x), \qquad f \in L_2(E^n). \tag{16}$$

Thus, by Lemma XII.1.2, S is closed. It follows that $\bar{S}_0 = S$. Moreover, the operator S is plainly symmetric, so that it follows from Corollary XII.4.13(b) that S is self adjoint.

Since $S_0 \subseteq WT_0W^{-1} \subseteq S$, while $\bar{S}_0 = S$, it follows by Lemma XII.4.8(a), and using the fact that W is an isometry of $L_2(E^n)$ onto itself, that $WTW^{-1} = S$. Hence T is a self adjoint operator.

It is apparent from expression (16) for the inverse $(kI-S)^{-1}$ that the spectrum $\sigma(S)$ is a subset of $[0, \infty]$. It is also clear from (16) that $|(kI-S)^{-1}|$ approaches ∞ if k approaches any point in the interval $[0, \infty]$. Thus, by Theorem XII.2.9(a), $\sigma(S) = [0, \infty]$. Since $WTW^{-1} = S$, we must have $\sigma(T) = [0, \infty]$ also, and the proof of part (a) of the present theorem is complete.

It follows from (4) that the function $V(\cdot)$ belongs to $L_1(E^n)$. In virtue of (4) the formula

$$\hat{V}(k) = (2\pi)^{-n/2} \int_{E^n} V(x)e^{ik\cdot x}\, dx \tag{17}$$

may be integrated repeatedly by parts to give

$$(-ik)^J\hat{V}(k) = (2\pi)^{-n/2}\int_{E^n} (\partial^J V)(x)e^{ik\cdot x}\, dx, \qquad |J| \leqq n+1 \tag{18}$$

(cf. the first paragraph of Section XIV.2 for an explanation of the notations employed). It follows in particular that $|\hat{V}(k)| = O(|k|^{-n-1})$ as $|k| \to \infty$. Therefore, $\hat{V}(k)$ is integrable, and we have

$$V(x) = (2\pi)^{-n/2}\int_{E^n} \hat{V}(k)e^{-ik\cdot x}\, dk. \tag{19}$$

Let V_1 be the operator in $L_2(E^n)$ defined by

$$(20) \qquad (V_1 f)(k) = (2\pi)^{-n/2} \int_{E^n} \hat{V}(k - k_1) f(k_1)\, dk_1, \qquad f \in L_2(E^n).$$

Then, by the n-dimensional analogue of Theorem XI.3.21(d) (cf. the discussion following XI.3.22) we have

$$(21) \qquad W V W^{-1} f = V_1 f, \qquad f \in L_2(E^n).$$

To prove part (b) of our theorem, we must therefore show that, if condition (4) is satisfied, the operators S and $S + V_1$ are similar.

We proceed toward this goal as follows. Suppose, as in Section XI.7, that μ is the measure of hypersurface on the unit sphere Σ in E^n, so that

$$(22) \qquad \int_{E^n} f(x)\, dx = \int_0^\infty \int_\Sigma \left\{ f(r\omega) r^{n-1} \right\} \mu(d\omega)\, dr$$

for each Borel measurable function f defined on E^n which is either integrable over E^n or non-negative. If f is a Borel measurable function on E^n, let $[f]$ be the vector valued function defined on $[0, \infty]$, and having values in the set of functions on the sphere Σ, such that

$$(23) \qquad \{[f](\rho)\}(\omega) = 2^{-1/2} \rho^{(n-2)/4} f(\rho^{1/2}\omega), \qquad 0 \leqq \rho < \infty,\ w \in \Sigma.$$

It follows from (22), using Theorems III.11.17, III.11.14, and the Radon–Nikodým theorem (cf. Corollary III.10.6) that the transform $Z : f \to [f]$ maps $L_2(E^n)$ isometrically onto $\mathscr{L}_2 = L_2([0, \infty), L_2(\Sigma))$. From (23) we have

$$(24) \qquad \{\rho[f](\rho)\}(\omega) = \{[g](\rho)\}(\omega),$$

where $g(k) = |k|^2 f(k)$, and thus it follows in the same way that the operator $ZSZ^{-1} = \mathscr{S}$ in $\mathscr{L}_2$ has the domain

$$(25) \qquad \mathfrak{D}(\mathscr{S}) = \left\{ f \in \mathscr{L}_2 \,\middle|\, \int_0^\infty \rho^2 |f(\rho)|^2\, d\rho < \infty \right\}$$

and satisfies

$$(26) \qquad (\mathscr{S} f)(\rho) = \rho f(\rho), \qquad f \in \mathscr{L}_2.$$

Arguing once more in the same way, we find the following description of the operator $Z V_1 Z^{-1} = \mathscr{V}$. Write

$$(27) \qquad V_2(\rho, \rho'; \omega, \omega') = \tfrac{1}{2}(2\pi)^{-n/2} (\rho\rho')^{(n-2)/4} \hat{V}(\rho^{1/2}\omega - (\rho')^{1/2}\omega').$$

For all $0 \leqq \rho, \rho' < \infty$, let $V_3(\rho, \rho')$ be the integral operator in $L_2(\Sigma, \mu)$ defined by

$$(28)\quad (V_3(\rho, \rho')\varphi)(\omega) = \int_\Sigma V_2(\rho, \rho'; \omega, \omega')\varphi(\omega')\mu(d\omega'), \qquad \varphi \in L_2(\Sigma, \mu).$$

Then we have

$$(29)\quad (\mathscr{V}f)(\rho) = \int_0^\infty V_3(\rho, \rho')f(\rho')\, d\rho', \qquad f \in L_2([0, \infty), L_2(\Sigma, \mu)).$$

Since Z is an isometry of $L_2(E^n)$ onto $\mathscr{L}_2$, it follows that to establish part (b) of our theorem, we have only to show that, if condition (4) is satisfied, the operators $\mathscr{S}$ and $\mathscr{S} + \mathscr{V}$ are similar.

This we do as follows: in virtue of (4), equation (18) may be differentiated once and twice under the sign of integration, and we conclude that, for a certain finite constant M independent of ε, we have

$$(30)\quad |\hat{V}(k)| + \sum_{i=1}^n \left|\frac{\partial \hat{V}}{\partial k_i}(k)\right| + \sum_{i,j=1}^n \left|\frac{\partial^2 \hat{V}}{\partial k_i\, \partial k_j}(k)\right| \leqq M\varepsilon(1 + |k|^{n+1})^{-1}.$$

Let us now note that there exists a positive absolute constant c such that

$$(31)\quad |\omega - \alpha\omega'| \geqq c\,|\omega - \omega'|$$

for all $0 < \alpha < 1$ and all $\omega, \omega' \in \Sigma$. Indeed, if θ is the angle between ω and ω', it is clear that $|\omega - \alpha\omega'| > 1$ if $\pi/2 \leqq |\theta| \leqq \pi$, while

$$(32)\quad |\omega - \alpha\omega'| \geqq |\sin\theta|, \qquad 0 \leqq |\theta| \leqq \frac{\pi}{2},$$

while $|\omega - \omega'| = 2\,|\sin(\theta/2)|$. Since there exists a positive constant c_0 such that $|\sin\theta| \geqq c_0\,|\sin(\theta/2)|$ for $0 \leqq |\theta| \leqq \pi/2$, formula (30) follows at once. Using (27), (30), and (31), we see that there exists a finite constant M' independent of ε such that

$$(33)\quad |\hat{V}(r\omega - r'\omega')| \leqq M'\varepsilon(1 + \max(r, r')|\omega - \omega'|)^{-n-1}.$$

Similarly, differentiating (27) once and twice, and using (30) and (31), we find

$$(34)\quad \left|\frac{\partial \hat{V}}{\partial r}(r\omega - r'\omega')\right| + \left|\frac{\partial \hat{V}}{\partial r'}(r\omega - r'\omega')\right| + \left|\frac{\partial^2 \hat{V}}{\partial r\, \partial r'}(r\omega - r'\omega')\right|$$
$$\leqq M'\varepsilon(1 + \max(r, r')|\omega - \omega'|)^{-n-1}.$$

For $0 \leqq r, r' < \infty$, let $V_4(r, r')$ be the integral operator in $L_2(\Sigma, \mu)$ defined by

$$(35) \qquad (V_4(r, r')\varphi)(\omega) = \int_\Sigma \hat{V}(r\omega - r'\omega')\varphi(\omega')\mu(d\omega'), \qquad \varphi \in L_2(\Sigma, \mu),$$

so that

$$(36) \qquad V_3(\rho, \rho') = \tfrac{1}{2}(2\pi)^{-n/2}(\rho\rho')^{(n-2)/4} V_4(\rho^{1/2}, (\rho')^{1/2}).$$

By (33) and 34), and using Lemma 5, we have

$$(37) \qquad |V_4(r, r')| + \left|\frac{\partial V_4}{\partial r}(r, r')\right| + \left|\frac{\partial V_4}{\partial r'}(r, r')\right| + \left|\frac{\partial^2 V_4}{\partial r\,\partial r'}(r, r')\right|$$
$$\leqq M'\varepsilon \int_\Sigma \frac{1}{(1 + \max(r, r')|\omega - \omega_0|)^{n+1}}\,\mu(d\omega),$$

where $\omega_0 = [1, 0, 0, \ldots, 0]$. Put

$$(38) \qquad I(\alpha) = \int_\Sigma \frac{1}{(1 + \alpha|\omega - \omega_0|)^{n+1}}\,\mu(d\omega), \qquad \alpha \geqq 0.$$

The integral $I(\alpha)$ plainly remains bounded if the parameter α remains bounded. Let Σ_0 be a small spherical neighborhood in Σ of the point ω_0, and let $\Sigma_1 = \Sigma - \Sigma_0$. Put

$$(39) \qquad I_0(\alpha) = \int_{\Sigma_0} \frac{1}{(1 + \alpha|\omega - \omega_0|)^{n+1}}\,\mu(d\omega),$$

and

$$(40) \qquad I_1(\alpha) = \int_{\Sigma_1} \frac{1}{(1 + \alpha|\omega - \omega_0|)^{n+1}}\,\mu(d\omega),$$

so that $I(\alpha) = I_0(\alpha) + I_1(\alpha)$. It is clear that $I_1(\alpha) = O(\alpha^{-(n+1)})$ as $\alpha \to \infty$. Moreover, by introducing coordinates in Σ_0, and using the ordinary formula for change of variables in a multiple integral, the reader will readily verify that $I_0(\alpha)$ may be written in the form

$$(41) \qquad I_0(\alpha) = \int_{\substack{|y| \leqq 1 \\ y \in E^{n-1}}} \frac{D(y)}{(1 + \alpha|y|)^{n+1}}\,dy,$$

where $D(y)$ is a certain positive infinitely often differentiable "density" function of the variable y. Making a change of variable in the integral (41), we find that

$$I_0(\alpha) = \alpha^{-(n-1)} \int_{\substack{|y| \leqq \alpha \\ y \in E^{n-1}}} \frac{D(y/\alpha)}{(1+|y|)^{n+1}}\, dy; \tag{42}$$

thus $I_0(\alpha) = O(\alpha^{-(n-1)})$ as $\alpha \to \infty$, and it follows that $I(\alpha) = O(\alpha^{-(n-1)})$ as $\alpha \to \infty$. This shows that there exists a finite absolute constant c' such that $I(\alpha)| \leqq c'(1+|\alpha|)^{-n+1}$. Using this inequality and using (37), we see that there exists a finite constant M'' independent of ε such that

$$|V_4(r, r')| + \left|\frac{\partial V_4}{\partial r}(r, r')\right| + \left|\frac{\partial V_4}{\partial r'}(r, r')\right| + \left|\frac{\partial^2 V_4}{\partial r\, \partial r'}(r, r')\right| \leqq M''\varepsilon(1+\max(r, r'))^{-n+1}. \tag{43}$$

Put $V_5(r,r') = (r\, r')^{(n-2)/2} V_4(r, r')$. Using (43) and Definitions 13 and 17, we find that there exists a finite constant M''' independent of ε such that

$$|V_4(\cdot, \cdot)|_{1/2,\, 1/2,\, 1} \leqq M'''\varepsilon. \tag{44}$$

Since $\frac{1}{2}(2\pi)^{-n/2} V_5(\rho^{1/2}, (\rho')^{1/2}) = V_3(\rho, \rho')$ by (36), it follows by a second use of Definitions 13 and 17 that there exists a finite constant M_0 independent of ε such that

$$|V_3(\cdot, \cdot)|_{1/4,\, 1/4,\, 1/2} \leqq M_0\, \varepsilon. \tag{45}$$

Therefore, using Theorem 21, we conclude that if $\varepsilon = \varepsilon(n)$ is sufficiently small, the operators $\mathscr{S}$ and $\mathscr{S} + \mathscr{V}$ are similar. Q.E.D.

The information established incidentally in the course of the above proof is sometimes useful. We record this information in the following corollary.

23 Corollary. *Let $n \geqq 3$, T, Σ, and μ be as in the preceding theorem, and let V be a complex valued function defined on E^n for which the quantity*

$$|V|^{\dagger} = \sum_{|J| \leqq n+1} \int_{E^n} (1+|x|^2)|\partial^J V(x)|\, dx \tag{46}$$

is finite. Then:

(a) *If $\hat{f}$ denotes the Fourier transform of the function $f \in L_2(E^n)$, so that*

$$\hat{f}(k) = (2\pi)^{-n/2} \operatorname*{l.i.m.}_{N \to \infty} \int_{|x| \leqq N} f(x) e^{ikx}\, dx, \tag{47}$$

and if we put $(Z_1 f)(\rho, \omega) = 2^{-1/2}\rho^{(n-2)/4} f(\rho^{1/2}\omega)$ *and* $(Z_2 f)(\rho) = (Z_1 f)(\rho, \cdot)$ *for* $f \in L_2(E^n)$, *then* Z_2 *is an isometry of* $L_2(E^n)$ *onto* $L_2([0, \infty), L_2(\Sigma, \mu)) = \mathscr{L}_2$.

(b) *The operator* $\mathscr{S} = Z_2 T Z_2^{-1}$ *has the domain*

$$\mathfrak{D}(\mathscr{S}) = \left\{ f \in \mathscr{L}_2 \,\middle|\, \int_0^\infty \rho^2 |f(\rho)|^2 \, d\rho < \infty \right\} \tag{48}$$

and satisfies

$$(\mathscr{S} f)(\rho) = \rho f(\rho), \qquad f \in \mathfrak{D}(\mathscr{S}). \tag{49}$$

(c) *If* V_1 *denotes the operator in* $L_2(E^n)$ *defined by*

$$(V_1 f)(x) = V(x) f(x), \tag{50}$$

then $\mathscr{V} = Z_2 V_1 Z_2^{-1}$ *is the operator in* $\mathscr{L}_2$ *defined by*

$$(\mathscr{V} f)(\rho) = \int_0^\infty V_2(\rho, \sigma) f(\sigma) \, d\sigma, \qquad f \in \mathscr{L}_2, \tag{51}$$

where $V_2(\rho, \sigma)$ *is the operator in* $L_2(\Sigma, \mu)$ *defined by*

$$(V_2(\rho, \sigma) g)(\omega) = \tfrac{1}{2}(2\pi)^{-n/2}(\rho\rho')^{(n-2)/4} \int_\Sigma \hat{V}(\rho^{1/2}\omega - (\rho')^{1/2}\omega') \times g(\omega')\mu(d\omega'), \qquad g \in L_2(\Sigma, \mu). \tag{52}$$

(d) *If* $V_3(r, r')$ *denotes the operator in* $L_2(\Sigma, \mu)$ *defined by*

$$(V_3(r, r') g)(\omega) = \int_\Sigma \hat{V}(r\omega - r'\omega') g(\omega') \mu(d\omega'), \qquad g \in L_2(\Sigma, \mu), \tag{53}$$

then there exists a finite constant M_n *depending only on* n *such that*

$$|V_3(r, r')| + \left| \frac{\partial V_3}{\partial r}(r, r') \right| + \left| \frac{\partial V_3}{\partial r'}(r, r') \right| + \left| \frac{\partial^2 V_3}{\partial r \, \partial r'}(r, r') \right| \leq M_n |V|^\dagger (1 + \max(r, r'))^{-n+1}. \tag{54}$$

We conclude the present section with the following theorem, due to John M. Freeman, which applies Corollary 3 to an interesting class of quasi-nilpotent operators.

24 THEOREM. *Let* $1 \leqq p < \infty$, *and let* $L_p[0, 1]$ *be the space of all complex valued, Borel-Lebesgue measurable functions defined in* $[0, 1]$ *and satisfying*

$$\left\{\int_0^1 |f(x)|^p \, dx\right\}^{1/p} = |f| < \infty. \tag{1}$$

Let $g(x, y)$ *be defined and continuous on the triangle* $D = \{[x, y] \,|\, 0 \leqq y \leqq x \leqq 1\}$, *and let* G *be the integral operator (of "Volterra type") defined by*

$$(Gf)(x) = \int_0^x g(x, y) f(y) \, dy, \qquad f \in L_p[0, 1]. \tag{2}$$

Suppose that

(i) g *has continuous partial derivatives of all orders up to the second everywhere in the interior of* D, *and that these derivatives may be extended continuously to the whole of* D;

(ii) $g(x, x) > 0$ *for* $0 \leqq x \leqq 1$, *and* $\int_0^1 g(x, x) \, dx = c$. *Then, if* J *denotes the integral operator defined by*

$$(Jf)(x) = \int_0^x f(y) \, dy, \qquad f \in L_p[0, 1], \tag{3}$$

it follows that the operators G *and* cJ *are similar.*

PROOF. We shall first prove Theorem 24 in the special case in which $g(x, x) = 1$, and then show, by an easy supplementary argument, that the general case follows from this special case.

Let $\mathfrak{A}$ be the class of functions A defined in the triangle D, and satisfying the following conditions:

(a) A is continuous in D, and has continuous first and second partial derivatives with respect to x in the interior of D, which can be extended continuously to the whole of D.

(b) $A(x, x) = \dfrac{\partial A}{\partial x}(x, x) = 0, \qquad 0 \leqq x \leqq 1.$

If $A \in \mathfrak{A}$, let

$$|A| = \sup_{[x, y] \in D} \left\{ |A(x, y)| + \left| \frac{\partial A}{\partial x}(x, y) \right| + \left| \frac{\partial^2 A}{\partial x^2}(x, y) \right| \right\}, \tag{4}$$

so that $\mathfrak{A}$ becomes a B-space. Put

$$(\varphi(A)f)(x) = \int_0^x A(x, y) f(y) \, dy, \qquad f \in L_p[0, 1] \tag{5}$$

for $A \in \mathfrak{A}$, so that $\varphi(A)$ is a bounded linear operator in $L_p[0, 1]$. Plainly $|\varphi(A)| \leqq K_1|A|$, where K_1 is a certain finite constant. Put

$$(6)\quad (\Gamma(A)f)(x) = \int_0^x \left\{ \frac{\partial^2}{\partial x \partial y} \int_0^y A(\xi + x - y, \xi)\, d\xi \right\} f(y)\, dy, \qquad f \in L_p[0, 1].$$

Since

$$(7)\quad \frac{\partial^2}{\partial x\, \partial y} \int_0^y A_1(\xi + x - y, \xi)\, d\xi = \frac{\partial A_1}{\partial x}(x, y) - \int_0^y \frac{\partial^2 A_1}{\partial x^2}(\xi + x - y, \xi)\, d\xi$$

for $A_1 \in \mathfrak{A}$, the function on the left is continuous, and thus $\Gamma(A)$ is a bounded linear operator in $L_p[0, 1]$, and $|\Gamma(A)| \leqq K_2|A|$, where K_2 is a certain finite constant.

If $A \in \mathfrak{A}$ and $f \in L_p$ [0, 1], we have, integrating by parts,

$$(8)\quad \begin{aligned} (\Gamma(A)Jf)(x) &= \int_0^x \left\{ \frac{\partial}{\partial y} \int_0^y \frac{\partial A}{\partial x}(\xi + x - y, \xi)\, d\xi \right\} \left\{ \int_0^y f(\eta)\, d\eta \right\} dy \\ &= -\int_0^x \left\{ \int_0^y \frac{\partial A}{\partial x}(\xi + x - y, \xi)\, d\xi \right\} f(y)\, dy \\ &\quad + \left\{ \int_0^x \frac{\partial A}{\partial x}(\xi, \xi)\, d\xi \right\} \left\{ \int_0^x f(x)\, dx \right\} \\ &= -\int_0^x \left\{ \int_0^y \frac{\partial A}{\partial x}(\xi + x - y, \xi)\, d\xi \right\} f(y)\, dy. \end{aligned}$$

Moreover, by Fubini's theorem we have

$$(9)\quad \begin{aligned} (J\Gamma(A)f)(x) &= \int_0^x \left\{ \int_0^\eta \left\{ \frac{\partial^2}{\partial \eta\, \partial y} \int_0^y A(\xi + \eta - y, \xi)\, d\xi \right\} f(y)\, dy \right\} d\eta \\ &= \int_0^x \left\{ \int_y^x \frac{\partial^2}{\partial \eta\, \partial y} \left\{ \int_0^y A(\xi + \eta - y, \xi)\, d\xi \right\} d\eta \right\} f(y)\, dy \\ &= \int_0^x \left\{ \frac{\partial}{\partial y} \left[\int_0^y A(\xi + x - y, \xi)\, d\xi - \int_0^y A(\xi, \xi)\, d\xi \right] \right\} f(y)\, dy \\ &= \int_0^x \left\{ \frac{\partial}{\partial y} \left[\int_0^y A(\xi + x - y, \xi)\, d\xi \right] \right\} f(y)\, dy \\ &= \int_0^x A(x, y) f(y) dy - \int_0^x \left\{ \frac{\partial A}{\partial x}(\xi + x - y, \xi)\, d\xi \right\} f(y)\, dy. \end{aligned}$$

Thus

$$J\Gamma(A) - \Gamma(A)J = \varphi(A), \qquad A \in \mathfrak{A}. \tag{10}$$

If we write $B_1(x, y)$ for the kernel of A_1, that is, for the continuous function appearing on the right-hand side of (7), then $\sup_{[x,y] \in D} |B_1(x, y)| \leqq |A_1|$, and we have

$$|(\Gamma(A_1)f)(x)| \leqq |A_1| \int_0^x |f(y)|\, dy, \qquad f \in L_p[0, 1], \tag{11}$$

for $A_1 \in \mathfrak{A}$, and thus, inductively, we have

$$|((\Gamma(A_1))^n f)(x)| \leqq \frac{|A_1|^n}{(n-1)!} \int_0^x (x-y)^{n-1}|f(y)|\, dy. \tag{12}$$

Thus

$$|(\Gamma(A_1))^n| = O\left(\frac{|A_1|^n}{(n-1)!}\right),$$

so that $\Gamma(A_1)$ is a quasi-nilpotent operator for each $A_1 \in \mathfrak{A}$.

Next, for $A, A_1 \in \mathfrak{A}$, put

$$(\psi(A, A_1))(x, y) = \int_y^x A(x, \eta)\left\{\frac{\partial^2}{\partial\eta\, \partial y} \int_0^y A_1(\xi + \eta - y, \xi)\, d\xi\right\} d\eta. \tag{13}$$

Then plainly

$$\varphi(\psi(A, A_1)) = \varphi(A)\Gamma(A_1). \tag{14}$$

Writing B_1 for the right-hand side of (7) once more, we have

$$(\psi(A, A_1))(x, y) = \int_y^x A(x, \eta)B_1(\eta, y)\, d\eta. \tag{15}$$

It is then plain that $(\psi(A, A_1))(x, y)$ is continuous. Moreover,

$$\frac{\partial\psi(A, A_1)}{\partial x}(x, y) = \int_y^x \frac{\partial A}{\partial x}(x, \eta)B_1(\eta, y)\, d\eta \tag{16}$$

and

$$\frac{\partial^2\psi(A, A_1)}{\partial x^2}(x, y) = \int_y^x \frac{\partial^2 A}{\partial x^2}(x, \eta)B_1(\eta, y)\, d\eta, \tag{17}$$

so that $\psi(A, A_1)(x, y)$ has two continuous partial derivatives with respect to x. Furthermore $\psi(A, A_1)(x, x) = (\partial\psi(A, A_1)/\partial x)(x, x) = 0$ for $0 \leqq x \leqq 1$, and therefore $\psi(A, A_1) \in \mathfrak{A}$. Taking (16) and (17) into account, we see at once that there exists a finite constant K_3 such that $|\psi(A, A_1)| \leqq K_3|A||A_1|$.

Let $A \in \mathfrak{A}$, and let σ_A be the transformation of $\mathfrak{A}$ into itself defined by $\sigma_A(A_1) = \psi(A, A_1)$. We shall show that σ_A is a quasi-nilpotent transformation. Indeed, from (13), (16), (17), and the inequality preceding (11), we conclude at once that

$$(18)\quad |\sigma_A(A_1)[x, y]| + \left|\frac{\partial \sigma_A(A_1)[x, y]}{\partial x}\right| + \left|\frac{\partial^2 \sigma_A(A_1)[x, y]}{\partial x^2}\right|$$
$$\leqq |A|\,|A_1|\,|x-y|; \qquad [x, y] \in D;\ A, A_1 \in \mathfrak{A}.$$

Let us assume, inductively, that

$$(19)\quad |\sigma_A^n(A_1)[x, y]| + \left|\frac{\partial \sigma_A^n(A_1)}{\partial x}[x, y]\right| + \left|\frac{\partial^2 \sigma_A^n(A_1)}{\partial x^2}[x, y]\right|$$
$$\leqq \frac{|A|^n}{n\,!}\,|A_1|[2\,|x-y|]^n, \qquad [x, y] \in D;\ A, A_1 \in \mathfrak{A}.$$

Then, using (13), (16), (17), and (7), we find that

$$(20)\quad |\sigma_A^{n+1}(A_1)[x, y]| + \left|\frac{\partial \sigma_1^{n+1}(A_1)}{\partial x}[x, y]\right| + \left|\frac{\partial^2 \sigma_A^{n+1}(A_1)}{\partial x^2}[x, y]\right|$$
$$\leqq |A| \int_y^x \left\{\left|\frac{\partial \sigma_A^n(A_1)}{\partial \eta}(\eta, y)\right| + \int_0^y \left|\frac{\partial^2 \sigma_A^n(A_1)}{\partial \eta^2}(\xi + \eta - y, \xi)\right| d\xi\right\} d\eta$$
$$\leqq \frac{|A|^{n+1}}{n\,!}\,|A_1|\,2^n(1+y)\int_y^x (\eta - y)^n d\eta$$
$$\leqq \frac{|A|^{n+1}}{(n+1)!}\,|A_1|[2(x-y)]^{n+1}.$$

The inequality (19) must therefore hold for all n, so that

$$(21)\qquad |\sigma_A^n| = O\left(\frac{(2\,|A|)^n}{n!}\right) \qquad \text{as } n \to \infty,$$

which proves that σ_A is a quasi-nilpotent transformation for each $A \in \mathfrak{A}$.

Applying Corollary 4, we find that $J + A$ is similar to J. This proves our theorem in the special case in which $g(x, x) = 1$ and $(\partial g/\partial x)(x, x) = 0$, $0 \leqq x \leqq 1$. To reduce the general case of Theorem 24 to this special case, we argue as follows. Let ψ be a monotone increasing function with two continuous derivatives, mapping the interval $[0, 1]$ into itself. Let $\hat{\psi}$ be the inverse of the mapping ψ. Let $a(x)$ be a complex valued function with two continuous derivatives defined in $[0, 1]$. Put $(\bar{\psi}f)(x) = \exp(a(x)) f(\psi(x))$ for each $f \in L_p[0, 1]$, so that $\bar{\psi}$ is a bounded linear transformation of $L_p[0, 1]$ into itself, whose bounded inverse $\bar{\psi}^{-1}$ is given by the formula $(\bar{\psi}^{-1}f)(x) = \exp(-a(\hat{\psi}(x)))f(\hat{\psi}(x))$. From the definition (2) of the transformation G, we find that

$$(22) \qquad (\bar{\psi}^{-1}G\bar{\psi}f)(x) = \int_0^{\hat{\psi}(x)} g(\hat{\psi}(x), y)\exp(a(y) - a(\hat{\psi}(x))f(\psi(y))\, dy$$

$$= \int_0^x g(\hat{\psi}(x), \hat{\psi}(y))\exp(a(\hat{\psi}(y)) - a(\hat{\psi}(x)))\hat{\psi}'(y)f(y)\, dy.$$

We now choose $\psi(x)$ so that

$$(23) \qquad \hat{\psi}'(x) = cg(\hat{\psi}(x), \hat{\psi}(x))^{-1},$$

that is,

$$(24) \qquad \psi'(x) = c^{-1}g(x, x),$$

where, as in (ii), $c = \int_0^1 g(x, x)\, dx$. Putting $\psi(0) = 0$ we have $\psi(1) = 1$ from (24), so that ψ is indeed a function with two continuous derivatives mapping the unit interval onto itself. Putting

$$g_1(x, y) = g(\hat{\psi}(x), \hat{\psi}(y))\hat{\psi}'(y)\exp(a(\hat{\psi}(y)) - a(\hat{\psi}(x))),$$

we have $g_1(x, x) = 1$, $0 \leqq x \leqq 1$. If we choose the function a so that

$$(25) \qquad \left\{\frac{\partial g}{\partial x}(\hat{\psi}(x), \hat{\psi}(x))(\hat{\psi}'(x))^2\right\}\left\{g(\hat{\psi}(x), \hat{\psi}(x))\hat{\psi}'(x)\right\}^{-1} = a(x),$$

we shall have $(\partial g_1/\partial x)(x, x) = 0$, $0 \leqq x \leqq 1$. Moreover, by (22), we have

$$(26) \qquad (\bar{\psi}^{-1}G\bar{\psi}f)(x) = \int_0^x g_1(x, y)f(y)\, dy.$$

Applying the special case of Theorem 23, which has already been proved, to the mapping $(1/c)\bar{\psi}^{-1}G\bar{\psi}$, we find that $(1/c)\bar{\psi}^{-1}G\bar{\psi}$ is similar to J, so that G is similar to cJ. Q.E.D.

3. The Friedrichs' Method for the Discrete Spectrum

The perturbation-theoretic methods developed in the preceding section can be adapted for application to operators with discrete spectrum, and even for application to operators with mixed discrete and continuous spectrum. In the present section we shall illustrate this assertion by proving a number of results, due to Robert E. L. Turner, concerning the spectral character of certain classes of compact operators.

We begin by generalizing Theorem 2.1.

→ 1 THEOREM. *Let $\mathfrak{X}$ be a B-space, and let $T \in B(\mathfrak{X})$. Let $\mathfrak{A}$ be an "auxiliary" B-space, with the norm $|||A|||$, $A \in \mathfrak{A}$. Let M_1 and M_2 be real numbers greater than zero. Suppose that:*

(a) *a continuous linear mapping $\varphi : \mathfrak{A} \to B(\mathfrak{X})$, of norm at most M_1, is given;*

(b) *a continuous linear mapping $\eta : \mathfrak{A} \to \mathfrak{A}$, of norm at most M_1, is given;*

(c) *a continuous linear mapping $\Gamma : \mathfrak{A} \to B(\mathfrak{X})$, of norm at most M_1, such that*

$$T\Gamma(A) - \Gamma(A)T = \varphi(A - \eta(A)), \qquad A \in \mathfrak{A},$$

is defined;

(d) *a continuous bilinear mapping $\psi(A, A_1)$ of $\mathfrak{A} \times \mathfrak{A}$ into $\mathfrak{A}$ is defined and satisfies*

(i) $$\varphi(\psi(A, A_1)) = \Gamma(A)\varphi(A_1), \qquad A, A_1 \in \mathfrak{A};$$

(ii) $$|||\psi(A, A_1)||| \leqq M_2 \, |||A||| \, |||A_1||| , \qquad A, A_1 \in \mathfrak{A};$$

(e) *a continuous bilinear mapping $\psi_0(A, A_1)$ of $\mathfrak{A} \times \mathfrak{A}$ into $\mathfrak{A}$ is defined and satisfies*

(i) $$\varphi(\psi_0(A, A_1)) = \varphi(\eta(A))\Gamma(A_1), \qquad A, A_1 \in \mathfrak{A},$$

(ii) $$|||\psi_0(A, A_1)||| \leqq M_2 \, |||A||| \, |||A_1||| , \qquad A, A_1 \in \mathfrak{A}.$$

Then for each $A_1 \in \mathfrak{A}$ such that $|||A_1||| < (6M_2)^{-1} + (2M_1)^{-1}$, there exists an operator $A \in \mathfrak{A}$ such that the operators $T + \varphi(A_1)$ and $T + \varphi(\eta(A))$ are similar.

REMARK. Theorem 2.1 is the special case of our present theorem in which $\eta(A_0) = 0$ for each $A_0 \in \mathfrak{A}$. For application to compact operators later in the present section, we shall define $\eta(A_0)$ in such a way that $\varphi(\eta(A_0))$ is the "diagonal part" of $\varphi(A_0)$ relative to a base for the space $\mathfrak{X}$ in which the operator T is diagonalized.

PROOF. We shall search for an operator U and an $A \in \mathfrak{A}$ satisfying the equation

$$(1) \qquad U(T + \varphi(A_1)) = (T + \varphi(\eta(A)))U,$$

assuming that U has the form $U = I + \Gamma(A)$. Taking U to be of this form, we see that equation (1) is equivalent to the equation

$$(2) \qquad (I + \Gamma(A))(T + \varphi(A_1)) = (T + \varphi(\eta(A)))(I + \Gamma(A)),$$

that is, to

$$(3) \qquad \varphi(A - \eta(A)) + \varphi(\eta(A)) + \varphi(\eta(A))\Gamma(A) - \Gamma(A)\varphi(A_1) = \varphi(A_1).$$

By hypotheses (d) and (e), this last equation would follow from

$$(4) \qquad A + \psi_0(A, A) - \psi(A, A_1) = A_1;$$

thus, to solve (1), we have only to solve (4). We solve equation (4) for A in terms of A_1 by an iterative procedure. Put $A^{(1)} = A_1$, and, inductively, put

$$(5) \qquad A^{(n+1)} = A_1 - \psi_0(A^{(n)}, A^{(n)}) + \psi(A^{(n)}, A_1).$$

Put $t_n = |||A^{(n)}|||$; then, noting that $A^{(1)} = A_1$, we find at once from (5) that

$$(6) \qquad t_{n+1} \leqq t_1 + M_2 t_n^2 + M_2 t_1 t_n .$$

We shall prove inductively that $t_n \leqq 2t_1$. For $n = 1$, this is clear. On the other hand, assuming the truth of our assertion for a given n, it follows from (6) that $t_{n+1} \leqq t_1 + 6M_2 t_1^2$. Since, by hypothesis, $6M_2 t_1 \leqq 1$, we have $t_{n+1} \leqq 2t_1$ and our assertion follows. It also follows from (5) and hypotheses (d) and (e) that

$$(7) \qquad |||A^{(n+1)} - A^{(n)}|||$$
$$\leqq M_2(|||A^{(n)}||| + |||A^{(n-1)}||| + |||A_1|||)\, |||A^{(n)} - A^{n(-1)}||| .$$

Since, by what we have proved, $|||A^{(n)}||| \leqq 2t_1$, it follows from (7) that

$$(8) \qquad |||A^{(n+1)} - A^{(n)}||| \leqq 5M_2 t_1 \, |||A^{(n)} - A^{(n-1)}||| .$$

Since $5M_2 t_1 < 1$ by hypothesis, it follows from (8) that the sequence $\{A^{(n)}\}$ converges to a limit A, which necessarily satisfies $|||A||| \leqq 2\, |||A_1|||$. Thus, by hypothesis, $|\Gamma(A)| \leqq 2M_1 |||A|||_1 < 1$, so that by VII.6.1 the operator $I + \Gamma(A)$ has a bounded inverse.

We have now shown that equation (4) has a solution A such that $U = I + \Gamma(A)$ has a bounded inverse. Since equation (4) implies equation (1), we see that $T + \varphi(A_1)$ and $T + \varphi(\eta(A))$ are similar operators, and the proof of our theorem is complete. Q.E.D.

We shall apply Theorem 1 as follows. Let $\mathfrak{X}$ be a separable Hilbert space, and let T be a compact normal operator. By Theorem X.3.4 there exists a complete orthonormal basis $\{x_n\}$ consisting of eigenvectors of T. We let $Tx_n = \lambda_n x_n$, and suppose that none of the eigenvalues λ_n is multiple (that is, that the sequence $\{\lambda_n\}$ contains no repetitions) and that $\lambda_n \neq 0$ for $n \geqq 1$. By Corollary X.3.5 we have $\lambda_n \to \infty$. Let r_n be the distance from the eigenvalue λ_n to the other points on the spectrum of T, so that $r_n > 0$, while, of course, the sequence $\{r_n\}$, like the sequence λ_n, is bounded. Let R be the closed, densely defined, unbounded operator in $\mathfrak{X}$ whose domain $\mathfrak{D}(R)$ consists of all the elements $x = \sum_{n=1}^{\infty} \alpha_n x_n$ in $\mathfrak{X}$ for which $\sum_{n=1}^{\infty} r_n^{-2} |\alpha_n|^2$ is finite, and which is defined by

$$(9) \qquad R\left(\sum_{n=1}^{\infty} \alpha_n x_n\right) = \sum_{n=1}^{\infty} r_n^{-1} \alpha_n x_n .$$

It is plain from the definition of R and the boundedness of the sequence $\{r_n\}$ that the operator R^{-1} is bounded. Let $\mathfrak{A}$ be the set of all operators $A \in B(\mathfrak{X})$ for which $A\mathfrak{X} \subseteq \mathfrak{D}(R)$ and for which RA belongs to the Hilbert-Schmidt class HS, and let $|||A||| = \|RA\|$, so that the norm of an element $A \in \mathfrak{A}$ is simply the Hilbert-Schmidt norm of the operator RA.

If $A_n \in \mathfrak{A}$ and $\{A_n\}$ is a Cauchy sequence, then, since R^{-1} is bounded, $|(A_n - A_m)x| \leqq \|A_n - A_m\| \leqq |R^{-1}| \, \|R(A_n - A_m)\| \leqq |R^{-1}| \, |||A_n - A_m|||$ by Corollary XI.6.5. It follows that $\{A_n x\}$ is a Cauchy sequence of vectors for each $x \in \mathfrak{X}$. Moreover, by definition of the norm in $\mathfrak{A}$, $\{RA_n x\}$ is also a Cauchy sequence. If $A = \lim_{n\to\infty} A_n$, then, since R is a closed operator, we have $Ax \in \mathfrak{D}(R)$ for each $x \in \mathfrak{X}$, and $RAx = \lim_{n\to\infty} RA_n x$. Thus, using Theorem XI.6.4 and the definition of the norm in $\mathfrak{A}$, we find that $\lim_{n\to\infty} |||A_n - A||| = 0$, proving that $\mathfrak{A}$ is a complete B-space.

Let φ be the identity mapping of $\mathfrak{A}$ into $B(\mathfrak{X})$. Since R^{-1} is bounded, it follows from Corollary XI.5.6 that

$$(10) \qquad |A| = |\varphi(A)| \leqq \|\varphi(A)\| \leqq |R^{-1}| \, \|RA\| = |R^{-1}| \, |||A||| .$$

Thus the norm of the mapping φ is at most $|R^{-1}|$. Let $\eta(A)$ be defined by

$$\eta(A)\left(\sum_{n=1}^{\infty}\alpha_n x_n\right)=\sum_{n=1}^{\infty}\alpha_n(Ax_n, x_n)x_n \tag{11}$$

for each $x=\sum_{n=1}^{\infty}\alpha_n x_n$ in $\mathfrak{X}$. Then, by Corollary XI.6.3, we have

$$\begin{aligned}|||\eta(A)|||^2 &= \sum_{n=1}^{\infty}|(Ax_n, x_n)|^2 r_n^{-2}=\sum_{n=1}^{\infty}|(RAx_n, x_n)|^2\\ &\leqq \|RA\|^2=|||A|||^2.\end{aligned} \tag{12}$$

Put

$$\Gamma(A)x=\sum_{\substack{n,m=1\\ n\neq m}}^{\infty}\frac{(x_m, A^*x_n)}{\lambda_n-\lambda_m}(x, x_m)x_n \tag{13}$$

for $x\in\mathfrak{X}$ and $A\in\mathfrak{A}$. By Schwarz' inequality, and since $\{x_n\}$ is an orthonormal basis, we have

$$\begin{aligned}&\sum_{n=1}^{\infty}\left(\sum_{\substack{m=1\\ m\neq n}}^{\infty}\left|\frac{(x_m, A^*x_n)}{\lambda_n-\lambda_m}\right||(x, x_m)|\right)^2\\ &\qquad\leqq\sum_{n=1}^{\infty}\left\{\left(\sum_{\substack{m=1\\ m\neq n}}^{\infty}r_n^{-2}|(Ax_m, x_n)|^2\right)\left(\sum_{\substack{m=1\\ n\neq m}}^{\infty}|(x, x_m)|^2\right)\right\}\\ &\qquad\leqq |x|^2\sum_{n,m=1}^{\infty}r_n^{-2}|(Ax_m, x_n)|^2.\end{aligned} \tag{14}$$

Therefore, by Lemma IV.4.9 and Corollary XI.6.3, the series (13) converges and $|\Gamma(A)|\leqq\|RA\|=|||A|||$. By (13) we have

$$\begin{aligned}(T\Gamma(A)-\Gamma(A)T)x &= \sum_{\substack{n,m=1\\ n\neq m}}^{\infty}(\lambda_n-\lambda_m)\frac{(x_m, A^*x_n)}{\lambda_n-\lambda_m}(x, x_m)x_n\\ &= \sum_{\substack{n,m=1\\ n\neq m}}^{\infty}(x_m, A^*x_n)(x, x_m)x_n\\ &= \sum_{n,m=1}^{\infty}(x_m, A^*x_n)(x, x_m)x_n-\sum_{n=1}^{\infty}(Ax_n, x_n)(x, x_n)x_n\end{aligned} \tag{15}$$

for $x\in\mathfrak{X}$. Thus, since $\{x_n\}$ is an orthonormal basis, and by (11) we have

$$\begin{aligned}(T\Gamma(A)-\Gamma(A)T)x &= \sum_{n=1}^{\infty}(Ax, x_n)x_n-\eta(A)x\\ &= Ax-\eta(A)x.\end{aligned} \tag{16}$$

This shows that hypotheses (a), (b), and (c) of Theorem 1 are verified with $M_1 = \max(1, |R^{-1}|)$.

If $A, A_1 \in \mathfrak{A}$, define $\psi(A, A_1)$ by $\psi(A, A_1) = \Gamma(A)A_1$, and $\psi_0(A, A_1)$ by $\psi_0(A, A_1) = \eta(A)\Gamma(A_1)$. We have already seen that $\eta(A) \in \mathfrak{A}$, so that $\eta(A)\mathfrak{X} \subseteq \mathfrak{D}(R)$. Thus, using the inequality $|\Gamma(A_1)| \leqq |||A_1|||$, and using (12) and Corollary XI.6.5, we have $|||\eta(A)\Gamma(A_1)||| = \|R\eta(A)\Gamma(A_1)\| \leqq |||\eta(A)||| \times |\Gamma(A_1)| \leqq |||A||| \; |||A_1|||$.

Next, let $x \in \mathfrak{X}$, and let $x = \sum_{n=1}^{\infty} \alpha_n x_n$. Then, by the completeness of the orthonormal set $\{x_n\}$, we have

$$A_1 x = \sum_{n,m=1}^{\infty} \alpha_n (A_1 x_n, x_m) x_m, \tag{17}$$

and therefore by (13) we have

$$\Gamma(A)A_1 x = \sum_{\substack{n,m,l=1 \\ m \neq l}}^{\infty} \alpha_n (A_1 x_n, x_m) \frac{(Ax_m, x_l)}{\lambda_m - \lambda_l} x_l. \tag{18}$$

The "matrix elements" $(\Gamma(A)A_1 x_n, x_l)$ are given by the formula

$$(\Gamma(A)A_1 x_n, x_l) = \sum_{\substack{m=1 \\ m \neq l}}^{\infty} \frac{(A_1 x_n, x_m)(Ax_m, x_l)}{\lambda_m - \lambda_l}. \tag{19}$$

Using this fact, Schwarz's inequality, and Corollary XI.6.3, we may estimate the norm $\|R\Gamma(A)A_1\|$ as follows:

$$\begin{aligned} \|R\Gamma(A)A_1\| &= \sum_{n,l=1}^{\infty} \left\{ r_l^{-2} \left| \sum_{\substack{m=l \\ m \neq l}}^{\infty} (A_1 x_n, x_m) \frac{(Ax_m, x_l)}{\lambda_m - \lambda_l} \right| \right\} \\ &\leqq \sum_{n,l=1}^{\infty} \left(\sum_{\substack{m=1 \\ m \neq l}}^{\infty} r_l^{-1} |A_1 x_n, x_m) \, |r_m^{-1}| \, (Ax_m, x_l)| \right)^2 \\ &\leqq \sum_{n,l=1}^{\infty} \left(\sum_{m=1}^{\infty} |(RA_1 x_n, x_m)| \, |(RAx_m, x_l)| \right)^2 \\ &\leqq \sum_{n,l=1}^{\infty} \left\{ \sum_{m=1}^{\infty} |(RA_1 x_n, x_m)|^2 \right\} \left\{ \sum_{m=1}^{\infty} |(RAx_m, x_l)|^2 \right\} \\ &= \|RA_1\| \, \|RA\| = |||A||| \; |||A_1|||. \end{aligned} \tag{20}$$

It follows by these same estimates that $\Gamma(A)A_1 x \in \mathfrak{D}(R)$ for all $x \in \mathfrak{X}$. Thus hypotheses (d) and (e) of Theorem 1 are satisfied with the constant $M_2 = 1$.

Therefore, if we apply Theorem 1, we obtain the following result.

2 THEOREM. *Let $\mathfrak{H}$ be a separable Hilbert space, and let T be a compact normal operator in $\mathfrak{H}$. Let $\{x_n\}$ be a complete orthonormal set of eigenvectors of T, and let $Tx_n = \lambda_n x_n$. Suppose that the sequence $\{\lambda_n\}$ contains no repetitions, and that all the eigenvalues λ_n are different from zero. Let R be the closed, densely defined, unbounded operator in $\mathfrak{H}$ whose domain $\mathfrak{D}(R)$ consists of all the elements $x = \sum_{n=1}^{\infty} \alpha_n x_n$ in $\mathfrak{H}$ for which $\sum_{n=1}^{\infty} r_n^{-2} |\alpha_n|^2$ is finite, and which is defined by $R(\sum_{n=1}^{\infty} \alpha_n x_n) = \sum_{n=1}^{\infty} r_n^{-1} \alpha_n x_n$. Let $\mathfrak{A}$ be the set of all operators $A \in B(\mathfrak{H})$ for which $A\mathfrak{H} \subseteq \mathfrak{D}(R)$, and for which RA belongs to the Hilbert-Schmidt class HS. Then there exists a positive constant $\varepsilon(T)$ depending only on T such that if $A \in \mathfrak{A}$ and $\|RA\| < \varepsilon(T)$, then $T + A$ is similar to an operator T_1 commuting with T.*

3 COROLLARY. *If, under the hypotheses of the preceding theorem, we have $\|RA\| < \varepsilon(T)$, then $T + A$ is a spectral operator of scalar type.*

PROOF. Since the operator T_1 of the preceding theorem commutes with T, it must map every eigenspace $\mathfrak{H}(\lambda_n)$ of T into $\mathfrak{H}(\lambda_n)$. Thus we must have $T_1 x_n = \mu_n x_n$, where μ_n is some complex constant. Then $(T_1^* x_n, x_m) = (x_n, T_1 x_m) = \bar{\mu}_m \delta_{n,m}$, proving that $T_1^* x_n = \bar{\mu}_n x_n$. Therefore T_1 and T_1^* commute, so that T_1 is normal. Since $T + A$ is similar to T_1, $T + A$ is a spectral operator of scalar type. Q.E.D.

Modern operator methods will next be used to discuss classical wave theory.

4. The Wave Operator Method

In the present section we shall develop a circle of ideas which have played a great role in recent theoretical physics. The central formal notion of this group of ideas is most easily seen in the proof of the following interesting (even if false!) pseudo-theorem.

PSEUDO-THEOREM. *Any two self adjoint operators in Hilbert space are unitarily equivalent.*

PSEUDO-PROOF. Let the two operators be H_1 and H_2; put $U_t = e^{itH_1} e^{-itH_2}$ for all real t, so that by Theorem XII.2.6 U_t is the product of two unitary operators and is therefore a unitary operator. Write $U_\infty = \lim_{t \to \infty} U_t$, so that U_∞, as the limit of unitary operators, is itself unitary. We have $e^{isH_1} U_t e^{-isH_2} = U_{t+s}$; thus $e^{isH_1} U_\infty e^{-isH_2} = U_\infty$, or $e^{isH_1} U_\infty = U_\infty e^{isH_2}$. Differentiating this last equation with respect to s and setting

$s=0$ we find $H_1U_\infty = U_\infty H_2$, so that $U_\infty^* H_1 U_\infty = H_2$, and therefore H_1 and H_2 are unitarily equivalent. Q.E.D.

The error enters, of course, in the implicit assumption that $U_t v$ has a limit $U_\infty v$ for all v in Hilbert space. However, by adding suitable hypotheses and by arguing more carefully, we can extract a kernel of truth from the erroneous proof above. Let us begin to retrace our steps, more carefully this time!

Throughout the remainder of the present section, $\mathfrak{H}$ will denote a complex Hilbert space.

1 DEFINITION. Let H_1 and H_2 be two self adjoint (possibly unbounded) operators in $\mathfrak{H}$. Put

$$\sum (H_1, H_2) = \{x \in \mathfrak{H} \mid \lim_{t\to\infty} e^{itH_1} e^{-itH_2} x \text{ exists}\}, \tag{1}$$

and

$$\mathscr{U}(H_1, H_2)x = \lim_{t\to\infty} e^{itH_1} e^{-itH_2} x, \qquad x \in \sum (H_1, H_2). \tag{2}$$

2 LEMMA. *Let H_1 and H_2 be two self adjoint operators in $\mathfrak{H}$. Then $\sum (H_1, H_2)$ is a closed subspace of $\mathfrak{H}$. Moreover, $(\mathscr{U}(H_1, H_2)v, \mathscr{U}(H_1, H_2)w) = (v, w)$ for all $v, w \in \sum (H_1, H_2)$.*

PROOF. The operators $\mathscr{U}_t = e^{itH_1}e^{-itH_2}$ are unitary by Theorem XII.2.6, so that our first assertion follows immediately from Theorem II.3.6. We have $(\mathscr{U}_t v, \mathscr{U}_t w) = (v, w)$ for all $v, w \in \sum (H_1, H_2)$, and, letting $t \to \infty$, we obtain our second assertion. Q.E.D.

3 LEMMA. *Let H_1, H_2, and H_3 be three self adjoint operators in $\mathfrak{H}$. Let $x \in \sum (H_2, H_1)$. Then $\mathscr{U}(H_2, H_1)x \in \sum (H_3 H_2)$ if and only if $x \in \sum (H_3, H_1)$, and, in this case, $\mathscr{U}(H_3, H_2)\mathscr{U}(H_2, H_1)x = \mathscr{U}(H_3, H_1)x$.*

PROOF. Put $\mathscr{U}_t^{(2,1)} = e^{itH_2}e^{-itH_1}$, $\mathscr{U}_t^{(3,2)} = e^{itH_3}e^{-itH_2}$, and $\mathscr{U}_t^{(3,1)} = e^{itH_3}e^{-itH_1}$ for all real t, so that the operators $\mathscr{U}_t^{(i,j)}$ are unitary by Theorem XII.2.6, and so that $\mathscr{U}_t^{(3,2)}\mathscr{U}_t^{(2,1)} = \mathscr{U}_t^{(3,1)}$. If $x \in \sum (H_2, H_1)$ and we put $x' = \mathscr{U}(H_2, H_1)x_1$, then

$$\mathscr{U}_t^{(3,1)}x = \mathscr{U}_t^{(3,2)}\mathscr{U}_t^{(2,1)}x = \mathscr{U}_t^{(3,2)}x' + \mathscr{U}_t^{(3,2)}(x' - \mathscr{U}_t^{(2,1)}x). \tag{3}$$

Since the second term on the right of this last equation converges to zero, it follows that $\mathscr{U}_t^{(3,1)}x$ has a limit as $t \to \infty$ if and only if $\mathscr{U}_t^{(3,2)}x'$ has a limit as $t \to \infty$, and that, if either limit exists, both limits are equal. Q.E.D.

4 Corollary. *Let* H_1 *and* H_2 *be two self adjoint operators in* $\mathfrak{H}$. *Then* $\mathscr{U}(H_1, H_2) \sum (H_1, H_2) = \sum (H_2, H_1)$, *and, if* $x \in \sum (H_1, H_2)$, *we have* $\mathscr{U}(H_2, H_1)\mathscr{U}(H_1, H_2)x = x$.

Proof. It is plain from Definition 1 that $\sum (H_1, H_1) = \mathfrak{H}$, and that $\mathscr{U}(H_1, H_1) = I$. Therefore, if we let the three operators of the preceding lemma be H_2, H_1, H_1, we obtain the present corollary. Q.E.D.

5 Corollary. *Under the hypotheses of the preceding corollary,* $\mathscr{U}(H_1, H_2)$ *is an isometric mapping of* $\sum (H_1, H_2)$ *onto* $\sum (H_2, H_1)$.

Proof. By the preceding corollary, $\mathscr{U}(H_1, H_2) \sum (H_1, H_2) = \sum(H_2, H_1)$; the isometry of $\mathscr{U}(H_1, H_2)$ is stated in Lemma 2. Q.E.D.

6 Lemma. *Let* H_1 *and* H_2 *be two self adjoint operators in* $\mathfrak{H}$. *Then*:

(a) *If* $-\infty < s < +\infty$, *and* $x \in \sum (H_1, H_2)$, *we have* $e^{isH_2} x \in \sum (H_1, H_2)$ *and* $\mathscr{U}(H_1, H_2)e^{isH_2} x = e^{isH_1} \mathscr{U}(H_1, H_2)x$.

(b) *If* $F(\cdot)$ *is any bounded Borel function defined on the real axis, then* $F(H_2) \sum (H_1, H_2) \subseteq \sum (H_1, H_2)$ *and*

$$(4) \qquad \mathscr{U}(H_1, H_2)F(H_2)x = F(H_1)\mathscr{U}(H_1, H_2)x, \qquad x \in \sum (H_1, H_2).$$

(c) *Let* $\mathfrak{D}(H_i)$ *be the domain of* H_i, $i = 1, 2$. *The restrictions* $H_1 | (\sum (H_2, H_1) \cap \mathfrak{D}(H_1))$ *and* $H_2 | (\sum (H_1, H_2) \cap \mathfrak{D}(H_2))$ *are self adjoint operators in* $\sum (H_2, H_1)$ *and* $\sum (H_1, H_2)$, *respectively. Moreover,*

$$(5) \quad H_1 | (\sum (H_2, H_1) \cap \mathfrak{D}(H_1)) \\ = \mathscr{U}(H_1, H_2)\{H_2 \Big| \sum (H_1, H_2) \cap \mathfrak{D}(H_2)\}\mathscr{U}(H_1, H_2)^{-1}.$$

Proof. Let $x \in \sum (H_1, H_2)$, and let s be real. Then $e^{-isH_2} e^{itH_1} e^{-itH_1} e^{isH_2} x = e^{i(t-s)H_1} e^{-i(t-s)H_2} x$, and therefore $e^{isH_2}x \in \sum (H_1, H_2)$ and $e^{-isH_1} \mathscr{U}(H_1, H_2)e^{isH_2} x = \sum (H_1, H_2)$, proving (a).

From (a) it follows that, if F_0 is a finite linear combination of the exponential functions $e^{it\lambda}$ of λ, then $F_0(H_2)x \in \sum (H_1, H_2)$, and $F_0(H_1) \mathscr{U}(H_1, H_2)x = \mathscr{U}(H_1, H_2)F_0(H_1)x$. By the Stone-Weierstrass theorem (IV.6.16), any continuous function F vanishing outside a compact subset of the real axis is the uniform limit of finite linear combinations of such functions. Thus, by Theorem XII.2.9(a), we can find a sequence F_n of functions which are finite linear combinations of exponential functions, such that $\lim_{n\to\infty} F_n(H_i) = F(H_i)$ in the uniform topology of operators,

$i = 1, 2$. We therefore find that equation (4) is valid whenever F is a continuous function vanishing outside a compact subset of the real axis.

Next let F be the characteristic function of an interval of the real axis. Using Theorem XII.2.6, it follows that we can find a sequence F_n of continuous functions, each vanishing outside a compact subset of the real axis, such that $\lim_{n\to\infty} F_n(H_i) = F(H_i)$ in the strong topology of operators, $i = 1, 2$. Therefore equation (4) is valid whenever F is the characteristic function of an interval of the real axis.

Next, let $E_i(\cdot)$ be the spectral resolution of the operator H_i, $i = 1, 2$. Put $x' = \mathscr{U}(H_1, H_2)x$. Let F be an arbitrary bounded Borel measurable function. By Lemma III.8.3, we may find a sequence $\{F_n^\dagger\}$ of functions, each of which is a finite linear combination of characteristic functions of subintervals of the real axis, such that

$$\lim_{n\to\infty} \int_{-\infty}^{+\infty} |F_n^\dagger(\lambda) - F(\lambda)|^2 \nu(d\lambda) = 0, \tag{6}$$

where, for each Borel set e,

$$\nu(e) = |E_1(e)x|^2 + |E_2(e)x'|^2. \tag{7}$$

If we put $F_n(\lambda) = F_n^\dagger(\lambda)$ if $|F_n^\dagger(\lambda)| \leqq C = \sup_{-\infty<\lambda<+\infty} |F(\lambda)|$, and $F_n(\lambda) = C^{-1}F_n^\dagger(\lambda)$ if $|F_n^\dagger(\lambda)| > C$, then $\{F_n(\lambda)\}$ is uniformly bounded, and we have $|F_n(\lambda) - F(\lambda)| \leqq |F_n^\dagger(\lambda) - F(\lambda)|$ for all n and λ, so that

$$\lim_{n\to\infty} \int_{-\infty}^{+\infty} |F_n(\lambda) - F(\lambda)|^2 \nu(d\lambda) = 0. \tag{8}$$

Thus, by Theorem XII.2.6, we have $\lim_{n\to\infty} F_n(H_1)x' = F(H_1)x'$ and $\lim_{n\to\infty} F_n(H_2)x = F_n(H_2)x$. Using what we have already proved and using Lemma 2, we find that $F(H_2)x \in \sum (H_1, H_2)$ and that

$$\begin{aligned} F(H_1)\mathscr{U}(H_1, H_2)x &= \lim_{n\to\infty} F_n(H_1)\mathscr{U}(H_1, H_2)x \\ &= \lim_{n\to\infty} \mathscr{U}(H_1, H_2)F_n(H_2)x \\ &= \mathscr{U}(H_1, H_2)F(H_2)x, \end{aligned} \tag{9}$$

proving (b).

To prove (c), note that by (b) the two operators $(\pm iI - H_2)^{-1}$ map $\sum (H_1, H_2)$ into itself. If $(iI - H_2)^{-1} \sum (H_1, H_2)$ were not dense in $\sum (H_1, H_2)$, it would follow by the Hahn-Banach theorem (cf. Corollary II.3.13) that for some $0 \neq x \in \sum (H_1, H_2)$ we have $(x, (iI - H_2)^{-1}y) = 0$,

$y \in \sum (H_1, H_2)$. But then $((-iI - H_2)^{-1}x, y) = 0$, $y \in \sum (H_1, H_2)$, so that $(-iI - H_2)^{-1}x = 0$, which is impossible. It follows that

$$(iI - H_2)^{-1} \sum (H_1, H_2) \subseteq \sum (H_1, H_2) \cap \mathfrak{D}(H_2)$$

is dense in $\sum (H_1, H_2)$.

Let $x \in \sum (H_1, H_2) \cap \mathfrak{D}(H_2)$. We shall show that $H_2 x \in \sum (H_1, H_2)$. To prove this, we reason as follows. We may write $H_2 x = y_1 + y_2$, where $y_1 \in \sum (H_1, H_2)$ and $y_2 \in \mathfrak{H} \ominus \sum (H_1, H_2)$. By Corollary XII.2.7, we have $(I + H_2^2)^{-1} = (iI - H_2)^{-1}(-iI - H_2)^{-1}$, and therefore, using (b), we have $(I + H_2^2)^{-1}H_2 x = (iI - H_2)^{-1}(i(iI + H_2)^{-1}x - x) \in \sum (H_1, H_2)$. Therefore $(I + H_2^2)^{-1}y_1 + (I + H_2^2)^{-1}y_2 \in \sum (H_1, H_2)$, and we see that $(I + H_2^2)^{-1} y_2 \in \sum (H_1, H_2)$. But since $(I + H_2^2)^{-1} \sum (H_1, H_2) \subseteq \sum (H_1, H_2)$, it follows that $(I + H_2^2)^{-1}(\mathfrak{H} \ominus \sum (H_1, H_2)) \subseteq \mathfrak{H} \ominus \sum (H_1, H_2)$. Thus $(I + H_2^2)^{-1}y_2 = 0$. But then $y_2 = 0$, which proves that $H_2 x \in \sum (H_1, H_2)$. It is immediate that $\hat{H}_2 = H_2 \mid (\sum (H_1, H_2) \cap \mathfrak{D}(H_2))$ is a symmetric operator in $\sum (H_1, H_2)$, and it follows that H_2 is closed. Since

$$(10) \qquad (\pm iI - \hat{H}_2)\{(\pm iI - H_2)^{-1} \mid \sum (H_1, H_2)\} = I,$$

it follows from Corollary XII.4.13(b) that $\hat{H}_2$ is self adjoint. Then, plainly, $(\pm iI - \hat{H}_2)^{-1} = (\pm iI - H_2)^{-1} \mid \sum (H_1, H_2)$. Arguing in the same way, we find that $\hat{H}_1 = H_1 \mid (\sum (H_2, H_1) \cap \mathfrak{D}(H_1))$ is a densely defined, self adjoint operator in $\sum (H_2, H_1)$, and that $(iI - \hat{H}_1)^{-1} = (iI - H_1)^{-1} \mid \sum (H_2, H_1)$. By (b), and by Corollary 4 and Corollary 5, we have

$$(11) \qquad (iI - \hat{H}_1)^{-1} = \mathscr{U}(H_1, H_2)(iI - \hat{H}_2)^{-1}\mathscr{U}(H_1, H_2)^{-1},$$

so that

$$(12) \qquad \hat{H}_1 = \mathscr{U}(H_1, H_2)\hat{H}_2\mathscr{U}(H_1, H_2)^{-1},$$

proving (c). Q.E.D.

We now wish to discuss a certain direct sum decomposition of Hilbert space, relative to a given self adjoint operator H, which is intimately related to the Lebesgue decomposition of a measure given by Theorem III.4.14.

7 Definition. Let H be a self adjoint operator, possibly unbounded, in the Hilbert space $\mathfrak{H}$ and let $E(e)$ be the spectral resolution of H. Let λ denote the Lebesgue measure on the real axis. R. Put

(13) $\sum_{ac}(H) = \{x \in \mathfrak{H} \mid |E(e)x|^2$ is a countably additive λ-continuous measure$\}$;

(14) $\sum_{sing}(H) = \{x \in \mathfrak{H} \mid |E(e)x|^2$ is a countably additive, λ-singular measure, and $|E(e)x|^2 = 0$ if e consists of a single point$\}$;

(15) $\sum_{p}(H) = \{x \in \mathfrak{H} \mid |E(R - e_0)x|^2 = 0$, where $e_0 = e_0(x)$ is a countable set of points$\}$.

8 LEMMA. *Let H be a self adjoint operator, possibly unbounded, in the Hilbert space $\mathfrak{H}$, and let $\mathfrak{D}(H)$ be the domain of H. Then*

(a) *The three spaces $\sum_{ac}(H)$, $\sum_{sing}(H)$, and $\sum_{p}(H)$ are closed and mutually orthogonal, and $\mathfrak{H} = \sum_{ac}(H) \oplus \sum_{sing}(H) \oplus \sum_{p}(H)$.*

(b) *Every $v \in \mathfrak{D}(H)$ can be written uniquely as $v = v_1 + v_2 + v_3$, where $v_1 \in \sum_{ac}(H) \cap \mathfrak{D}(H)$, $v_2 \in \sum_{sing}(H) \cap \mathfrak{D}(H)$, and $v_3 \in \sum_{p}(H) \cap \mathfrak{D}(H)$.*

(c) *The restriction $H \mid (\sum_{ac}(H) \cap \mathfrak{D}(H))$ is a self adjoint operator in $\sum_{ac}(H)$; the restriction $H \mid (\sum_{sing}(H) \cap \mathfrak{D}(H))$ is a self adjoint operator in $\sum_{sing}(H)$; the restriction $H \mid (\sum_{p}(H) \cap \mathfrak{D}(H))$ is a self adjoint operator in $\sum_{p}(H)$.*

(d) *$\sum_{p}(H)$ is the closed subspace spanned by all the eigenvectors of the operator H.*

PROOF. Let $x \in \mathfrak{H}$, and denote the real axis by R. By Theorem III.4.14, if we put $\nu(e) = |E(e)x|^2$, we may write $\nu = \alpha + \beta$ where α is λ-continuous and where there exists a λ-null set e_0 such that $\beta(e \cap (R - e_0)) = 0$ for each Borel set e. But then evidently $\alpha(e) = |E(e \cap (R - e_0))x|^2$ for each Borel set e. Let e_1 be the set of all points $s \in R$ such that $|E(\{s\})\, x|^2 > 0$. Since $|E(R)x|^2 = |x|^2$ is finite, e_1 is countable. Let $e_2 = e_0 - e_1$. If $x_1 = E(R - e_0)x$, $x_2 = E(e_2)x$, and $x_3 = E(e_1)x$. Then $x = x_1 + x_2 + x_3$, $|E(e)x_1|^2 = |E(e \cap (R - e_0))x|^2$, $|E(e)x_2|^2 = |E(ee_2)x|^2$, and $|E(e)x_3| = |E(ee_1)x|^2$. It is therefore clear that $x_1 \in \sum_{ac}(H)$, $x_2 \in \sum_{sing}(H)$, and $x_3 = \sum_{p}(H)$. If $x \in \mathfrak{D}(H)$, it follows from Theorem XII.2.6 that $x_i \in \mathfrak{D}(H)$, $i = 1, 2, 3$.

Next, let $v_1 \in \sum_{ac}(H)$, $v_2 = \sum_{sing}(H)$, and $v_3 \in \sum_{p}(H)$. Since the measure $|E(\cdot)v_2|^2$ is λ-singular, there exists a λ-null set e_0 such that $|E(R - e_0)v_2|^2 = 0$, so that $E(e_0)v_2 = v_2$. But then, since $|E(\cdot)v_1|^2$ is λ-continuous, we have $E(e_0)v_1 = 0$, so that

$$(v_1, v_2) = (v_1, E(e_0)v_2) = (E(e_0)v_1, v_2) = 0.$$

We leave the details of the similar proofs that $(v_1, v_3) = 0 = (v_2, v_3)$ to the reader, and conclude without further ado that $\sum_{ac}(H)$, $\sum_{sing}(H)$, and $\sum_p(H)$ are mutually orthogonal. Statement (b) of our lemma follows at once.

If $x_n \in \sum_{ac}(H)$ and $\lim_{n\to\infty} x_n = x$, then, by what we have already proved, we may write $x = y_1 + y_2 + y_3$, where $y_1 \in \sum_{ac}(H)$ and y_2, y_3 are orthogonal to $\sum_{ac}(H)$. But, since $x_n \in \sum_{ac}(H)$ we have $(x_n, y) = 0$ for $y \in \mathfrak{H} \ominus \sum_{ac}(H)$, and therefore $(x, y) = 0$ for $y \in \mathfrak{H} \ominus \sum_{ac}(H)$, which implies $(y_2 + y_3, y) = 0$ for $y \in \mathfrak{H} \ominus \sum_{ac}(H)$, or $y_2 + y_3 = 0$. Therefore $x = y_1 \in \sum_{ac}(H)$, proving that $\sum_{ac}(H)$ is closed. We may show similarly that $\sum_{sing}(H)$ and $\sum_p(H)$ are closed. Thus statement (a) of our theorem follows.

If F is a bounded Borel function defined on the real axis and $x \in \sum_{ac}(H)$, then, by Theorem XII.2.6 and Lemma III.4.13,

$$|E(e)F(H)x|^2 = \int_e |f(x)|^2 |E(d\lambda)x|^2 \leq \sup_{s\in R} |f(s)|^2 |E(e)x|^2 \tag{16}$$

is λ-continuous. This shows that $F(H)\sum_{ac}(H) \subseteq \sum_{ac}(H)$. It may be shown similarly that $F(H)\sum_{sing}(H) \subseteq \sum_{sing}(H)$, and $F(H)\sum_p(H) \subseteq \sum_p(H)$; we leave the details to the reader.

Using this last fact, it is easy to prove assertion (c) of the present lemma. We argue as follows. If $(iI - H)^{-1}\sum_{ac}(H)$ were not dense in $\sum_{ac}(H)$, it would follow by the Hahn-Banach theorem (cf. Corollary II.3.13) that for some $x \neq 0$ in $\sum_{ac}(H)$ we have $(x, (iI - H)^{-1}y) = 0$, $y \in \sum_{ac}(H)$. But then $((-iI - H)^{-1}x, y) = 0$, $y \in \sum_{ac}(H)$, so that $(-iI - H)^{-1}x = 0$, which is impossible. It follows that $(iI - H)^{-1}\sum_{ac}(H) \subseteq \sum_{ac}(H) \cap \mathfrak{D}(H)$ is dense in $\sum_{ac}(H)$.

Let $x \in \sum_{ac}(H) \cap \mathfrak{D}(H)$. We shall show that $Hx \in \sum_{ac}(H)$. To prove this, we reason as follows. We may write $Hx = y_1 + y_2$, where $y_1 \in \sum_{ac}(H)$, $y_2 \in \mathfrak{H} \ominus \sum_{ac}(H)$. We have $(I + H^2)^{-1} = (iI - H)^{-1}(iI - H)^{-1}$, by Theorem XII.2.6, and therefore, using what we have already proved, we have $(I + H^2)^{-1}Hx = (iI - H_2)^{-1}(i(iI + H)^{-1}x - x) \in \sum_{ac}(H)$. Therefore $(I + H^2)^{-1}y_1 + (I + H^2)^{-1}y_2 \in \sum_{ac}(H)$, and it follows that $(I + H^2)^{-1}y_2 \in \sum_{ac}(H)$. But, since $(I + H^2)\sum_{ac}(H) \subseteq \sum_{ac}(H)$, it follows that $(I + H^2)(\mathfrak{H} \ominus \sum_{ac}(H)) \subseteq \mathfrak{H} \ominus \sum_{ac}(H)$. Thus $(I + H^2)^{-1}y_2 = 0$, so that $y_2 = 0$, which proves that $Hx \in \sum_{ac}(H)$. It is immediate that $\hat{H} = H|(\sum_{ac}(H) \cap \mathfrak{D}(H))$

is a symmetric operator in $\sum_{ac}(H)$, and it follows easily that $\hat{H}$ is closed. Since

$$(\pm iI - \hat{H})\{(\pm iI - H)^{-1} | \textstyle\sum_{ac}(H)\} = I, \tag{17}$$

it follows from Corollary XII.4.13(b) that $\hat{H}$ is self adjoint. We may show similarly that $H | (\sum_{sing}(H) \cap \mathfrak{D}(H))$ is a densely defined self adjoint operator in $\sum_{sing}(H)$, and that $H | (\sum_{p}(H) \cap \mathfrak{D}(H))$ is a densely defined self adjoint operator in $\sum_{p}(H)$; details are left to the reader. Statement (c) of the present lemma follows at once.

Finally, to prove (d), let $x \in \sum_{p}(H)$. Then by Definition 7 there exists a countable subset $e_0 = \{\lambda_j\} \subseteq R$ such that $x = E(e_0)x$. But then $x = \sum_{j=1}^{\infty} E(\lambda_j)x$, and by Theorem XII.2.6, $HE(\lambda_j)x = \lambda_j E(\lambda_j)x$. This shows that each $x \in \sum_{p}(H)$ is the sum of a series of eigenvectors of H. Conversely, let $y \in \mathfrak{D}(H)$, and let $Hy = \lambda_0 y$, where $\lambda_0 \beta$ is real. Then by Theorem XII.2.6 we have

$$0 = |Hy - \lambda_0 y|^2 = \int_{-\infty}^{+\infty} |\lambda - \lambda_0|^2 \, |E(d\lambda)y|^2. \tag{18}$$

Thus $E(R - \{\lambda_0\})y = 0$, so that $y \in \sum_{p}(H)$. This shows that every eigenvector of H belongs to $\sum_{p}(H)$, so that assertion (d) follows at once. Q.E.D.

Our main aim in the present section is to prove the following theorem, due to Kato and Kuroda, which shows that the preceding corollaries and lemmas numbered 2 through 6 have non-trivial application to a wide class of self adjoint operators.

→ 9 THEOREM. *Let H be a self adjoint operator in $\mathfrak{H}$, with domain $\mathfrak{D}(H)$ and resolution of the identity $E(\cdot)$. Let V be a symmetric operator in $\mathfrak{H}$, with domain $\mathfrak{D}(V)$. Suppose that $\mathfrak{D}(V) \supseteq \mathfrak{D}(H)$, that the operator $V(iI - H)^{-1}$ is compact, and that $(iI - H)^{-1}V(iI - H)^{-1}$ is of trace class. Then*

(a) *$H_1 = H + V$ is a self adjoint operator;*
(b) *$\sum(H_1, H) \supseteq \sum_{ac}(H)$;*
(c) *$\mathscr{U}(H_1, H)\sum_{ac}(H) = \sum_{ac}(H_1)$.*

To the proof of Theorem 9 we must prefix a number of lemmas.

10 LEMMA. *Let V be a bounded operator in $\mathfrak{H}$, and let $V = QR$ be its canonical factorization into the product of a partial isometry Q and a positive Hermitian operator R (cf. Theorem XII.7.7). Then V belongs to one of the*

classes $\mathscr{C}_p$ *of compact operators (cf. Definition* XI.9.1) *if and only if R belongs to this same class* $\mathscr{C}_p$; *moreover, V and R have the same norm as elements of* $\mathscr{C}_p$.

11 Corollary. *An operator V belonging to the trace class* $\mathscr{C}_1$ *can be factored as* $V = AB$, *where A and B belong to the Hilbert-Schmidt class* $HS = \mathscr{C}_2$, *and where the Hilbert-Schmidt norms* $\|A\|$ *and* $\|B\|$ *are both equal to the square root* $|V|_1^{1/2}$ *of the trace norm* $|V|_1$ *of the operator V (cf. Definition* XI.9.1 *and Lemma* XI.9.9(e)). *The operator V may also be factored as* $V = CD$, *where C is compact, and where D is of trace class.*

Proof of Lemma 10. We note from Theorem XII.7.7 that in the canonical factorization $V = QR$ the initial domain of Q is the closure of the range of the operator R, and thus (cf. Definition XII.7.4) $V^*V = RQQ^*R = R^2$. Therefore, by Lemma XII.7.3, R is the unique positive square root $(V^*V)^{1/2}$ of the bounded self adjoint operator V^*V. As is observed in the first sentence of Section XI.9, $(V^*V)^{1/2}$ is compact. Lemma 10 now follows immediately from Definition XI.9.1. Q.E.D.

Proof of Corollary 11. Let Q and R be as in Lemma 10. If the positive eigenvalues of R, arranged in decreasing order and repeated according to multiplicity, are $\mu_1, \mu_2, \ldots$, then the positive eigenvalues of $R^{1/2}$, arranged in decreasing order and repeated according to multiplicity, are $\mu_1^{1/2}, \mu_2^{1/2}, \ldots$. It follows by Definition XI.9.1 that $|R|_1 = |R^{1/2}|_2^2$. Put $A = QR^{1/2}$, $B = R$. Since Q is a partial isometry, $|Q| \leqq 1$, and thus, using Lemma 10 and Lemma XI.9.9, we have $|A|_2 \leqq |V|_1^{1/2}$ and $|B|_2 = |V|_1^{1/2}$. But then, by Lemma XI.9.14, we must also have $|A|_2 \geqq |V|^{1/2}$. Since $|A|_2 = \|A\|$ and $|B|_2 = \|B\|$, the first assertion of Corollary 11 is proved.

To prove the second assertion of Corollary 11, let x_i be an eigenvector of R belonging to the eigenvalue μ_i. Since $\sum_{i=1}^{\infty} \mu_i = |R|_1 < \infty$, we can find an increasing sequence of integers $\{n_j\}$ such that $\sum_{n=n_j}^{\infty} \mu_i < 2^{-j}$, and then, putting $c_i = 2^{j/2}$ for $n_j \leqq i < n_{j+1}$, we have $\sum_{i=1}^{\infty} c_i \mu_i < \infty$ while $\lim_{i\to\infty} c_i = \infty$. Define operators R_1 and R_2 by requiring that $R_1 x_i = c_j^{-1} x_i$, that $R_2 x_i = c_i \mu_i x_i$, and that $R_1 x = R_2 x = 0$ if $x \in \mathfrak{H}$ is orthogonal to all the vectors x_i. Plainly $R_1 R_2 = R$, whereas $R_2 \in \mathscr{C}_1$. If E_i is the projection on the one-dimensional space spanned by x_i, then we have

$$\lim_{n\to\infty} \sum_{i=1}^{n} c_i E_i = R_1$$

in the uniform topology of operators. Hence, by Corollary VI.5.5, R_1

is compact. Put $C = QR_1$, and $D = R_2$, so that $V = CD$. The operator C is compact by Corollary VI.5.5, and thus proof of Corollary 11 is complete. Q.E.D.

12 LEMMA. *If C is a compact operator in $\mathfrak{H}$, and $\{T_n\}$ is a uniformly bounded sequence of operators in $\mathfrak{H}$ converging strongly to zero, then $\{T_n C\}$ converges uniformly to zero and $\{CT_n^*\}$ converges uniformly to zero. Moreover, if C belongs to the trace class $\mathscr{C}_1$, then $T_n C$ converges to zero in trace norm, and CT_n^* converges to zero in trace norm.*

PROOF. The set $K = C(\{x \in \mathfrak{H} \mid |x| \leqq 1\})$ is conditionally compact, and thus for each $\varepsilon > 0$ there exists a finite set $x_1, \ldots, x_m$ of elements of K such that each $x \in K$ satisfies $|x - x_i| < \varepsilon$ for some $1 \leqq i \leqq m$. Thus, if we let $M = \sup_n |T_n|$, and choose n_0 so large that $|T_n x_i| < \varepsilon$ for $n \geqq n_0$ and $1 \leqq i \leqq m$, we have $|T_n x| \leqq |T_n x_i| + |T_n(x - x_i)| \leqq \varepsilon + M\varepsilon$ for $x \in K$ and $n \geqq n_0$. Therefore $|T_n C| \leqq \varepsilon(M+1)$ for $n \geqq n_0$, proving $|T_n C| \to 0$ as $n \to \infty$.

By Theorem VI.5.2, C^* is a compact operator. Thus, by what we have already proved, $|T_n C^*| \to 0$ as $n \to \infty$. But $|CT_n^*| = |(T_n C^*)^*| = |T_n C^*|$, so $|CT_n^*| \to 0$ as $n \to \infty$ also.

This proves the first assertion of the lemma. To prove the second, we use Corollary 11 to write $C = AB$, where A is compact and B belongs to the trace class. Then, by what we have already proved, $T_n A$ converges to zero in norm, and thus, by Lemma XI.9.9, $T_n C = (T_n A)B$ converges to zero in trace norm.

By Lemma XI.9.6(c) and Definition XI.9.1, C^* belongs to the trace class. By what we have already proved, it follows that $T_n C^*$ converges to zero in trace norm. Thus, using Lemma XI.9.6(c) and Definition XI.9.1 once more, we can see that $CT_n^* = (T_n C^*)^*$ converges to zero in trace norm also. Q.E.D.

13 LEMMA. *Let H be a self adjoint operator in $\mathfrak{H}$, with domain $\mathfrak{D}(H)$, and let V be a symmetric operator in $\mathfrak{H}$, with domain $\mathfrak{D}(V)$. Suppose that $\mathfrak{D}(V) \supseteq \mathfrak{D}(H)$, and that the operator $V(iI - H)^{-1}$ is compact, Then*

(a) *$H + V$ is self adjoint;*

(b) *we have $|V(\lambda_0 I - H)^{-1}| < \frac{1}{2}$ if $\mathscr{I}(\lambda_0)$ is sufficiently large;*

(c) *if $\mathscr{I}\lambda_0 \neq 0$, and $I - V(\lambda_0 I - H)^{-1}$ has a bounded inverse, then we have*

$$(\lambda_0 I - H - V)^{-1} = (\lambda_0 I - H)^{-1}(I - V(\lambda_0 I - H)^{-1})^{-1}.$$

PROOF. We may write

$$V(\lambda_0 I - H)^{-1} = V(iI - H)^{-1}(iI - H)(\lambda_0 I - H)^{-1} \tag{19}$$
$$= V(iI - H)^{-1}(I + (i - \lambda_0)(\lambda_0 I - H)^{-1}).$$

If we set $\varphi_{\lambda_0}(\lambda) = -(i + \lambda)(\lambda_0 - \lambda)^{-1}$, then we have $\varphi_{\lambda_0}(H) = \{I + (i - \lambda_0)(\lambda_0 I - H)^{-1}\}^*$. Using Theorem XII.2.6, we see that $\lim_{|\mathscr{I}\lambda_0| \to \infty} \{I + (i - \lambda_0)(\lambda_0 I - H)^{-1}\}^* g = 0$ for each $g \in \mathfrak{H}$. Statement (b) now follows from Lemma 12.

If $|V(\lambda_0 I - H)^{-1}| < 1$, then $I - V(\lambda_0 I - H)^{-1}$ has a bounded inverse by Lemma VII.3.4. Let us now suppose that $\mathscr{I}\lambda_0 \neq 0$, and that $I - V(\lambda_0 I - H)^{-1}$ has a bounded inverse. Write $R(\lambda_0 ; H) = (\lambda_0 I - H)^{-1}$. Then, if $x \in \mathfrak{H}$, we have

$$(\lambda_0 I - H - V)[R(\lambda_0 ; H)(I - VR(\lambda_0 ; H))^{-1}]x \tag{20}$$
$$= (I - VR(\lambda_0 ; H))(I - VR(\lambda_0 ; H))^{-1}x = x.$$

Moreover, if $x \in \mathfrak{D}(H) = \mathfrak{D}(H + V)$, we have

$$[R(\lambda_0 ; H)(I - VR(\lambda_0 ; H))^{-1}](\lambda_0 I - H - V)x \tag{21}$$
$$= R(\lambda_0 ; H)(I - VR(\lambda_0 ; H))^{-1}(\lambda_0 I - H - V)R(\lambda_0 ; H)(\lambda_0 I - H)x$$
$$= R(\lambda_0 ; H)(I - VR(\lambda_0 ; H))^{-1}(I - VR(\lambda_0 ; H))(\lambda_0 I - H)x$$
$$= R(\lambda_0 ; H)(\lambda_0 I - H)x = x.$$

This shows that $(\lambda_0 I - H - V)^{-1}$ exists and is an everywhere defined, bounded operator, proving assertion (c) of our lemma. By Lemma XII.1.2, $H + V$ is a closed operator. By Theorem XII.4.19 the deficiency indices of $H + V$ are both zero. Therefore, by Corollary XII.4.13(b), $H + V$ is self adjoint, proving assertion (a) of our lemma. Q.E.D.

14 LEMMA. *Let H be a self adjoint operator in $\mathfrak{H}$, with domain $\mathfrak{D}(H)$. Let V and V_n, $n \geq 1$, be symmetric operators in $\mathfrak{H}$, with domains $\mathfrak{D}(V)$ and $\mathfrak{D}(V_n)$. Suppose that $\mathfrak{D}(V) \supseteq \mathfrak{D}(H)$, that $\mathfrak{D}(V_n) \supseteq \mathfrak{D}(H)$, and that the operators $V(iI - H)^{-1}$ and $V_n(iI - H)^{-1}$ are all compact. Suppose finally that $\lim_{n \to \infty} V_n(iI - H)^{-1} = V(iI - H)^{-1}$ in the uniform operator topology. Then for each real t, we have*

$$\lim_{n \to \infty} e^{it(H + V_n)} = e^{it(H + V)} \tag{22}$$

in the strong operator topology.

PROOF. Write $R(\lambda; T) = (\lambda I - T)^{-1}$ for each self adjoint operator T and all non-real λ. By Lemma 13, there exists a finite constant $M \geqq 1$ such that $|VR(\lambda; H)| < \frac{1}{2}$ if $|\mathscr{I}\lambda| \geqq M$. By Theorem XII.2.6 we have

$$
\begin{aligned}
(23)\qquad |(V - V_n)R(\lambda; H)| &= |(V - V_n)R(i; H)(iI - T)R(\lambda; H)| \\
&\leqq |(V - V_n)R(i; H)| \sup_{< \infty < \mu < +\infty} \left| \frac{i - \mu}{\lambda - \mu} \right| \\
&\leqq |(V - V_n)R(i; H)|
\end{aligned}
$$

if $|\mathscr{I}\lambda| \geqq M \geqq 1$. This shows that $\lim_{n\to\infty} V_n R(\lambda; H) = VR(\lambda; H)$ in the uniform topology of operators, uniformly for $|\mathscr{I}\lambda| \geqq M$. It therefore follows, using Lemma VII.3.4, that for all λ with $|\mathscr{I}\lambda| \geqq M$ and for all sufficiently large n, the inverse of $I - V_n R(\lambda; H)$ exists. Thus, by Lemma 13, we have

$$
(24)\qquad R(\lambda; H + V) = R(\lambda; H)(I - VR(\lambda; H))^{-1}, \qquad |\mathscr{I}\lambda| \geqq M,
$$

and

$$
(25)\qquad R(\lambda; H + V_n) = R(\lambda; H)(I - V_n R(\lambda; H))^{-1}, \qquad |\mathscr{I}\lambda| \geqq M,
$$

for all sufficiently large n. Using Lemma VII.3.4, we have

$$
|(I - VR(\lambda; H))^{-1}| \leqq 2, \qquad |\mathscr{I}\lambda| \geqq M.
$$

Therefore, by Corollary VII.6.2 and by (23) we have

$$
\begin{aligned}
(26)\qquad & |(I - VR(\lambda; H))^{-1} - (I - V_n R(\lambda; H))^{-1}| \\
&\leqq 2(1 - |(V - V_n)R(i; H)|)^{-1}, \qquad |\mathscr{I}\lambda| \geqq M,
\end{aligned}
$$

for sufficiently large n. It follows, using (24) and (25), that

$$
(27)\qquad \lim_{n\to\infty} R(\lambda; H + V_n) = R(\lambda; H + V)
$$

as $n \to \infty$, uniformly for $|\mathscr{I}\lambda| \geqq M$.

Using Lemma VIII.2.7, we may write the Cauchy integral formula

$$
(28)\qquad \exp(it(H + V))x = \frac{1}{2\pi i} \int_{\Gamma_M} \frac{e^{it\lambda} R(\lambda; H + V)}{(\lambda - \alpha)^2} (\alpha I - H - V)^2 x \, d\lambda;
$$

here the contour Γ_M of integration consists of the line $s - iM$, s real, traversed from $s = -\infty$ to $s = +\infty$, together with the line $s + iM$, s real, traversed from $s = +\infty$ to $s = -\infty$; α is any complex number such that

$|\mathscr{I}\alpha| > M$; and we require that $x \in \mathfrak{D}((H+V)^2)$. Formula (28) may be written as

$$(29) \quad \exp(it(H+V))(R(\alpha; H+V))^2 = \frac{1}{2\pi i}\int_{\Gamma_M} \frac{e^{it\lambda}R(\lambda; H+V)}{(\lambda-\alpha)^2}\,d\lambda.$$

Similarly, we have

$$(30) \quad \exp(it(H+V_n))(R(\alpha; H+V_n))^2 = \frac{1}{2\pi i}\int_{\Gamma_M} \frac{e^{it\lambda}R(\lambda; H+V_n)}{(\lambda-\alpha)^2}\,d\lambda$$

for all n. Using (27), we find from (29) and (30) that

$$(31) \quad \lim_{n\to\infty} \exp(it(H+V_n))(R(\alpha; H+V_n))^2 = \exp(it(H+V))(R(\alpha; H+V))^2$$

in the uniform topology of operators. But since $|\exp(it(H+V_n))| \leqq 1$, and since

$$(32) \quad |\exp(it(H+V_n))(R(\alpha; H+V))^2 - \exp(it(H+V))(R(\alpha; H+V))^2|$$
$$\leqq |\exp(it(H+V_n))\{(R(\alpha; H+V))^2 - (R(\alpha; H+V_n))^2\}|$$
$$+ |\exp(it(H+V_n))(R(\alpha; H+V_n))^2 - \exp(it(H+V))(R(\alpha; H+V))^2|,$$

we find from (31) and (27) that

$$\lim_{n\to\infty} (\exp(it(H+V_n)) - \exp(it(H+V)))(R(\alpha; H+V))^2 = 0.$$

Therefore

$$\lim_{n\to\infty} \exp(it(H+V_n))x = \exp(it(H+V))x, \qquad x \in \mathfrak{D}((H+V)^2).$$

Since $\mathfrak{D}((H+V)^2)$ is dense in $\mathfrak{H}$, it follows by Theorem II.3.6 that $\lim_{n\to\infty} \exp(it(H+V_n))x = \exp(it(H+V))x$ for all $x \in \mathfrak{H}$, and our lemma is proved. Q.E.D.

We are now in a position to begin the main part of the proof of Theorem 9. Our method will be as follows. First we shall establish Theorem 9 in a special case in which the Hermitian operator H is a multiplication operator and the symmetric operator V is an integral operator. Then, using the spectral representation theory of Section XII.3 (especially Theorem XII.3.16), we shall show that this special case of Theorem 9 actually implies the general case.

To describe the special case of Theorem 9 which we first wish to consider, we must introduce certain notations. Let R denote the real axis, λ the Lebesgue measure on R, and ν a finite positive λ-singular measure on R, so that there exists a λ-null set e_ν such that $\nu(R - e_\nu) = 0$. Put $\mu = \nu + \lambda$. Let $\mathfrak{H}'$ denote the set of all sequences $\tilde{f} = \{f_i(\cdot)\}$ of μ-measurable functions defined on R such that

$$|\tilde{f}|^2 = \sum_{i=1}^{\infty} \int_R |f_i(a)|^2 \mu(da) < \infty. \tag{33}$$

The set $\mathfrak{H}'$ is a complete Hilbert space with the inner product

$$(\tilde{f}, \tilde{g}) = \sum_{i=1}^{\infty} \int_R f_i(a)\overline{g_i(a)}\mu(da). \tag{34}$$

Let H be the unbounded operator in $\mathfrak{H}'$ defined as follows:

$$\mathfrak{D}(H) = \{\tilde{f} \in \mathfrak{H}' \,\Big|\, \sum_{i=1}^{\infty} \int_R a^2 |f_i(a)|^2 \mu(da) < \infty\}; \tag{35}$$

$$H\tilde{f} = \tilde{g}, \quad \text{where} \quad g_i(a) = af_i(a). \tag{36}$$

It is clear that H is symmetric, and that, for $\mathscr{I}\lambda \neq 0$, $\lambda I - H$ has a bounded inverse defined by

$$(\lambda I - H)^{-1}\tilde{f} = \tilde{h}, \quad \text{where} \quad h_i(a) = (\lambda - a)^{-1} f_i(a). \tag{37}$$

Thus, by Lemma XII.1.2, H is closed, and, by XII.4.13(b), H is self adjoint. We leave it to the reader to show that the spectral resolution $E(\cdot)$ of H is defined by

$$E(e)\tilde{f} = \tilde{g}, \quad \text{where} \quad g_i(a) = \chi_e(a) f_i(a), \tag{38}$$

χ_e denoting the characteristic function of the Borel set e; and that if F is a bounded Borel function, the bounded operator $F(H)$ is given by

$$F(H)\tilde{f} = \tilde{g}, \quad \text{where} \quad g_i(a) = F(a) f_i(a). \tag{39}$$

Let $\mathscr{V}_0$ denote the set of all symmetric operators V in $\mathfrak{H}'$ such that the operator $V(iI - H)^{-1}$ is compact, and such that the operator $(iI - H)^{-1}VE(e)$ belongs to the trace class whenever e is a finite subinterval of R. We introduce a norm into $\mathscr{V}_0$ by writing

$$|V|^* = |V(iI - H)^{-1}| + \sum_{n=1}^{\infty} \frac{1}{2^n} \frac{|(iI - H)^{-1}VE([-n, n])|_1}{1 + |(iI - H)^{-1}VE([-n, n])|_1}; \tag{40}$$

here as previously, $|S|_1$ denotes the trace norm of the operator S of trace class; cf. Definition XI.9.1.

Using the notations just introduced, we can state the following lemma, which describes the special case of Theorem 9 which we will prove first, and to which we will reduce the general case of Theorem 9.

15 LEMMA. *Let H be as in* (35)–(36), *and let $\mathfrak{H}'$ and $\mathscr{V}_0$ be as above. Suppose that $V \in \mathscr{V}_0$. Then*

(a) *$H + V$ is a self adjoint operator;*

(b) *if $\tilde{f} = \{f_i(\cdot)\} \in \mathfrak{H}'$, and $f_i(a) = 0$ for $i \geqq 1$ and $a \in e_\nu$, then $\tilde{f} \in \sum (H + V, H)$.*

The proof of Lemma 15 will be accomplished in several steps. First we shall show that any $V \in \mathscr{V}_0$ can be approximated by another element $V' \in \mathscr{V}_0$ of a particularly convenient special form. Next we establish a certain basic inequality, first on the hypothesis that we deal with an operator $H + V'$ for which V' has the above-mentioned special form, and then, by an approximation argument, generally. Once this inequality is established, Lemma 15 will follow generally.

The following lemma describes the way in which any element $V \in \mathscr{V}_0$ is to be approximated by elements of special form.

16 LEMMA. *Suppose that we call an element $V_0 \in \mathscr{V}_0$ smooth and finite if*

(i) *there exists an integer n and functions $K_{ij}(\cdot, \cdot) \in L_2(\mu \times \mu)$, vanishing for $i > n$, for $j > n$, and vanishing outside $[-n, n] \times [-n, n]$ in any case, such that*

$$V_0 \tilde{f} = \tilde{g}, \quad \textit{where} \quad g_i(a) = \sum_{j=1}^{n} \int_R K_{i,j}(a, b) f_j(b) \mu(db), \tag{41}$$

and

(ii) *the functions $K_{i,j}$ agree for $b \notin e_\nu$ with a function $\mathring{K}_{i,j}$ which is infinitely often differentiable in b, and which is such that $\partial^m \mathring{K}_{i,j}/\partial b^m$ belongs to $L_2(\mu \times \lambda)$ for all m, i, j.*

Then, for each $V \in \mathscr{V}_0$ and $\varepsilon > 0$, there exists an element $V^{(\varepsilon)} \in \mathscr{V}_0$, such that $|V^{(\varepsilon)}|^* \leqq \varepsilon$, and such that $V + V^{(\varepsilon)}$ is smooth and finite.

PROOF. Let P_n denote the projection in $\mathfrak{H}'$ defined by

$$P_n \tilde{f} = \tilde{g}, \quad \text{where} \quad \begin{aligned} g_i(a) &= f_i(a), & 1 \leqq i \leqq n, \\ g_i(a) &= 0, & i > n. \end{aligned} \tag{42}$$

Plainly $P_n\tilde{f}\to\tilde{f}$ as $n\to\infty$ for each $\tilde{f}\in\mathfrak{H}'$. It is no less plain that P_n commutes with every bounded Borel function of H. Thus, if we put $V_{(n)}=P_nE([-n,n])VE([-n,n])P_n$ for each $V\in\mathscr{V}_0$, the operator $V_{(n)}$ is an everywhere defined, bounded operator belonging to the trace class. By Lemma 12, $\lim_{n\to\infty}|V_{(n)}-V|^*=0$.

Since $V_{(n)}$ belongs to the trace class $\mathscr{C}_1$, it also (cf. Lemma XI.9.9(a) and (e)) belongs to the Hilbert-Schmidt class $\mathscr{C}_2=HS$. Thus, using Lemma XI.10.5 and taking note of the fact that $V_{(n)}=P_nV_{(n)}P_n=E([-n,n])V_{(n)}E([-n,n])$, we find that there exist functions $K_{ij}^{(n)}(\cdot,\cdot)\in L_2(\mu\times\mu)$, vanishing for $i>n$ and for $j>n$, and vanishing outside the set $[-n,n]\times[-n,n]$ for all i,j, such that

$$(43)\qquad V_{(n)}\tilde{f}=\tilde{g},\quad\text{where}\quad g_i(a)=\sum_{j=1}^{n}\int_R K_{ij}^{(n)}(a,b)f_i(b)\mu(db),\quad \tilde{f}\in\mathfrak{H}'.$$

Let φ be a non-negative function belonging to $C^\infty(R)$, vanishing outside $[-1,+1]$, and satisfying $\varphi(x)=\varphi(-x)$ and $\int_R\varphi(a)\,da=1$. For each $\varepsilon>0$, and each $\tilde{f}\in\mathfrak{H}'$, define $\Phi_\varepsilon\tilde{f}$ by

$$(44)\qquad \Phi_\varepsilon\tilde{f}=\tilde{g},\quad\text{where}\quad g_i(a)=f_i(a),\qquad a\in e_\nu,$$

$$g_i(a)=\varepsilon^{-1}\int_R\varphi(\varepsilon^{-1}(a-b))f_i(b)\,db,\qquad a\notin e_\nu.$$

Using Lemma XI.3.1, we see that Φ_ε is a mapping of $\mathfrak{H}'$ into itself of norm at most 1. If $\tilde{f}$ is such that $f_i(a)=0$ for all $i\neq i_0$, and $f_{i_0}(a)$ agrees for all $a\notin e_\nu$ with a continuous function h vanishing outside a bounded interval of R, then we have $\Phi_\varepsilon\tilde{f}-\tilde{f}=\tilde{g}^\varepsilon$, where $g_i^\varepsilon(a)=0$ for $i\neq i_0$, $g_{i_0}(a)=0$ for $a\in e_\nu$, and

$$(45)\qquad g_{i_0}^\varepsilon(a)=\varepsilon^{-1}\int_R\varphi(\varepsilon^{-1}(a-b))(h(b)-h(a))\,da,\qquad a\notin e_\nu.$$

It follows easily that $\int_R|g_{i_0}^\varepsilon(a)|^2\mu(da)\to0$ as $\varepsilon\to0$. Thus, using Theorem II.3.6, we conclude that $\lim_{\varepsilon\to0}\Phi_\varepsilon=I$ in the strong topology of operators.

Therefore, putting $V_{(n,\varepsilon)}=\Phi_\varepsilon V_{(n)}\Phi_\varepsilon$, it follows by Lemma 12 that $\lim_{\varepsilon\to0}|V_{(n,\varepsilon)}-V_{(n)}|^*=0$. Using (43) and (44), we see that if we define the functions $K_{i,j}^{(n,\varepsilon)}(\cdot,\cdot)\in L_2(\mu\times\mu)$ by

(46)

$$K_{i,j}^{(n,\varepsilon)}(a,b)=K_{i,j}^{(n)}(a,b),\qquad a\in e_\nu,\ b\in e_\nu;$$

$$K_{i,j}^{(n,\varepsilon)}(a,b)=\varepsilon^{-1}\int_R\varphi(\varepsilon^{-1}(a-a'))K_{i,j}^{(n)}(a',b)\,da',\qquad a\in R-e_\nu,\ b\in e_\nu;$$

$$K_{i,j}^{(n,\varepsilon)}(a, b) = \varepsilon^{-1}\int_R \varphi(\varepsilon^{-1}(b-b'))K_{i,j}^{(n)}(a, b')\,db', \qquad a \in e_\nu,\ b \in R - e_\nu;$$

$$K_{i,j}^{(n,\varepsilon)}(a, b) = \varepsilon^{-2}\int_R\int_R \varphi(\varepsilon^{-1}(a-a'))\varphi(\varepsilon^{-1}(b-b'))K_{i,j}^{(n)}(a', b')\,da'\,db',$$

if $a \in R - e_\nu$ and $b \in R - e_\nu$, we have $V_{(n,\varepsilon)}\tilde{g} = \tilde{h}$ for each $\tilde{g} \in \mathfrak{H}'$, where

$$h_i(a) = \sum_{j=1}^{n}\int_R K_{i,j}^{(n,\varepsilon)}(a, b)g(b)\mu(db).$$

It is plain from (45) that for each $a \in R$, $K_{i,j}^{(n,\varepsilon)}(a, b)$ agrees for $b \notin e_\nu$ with a function $\mathring{K}_{i,j}^{(n,\varepsilon)}(a, b)$ which is infinitely often differentiable in b, and such that $\partial^m \mathring{K}_{i,j}^{(n,\varepsilon)}/\partial b^m$ belongs to $L_2(\mu \times \lambda)$ for all m, i, and j. Therefore, for each $n \geqq 1$ and each $\varepsilon > 0$, the operator $V_{(n,\varepsilon)}$ is very smooth and finite. If we now choose $n = n(\varepsilon)$ so that $|V_{(n)} - V|^* < \varepsilon/2$, then choose $\delta = \delta(\varepsilon)$ so that $|V_{(n,\delta)} - V_{(n)}|^* < \varepsilon/2$, and put $V^{(\varepsilon)} = V_{(n,\delta)} - V$, it follows that $V + V^{(\varepsilon)}$ is very smooth and finite, and that $|V^{(\varepsilon)}|^* < \varepsilon$. Thus the proof of our lemma is complete. Q.E.D.

Next we prove an important inequality.

17 LEMMA. *Suppose that we call on element $\tilde{f} \in \mathfrak{H}'$ very smooth and finite if*

(i) *there exists an integer n such that $f_i(a) = 0$ for $i > n$;*

(ii) *for $1 \leqq i \leqq n$, we have $f_i(a) = 0$ if $a \in e_\nu$. Moreover, there exists a function $\mathring{f}_i$ defined on R which is infinitely often differentiable and vanishes outside a bounded subset of R, such that $f_i(a) = \mathring{f}_i(a)$ for $a \notin e_\nu$.*

Then, if $\tilde{f} \in \mathfrak{H}'$ is very smooth and finite, there exists a finite constant $C(\tilde{f})$ depending only on $\tilde{f}$, such that

$$\int_{-\infty}^{+\infty} |Te^{itH}\tilde{f}|^2\,dt \leqq \|T\|^2 C(\tilde{f}) \tag{47}$$

for each operator T in $\mathfrak{H}'$ of the Hilbert-Schmidt class HS. (Here, as previously, $\|T\|$ denotes the Hilbert-Schmidt norm of T.)

PROOF. Let $\tilde{f}$, f_i, $\mathring{f}_i$, and T be as above. By Lemma XI.10.5, there exists a double sequence $L_{ij}(\cdot\,,\cdot)$ of elements of $L_2(\mu \times \mu)$, $1 \leqq i, j < \infty$, such that

$$Te^{itH}\tilde{f} = \tilde{g}^{(t)}, \quad \text{where} \quad g_i^{(t)}(a) = \sum_{j=1}^{n}\int_R L_{i,j}(a, b)e^{itb}f_j(b)\mu(db). \tag{48}$$

Moreover, by this same lemma, we have

$$(49) \qquad \|T\|^2 = \sum_{i,j=1}^{\infty} \int\int_R |L_{ij}(a, b)|^2 \mu(da)\mu(db).$$

Using (48) and the fact that $f_i(a) = 0$ for $a \in e_\nu$, we have

$$(50) \quad \int_{-\infty}^{+\infty} |Te^{itH}\tilde{f}|^2 \, dt = \sum_{i=1}^{\infty} \int_{-\infty}^{+\infty} \int_R \left| \sum_{j=1}^{n} \int_R L_{i,j}(a, b) \mathring{f}_j(b) e^{itb} \, db \right|^2 \mu(da) \, dt,$$

and therefore, using Plancherel's theorem (XV.11.3) and Fubini's theorem, we have

$$(51) \quad \int_{-\infty}^{+\infty} |Te^{itH}\tilde{f}|^2 \, dt$$

$$= \sum_{i=1}^{\infty} \int_R \int_{-\infty}^{+\infty} \left| \int_R \sum_{j=1}^{n} L_{i,j}(a, b) \mathring{f}_j(b) e^{itb} \, db \right|^2 dt \, \mu(da)$$

$$= 2\pi \sum_{i=1}^{\infty} \int_R \int_R \left| \sum_{j=1}^{n} L_{i,j}(a, b) \mathring{f}_j(b) \right|^2 db \, \mu(da)$$

$$\leqq 2\pi \sup_{b \in R} \left\{ \sum_{j=1}^{n} |\mathring{f}_j(b)|^2 \right\} \sum_{i=1}^{\infty} \int_R \int_R \sum_{j=1}^{n} |L_{i,j}(a, b)|^2 \mu(da)\mu(db)$$

$$\leqq 2\pi \sup_{b \in R} \left\{ \sum_{j=1}^{n} |\mathring{f}_j(b)|^2 \right\} \|T\|^2,$$

and our lemma is proved. Q.E.D.

Now we are able to give the proof of the fundamental Lemma 15.

PROOF OF LEMMA 15. Since the proof is somewhat lengthy, we summarize it in general outline before proceeding to explicit details. Our first aim is to prove that the following identity is valid for each $V_0 \in \mathscr{V}$:

$$\frac{d}{dt} e^{it(H+V_0)} e^{-itH} g = i e^{it(H+V_0)} V_0 e^{-itH} g, \qquad g \in \mathfrak{D}(H);$$

this follows by quite straightforward functional calculus arguments. Integrating this identity, it follows easily that, if both f and V_0 satisfy appropriate smoothness conditions, then $f \in \sum (H + V, H)$. Moreover, transforming the integrated identity, we are able to obtain an *a priori* estimate (Rosenblum's inequality) for the rate of convergence of

$$e^{it(H+V_0)}e^{itH}f$$

to its limit. This estimate (formula (74) below) turns out to involve only an integral for which Lemma 17 gives a convenient bound. Once this estimate is available for smooth V_0 and f, we are readily able to show by a limit argument of standard form, that the same estimate holds for all V and f satisfying the hypothesis of Lemma 15. This fact leads at once to the convergence assertion of Lemma 15.

With this prospectus, we pass on to details.

That $H + V$ is self adjoint follows immediately from Lemma 13. This proves (a) of Lemma 15.

Let $\tilde{f} \in \mathfrak{H}'$ be very smooth and finite, so that there exists an integer n and functions $\mathring{f}_1, \ldots, \mathring{f}_n \in C^\infty(R)$, vanishing outside $[-n, n]$, such that $f_i(a) = 0$ if either $i > n$ or $a \in e_\nu$, and such that $f_i(a) = \mathring{f}_i(a)$ if $a \notin e_\nu$, $1 \leqq i \leqq n$. Let $V \in \mathscr{V}_0$. Let $V^{(\varepsilon)}$ be as in Lemma 16, so that for each $\varepsilon > 0$ there exists a finite integer $n(\varepsilon)$, functions $K_{i,j}^{(\varepsilon)}(\cdot\,,\,\cdot) \in L_2(\mu \times \mu)$, and functions $\mathring{K}_{i,j}^{(\varepsilon)} \in L_2(\mu \times \lambda)$, such that $K_{i,j}^{(\varepsilon)}(a, b) = 0$ if either $i > n$, $j > n$, $|a| > n$, or $|b| > n$, such that

(i) $(V + V^{(\varepsilon)})\tilde{g} = \tilde{h}$, where $h_i(a) = \sum_{j=1}^{n(\varepsilon)} \int_R K_{i,j}^{(\varepsilon)}(a, b) g_j(b)\mu(db)$, $\tilde{g} \in \mathfrak{H}'$;

(ii) $K_{i,j}^{(\varepsilon)}(a, b) = \mathring{K}_{i,j}^{(\varepsilon)}(a, b)$ if $b \notin e_\nu$; $\mathring{K}_{i,j}^{(\varepsilon)}$ is infinitely often differentiable in b, and $\partial^m \mathring{K}_{i,j}^{(\varepsilon)}/\partial b^m \in L_2(\mu \times \lambda)$ for all m, i, and j.

If $\tilde{g} \in \mathfrak{D}(H)$, then by Theorem XII.2.6 and the Lebesgue dominated convergence theorem we have

$$(52)\quad \lim_{h\to 0} \left| \frac{e^{ihH}\tilde{g} - \tilde{g}}{h} - iH\tilde{g} \right|^2 = \lim_{h\to 0} \int_{-\infty}^{+\infty} \left| \frac{e^{ih\lambda} - 1 - ih\lambda}{h\lambda} \right|^2 |\lambda|^2 |E(d\lambda)\tilde{g}|^2 = 0.$$

This shows that $\lim_{h\to 0} ((e^{ihH} - I)/h - iH)(iI - H)^{-1} = 0$ strongly. Similarly, if $V_0 \in \mathscr{V}$, then since $\mathfrak{D}(H) = \mathfrak{D}(H + V_0)$, we find that

$$(53)\quad \lim_{h\to 0} \left(\frac{e^{ih(H+V_0)} - I}{h} - i(H + V_0) \right) (iI - H)^{-1} = 0 \text{ strongly.}$$

Consequently

$$(54)\quad \lim_{h\to 0} \frac{e^{i(t+h)(H+V_0)}e^{-i(t+h)H} - e^{it(H+V_0)}e^{-itH}}{h}\tilde{g}$$

$$= \lim_{h\to 0} e^{it(H+V_0)}\left(\frac{e^{ih(H+V_0)}e^{-ihH} - I}{h}\right)e^{-itH}\tilde{g}$$

$$= \lim_{h\to 0} e^{it(H+V_0)}\left(\frac{(e^{ih(H+V_0)} - I)e^{-ihH} + e^{-ihH} - I}{h}\right)e^{-itH}\tilde{g}$$

$$= \lim_{h\to 0} e^{it(H+V_0)}\left(\frac{(e^{ih(H+V_0)} - I)(iI - H)^{-1}}{h}\right)e^{-ihH}e^{-itH}(iI - H)\tilde{g}$$

$$+ \lim_{h\to 0} e^{it(H+V_0)}\left(\frac{e^{-ihH} - I}{h}\right)e^{-itH}\tilde{g}$$

$$= e^{it(H+V_0)}(i(H+V_0)(iI-H)^{-1})e^{-itH}(iI-H)\tilde{g}$$

$$- e^{it(H+V_0)}(iH)e^{-itH}\tilde{g}$$

$$= e^{it(H+V_0)}(i(H+V_0) - iH)e^{-itH}\tilde{g}$$

$$= ie^{it(H+V_0)}V_0e^{-itH}\tilde{g}.$$

That is,

$$(55)\qquad \frac{d}{dt}e^{it(H+V_0)}e^{-itH}\tilde{g} = ie^{it(H+V_0)}V_0e^{-itH}\tilde{g}, \qquad \tilde{g}\in\mathfrak{D}(H).$$

We leave to the reader the similar but even easier proof that the right-hand side of this last equation is continuous in t. Using this fact we may integrate both sides of equation (55) to obtain

$$(56)\quad (e^{it(H+V_0)}e^{-itH} - e^{is(H+V_0)}e^{-isH})\tilde{g}$$

$$= i\int_s^t e^{iu(H+V_0)}V_0e^{-iuH}\tilde{g}\,du, \qquad \tilde{g}\in\mathfrak{D}(H).$$

As a special case of this last equation we have

$$(57)\quad \exp(it(H+V+V^{(\varepsilon)}))\exp(-itH)\tilde{f}$$

$$- \exp(is(H+V+V^{(\varepsilon)}))\exp(-isH)\tilde{f}$$

$$= i\int_s^t \exp(iu(H+V+V^{(\varepsilon)}))(V+V^{(\varepsilon)})\exp(-iuH)\tilde{f}\,du.$$

On the other hand, we have

$$(V + V^{(\varepsilon)})\exp(-iuH)\tilde{f} = \tilde{g}^{(u)}, \tag{58}$$

where

$$g_i^{(u)}(a) = \sum_{j=1}^{n(\varepsilon)} \int_R \mathring{K}_{i,j}^{(\varepsilon)}(a, b)e^{-iub}\mathring{f}(b)\, db.$$

If we integrate twice by parts, we find that

$$g_i^{(u)}(a) = -u^{-2}\sum_{j=1}^{n(\varepsilon)} \int_R \left\{\frac{\partial^2}{db^2}[\mathring{K}_{i,j}^{(\varepsilon)}(a, b)\mathring{f}(b)]\right\}e^{-iub}\, db, \qquad 1 \leqq i \leqq n, \tag{59}$$

$$g_i^{(u)}(a) = 0 \qquad i > n,$$

from which it is clear that there exists a finite constant C such that $\tilde{g}^{(u)} \leqq C(1 + |u|^2)$. Therefore

$$\int_{-\infty}^{+\infty} |(V + V^{(\varepsilon)})\exp(-iuH)\tilde{f}|\, du < \infty. \tag{60}$$

Since by (57) we have

$$\begin{aligned} &|\exp(it(H + V + V^{(\varepsilon)}))\exp(-itH)\tilde{f} \\ &\qquad - \exp(is(H + V + V^{(\varepsilon)}))\exp(-isH)\tilde{f}| \\ &\qquad\qquad \leqq \int_s^t |(V + V^{(\varepsilon)})(\exp(-iuH))\tilde{f}|\, du, \end{aligned} \tag{61}$$

it follows that $\tilde{f} \in \sum (H + V + V^{(\varepsilon)}, H)$. Let

$$\mathscr{U}_t = \exp(it(H + V))\exp(-itH) \tag{62}$$

and

$$\mathscr{U}_t^{(\varepsilon)} = \exp(it(H + V + V^{(\varepsilon)}))\exp(itH). \tag{63}$$

We have

$$|(\mathscr{U}_t - \mathscr{U}_s)\tilde{f}|^2 = 2|\tilde{f}|^2 - 2\mathscr{R}(\mathscr{U}_t\tilde{f}, \mathscr{U}_s\tilde{f}) = 2\mathscr{R}((\mathscr{U}_t - \mathscr{U}_s)\tilde{f}, \mathscr{U}_t\tilde{f}). \tag{64}$$

Similarly,

$$|(\mathscr{U}_t^{(\varepsilon)} - \mathscr{U}_s^{(\varepsilon)})\tilde{f}|^2 = 2\mathscr{R}(\mathscr{U}_t^{(\varepsilon)} - \mathscr{U}_s^{(\varepsilon)}\tilde{f}, \mathscr{U}_t^{(\varepsilon)}\tilde{f}). \tag{65}$$

Write $\mathscr{U}_\infty^{(\varepsilon)} = \mathscr{U}(H + V + V^{(\varepsilon)}, H)$. Taking the inner product of both sides of (57) with $\mathscr{U}_t^{(\varepsilon)}\tilde{f}$ and letting $t \to \infty$, we find that

$$(66) \quad |(\mathscr{U}_\infty^{(\varepsilon)} - \mathscr{U}_s^{(\varepsilon)})\tilde{f}|^2$$

$$= -2\mathscr{I} \int_s^\infty (\exp(iu(H + V + V^{(\varepsilon)}))(V + V^{(\varepsilon)})\exp(-iuH)\tilde{f}, \mathscr{U}_\infty^{(\varepsilon)}\tilde{f})\, du$$

$$\leqq 2 \int_s^\infty |(\exp(iu(H + V + V^{(\varepsilon)}))(V + V^{(\varepsilon)})\exp(-iuH)\tilde{f}, \mathscr{U}_\infty^{(\varepsilon)}\tilde{f})|\, du.$$

Using Lemma 6(b), we may rewrite this last inequality as

$$(67) \quad |(\mathscr{U}_\infty^{(\varepsilon)} - \mathscr{U}_s^{(\varepsilon)})\tilde{f}|^2$$

$$\leqq 2 \int_s^\infty |((V + V^{(\varepsilon)})\exp(-iuH)\tilde{f}, \mathscr{U}_\infty^{(\varepsilon)}\exp(-iuH)\tilde{f})|\, du.$$

Now, since $\tilde{f}$ is very smooth and finite, there exists a bounded subinterval e of R such that $E(e)\tilde{f} = \tilde{f}$. Let $E^{(\varepsilon)}(\cdot)$ be the resolution of the identity of the self adjoint operator $H^{(\varepsilon)} = H + V + V^{(\varepsilon)}$. Then it follows from (67), using Lemma 6(b) once more, that

$$(68) \quad |(\mathscr{U}_t^{(\varepsilon)} - \mathscr{U}_s^{(\varepsilon)})\tilde{f}|^2$$

$$\leqq 8 \int_{\min(t,s)}^\infty |(E^{(\varepsilon)}(e)(V + V^{(\varepsilon)})E(e)\exp(-iuH)\tilde{f}, \mathscr{U}_\infty^{(\varepsilon)} \exp(-iuH)\tilde{f})|\, du.$$

Using Lemma 13, choose λ_0 with $\mathscr{I}\lambda_0 \neq 0$ such that $|VR(\lambda_0; H)| < \frac{1}{2}$. Then since $|V^{(\varepsilon)}|^* \leqq \varepsilon$, we have $|(V + V^{(\varepsilon)})R(\lambda_0; H)| < \frac{1}{2}$ for all sufficiently small ε. Hence, by Lemma 13, we have

$$(69) \qquad R(\lambda_0; H^{(\varepsilon)}) = R(\lambda_0; H)(I - (V + V^{(\varepsilon)})R(\lambda_0; H))^{-1}$$

for ε sufficiently small; thus, taking conjugates, we have

$$(70) \qquad R(\bar{\lambda}_0; H^{(\varepsilon)}) = (I - ((V + V^{(\varepsilon)})R(\lambda_0; H))^*)^{-1}(\bar{\lambda}_0; H).$$

Note also that it follows from Lemma VII.3.4 that the operator $(I - ((V + V^{(\varepsilon)})R(\lambda_0; H))^*)^{-1}$ is of norm at most 2 for all sufficiently small ε.

By Theorem XII.2.9, we see that we may write $E^{(\varepsilon)}(e) = M^{(\varepsilon)}R(\bar{\lambda}_0; H^{(\varepsilon)})$, where $M^{(\varepsilon)}$ is a bounded operator satisfying $|M^{(\varepsilon)}| \leqq C(e, \lambda_0)$, $C(e, \lambda_0)$ being a constant depending only on the bounded interval e and on λ_0, but not on ε. If we write $P_\infty^{(\varepsilon)}$ for the orthogonal projection on the domain $\sum(H + V + V^{(\varepsilon)}, H)$ of $\mathscr{U}_\infty^{(\varepsilon)}$, and put

$$(71) \qquad \hat{M}^{(\varepsilon)} = (M^{(\varepsilon)}(I - ((V + V^{(\varepsilon)})R(\lambda_0; H))^*)^{-1})^* \mathscr{U}_\infty^{(\varepsilon)} P_\infty^{(\varepsilon)},$$

we have $|\hat{M}^{(\varepsilon)}| \leqq 2C(e, \lambda_0)$ for sufficiently small ε, and also, using (70), can rewrite (70) in the form

$$(72)\quad |(\mathscr{U}_t^{(\varepsilon)} - \mathscr{U}_s^{(\varepsilon)})\tilde{f}|^2$$
$$\leqq 8 \int_{\min(t,s)}^{\infty} |(R(\bar{\lambda}_0; H)(V + V^{(\varepsilon)})E(e)\exp(-iuH)\tilde{f}, \hat{M}^{(\varepsilon)}\exp(-iuH)\tilde{f})|\, du.$$

From Corollary 11, we see that we may write $R(\bar{\lambda}_0; H)VE(e) = BA$, where A and B are operators of the Hilbert-Schmidt class. Similarly, we may write $R(\bar{\lambda}_0; H)V^{(\varepsilon)}E(e) = B^{(\varepsilon)}A^{(\varepsilon)}$, where $A^{(\varepsilon)}$ and $B^{(\varepsilon)}$ are operators of the Hilbert-Schmidt class such that $\|A^{(\varepsilon)}\| \to 0$ and $\|B^{(\varepsilon)}\| \to 0$ as $\varepsilon \to 0$. Put $\hat{B}^{(\varepsilon)} = B^*\hat{M}^{(\varepsilon)}$ and $\tilde{B}^{(\varepsilon)} = (B^{(\varepsilon)})^*\hat{M}^{(\varepsilon)}$. Then, by Lemma XI.9.9(d), we have $\|\hat{B}^{(\varepsilon)}\| \leqq 2C(e, \lambda_0)\|B\|$ for sufficiently small ε. For the same reasons, we have $\|\tilde{B}^{(\varepsilon)}\| \to 0$ as $\varepsilon \to 0$. From (72) we have

$$(73)\quad |(\mathscr{U}_t^{(\varepsilon)} - \mathscr{U}_s^{(\varepsilon)})\tilde{f}|^2 \leqq 8 \int_{\min(t,s)}^{\infty} |(A \exp(-iuH)\tilde{f}, \hat{B}^{(\varepsilon)}\exp(-iuH)\tilde{f})|\, du$$
$$+ 8 \int_{\min(t,s)}^{\infty} |(A^{(\varepsilon)}\exp(-iuH)\tilde{f}, \tilde{B}^{(\varepsilon)}\exp(-iuH)\tilde{f})|\, du.$$

Therefore, using Lemma 17 and Schwarz' inequality, we have

$$(74)\quad |(\mathscr{U}_t^{(\varepsilon)} - \mathscr{U}_s^{(\varepsilon)})\tilde{f}|^2$$
$$\leqq 16C(e, \lambda_0)\|B\|C(\tilde{f})\left\{\int_{\min(t,s)}^{\infty} |A \exp(-iuH)\tilde{f}|^2\, du\right\}^{1/2}$$
$$+ 8\, \|A^{(\varepsilon)}\|\, \|\tilde{B}^{(\varepsilon)}\| C^2(\tilde{f}).$$

We may now use Lemma 14 to let $\varepsilon \to 0$, and find that

$$(75)\quad |(\mathscr{U}_t - \mathscr{U}_s)\tilde{f}|^2 \leqq 16C(e, \lambda_0)\|B\|C(\tilde{f})\left\{\int_{\min(t,s)}^{\infty} |A(\exp(-iuH))\tilde{f}|^2\, du\right\}^{1/2}.$$

It now follows immediately from Lemma 17 that $\lim_{t,s\to\infty} |(\mathscr{U}_t - \mathscr{U}_s)\tilde{f}| = 0$. Hence $\tilde{f} \in \sum (H + V, H)$. Since $\tilde{f}$ was any very smooth, finite element of $\mathfrak{H}'$, we see that $\sum (H + V, H)$ contains any very smooth, finite element of $\mathfrak{H}'$.

Next, note that if $\tilde{g}$ is any element of $\mathfrak{H}'$ such that $g_i(a) = 0$ for all $a \in e_\nu$, then

(76) there exists a sequence $\{\tilde{f}^{(n)}\}$ of very smooth, finite elements of $\mathfrak{H}'$ such that $\lim_{n\to\infty} |\tilde{f}^{(n)} - \tilde{g}| = 0$.

To prove (76), we first let $\tilde{g}$ be such that $g_i(a) = 0$ for all i except some certain i_0, while $g_{i_0}(a) = \chi_{[c,d]-e_\nu}(a)$, $[c, d]$ being a finite closed interval and χ_e denoting the characteristic function of the set e. Since there is a sequence $\{f^{(n)}\}$ of functions belonging to $C^\infty(R)$ and vanishing outside a bounded subinterval of R such that $\lim_{n\to\infty} \int_R |f^{(n)}(a) - \chi_{[c,d]}(a)|^2\, da = 0$, we have

$$\lim_{n\to\infty} \int_R |f^{(n)}(a)\chi_{R-e_\nu}(a) - \chi_{[c,d]-e_\nu}(a)|^2\, \mu(da) = 0, \tag{77}$$

proving (76) for $\tilde{g}$ of the special form described above. Next, let $\tilde{g}$ be any element of $\mathfrak{H}'$ such that $g_i(a) = 0$ for all $a \in e_\nu$. Using Lemma III.8.3, we see that there exists a sequence $\tilde{g}^{(n)}$ of elements of $\mathfrak{H}'$, each of which is a finite linear combination of elements of the special form described above, such that $\lim_{n\to\infty} |\tilde{g} - \tilde{g}^{(n)}| = 0$. We therefore see that, as asserted, statement (76) holds for each $\tilde{g} \in \mathfrak{H}'$ such that $g_i(a) = 0$ for all $a \in e_\nu$.

Since $\sum (H + V, H)$ is a closed subspace of $\mathfrak{H}'$ (cf. Lemma 2) it follows that every $\tilde{g} \in \mathfrak{H}'$ such that $g_i(a) = 0$ for all $a \in e_\nu$ belongs to $\sum (H + V, H)$. This proves statement (b) of Lemma 15. Q.E.D.

Next we show that the result given by Lemma 15 for the full space $\mathfrak{H}'$ carries over without difficulty to certain subspaces of $\mathfrak{H}'$.

18 LEMMA. *Let $\mathfrak{H}'$, H, $\mathscr{V}_0$, and so on, be as in Lemma* 15 *and in the formulas preceding Lemma* 15. *Let a sequence $\hat{e}_1, \hat{e}_2, \ldots$ of Borel subsets of R be given, and let $\mathfrak{H}_1$ be the set of all $\tilde{f} \in \mathfrak{H}'$ such that $f_i(a) = 0$ for all $a \notin \hat{e}_i$. Let $V \in \mathscr{V}_0$, and let $V(\mathfrak{D}(V) \cap \mathfrak{H}_1) \subseteq \mathfrak{H}_1$. Then*

(a) *$\mathfrak{H}_1$ is a closed subspace of $\mathfrak{H}$;*

(b) *the restriction $H_2 = H \,|\, (\mathfrak{D}(H) \cap \mathfrak{H}_1)$ is a self adjoint operator in $\mathfrak{H}_1$;*

(c) *the restriction $H_3 = (H + V) \,|\, (\mathfrak{D}(H) \cap \mathfrak{H}_1)$ is a self adjoint operator in $\mathfrak{H}_1$;*

(d) *if $\tilde{f} \in \mathfrak{H}_1$ belongs to $\sum_{ac} (H_2)$, then $\tilde{f} \in \sum (H_3, H_2)$.*

PROOF. The routine proof of statement (a) is left to the reader.

Plainly, $H_2\mathfrak{D}(H_2) \subseteq \mathfrak{H}_1$. It is clear that H_2 is symmetric, and that, for $\mathscr{I}\lambda \neq 0$, $\lambda I - H_2$ has a bounded inverse defined by $(\lambda I - H_2)^{-1} = (\lambda I - H)^{-1} \,|\, \mathfrak{H}_1$. Thus, by XII.1.2, H_2 is closed, and, by Corollary XII.4.13(b), H_2 is self adjoint. This proves statement (b). We leave it to the reader to show that if F is a bounded Borel function, then $F(H_2) = F(H) \,|\, \mathfrak{H}_1$.

Suppose next that $V \in \mathscr{V}_0$, and that $V(\mathfrak{D}(V) \cap \mathfrak{H}_1) \subseteq \mathfrak{H}_1$. By Lemma 13, there exists a finite constant M such that $|V(\lambda_0 I - H)^{-1}| < \frac{1}{2}$ and

$$(78) \qquad (\lambda_0 I - H - V)^{-1} = (\lambda_0 I - H)^{-1}(I - V(\lambda_0 I - H)^{-1})^{-1}$$

if $|\mathscr{I}\lambda_0| \geqq M$. This makes it plain that for $|\mathscr{I}\lambda_0| \geqq M$, $(\lambda_0 I - H - V)^{-1}$ maps $\mathfrak{H}_1$ into itself. If we put $H_3 = (H + V) \mid (\mathfrak{D}(H) \cap \mathfrak{H}_1)$, we see that for $|\mathscr{I}\lambda_0|$ sufficiently large, $\lambda_0 I - H_3$ has a bounded inverse defined by $(\lambda_0 I - H_3)^{-1} = (\lambda I - H - V)^{-1} \mid \mathfrak{H}_1$. Thus, by Lemma XII.1.2, H_3 is closed, and, by Theorem XII.4.19 and Corollary XII.4.13(b), H_3 is self adjoint. This proves statement (c). By the Stone-Weierstrass theorem (IV.6.16), every continuous function F defined on R and vanishing at $\pm\infty$ can be approximated uniformly by linear combinations or products of functions of the form $G(\lambda) = (\lambda_0 - \lambda)^{-1}$, where $|\mathscr{I}\lambda_0| \geqq M$. Hence, by Theorem XII.2.6, $F(H + V)$ can be approximated arbitrarily closely, in the uniform topology of operators, by linear combinations of products of the operators $(\lambda_0 I - H - V)^{-1}$, and $F(H_3)$ can be approximated arbitrarily closely in the same topology by the corresponding linear combinations of products of the operators $(\lambda_0 I - H_3)$. It follows that

$$(79) \qquad F(H + V)\mathfrak{H}_1 \subseteq \mathfrak{H}_1 \quad \text{and} \quad F(H_3) = F(H + V) \mid \mathfrak{H}_1$$

for each continuous function F defined on R and vanishing at $\pm\infty$. Since every bounded continuous function F defined on R is the limit of a uniformly bounded sequence of continuous functions F_n, each vanishing at $\pm\infty$, it follows by Theorem XII.2.6 and the Lebesgue dominated convergence theorem that (79) holds for each such function F, and, in particular, for the exponential function $F(\lambda) = \exp(it\lambda)$. Therefore, $\exp(itH_3)\exp(-itH_2) = \exp(it(H + V))\exp(-itH) \mid \mathfrak{H}_1$. Thus, by what has been proved above, every $\tilde{f} \in \mathfrak{H}_1$ such that $f_i(a) = 0$ for $a \in e_\nu$ belongs to $\sum (H_3, H_2)$.

Let us now note that if $\tilde{f} \in \mathfrak{H}_1$ belongs to $\sum_{ac} (H_2)$, then, since $\lambda(e_\nu) = 0$, we have $E(e_\nu)\tilde{f} = 0$, so that $f_i(a) = 0$ μ-almost everywhere in e_ν for each $i \geqq 1$. This shows that $\sum_{ac} (H_2) \subseteq \sum (H_3, H_2)$, which proves statement (d). Q.E.D.

The following lemma carries the statement of Lemma 15 over from the concretely defined pair H_2, H_3 of operators to an abstractly defined pair H_4, H_5 of operators.

19 LEMMA. *Let H_4 be a self adjoint operator in $\mathfrak{H}_1$ with domain $\mathfrak{D}(H_4)$ and resolution of the identity $E_4(\cdot)$. Suppose that V_4 is a symmetric operator in $\mathfrak{H}$, that $\mathfrak{D}(V_4) \supseteq \mathfrak{D}(H_4)$, that the operator $V_4(iI - H_4)^{-1}$ is compact, and that for each bounded interval e of the real axis the operator $(iI - H_4)^{-1} V_4 E_4(e)$ is of trace class. Then*

(a) *$H_5 = H_4 + V_4$ is a self adjoint operator;*

(b) $\sum (H_5, H_4) \supseteq \sum_{ac} (H_4)$.

PROOF. We use the spectral representation theorem, XII.3.16. By that theorem, $\mathfrak{H}$ has an ordered representation relative to the self adjoint operator H_4. Let the measure of this ordered representation be μ_1, and let the multiplicity sets be $e_1, e_2, \ldots$. By Theorem III.4.14, we may write $\mu_1 = \nu + \lambda_1$, where ν is λ-singular, and λ_1 is λ-continuous. As before, let e_ν be a λ-null set such that $\nu(R - e_\nu) = 0$. By the Radon-Nikodým theorem (III.10.2) there exists a λ-integrable function ψ such that $\lambda_1(e) = \int_e \psi(a)\lambda(da)$ for each Borel set e. Since λ_1 is non-negative, it is easily seen that ψ is non-negative. Let $\tilde{e} = \{a \mid \psi(a) > 0\}$. Put $\lambda_2(e) = \lambda(\tilde{e} \cap e)$ for each Borel set e. Then plainly $\lambda_2(e) > 0$ if and only if $\lambda_1(e) > 0$, so $\lambda_1 \cong \lambda_2$, that is, λ_1 and λ_2 are equivalent measures. It follows that if we put $\mu = \nu + \lambda_2$, we have $\mu \cong \mu_1$. Let $\hat{e}_i = e_i \cap R$, $i = 1, 2, \ldots$, let $\mathfrak{H}_1$ be the space of sequences $\tilde{f}$ of μ_2-measurable functions f_i defined on R such that $f_i(a) = 0$ for $a \notin \hat{e}_i$, and such that

$$\sum_{i=1}^{\infty} \int_R |f_i(a)|^2 \mu(da) < \infty; \tag{80}$$

and let H_2 be the self adjoint operator in $\mathfrak{H}_1$ defined by

$$\mathfrak{D}(H_2) = \{\tilde{f} \,\Big|\, \sum_{i=1}^{\infty} \int_R a^2 |f_i(a)|^2 \mu_2(da) < \infty\}, \tag{81}$$

$$H_2 \tilde{f} = \tilde{g}, \quad \text{where} \quad g_i(a) = a f_i(a). \tag{82}$$

Then H_4 and H_2 have spectral representation with equivalent measures and the same multiplicity sets, so that by Theorem XII.3.16, there exists an isometric mapping U of $\mathfrak{H}$ onto $\mathfrak{H}_1$, such that $U^{-1} H_2 U = H_4$.

If we put $V_2 = U V_4 U^{-1}$, V_2 is a symmetric operator in $\mathfrak{H}_1$ such that $\mathfrak{D}(V_2) \supseteq \mathfrak{D}(H_2)$, such that $V_2(iI - H_2)^{-1}$ is compact, and such that $(iI - H_2)^{-1} V_2(E(e) \mid \mathfrak{H}_1)$ is of trace class for each bounded subinterval e of R, $E(\cdot)$ being the spectral resolution of H_2. The space $\mathfrak{H}_1$ we have defined is obviously the same as the space $\mathfrak{H}_1$ in Lemma 18, and may be regarded as a subspace of the larger space $\mathfrak{H}'$ of Lemma 15, while equally plainly H_2 may be regarded as the restriction to $\mathfrak{H}_1$ of the operator H of Lemma 15 (cf. (33)–(36) above). Let Q be the projection of $\mathfrak{H}'$ onto its subspace $\mathfrak{H}_1$, and put $V = V_2 Q$. Plainly, V is symmetric. The operator $V(iI - H_2)^{-1} = V_2 Q(iI - H_2)^{-1} = V_2(iI - H_2)^{-1}Q$ is seen to be compact by use of Corollary VI.5.5, and the operator $(iI - H)^{-1}VE(e) = (iI - H)^{-1}V_2 QE(e) = (iI - H_2)^{-1}V_2(E(e) \mid \mathfrak{H}_1)Q$ is seen to belong to the trace class by Lemma XI.9.9(d). Finally, we have $V_2 = V \mid (\mathfrak{D}(V) \cap \mathfrak{H}_1)$. Thus, by Lemma 18, $\sum (H_1 + V_1, H_1) \supseteq \sum_{ac} (H_1)$. Using the isometric equivalences $U^{-1}H_2 U = H_4$ and $U^{-1}V_2 U = V_4$, we conclude at once that $H_4 + V_4$ is self adjoint and that $\sum (H_4 + V_4, H_4) \supseteq \sum_{ac} (H_4)$. Q.E.D.

Using Lemma 19, it is easy for us to complete the proof of Theorem 9.

Proof of Theorem 9. Let e be a bounded interval of the real axis. By Theorem XII.2.6, we may write $E(e) = (iI - H)^{-1}F(H)$, where F is a bounded Borel function. Thus the hypothesis of Theorem 9 implies the hypothesis of Lemma 19, so that assertions (a) and (b) of Theorem 9 follow immediately from Lemma 19.

Let us now observe that the self adjoint operators H and $H_1 = H + V$ of Theorem 9 are symmetrically related. Indeed, it follows by Lemma 13 that, if $|\mathscr{I}\lambda_0|$ is sufficiently large, we have

$$(83) \qquad (\lambda_0 I - H_1)^{-1} = (\lambda_0 I - H)^{-1}(I - V(\lambda_0 I - H))^{-1},$$

and thus, taking adjoints,

$$(84) \qquad (\lambda_0 I - H_1)^{-1} = (I - (V(\bar{\lambda}_0 I - H))^*)^{-1}(\lambda_0 I - H)^{-1}.$$

Then, by Corollary VI.5.5, $V(\lambda_0 I - H_1)^{-1}$ is compact, and, by Lemma XI.9.9(d), $(\lambda_0 I H -_1)^{-1}V(\lambda_0 I - H_1)^{-1}$ belongs to the trace class. It follows from Theorem XII.2.6 that $(\lambda_0 I - H)^{-1} = (iI - H)^{-1}G(H) = G(H)(iI - H)^{-1}$, where G is a bounded Borel function. Therefore, using Corollary VI.5.5 and Lemma XI.9.9(d) once more, we see that $V(iI - H_1)^{-1}$ is compact, and that $(iI - H_1)^{-1}V(iI - H_1)^{-1}$ belongs to the trace class, verifying our assertion that H and H_1 are symmetrically related.

Hence, using Lemma 19, we see that $\sum (H, H_1) \supseteq \sum_{ac} (H_1)$. Now let $x \in \sum_{ac} (H)$, and let $E_1(\cdot)$ be the resolution of the identity for H_1. If e_0 is a Borel set of Lebesgue measure zero, then $E(e_0)x = 0$, and thus, by Lemma 6, we have $0 = \mathscr{U}(H_1, H)E(e_0)x = E_1(e_0)\mathscr{U}(H_1, H)x$. This shows that

$\mathscr{U}(H_1, H) \sum_{ac}(H) \subseteqq \sum_{ac}(H_1)$. Similarly, $\mathscr{U}(H, H_1) \sum_{ac}(H_1) \subseteqq \sum_{ac}(H)$. Using Corollary 4, we find that $\mathscr{U}(H_1, H) \sum_{ac}(H) = \sum_{ac}(H_1)$, and the proof of Theorem 9 is complete. Q.E.D.

Theorem 9 has a number of interesting applications to the spectral theory of particular operators, references to which are given in the section of Notes and Remarks appended to the present chapter. This theorem also permits a number of interesting extensions, some of which are given in the Notes and Remarks, and others of which are developed as exercises in the immediately following section.

5. Exercises

1 Let H_1 and H_2 be unbounded self adjoint operators in a Hilbert space $\mathfrak{X}$; suppose that the intersection $\mathfrak{Y}$ of the domains of H_1 and H_2 is dense. Show that for every vector $x \in \mathfrak{Y}$ for which

$$\int_0^\infty |(H_2 - H_1)e^{-itH_1}x| \, dt < \infty$$

the limit $\lim_{t\to\infty} e^{itH_2}e^{-itH_1}x$ exists strongly.

2 Let H_1 be the operator $f(x) \to xf(x)$ in the Hilbert space $L_2(0, 1)$, and let V be an integral operator

$$(Vf)(x) = \int_0^1 V(x, y)f(y) \, dy$$

with a measurable kernel satisfying $V(x, y) = \overline{V(y, x)}$ and

$$\int_0^1 \int_0^1 |V(x, y)|^2 \, dx \, dy < \infty.$$

Put $H_2 = H_1 + V$. Show that if V is absolutely continuous in y for each fixed x and if

$$\int_0^1 \int_0^1 \left| \frac{\partial}{\partial y} V(x, y) \right|^2 dx \, dy < \infty,$$

then $\lim_{t\to\infty} e^{itH_2}e^{-itH_1}$ exists strongly. (Hint: Use Exercise 1.)

3 In the following exercise, if τ denotes a formal partial differential operator defined in the Euclidean space E^n, then $T_1(\tau)$ denotes the closed operator in Hilbert space defined in the semi-final paragraph of Section XIV.3.

(a) Let $\nabla = \partial^2/\partial x_1^2 + \cdots + \partial^2/\partial x_n^2$ be the Laplacian operator in E^n. Show that $T_1(\nabla)$ is a self adjoint operator, and that if $f \in \mathfrak{D}(T_1(\nabla))$ is a function in its domain, then $\partial f/\partial x_i$ and $\partial^2 f/\partial x_i\, \partial x_j$ are square integrable for all $1 \leqq i, j \leqq n$.

(b) Suppose that the coefficient functions a_{ij}, a_i, and a are defined in E^n, bounded, and approach zero as $|x| \to 0$ in E^n. Let

$$\tau_1 = \sum_{i,j=1}^{n} a_{ij}(x) \frac{\partial^2}{\partial x_i\, \partial x_j} + \sum_{j=1}^{n} a_i(x) \frac{\partial}{\partial x_i} + a(x).$$

Show that, for $\lambda < 0$, $T_1(\tau_1)(T_1(\nabla) + \lambda I)^{-1}$ is a compact operator, and that, as $\lambda \to -\infty$, $T_1(\tau_1)(T_1(\nabla) + \lambda I)^{-1}$ converges uniformly to zero.

(c) Show, under the hypotheses of (b), that if τ_1 is formally self adjoint, that $T_1(\nabla + \tau_1)$ is a self adjoint operator in $L_2(E^n)$.

(d) Put $f_{a,k}(x) = \exp(-|x - a|^2/2k)$. Show that the functions $f_{a,k}$ span all of $L_2(E^n)$, and that

$$\exp(-itT_1(\nabla))f_{a,k} = (1 + 4ikt)^{-n/2} \exp(-k|x - a|^2/(1 + 4ikt)).$$

(e) Show, under the hypotheses of (b), and under the additional hypotheses that

$$\int_{E^n} \left\{ |x|^4 \sum_{i,j=1}^{n} |a_{ij}(x)|^2 + |x|^2 \sum_{i=1}^{n} |a_i(x)|^2 + |a(x)|^2 \right\} dx < \infty,$$

that the strong limit $\lim_{t\to\infty} \exp(itT_1(\nabla + \tau_1))\exp(-itT_1(\nabla))$ exists provided that $n \geqq 3$; and hence that $T_1(\nabla + \tau_1)$, restricted to an appropriate closed invariant subspace, yields an operator unitarily equivalent to $T_1(\nabla)$. (Hint: Use (d) and Exercise 1.)

4 (Putnam) Let τ be a formally self adjoint formal differential operator on an interval (a, b), and let H_1 and H_2 be self adjoint operators in the Hilbert space $L_2(a, b)$, both defined by the formal differential operator τ, but by different sets of linear boundary conditions. Let λ be a real number not in the spectrum of H_1.

(a) Show that there exists a real number μ which is not in the spectrum either of H_1 or of H_2.

(b) Show that, if μ is as above, then

$$(\mu I - H_1)^{-1} - (\mu I - H_2)^{-1}$$

is a self adjoint operator with finite dimensional range.

(c) Show that $H_1 | \sum_{ac}(H_1)$ and $H_2 | \sum_{ac}(H_2)$ are isometrically equivalent.

5 Let l_p denote the B-space of sequences $z = [z_0, z_1, \ldots]$ with the norm $|z| = \{\sum_{j=0}^{\infty} |z_j|^p\}^{1/p}$, and let $\mathfrak{A}$ be the B-space of infinite matrices $a = \{a_{ij}, i, j \geqq 0\}$ such that

$$\|a\| = \sum_{i,j=0}^{\infty} |a_{ij}| < \infty.$$

Let $\mathfrak{B}$ be the B-space of infinite matrices $b = \{b_{ij}, i, j \geqq 0\}$ such that

$$|b| = \max\left(\max_{i \geqq 0} \sum_{j=0}^{\infty} |a_{ij}|, \quad \max_{j \geqq 0} \sum_{i=0}^{\infty} |a_{ij}|\right) < \infty.$$

Show that

(a) $\mathfrak{A} \subseteqq \mathfrak{B}$, and $|a| \leqq \|a\|$, $\qquad a \in \mathfrak{A}$.

(b) If $a \in \mathfrak{A}$ and $b \in \mathfrak{B}$, the product matrices ab and ba both belong to $\mathfrak{B}$, and

$$|ab| \leqq \|a\| \, |b|, \qquad |ba| \leqq \|a\| \, |b|.$$

(c) If $b \in \mathfrak{B}$, then $z \to bz$ defines a bounded linear transformation in the space l_p, and, in fact, $|bz| \leqq |b| \, |z|$.

6 (Freeman) Let S be the unilateral left shift transformation

$$z \to Sz = [z_1, z_2, \ldots]$$

in the space l_p of the preceding exercise, and use the notations $\mathfrak{A}$, $\mathfrak{B}$, and so on, of the preceding exercise. If $a \in \mathfrak{A}$, let $b = \Gamma a$ be the infinite matrix whose elements are defined by

$$b_{i,j} = a_{i-1,j} + a_{i-2,j-1} + \cdots$$

(where, for convenience of notation, we write $a_{i,j} = 0$ if either one of the indices i, j is negative). Show that

(a) $S(\Gamma a) - (\Gamma a)S = a, \qquad a \in \mathfrak{A}$.

(b) $|\Gamma a| \leqq \|a\|$.

(c) If $\|a\| < \frac{1}{2}$, then the operator $z \to Sz + az$ in l_p is similar to the operator S in l_p.

7 (Freeman) Let S, $\mathfrak{A}$, l_p, and so on, be as in Exercises 5 and 6. Show that

(a) If $a_n, a \in \mathfrak{A}$ and $\|a_n - a\| \to 0$, then the transformations $b \to \Gamma(b)a_n$ in the B-space $\mathfrak{A}$ converge uniformly to the transformation $b \to \Gamma(b)a$.

(b) If $a \in \mathfrak{A}$ is purely superdiagonal, that is, if the matrix elements $a_{i,j}$ of a vanish whenever $i \geqq j$, then the transformation $b \to \Gamma(b)a$ in the space $\mathfrak{A}$ is quasinilpotent.

(Hint: Use the preceding part to approximate a by a finite superdiagonal matrix; show in this case that $b \to \Gamma(b)a$ is nilpotent.)

(c) If $a \in \mathfrak{A}$ is purely superdiagonal, the equation $b - \Gamma(b)a = a$ has a unique superdiagonal solution $b \in \mathfrak{A}$; the elements $b_{i-1,i}$ of this solution are determined inductively by the equations

$$b_{0,1} = a_{0,1}; \qquad b_{1,2} = a_{1,2}(1 + b_{0,1}); \qquad b_{2,3} = a_{2,3}(1 + b_{0,1} + b_{1,2}), \ldots.$$

(e) If $a \in \mathfrak{A}$ is purely superdiagonal, and $a_{n,n+1} \neq -1$ for all $n \geqq 1$, then the transformation $z \to Sz + az$ in the space l_p is similar to the transformation S.

8 Let S, $\mathfrak{A}$, l_p, and so on, be as in Exercises 5, 6, and 7. Show that

(a) If $a \in \mathfrak{A}$ is purely subdiagonal, that is, if the matrix elements $a_{i,j}$ of a vanish if $j \geqq i$, then the transformation $b \to a\Gamma(b)$ in the space $\mathfrak{A}$ is quasinilpotent.

(Hint: Use the procedure of the corresponding part of Exercise 7.)

(b) If $a \in \mathfrak{A}$ is purely subdiagonal, then the transformation $z \to Sz + az$ in the space l_p is similar to the transformation S.

9 Consider the Hilbert space $L_2 = L_2(0, 1)$, and the operator T defined by $(T_0 f)(x) = xf(x)$, $f \in L_2(0, 1)$. Let $\varphi \in L_2$ be a continuously differentiable function, let c be real, and consider the bounded self adjoint operator T defined by $Tf = T_0 f + c\varphi(f, \varphi)$.

(a) Show that the eigenvalues of T in the interval $[0, 1]$ are the points λ such that $\varphi(\lambda) = 0$ and such that

$$c \int_0^1 \frac{|\varphi(x)|^2}{x - \lambda}\, dx = 1.$$

Show that the continuous spectrum of T is $[0, 1] - \sigma_p(T)$, where $\sigma_p(T)$ is the point spectrum of T.

(b) Construct a function φ belonging to L_2 and continuously differentiable in the interior of $[0, 1]$ for which the operator T of (a) has infinitely many eigenvalues within the interval $[0, 1]$.

10 Let $\mathfrak{X}$ be a B-space. Let $A(s)$ be an $\mathfrak{X}$-valued function defined on the real axis R. Suppose that $\gamma > 0$, that $1 > \beta > 0$, and that $\|A\|_{\gamma,\beta} < \infty$, where the norm is as specified in Definition 2.11.

(a) Show that

$$\lim_{\varepsilon \to 0+} \int_{-\infty}^{+\infty} \frac{A(\sigma)}{s-\sigma+i\varepsilon}\, d\sigma = \mathscr{P} \int_{-\infty}^{+\infty} \frac{A(\sigma)}{s-\sigma}\, d\sigma - i\pi A(s).$$

(b) Show that

$$\lim_{\varepsilon \to 0+} \int_{-\infty}^{+\infty} \frac{A(\sigma)}{s-\sigma-i\varepsilon}\, d\sigma = \mathscr{P} \int_{-\infty}^{+\infty} \frac{A(\sigma)}{s-\sigma}\, d\sigma + i\pi A(s).$$

Let $B(s, t)$ be an $\mathfrak{X}$-valued function defined for all real s and t, and suppose that $\|B\|_{\gamma,\beta} < \infty$, where the norm is as specified in Definition 2.13.

(c) Show that

$$\lim_{\varepsilon \to 0+} \int_{-\infty}^{+\infty} \frac{B(s, \sigma)}{s-\sigma\pm i\varepsilon}\, d\sigma = \mathscr{P} \int_{-\infty}^{+\infty} \frac{B(s, \sigma)}{s-\sigma}\, d\sigma \mp i\pi A(s, s).$$

Let $\mathfrak{H}$ be a Hilbert space, and let $B(\mathfrak{H})$ be the B-space of all bounded operators in $\mathfrak{H}$. Let $C(s, t)$ be a $B(\mathfrak{H})$-valued function defined for all s and t, and let $f \in L_2(R, \mathfrak{H})$. Suppose that $\|C\|_{\gamma,\beta} < \infty$, where the norm is as specified in Definition 2.13.

(d) Show that

$$\lim_{\varepsilon \to 0+} \int_{-\infty}^{+\infty} \frac{C(s, \sigma)f(\sigma)}{s-\sigma\pm i\varepsilon}\, d\sigma = \mathscr{P} \int_{-\infty}^{+\infty} \frac{C(s, \sigma)f(\sigma)}{s-\sigma}\, d\sigma \mp i\pi C(s, s)f(s)$$

almost everywhere, the limit on the left and the proper value integral on the right existing almost everywhere.

11 (Friedrichs' Identity) Let $\mathfrak{H}$ be a Hilbert space, and let $B(\mathfrak{H})$ be the B-space of all bounded operators in $\mathfrak{H}$. Let $\gamma > 0$ and $1 > \beta > 0$ be two real numbers, and let $\mathfrak{A}_{\gamma,\beta}(B(\mathfrak{H}))$ be the space of Definition 2.13. For each $A \in \mathfrak{A}_{\gamma,\beta}(B(\mathfrak{H}))$, let $\Gamma_+(A)$ be the bounded operator in the Hilbert space $L_2(R, \mathfrak{H})$ defined by

$$(\Gamma_+(A)f)(s) = \lim_{\varepsilon \to 0+} \int_{-\infty}^{+\infty} \frac{A(s, \sigma)}{s-\sigma+i\varepsilon} f(\sigma)\, d\sigma, \qquad f \in L_2(R, \mathfrak{H})$$

(cf. part (d) of the preceding exercise). Show that, if $A, B \in \mathfrak{A}_{\gamma,\beta}(B(\mathfrak{H}))$, then

$$\Gamma_+(A)\Gamma_+(B) = \Gamma_+(A\Gamma_+(B) + \Gamma_+(A)B).$$

Here, we write $A\Gamma_+(B)$ for the limit integral

$$\lim_{\varepsilon \to 0+} \int_{-\infty}^{+\infty} A(s, \sigma) \frac{B(\sigma, t)}{\sigma - t + i\varepsilon}\, d\sigma,$$

and $\Gamma_+(A)B$ for the limit integral

$$\lim_{\varepsilon \to 0+} \int_{-\infty}^{+\infty} \frac{A(s, \sigma)}{s - \sigma + i\varepsilon} B(\sigma, t)\, d\sigma.$$

12 (Friedrichs) Let $\mathfrak{H}$, $B(\mathfrak{H})$, γ, β, $\mathfrak{A}_{\gamma,\beta}$, and A be as the preceding exercise. Let r and l be elements of $\mathfrak{A}_{\gamma,\beta}(B(\mathfrak{H}))$ such that

$$A + \Gamma_+(l)A - l = 0; \qquad A + A\Gamma_+(r) + r = 0.$$

(a) Show that

$$(I + \Gamma_+(l))r = -l(I + \Gamma_+(r)).$$

(b) Show that

$$(I + \Gamma_+(l))(I + \Gamma_+(r)) = I.$$

(c) Show that

$$P = (I + \Gamma_+(r))(I + \Gamma_+(l))$$

is a projection.

(d) Show that, if $\mathfrak{H}$ is one-dimensional, the complementary projection I–P of the projection P has a finite dimensional range.

13 Let $\mathfrak{H}$, $B(\mathfrak{H})$, γ, β, $\mathfrak{A}_{\gamma,\beta}$, and A be as in Exercise 11.

(a) Show that if we define A^* by the equation

$$A^*(s, t) = (A(t, s))^*,$$

then

$$\Gamma_+(A^*) = -\Gamma_+(A)^*.$$

(b) Show that, if $l \in \mathfrak{A}_{\gamma,\beta}(B(\mathfrak{H}))$ is a solution of the equation

$$A + \Gamma_+(l)A - l = 0,$$

then $r = -l^*$ is a solution of

$$A^* + A^*\Gamma_+(r) + r = 0.$$

(c) Show that, if $A = A^*$ and the equation

$$A + \Gamma_+(l)A - l = 0,$$

has a solution $l \in \mathfrak{A}_{\gamma,\beta}(B(\mathfrak{H}))$, then $I + \Gamma(l)$ is a partially isometric operator, and that if in addition $\mathfrak{H}$ is one-dimensional, then the range of $I + \Gamma(l)$ has a finite dimensional orthocomplement.

14 Let $\mathfrak{H}$ be a Hilbert space, R the real axis, and consider the space $L_2 = L_2(R, \mathfrak{H})$. Let H be the self adjoint operator in L_2 defined by $(Hf)(x) = xf(x)$. For each $\varphi \in L_2$, let T_φ be the self adjoint operator with one-dimensional range defined by $T_\varphi f = \varphi(f, \varphi)$. Let c be real.

(a) Show that the equation $W(H + cT_\varphi) = HW$ has the formal solution $W = I + \Gamma_+(A)$, where A is the integral operator whose kernel $A(x, y)$ is defined by

$$A(x, y)v = a(x)(v, \varphi(y)), \qquad v \in \mathfrak{H},\ x, y \in R,$$

and where

$$a(x) = \frac{c\varphi(x)}{1 - c\displaystyle\int_{-\infty}^{+\infty} \frac{|\varphi(t)|^2}{x - t + i0}\, dt}.$$

(b) Show that, if $\gamma > 0$ and $1 > \beta > 0$, while $\|\varphi\|_{\gamma, \beta} < \infty$, the norm being as in Definition 2.13, then $W(H + cT_\varphi)f = HWf$ holds rigorously for all f in the domain of the operator H.

(c) Show that the denominator of the fraction defining $a(x)$ vanishes only at points x_0 at which $\varphi(x_0) = 0$, and that this denominator is asymptotic to 1 as $x \to \pm\infty$.

(d) Show that, if φ has a Hölder-continuous first derivative, then the derivative of the denominator of the fraction defining $a(x)$ has the value

$$c\int_{-\infty}^{+\infty} \frac{|\varphi(t)|^2}{(x_0 - t)^2}\, dt$$

at each of its zeros x_0, so that each such zero is simple.

(e) Show that, if φ has a Hölder-continuous first derivative, the vector valued function $a(x)$ is Hölder-continuous, so that the equation $W(H + cT_\varphi) = HW$ always has a solution W which is an isometric operator in the Hilbert space L_2.

(f) Show that, for the isometric operator W constructed above, the projection $P = WW^*$ has a complement $I - P$ which is compact, so that the range of W has a finite dimensional complement.

15 Let $\mathfrak{H}$ be a Hilbert space, R the real axis, and consider the space $L_2 = L_2(R, \mathfrak{H})$. Let H be the self adjoint operator in L_2 defined by $(Hf)(x) = xf(x)$. Let $\varphi_i \in L_2(R, \mathfrak{H})$, $i = 1, \ldots, n$; let $\{k_{ij}, i, j = 1, \ldots, n\}$

be an Hermitian matrix, and let T be the self adjoint operator in L_2 defined by

$$Tf = \sum_{i,j=1}^{n} k_{ij}\varphi_i(f, \varphi_j).$$

Suppose that $1 > \beta > 0$, that $(1+|x|)^\beta \varphi_j(x)$ is bounded, $1 \leqq j \leqq n$; and that $\varphi_j \in C^\infty(R)$, $1 \leqq j \leqq n$.

(a) Show that there exists an isometry W in $L_2(R, \mathfrak{H})$, whose range has a finite dimensional complement, such that $W(H+T)f = HWf$ for all f in the domain of H.
(Hint: Use Exercise 14, and induction on the dimension of the range of T.)

(b) Show that $\sum_{ac}(H+T)$ has a finite dimensional complement, so that $H+T$ reduces to the direct sum of an operator equivalent to H and of a finite dimensional self adjoint operator.

16 (a) Let H be a self adjoint operator in a Hilbert space $\mathfrak{X}$. Let $\mathfrak{Y}$ be an auxiliary Hilbert space, and let λ denote the Lebesgue measure on the real axis R. Show that there exist a positive λ-singular Borel measure ν on R and an isometric embedding of $\mathfrak{X}$ in the Hilbert space $L_2(\mu, \mathfrak{Y})$ of vector valued functions, such that $H = H_0 \mid \mathfrak{X}$, where H_0 is the multiplication operator $f(x) \to xf(x)$. Here $\mu = \lambda + \nu$.

(b) Let H, $\mathfrak{X}$, $\mathfrak{Y}$, and μ be as in (a), and let $\{E(\cdot)\}$ be the spectral resolution of H. For each $w \in \mathfrak{X}$, let

$$\frac{d(E(\cdot)w, w)}{d\lambda} = d_w(\cdot)$$

denote the Radon-Nikodým derivative with respect to the Lebesgue measure λ of the measure $(E(e)w, w)$. Write

$$\|w\|_H = \lambda\text{-ess}\sup_{-\infty < x < +\infty} |d_w(x)|^{1/2}.$$

Show that if using the embedding of $\mathfrak{X}$ in $L_2(\mu, \mathfrak{Y})$ we regard w as a vector valued function defined on the real axis R, then $\|w\|_H$ is identical with the λ-essential supremum of the function $|w(x)|$.

(c) Prove that $\|w\|_H < \infty$ for all w in a dense subspace of $\sum_{ac}(H)$.

(d) Prove that if $w \in \sum_{ac}(H)$ and $\|w\|_H < \infty$, while v is any vector in Hilbert space, then

$$\int_{-\infty}^{+\infty} |(e^{itH}w, v)|^2\, dt \leqq 2\pi \|w\|_H^2 \|v\|^2.$$

(e) Prove that if A is an operator of the Hilbert-Schmidt class, while $w \in \sum_{ac}(H)$ and $\|w\|_H < \infty$, then

$$\int_{-\infty}^{+\infty} |Ae^{itH}w|^2 \, dt \leqq 2\pi \, \|A\|_2^2 \, \|w\|_H^2 .$$

(Hint: Consider A self adjoint, and expand in the eigenvectors of A.)

17 (a) Let f be a continuous function of bounded variation on an interval $[a, b]$, and let $V(f)$ denote its total variation. Show that

$$\left| \int_a^b e^{ixf}(x) \, dx \right| \leqq 2 \max_{s \leqq x \leqq b} |f(x)| + V(f).$$

(Hint: Integrate by parts.)

(b) Let f be a strictly increasing function on an interval $[a, b]$. Suppose that its derivative f' is positive, continuous, and of bounded variation, and let $V((f')^{-1})$ be the total variation of the reciprocal of f'. Prove that

$$\left| \int_a^b e^{if(x)} dx \right| \leqq 2 \max_{a \leqq x \leqq b} (|f'(x)|^{-1}) + V((f')^{-1}).$$

(Hint: Make the change of variables $y = f(x)$.)

(c) Let f be a strictly increasing function on an interval $[a, b]$. Suppose that its derivative is positive, continuous, and of bounded variation. Show that

$$\lim_{s \to \infty} \int_0^\infty \left| \int_a^b e^{i(tx + sf(x))} \, dx \right|^2 dt = 0.$$

(Hint: Use (b).)

(d) Let A be an operator of the Hilbert-Schmidt class. Show, under the hypotheses of part (c) of the present exercise, that if H is a self adjoint operator in Hilbert space,

$$\lim_{s \to \infty} \int_0^\infty |A \exp(i(tH + sf(H)))w|^2 \, dt = 0$$

for all $w \in \sum_{ac}(H)$ such that $\|w\|_H < \infty$. (Cf. Exercise 16, parts (b) and (c).)

18 Let $\mathfrak{Y}$ be a Hilbert space, and let λ denote the Lebesgue measure on the real axis R. Let ν be a positive λ-singular Borel measure on R, and put $\mu = \lambda + \nu$. Consider the Hilbert space $L_2(\mu, \mathfrak{Y})$ of vector valued functions, and let H_1 be the multiplication operator $f(x) \to xf(x)$ in this space. Let V be a self adjoint operator of trace class. Put $H_2 = H_1 + V$, and $U_t = \exp(itH_2)\exp(-itH_1)$.

(a) Show that

$$|(U_t - U_s)w|^2 = 2R((U_t - U_s)w,\ U_t w), \qquad w \in L_2(\mu,\ \mathfrak{Y}).$$

(b) Show that if $w \in L_2(\mu, \mathfrak{Y})$ and vanishes outside a bounded subset of R, then

$$|(U_t - U_s)w|^2 = \int_s^t (\exp(ixH_2)V \exp(-ixH_1)w,\ U_t w)\, dx.$$

(c) Find a dense subset D of the space $\sum_{ac}(H_1)$ such that if $w \in D$ the limit $W_+ w = \lim_{t\to\infty} U_t w$ exists and

$$|(W_+ - U_s)w|^2 = \int_s^\infty (V \exp(-ixH_1)w, W_+ \exp(-ixH_1)w)\, dx.$$

(Hint: Use Theorem 4.9.)

(d) (Rosenblum's Inequality) Show that there exist operators A, B of Hilbert-Schmidt class depending only on V and μ such that if $w \in \sum_{ac}(H_1)$ and $\|w\|_{H_1} < \infty$ then

$$|(W_+ - U_s)w|^2 \leqq \left(\int_s^\infty |A \exp(-ixH_1)w|^2\, dx\right)^{1/2} \times \left(\int_s^\infty |B \exp(-ixH_1)w|^2\, dx\right)^{1/2}.$$

(Hint: Use (c) and (e) of the preceding exercise.)

19 Let H be a self adjoint operator in Hilbert space. Let V be an operator of trace class, and put $H_1 = H + V$. Put $U_t = \exp(itH_1)\exp(-itH)$. Let $w \in \sum_{ac}(H)$ and $\|w\|_H < \infty$ (cf. Exercise 17.)

(a) Show that the limit $W_+ w = \lim_{t\to\infty} U_t w$ exists, and that there exist operators A, B of the Hilbert-Schmidt class, depending only on V and H, such that

$$|(W_+ - U_s)w|^2 \leqq \left(\int_s^\infty |A \exp(-ixH)w|^2\, dx\right)^{1/2} \times \left\{\int_s |B \exp(-ixH)w|^2\, dx\right\}^{1/2}.$$

(b) Show that, if f is any strictly increasing function on the real axis, having a positive, continuous derivative of bounded variation, then

$$\lim_{s\to\infty} (W_+ - I)\exp(-isf(H))w = 0$$

for all $w \in \sum_{ac}(H)$ such that $\|w\|_H < \infty$.
(Hint: Use Exercise 17.)

(c) (Wave operator invariance theorem) Show, under the preceding hypotheses, that

$$\lim_{s\to\infty} \exp(isf(H_1))\exp(-isf(H))w = w = W_+ w$$

for all $w \in \sum_{ac}(H)$.

20 Let H_1 and H_2 be strictly positive self adjoint operators in Hilbert space. Suppose that for some positive real number x the difference $H_1^{-x} - H_2^{-x}$ belongs to the trace class. Then

(a) $\lim_{t\to\infty} e^{itH_2}e^{-itH_1} w$ exists for all $w \in \sum_{ac}(H_1)$.

(b) $H_1 | \sum_{ac}(H_1)$ and $H_2 | \sum_{ac}(H_2)$ are isometrically equivalent.

21 Let H_1 and H_2 be self adjoint operators in Hilbert space. Let V_1, V_2 denote any two operators of trace class, and $\|V_1\|_1$, $\|V_2\|_1$ their trace norms. Write

$$W_+(H_2 + V_2, H_1 + V_1)w = \lim_{t\to\infty} \exp(it(H_2 + V_2))\exp(-it(H_1 + V_1))w$$

if the limit on the right exists. Then

(a) if $W_+(H_2, H_1)w$ exists so does $W_+(H_2, H_1 + V)w$ and

$$\lim_{\|V\|_1\to 0} W_+(H_2, H_1 + V)w = W_+(H_2, H_1)w$$

in the weak topology;

(b) if $W_+(H_2, H_1)w$ exists, so does $W_+(H_2 + V_1, H_1)\, w$ and

$$\lim_{\|V\|_1\to 0} W_+(H_2 + V, H_1)w = W_+(H_2, H_1)w$$

in the strong topology.

6. Notes and Remarks

The method used in Section 1 for the spectral analysis of the second order operator is due essentially to Naĭmark [10, 11, 12]. Of course, this method is simply a generalization to nonselfadjoint operators of an idea that has long been known (see Weyl [5]) for the self adjoint case; but in the nonselfadjoint case, the arguments must be arranged in such a way as to avoid all dependence on the spectral theorem, which makes the whole theory fundamentally more difficult. Naĭmark also remarks that his methods can be generalized to equations of higher order. Such a generalization is carried out in G. E. Huige [1]. See also E. Balslev and T. W. Gamelin

[1], E. Balslev [1], as well as V. E. Ljance [4]. Some of Huige's arguments make use of the "factorization method" described below.

The work of Moser [1] is closely related to the theory developed in Section 1. Moser considers a formally self adjoint formal differential operator of the form

$$\tau_\varepsilon = -\frac{d}{dt}p(t)\frac{d}{dt} + q_0(t) + \sum_{i=1}^{\infty}\varepsilon^i q_i(t),$$

the coefficients p and q_i being real, and the dependence on the parameter ε being analytic "uniformly in t" in a sense to be specified more exactly below. The operators τ_ε are supposed to be defined on an interval (a, b), to have two boundary values at a and no boundary values at b. A common boundary condition $A(f)=0$ for all the operators τ_ε is imposed; in this way, a family T_ε of self adjoint operators is defined. It is then established that

(1) under suitable analytic conditions, the spectral resolution $E(T_\varepsilon; \Lambda)$, Λ being a certain interval of $\sigma(T_\varepsilon)$, depends analytically on ε (for ε real and small);

(2) under still more stringent analytic conditions, there is an analytically varying unitary operator U_ε such that $T_\varepsilon = U_\varepsilon^{-1}T_0U_\varepsilon$ for ε small.

Precise forms of Moser's theorems, in the somewhat special case in which τ_ε is defined on a half-open interval $[a, b)$, may be stated as follows.

THEOREM. *Let T_ε be the operator defined by the formal differential operator*

$$\tau_\varepsilon = -\frac{d}{dt}p(t)\frac{d}{dt} + q_0(\tau) + \sum_{j=1}^{\infty}\varepsilon^j q_j(t),$$

on the interval $[a, b)$, all the coefficients p and q_j being real, and by a boundary condition

$$f(a) + kf'(a) = 0, \qquad -\infty < k \leqq +\infty.$$

Let Δ be a subinterval of $\sigma(T_0)$. Suppose that there exist

(a) *a non-negative continuous function Φ defined for $t \in [a, b)$, and constants K and α such that*

$$\int_a^b \Phi^2(t)|q_j(t)|\,dt < \gamma K^{j-1}, \qquad j \geqq 1,$$

and such that the solutions φ *and* ψ *of the equation* $\tau\sigma = \lambda\sigma$ *defined by the respective boundary conditions* $\varphi(a) = 0$, $\varphi'(a) = 1$ *and* $\psi(a) = 1$, $\psi'(a) = 0$ *satisfy*

$$[*] \qquad |\varphi(t, \lambda)|^2 + |\psi(t, \lambda)|^2 \leqq \Phi(t)^2, \qquad t \in [a, b),\ \lambda \in \Delta;$$

(b) *two constants* $C \geqq 1$, $\delta > 0$ *such that the solution* χ *of the equation* $\tau\sigma = \lambda\sigma$ *defined by the boundary conditions*

$$\chi(a) + k\chi'(a) = 0,$$

$$k\chi(a) - \chi'(a) = 1$$

satisfies

$$[**] \qquad |\chi(t, \lambda + i\delta)| \leqq C\Phi(t), \qquad 0 < \delta < \delta_0, \qquad t \in [a, b),\ \lambda \in \Delta.$$

Then, if Λ *is a subinterval of* Δ, *the projection* $E(T_\varepsilon, \Lambda)$ *depends analytically on* ε *for* ε *real and small.*

THEOREM. *Let the hypotheses of the previous theorem be satisfied, and suppose in fact that conditions* [*] *and* [**] *are satisfied for all* $\lambda \in \sigma(T_0)$, *and not merely for all* $\lambda \in \Delta$. *Then there exists an analytically varying unitary operator defined for all small real* ε, *such that* $T_\varepsilon = U_\varepsilon^{-1} T_0 U_\varepsilon$.

Even though Moser's theorems are stated for formally self adjoint formal operators, the fact that $E(T_\varepsilon, \Lambda)$ and U_ε on ε depend analytically on ε allows his results to be continued into the complex plane, and thus allows the deduction of results which apply to nonselfadjoint operators. By this method, Moser is able to deduce the following theorem as a particular case of his general theory.

THEOREM. *Let* τ_ε *be the formal differential operator*

$$-\left(\frac{d}{dt}\right)^2 + \varepsilon q(t), \qquad 0 \leqq t < \infty, \qquad q \text{ real}.$$

Suppose that

$$\int_0^\infty (1 + t^2)|q(t)|\, dt < \infty.$$

Let T_ε *be the operator defined by* τ_ε *and by a boundary condition*

$$f(0) + kf'(0) = 0, \qquad -\infty < k \leqq +\infty.$$

Then, for $|\varepsilon|$ sufficiently small, T_ε is similar to the self adjoint operator T_0; that is, there exists an (analytically varying) operator U_ε with a bounded inverse, such that $T_\varepsilon = U_\varepsilon^{-1}T_0U_\varepsilon$.

This is a theorem almost including the main theorem of Section 1.

In this connection, see also Phillips [13]. Moser's paper [2] generalizes the results of Moser [1] to operators of the form $\tau + \varepsilon q(x)$, where τ is a self adjoint ordinary differential operator of even order. Extension in a similar direction of the theorems of Moser [1] are given by Butler [2, 3].

In Schwartz [4], the following general principle permitting the application of Theorem XVIII.2.34 is developed. Let $H_1 = H + V$, H being assumed self adjoint for the sake of simplicity in exposition. Then the ordinary perturbation series for the resolvent $R_1(\lambda) = (\lambda I - H_1)^{-1}$ in terms of $R(\lambda) = (\lambda I - H)^{-1}$ is

$$R_1(\lambda) = R(\lambda) + R(\lambda)VR(\lambda) + R(\lambda)VR(\lambda)VR(\lambda) + \cdots.$$

If V is assumed to be a product AB, we may write this series as

$$\begin{aligned} R_1(\lambda) &= R(\lambda)A(I + BR(\lambda)A + BR(\lambda)ABR(\lambda)A + \cdots)BR(\lambda) \\ &= R(\lambda) + R(\lambda)A(I - BR(\lambda)A)^{-1}BR(\lambda). \end{aligned}$$

It may now be observed that it is entirely possible that the operator $BR(\lambda)A$ should depend continuously on λ, even when λ approaches the continuous spectrum of H. Consider, for example, the case in which our Hilbert space is the space $L_2[a, b]$ of (possibly vector valued) functions, and $Hf(x) = xf(x)$. Suppose also that A and B are integral operators with Hölder-continuous kernels $\alpha(x, y)$ and $\beta(x, y)$. Then $BR(\lambda)A$ is the integral operator with the kernel

$$\int_a^b \frac{b(x, t)a(t, y)}{\lambda - t}\, dt;$$

this kernel is Hölder-continuous in all the three parameters λ, x, y.

If we let $\delta(\lambda) = \det_2(I - BR(\lambda)A)$ be the Hilbert-Fredholm determinant of the Hilbert-Schmidt operator $I - BR(\lambda)A$ (cf. Definition XI.9.21) and write $C(\lambda) = \delta(\lambda)(I - BR(\lambda)A)^{-1}$, then by Theorem XI.9.26 (cf. also Lemma XI.9.23) it follows that $C(\lambda)$ and $\delta(\lambda)$ have the same continuity properties as $BR(\lambda)A$, and hence, in the case in which we are interested, depend Hölder-continuously on λ. Since

$$R_1(\lambda) = R(\lambda) + (\delta(\lambda))^{-1}R(\lambda)AC(\lambda)BR(\lambda),$$

Theorem XVIII.2.34 is applicable and yields a spectral analysis of H_1 whenever $\delta(\lambda)$ has no zeros in the continuous spectrum of H. The cited paper of Schwartz may be consulted for a detailed development of this line of argument.

The "Friedrichs method" described in Section 2 was developed by K. O. Friedrichs in Friedrichs [1]; cf. Friedrichs [2] for an expository account of this work and for some extensions.

The application of the Friedrichs method to quasi-nilpotent integral operators which we have developed at the end of Section 2 is due to J. M. Freeman [3]. Freeman's results have been subsequently extended by A. Dupras [1] who studies integral operators of the form

$$f(x) \to g(x) = \int_0^x (x-y)^\alpha A(x, y) f(y)\, dy, \qquad 0 \leqq x \leqq 1,$$

where $A(x, x) \neq 0$, $0 \leqq x \leqq 1$, and gives conditions under which these operators are equivalent to the operator

$$f(x) \to g(x) = \int_0^x (x-y)^\alpha f(y)\, dy$$

of fractional integration.

Additional theorems belonging to the similarity theory of Volterra operators are to be found in S. J. Osher [1]. Osher studies Volterra operators of the form $\int_0^x A(x, y) f(y)\, dy$, in which the function $A(x, x)$ has a single zero on the unit interval, and finds similarity invariants for this class of operators which are related to the position of the isolated zero of $A(x, x)$. See also G. K. Kalisch [2, 3]. Freeman [2] applies his "quasi-nilpotent" version of the Friedrichs method to operators obtained by perturbing the unilateral shift operator

$$S\colon (x_0, x_1, \ldots) \to (x_1, x_2, \ldots)$$

in the sequence Hilbert space l_2 by upper-triangular infinite matrices.

Applications of the Friedrichs method to the study of various phenomena of spectral perturbation have been made by Friedrichs and others; see Friedrichs and Rejto [1]. Friedrichs and Rejto study the perturbation problem for the family $f(x) \to \varphi(x) \int_{-\infty}^{+\infty} f(y)\varphi(y)\, dy + \varepsilon x f(x)$

of operators, which is, of course, identical with the asymptotic theory of the perturbation problem for the operators

$$f(x) \to xf(x) + \lambda\varphi(x) \int_{-\infty}^{+\infty} f(y)\varphi(y)\, dy$$

as λ becomes large. The phenomena which appear have been described by Friedrichs and Rejto in the suggestive phrase "spectral concentration," and studied in a more general setting by Conley and Rejto [1]. See also J. T. Schwartz [8]. The class of singular integral operators studied in this last paper have also been analyzed by other methods, some of which give much more precise results in the self adjoint case. Cf. W. Koppelman and J. Pincus [1], W. Koppelman [2], J. D. Pincus [1, 2, 3, 4], and Pincus and Rovnyak [1].

Friedrichs [1] gives an important extension of his general method to the case of perturbations which are quantitatively large. Call an element $K \in F_2$ *finitary* if K is of the form

$$K(x_1, x_2) = \sum_{i=1}^{N} a_i(x) b_i(x_2), \tag{1}$$

where a_i and b_i are Hölder-continuous, and where $a_i(x) = b_i(x) = 0$ if $|x| \geqq 1$; or, more generally, if K is approximable arbitrarily closely in Hölder norm, by elements of the form (1). Then, if K is finitary, it follows that for each $\varepsilon > 0$ there exists a sum K_0 of the form (1) such that

$$\|K_0 - K'\| < \varepsilon. \tag{2}$$

Let $K_0 - K' = K_1$, so that $K' = K_0 + K_1$. Then, if ε is taken to be small enough, and if T is the operation of multiplication by x in the space of functions $L_2(-1, +1)$, it follows by the theorems of Friedrichs proved in Section 2 that there exists an operator U such that

$$T + K_1 = UTU^{-1}.$$

Moreover, an examination of the method of proof of the theorem of Friedrichs cited shows that the operator U and its inverse both are of the form $I + \Gamma$, where Γ is a (singular) integral operator. Thus, since K_0 is of the form (1), $U^{-1}K_0U$ is of this same form. Consequently, the operator $T + K'$ is equivalent to an operator $T + K$, where K is of the form (1), and where we may evidently suppose in addition that the functions a_i are linearly independent.

To analyze the operator $T+K$, one attempts to find a pair of solutions U, V of the equations

$$(T+K)U = UT, \qquad V(T+K) = TV, \tag{3}$$

where we make the *ansatz*

$$U = I + \Gamma R, \tag{4}$$

the integral kernel R being taken to be of the form

$$R(x_1, x_2) = \sum_{i=1}^{N} a_i(x_1) r_i(x_2). \tag{5}$$

The first equation (3) is then easily seen to be equivalent to the following equation for the functions r_i:

$$r_i(x) - \sum_{j=1}^{N} \left(\int \frac{b_i(y)a_j(y)}{x-y} r_j(x)\, dy \right) = b_i(x). \tag{6}$$

It is evident that this system of equations will have a solution if the following system has no non-zero solution for any y_0 in $[-1, +1]$:

$$\rho_i = \sum_{j=1}^{N} \left(\int \frac{b_i(y)a_j(y)}{y_0 - y}\, dy \right) \rho_j. \tag{7}$$

If equations (7) hold, then clearly the function

$$\alpha(x) = \sum_{j=1}^{N} a_j(x)\rho_j \tag{8}$$

satisfies

$$\alpha(x) = \int \frac{K(x, y)}{y_0 - y} \alpha(y)\, dy, \tag{9}$$

or, putting $\alpha(x) = (x - y_0)\beta_{y_0}(x)$,

$$x\beta_{y_0}(x) + \int K(x, y)\beta_{y_0}(y)\, dy = y_0 \beta_{y_0}(x). \tag{10}$$

Thus,

ASSUMPTION A. If the operator $T+K$ has no singular eigenfunction β of the form

$$\beta(x) = \frac{\alpha(x)}{x - y_0}$$

satisfying the equation $(T+K)\beta = y_0\beta$, for $-1 \leqq y_0 \leqq +1$, then the first equation (3) has a solution U.

Assumption A may be stated, roughly and heuristically, as the assumption that no point eigenvalue of $T+K$ lies in the continuous spectrum of T.

It is readily seen that Assumption A also implies that the second equation (3) has a solution. Having carried the analysis in the cited paper [1] to this point, Friedrichs then proceeds to show, on the basis of certain general identities, that the operators UV and VU differ from each other by a compact operator, and that VU (which commutes with the multiplication operator T) satisfies $VU = I$. Friedrichs [1] then uses this identity to show that the space $L_2(I)$ decomposes into the direct sum of two spaces $L_2 + H_1 \oplus H_2$, each invariant under the operator $T+K$. The space H_2 is finite dimensional, and the restriction of $T+K$ to the Hilbert space H_1 is an operator equivalent to the operator T in the space $L_2(I)$. Thus, large but finitary perturbations can add a finite number of additional point eigenvalues to the original continuous spectrum of T, but cannot, under Assumption A, change the spectrum of T in any more drastic sense.

For a detailed development of the line of argument sketched above, the reader should refer to the cited work of Friedrichs.

A related argument is employed in Ladyženskaya and Faddeev [1]. The Ladyženskaya-Faddeev argument involves no quantitative restriction on the size of the perturbing operator, but is restricted to self adjoint operators. The cited paper is the first in which the general Friedrichs similarity method is successfully applied to the spectral analysis of an operator of the form $-\nabla + V(x)$.

A variant, rather similar, abstract argument is given in Rejto [1, I].

The conditions on the potential function V which are imposed in Theorem 2.22 can be very considerably relaxed by various devices, especially if V is assumed to be real, so that the wave operator method is available. For improvements of this sort, together with the more basic improvements in argument which make them possible, see Brownell [1], Kato [12] and Jauch and Zinnes [1]. Ikebe [1] expresses the spectral resolution of the operator $-\nabla + V(x)$ directly in terms of a modified n-dimensional Fourier integral $f(k) = \int \psi_k(y) f(y)\, dy$, in which $\psi_k(x)$ is a solution of the equation $-(\nabla + V(x))\psi_k(x) = k^2\psi_k(x)$ which is asymptotic to the "free wave" function e^{ikx} as $|x| \to \infty$. Cf. also Povzner [1] and Povzner [9] for an earlier result of a very similar sort. Ikebe [2] gives a detailed analysis of the scattering operator S associated with the perturbed Laplacian $-\nabla + V(x)$. Greiner [1] establishes a related result for operators of the more general form $P(D) + V(x)$, in which D is an elliptic operator defined in all of Euclidean n-space, and P is a polynomial.

Theorems concerning the related situation in which the free space Laplacian operator $-\nabla$ is perturbed, not by the addition of a potential V, but by the excision of a finite body from Euclidean space, are given in Ikebe [3, 4]; see also Y. Shizuta [1]. A similar problem is treated for a more general class of elliptic partial differential operators in an exterior domain in Birman [5]; cf. also Birman [6] and [1]. Birman's argument depends on noting that, if S_1 and S_2 are self adjoint operators in Hilbert space defined in an exterior domain by a common elliptic formal differential operator but by differing boundary conditions, then for some real λ, $(\lambda I + S_1)^{-1} - (\lambda I + S_2)^{-1}$ is a compact operator. By estimating the eigenvalues of this compact operator, Birman is able to show that in certain cases it is not only compact but of trace class, in which case it is possible to apply the Kato-Kuroda theorem to relate the spectra of S and S_1.

The circle of ideas described in Section 4 was developed first among theoretical physicists; cf. C. Møller [1]; cf. also Jauch [1]. The fundamental inequality which we have given as Lemma 4.17 is due to M. Rosenblum [2]. The application of this inequality to prove the general Theorem 4.9 is due to Kato [9, 10]. The extension to perturbation of unbounded self adjoint operators was given by S. T. Kuroda [3, I]. Kuroda [3, II] develops variants of the wave operator theorems expressed in terms of quadratic forms rather than in terms of unbounded operators.

If the two wave operators $W_\pm = \lim_{t\to\pm\infty} e^{itH_2}e^{-itH_1}$ exist, we may regard these operators as being defined by asymptotic correspondences

$$e^{itH_1}x_+ = e^{itH_2}y + o(1), \qquad \text{as } t \to \infty,$$
$$e^{itH_1}x_- = e^{itH_2}y + o(1), \qquad \text{as } t \to -\infty,$$

where, of course, the first correspondence defines $y = W_- x_+$ and the second correspondence defines $y = W_+ x_-$. If W_+ and W_- have the same range, this pair of asymptotic relationships defines a unitary correspondence between x_+ and x_-: $x_+ = W_-^* W_+ x_-$. The operator $S = W_-^* W_+$

plays an important role in quantum mechanical and other applications and is called the *scattering operator*. Heuristically, a wave function x propagating according to a "free particle equation" $(d/dt)x = iH_1 x$ but transformed by S at some time between $t = -\infty$ and $t = +\infty$ into Sx, will be asymptotic to a wave function y propagating according to the "interacting particle equation" $(d/dt)y = iH_2 y$ both at $t = -\infty$ and at $t = +\infty$. Thus the scattering operator S summarizes the overall difference between the two propagation equations as an "instantaneous" transformation. Since, as observed in the early paragraphs of Section 4, we have $H_2 W_\pm = W_\pm H_1$, it follows that $W_-^* H_2 = H_1 W_-^*$; thus $SH_1 = W_-^* W_+ H_1 = W_-^* H_2 W_+ = H_1 W_-^* W_+ = H_1 S$; that is, the scattering operator S commutes with the "unperturbed Hamiltonian operator" H_1.

An insight into the principal properties of the scattering operator (in the simplest case) can be obtained by considering the special situation in which H_1 is a multiplication operator $F(x) \to xf(x)$ in a space of vector valued functions, in which $H_2 = H_1 + V$ where V is an integral operator with a smooth kernel $V(x, y)$, and in which the Friedrichs similarity method of Section 2 is assumed to apply. Then, as shown in Section 2, $e^{itH_2} = (I + \Gamma)^{-1} e^{itH_1}(I + \Gamma)$, where Γ is a singular integral operator

$$(\Gamma f)(x) = \mathscr{P} \int_{-\infty}^{+\infty} \frac{G(x, y)}{x - y} f(y)\, dy,$$

the kernel G being smooth. Thus we have

$$W_\pm = (I + \Gamma)^{-1}(I + \lim_{t \to \pm\infty} e^{itH_1} \Gamma e^{-itH_1}).$$

Writing $G_0(x, y) = (G(x, y) - G(x, x))(x - y)^{-1}$, we find that

$$(\Gamma f)(x) = \int_{-\infty}^{+\infty} G_0(x, y) f(y)\, dy + G(x, x) \int_{-\infty}^{+\infty} \frac{f(y)}{x - y}\, dy;$$

and since only the second integral is singular we have

$$\lim_{t \to \pm\infty} (e^{itH_1} \Gamma e^{-itH_1} f)(x) = G(x, x) \lim_{t \to \pm\infty} \int_{-\infty}^{+\infty} \frac{e^{it(x-y)}}{x - y} f(y)\, dy.$$

The integral on the right of the above formula is easily evaluated (for example, by use of the Fourier transformation), and we find that

$$\lim_{t \to \pm\infty} (e^{itH_1} \Gamma e^{-itH_1} f)(x) = \pm\, i\pi G(x, x) f(x).$$

We thus have

$$W_{\pm} = (I + \Gamma)^{-1}(I \pm i\pi G),$$

where $(Gf)(x) = G(x, x)f(x)$; so that

$$S = (W_{-})^{-1}W_{+} = (I - i\pi G)^{-1}(I + i\pi G).$$

It is plain from this expression that the scattering operator S commutes with the multiplication operator $x \to xf(x)$. The scattering operator is clearly determined entirely by the diagonal singularity of the Friedrichs similarity operator Γ; in a representation in which H_1 is diagonal, S appears as an operation of multiplication by an operator valued function $I + T(x)$, where $T(x)$ is an operator of trace class for each x.

Results concerning the scattering operator of the sort described above are proved rigorously on the basis of rather general hypotheses in Birman and Kreĭn [1].

In that paper, the scattering operator S for a pair of Hermitian operators H_2 and H_1 is studied under the hypothesis that $H_2^{-1} - H_1^{-1}$ belongs to the trace class. Taking an ordered representation (relative to H_1) of the Hilbert space $\mathfrak{X}$ in which these operators act, one may represent $\mathfrak{X}$ as a space of vector valued functions $f(\lambda)$ in which H_1 is represented as a multiplication operator $f(\lambda) \to \lambda f(\lambda)$. Since S commutes with H_1, it must have the form $f(\lambda) \to S(\lambda)f(\lambda)$, where $S(\lambda)$ is a measurable operator valued function. Birman and Kreĭn show that $S(\lambda)$ has the form $S(\lambda) = I + T(\lambda)$ where $T(\lambda)$ is of trace class for all λ, and that $S(\lambda)$ may also be represented in the form $S(\lambda) = \exp(-2\pi i K(\lambda))$, where $K(\lambda)$ is a Hermitian operator of trace class for all λ, and where $K(\lambda)$ is positive for almost all λ if $H_2 - H_1$ is a positive operator, or more generally if there exists a Hermitian operator T_0 with finite range such that $H_2 - H_1 + T_0$ is positive. Similar results are proved by Birman and Kreĭn for a pair U_2, U_1 of unitary operators, under the hypothesis that $U_2 - U_1$ belongs to the trace class.

Birman [4] gives a version of the fundamental Theorem 4.9 which is "local" in the sense that it applies to portions of the spectra of a pair of operators rather than to the entire spectra. Birman's result may be stated as follows. Let H_2 and H_1 be a pair of Hermitian operators with spectral resolutions $E_2(\cdot)$ and $E_1(\cdot)$, respectively. Let G be a Borel subset of the real axis R, and let G_n be an increasing sequence of subsets of G, such that $\bigcup_{n \geq 1} G_n = G$. Suppose that

(i) the difference $H_2E_2(G)E_1(G) - E_2(G)H_1E_1(G)$ is an operator of the trace class;

(ii) the operators $E(R-G)E_0(G_n)$ are all compact. Then the strong limits $\lim_{t\to\pm\infty} e^{itH_2}e^{-itH_1}E_1(G)$ exist.

As a consequence of this result, it follows that if H_1 and H_2 are a pair of Hermitian operators, if $j \geqq 0$, $k \geqq 0$, and $j+k>0$, and if $(\lambda I - H_1)^{-j}(H_1 - H_2)(\lambda I - H_2)^{-k}$ belongs to the trace class, then the wave operator limits $W_\pm = \lim_{t\to\pm\infty} e^{itH_2}e^{-itH_1}$ exist and define unitary equivalences between the absolutely continuous parts of H_1 and H_2.

Birman also studies the properties of the partial scattering operator $S = (\lim_{t\to-\infty} e^{itH_2}e^{-itH_1}E(G))(\lim_{t\to+\infty} e^{itH_2}e^{-itH_1}E_1(G))$ under the above hypotheses (i) and (ii). This operator commutes with H_1. If the subspace $E_1(G)\mathfrak{X}$ of the Hilbert space $\mathfrak{X}$ in which H_1 and H_2 act is given an ordered representation relative to H_1, then, as above, the restriction of S to $E_1(G)\mathfrak{X}$ will be represented in the form $f(\lambda) \to S(\lambda)f(\lambda)$, where $S(\lambda)$ is a measurable operator valued function. Birman shows that, under hypotheses (i) and (ii), $S(\lambda)$ has the form $I + T(\lambda)$, where $T(\lambda)$ is of trace class for almost all $\lambda \in G$.

The existence of the pair of limits $W_\pm = \lim_{t\to\pm\infty} e^{itH_2}e^{-itH_1}$ implies a close relationship between the Hermitian operators H_2 and H_1. If, in particular, the "perturbation" $V = H_2 - H_1$ is not of trace class, various interesting new phenomena related to these limits make their appearance. In the simplest case, even if H_1 has an absolutely continuous spectrum, the limits $W_\pm$ may still exist, but the orthocomplement of the range of the wave operators $W_\pm$ may be infinite rather than finite dimensional. This is the situation for the quantum mechanical three-body problem studied by Faddeev in the papers cited below. In other cases, the limits $W_\pm$ may fail to exist, but the modified limits $\lim_{t\to\pm\infty} e^{itH_2}e^{-it\varphi(H_1)}$ may exist. This is the case in the quantum field theoretic situations discussed in Friedrichs' monograph [17]. In such cases, H_2 is similar not to H_1 but to the operator $\varphi(H_1)$; the necessary occurrence of a function φ in this case constitutes an elementary version of the phenomenon of "renormalization" which is so significant in the quantum theory of fields. In Schwartz [8] the situation in which H_1 is a multiplication operator and V is a singular integral operator is studied. In this case, limits $\lim_{t\to\infty} e^{itH_2}e^{-it\varphi_+(H_1)}$ and $\lim_{t\to\infty} e^{itH_2}e^{-it\varphi_-(H_1)}$ exist, but the functions φ_+ and φ_- are distinct. Such a case may be regarded as exhibiting a curious type of "unsymmetric renormalization."

In Kreĭn [8, 24] the infinite determinant

$$\Delta(z) = \det((H_2 - zI)(H_1 - zI)^{-1})$$

is studied under the hypothesis that $H_2 - H_1$ belongs to the trace class of operators. Kreĭn proves that the function $\log \Delta(z)$ has the integral representation

$$\log \Delta(z) = \int_{-\infty}^{+\infty} \frac{\delta(\lambda)}{\lambda - z}\, d\lambda,$$

where the kernel function δ satisfies the integral inequality

$$\int |\delta(\lambda)|\, d\delta \leqq \|H_2 - H_1\|_1,$$

the norm on the right being the trace norm of Chapter XI. Moreover, Kreĭn establishes the trace formula

$$\operatorname{tr}(\varphi(H_2) - \varphi(H_1)) = \int_{-\infty}^{+\infty} \delta(\lambda)\varphi'(\lambda)\, d\lambda$$

for an extensive class of functions φ, from which the formula

$$\operatorname{tr}(H_2 - H_1) = \int_{-\infty}^{+\infty} \delta(\lambda)\, d\lambda$$

follows as a special case. Corresponding results are proved for a pair U_2, U_1 of unitary operators under the hypothesis that $U_2 - U_1$ belongs to the trace class of operators. Similar results are given in the earlier paper of Kreĭn cited above, in which detailed proofs may be found.

Assume once more that we have two Hermitian operators H_2, H_1, and that H_1 is a multiplication operator $f(x) \to xf(x)$ in a space of vector valued functions, while $H_2 = H_1 + V$, where V is an integral operator with smooth kernel to which the Friedrichs similarity method of Section 2 applies. Then, as shown in Section 2, $H_2 = (I + \Gamma)^{-1}H(I + \Gamma)$, where, as we have seen above, the operator Γ may be written as a singular integral operator of the form

$$(\Gamma f)(x) = \int_{-\infty}^{+\infty} \frac{G(x, y)}{x - y} f(y)\, dy.$$

If φ is any Borel function, we have

$$e^{it\varphi(H_2)}e^{-it\varphi(H_1)} = (I + \Gamma)^{-1}(I + e^{it\varphi(H_1)}\Gamma e^{-it\varphi(H_1)}).$$

It is plain that

$$(e^{it\varphi(H_1)}\Gamma e^{-it\varphi(H_1)}f)(x) = \mathscr{P}\int_{-\infty}^{+\infty}\frac{G(x, y)}{x-y}e^{it(\varphi(x)-\varphi(y))}f(y)\,dy.$$

If φ is smooth and monotone increasing, we may therefore write

$$\begin{aligned}(e^{it\varphi(H_1)}&\Gamma e^{-it\varphi(H_1)}f)(\varphi^{-1}(x))\\ &= \mathscr{P}\int_{-\infty}^{+\infty}\frac{G(\varphi^{-1}(x), \varphi^{-1}(y))}{\varphi^{-1}(x)-\varphi^{-1}(y)}e^{it(x-y)}f(\varphi^{-1}(y))(\varphi'(\varphi^{-1}(y)))^{-1}\,dy\\ &= \mathscr{P}\int_{-\infty}^{+\infty}\frac{G(\varphi^{-1}(x), \varphi^{-1}(y))}{x-y}\,\frac{x-y}{\varphi^{-1}(x)-\varphi^{-1}(y)}\\ &\qquad\times(\varphi'(\varphi^{-1}(y)))^{-1}e^{it(x-y)}f(\varphi^{-1}(y))\,dy.\end{aligned}$$

Arguing as in the preceding discussion of the scattering operator, we find that

$$\begin{aligned}\lim_{t\to\pm\infty}\,&(e^{it\varphi(H_1)}\Gamma e^{-it\varphi(H_1)}f)(\varphi^{-1}(x))\\ &= \pm i\pi G(\varphi^{-1}(x), \varphi^{-1}(x))((\varphi^{-1}(x))')^{-1}(\varphi'(\varphi^{-1}(x)))^{-1}f(\varphi^{-1}(x)).\end{aligned}$$

Since $\varphi(\varphi^{-1}(x)) = x$, we have $\varphi'(\varphi^{-1}(x))(\varphi^{-1}(x))' = 1$; and thus it follows that

$$\lim_{t\to\pm\infty}\,(e^{it\varphi(H_1)}\Gamma e^{-it\varphi(H_1)}f)(x) = \pm i\pi G(x, x)f(x)$$

is independent of the function φ. We conclude that the generalized "wave operator" limits

$$\lim_{t\to\pm\infty}\,e^{it\varphi(H_2)}e^{-it\varphi(H_1)} = W_\pm$$

are also independent of φ. This interesting and significant result is called the *theorem of invariance of wave operators* by T. Kato. In Kato [14] this result is proved rigorously on the basis of rather general hypotheses. Related results are given in Birman [3]. In this paper, theorems of the nature of the wave operator invariance theorem are used to establish the existence of the limits $\lim_{t\to\pm\infty} e^{itH_2}e^{-itH_1}$ on the basis of hypotheses of the general form $\varphi(H_2) - \varphi(H_1) \in C_1$, C_1 denoting the trace class of operators. For other results of this same kind, see Birman [2].

Birman and Kreĭn [1] (cited above) give interesting variants of the Kato-Kuroda wave operator theorems, in forms applying directly to pairs of unitary operators. In particular, they prove the following results.

THEOREM. *Let U_1 and U_2 be two unitary operators, and suppose that $U_1 - U_2$ is of trace class. Then the limit $W_\infty(U_1, U_2)x = \lim_{n\to\infty} U_1^n U_2^{-n}x$ exists for all x such that $E_2(e)x = 0$ whenever e is a subset of the unit circle of Lebesgue measure zero; here $E_2(\cdot)$ is the spectral measure of U_2.*

THEOREM. *Let H_1 and H_2 be two invertible Hermitian operators, and let $H_1^{-1} - H_2^{-1}$ be of trace class. Let x be an element of Hilbert space such that $E_2(e)x = 0$ whenever e is a set of Lebesgue measure 0, where $E_2(\cdot)$ is the spectral measure of H_2. Let $U_1 = (iI - H_1)(iI + H_1)^{-1}$ and $U_2 = (iI - H_2)(iI + H_2)^{-1}$. Then the limits*

$$\lim_{t\to\infty} U_1^n U_2^{-n}x,$$

$$\lim_{t\to\infty} e^{itH_1}e^{-itH_2}x,$$

and

$$\lim_{t\to\infty} e^{itH_1^{-1}}e^{-itH_2^{-1}}x$$

all exist and are equal.

Kuroda [5, 6] develops a "stationary" approach to the definition of the wave operator W, which in Section 4 we defined in "time dependent" terms as $W = \lim_{t\to\infty} e^{itH_1}e^{-itH_2}$. The formal basis of this alternate approach may be derived (albeit in completely non-rigorous fashion) as follows. Let H_1 and H_2 be two self adjoint operators, and let $H_2 = H_1 + V$, so that $H_1 = H_2 - V$. Consider the functions

$$W_1(z) = (H_2 - zI)(H_1 - zI)^{-1} = I + V(H_1 - zI)^{-1}$$

and

$$W_2(z) = (H_1 - zI)(H_2 - zI)^{-1} = I - V(H_2 - zI)^{-1},$$

which are each other's inverse. Under suitable hypotheses, there will be a dense subspace S of Hilbert space with the property that $W_j(z): S \to S$ for $\mathscr{I}z \neq 0$, $j = 1, 2$; and that if we let $z \to x + i0$ or $x - i0$, $W_j(z)$ converges to a limit $W_j(x \pm i0): S \to S$, $j = 1, 2$. For example, if H_1 is a multiplication operator $f(x) \to xf(x)$, if V is an integral operator $f(x) \to \int V(x, y)f(y)\,dy$ with a differentiable or even a Hölder-continuous kernel, and if S is the set of Hölder-continuous functions, then, for each real t,

$$f(x) \to \int \frac{V(x, y)}{t \pm i0 - y} f(y)\,dy$$

defines a mapping of S into itself. This fact forms a basis for the following calculations. On the one hand, since $W_1(z) = (W_2(z))^{-1}$, we have $W_1(x \pm i0) = (W_2(x \pm i0))^{-1}$. Thus

$$\begin{aligned} W_2(x+i0) - W_2(x-i0) &= (W_1(x+i0))^{-1} - (W_1(x-i0))^{-1} \\ &= (W_1(x+i0))^{-1}(W_1(x-i0) - W_1(x+i0))(W_1(x-i0))^{-1} \\ &= W_2(x+i0)(W_1(x-i0) - W_1(x+i0))W_2(x-i0). \end{aligned}$$

Let $E_1(\cdot)$ and $E_2(\cdot)$ be the spectral resolutions of H_1 and H_2, respectively. By Theorem X.6.1, $VE_j(e) = \pm(2\pi i)^{-1} \int_e (W_j(x+i0) - W_j(x-i0))\, dx$. If we write $\mathscr{R}(T)$ and $\mathscr{L}(T)$ for the operators of right and left multiplication, respectively, by an operator T, we may write

$$\begin{aligned} -2\pi i VE_2(e) &= \int_e W_2(x+i0)(W_1(x-i0) - W_1(x+i0))W_2(x-i0)\, dx \\ &= \int_e \mathscr{L}(W_2(x+i))\mathscr{R}0(W_2(x-i0))(W_1(x-i0) - W_1(x+i0))\, dx, \end{aligned}$$

so that by the above formula and the Radon-Nikodým theorem we have

$$\begin{aligned} VE_2(e) &= \int_e \mathscr{L}(W_2(x+i0))\mathscr{R}(W_2(x-i0))VE_1(dx), \\ &= \int_e \mathscr{L}(W_2(x+i0)V)\mathscr{R}(W_2(x-i0))E_1(dx), \end{aligned}$$

at least formally. For each collection of disjoint sets e_j we have

$$\begin{aligned} &(\sum_j \mathscr{L}(W_2(x_j+i0))VE_1(e_j))(\sum_j \mathscr{R}(W_2(x-i0))E_1(e_j)) \\ &\qquad = \sum_j \mathscr{L}(W_2(x_j+i0)V)\mathscr{R}(W_2(x-i0))E_1(e_j) \end{aligned}$$

by the orthogonality of projections in a spectral resolution and the commutativity of right with left multiplication; thus in the limit we have

$$\begin{aligned} &\left(\int_e \mathscr{L}(W_2(x+i0)V)E_1(dx)\right)\left(\int_e \mathscr{R}(W_2(x-i0))E_1(dx)\right) \\ &\qquad = \int_e \mathscr{L}(W_2(x+i0)V)\, \mathscr{R}(W_2(x-i0))E_1(dx). \end{aligned}$$

By a similar argument we have

$$\int_e \mathscr{L}(W_2(x+i0)V)E_1(dx) = \left(\int_{-\infty}^{+\infty} \mathscr{L}(W_2(x+i0)V)E_1(dx)\right)E_1(e)$$

and

$$\int_e \mathscr{R}(W_2(x-i0))E_1(dx) = \left(\int_{-\infty}^{+\infty} \mathscr{R}(W_2(x-i0))E_1(dx)\right)E_1(e).$$

Combining the preceding formulas, we may write

$$VE_2(e) = \left(\int_{-\infty}^{+\infty} W_2(x+i0)VE_1(dx)\right)E_1(e)\left(\int_e E_1(dx)W_2(x-i0)\right).$$

Since $W_2(z)V = V - V(H_2 - zI)^{-1}V$, we have $W_2(z)V = V(W_2(\bar{z}))^*$, and, taking limits as z approaches the x-axis from above, we have $W_2(x+i0)V = V(W(x-i0))^*$. We may therefore write the preceding formula as

$$E_2(e) = \left(\int_{-\infty}^{+\infty} E_1(dx)W_2(x-i0)\right)^* E_1(e)\left(\int_{-\infty}^{+\infty} E_1(dx)W_2(x-i0)\right)$$

or

$$E_2(e) = W^*E_1(e)W, \tag{1}$$

where

$$W = \int_{-\infty}^{+\infty} E_1(dx)W_2(x-i0). \tag{2}$$

Putting $e = (-\infty, +\infty)$, we find that $W^*W = I$; thus formula (1) shows that formula (2) defines a wave operator establishing a unitary equivalence between H_2 and H_1. Using the analyticity in the lower complex half plane of the operator $W_2(z)$, the operator W may be identified under suitable hypotheses with the limit operator studied in Section 4.

Birman and Entina [1] develop formulas for the wave and scattering operators along similar lines.

For another discussion of this approach to the theory, cf. Louis de Branges [1]. The stationary approach to the wave operator theory has the advantage of yielding formulas which are somewhat more explicit than those obtained from the "time-dependent" approach which we have employed in Section 4.

The formal device of factoring the "perturbation" $V = H_2 - H_1$ into the product of two operators, described above in connection with Schwartz' note [4, 5], is used in Kato's very interesting paper [12]. Kato assumes that $H_2 - H_1$ may be written as a product AB, and that $B(\lambda I - H_1)^{-1}A$ is uniformly bounded by a constant K in the neighborhood of the spectrum of H, while the integrals

$$\int_{-\infty}^{+\infty} |A((\lambda \pm i\varepsilon)I - H_1)^{-1}v|^2 \, d\lambda$$

and

$$\int_{-\infty}^{+\infty} |B^*((\lambda \pm i\varepsilon)I - H_1)^{-1}v|^2 \, d\lambda$$

are bounded as $\varepsilon \to 0$. In this case, and under the additional assumption that the constant K is small, Kato uses an appropriately modified version of the stationary wave operator method to construct operators $W_\pm$ such that $H_2 = W_\pm H_1 (W_\pm)^{-1}$. The wave operators $W_\pm$ have the expression

$$(W_\pm x, y) = (x, y) \mp \frac{1}{2\pi i} \int_{-\infty}^{+\infty} (BR(\lambda \pm i0)x, A^*R(\lambda \pm i0)y) \, d\lambda,$$

where $R(\lambda) = (\lambda I - H_1)^{-1}$, and where x, y denote arbitrary vectors in Hilbert space.

When a self adjoint operator H is perturbed by the addition of a self adjoint term V which is not small relative to H in the sense of Theorem 4.9, the spectrum of H can change drastically. In particular, the "absolutely continuous spectrum" $\sigma(H \mid \sum_{ac}(h))$ (cf. Definition 4.7) can disappear, grow, or be shifted. The study of these phenomena is of great interest in connection with the quantum theory of fields and the question of "renormlization." Cf. K. O. Friedrichs' monograph [17]. Similar phenomena arise in the quantum mechanical three-body problem, studied in L. D. Faddeev [2, 3, 4]. Faddeev shows that if H is the six-dimensional Laplacian, and V is a sum of three multiplication operators (each corresponding to a two-body force in a three-body system), then the spectrum of $H + V$ consists of the purely continuous spectrum corresponding to the spectrum of H, a point spectrum corresponding to the "bound states" of the three-body system, and three additional branches of continuous spectrum, corresponding to the states of the three-body system in which two particles are "bound" together, but the third is "ionized".

Physicists customarily discuss the spectral theory of operators of this sort, and even of more general classes of operators, on the basis of semi-formal heuristic principles inferred inductively from illustrative examples. For a typical example of such a discussion, including an analysis of the quantum mechanical three-body problem, cf. Hack [1] and Jauch [2].

From the identity

$$\frac{d}{dt}(e^{itH_2}e^{-itH_1}) = ie^{itH_2}(H_2 - H_1)e^{-itH_1}$$

it follows very easily that the condition

$$\int_{-\infty}^{\infty} |(H_2 - H_1)e^{-itH_1}x| \, dt < \infty$$

is sufficient for the existence of the limits $\lim_{t\to\pm\infty} e^{itH_2}e^{-itH_1}x$. In particular, if the integral displayed above is finite for a dense set of vectors x, then the wave operator $W_\pm = \lim_{t\to\pm\infty} e^{itH_2}e^{-itH_1}$ will exist as a strong limit. If the operator H_1 has simple form and a known spectral resolution, the finiteness of the above integral may often be verified for suitable x by direct calculation. This line of argument, introduced by J. M. Cook [1], is often called Cook's method, and often provides the simplest method for establishing the existence of the wave operators $W_\pm$. Note, however, that this argument by itself cannot be used to determine the range of the operators $W_\pm$, and thus yields only fragmentary information concerning the spectrum of the operator H_2. For a partial discussion of the quantum mechanical n-body problem using this method of Cook, see Kuroda [2].

In connection with the wave operator theorems of Section 4, the early work of von Neumann, von Neumann [6], is to be noted. Von Neumann shows that, by adding a Hilbert-Schmidt class self adjoint operator V to a bounded self adjoint operator H such that $\sum_p (H) = \sum_{sing} (H) = \{0\}$ Definition (4.7), one can produce a sum $H + V$ for which $\sum_{ac} (H + V) = 0$. Thus, in the hypotheses of Theorem 4.9, the reference to operators of trace class is essential for the validity of the conclusion.

An interesting series of papers by Calvin R. Putnam is devoted to analyses of the existence and properties of the wave operator. In Putnam [27] it is shown that, under fairly general hypotheses, the two self adjoint operators defined by a second order ordinary differential operator on a half-axis and two different sets of end conditions are unitarily equivalent. In [28] Putnam studies a pair of self adjoint operators H, V, where V is bounded and positive, and where the existence of a unitary operator U such that $UHU^* = H + V$ is assumed, and proves various interesting relations between the norms and spectra of U, H, and V. In Putnam's papers [24, 29] it is shown that the wave operators corresponding to the pair of differential operators $-(d/dx)^2$ and $-(d/dx)^2 + V(x)$, where V is integrable, non-negative, and bounded, have continuous spectra covering the unit circle, and, under slightly stronger hypotheses, that these spectra are absolutely continuous. In the second of these papers, the following more general result is proved.

THEOREM. *Let H and V be self adjoint operators; V bounded and positive. Suppose that the domain of H intersects the range of $V^{1/2}$ in a dense set. Then, if U is unitary and $UHU^* = H + V$, U has an absolutely continuous spectrum.*

Another line of thought closely related to the problem studied in the present chapter was initiated by E. R. Lorch [2], and continued by Wermer [1, 2], by F. Wolf [4], and finally by Ciprian Foiaş in a series of papers and in a joint monograph with Colojoară (see Colojoară-Foiaş [4]). The initial idea in this development is as follows. Let T be an operator in a B-space such that, for some A and $k > 0$,

$$(*) \qquad |T^k| < A(1 + |n|)^k, \qquad -\infty < n < +\infty.$$

Then it is easy to see that $\sigma(T)$ is a subset of the unit circle in the complex plane. An easy argument, which may be based simply on repeated integration by parts of the fundamental Cauchy formula for functions of T, will show that

$$|f(T)| \leqq B \sup_{0 \leqq \theta \leqq 2\pi} \sup_{0 \leqq j \leqq k+1} |f^{(j)}(e^{i\theta})|,$$

for some constant B. Thus the calculus of analytic functions of T may be extended to a calculus of functions $f(T)$, f being required to have $k + 1$ continuous derivatives on the unit circle. This fact, taken together with the known form for the conjugate space of C^{k+1}, immediately implies the existence of an integral formula

$$f(T) = A_0 f(1) + \cdots + A_k f^{(k)}(1) + \int_0^{2\pi} f^{(k+1)}(e^{i\theta}) A(d\theta),$$

$A(e)$ being a strongly countably additive operator valued measure. The functional calculus which T possesses in view of this formula may be used to prove a number of interesting properties of T.

This rudimentary idea has been developed in a sophisticated manner by Foiaş [9, 10, 12]. The second of the above cited papers of Foiaş discusses the connection between the theory under consideration and the general theory of spectral operators. These matters are systematically discussed in the monograph by Colojoară and Foiaş [4].

A variety of more or less related results on the existence of invariant subspaces for an operator T are given by Wermer [1, 2].

Sz.-Nagy has shown that if the constant k in formula $(*)$ above is 0, and the B-space in question is Hilbert space, then T is equivalent to a unitary operator. For a proof of this and a number of similar theorems, see the Appendix in Riesz and Sz.-Nagy [1].

Various of the perturbation-theoretic ideas touched on explicitly or by indirection in the present chapter are developed systematically by T. Kato in his treatise [13]. This treatise also contains an extensive bibliography of perturbation theory. A lively and enlightening survey of spectral perturbation theory from the hand of Friedrichs himself is available to us in K. O. Friedrichs [17]. Friedrichs' monograph contains, among other things, an especially interesting discussion of the spectral perturbation problems of quantum field theory, and the associated phenomena of renormalization. The monograph of Peter Lax and Ralph Phillips [2] studies a set of questions related to the wave operator methods of Section 4, in connection with an analysis of various classes of hyperbolic partial differential equations. Related work of Morawetz on the wave equation in exterior domains is discussed in this monograph. The Lax-Phillips analysis is based in part on the harmonic analysis of contraction semi-groups in Hilbert space, to which subject the monograph of B. Sz.-Nagy and Ciprian Foiaş is also devoted. C. Dolph [1] gives a survey of the theory of nonselfadjoint problems, with emphasis on perturbation theory and scattering theory, with a view toward the physical applications of these theories. Another survey of related areas, emphasizing perturbation theory, is to be found in Naĭmark [15]. See also Høegh-Krohn [1] as well as the monograph by F. A. Berezhin [1] which gives an account of the mathematical background of this interesting subject.

PERTURBATION THEORY. For additional references of articles dealing with the perturbation of operators, we add Apostol [13], Balabanov [1], Balslev [1], Balslev and Gamelin [1], Baumgärtel [1 through 5], Beals [1], Birman [5, 6], Bisshopp [1], de Branges [2], Brownell [1], Butler [1 through 3], Coburn [1], Conley [1], Conley and Rejto [1], Davis [2, 3], Donoghue [3], Dupras [1], Faddeev [1], Foguel [4, 6], J. M. Freeman

[1, 2], Friedrichs [2, 17, 18], Friedrichs and Rejto [1], Gehtman and Stankevič [1], Gilbert and Kramer [1], Goldberg [2], Gol'dman and Kračkovskiĭ [1 through 4], Greiner [1], Gustafson [1], Hadeler [2], Hasegawa [1], Huige [1], Javrjan [1], Kaashoek [1, 2], Kato [9 through 14], Konno and Kuroda [1], Kreĭn [24, 27], Kuroda [1, 3 through 9], Ladyženskaya and Faddeev [1], Langer [4], Loginov [1], Markus [1], Martirosjan [3, 4], Miyadera [2], Mochizuki [1], Moser [1, 2], Newburgh [1, 2], Nižnik [1], Osborn [1], Osher [1], Paraska [1], Porath [1], Przeworska-Rolewicz [1], Putnam [22, 31, 32], Rejto [1 through 3], Rosenblum [1, 2], Sahnovič [2, 3], Schechter [1, 2], J. Schwartz [2, 3, 9], Sigalov [1, 2], Simpson [3], Stampfli [5], Stankevič [1], Turner [1, 2], Tzafriri [3], Viziteĭ [1], Yosida [13], and Zaanen [10].

Scattering and Wave Operators. The following references deal with scattering and wave operators: Adamjan and Arov [1, 2], Birman [1 through 4], Birman and Entina [1], Birman and Kreĭn [1, 2], Brownell [1], Cook [1], Dolph and Penzlin [1], Faddeev [2 through 4], Greiner [1], Hack [1], Ikebe [1 through 3], Jauch [1, 2], Jauch and Zinnes [1], Kato [12 through 14], Kreĭn [27], Kuroda [2, 5 through 9], Lax and Phillips [1 through 3], Putnam [24, 29, 31], Stankevič [1, 2], and Thoe [1].

REFERENCES

Adamjan, V. M., and Arov, D. Z.

1. *On a class of scattering operators and characteristic operator-functions of contractions.* Doklady Akad. Nauk SSSR 160, 9–12 (1965). (Russian) Math. Rev. 30, #5169, 961 (1965).
2. *On scattering operators and contraction semi-groups in Hilbert space.* Doklady Akad. Nauk SSSR 165, 9–12 (1965). (Russian) Math. Rev. 32 #6240, 1068 (1966). Soviet Math. Dokl. 6, 1377–1380 (1965).

Aleksandrjan, R. A.

1. *Spectral expansion of arbitrary self-adjoint operators in eigenfunctionals.* Doklady Akad. Nauk SSSR 162, 11–14 (1965). (Russian) Math. Rev. 31 #5089, 927 (1966). Soviet Math. Dokl. 6, 607–611 (1965).

Aleksandrjan, R. A., and Mkrtčjan, R. Z.

1. *Certain criteria characterizing the spectrum of a selfadjoint operator in an abstract Hilbert space.* Izv. Akad. Nauk Armjan. SSR Ser. Mat 1, no. 1, 25–34 (1966). (Russian. Armenian and English summaries) Math. Rev. 34 #8180, 1504 (1967).

Allahverdiev, Dž. È.

1. *On the completeness of a system of eigen-elements and adjoined elements of non-selfadjoint operators close to normal ones.* Doklady Akad. Nauk SSSR 115, 207–210 (1957). (Russian) Math. Rev. 20 #1227, 205 (1959).
2. *On the completeness of systems of eigen-elements and adjoint elements of non-selfadjoint operators.* Akad. Nauk Azerbaĭdžan SSR Dokl. 18, no. 7, 3-7(1962) (Russian. Azerbaĭjani summary) Math. Rev. 27#609, 131 (1964).
3. *On the completeness of a system of characteristic and adjoint elements of operators which are rational functions of a parameter.* Doklady Akad. Nauk SSSR 159, 951–954 (1964). (Russian) Math. Rev. 29 #6273, 1176 (1965).
4. *On the completeness of the system of eigenelements and adjoint elements of non-selfadjoint operators.* Doklady Akad. Nauk SSSR 160, 503–506 (1965). (Russian) Math. Rev. 30 #3370a, 638 (1965).
5. *On the completeness of the system of eigenelements and adjoint elements of a class of non-selfadjoint operators depending on a parameter* λ. Doklady Akad. Nauk SSSR 160, 1231–1234 (1965) (Russian) Math. Rev. 30 #3370b, 638 (1965).

Allan, G. R.

1. *A spectral theory for locally convex algebras.* Proc. London Math. Soc. (3) 15, 399–421 (1965).

2. *On a class of locally convex algebras.* Proc. London Math. Soc. (3) 17, 91–114 (1967).

Altman, M.

6. *On the Riesz-Schauder theory of linear operator equations in spaces of type* (B_0). Studia Math. 15, 136–143 (1956). (Russian) Math. Rev. 17, 1226 (1956).

Andersen, E. Sparre

1. *On the fluctuations of sums of random variables.* Math. Scand. 1, 263–265 (1953).
2. *Remarks to the paper: On the fluctuations of sums of random variables.* Math. Scand. 2, 193–223 (1954).

Andô, Tsuyoshi

1. *Positive linear operators in semi-ordered linear spaces.* J. Fac. Sci. Hokkaido Univ. Ser. I.13, 214–228 (1957).
2. *On a pair of commutative contractions.* Acta. Sci. Math. Szeged 24, 88–90 (1963).
3. *Matrices of normal extensions of subnormal operators.* Acta Sci. Math. Szeged 24, 91–96 (1963).
4. *Note on invariant subspaces of a compact normal operator.* Arch. Math. 14, 337–340 (1963).

Apostol, C.

1. *Propriétés de certains opérateurs bornés des espaces de Hilbert,* I, II.
   I. Rev. Roumaine Math. Pures Appl. 10, 643–644 (1965).
   II. ibid. 12, 759–762 (1967).
2. *Sur la partie normale d'un ensemble d'opérateurs de l'espace de Hilbert.* Acta Math. Acad. Sci. Hung. 17, 1–4 (1966).
3. *Sur les opérateurs scalaires généralisés.* Bull. Sci. Math. 91, 57–61 (1967).
4. *Restrictions and quotients of decomposable operators in a Banach space.* Rev. Roumaine Math. Pures Appl. 13, 147–150 (1968).
5. *Roots of decomposable operator-valued functions.* Rev. Roumaine Math. Pures Appl. 13, 433–438 (1968).
6. *On the roots of spectral operator-valued analytic functions.* Rev. Roumaine Math. Pures Appl. 13, 587–589 (1968).
7. *Roots of scalar operator-valued analytic functions and their functional calculus.* J. Sci. Hiroshima Univ. Ser. A–I 32, 173–180 (1968).
8. *On the roots of spectral operators.* Proc. Amer. Math. Soc. 19, 811–814 (1968).
9. *On the roots of generalized spectral operator-valued analytic functions.* Glasnik Mat. 3 (23), 247–252 (1968).
10. *Teorie spectrală şi calcul functional.* Stud. Cerc. Mat. 20, 635–668 (1968).
11. *Spectral decompositions and functional calculus.* Rev. Roumaine Math. Pures Appl. 13, 1481–1528 (1968).
12. *A theorem on invariant subspaces.* Bull. Acad. Polon. Sci. 16, 181–183 (1968).
13. *Remarks on the perturbation and a topology of operators.* J. Funct. Anal. 2, 395–408 (1968).

14. *Sur l'équivalence asymptotique des opérateurs.* Rev. Roumaine Math. Pures Appl. 12, 601–606 (1967).
15. *Some properties of spectral maximal spaces and decomposable operators.* Rev. Roumaine Math. Pures Appl. 12, 607–610 (1967).
16. *Some properties of a couple of operators on a Banach space.* Rev. Roumaine Math. Pures Appl. 12, 1005–1010 (1967).
17. *On some multiplication operators.* Rev. Roumaine Math. Pures Appl. 13, 911–913 (1968).

Aronszajn, N., and Smith, K. T.
1. *Invariant subspaces of completely continuous operators.* Ann. of Math. (2) 60, 345–350 (1954).

Arov, D. Z. (see Adamjan, V. M.)

Arveson, W. B., and Feldman, J.
1. *A note on invariant subspaces.* Michigan Math. J. 15, 61–64 (1968).

Askerov, N. G., Kreĭn, S. G., and Laptev, G. I.
1. *On a class of non-selfadjoint boundary-value problems.* Doklady Akad. Nauk SSSR 155, 499–502 (1964). (Russian) Math. Rev. 28 #3347, 656 (1964).

Atkinson, F. V.
2. *The normal solubility of linear equations in normed spaces.* Mat. Sbornik N. S. 28 (70), 3–14 (1951). (Russian) Math. Rev. 13, 46 (1952).
4. *On relatively regular operators.* Acta Sci. Math. Szeged 15, 38–56 (1953).
6. *Some aspects of Baxter's functional equation.* J. Math. Anal. Appl. 7, 1–30 (1963).

Bade, W. G.
2. *Unbounded spectral operators.* Pacific J. Math. 4, 373–392 (1954).
3. *Weak and strong limits of spectral operators.* Pacific J. Math. 4, 393–413 (1954).
4. *On Boolean algebras of projections and algebras of operators.* Trans. Amer. Math. Soc. 80, 345–360 (1955).
5. *A multiplicity theory for Boolean algebras of projections in Banach spaces.* Trans. Amer. Math. Soc. 92, 508–530 (1959).

Bade, W. G., and Curtis, P. C., Jr.
1. *The Wedderburn decomposition of commutative Banach algebras.* Amer. J. Math. 82, 851–866 (1960).
2. *Embedding theorems for commutative Banach algebras.* Pacific J. Math. 18, 391–409 (1966).
3. Banach algebras on F-spaces. *Function algebras.* Scott-Foresman, Chicago, 1966.

Bahtin, I. A.
1. *On the existence of eigenvectors of positive linear operators which are not completely continuous.* Mat. Sbornik N. S. 64 (106), 102–114 (1964). (Russian) Math. Rev. 29 #6286, 1178 (1965).
2. *On positive linear operators and hypercomplex systems.* Ukrain. Mat. Ž., 17, no. 4, 3–11 (1965). (Russian) Math. Rev. 34 #4915, 891 (1967).

Bahtin, I. A., Krasnosel'skiĭ, M.A., and Stečenko, V. Ja.

1. *On the continuity of positive linear operators.* Sibirsk. Mat. Ž. 3, 156–160 (1962). (Russian) Math. Rev. 25 #2451, 475 (1963).

Balabanov, V. A.

1. *The problem of stability of the eigenelements of non-linear operators.* Naučn. Dokl. Vysš. Skoly Fiz.-Mat. Nauki 1958, 3–7. (Russian) Math. Rev. 28 #470, 98 (1964).

Balslev, E.

1. *Perturbation of ordinary differential operators.* Math. Scand. 11, 131–148 (1962).

Balslev, E., and Gamelin, T. W.

1. *The essential spectrum of a class of ordinary differential operators.* Pacific J. Math. 14, 755–776 (1964).

Barry, J. Y.

1. *On the convergence of ordered sets of projections.* Proc. Amer. Math. Soc. 5, 313–314 (1954).

Bartle, R. G.

6. *Spectral localization of operators in Banach spaces.* Math. Ann. 153, 261–269 (1964).
7. *Spectral decomposition of operators in Banach spaces.* Proc. London Math. Soc. (3) 20, (1970).
8. *The Elements of Real Analysis.* John Wiley & Sons, New York, 1964.

Baumgärtel, H.

1. *Zur Störungstheorie beschränkter linearer Operatoren eines Banachschen Raumes.* Math. Nachr. 26, 361–379 (1963/64).
2. *Eindimensionale Störung eines selbstadjungierten Operators mit reinen Punktspektrum.* Monatsb. Deutsch. Akad. Wiss. Berlin 7, 245–251 (1965).
3. *Analytische Störung isolierter Eigenwerte endlicher algebraischer Vielfachheit von nichtselbstadjungierten Operatoren.* Monatsb. Deutsch. Akad. Wiss. Berlin 10, 250–258 (1968).
4. *Jordansche Normalform holomorpher Matrizen.* Monatsb. Deutsch. Akad. Wiss. Berlin 11, 23–24 (1969).
5. *Ein Reduktionsprozess für analytische Störungen nichthalbeinfacher Eigenwerte.* Monatsb. Deutsch. Akad. Wiss. Berlin 11, 81–88 (1969).

Baxter, G.

1. *An operator identity.* Pacific J. Math. 8, 649–663 (1958).
2. *An analytic problem whose solution follows from a simple algebraic identity.* Pacific J. Math. 10, 731–742 (1960).

Beals, R. W.

1. *A note on the adjoint of a perturbed operator.* Bull. Amer. Math. Soc. 70, 314—315 (1964).

Berberian, S. K.

1. *The numerical range of a normal operator.* Duke Math. J. 31, 479–483 (1964).

2. *The spectral mapping theorem for a Hermitian operator.* Amer. Math. Monthly 70, 1049–1051 (1963).

3. *A note on operators whose spectrum is a spectral set.* Acta Sci. Math. Szeged 27, 201–203 (1966).

4. *Notes on spectral theory.* Van Nostrand Math. Studies 5, Princeton, 1966.

5. *Naĭmark's moment theorem.* Michigan Math. J. 13, 171–184 (1966).

Berberian, S. K., and Orland, G. H.

1. *On the closure of the numerical range of an operator.* Proc. Amer. Math. Soc. 18, 499–503 (1967).

Berezanskiĭ, Ju. M. (Yu. M.)

3. *On an eigenfunction expansion for self-adjoint operators.* Ukrain. Math Ž. 11, 16–24 (1959). (Russian. English summary) Math. Rev. 23 #A518, 88 (1962).

4. *Some questions of spectral theory of self-adjoint partial differential operators.* Outlines Joint Sympos. Partial Diff. Equations (Novosibirsk, 1963), pp. 26–32. Acad. Sci. USSR Siberian Branch, Moscow, 1963.

5. *Expansions in eigenfunctions of self-adjoint operators.* Amer. Math. Soc. Transl. of Math. Monographs 17, Providence, 1968.

Berezhin, F. A.

1. *The method of second quantization.* Academic Press, New York, 1966.

Berkson, E.

1. *Sequel to a paper of A. E. Taylor.* Pacific J. Math. 10, 767–776 (1960).

2. *A characterization of scalar type operators on reflexive Banach spaces.* Pacific J. Math. 13, 365–373 (1963).

3. *Some types of Banach spaces, Hermitian operators and Bade functionals.* Trans. Amer. Math. Soc. 116, 376–385 (1965).

4. *Some characterizations of C^*-algebras.* Illinois J. Math. 10, 1–8 (1966).

5. *Semi-groups of scalar type operators and a theorem of Stone.* Illinois J. Math. 10, 345–352 (1966).

Berkson, E., and Dowson, H. R.

1. *Prespectral operators.* Illinois J. Math. 13, 291–315 (1969).

2. *On uniquely decomposable well-bounded operators.* Proc. London Math. Soc. (to appear).

Bernau, S. J.

1. *The spectral theorem for normal operators.* J. London Math. Soc. 40, 478–486 (1965).

2. *The spectral theorem for unbounded normal operators.* Pacific J. Math. 19, 391–406 (1966).

3. *Extreme eigenvectors of a normal operator.* Proc. Amer. Math. Soc. 18, 127–128 (1967).

Bernau, S. J., and Smithies, F.

1. *A note on normal operators.* Proc. Cambridge Philos. Soc. 59, 727–729 (1963).

Bernstein, A. R., and Robinson, A.

1. *Solution of an invariant subspace problem of K. T. Smith and P. R. Halmos.* Pacific J. Math. 16, 421–432 (1966).

Bianchi, L., and Favella, L.

1. *A convolution integral for the resolvent of the sum of two commuting operators.* Nuovo Cimento (10) 34, 1825–1828 (1964).

Biriuk, G., and Coddington, E. A.

1. *Normal extensions of unbounded formally normal operators.* J. Math. Mech. 13, 617–634 (1964).

Birkhoff, G.

9. *Note on positive linear operators.* Proc. Amer. Math. Soc. 16, 14–16 (1965).

Birkhoff, G. D.

3. *Boundary value and expansion problems of ordinary linear differential equations.* Trans. Amer. Math. Soc. 9, 373–395 (1908).
6. *Note on the expansion of the Green's function.* Math. Ann. 72, 292–294 (1912).
7. *Note on the expansion problems of ordinary linear differential equations.* Rend. Circ. Mat. Palermo 36, 115–126 (1913).

Birkhoff, G. D., and Langer, R. E.

1. *The boundary problems and developments associated with a system of ordinary differential equations of the first order.* Proc. Amer. Acad. Arts Sci. (2) 58, 51–128 (1923).

Birman, M. Š.

1. *On existence conditions for wave operators.* Doklady Akad. Nauk SSSR 143, 506–509 (1962). (Russian) Math. Rev. 26 #6823, 1295 (1963). Soviet Math. Dokl. 3, 408 (1962).
2. *On a test for the existence of wave operators.* Doklady Akad. Nauk SSSR 147, 1008–1009 (1962). (Russian) Math. Rev. 29 #5107, 963 (1965). Soviet Math. Dokl. 3, 1747 (1962).
3. *Existence conditions for wave operators.* Izv. Akad. Nauk SSSR Ser. Mat. 27, 883–906 (1963). (Russian) Math. Rev. 28 #4359, 848 (1964).
4. *A local criterion for the existence of wave operators.* Doklady Akad. Nauk SSSR 159, 485–488 (1964). (Russian) Math. Rev. 30 #4171, 786 (1965). Soviet Math. Dokl. 5, 1505 (1964).
5. *Perturbation of the spectrum of a singular elliptic operator under variation of the boundary and boundary conditions.* Doklady Akad. Nauk SSSR 137, 761–763 (1961). (Russian) Math. Rev. 31 #1574, 284 (1966). Soviet Math. Dokl. 2, 326 (1961).
6. *Perturbation of the continuous spectrum of a singular elliptic operator by varying the boundary and the boundary conditions.* Vestnik Leningrad Univ. 17, 22–55 (1962) (Russian. English summary). Math. Rev. 25 #2314, 450 (1963).

Birman, M. Š., and Entina, S. B.

1. *Stationary approach to the abstract theory of scattering*. Doklady Akad. Nauk SSSR 155, 506–508 (1964). (Russian) Math. Rev. 29 #1887, 367 (1965). Soviet Math. Dokl. 5, 432 (1964).

Birman, M. Š., and Kreĭn, M. G.

1. *On the theory of wave operators and scattering operators*. Doklady Akad. Nauk SSSR 144, 475–478 (1962). (Russian) Math. Rev. 25 #2447, 475 (1963) Soviet Math. Dokl. 3, 740–747 (1962).
2. *Some topics on the theory of the wave and scattering operators*. Outlines Joint Sympos. Partial Differential Equations (Novosibirsk, 1963), pp. 39–45. Acad. Sci. USSR Siberian Branch, Moscow 1963. (Russian) Math. Rev. 34 #637, 109 (1967).

Bishop, E.

1. *Spectral theory for operators on a Banach space*. Trans. Amer. Math. Soc. 86, 414–445 (1957).
2. *A duality theorem for an arbitrary operator*. Pacific J. Math. 9, 379–397 (1959).

Bisshopp, F. E.

1. *A note on regular perturbation theories*. J. Math. Anal. Appl. 12, 71–86 (1965).

Bognár, J.

1. *On the existence of square roots of an operator which is self-adjoint with respect to an indefinite metric*. Magyar Tud. Akad. Mat. Kutató Int. Közl. 6, 351–363 (1961). (Russian. English summary) Math. Rev. 27 #6134, 1172 (1964).
2. *Some relations among the negativity properties of operators in spaces with indefinite metric*. Magyar Tud. Akad. Mat. Kutató Int. Közl. 8, 201–212 (1963). (Russian. English summary) Math. Rev. 29 #2644, 513 (1965).
3. *Non-negativity properties of operators in spaces with indefinite metric*. Ann. Acad. Sci. Fenn. Ser. AI No. 336/10 (1963), 9 pp.

Bonsall, F. F.

2. *Endomorphisms of partially ordered vector spaces*. J. London Math. Soc. 30, 133–144 (1955).
3. *Endomorphisms of a partially ordered vector space without order unit*. J. London Math. Soc. 30, 144–153 (1955).
4. *Linear operators in complete positive cones*. Proc. London Math. Soc. (3) 8, 53–75 (1958).
5. *The iteration of operators mapping a positive cone into itself*. J. London Math. Soc. 34, 364–366 (1959).
6. *A formula for the spectral family of an operator*. J. London Math. Soc. 35, 321–333 (1960).
7. *Positive operators compact in an auxiliary topology*. Pacific J. Math. 10, 1131–1138 (1960).
8. *A polynomial iteration for the spectral family of an operator*. Proc. Glasgow Math. Assoc. 6, 65–69 (1963).

9. *Compact linear operators from an algebraic viewpoint.* Glasgow Math. J. 8, 41–49 (1967).

Bonsall, F. F., and Duncan, J.

1. *Numerical range.* London Math. Soc. Lecture Note Series, no. 2, 1971.

Bonsall, F. F., and Tomiuk, B. J.

1. *The semi-algebra generated by a compact linear operator.* Proc. Edinburgh Math. Soc. (2) 14, 177–196 (1964/65).

Bonsall, F. F., Lindenstrauss, J., and Phelps, R. R.

1. *Extreme positive operators on algebras of functions.* Math. Scand. 18, 161–182 (1966).

Bos, W.

1. *Zur Abschätzung der Eigenwerte einer beschränkten linearen Transformation mit Hilfe der singulären Werte.* Math. Ann. 157, 276–277 (1964).

Bourbaki, N.

6. *Éléments de mathématique. Fasc.* 32 *Théories spectrales.* Chapitre I: Algèbres normées. Chapitre II: Groupes localement compacts commutatifs. Hermann et Cie., Act. Sci. et Ind., no. 1332, Paris, 1967.

Brainerd, B.

1. *Averaging operators on the ring of continuous functions on a compact space.* J. Austral. Math. Soc. 4, 293–298 (1964).

Bram, J.

1. *Subnormal operators.* Duke Math. J. 22, 75–94 (1955).

de Branges, L.

1. *Some Hilbert spaces of entire functions,* I–IV.
   I. Trans. Amer. Math. Soc. 96, 259–295 (1960).
   II. ibid., 99, 118–152 (1961).
   III. ibid., 100, 73–115 (1961).
   IV. ibid., 105, 43–83 (1962).
2. *Perturbations of self-adjoint transformations.* Amer. J. Math. 84, 543–560 (1962).
3. *Invariant subspaces of non-self-adjoint transformations.* Bull. Amer. Math. Soc. 69, 587–590 (1963).
4. *Some Hilbert spaces of analytic functions.*
   I. Trans. Amer. Math. Soc. 106, 445–468 (1963).

de Branges, L., and Rovnyak, J.

1. *The existence of invariant subspaces.* Bull. Amer. Math. Soc. 70, 718–721 (1964).
2. *Correction to "The existence of invariant subspaces."* Bull. Amer. Math. Soc. 71, 396 (1965).

Brehmer, S.

1. *Über vertauschbare Kontraktionen des Hilbertschen Raumes.* Acta Sci. Math. Szeged 22, 106–111 (1961).

Breuer, M.

1. *Banachalgebren mit Anwendungen auf Fredholmoperatoren und singuläre Integralgleichungen*. Bonn. Math. Schr. No. 24, (1965), 108 pp.

Breuer, M., and Cordes, H.-O.

1. *On Banach algebras with σ-symbol.* I, II.
I. J. Math. Mech. 13, 313–323 (1964).
II. ibid. 14, 299–313 (1965).

Brodskiĭ, M. S.

1. *Integral representations of bounded non-selfadjoint operators with a real spectrum.* Doklady Akad. Nauk SSSR 126, 1166–1169 (1959). (Russian) Math. Rev. 21 #7438, 1380 (1960).
2. *Triangular representation of some operators with completely continuous imaginary part.* Doklady Akad. Nauk SSSR 133, 1271–1274 (1960). (Russian) Math. Rev. 26 #6778, 1285 (1963). Soviet Math. Dokl. 1, 952–955 (1961).
3. *Unicellularity criteria for Volterra operators.* Doklady Akad. Nauk SSSR 138, 512–514 (1961). (Russian) Math. Rev. 24 #A1015, 187 (1962).
4. *Unicellularity of real Volterra operators.* Doklady Akad. Nauk SSSR 147, 1010–1012 (1962). Math. Rev. 26 #6779, 1286 (1963).
5. *Operators with nuclear imaginary components.* Acta Sci. Math. Szeged 27, 147–155 (1966). (Russian) Math. Rev. 34#1843, 319 (1967).
6. *On the triangular representation of completely continuous operators with one-point spectrum.* Uspehi Mat. Nauk 16, no. 1 (97), 135–141 (1961) (Russian) Math. Rev. 24 #A426, 80 (1962). Amer. Math. Soc. Transl. (2) 47, 59–65.

Brodskiĭ, M. S., Gohberg, I. C., Kreĭn, M. G., and Macaev, V. I.

1. *Some new investigations in the theory of nonselfadjoint operators.* Proc. Fourth All-Union Math. Congress, Leningrad, 1964. Amer. Math. Soc. Transl. (2) 65, 237–251.

Brodskiĭ, M. S., and Kisilevskiĭ, G. È.

1. *Criterion for unicellularity of dissipative Volterra operators with nuclear imaginary components.* Izv. Akad. Nauk SSSR Ser. Mat. 30, 1213–1228 (1966). (Russian) Math. Rev. 34 #3310, 597 (1967).

Brodskiĭ, M. S., and Livšic, M. S.

2. *Spectral analysis of nonselfadjoint operators and intermediate systems.* Uspehi Mat. Nauk (N. S.) 13, no. 1 (79), 3–85 (1958). (Russian) Math. Rev. 20 #7221, 1883 (1959). Amer. Math. Soc. Transl. (2) 13, 265–346.

Brodskiĭ, M. S., and Šmul′jan, Ju. L.

1. *Invariant subspaces of a linear operator and divisors of its characteristic function.* Uspehi Mat. Nauk 19, no. 1 (115), 143–149 (1964). (Russian) Math. Rev. 29 #2645, 514 (1965).

Brodskiĭ, V. M.

1. *Eigenvectors of completely continuous linear operators defined in partially ordered non-normed spaces.* Sibirsk. Mat. Ž. 5, 468–471 (1964). (Russian) Math. Rev. 28 #5339, 1033 (1964).

Broido, M. M.

1. *Spectral representations for families of self-adjoint operators.* Proc. Cambridge Philos. Soc. 62, 209–213 (1966).

Browder, F. E.

6. *On the eigenfunctions and eigenvalues of the general linear elliptic differential operator.* Proc. Nat. Acad. Sci. 39, 433–439 (1953).

Brown, A.

1. *On the adjoint of a closed transformation.* Proc. Amer. Math. Soc. 15, 239–240 (1964).

Brown, A., and Pearcy, C.

1. *Spectra of tensor products of operators.* Proc. Amer. Math. Soc. 17, 162–166 (1966).

Brown, C. C.

1. *Über schwach-kompakte Operatoren im Banachraum.* Math. Scand. 14, 45–64 (1964).

Brownell, F. H.

1. *A note on Cook's wave-matrix theorem.* Pacific J. Math. 12, 47–52 (1962).

de Bruijn, N. G.

1. *On unitary equivalence of unitary dilations of contractions in Hilbert space.* Acta Sci. Math. Szeged 23, 100–105 (1963).

Buraczewski, A.

1. *Determinant systems for generalized Fredholm operators.* Bull. Acad. Polon. Sci. 9, 435–440 (1961).
2. *The determinant theory of generalized Fredholm operators.* Studia Math. 22, 265–307 (1963).

Butler, J. B.

1. *Perturbation series for eigenvalues of analytic non-symmetric operators.* Arch. Math. 10, 21–27 (1959).
2. *Perturbation of the continuous spectrum of even order differential operators.* Canadian J. Math. 12, 304–323 (1960).
3. *Perturbation of the continuous spectrum of systems of ordinary differential operators.* Canadian J. Math. 14, 359–379 (1962).

Caradus, S. R.

1. *Operators of Riesz type.* Pacific J. Math. 18, 61–71 (1966).
2. *Operators with finite ascent and descent.* Pacific J. Math. 18, 437–449 (1966).
3. *On meromorphic operators,* I, II.
   I. Canadian J. Math. 19, 723–736 (1967).
   II. ibid. 19, 737–748 (1967).

Cartan, H.

2. *Théorie spectrale des C-algèbres commutatives.* Séminaire Bourbaki, Exposé 125 (1955/56).

Cekanovskiĭ, È. R.

1. *Model elements of non-self-adjoint operators.* Doklady Akad. Nauk SSSR 142, 1043–1046 (1962). (Russian) Math. Rev. 25 #455, 96 (1963).

Chow, T. R.

1. *A spectral theory for direct integrals of operators.* Math. Ann. 188, 285–303 (1970).

Chung, Kai Lai

1. *On the exponential formulas of semi-group theory.* Math. Scand. 10, 153–162 (1962).

Ciorănescu, I.

1. *Sous-espaces invariants dans les espaces localement convexes.* Bull. Acad. Polon. Sci. 16, 721–725 (1968).

Clark, C.

1. *On relatively bounded perturbations of ordinary differential operators.* Pacific J. Math. 25, 59–70 (1968).

Coburn, L. A.

1. *Weyl's theorem for non-normal operators.* Michigan Math. J. 13, 285–288 (1966).

Coburn, L. A., and Lebow, A.

1. *Algebraic theory of Fredholm operators.* J. Math. Mech. 15, 577–584 (1966).
2. *Approximation by Fredholm operators in the metric space of closed operators.* Rend. Sem. Mat. Univ. Padova 36, 217–222 (1966).

Coddington, E. A. (see also Biriuk, G.)

5. *Formally normal operators having no normal extensions.* Canadian J. Math. 17, 1030–1040 (1965).

Coddington, E. A., and Gilbert, R. C.

1. *Generalized resolvents of ordinary differential operators.* Trans. Amer. Math. Soc. 93, 216–241 (1959).

Coddington, E. A., and Levinson, N.

1. *Theory of differential equations.* McGraw-Hill, New York, 1955.

Colojoară, I.

1. *Generalized spectral operators.* Rev. Roumaine Math. Pures Appl. 7, 459–465 (1962).
2. *Operatori spectrali generalizaţi,* II. Com. Acad. R. P. Romîne 12, 973–977 (1962).
3. *Operatori spectrali generalizaţi.* Stud. Cerc. Mat. 15, 499–536 (1964).
4. *Logarithms of generalized spectral operators.* Rev. Roumaine Math. Pures Appl. 10, 319–322 (1965).
5. *Elemente de teorie spectrală.* (Romanian) Editura Academiei Rep. Soc. România, Bucarest, 1968.

Colojoară, I., and Foiaş, C.

1. *Quasi-nilpotent equivalence of not necessarily commuting operators.* J. Math. Mech. 15, 521–540 (1966).
2. *The Riesz-Dunford functional calculus with decomposable operators.* Rev. Roumaine Math. Pures Appl. 12, 627–641 (1967).
3. *Spectral distribution of finite multiplicity.* Rev. Roumaine Math. Pures Appl. 12, 1039–1042 (1967).
4. *Theory of generalized spectral operators.* Gordon and Breach, New York, 1968.

5. *Commutators of decomposable operators.* Rev. Roumaine Math. Pures Appl. 12, 807–815 (1967).

Conley, C. C.

1. *A note on perturbations which create new point eigenvalues.* J. Math. Anal. Appl. 15, 421–433 (1966).

Conley, C. C., and Rejto, P. A.

1. *On spectral concentration.* Technical Report IMM-NYU 293, New York Univ. 1962.

Cook, J. M.

1. *Convergence to the Møller wave-matrix.* J. Math. Phys. 36, 82–87 (1957).

Cordes, H.-O. (see also Breuer, M.)

3. *The algebra of singular integral operators in R^n.* J. Math. Mech. 14, 1007–1032 (1965).
4. *Über eine nicht algebraische Charakterisierung von $\mathscr{I}$-Fredholm-Operatoren.* Math. Ann. 163, 212–229 (1966).

Cordes, H.-O., and Labrousse, J. P.

1. *The invariance of the index in the metric of closed operators.* J. Math. Mech. 12, 693–719 (1963).

Crimmins, T., and Rosenthal, P.

1. *On the decomposition of invariant subspaces.* Bull. Amer. Math. Soc. 73, 97–99 (1967).

Cuculescu, I., and Foiaş, C.

1. *An individual ergodic theorem for positive operators.* Rev. Roumaine Math. Pures Appl. 11, 581–594 (1966).

Čumakin, M. E.

1. *Generalized resolvents of an isometric operator.* Doklady Akad. Nauk SSSR 154, 791–794 (1964). (Russian) Math. Rev. 29 #479, 97 (1965).

Curtis, P. C., Jr. (see Bade, W. G.)

Davis, Chandler

1. *Various averaging operators onto subalgebras.* Illinois J. Math. 3, 538–553 (1959).
2. *The rotation of eigenvectors by a perturbation.* J. Math. Anal. Appl. 6, 159–173 (1963).
3. *The rotation of eigenvectors by a perturbation,* I, II.
   I. J. Math. Anal. Appl. 6, 159–173 (1963).
   II. ibid. 11, 20–27 (1965).

Davis, C., and Rider, D. G.

1. *Spectral sets and numerical range.* Rev. Roumaine Math. Pure Appl. 10, 125–131 (1965).

Day, M. M.

8. *Means for the bounded functions and ergodicity of the bounded representations of semi-groups.* Trans. Amer. Math. Soc. 69, 276–291 (1950).
10. *Ergodic theorems for abelian semi-groups.* Trans. Amer. Math. Soc. 51, 399–412 (1942).

12. *Normed linear spaces.* Ergebnisse der Math. N. F., Heft 21. Springer-Verlag, Berlin-Göttingen-Heidelberg, 1958.

Deal, E. R.

1. *Quasi-spectral theory.* Math. Scand. 13, 188–198 (1963).
2. *A quasi-spectral operator.* Math. Scand. 16, 29–32 (1965).

Dean, D. W.

1. *Schauder decompositions in (m).* Proc. Amer. Math. Soc. 18, 619–623 (1967).

Deckard, D., and Pearcy, C.

1. *On unitary equivalence of Hilbert-Schmidt operators.* Proc. Amer. Math. Soc. 16, 671–675 (1965).
2. *On rootless operators and operators without logarithms.* Acta Sci. Math. Szeged 28, 1–7 (1967).

Deprit, A.

1. *Endomorphismes de Riesz.* Ann. Soc. Sci. Bruxelles. Sér. I. 70, 165–183 (1956).
2. *Contribution à l'étude de l'algèbre des applications linéaires continues d'un espace localement convexe séparé: Théorie de Riesz-théorie spectrale.* Acad. Roy. Belg. Cl. Sci. Mém. Coll. in 8° 31, no. 2., 170 pp. (1959).

Derr, J., and Taylor, A. E.

1. *Operators of meromorphic type with multiple poles of the resolvent.* Pacific J. Math. 12, 85–111 (1962).

De Wilde, M.

1. *Opérateurs semi-compacts.* Bull. Soc. Roy. Sci. Liège 34, 194–208 (1965).
2. *Sur les opérateurs prénucleaires et intégraux.* Bull. Soc. Roy. Sci. Liège 35, 22–39 (1966).

Dieudonné, J.

19. *Sur la bicommutante d'un algèbre d'opérateurs.* Portugaliae Math. 14, 35–38 (1955).
20. *Sur la théorie spectrale.* J. Math. Pures Appl. (9) 35, 175–187 (1956).
21. *Champs de vecteurs non localement triviaux.* Archiv des Math. 7, 6–10 (1956).
22. *Sur les homomorphismes d'espaces normés.* Bull. Sci. Math. (2) 67, 72–84 (1943).

Dixmier, J.

1. *Les moyennes invariantes dans les semi-groupes et leurs applications.* Acta Sci. Math. Szeged 12 Pars A, 213–227 (1950).

Dollinger, M. B.

1. *Some aspects of spectral theory on Banach spaces.* Dissertation, Univ. of Illinois, 1968.
2. *A type of spectral decomposition for a class of operators.* J. Math. Mech. 18, 1059–1066 (1969).

Dollinger, M. B., and Oberai, K. K.

1. *Perturbation of local spectra.* (to appear).

Dolph, C. L.

1. *Recent developments in some non-self-adjoint problems of mathematical physics.* Bull. Amer. Math. Soc. 67, 1–69 (1961).

2. *Positive real resolvents and linear passive Hilbert systems.* Ann. Acad. Sci. Fenn. Ser. A.I no. 336 (1963), 39 pp.

Dolph, C. L., and Penzlin, F.

1. *On the theory of a class of non-self-adjoint operators and its applications to quantum scattering theory.* Ann. Acad. Sci. Fenn. Ser. A. I. no. 263 (1959), 36 pp.

Domar, Y.

1. *Harmonic analysis based on certain commutative Banach algebras.* Acta Math. 9, 1–66 (1956).

Donoghue, W. F., Jr.

1. *The lattice of invariant subspaces of a completely continuous quasi-nilpotent transformation.* Pacific J. Math. 7, 1031–1035 (1957).
2. *On a problem of Nieminen.* Inst. Hautes Études Sci. Publ. Math. no. 16, 31–33 (1963).
3. *On the perturbation of spectra.* Comm. Pure Appl. Math. 18, 559–579 (1965).

Dowson, H. R. (see also Berkson, E.)

1. *Restrictions of spectral operators.* Proc. London Math. Soc. (3) 15, 437–457 (1965).
2. *On some algebras of operators generated by a scalar-type spectral operator.* J. London Math. Soc. 40, 589–593 (1965).
3. *Operators induced on quotient spaces by spectral operators.* J. London Math. Soc. 42, 666–671 (1967).
4. *On the commutant of a complete Boolean algebra of projections.* Proc. Amer. Math. Soc. 19, 1448–1452 (1968).
5. *Restrictions of prespectral operators.* J. London Math. Soc. (2) 1,633–642. (1969).
6. *On a Boolean algebra of projections constructed by Dieudonné.* Proc. Edinburgh Math. Soc. 16 (2), 259–262 (1969).

Dunford, N.

2. *Direct decompositions of Banach spaces.* Bol. Soc. Mat. Mexicana 3, 1–12 (1946).
6. *Spectral theory.* Bull. Amer. Math. Soc. 49, 639–661 (1940).
7. *Spectral theory,* I. *Convergence to projections.* Trans. Amer. Math. Soc. 54, 185–217 (1943).
14. *Spectral theory in abstract spaces and Banach algebras.* Proc. Symposium on Spectral Theory and Differential Problems, 1–65 (1951). Oklahoma Agricultural and Mechanical College, Stillwater, Oklahoma.
15. *Spectral theory.* Proc. Symposium on Spectral Theory and Differential Problems, 203–208 (1951).
16. *The reduction problem in spectral theory.* Proc. International Congress Math., Cambridge, Mass., 1950, vol. 2, 115–122.
17. *Spectral theory,* II. *Resolutions of the identity.* Pacific J. Math. 2, 559–614 (1952).

18. *Spectral operators.* Pacific J. Math. 4, 321–354 (1954).
19. *A survey of the theory of spectral operators.* Bull. Amer. Math. Soc. 64, 217–274 (1958).
20. *Spectral operators in a direct sum of Hilbert spaces.* Proc. Nat. Acad. Sci. U.S.A. 50, 1041–1043 (1963).
21. *A spectral theory for certain operators on a direct sum of Hilbert spaces.* Math. Ann. 162, 294–330 (1966).

Dupras, A.
1. *Similarity of certain Volterra operators.* Dissertation, New York Univ., 1965.

Duren, P. L.
1. *Invariant subspaces of tridiagonal operators.* Duke Math. J. 30, 239–248 (1963).

Durszt, E.
1. *On the numerical range of normal operators.* Acta Sci. Math. Szeged 25, 262–265 (1964).
2. *On unitary ρ-dilations of operators.* Acta Sci. Math. Szeged 27, 247–250 (1966).

Eberly, W. S.
1. *A convergence theorem for bounded operators.* J. London Math. Soc. 40, 533–539 (1965).

Edwards, D. A., and Ionescu Tulcea, C.
1. *Some remarks on commutative algebras of operators on Banach spaces.* Trans. Amer. Math. Soc. 93, 541–551 (1959).

Elliott, J.
1. *The boundary value problems and semi-groups associated with certain integro-differential operators.* Trans. Amer. Math. Soc. 76, 300–331 (1954).
2. *Eigenfunction expansions associated with singular differential operators.* Trans. Amer. Math. Soc. 78, 406–425 (1955).

Ellis, A. J.
1. *Extreme positive operators.* Quart. J. Math. Oxford Ser. (2) 15, 342–344 (1964).

Ellis, R. J.
1. *The Fredholm alternative for non-Archimedean fields.* J. London Math. Soc. 42, 701–705 (1967).

Embry, M. R.
1. *Conditions implying normality in Hilbert space.* Pacific J. Math. 18, 457–460 (1966).

Entina, S. B. (see Birman, M. Š.)

Ercolano, J., and Schechter, M.
1. *Spectral theory for operators generated by elliptic boundary problems with eigenvalue parameter in boundary conditions,* I. Comm. Pure Appl. Math. 18, 83–105 (1965).

Esajan, A. R., and Stečenko, V. Ja.
1. *Estimates of the spectrum of integral operators and infinite matrices.* Doklady Akad. Nauk SSSR 157, 254–257 (1964). (Russian) Math. Rev. 29 #3899, 748 (1965).

Etienne, J.

1. *Opérateurs scalaires dans un espace linéaire semi-norme*. Bull. Soc. Roy. Sci. Liège 325–326 (supplement), 419–429 (1963).

Faddeev, L. D. (see also Ladyženskaya, O. A.)

1. *On a model of Friedrichs in the theory of perturbations of the continuous spectrum*. Trudy Mat. Inst. Steklov. 73, 292–313 (1964). (Russian) Math. Rev. 31 #2620, 473 (1966). Amer. Math. Soc. Transl. (2) 62, 177–203.
2. *The resolvent of the Schrödinger operator for a system of three particles interacting in pairs*. Doklady Akad. Nauk SSSR 138, 565–567 (1961) (Russian) Math. Rev. 24 #B 1568, 245 (1962). Soviet Physics Dokl. 6, 384–386 (1961).
3. *The construction of the resolvent of the Schrödinger operator for a three-particle system and the scattering problem*. Doklady Akad. Nauk SSSR 145, 301–304 (1962). (Russian) Math. Rev. 27 #3285, 639 (1964). Soviet Physics Dokl. 7 600–602 (1963).
4. *Mathematical questions in the quantum theory of scattering for three-particle systems*. Trudy Mat. Inst. Steklov, 69, 1–122 (1963). (Russian) Math. Rev. 29 #995, 189 (1965).

Fan, Ky

6. *Invariant subspaces of certain linear operators*. Bull. Amer. Math. Soc. 69, 773–777 (1963).

Favella, L. (see Bianchi, L.)

Feldman, Jacob

1. *On the functional calculus of an operator measure*. Acta. Sci. Math. Szeged 23 268–271 (1962).

Feldzamen, A. N.

1. *A generalized Weyr characteristic*. Bull. Amer. Math. Soc. 65, 79–83 (1959).
2. *Semi-similarity invariants for spectral operators on Hilbert space*. Trans. Amer. Math. Soc. 100, 277–324 (1961).

Fišman, K. M., and Valickiĭ, Yu. N.

1. *The applicability of Fredholm's theory to certain linear topological spaces*. Doklady Akad. Nauk SSSR 117, 943–946 (1957). (Russian) Math. Rev. 20 #230, 41 (1959).

Fixman, U.

1. *Problems in spectral operators*. Pacific J. Math. 9, 1029–1051 (1959).

Foguel, S. R.

1. *Sums and products of commuting spectral operators*. Ark. Mat. 3, 449–461 (1958).
2. *The relations between a spectral operator and its scalar part*. Pacific J. Math. 8, 51–65 (1958).
3. *Normal operators of finite multiplicity*. Comm. Pure Appl. Math. 11, 297–313 (1958).
4. *A perturbation theorem for scalar operators*. Comm. Pure Appl. Math. 11, 293–295 (1958).

5. *Boolean algebras of projections of finite multiplicity.* Pacific J. Math. 9, 681–693 (1959).
6. *Finite dimensional perturbations in Banach spaces.* Amer. J. Math. 82, 260–270 (1960).
7. *Computations of the multiplicity function.* Pacific J. Math. 10, 539–546 (1960).
8. *On a paper of A. Feldzamen.* Israel J. Math. 1, 133–138 (1963).
9. *Powers of a contraction in Hilbert space.* Pacific J. Math. 13, 551–562 (1963).
10. *A counterexample to a problem of Sz.-Nagy.* Proc. Amer. Math. Soc. 15, 788–790 (1964).
11. *Weak limits of powers of a contraction in Hilbert space.* Proc. Amer. Math. Soc. 16, 659–661 (1965).
12. *On spectrality criterion for operators on a direct sum of Hilbert spaces.* Israel J. Math. 3, 248–250 (1965).

Foiaș, C. (see also Colojoară, I., Cuculescu, I., and Sz.-Nagy, B.)

1. *La mesure harmonique-spectrale et la théorie spectrale des opérateurs généraux d'un espace de Hilbert.* Bull. Soc. Math. France 85, 263–282 (1957).
2. *Sur certains théorèmes de J. von Neumann concernant les ensembles spectraux.* Acta Sci. Math. Szeged 18, 15–20 (1957).
3. *On strongly continuous semigroups of spectral operators in Hilbert space.* Acta Sci. Math. Szeged 19, 188–191 (1958).
4. *Décompositions intégrales des familles spectrales et semi-spectrales en opérateurs qui sortent de l'espace hilbertien.* Acta Sci. Math. Szeged 20, 117–155 (1959).
5. *Sur la décomposition intégrale des familles semi-spectrales en opérateurs qui sortent de l'espace de Hilbert.* C. R. Acad. Sci. Paris 248, 904–906 (1959).
6. *Sur la décomposition spectrale en opérateurs propres des opérateurs linéaires dans les espaces nucléaires.* C. R. Acad. Sci. Paris 248, 1105–1108 (1959).
7. *Certaines applications des ensembles spectraux.* I. *Mesure harmonique-spectrale.* Acad. R. P. Romine Stud. Cerc. Mat. 10, 365–401 (1959). (Romanian. Russian and French summaries) Math. Rev. 22 #8340, 1421 (1961).
8. *On Hille's spectral theory and operational calculus for semi-groups of operators in Hilbert space.* Compositio Math. 14, 71–73 (1959).
9. *Une application des distributions vectorielles à la théorie spectrale.* Bull. Sci. Math. (2) 84, 147–158 (1960).
10. *Relation entre opérateurs spectraux et scalaires généralisés.* Com. Acad. R. P. Romîne 11, 1427–1429 (1961). (Romanian. Russian and French summaries) Math. Rev. 24 #A2852, 532 (1962).
11. *Relația dintre operatori spectrali și scalari generalizați.* Com. Acad. R. P. Romîne 11, 1427–1430 (1961).
12. *Spectral maximal spaces and decomposable operators in Banach space.* Arch. Math. 14, 341–349 (1963).
13. *Asupra unei probleme de teorie spectrală.* Stud. Cerc. Mat. 17, 921–923 (1965).

14. *Modèles fonctionnels, liason entre les théories de la prédiction, de la fonction caractéristique et de la dilation unitaire.* Deuxième Colloq. l'Analyse Fonctionnelle, Liège, 1964, pp. 63–76.

15. *Sur les mesures spectrales qui interviennent dans la théorie ergodique.* J. Math. Mech. 13, 639–658 (1964).

16. *Măsuri spectrale și semispectrale.* Stud. Cerc. Mat. 18, 7–56 (1966).

17. *Spectral capacities and decomposable operators.* Rev. Roumaine Math. Pures Appl. 13, 1539–1545 (1968).

18. *Décompositions en opérateurs et vecteurs propres.* I, II.
    I. *Études de ces décompositions et leurs rapports avec les prolongements des opérateurs.* Rev. Roumaine Math. Pures Appl. 7, 241–282 (1962) [errata ibid. 9, 805–809 (1964)].
    II. *Éléments de théorie spectrale dans les espaces nucléaires.* Ibid. 7, 571–602 (1962).

Foiaș, C., and Gehér, L.

1. *Über die Weylsche Vertauschungsrelation.* Acta. Sci. Math. Szeged 24, 97–102 (1963).

Foiaș, C., and Lions, J. L.

1. *Sur certains théorèmes d'interpolation.* Acta Sci. Math. Szeged 22, 269–282 (1961).

Foiaș, C., and Mlak, W.

1. *The extended spectrum of completely non-unitary contractions and the spectral mapping theorem.* Studia Math. 26, 239–245 (1966).

Foiaș, C., and Suciu, I.

1. *Szegö-measures and spectral theory in Hilbert spaces.* Rev. Roumaine Math. Pures Appl. 11, 147–159 (1966).
2. *On operator representation of log-modular algebras.* Bull. Acad. Polon. Sci. 16, 505–509 (1968).

Freeman, J. M.

1. *The perturbation of some Volterra operators.* Dissertation, Mass. Inst. of Tech., 1963.
2. *Perturbations of the shift operator.* Trans. Amer. Math. Soc. 114, 251–260 (1965).
3. *Volterra operators similar to $J: f \to \int_0^x f(t)\,dt$.* Trans. Amer. Math. Soc. 116, 181–192 (1965).

Freeman, R. S.

1. *Closed operators and their adjoints associated with elliptic differential operators.* Pacific J. Math. 22, 71–97 (1967).

Friedman, B., and Mishoe, L. I.

1. *Eigenfunction expansions associated with a non-self adjoint differential equation.* Pacific J. Math. 5. 249–270 (1956).

Friedrichs, K. O.

2. *On the perturbation of continuous spectra.* Comm. Pure Appl. Math. 1, 361–406 (1948).

17. *Perturbation of spectra in Hilbert space.* Lectures in Applied Math., Vol. III, Amer. Math. Soc., Providence, R. I., 1965.

18. *Spectral perturbation phenomena.* Perturbation Theory and its Applications in Quantum Mechanics, Wiley, New York, 1966.

Friedrichs, K. O., and Rejto, P. A.

1. *On a perturbation through which a discrete spectrum becomes continuous.* Comm. Pure Appl. Math. 15, 219–235 (1962).

Fuglede, B.

1. *A commutativity theorem for normal operators.* Proc. Nat. Acad. Sci. U.S.A. 36, 35–40 (1950).

Galindo, A.

1. *On the existence of J-selfadjoint extensions of J-symmetric operators with adjoint.* Comm. Pure Appl. Math. 15, 423–425, (1962).

Gamelin, T. W. (see also Balslev, E.)

1. *Decomposition theorems for Fredholm operators.* Pacific J. Math. 15, 97–106 (1965).

Gamlen, J. L. B., and Miller, J. B.

1. *Averaging and Reynolds operators on Banach algebras.* II. *Spectral properties of averaging operators.* J. Math. Anal. Appl. 23, 183–197 (1968).

Gehér, L. (see Foiaş, C.)

Gehtman, M. M., and Stankevič, I. V.

1. *On the spectrum of non-selfadjoint differential operators.* Doklady Akad. Nauk SSSR 158, 29–32 (1964). (Russian) Math. Rev. 29 #6343, 1190 (1965).

Gelfand, I. M., and Neumark, M. A.

3. *Unitäre Darstellungen der klassischen Gruppen.* Akademie Verlag, Berlin, 1957.

Gelfand, I. M., and Raĭkov, D. A.

1. *On the theory of characters of commutative topological groups.* Doklady Akad. Nauk SSSR (N. S.) 28, 195–198 (1940).

Gelfand, I. M., and Šilov, G. E.

2. *Generalized Functions,* Vol. 3, *Theory of differential equations.* Academic Press, 1967.

George, M. D.

1. *The spectrum of an operator in Banach space.* Proc. Amer. Math. Soc. 16, 980–982 (1965).

Gerlach, E.

1. *On spectral representation for self adjoint operators. Expansion in generalized eigenelements.* Ann. Inst. Fourier (Grenoble) 15, fasc. 2, 537–574 (1965).

Ghika, A.

1. *Décompositions spectrales généralisées des transformations linéaires d'un espace hilbertien dans un autre.* Rev. Math. Pures Appl. 2, 61–109 (1957).

Gide, A.

1. *Self-Portraits, The Gide/Valéry Letters.* The University of Chicago Press, 1966.

Giertz, M.

1. *On the expansion of certain generalized functions in series of orthogonal functions.* Proc. London Math. Soc. (3) 14, 45-52 (1964).

Gilbert, R. C. (see also Coddington, E. A.)

1. *Extremal spectral functions of a symmetric operator.* Pacific J. Math. 14, 75–84 (1964).

Gilbert, R. C., and Kramer, V. A.

1. *Trace formulas for powers of a Sturm-Liouville operator.* Canadian J. Math. 16, 412–422 (1964).

Gil'derman, Ju. I., and Korotkov, V. B.

1. *On the general form of completely continuous operators acting from L_p into a B-space X.* Sibirsk. Mat. Ž. 4, 1426–1430 (1963). (Russian) Math. Rev. 28 #2431, 480 (1964).

Gillespie, T. A., and West, T. T.

1. *A characterization and two examples of Riesz operators.* Glasgow Math. J. 9, 106–110 (1968).

Gindler, H. A.

1. *An operational calculus for meromorphic functions.* Nagoya Math. J. 26, 31–38 (1966).

Gindler, H. A., and Taylor, A. E.

1. *The minimum modulus of a linear operator and its use in spectral theory.* Studia Math. 22, 15–41 (1962/63).

Ginzburg, Ju. P.

1. *Multiplicative representations of bounded analytic operator-functions.* Doklady Akad. Nauk SSSR 170, 23–26 (1966). (Russian) Math. Rev. 34 #611, 104 (1967). Soviet Math Dokl. 7, 1125–1128 (1966).

Ginzburg, Ju. P., and Iohvidov, I. S.

1. *A study of the geometry of infinite-dimensional spaces with bilinear metric.* Uspehi Mat. Nauk 17, no. 4 (106), 3–56 (1962). (Russian) Math. Rev. 26 #2850, 551 (1963).

Glazman, I. M.

5. *Direct methods of qualitative spectral analysis of singular differential operators.* (Russian) Moscow, 1963. (Transl. Israel Program for Sci. Transl., Jerusalem 1965, Daniel Davey and Co., New York, 1966.)

Glickfield, B. W.

1. *A metric characterization of C(X) and its generalization to C*-algebras,* Illinois J. Math. 10, 547–556 (1966).

Godič, V. I.

1. *On invariant subspaces of completely continuous bisymmetric operators.* Ukrain. Mat. Ž. 18, no. 3, 103–107 (1966). Math. Rev. 34 #612, 104 (1967).

Gohberg, I. C. (see also Brodskiĭ, M. S.)

4. *On the index of an unbounded operator.* Mat. Sbornik N. S. 33 (75), 193–198 (1953). (Russian) Math. Rev. 15, 233 (1954).

5. *Criteria for one-sided reversibility of elements of normed rings and their applications.* Doklady Akad. Nauk SSSR 145, 971–974 (1962). (Russian) Math. Rev. 27 #6147, 1175 (1964).

6. *The factorization problem for operator functions.* Izv. Akad. Nauk SSSR Ser. Mat. 28, 1055–1082 (1964). (Russian) Math. Rev. 30 #5182, 964 (1965).

7. *On linear equations in normed spaces.* Doklady Akad. Nauk SSSR 76, 477–480 (1951). (Russian) Math. Rev. 13, 46 (1952).

Gohberg, I. C., and Kreĭn, M. G.

1. *On completely continuous quasinilpotent operators.* Doklady Akad. Nauk SSSR 128, 227–230 (1959). (Russian) Math. Rev. 24 #A1022, 189 (1962).

2. *The basic propositions on defect numbers, root numbers and indices of linear operators.* Uspehi Mat. Nauk (N.S.) 12, no. 2(74), 43–118 (1957). (Russian) Math. Rev. 20 #3459, 572 (1959). Amer. Math. Soc. Transl. (2)13, 185–264.

3. *Systems of integral equations on the half-line with kernels depending on the difference of the arguments.* Uspehi Mat. Nauk 13, no. 2 (80), 3–72 (1958). (Russian) Math. Rev. 21 #1506, 286 (1960). Amer. Math. Soc. Transl. (2) 14, 217–287.

4. *On the theory of triangular representations of non-selfadjoint operators.* Doklady Akad. Nauk SSSR 137, 1034–1037 (1961). (Russian) Math. Rev. 25 #3370, 656 (1963).

5. *Volterra operators with imaginary component in one class or another.* Doklady Akad. Nauk SSSR 139, 779–782 (1961). (Russian) Math. Rev. 25 #3371, 656 (1963).

6. *Completeness criteria for the system of root vectors of a contraction.* Ukrain. Mat. Ž. 16, 78–82 (1964). (Russian) Math. Rev. 29 #2651, 515 (1965). Amer. Math. Soc. Transl. (2) 54, 119–124.

7. *Introduction to the theory of linear non-self-adjoint operators.* Izdat. "Nauka", Moscow 1965, (Russian). Amer. Math. Soc. Transl. of Math. Monographs, 18, Providence, 1969.

8. *Theory of Volterra operators in Hilbert space and its application.* Moscow 1967. (Russian).

9. *Triangular representations of linear operators and multiplicative representions of their characteristic functions.* Doklady Akad. Nauk SSSR 175, 272–275 (1967). (Russian) Math. Rev. 35 #7157, 1330 (1968). Soviet Math. Dokl. 8, 831–834 (1967).

10. *Factorization of operators in Hilbert space.* Acta Sci. Math. Szeged 25, 90–123 (1964) (Russian) Math. Rev. 29 #6313, 1183 (1965). Amer. Math. Soc. Transl. (2) 51, 155–188.

Gohberg, I. C., and Markus, A. S.

1. *Some relations between eigenvalues and matrix elements of linear operators.* Mat. Sbornik (N.S.) 64 (106), 481–496 (1964). (Russian) Math. Rev. 30 #457, 94 (1965).

Goldberg, S.

1. *Closed linear operators and associated continuous linear operators.* Pacific J. Math. 12, 183–186 (1962).
2. *Unbounded linear operators: Theory and applications.* McGraw-Hill, New York, 1966.

Goldberg, S., and Schubert, C.

1. *Some applications of the theory of unbounded operators to ordinary differential equations.* J. Math. Anal. Appl. 19, 78–92 (1967).

Goldberg, S., and Thorp, E. O.

1. *The range as range space for compact operators.* J. Reine Angew. Math. 211, 113–115 (1962).
2. *On some open questions concerning strictly singular operators.* Proc. Amer. Math. Soc. 14, 334–336 (1963).

Gol'dengeršel', È. I.

1. *On the spectrum of a certain class of nonselfadjoint operators.* Sibirsk. Mat. Ž.6, 1420–1422 (1965). (Russian) Math. Rev. 33 #7885, 1372 (1967).
2. *On the resolvent of a Volterra operator with kernel depending on the difference.* Sibirsk. Mat. Ž.6, 1423–1434 (1965). (Russian) Math. Rev. 33 #7886, 1372 (1967).

Gol'dman, M. A.

1. *On the stability of the property of normal solvability of linear equations.* Doklady Akad. Nauk SSSR 100, 201–204 (1955). (Russian) Math. Rev. 17, 284 (1956).

Gol'dman, M. A., and Kračkovskiĭ, S. N.

1. *Invariance of certain spaces associated with the operator $A - \lambda I$.* Doklady Akad. Nauk SSSR 154, 500–502 (1964). (Russian) Math. Rev. 29 #467, 95 (1965).
2. *On certain perturbations of a closed linear operator.* Doklady Akad. Nauk SSSR 158, 507–509 (1964). (Russian) Math. Rev. 31 #5094, 927 (1966). Soviet Math. Dokl. 5, 1243–1245 (1964).
3. *The d-characteristic of a linear operator.* Doklady Akad. Nauk SSSR 165, 476–478 (1965). (Russian) Math. Rev. 32 #6236, 1067 (1966). Soviet Math. Dokl. 6, 1455–1457 (1965).
4. *Perturbation of homomorphisms by operators of finite rank.* Doklady Akad. Nauk SSSR 174, 743–746 (1967). (Russian) Math. Rev. 34 #7164, 1331 (1968). Soviet Math. Dokl. 8, 670–673 (1967).

Gol'dman, M. A., and Levič, E. M.

1. *On invariant supplementability of certain spaces generated by a linear operator.* Doklady Akad. Nauk SSSR 166, 267–270 (1966). (Russian) Math. Rev. 32 #6224, 1064 (1966). Soviet Math. Dokl. 7, 53–55 (1966).

Gonshor, H.

1. *Spectral theory for a class of non-normal operators*, I, II.
   I. Canadian J. Math. 8, 449–461 (1956).
   II. ibid. 10, 97–102 (1958).

Gramsch, B.

1. *σ-Transformationen in lokalbeschränkten Vektorräumen.* Math. Ann. 165, 135–151 (1966).
2. *Ein Schema zur Theorie Fredholmscher Endomorphismen und eine Anwendung auf die Idealkette der Hilberträume.* Math. Ann. 171, 263–272 (1967).

Graves, L. M.

6. *A generalization of the Riesz theory of completely continuous transformations.* Trans. Amer. Math. Soc. 79, 141–149 (1955).

Gray, J. D.

1. *Local analytic extensions of the resolvent.* Pacific J. Math. 27, 305–324 (1968).

Greiner, P. C.

1. *Eigenfunction expansions and scattering theory for perturbed elliptic partial differential operators.* Bull. Amer. Math. Soc. 70, 517–521 (1964).

Grothendieck, A.

3. *Produits tensoriels topologiques et espaces nucléaires.* Memoirs Amer. Math. Soc. no. 16, 1955.
6. *The trace of certain operators.* Studia Math. 20, 141–143 (1961).

Guenin, M., and Misra, B.

1. *De la permutabilité des opérateurs non bornes.* (English summary) Helv. Phys. Acta 37, 233–240 (1964).

Gustafson, K.

1. *A perturbation lemma.* Bull. Amer. Math. Soc. 72, 334–338 (1966).

Haahti, H.

1. *Zur Verallgemeinerung des Spur-Operators.* Ann. Acad. Sci. Fenn. Ser. AI No. 369 (1965), 14 pp.

Haar, A.

3. *Zur Theorie der orthogonalen Funktionensysteme,* I, II.
   I. Math. Ann. 69, 331–371 (1910).
   II. ibid. 71, 38–53 (1911).

Hack, M. N.

1. *Wave operators in multichannel scattering.* Nuovo Cimento 13, 231–236 (1959).

Hadamard, J.

2. *Lectures on Cauchy's problem in linear partial differential equations.* Yale University Press, 1923. Dover edition, 1952.

Hadeler, K.-P.

1. *Estimates for the spectrum of normal operators.* Doklady Akad. Nauk SSSR 157, 284–287 (1964). (Russian) Math. Rev. 29 #2648, 514 (1965).
2. *On the spectrum of normal operators and perturbations of them.* Doklady Akad. Nauk SSSR 158, 1042–1043 (1964). (Russian) Math. Rev. 31 #606, 110 (1966).

Halberg, C. J. A., Jr.

1. *The spectra of bounded linear operators on the sequence spaces.* Proc. Amer. Math. Soc. 8, 728–732 (1957).

2. *Semigroups of matrices defining linked operators with different spectra.* Pacific J. Math. 13, 1187–1191 (1963).

Halberg, C. J. A., Jr., and Taylor, A. E.

1. *On the spectra of linked operators.* Pacific J. Math. 6, 283–290 (1956).

Halmos, P. R.

3. *Commutativity and spectral properties of normal operators.* Acta Sci. Math. Szeged 12 Pars B, 153–156 (1950).
6. *Introduction to Hilbert space and the theory of spectral multiplicity.* Chelsea, New York, 1951.
11. *Shifts on Hilbert spaces.* J. Reine Angew. Math. 208, 102–112 (1961).
12. *What does the spectral theorem say?* Amer. Math. Monthly 70, 241–247 (1963).
13. *A glimpse into Hilbert space.* Lectures on Modern Mathematics, Vol. 1, pp. 1–22. Wiley, New York, 1963.
14. *Numerical ranges and normal dilations.* Acta Sci. Math. Szeged 25, 1–5 (1964).
15. *On Foguel's answer to Nagy's question.* Proc. Amer. Math. Soc. 15, 791–793 (1964).
16. *Invariant subspaces of polynomially compact operators.* Pacific J. Math. 16, 433–438 (1966).
17. *A Hilbert space problem book.* D. Van Nostrand Co., Princeton, 1967.

Halmos, P. R., and Lumer, G.

1. *Square roots of operators,* II. Proc. Amer. Math. Soc. 5, 589–595 (1954).

Halmos, P. R., Lumer, G., and Schäffer, J. J.

1. *Square roots of operators.* Proc. Amer. Math. Soc. 4, 142–149 (1953).

Halmos, P. R., and McLaughlin, J. E.

1. *Partial isometries.* Pacific J. Math. 13, 585–596 (1963).

Halperin, Israel

6. *The unitary dilation of a contraction operator.* Duke Math. J. 28, 563–571 (1961).
7. *Unitary dilations which are orthogonal bilateral shift operators.* Duke Math. J. 29, 573–580 (1962).
8. *Sz.-Nagy-Brehmer dilations.* Acta Sci. Math. Szeged 23, 279–289 (1962).
9. *Intrinsic description of the Sz.-Nagy-Brehmer unitary dilation.* Studia Math. 22, 211–219 (1962/63).
10. *Interlocking dilations.* Duke Math. J. 30, 475–484 (1963).
11. *The spectral theorem.* Amer. Math. Monthly 71, 408–410 (1964).

Hamburger, H. L.

1. *Five notes on a generalization of quasi-nilpotent transformations in Hilbert space.* Proc. London Math. Soc. (3) 1, 494–512 (1951).

Harazov, D. F.

4. *On the spectral theory of semi-bounded operators.* Akad. Nauk Gruzin. SSR Trudy Tbiliss. Math. Inst. Razmadze 26, 153–170 (1959). (Russian) Math. Rev. 25 #2457, 477 (1963).

5. *Spectral theory of some linear operators depending meromorphically on a parameter.* Studia Math. 20, 19–45 (1961). (Russian) Math. Rev. 25 #1459, 290 (1963).
6. *On the spectrum of completely continuous operators depending analytically on a parameter, in topological linear spaces.* Acta. Sci. Math. Szeged 23, 38–45 (1962). (Russian) Math. Rev. 25 #3374, 657 (1963).
7. *On the separation of the eigenvalues of operators with discrete spectrum.* Mathematica (Cluj) 4(27), 253–260 (1962). (Russian) Math. Rev. 29 #6311, 1182 (1965).
8. *Theorems of comparison type for eigenvalues of certain operators with discrete spectrum.* Akad. Nauk Gruzin. SSR Trudy Tbiliss. Mat. Inst. Razmadze 29, 219–227 (1964). (Russian, Georgian summary) Math. Rev. 31 #3870, 699 (1966).

Hartogs, F., and Rosenthal, A.
1. *Über Folgen analytischer Funktionen.* Math. Ann. 104, 606–610 (1931).

Hasegawa, M.
1. *On the convergence of resolvents of operators.* Pacific J. Math. 21, 35–47 (1967).

Hasumi, M., and Srinivasan, T. P.
1. *Doubly invariant subspaces.* II. Pacific J. Math. 14, 525–535 (1964).
2. *Invariant subspaces of continuous functions.* Canadian J. Math. 17, 643–651 (1965).

Hellwig, G.
1. *Differentialoperatoren der mathematischen Physik. Eine Einführung.* Springer-Verlag, Berlin, 1964.

Helson, H.
1. *Lectures on invariant subspaces.* Academic Press, New York-London, 1964.

Helson, H., and Lowdenslager, D.
1. *Invariant subspaces.* Proc. Internat. Sympos. Linear Spaces (Jerusalem, 1960), 251–262.

Hempel, P.
1. *Einschliessungsaussagen für das Spektrum selbstadjungierter und normaler Transformationen im Hilbert-Raum durch Abschätzung der Norm der Resolvente.* Arch. Rational Mech. Anal. 13, 147–156 (1963).

Hestenes, M. R.
1. *Relative self-adjoint operators in Hilbert space.* Pacific J. Math. 11, 1315–1357 (1961).

Heuser, H.
1. *Über die Iteration Rieszscher Operatoren.* Arch. Math. 9, 202–210 (1958).
2. *Zur Eigenwerttheorie einer Klasse Rieszscher Operatoren.* Arch. Math. 14, 39–46 (1963).
3. *Über Eigenwerte and Eigenlösungen symmetrisierbarer finiter Operatoren.* Arch. Math. 10, 12–20 (1959).

Heyn, E.

1. *Die Differentialgleichung $dT/dt = P(t)T$ für Operatorfunktionen.* Math. Nachrichten 24, 281–330 (1962).
2. *Skalare Spektraloperatoren im reflexiven Banachraum.* Math. Nachrichten 31, 169–177 (1966).

Hildebrandt, Stefan

1. *The closure of the numerical range of an operator as spectral set.* Comm. Pure Appl. Math. 17, 415–421 (1964).
2. *Über den numerischen Wertebereich eines Operators.* Math. Ann. 163, 230–247 (1966).

Hille, E.

6. *On roots and logarithms of elements of a complex Banach algebra.* Math. Ann. 136, 46–57 (1958).
7. *Some aspects of Cauchy's problem.* Proc. Intern. Cong. Math. 1954. Amsterdam, 3, 109–116 (1956).

Hille, E., and Phillips, R. S.

1. *Functional analysis and semi-groups.* Amer. Math. Soc. Coll. Publ. 31, 1957, 808 pp.

Hirschfeld, R. A.

1. *Expansion in eigenfunctionals.* Nederl. Akad. Wetensch. Proc. Ser. A68, 513–520 (1965).

Høegh-Krohn, J. R.

1. *Partly gentle perturbation with application to perturbation by annihilation-creation operators.* Proc. Nat. Acad. Sci. USA 56, 2187–2192 (1967).

Hoffman, S. P., Jr.

1. *Second order linear differential operators defined by irregular boundary conditions.* Dissertation, Yale University (1957).

Hopf, E.

1. *Ergodentheorie.* Ergebnisse der Math., V.2, J. Springer, Berlin, 1937. Reprinted by Chelsea Publ. Co., New York, 1948.
4. *Mathematical problems of radiative equilibrium.* Cambridge Univ. Press, 1934.

Hörmander, L.

1. *Translation invariant operators.* Acta Math. 104, 93–139 (1960).
2. *Linear partial differential operators.* Springer-Verlag, Berlin, 1963.

Huige, G. E.

1. *The spectral theory of some non-selfadjoint differential operators.* Comm. Pure Appl. Math. 21, 25–49 (1968).

Hukuhara, M., and Sibuya, Y.

1. *Sur l'endomorphisme complètement continu.* Proc. Japan Acad. 31, 595–599 (1955).
2. *Théorie des endomorphismes complètement continus*, I, II.
   I. J. Fac. Sci. Univ. Tokyo Sect. I. 7, 391–405 (1957).
   II. ibid. 7, 511–525 (1958).

Ikebe, T.

1. *Eigenfunction expansions associated with the Schrödinger operators and their applications to scattering theory.* Arch. Rat. Mech. Anal. 5, 1–34 (1960).
2. *On the phase-shift formula for the scattering operator.* Pacific J. Math. 15, 511–523 (1965).
3. *Orthogonality of the eigenfunctions for the exterior problem connected with* $-\Delta$. Arch. Rat. Mech. Anal. 19, 71–73 (1965).
4. *On the eigenfunction expansion connected with the exterior problem for the Schrödinger equation.* Japanese J. Math. 36, 33–55 (1967).

Inoue, Sakuji

1. *On the distribution of the spectra of normal operators in Hilbert spaces.* Proc. Japan Acad. 37, 464–468 (1961).
2. *Simplification of the canonical spectral representation of a normal operator in Hilbert space and its applications.* Mem. Fac. Educ. Kumamoto Univ. 3, Supplement 1, 1–50 (1955).
3. *Some analytical properties of the spectra of normal operators in Hilbert spaces.* Proc. Japan Acad. 37, 566–570 (1961).
4. *Functional-representations of normal operators in Hilbert spaces and their applications.* Proc. Japan Acad. 37, 614–618 (1961).
5. *On the functional-representations of normal operators in Hilbert spaces.* Proc. Japan Acad. 38, 18–22 (1962).
6. *Some applications of the functional-representations of normal operators in Hilbert spaces,* I–XXIV.

I. Proc. Japan Acad. 38, 263–268 (1962).
II. ibid. 38, 452–456 (1962).
III. ibid. 38, 641–645 (1962).
IV. ibid. 38, 646–650 (1962).
V. ibid. 38, 706–710 (1962).
VI. ibid. 39, 109–113 (1963).
VII. ibid. 39, 338–341 (1963).
VIII. ibid. 39, 455–460 (1963).
IX. ibid. 39, 566–568 (1963).
X. ibid. 40, 317–322 (1964).
XI. ibid. 40, 391–395 (1964).
XII. ibid. 40, 487–491 (1964).
XIII. ibid. 40, 492–497 (1964).
XIV. ibid. 40, 654–659 (1964).
XV. ibid. 41, 150–154 (1965).
XVI. ibid. 41, 541–546 (1965).
XVII. ibid. 41, 702–705 (1965).
XVIII. ibid. 41, 911–914 (1965).
XIX. ibid. 41, 915–918 (1965).
XX. ibid. 42, 364–369 (1966).

XXI. ibid. 42, 583–588 (1966).
XXII. ibid. 42, 743–748 (1966).
XXIII. ibid. 42, 749–754 (1966).
XXIV. ibid. 42, 901–906 (1966).

Iohvidov, I. S.

1. *Singular linear manifolds in spaces with an arbitrary Hermitian bilinear metric.* Uspehi Mat. Nauk 17 (106), 127–133 (1962). (Russian) Math. Rev. 28 #468, 98 (1964).
2. *Operators with completely continuous iterations.* Doklady Akad. Nauk SSSR 153, 258–261 (1963). (Russian) Math. Rev. 28 #469, 98 (1964).
3. *Singular linear manifolds in the spaces* Π_κ. Ukrain. Mat. Ž. 16, 300–308 (1964). (Russian. English summary) Math. Rev. 29 #2649, 514 (1965).

Iohvidov, I. S., and Kreĭn, M. G.

1. *Spectral theory of operators in spaces with indefinite metric,* I, II.
   I. Trudy Moskov. Mat. Obšč. 5, 367–432 (1956). (Russian) Math. Rev. 18, 320 (1957). Amer. Math. Soc. Transl. (2) 13, 105–175.
   II. ibid. 8, 413–496 (1959). (Russian) Math Rev. 21 #6543, 1221 (1960). Amer. Math. Soc. Transl. (2) 34, 283–373.

Ionescu Tulcea, C. T. (see also Edwards, D. A.)

1. *Spaţii Hilberti.* Editura Acad. Rep. Populare Romane, 1956.
2. *Spectral representations of certain semi-groups of operators.* J. Math. Mech. 8, 95–110 (1959).
3. *Spectral operators on locally convex spaces.* Bull. Amer. Math. Soc. 67, 125–128 (1961).
4. *Scalar dilations and scalar extensions of operators on Banach spaces* (I). J. Math. Mech. 14, 841–856 (1965).
5. *Notes on spectral theory.* Technical report, U.S. Army Research Office (Durham), 1964.

Ionescu Tulcea, C., and Plafker, S.

1. *Dilations et extensions scalaires sur les espaces de Banach.* C. R. Acad. Sci. Paris 265, 734–735 (1967).

Ionescu Tulcea, C., and Simon, A. B.

1. *Spectral representations and unbounded convolution operators.* Proc. Nat. Acad. Sci. USA 45, 1765–1767 (1959).

Istrăteşcu, V.

1. *On some hyponormal operators.* Pacific J. Math. 22, 413–417 (1967).

Itô, Takasi

1. *On the commutative family of subnormal operators.* J. Fac. Sci. Hokkaido Univ. Ser. I 14, 1–15 (1958).

Jakubov, S. Ja.

1. *Hilbert-Schmidt theory for J-symmetrizable operators acting in a Banach space.* Izv. Akad. Nauk Azerbaĭdžan. SSR Ser. Fiz.-Mat. Tehn. Nauk 1961, 39–48. (Russian. Azerbaijani summary) Math. Rev. 29 #1540, 305 (1965).

Jauch, J. M.

1. *Theory of the scattering operator.* Helv. Phys. Acta 31, 127–158 (1958).
2. *Theory of the scattering operator.* II. *Multichannel scattering.* Helv. Phys. Acta 31, 661–684 (1958).

Jauch, J. M., and Misra, B.

1. *The spectral representation.* Helv. Phys. Acta 38, 30–52 (1965).

Jauch, J. M., and Zinnes, I. I.

1. *The asymptotic condition for simple scattering systems.* Nuovo Cimento 11, 553–567 (1959).

Javrjan, V. A.

1. *Some perturbations of self-adjoint operators.* Akad. Nauk Armjan. SSR Dokl. 38, 3–7 (1964). (Russian. Armenian summary) Math. Rev. 28 #5340, 1033 (1964).

Kaashoek, M. A.

1. *Closed linear operators on Banach spaces.* Nederl. Akad. Wetensch. Proc. Ser. A 68, 405–414 (1965).
2. *Stability theorems for closed linear operators.* Nederl. Akad. Wetensch. Proc. Ser. A 68, 452–466 (1965).

Kaashoek, M. A., and Lay, D. C.

1. *On operators whose Fredholm set is the complex plane.* Pacific J. Math. 21, 275–278 (1967).

Kac, G. I.

1. *Spectral decompositions of self-adjoint operators in terms of generalized elements of a Hilbert space.* Ukrain. Math. Ž, 13, no. 4, 13–33 (1961). (Russian. English summary) Math. Rev. 26 #1764, 341 (1963).

Kac, I. S.

1. *Spectral multiplicity of a second-order differential operator and expansion in eigenfunctions.* Izv. Akad. Nauk SSSR Ser. Mat. 27, 1081–1112 (1963). (Russian) Math. Rev. 28 #3196, 623 (1964). Corrections, same Izv. 28, 951–952 (1964).

Kacnel'son, V. È.

1. *Conditions for a system of root vectors of certain classes of operators to be a basis.* Teor. Funkciĭ Funkcional. Anal. i. Priložen. 1., no. 2, 39–51 (1967). (Russian) Math. Rev. 35 #4757, 876 (1968).

Kacnel'son, V. È., and Macaev, V. I.

1. *Spectral sets for operators in a Banach space and estimates of functions of finite-dimensional operators.* Teor. Funkciĭ Funkcional. Anal. i. Priložen. Vyp. 3, 3–10 (1966). (Russian) Math. Rev. 34 #6532, 1198 (1967).

Kakutani, S.

15. *An example concerning uniform boundedness of spectral measures.* Pacific J. Math. 4, 363–372 (1954).

Kalisch, G. K.

1. *On similarity, reducing manifolds, and unitary equivalence of certain Volterra operators.* Ann. of Math. (2) 66, 481–494 (1957).

2. *On similarity invariants of certain operators in L_p*. Pacific J. Math. 11, 247–252 (1961).
3. *On isometric equivalence of certain Volterra operators*. Proc. Amer. Math. Soc. 12, 93–98 (1961).
4. *Théorème de Titchmarsh sur la convolution et opérateurs de Volterra*. Séminare d'Analyse, dirigé par P. Lelong 1962/63. Paris, 1963.
5. *Direct proofs of spectral representation theorems*. J. Math. Anal. Appl. 8, 351–363 (1964).
6. *Characterizations of direct sums and commuting sets of Volterra operators*. Pacific J. Math. 18, 545–552 (1966).
7. *On fractional integrals of pure imaginary order in L_p*. Proc. Amer. Math. Soc. 18, 136–139 (1967).

Kal'muševskiĭ, I. I.
1. *On the linear equivalence of Volterra operators*. Uspehi Mat. Nauk 20, no. 6 (126) 93–97 (1965). (Russian) Math. Rev. 32 #8161, 1405 (1966).

Kamowitz, H.
1. *On operators whose spectrum lies on a circle or a line*. Pacific J. Math. 20, 65–68 (1967).

Kaniel, S.
1. *Unbounded normal operators in Hilbert space*. Trans. Amer. Math. Soc. 113, 488–511 (1964).

Kaniel, S., and Schechter, M.
1. *Spectral theory for Fredholm operators*. Comm. Pure Appl. Math. 16, 423–448 (1963).

Kantorovitz, Shmuel
1. *An operational calculus and spectral operators*. Dissertation, University of Minnesota, 1962.
2. *Operational calculus in Banach algebras for algebra-valued functions*. Trans. Amer. Math. Soc. 110, 519–537 (1964).
3. *Classification of operators by means of the operational calculus*. Bull. Amer. Math. Soc. 70, 316–320 (1964).
4. *On the characterization of spectral operators*. Trans. Amer. Math. Soc. 111, 152–181 (1964).
5. *Classification of operators by means of their operational calculus*. Trans. Amer. Soc. 115, 194–224 (1965).
6. *A Jordan decomposition for operators in Banach space*. Bull. Amer. Math. Soc. 71, 891–893 (1965).
7. *A Jordan decomposition for operators in Banach space*. Trans. Amer. Math. Soc. 120, 526–550 (1965).
8. *The semi-simplicity manifold of arbitrary operators*. Trans. Amer. Math. Soc. 123, 241–252 (1966).
9. *The C^k-classification of certain operators on L_p*. Trans. Amer. Math. Soc. 132, 323–333 (1968).

10. *Local C^n-operational calculus.* J. Math. Mech. 17, 181–188 (1967).

Kaplansky, I.

7. *A theorem on rings of operators.* Pacific J. Math. 1, 227–232 (1951).

Kariotis, C. A.

1. *Spectral properties of certain classes of operators.* Dissertation, Univ. of Illinois, 1966.

Karlin, S.

3. *Positive operators.* J. Math. Mech. 8, 907–937 (1959).

Kato, T.

9. *Perturbation of continuous spectra by trace class operators.* Proc. Japan Acad. 33, 260–264 (1957).
10. *On finite-dimensional perturbations of self-adjoint operators.* J. Math. Soc. Japan 9, 239–249 (1957).
11. *Perturbation theory for nullity, deficiency and other quantities of linear operators.* J. Analyse Math. 6, 261–322 (1958).
12. *Wave operators and similarity for non-selfadjoint operators.* Math. Ann. 162, 258–279 (1966).
13. *Perturbation theory for linear operators.* Springer-Verlag, New York, 1966.
14. *Wave operators and unitary equivalence.* Pacific J. Math. 15, 171–180 (1965).

Keldyš, M. V.

1. *On the characteristic values and characteristic functions of certain classes of non-self-adjoint equations.* Doklady Akad. Nauk SSSR (N. S.) 77, 11–14 (1951). (Russian) Math. Rev. 12, 835 (1951).

Keldyš, M., and Lidskiĭ, V. B.

1. *On the spectral theory of non-selfadjoint operators.* Proc. Fourth All-Union Math. Congr. (Leningrad, 1961), Vol. I, pp. 101-120. Izdat. Akad. Nauk SSSR, Leningrad, 1963. (Russian) Math. Rev. 30 #1414, 278 (1965).

Kelley, J. L., and Namioka, I., *et al.*

1. *Linear topological spaces.* D. van Nostrand Inc., Princeton, 1963.

Kemp, R. R. D.

1. *On a class of singular differential operators.* Canadian J. Math. 13, 316–330 (1961).

Kesel'man, G. M.

1. *On the single-valued analytic continuability of the resolvent of a bounded linear operator.* Uspehi Mat. Nauk 17, no. 4 (106), 135–139 (1962). (Russian) Math. Rev. 26 #600, 116 (1963).
2. *On the structure of a non-selfadjoint second-order differential operator on a semi-axis.* Dopovidi Akad. Nauk Ukraïn. RSR 1963, 588–591. (Ukrainian. Russian and English summaries). Math. Rev. 30 #3386, 642 (1965).

Kilpi, Y.

2. *Über die Anzahl der hypermaximalen normalen Fortsetzungen normaler Transformationen.* Ann. Univ. Turku. Ser. AI No. 65 (1963), 12 pp.

Kisilevskiĭ, G. È. (see also Brodskiĭ, M. S.)

1. *Conditions for unicellularity of dissipative Volterra operators with finite-dimensional imaginary component.* Doklady Akad. Nauk SSSR 159, 505–508 (1964). (Russian) Math. Rev. 30 # 5162, 960 (1965).
2. *On the order of the characteristic matrix functions of dissipative Volterra operators.* Doklady Akad. Nauk SSSR 159, 730–733 (1964). (Russian) Math. Rev. 30 #5163, 960 (1965).

Kleinecke, D. C.

4. *On operator commutators.* Proc. Amer. Math. Soc. 8, 535–536 (1957).
5. *Almost-finite, compact, and inessential operators.* Proc. Amer. Math. Soc. 14, 863–868 (1963).

Kluvánek, I.

1. *Characterization of scalar-type spectral operators.* Arch. Math. (Brno) 2, 153–156 (1966).
2. *Characterization of Fourier-Stieltjes transformations of vector and operator valued measures.* Czech. Math. J. 17 (92), 261–277 (1967).

Kluvánek, I., and Kovářiková, M.

1. *Product of spectral operators.* Czech. Math. J. 17 (92), 248–256 (1967). (Russian summary)

Kocan, D.

1. *Spectral manifolds for a class of operators.* Illinois J. Math. 10, 605–622 (1966).

Komatsu, H.

1. *Semi-groups of operators in locally convex spaces.* J. Math. Soc. Japan 16, 230–262 (1964).

Konno, R., and Kuroda, S.-T.

1. *On the finiteness of perturbed eigenvalues.* J. Fac. Sci. Univ. Tokyo Sect. I 13, 55–63 (1966).

Koppelman, Walter

1. *Spectral multiplicity theory for a class of singular integral operators.* Trans. Amer. Math. Soc. 113, 87–100 (1964).
2. *On the spectral theory of singular integral operators.* Trans. Amer. Math. Soc. 97, 35–63 (1960).

Koppelman, W., and Pincus, J. D.

1. *Spectral representation for finite Hilbert transforms.* Math. Zeit. 71, 399–407 (1959).

Korotkov, V. B. (see also Gil'derman, Ju. I.)

1. *Abstract set functions and imbedding theorems.* Doklady Akad. Nauk SSSR 146, 531–534 (1962). (Russian) Math. Rev. 26 #1747, 338 (1963).

Köthe, G.

12. *Topologische lineare Räume*, I. Springer-Verlag, Berlin, 1960.
13. *General linear transformations of locally convex spaces.* Math. Ann. 159, 309–328 (1965).

Kovářiková, M. (see also Kluvánek, I.)

1. *On the polar decomposition of scalar-type operators.* Czech. Math. J. 17 (92), 313–316 (1967). (Russian summary).

Krabbe, G. L.

1. *On the logarithm of a uniformly bounded operator.* Trans. Amer. Math. Soc. 81, 155–166 (1956).
2. *On the spectra of certain Laurent matrices.* Proc. Amer. Math. Soc. 8, 894–897 (1957).
3. *Convolution operators which are not of scalar type.* Math Zeit. 69, 346–350 (1958).
4. *Spectral invariance of convolution operators on $L^p(-\infty, \infty)$.* Duke Math. J. 25, 131–141 (1958).
5. *Spectral isomorphisms for some rings of infinite matrices on a Banach space.* Amer. J. Math. 78, 42–50 (1956).
6. *Convolution operators that satisfy the spectral theorem.* Math. Zeit. 70, 446–462 (1959).
7. *Vaguely normal operators on a Banach space.* Arch. Rat. Mech. Anal. 3, 51–59 (1959).
8. *Normal operators on the Banach spaces $L^p(-\infty, \infty)$, I, II.*
   I. *Bounded operators.* Bull. Amer. Math. Soc. 65, 270–272 (1959).
   II. *Unbounded transformations.* Bull. Amer. Math. Soc. 66, 86–90 (1960).
9. *Normal operators on the Banach space $L^p(-\infty, \infty)$, I, II.*
   I. Canadian J. Math. 13, 505–518 (1961).
   II. *Unbounded operators.* J. Math. Mech. 10, 111–133 (1961).
10. *Integration with respect to operator-valued functions.* Bull. Amer. Math. Soc. 67, 214–218 (1961).
11. *Réfractions non-hilbertiennes d'une transformation symétrique bornée.* Studia Math. 20, 347–357 (1961).
12. *Sur la permanence spectrale.* C. R. Acad. Sci. Paris 255, 1326–1328 (1962).
13. *Spectrale permanence of scalar operators.* Pacific J. Math 13, 1289–1303 (1963).
14. *Generalized measures whose values are operators into an intermediate space.* Bull. Amer. Math. Soc. 68, 42–46 (1962).
15. *Generalized measures whose values are operators into an intermediate space.* Math. Ann. 151, 219–238 (1963).
16. *Stieltjes integration, spectral analysis, and the locally-convex algebra (BV).* Bull. Amer. Math. Soc. 71, 184–189 (1965).

Kračkovskiĭ, S. N. (see Gol'dman, M. A.)

Kramer, H. P.

1. *Perturbation of differential operators.* Dissertation, Univ. of California, Berkeley, 1954.
2. *Perturbation of differential operators.* Pacific J. Math. 7, 1405–1435 (1957).

Kramer, V. A. (see Gilbert, R. C.)

Krasnosel'skiĭ, M. A. (see also Bahtin, I. A., Kreĭn, M. G., and Zabreĭko, P. P.
5. *Positive solutions of operator equations*. Gosudarstv. Izdat. Fiz.-Math. Lit., Moscow, 1962.

Kreĭn, M. G. (see also Birman, M. Š., Brodskiĭ, M. S., Gohberg, I. C., and Iohvidov, I. S.)
8. *On the trace formula in perturbation theory*. Mat. Sbornik N. S. 33 (75), 597–626 (1953). (Russian) Math. Rev. 15, 720 (1954).
22. *Criteria for completeness of the system of root vectors of a dissipative operator*. Uspehi Mat. Nauk 14, no. 3 (87), 145–152 (1959). (Russian) Math. Rev. 22 #9856 (1961). Amer. Math. Soc. Transl. (2) 26, 221–229.
23. *A contribution to the theory of linear non-selfadjoint operators*. Doklady Akad. Nauk SSSR 130, 254–256 (1960). (Russian) Math. Rev. 24 #A1024, 189 (1962). Soviet Math. Dokl. 1, 38–40 (1960).
24. *On perturbation determinants and a trace formula for unitary and selfadjoint operators*. Doklady Akad. Nauk SSSR 144, 268-271 (1962). (Russian) Math. Rev. 25 #2446, 475 (1963). Soviet Math. Dokl. 3, 707 (1962).
25. *A new application of the fixed-point principle in the theory of operators in a space with indefinite metric*. Doklady Akad. Nauk SSSR 154, 1023–1026 (1964). (Russian) Math. Rev. 29 #6314, 1183 (1965).
26. *Integral equations on a half-line with kernel depending on the difference of the arguments*. Uspehi Mat. Nauk 13, no. 5 (83), 3–120 (1958). (Russian) Math. Rev. 21 #1507, 287 (1960) Amer. Math. Soc. Transl. (2) 22, 163–288.
27. *Some new studies in the theory of perturbations of selfadjoint operators*. First Math. Summer School, Part I, pp. 103–187. Izdat. "Naukova Dumka," Kiev, 1964. (Russian) Math. Rev. 32 #2919, 494 (1966).
28. *Introduction to the geometry of indefinite J-spaces and to the theory of operators in those spaces*. Second Math. Summer School, Part I, pp. 15–92. Izdat. "Naukova Dumka," Kiev, 1965. (Russian) Math. Rev. 33 #574, 102 (1967).

Kreĭn, M. G., and Krasnosel'skiĭ, M. A.
1. *Stability of the index of an unbounded operator*. Mat. Sbornik N. S. 30 (72) 219–224 (1952). (Russian) Math. Rev. 13, 849 (1952).

Kreĭn, M. G., and Langer, G. K. [Heinz]
1. *On the spectral function of a self-adjoint operator in a space with indefinite metric*. Doklady Akad. Nauk SSSR 152, 39–42 (1963). (Russian) Math. Rev. 29 #2650, 514 (1965).
2. *On the theory of quadratic pencils of self-adjoint operators*. Doklady Akad. Nauk SSSR 154, 1258–1261 (1964). (Russian) Math. Rev. 29 #6315, 1183 (1965).

Kreĭn, M. G., and Rutman, M. A.
1. *Linear operators leaving invariant a cone in a Banach space*. Uspehi Mat. Nauk 3, no. 1 (23), 3–95 (1948). (Russian) Amer. Math. Soc. Transl. 26 (1950).

Kreĭn, M. G., and Šmul'jan, Ju. L.
1. *A class of operators in a space with an indefinite metric*. Doklady Akad. Nauk SSSR 170, 34–37 (1966). (Russian) Math. Rev. 34 #615, 105 (1967). Soviet Math. Dokl. 7, 1137–1141 (1966).

Kreĭn, S. G. (editor) (see also Askerov, N. G.)

1. *Functional analysis*. Izdat. "Nauka," Moscow, 1964.

Kultze, R.

1. *Zur Theorie Fredholmscher Endomorphismen in nuklearen topologischen Vektorräumen*. J. Reine Angew. Math. 200, 112–124 (1958).

Kurepa, Svetozar

1. *A note on logarithms of normal operators*. Proc. Amer. Math. Soc. 13, 307–311 (1962).
2. *On n-th roots of normal operators*. Math. Zeit. 78, 285–292 (1962).
3. *On roots of an element of a Banach algebra*. Publ. Inst. Math. (Beograd) (N. S.) 1 (15), 5–10 (1962).
4. *Logarithms of spectral type operators*. Glasnik Mat.-Fiz. Astronom. Ser. II 18, 53–57 (1963).
5. *A theorem about similarity of operators*. Arch. Math. 14, 411–414 (1963).
6. *On operator-roots of an analytic function*. Glasnik Mat.-Fiz. Astronom. Ser. II 18, 49–51 (1963). (Serbo-Croatian summary).

Kuroda, S.-T. (see also Konno, R.)

1. *On a theorem of Weyl-von Neumann*. Proc. Japan Acad. 34, 11–15 (1958).
2. *On the existence and unitary properties of the scattering operator*. Il Nuovo Cimento 12, 431–454 (1959).
3. *Perturbation of continuous spectra by unbounded operators*, I, II.
   I. J. Math. Soc. Japan 11, 246–262 (1959).
   II. ibid. 12, 243–257 (1960).
4. *On a generalization of the Weinstein-Aronszajn formula and the infinite determinant*. Sci. Papers Coll. Gen. Ed. Univ. Tokyo 11, 1–12 (1961).
5. *Finite-dimensional perturbation and a representation of the scattering operator*. Pacific J. Math. 13, 1305–1318 (1963).
6. *On a stationary approach to scattering problems*. Bull. Amer. Math. Soc. 70, 556–560 (1964).
7. *Stationary methods in the theory of scattering*. Perturbation Theory and its Applications in Quantum Mechanics, pp. 185–214. Wiley, New York, 1966.
8. *Perturbation of eigenfunction expansions*. Proc. Nat. Acad. Sci. USA 57, 1213–1217 (1967).
9. *An abstract stationary approach to perturbation of continuous spectra and scattering theory*. J. Analyse Math. 20, 57–117 (1967).

Kužel', A. V. (=O.V.)

1. *Spectral analysis of unbounded non-selfadjoint operators*. Doklady Akad. Nauk SSSR 125, 35–37 (1959). (Russian) Math. Rev. 22 #2905, 493 (1961).
2. *Spectral decomposition of quasi-unitary operators of arbitrary rank in a space with indefinite metric*. Dopovidi Akad. Nauk. Ukraïn. RSR 1963, 430–433. (Ukrainian. Russian and English summaries.) Math. Rev. 29 #1544, 307 (1965).

3. *Spectral analysis of bounded non-selfadjoint operators in a space with indefinite metric.* Doklady Akad. Nauk SSSR 151, 727–774 (1963). (Russian) Math. Rev. 28 #1496, 302 (1964).

Labrousse, J. P. (see Cordes, H.-O.)

Ladyženskaya, O. A., and Faddeev, L. D.

1. *On continuous spectrum perturbation theory.* Doklady Akad. Nauk SSSR 120, 1187–1190 (1958). (Russian) Math. Rev. 21 #1536, 293 (1960).

Langer, Heinz (=G.) (see also Kreĭn, M. G.)

1. *Zur Spektraltheorie J-selbstadjungierter Operatoren.* Math. Ann. 146, 60–85 (1962).
2. *On J-Hermitian operators.* Doklady Akad. Nauk SSSR 134, 263–266 (1960). (Russian) Math. Rev. 25 #1457, 289 (1963).
3. *Eine Verallgemeinerung eines Satzes von L. S. Pontrjagin.* Math. Ann. 152, 434–436 (1963).
4. *Eine Erweiterung der Spurformel der Störungstheorie.* Math. Nachr. 30, 123–135 (1965).
5. *Spektralfunktionen einer Klasse J-selbstadjungierter Operatoren.* Math. Nachr. 33, 107–120 (1967).

Lanier, L. H., Jr.

1. *Semi-groups of spectral operators.* Dissertation, University of Illinois, Urbana, 1963.

Laptev, G. I. (see Askerov, N. G.)

Lavrentieff, M.

1. *Sur les fonctions d'une variable complexe représentables par des séries de polynomes.* Act. Sci. Ind. 441, Paris, 1936.

Lax, P. D., and Phillips, R. S.

1. *Scattering theory.* Bull. Amer. Math. Soc. 70, 130–142 (1964).
2. *Scattering theory.* Academic Press, New York, 1967.
3. *Analytic properties of the Schrödinger scattering matrix.* Perturbation Theory and its Applications in Quantum Mechanics, pp. 243–253. Wiley, New York, 1966.

Lay, D. C. (see Kaashoek, M. A.)

Leaf, G. K.

1. *A spectral theory for a class of linear operators.* Pacific J. Math. 13, 141–155 (1963).
2. *An approximation theorem for a class of operators.* Proc. Amer. Math. Soc. 16, 991–995 (1965). Errata ibid. 18, 1141–1142 (1967).

Lebow, A. (see also Coburn, L. A.)

1. *On von Neumann's theory of spectral sets.* J. Math. Anal. Appl. 7, 64–90 (1963).
2. *A note on normal dilations.* Proc. Amer. Math. Soc. 16, 995–998 (1965).

Levič, E. M. (see Gol'dman, M. A.)

Levin, B. Ja.

1. *Transformations of Fourier and Laplace types of means of solutions of differential equations of second order.* Doklady Akad. Nauk SSSR 106, 187–190 (1956). (Russian) Math. Rev. 18, 35 (1957).

Leżański, T.

1. *The Fredholm theory of linear equations in Banach spaces.* Studia Math. 13, 244–276 (1953).

Lidskiĭ, V. B.

1. *Conditions for completeness of a system of root subspaces for non-selfadjoint operators with discrete spectrum.* Trudy Moskov. Mat. Obšč. 8, 83–120 (1959). (Russian) Math. Rev. 21 #6539, 1219 (1960). Amer. Math. Soc. Transl. (2) 34, 241–281.
2. *Summability of series in terms of the principal vectors of non-selfadjoint operators.* Trudy Moskov. Mat. Obšč. 11, 3–35 (1962). (Russian) Math. Rev. 26 #1760, 340 (1963). Amer. Math. Soc. Transl. (2) 40, 193–228.

Lindenstrauss, J. (see also Bonsall, F. F.)

1. *Extension of compact operators.* Memoirs Amer. Math. Soc. 48, 1964.

Lindenstrauss, J., and Pełczyński, A.

1. *Absolutely summing operators in L_p-spaces and their applications.* Studia Math. 29, 275–326 (1968).

Lions, J. L. (see Foiaș, C.)

Littman, W., Mc Carthy, C., and Riviere, N.

1. *L^p multiplier theorems.* Studia Math. 30, 197–221 (1969).

Livšic, M. S.

6. *On spectral resolutions of linear non-selfadjoint operators.* Mat. Sbornik N. S. 34 (76), 145–199 (1954). (Russian) Math. Rev. 16, 48 (1955). Amer. Math. Soc. Transl. (2) 5, 67–114.
7. *On a class of linear operators in Hilbert space.* Mat. Sbornik N.S. 19 (61), 239–262 (1946). (Russian) Math. Rev. 8, 588 (1947). Amer. Math. Soc. Transl. (2) 13, 61–83.

Ljance, V. È.

1. *Rings of unbounded operators based on a resolution of the identity and their representations.* Doklady Akad. Nauk SSSR 121, 801–804 (1958). (Russian) Math. Rev. 21 #2913, 552 (1960).
2. *A generalization of the concept of spectral operator.* Mat. Sbornik N. S. 61 (103), 80–120 (1963). (Russian) Math. Rev. 27 #2874, 561 (1964). Amer. Math. Soc. Transl. (2) 51, 273–316.
3. *Unbounded operators commuting with the resolution of the identity.* Ukrain. Mat. Ž. 15, 376–384 (1963). (Russian. English summary) Math. Rev. 28 #3327, 652 (1964).
4. *On differential operators with spectral singularities*, I, II. (Russian).
   I. Mat. Sbornik N. S. 64 (106), 521–561 (1964). Amer. Math. Soc. Transl. (2) 60, 185–225.
   II. ibid. 65 (107), 47–103 (1964). Amer. Math. Soc. Transl. (2) 60, 227–283.

5. *Certain properties of idempotent operators*. Teoret. Prikl. Mat. Vyp. 1, 16–22 (1958). (Russian) Math. Rev. 33 #4680, 792 (1967).

6. *Expansion in characteristic functions of a non-selfadjoint differential operator with spectral singularities*. Doklady Akad. Nauk SSSR 149, 256–259 (1963). (Russian) Soviet Math. Dokl. 4, 363–366 (1963).

7. *The inverse problem for a nonselfadjoint operator*. Doklady Akad. Nauk SSSR 166, 30–33 (1966). (Russian) Math. Rev. 33 #4720, 800 (1967). Soviet Math. Dokl. 7, 27–30 (1966).

8. *Expansions in principal functions of an operator with spectral singularities*. Rev. Roumaine Math. Pures Appl. 11, 921–950, 1187–1224 (1966). (Russian).

9. *Nonselfadjoint one-dimensional perturbation of the operator of multiplication by the independent variable*. Doklady Akad. Nauk SSSR 182, 1010–1013 (1968). Math. Rev. 38 #1557, 292 (1969). Soviet Math Dokl. 9, 1241–1244 (1968).

10. *On the perturbation of a continuous spectrum*. Doklady Akad. Nauk SSSR 187, 514–517 (1969). Soviet Math Dokl. 10, 896–899 (1969).

Ljubič, Ju. I. (=Lyubič, Yu. I.)

1. *Almost periodic functions in the spectral analysis of operators*. Doklady Akad. Nauk SSSR 132, 518–520 (1960). (Russian) Soviet Math. Dokl. 1, 593–595 (1960).

2. *On a certain class of operators in Banach space*. Uspehi Mat. Nauk. 20, no. 6 (126), 131–133 (1965). (Russian) Math. Rev. 34 #3328, 601 (1967).

3. *Conservative operators*. Uspehi Mat. Nauk 20, no. 5 (125), 221–225 (1965). (Russian) Math. Rev. 34 #4925, 893 (1967).

Ljubič, Ju. I., and Macaev, V. I.

1. *On the spectral theory of linear operators in a Banach space*. Doklady Akad. Nauk SSSR 131, 21–23 (1960). (Russian) Math. Rev. 22 #3980, 673 (1961). Soviet Math. Dokl. 1, 184–186 (1960).

2. *Operators with separable [decomposable] spectrum*. Mat. Sbornik N. S. 56 (98), 433–468 (1962). Errata, ibid. 71 (113), 287–288 (1966). (Russian) Math. Rev. 25 #2450, 475 (1963). Amer. Math. Soc. Transl. (2) 47, 89–129.

Lloyd, S. P.

1. *On extreme averaging operators*. Proc. Amer. Math. Soc. 14, 305–310 (1963).

Loginov, B. V.

1. *An estimate of precision for the method of perturbations in the case of a bounded unperturbed operator*. Izv. Akad. Nauk UzSSR Ser.-Fiz.-Mat. Nauk 1963, no. 5, 21–25. (Russian. Uzbek summary) Math. Rev. 31 #608, 110 (1966).

Lorch, E. R.

1. *Bicontinuous linear transformations in certain vector spaces*. Bull. Amer. Math. Soc. 45, 564–569 (1939).

2. *On a calculus of operators in reflexive vector spaces*. Trans. Amer. Math. Soc. 45, 217–234 (1939).

7. *The integral representation of weakly almost-periodic transformations in reflexive vector spaces.* Trans. Amer. Math. Soc. 49, 18–40 (1941).
15. *Spectral theory.* Oxford Univ. Press, New York, 1962.

Lotz, H. P.
1. *Über das Spektrum positiver Operatoren.* Math. Zeit. 108, 15–32 (1968).

Lotz, H. P., and Schaefer, H. H.
1. *Über einen Satz von F. Niiro und I. Sawashima.* Math. Zeit. 108, 33–36 (1968).

Lowdenslager, D. (see Helson, H.)

Lumer, G. (see also Halmos, P. R.)
1. *Semi-inner product spaces.* Trans. Amer. Math. Soc. 100, 29–43 (1961).
2. *Spectral operators, Hermitian operators, and bounded groups.* Acta Sci. Math. Szeged 25, 75–85 (1964).
3. *Remarks on n-th roots of operators.* Acta Sci. Math. Szeged 25, 72–74 (1964).

Luxemburg, W. A. J., and Zaanen, A. C.
1. *Compactness of integral operators in Banach function spaces.* Math. Ann. 149, 150–180 (1962/63).

Macaev, V. I. (see also Brodskiĭ, M. S., Kacnel'son, V. È., and Ljubič, Ju. I.)
1. *A class of completely continuous operators.* Doklady Akad. Nauk SSSR 139, 548–551 (1961). (Russian) Math. Rev. 24 #A1617, 297 (1962).
2. *A method of estimation for resolvents of non-selfadjoint operators.* Doklady Akad. Nauk SSSR 154, 1034–1037 (1964). (Russian) Math. Rev. 28 #1495, 302 (1964).
3. *Several theorems on completeness of root subspaces of completely continuous operators.* Doklady Akad. Nauk SSSR 155, 273–276 (1964). (Russian) Math. Rev. 28 #2444, 483 (1964).

McCarthy, C. A. (see also Littman, W.)
1. *The nilpotent part of a spectral operator*, I, II.
   I. Pacific J. Math. 9, 1223–1231 (1959).
   II. ibid. 15, 557–559 (1965).
2. *Commuting Boolean algebras of projections*, I, II.
   I. Pacific J. Math. 11, 295–307 (1961).
   II. Proc. Amer. Math. Soc. 15, 781–787 (1964).
3. c_p. Israel J. Math. 5, 249–271 (1967).

McCarthy, C. A., and Schwartz, J.
1. *On the norm of a finite Boolean algebra of projections, and applications to theorems of Kreiss and Morton.* Comm. Pure Appl. Math. 18, 191–201 (1965).

McCarthy, C. A., and Stampfli, J. G.
1. *On one-parameter groups and semi-groups of operators in Hilbert space.* Acta Sci. Math. Szeged 25, 6–11 (1964).

McCarthy, C. A., and Tzafriri, L.
1. *Projections in $\mathscr{L}_1$ and $\mathscr{L}_\infty$-spaces.* Pacific J. Math. 26, 529–546 (1968).

MacCluer, C. R.
1. *On extreme points of the numerical range of normal operators.* Proc. Amer. Math. Soc. 16, 1183–1184 (1965).

McGarvey, D. C.

1. *Operators commuting with translation by one*, I–III.
   I. *Representation theorems*. J. Math. Anal. Appl. 4, 366–410 (1962).
   II. *Differential operators with periodic coefficients in* $L_p(-\infty, \infty)$. ibid. 11, 564–569 (1965).
   III. *Perturbation results for periodic differential operators*. ibid. 12, 187–234 (1965).

McKelvey, R.

1. *The spectra of minimal self-adjoint extensions of a symmetric operator*. Pacific J. Math. 12, 1003–1022 (1962).
2. *Spectral measures, generalized resolvents, and functions of positive type*. J. Math. Anal. Appl. 11, 447–477 (1965).

Mackey, G. W.

4. *Commutative Banach algebras*. Mimeographed lecture notes, Harvard Univ., 1952.

McLaughlin, J. E. (see Halmos, P. R.)

Maeda, F.-Y.

1. *Spectral theory on locally convex spaces*. Dissertation, Yale Univ., 1961.
2. *A characterization of spectral operators on locally convex spaces*. Math. Ann. 143, 59–74 (1961).
3. *Remarks on spectra of operators on a locally convex space*. Proc. Nat. Acad. Sci. USA 47, 1052–1055 (1961).
4. *Generalized spectral operators on locally convex spaces*. Pacific J. Math. 13, 177–192 (1963).
5. *Function of generalized scalar operators*. J. Sci. Hiroshima Univ. Ser. A-I 26, 71–76 (1962).
6. *On spectral representations of generalized spectral operators*. J. Sci. Hiroshima Univ. Ser. A-I 27, 137–149 (1963).
7. *Generalized unitary operators*. Bull. Amer. Math. Soc. 71, 631–633 (1965).
8. *Generalized scalar operators whose spectra are contained in a Jordan curve*. Illinois J. Math. 10, 431–459 (1966).

Maltese, G.

1. *Spectral representations for solutions of certain abstract functional equations*. Compositio Math. 15, 1–22 (1961).
2. *Spectral representations for some unbounded normal operators*. Trans. Amer. Math. Soc. 110, 79–87 (1964).

Marčenko, V. A.

3. *Expansion in eigenfunctions of a non-self-adjoint singular differential operator of second order*. Mat. Sbornik N. S. 52 (94), 739–788 (1960). (Russian) Math. Rev. 23 #A3313, 630 (1962). Amer. Math. Soc. Transl. (2) 25, 77–130.

Marčenko, V. A., and Rofe-Beketov, F. S.

1. *Expansion in characteristic functions of non-self-adjoint singular differential operators*. Doklady Akad. Nauk SSSR 120, 963–966 (1958). (Russian) Math. Rev. 21 #2092, 400 (1960).

Marek, Ivo

1. *On some spectral properties of Radon-Nicolski operators and their generalizations.* Comment. Math. Univ. Carolinae 3, 20–30 (1962).
2. *A note on K-positive operators.* Comment. Math. Univ. Carolinae 4, 137–146 (1963).
3. *On the minimax principle for K-positive operators.* Comment. Math. Univ. Carolinae 7, 109–112 (1966).

Markus, A. S. (see also Gohberg, I. C.)

1. *The root-vector expansion of a weakly perturbed self-adjoint operator.* Doklady Akad. Nauk SSSR 142, 538–541 (1962). (Russian) Math. Rev. 27 #1837, 361 (1964).
2. *Eigenvalues and singular values of the sum and product of linear operators.* Uspehi Mat. Nauk 19, no. 4 (118), 93–123 (1964). (Russian) Math. Rev. 29 #6318, 1184 (1965).
3. *Some tests of completeness of the system of root vectors of a linear operator and the summability of series in that system.* Doklady Akad. Nauk SSSR 155, 753–756 (1964). (Russian) Math. Rev. 28 #5345, 1034 (1964).
4. *Certain criteria for the completeness of a system of root-vectors of a linear operator in a Banach space.* Mat. Sbornik N.S. 70 (112), 526–561 (1966). (Russian).

Martirosjan, R. M.

1. *Defect indices and the spectrum of certain linear operators.* Akad. Nauk Armjan. SSR Dokl. 34, 49–55 (1962). (Russian. Armenian summary) Math. Rev. 27 #6126, 1170 (1964).
2. *On the invariance of the spectrum of small perturbations of the polyharmonic operator.* Izv. Akad. Nauk SSSR Ser. Mat. 28, 79–90 (1964). (Russian) Math. Rev. 28 #3343, 655 (1964).
3. *A method of studying the spectrum of perturbations of self-adjoint operators.* Akad. Nauk Armjan. SSR Dokl. 41, 257–263 (1965). (Russian, Armenian summary) Math. Rev. 32 #8175, 1408 (1966).
4. *The spectrum of certain nonselfadjoint perturbations of selfadjoint differential operators.* Izv. Akad. Nauk. Armjan. SSR Ser. Mat. 1, no. 3, 192–216 (1966). (Russian, Armenian and English summaries) Math. Rev. 34 #4952, 898 (1967).
5. *On the spectra of some non-self-adjoint operators.* Izv. Akad. Nauk SSSR Ser. Mat. 27, 677–700 (1963). (Russian) Math. Rev. 27 #378, 79 (1964). Amer. Math. Soc. Transl. (2) 53, 81–103.

Maurin, Krzysztof

1. *Allgemeine Eigenfunktionsentwicklungen. Spektraldarstellung abstrakter Kerne. Eine Verallgemeinerung der Distributionen auf Lie'schen Gruppen.* Bull. Acad. Polon. Sci. 7, 471–479 (1959).
2. *Eine Bemerkung zur allgemeinen Eigenfunktionsentwicklungen für vertauschbare Operatorensysteme beliebiger Mächtigkeit.* Bull. Acad. Polon. Sci. 8, 381–384 (1960). (Russian summary).

3. *Spektraldarstellung der Kerne. Eine Verallgemeinerung der Sätze von Källén-Lehmann und Herglotz-Bochner u. a.* Bull. Acad. Polon. Sci. 7, 461–470 (1959). (Russian summary).
4. *Abbildungen vom Hilbert-Schmidtschen Typus und ihre Anwendungen.* Math. Scand. 9, 359–371 (1961).
5. *Mappings of Hilbert-Schmidt-type. Their application to eigenfunction expansions and elliptic boundary problems.* Bull. Acad. Polon. Sci. 9, 7–11 (1961). (Russian summary).
6. *General eigenfunction expansions and unitary representations of topological groups.* Deuxième Colloq. d'Anal. Fonct. pp. 49–55. Louvain, 1964.
7. *Methods of Hilbert spaces.* Monografie Matematyczne 45, Warszawa, 1967.

Maurin, Lidia, and Maurin, Krzysztof
1. *Spektraltheorie separierbarer Operatoren.* Studia Math. 23, 1–29 (1963).
2. *Nuklearität gewisser Rellich-Sobolevschen Einbettungen. Anwendung auf Spektraltheorie der Differentialoperatoren.* Bull. Acad. Polon. Sci. 8, 621–624 (1960). (Russian summary).

Mejlbo, L. C.
1. *On the solution of the commutation relation* $PQ - QP = -iI$. Math. Scand. 13, 129–139 (1963).

Mergelyan, S. N.
1. *On the representation of functions by series of polynomials on closed sets.* Doklady Akad. Nauk SSSR (N.S.) 78, 405–408 (1951). (Russian) Math. Rev. 13, 23 (1952). Amer. Math. Soc. Translation, no. 85.

Mewborn, A. C.
1. *Generalizations of some theorems on positive matrices to completely continuous linear transformations on a normed linear space.* Duke Math. J. 27, 273–281 (1960).

Miller, J. B. (see also Gamlen, J. L. B.)
1. *Some properties of Baxter operators.* Acta Math. Acad. Sci. Hungar. 17, 387–400 (1966).
2. *Averaging and Reynolds operators on Banach algebras.*
   I. *Representation by derivations and antiderivations.* J. Math. Anal. Appl. 14, 527–548 (1966).

Milne, W. E.
2. *On the degree of convergence of expansions in an infinite interval.* Trans. Amer. Math. Soc. 31, 906–918 (1929).

Mimura, Y.
1. *Über Funktionen von Funktionaloperatoren in einem Hilbertschen Raum.* Jap. J. Math. 13, 119–128 (1936).

Mishoe, L. I., and Ford, G. C.
1. *Studies in the eigenfunction series associated with a non-self-adjoint differential system.* Tech. Report, Nat. Sci. Foundation, 1955.

Misra, B. (see Guenin, M., and Jauch, J. M.)

Mitjagin (Mitiagin), B. S., and Pełczyński, A.

1. *Nuclear operators and approximative dimension.* Proc. International Congress Math., Moscow 1966, pp. 366–372. (Also Amer. Math. Soc. Transl. (2) 70, 137–145.)

Miyadera, I.

2. *On perturbation theory for semi-groups of operators.* Tôhoku Math. J. (2) 18, 299–310 (1966).

Mkrtčjan, R. Z. (see Aleksandrjan, R. A.)

Mlak, W.

1. *Characterization of completely non-unitary contractions in Hilbert spaces.* Bull. Acad. Polon. Sci. 11, 111–113 (1963).
2. *Note on the unitary dilation of a contraction operator.* Bull. Acad. Polon. Sci. 11, 463–467 (1963).
3. *Some prediction theoretical properties of unitary dilations.* Bull. Acad. Polon. Sci. 12, 37–42 (1964).
4. *Unitary dilations of contraction operators.* Rozprawy Mat. 46 (1965), 91 pp.
5. *On semi-groups of contractions in Hilbert space.* Studia Math. 26, 263–272 (1966).

Mochizuki, K.

1. *On the large perturbation by a class of non-selfadjoint operators.* J. Math. Soc. Japan 19, 123–158 (1967).

Møller, C.

1. *General properties of the characteristic matrix in the theory of elementary particles,* I, II.
   I. Danske Vid. Selsk. Mat.-Fys. Medd. 21, no. 1 (1945).
   II. ibid. 22, no. 19 (1946).

Moser, J.

1. *Störungstheorie des kontinuierlichen Spektrums für gewöhnliche Differentialgleichungen zweiter Ordnung.* Math. Ann. 125, 366–393 (1953).
2. *Singular perturbation of eigenvalue problems for linear differential equations of even order.* Comm. Pure Appl. Math. 8, 251–278 (1955).

Moy, S.-T. C.

1. *Characterizations of conditional expectation as a transformation on function spaces.* Pacific J. Math. 4, 47–63 (1954).

Moyal, J. E.

1. *The theory of spectral and scalar algebras.* (to appear).

Murray, F. J.

3. *The analysis of linear transformations.* Bull. Amer. Math. Soc. 48, 76–93 (1942).

Muskhelishvili, N. I.

1. *Singular integral equations.* Moscow, 1946. Revised translation (ed. by J. R. M. Radok), P. Noordhoff N. V., Groningen, Holland, 1953.

Naĭmark, M. A. (see Neumark, M. A.)

Nakamura, M., and Yoshida, M.

1. *On Bückner's inclusion theorems for Hermitean operators.* Proc. Japan Acad. 40, 180–182 (1964).

Nakano, H.

8. *Über Abelsche Ringe von Projektionsoperatoren.* Proc. Phys.-Math. Soc. Japan (3) 21, 357–375 (1939).
9. *Unitärinvariante hypermaximale normale Operatoren.* Ann. of Math. (2) 42, 657–664 (1941)
12. *Modern spectral theory.* Maruzen Co., Tokyo, 1950.
13. *Spectral theory in the Hilbert space.* Japan Soc. for Promotion of Sci., Tokyo, 1953.
19. *On unitary dilations of bounded operators.* Acta Sci. Math. Szeged 22, 286–288 (1961).

Nel, L. D.

1. *A characterization of unbounded spectral operators.* J. London Math. Soc. 37, 317–319 (1962).

Nelson, E.

1. *Kernel functions and eigenfunction expansions.* Duke Math. J. 25, 15–27 (1958). ibid. 26, 697–698 (1959).

Neubauer, G.

1. *Zur Spektraltheorie in lokalkonvexen Algebren,* I, II.
   I. Math. Ann. 142, 131–164 (1960/61).
   II. ibid. 143, 251–263 (1961).
2. *Zu einem Satz von N. Dunford.* Arch. Math. 11, 366–367 (1960).
3. *Über den Index abgeschlossener Operatoren in Banachräumen,* I, II.
   I. Math. Ann. 160, 93–130 (1965).
   II. ibid. 162, 92–119 (1965/66).

Neugebauer, O.

1. *Astronomical cuneiform texts,* (3 vols.). Lund Humphries, London, 1955.
2. *The exact sciences in antiquity.* Brown University Press, 1957. Dover edition, 1969.

von Neumann, J.

6. *Charakterisierung des Spektrums eines Integraloperators.* Act. Sci. et Ind. 229, Paris, 1935.
24. *Approximative properties of matrices of high finite order.* Portugaliae Math. 3, 1–62 (1942).

Neumark (Naĭmark), M. A. (see also Gelfand, I. M.)

5. *Linear differential operators.* Gosudarstv. Izdat. Tehn.-Teo. Lit., Moscow, 1954.
10. *Investigation of the spectrum and expansion in eigenfunctions of singular non-self-adjoint differential operators of the second order.* Uspehi Matem. Nauk (N. S.) 8, no. 4 (56), 174–175 (1953). (Russian) Math. Rev. 15, 530 (1954).

11. *On expansion in characteristic functions of non-self-adjoint singular differential operators of the second order*. Doklady Akad. Nauk SSSR (N. S.) 89, 213–216 (1953). (Russian) Math. Rev. 15, 33 (1954).
12. *Investigation of the spectrum and the expansion in eigenfunctions of a non: self-adjoint operator of the second order on a semi-axis*. Trudy Moskov. Mat. Obšč. 3, 181–270 (1954). (Russian) Math. Rev. 15, 959 (1954). Amer. Math. Soc. Transl. (2) 16, 103–193 (1960).
13. *Normed rings*. Gosudarstv. Izdat. Tehn.-Teor. Lit., Moscow, 1956. English transl., Nordhoff, Groningen, 1959.
14. *Linear representation of the Lorentz group*. Uspehi Matem. Nauk (N. S.) 9, no. 4 (62), 19–93 (1954). (Russian) Amer. Math. Soc. Transl. (2) 6, 379–458 (1957).
15. *Spectral analysis of non-self-adjoint operators*. Uspehi Matem. Nauk (N. S.) 11, no. 6 (72), 183–202 (1956). (Russian) Math. Rev. 21 #3645, 677 (1960). Amer. Math. Soc. Transl. (2) 20, 55–75 (1962).
16. *On commuting unitary operators in spaces with indefinite metric*. Acta Sci. Math. Szeged 24, 177–189 (1963).

Nevanlinna, F., and Nieminen, T.

1. *Das Poisson-Stieltjes'sche Integral und seine Anwendung in der Spektral theorie des Hilbert'schen Raumes*. Ann. Acad. Sci. Fenn. Ser. AI no. 207, 1–38 (1955).

Newberger, S. M.

1. *The σ-symbol of the singular integral operators of Calderón and Zygmund*. Illinois J. Math. 9, 428–443 (1965).

Newburgh, J. D.

1. *The variation of spectra*. Duke Math. J. 18, 165–176 (1951).
2. *A topology for closed operators*. Ann. of Math. (2) 53, 250–255 (1957).

Nieminen, T. (see also Nevanlinna, F.)

1. *A condition for the self-adjointness of a linear operator*. Ann. Acad. Sci. Fenn. Ser. AI no. 316, 5pp. (1962).

Niiro, F.

1. *On indecomposable operators in $l_p(1 < p < \infty)$ and a problem of H. Schaefer*. Sci. Papers College Gen. Ed. Univ. Tokyo 14, 165–179 (1964).

Niiro, F., and Sawashima, I.

1. *On the spectral properties of positive irreducible operators in an arbitrary Banach lattice and problems of H. H. Schaefer*. Sci. Papers College Gen. Ed. Univ. Tokyo 16, 145–183 (1966).

Nikol'skiĭ, N. K.

1. *Invariant subspaces of unitary operators*. Vestnik Leningrad Univ. 21, no. 19, 36–43 (1966). (Russian. English summary) Math. Rev. 34 #4910, 890 (1967).

Nižnik, L. P.

1. *Spectral structure and the self-adjointness of perturbations of differential operators with constant coefficients*. Ukrain. Mat. Ž. 15, 385–399 (1963). (Russian. English summary) Math. Rev. 28 #1151, 305 (1964).

Novosel'skiĭ, I. A.

1. *On certain criteria of completeness of the system of root vectors of a completely continuous operator.* Bul. Akad. Štiince RSS Moldoven. 1965, no. 7, 47–54. (Russian. Moldavian summary) Math. Rev. 33 #7859, 1366 (1967).

Oberai, K. K.

1. *Sum and product of commuting spectral operators.* Pacific J. Math. 25, 129–146 (1968).
2. *Spectrum of a spectral operator.* Proc. Amer. Math. Soc. 19, 325–331 (1968).
3. *Spectral interpolation in L_p spaces.* Math. Zeit. 103, 122–128 (1968).
4. *Nilpotency and the rate of growth condition.* Math. Ann. 184, 233–287 (1970).
5. *On spectral permanence.* J. Math. Mech. 18, 553–558 (1968).

Olagunju, P. A., and West, T. T.

1. *The spectra of Fredholm operators in locally convex spaces.* Proc. Cambridge Philos. Soc. 60, 801–806 (1964).

Orland, G. H.

1. *On a class of operators.* Proc. Amer. Math. Soc. 15, 75–79 (1964).
2. *On some theorems of Bram for subnormal operators.* Amer. Math. Monthly 73, 377–378 (1966).

Osborn, J. E.

1. *Approximation of the eigenvalues of nonself-adjoint operators.* J. Math. and Phys. 45, 391–401 (1966).

Osher, S. J.

1. *Two papers on the similarity of certain Volterra integral operators.* Memoirs Amer. Math. Soc. 73, 1967.

Palant, Ju. A.

1. *A test for the completeness of a system of eigenvectors and adjoint vectors of a polynomial bundle of operators.* Doklady Akad. Nauk SSSR 141, 558–560 (1961). (Russian) Math. Rev. 26 #1737, 336 (1963).

Panchapagesan, T. V.

1. *Unitary operators in Banach spaces.* Pacific J. Math 22, 465–475 (1967).

Paraska, V. I.

1. *A metric in the space of linear closed operators and its application in perturbation theory.* Mat. Issled. 2, no. 1, 45–66 (1967). (Russian) Math. Rev. 35 #793, 154 (1968).

Pavlov, B. S.

1. *The non-selfadjoint operator $-y'' + q(x)y$ on a half-line.* Doklady Akad. Nauk SSSR 141, 807–810 (1961). (Russian) Math. Rev. 24 #A2085, 383 (1962). Soviet Math. Dokl. 2, 1565–1568 (1961).
2. *On the spectral theory of non-self-adjoint differential operators.* Doklady Akad. Nauk SSSR 146, 807–810 (1961). (Russian) Math. Rev. 28 #1341, 268 (1964). Soviet Math. Dokl. 3, 1483–1487 (1962).

Pearcy, C. (see Brown, A., Deckard, D.)

Pedersen, N. W.

1. *The resolutions of the identity for sums and products of commuting spectral operators.* Math. Scand. 11, 123–130 (1962).

Pełczyński, A. (see also Lindenstrauss, J., and Mitjagin, B. S.)

1. *A characterization of Hilbert-Schmidt operators.* Studia Math. 28, 355–360 (1967).
2. *Proof of Grothendieck's theorem on the characterization of nuclear spaces.* Prace Mat. 7, 155–167 (1962). (Russian) Math. Rev. 34 #8151, 1499 (1967).
3. *On strictly singular and strictly cosingular operators*, I, II.
   I. *Strictly singular and strictly cosingular operators in $C(S)$-spaces.* Bull. Acad. Polon. Sci. 13, 31–36 (1965).
   II. *Strictly singular and strictly cosingular operators in* $L(\nu)$-spaces. ibid. 13, 37–41 (1965).

Penzlin, F. (see Dolph, C. L.)

Peressini, A. L.

1. *Ordered topological vector spaces.* Harper and Row, New York, 1967.

Peressini, A. L. and Sherbert, D. R.

1. *Multiplicative operators and substochastic matrices.* J. London Math. Soc. 41, 605–611 (1966).
2. *Order properties of linear mappings on sequence spaces.* Math. Ann. 165, 318–332 (1966).
3. *Ordered topological tensor products.* Proc. Lond. Math. Soc. (3) 19, 177–190 (1969).

Pettineo, B.

1. *Equazioni funzionali negli spazi di Hilbert e teoria fredholmiana.* Atti. Accad. Sci. Lett. Arti Palermo Parte I (4) 20 (1959/60), 117–185 (1961).
2. *Teoreme dell'alternativa e teoria fredholmiana per talune equazioni negli spazi hilbertiani.* Celebrazioni Archimedee del Sec. XX (Siracusa, 1961), Vol. II, 91–105. Edizioni "Oderisi", Gubbio, 1962.

Pflüger, A.

1. *Verallgemeinerte Poisson-Stieltjes'sche Integraldarstellung und kontraktive Operatoren.* Ann. Acad. Sci. Fenn. Ser. AI No. 336/13 (1963), 14 pp.

Phelps, R. R. (see also Bonsall, F. F.)

1. *Extreme positive operators and homomorphisms.* Trans. Amer. Math. Soc. 108, 265–274 (1963).

Phillips, R. S. (see Hille, E., and Lax, P. D.)

Pietsch, A.

1. *Zur Theorie der σ-Transformationen in lokalkonvexen Vektorräumen.* Math. Nachr. 21, 347–369 (1960).
2. *Homomorphismen in lokalkonvexen Vektorräumen.* Math. Nachr. 22, 162–174 (1960).
3. *Unstetige lineare Abbildungen in lokalkonvexen Vektorräumen.* Math. Ann. 140, 153–164 (1960).

4. *Ein verallgemeinertes Spektralproblem für kompakte lineare Abbildungen in lokalkonvexen Vektorräumen.* Math. Ann. 140, 147–152 (1960).
5. *Quasi-präkompakte Endomorphismen und ein Ergodensatz in lokalkonvexen Vektorräumen.* J. Reine Angew. Math. 207, 16–30 (1961).
6. *Zur Fredholmschen Theorie in lokalkonvexen Räumen.* Studia Math. 22, 161–179 (1962/63).
7. *Einige neue Klassen von kompakten linearen Abbildungen.* Rev. Roumaine Math. Pures Appl. 8, 427–447 (1963).
8. *Eine neue Charakterisierung der nuklearen lokalkonvexen Räume,* I, II.
   I. Math. Nachr. 25, 31–36 (1963).
   II. ibid. 25, 49–58 (1963).
9. *Absolut summierende Abbildungen in lokalkonvexen Räumen.* Math. Nachr. 27, 77–103 (1963).
10. *Nukleare lokalkonvexe Räume.* Akad.-Verlag, Berlin, 1965.
11. *Über die Erzeugung von (F)-Räumen durch selbstadjungierte Operatoren.* Math. Ann. 164, 219–224 (1966).
12. *Absolut p-summierende Abbildungen in normierten Räumen.* Studia Math. 28, 333–353 (1967).

Pincus, J. D. (see also Koppelman, W.)
1. *On the spectral theory of singular integral operators.* Trans. Amer. Math. Soc. 113, 101–128 (1964).
2. *Commutators, generalized eigenfunction expansions and singular integral operators.* Trans. Amer. Math. Soc. 121, 358–377 (1966).
3. *On the explicit construction of generalized eigenfunction expansions.* Brookhaven Nat. Lab., Appl. Math. Report 343, April 1964.
4. *Commutators and systems of singular integral equations,* I. Brookhaven Nat. Lab., Appl. Math. Report 499, April 1967.
5. *The spectral theory for self-adjoint Wiener-Hopf operators.* Bull. Amer. Math. Soc. 72, 882–887 (1966).

Pincus, J. D., and Rovnyak, J.
1. *A spectral theory for some unbounded self-adjoint singular integral operators.* Brookhaven Nat. Lab., Appl. Math. Report 509, 1967.

Plafker, S. (see also Ionescu Tulcea, C.)
1. *Spectral representations for a general class of operators on a locally convex space.* Illinois J. Math. 13, 573–582 (1969).
2. *Generalized subscalar operators on Banach spaces.* J. Math. Anal. Appl. 24, 345–361 (1968).

Plesner, A. I.
3. *Spectral theory of linear operators.* Izdat. "Nauka", Moscow 1965, (Russian) Math. Rev. 33 #3106, 537 (1967).

Poincaré, H.
2. *Sur les équations de la physique mathématique.* Rend. Circ. Mat. Palermo 8, 57–156 (1894).

Porath, G.

1. *Störungstheorie für abgeschlossene lineare Transformationen im Banachschen Raum.* Math. Nachr. 17, 62–72 (1958).

Povzner, A.

1. *On some applications of a class of Hilbert spaces of functions.* Doklady Akad. Nauk SSSR (N. S.) 74, 13–16 (1950). (Russian) Math. Rev. 12, 343 (1951).
9. *On the expansion of arbitrary functions in terms of the eigenfunctions of the operator* $-\Delta u + cu$. Mat. Sbornik N. S. 32 (74), 109–156 (1953) (Russian). Math. Rev. 14, 755 (1953).

Przeworska-Rolewicz, D., and Rolewicz, S.

1. *On operators with a finite d-characteristic.* Studia Math. 24, 257–270 (1964).
2. *On operators preserving a conjugate space.* Studia Math. 25, 245–249 (1964/65)
3. *Remarks on* Φ*-operators in linear topological spaces.* Prace Mat. 9, 91–94 (1965).
4. *On quasi-Fredholm ideals.* Studia Math. 26, 67–71 (1965).
5. *Equations in linear spaces.* Monografie Matematyczne 47, Warszawa, 1968.

Pustyl'nik, E. I.

1. *On the convergence of series of eigenfunctions of a completely continuous operator in Banach spaces.* Sibirsk. Mat. Ž. 4 705–708 (1963). (Russian) Math. Rev. 28 #475, 99 (1964).

Putnam, C. R.

19. *On square roots of normal operators.* Proc. Amer. Math. Soc. 8, 768–769 (1957).
20. *On bounded matrices with non-negative elements.* Canadian J. Math. 10, 587–591 (1958).
21. *On square roots and logarithms of self-adjoint operators.* Proc. Glasgow Math. Assoc. 4, 1–2 (1958).
22. *Commutators, perturbations, and unitary spectra.* Acta Math. 106, 215–232 (1961).
23. *A note on non-negative matrices.* Canadian J. Math. 13, 59–62 (1961).
24. *Absolute continuity of certain unitary and half-scattering operators.* Proc. Amer. Math. Soc. 13, 844–846 (1962).
25. *Positive matrices and eigenvectors.* Proc. Glasgow Math. Assoc. 6, 27–30 (1963).
26. *On the structure of semi-normal operators.* Bull. Amer. Math. Soc. 69, 818–819 (1963).
27. *Continuous spectra and unitary equivalence.* Pacific J. Math. 7, 993–995 (1957).
28. *On differences of unitarily equivalent self-adjoint operators.* Proc. Glasgow Math. Assoc. 4, 103–107 (1960).
29. *On the spectra of unitary half-scattering operators.* Quart. Appl. Math. 20, 85–88 (1962/63).
30. *On the spectra of semi-normal operators.* Trans. Amer. Math. Soc. 119, 509–523 (1965).
31. *Commutation properties of Hilbert space operators and related topics.* Ergebnisse der Math., Band 36, Springer-Verlag, Berlin, 1967.

32. *Perturbations of bounded operators.* Nieuw Arch. Wisk. (3) 15, 146–152 (1967).
33. *Wiener-Hopf operators and absolutely continuous spectra.* Bull. Amer. Math. Soc. 73, 659–662 (1967).

Rejto, P. A. (see also Conley, C. C., and Friedrichs, K. O.)
1. *On gentle perturbations,* I, II.
   I. Comm. Pure Appl. Math. 16, 279–303 (1963).
   II. ibid. 17, 257–292 (1964).
2. *On gentle perturbations. Perturbation theory and its application in quantum mechanics.* Wiley, New York, 57–95, 1966.
3. *On partly gentle perturbations,* I-III.
   I. J. Math. Anal. Appl. 17. 435–462 (1967).
   II. ibid. 20, 145–187 (1967).
   III. ibid. 27, 21–67 (1969).

Rellich, F.
2. *Störungstheorie der Spektralzerlegung,* I-V.
   I. Math. Ann. 113, 600–619 (1936).
   II. ibid. 113, 677–685 (1936).
   III. ibid. 116, 555–570 (1939).
   IV. ibid. 117, 356–382 (1940–1941).
   V. ibid. 118, 462–484 (1941–1943).

Rider, D. G. (see Davis, C.)

Riedl, J.
1. *Partially ordered locally convex vector spaces and extensions of positive continuous linear mappings.* Math. Ann. 157, 95–124 (1964).

Riesz, F.
21. *Sur les fonctions des transformations hermitiennes dans l'espace de Hilbert.* Acta Sci. Math. Szeged 7, 147–159 (1935).

Riesz, F., and Sz.-Nagy, B.
1. *Leçons d'analyse fonctionnelle.* Akadémiai Kiadó, Budapest, 1952.

Ringrose, J. R.
1. *Precompact linear operators in locally convex spaces.* Proc. Cambridge Philos. Soc. 53, 581–591 (1957).
2. *Operators of Volterra type.* J. London Math. Soc. 33, 418–424 (1958).
3. *On well-bounded operators,* I, II.
   I. J. Austral. Math. Soc. 1, 334–343 (1959/60).
   II. Proc. London Math. Soc. (3) 13, 613–638 (1963).
4. *Super-diagonal forms for compact linear operators.* Proc. London Math. Soc. (3) 12, 367–384 (1962).
5. *On the triangular representation of integral operators.* Proc. London Math. Soc. (3) 12, 385–399 (1962).
6. *On the resolvent and the principal vectors of a compact linear operator.* Proc. Cambridge Philos. Soc. 60, 525–531 (1964).

Riviere, N. (see Littman, W.)

Robertson, A. P., and Robertson, W. J.

1. *Topological vector spaces.* Cambridge Univ. Press, London, 1964.

Robinson, A. (see Bernstein, A. R.)

Rofe-Beketov, F. S. (see also Marčenko, V. A.)

1. *Expansion in eigenfunctions of finite systems of differential equations in the non-self-adjoint and self-adjoint cases.* Mat. Sbornik N. S. 51 (93) 293–342 (1960). (Russian) Math. Rev. 23 #A2603, 492 (1962).

Rolewicz, S. (see Przeworska-Rolewicz, D.)

Rosenblum, M.

1. *On the operator equation* $BX - XA = Q$. Duke Math. J. 23, 263–270 (1956).
2. *Perturbations of the continuous spectrum and unitary equivalence.* Pacific J. Math. 7, 997–1010 (1957).
3. *On a theorem of Fuglede and Putnam.* J. London Math. Soc. 33, 376–377 (1958).
4. *A spectral theory for self-adjoint singular integral operators.* Amer. J. Math. 88, 314–328 (1966).

Rosenthal, A. (see Hartogs, F.)

Rosenthal, P. (see Crimmins, T.)

Rota, G.-C.

1. *Note on the invariant subspaces of linear operators.* Rend. Circ. Mat. Palermo (2) 8, 182–184 (1959).
2. *Spectral theory of smoothing operators.* Proc. Nat. Acad. Sci. U.S.A. 46, 863–868 (1960).
3. *On the representation of averaging operators.* Rend. Sem. Mat. Univ. Padova 30, 52–64 (1960).
4. *On models for linear operators.* Comm. Pure Appl. Math. 13, 469–472 (1960).
5. *On the eigenvalues of positive operators.* Bull. Amer. Math. Soc. 67, 556–558 (1961). addend. 68, 49 (1962).
6. *Reynolds operators.* Proc. Sympos. Appl. Math., Vol. XVI, pp. 70–83. Amer. Math. Soc., 1964.
7. *Baxter Algebras and combinatorial identities,* I, II.
   I. Bull Amer. Math. Sci. 75, 325–329 (1969).
   II. ibid. 330–334 (1969).

Rota, G.-C., and Strang, W. G.

1. *A note on the joint spectral radius.* Nederl. Akad. Wetensch. Proc. Ser. A63, 379–381 (1960).

Rovnyak, J. (see de Branges, L., and Pincus, J. D.)

Ruston, A. F.

2. *On the Fredholm theory of integral equations for operators belonging to the trace class of a general Banach space.* Proc. London Math. Soc. (2) 53, 109–124 (1951).
3. *Direct products of Banach spaces and linear functional equations.* Proc. London Math Soc. (3) 1, 327–384 (1951).

5. *Formulae of Fredholm type for compact linear operations on a general Banach space.* Proc. London Math. Soc. (3) 3, 368–377 (1953).
6. *Operators with a Fredholm theory.* J. London Math. Soc. 29, 318–326 (1954).

Rutman, M. A. (see Kreĭn, M. G.)

Saffern, W. W.
1. *Subscalar operators.* Dissertation, Columbia Univ., 1962.

Sahnovič, L. A.
1. *On reduction of Volterra operators to the simplest form and on inverse problems.* Izv. Akad. Nauk SSSR Ser. Mat. 21, 235–262 (1957). (Russian) Math. Rev. 19, 970 (1958).
2. *Reduction of a non-selfadjoint operator with continuous spectrum to diagonal form.* Uspehi Mat. Nauk 13, no. 4 (82), 193–196 (1958). (Russian) Math. Rev. 20 #7222, 1183 (1959).
3. *Reduction to diagonal form of non-selfadjoint operators with continuous spectrum.* Mat. Sbornik N. S. 44 (86), 509–548 (1958). (Russian) Math. Rev. 20 #7223, 1184 (1959).
4. *The reduction of non-selfadjoint operators to triangular form.* Izv. Vysš. Učebn. Zaved. Matematika, no. 1 (8), 180–186 (1959). (Russian) Math. Rev. 25 #460, 97 (1963).
5. *A study of the "triangular form" of non-selfadjoint operators.* Izv. Vysš. Učebn. Zaved. Matem., no. 4 (11), 141–149 (1959). (Russian) Math. Rev. 25 #461, 98 (1963). Amer. Math. Soc. Transl. (2) 54, 75–84.
6. *Spectral analysis of Volterra operators prescribed in the vector-function space* $L_m^2[0, l]$. Ukrain. Mat. Ž. 16, 259–268 (1964). (Russian) Math. Rev. 29 #2680, 521 (1965).

Saitô, T. and Yoshino, T.
1. *Note on the canonical decomposition of contraction.* Tôhoku Math. J. (2) 16, 309–312 (1964).
2. *On a conjecture of Berberian.* Tôhoku Math. J. (2) 17, 147–149 (1965).

Salehi, H.
1. *A transformation theorem on spectral measures.* Proc. Amer. Math. Soc. 18, 610–613 (1967).

Saphar, P.
1. *Sur les sous-espaces invariants d'un opérateur linéaire continu dans un espace vectoriel topologique.* C. R. Acad. Sci. Paris 250, 1165–1166 (1960).
2. *Sur le spectre d'un opérateur linéaire continu dans un espace de Banach.* C. R. Acad. Sci. Paris 255, 3107–3108 (1962).
3. *Sur quelques propriétés d'un opérateur linéaire continu dans un espace vectoriel topologique.* C. R. Acad. Sci. Paris 254, 3946–3948 (1962).
4. *Calcul fonctionnel et sous-espaces stables pour une application linéaire continue dans un espace de Banach.* C. R. Acad. Sci. Paris 258, 6055–6057 (1964).
5. *Contribution à l'étude des applications linéaires dans un espace de Banach.* Bull. Soc. Math. France 92, 363–384 (1964).

6. *Sur les applications linéaires dans un espace de Banach*, I, II.
I. Bull. Soc. Math. France 92, 363–384 (1964).
II. Ann. Sci. École Norm. Sup. (3) 82, 205–240 (1965).
7. *Applications à puissance nucléaire et applications de Hilbert-Schmidt dans les espaces de Banach*. C. R. Acad. Sci. Paris 261, 867–870 (1965).
8. *Applications à puissance nucléaire et applications de Hilbert-Schmidt dans les espaces de Banach*. Ann. Sci. École Norm. Sup. (3) 83, 113–151 (1966).

Sarason, D.
1. *On spectral sets having connected complement*. Acta. Sci. Math. Szeged 26, 289–299 (1965).
2. *A remark on the Volterra operator*. J. Math. Anal. Appl. 12, 244–246 (1965).
3. *Invariant subspaces and unstarred operator algebras*. Pacific J. Math. 17, 511–517 (1966).

Sasser, D. W.
1. *Quasi-positive operators*. Pacific J. Math. 14, 1029–1037 (1964).

Sawashima, I. (see also Niiro, F.)
1. *Some counter examples in the theory of positive operators*. Sci. Papers College Gen. Ed. Univ. Tokyo 14, 181–182 (1964).
2. *On spectral properties of some positive operators*. Natur. Sci. Rep. Ochanomizu Univ. 15, 53–64 (1964).
3. *On spectral properties of positive irreducible operators in $C(X)$ and a problem of H. H. Schaefer*. Natur. Sci. Rep. Ochanomizu Univ. 17, 1–15 (1966).

Schaefer, H. H. (see also Lotz, H.P.)
1. *Positive Transformationen in lokalkonvexen halbgeordneten Vektorräumen*. Math. Ann. 129, 323–329 (1955).
2. *Über singulare Integralgleichungen und eine Klasse von Homomorphismen in lokalkonvexen Räumen*. Math. Zeit. 66, 147–163 (1956).
3. *Halbgeordnete lokalkonvexe Vektorräume*, I–III.
I. Math. Ann. 135, 115–141 (1958).
II. ibid. 138, 259–286 (1959).
III. ibid. 141, 113–142 (1960).
4. *On nonlinear positive operators*. Pacific J. Math. 9, 847–860 (1959).
5. *On the Fredholm alternative in locally convex linear spaces*. Studia Math. 18, 229–245 (1959).
6. *Some spectral properties of positive linear operators*. Pacific J. Math. 10, 1009–1019 (1960).
7. *A new class of spectral operators*. Bull. Amer. Math. Soc. 67, 154–155 (1961).
8. *A generalized moment problem*. Math. Ann. 146, 326–330 (1962).
9. *Über die Additivität von Spektralmassen*. Math. Zeit. 79, 456–459 (1962).
10. *Spectral measures in locally convex algebras*. Acta Math. 107, 125–173 (1962).
11. *Convex cones and spectral theory*. Proc. Sympos. Pure Math., Vol. VII, 451–471. Amer. Math. Soc., Providence, 1963.
12. *Spektraleigenschaften positiver linearer Operatoren*. Math. Zeit. 82, 303–313 (1963).

13. *Eine Bemerkung zur Existenz invarianter Teilräume linearer Abbildungen.* Math. Zeit. 82, 90 (1963).
14. *On the point spectrum of positive operators.* Proc. Amer. Math. Soc. 15, 56–60 (1964).
15. *On the role of order structures in spectral theory.* Colloque sur l'Analyse Fonctionnelle, Louvain-Paris (1964).
16. *Über das Randspektrum positiver Operatoren.* Math. Ann. 162, 289–293 (1965/66).
17. *Eine Klasse irreduzibler positiver Operatoren.* Math. Ann. 165, 26–30 (1966).
18. *Topological vector spaces.* Macmillan, New York, 1966.
19. *Invariant ideals of positive operators in $C(X)$*, I, II.
    I. Illinois J. Math. 11, 703–715 (1967).
    II. ibid. 12, 525–538 (1968)
20. *Banach lattices and positive operators.* Springer-Verlag (to appear).

Schaefer, H. H., and Walsh, B. J.
1. *Spectral operators in spaces of distributions.* Bull. Amer. Math. Soc. 68, 509–511 (1962).

Schäffer, J. J. (see also Halmos, P. R.)
3. *More about invertible operators without roots.* Proc. Amer. Math. Soc. 16, 213–219 (1965).

Schatten, R.
2. *Norm ideals of completely continuous operators.* Ergebnisse der Math. N. F., Heft 27. Springer-Verlag, Berlin-Göttingen-Heidelberg, 1960.

Schechter, M. (see also Ercolano, J., and Kaniel, S.)
1. *Invariance of the essential spectrum.* Bull. Amer. Math. Soc. 71, 365–367 (1965).
2. *On the essential spectrum of an arbitrary operator* I. J. Math. Anal. Appl. 13, 205–215 (1966).

Schreiber, M.
2. *Unitary dilations of operators.* Duke Math. J. 23, 579–594 (1956).
3. *A functional calculus for general operators in Hilbert space.* Trans. Amer. Math. Soc. 87, 108–118 (1958).
4. *On the spectrum of a contraction.* Proc. Amer. Math. Soc. 12, 709–713 (1961).
5. *Absolutely continuous operators.* Duke Math. J. 29, 175–190 (1962).
6. *Numerical range and spectral sets.* Michigan Math. J. 10, 283–288 (1963).
7. *Semi-Carleman operators.* Acta Sci. Math. Szeged 24, 82–87 (1963).
8. *Remark on a paper of Kalisch.* J. Math. Anal. Appl. 7, 62–63 (1963).

Schubert, C. (see Goldberg, S.)

Schwartz, J. T. (see also McCarthy, C. A.)
2. *Perturbation of spectral operators, and applications*, I. Pacific J. Math. 4, 415–458 (1954).
3. *Two perturbation formulae.* Comm. Pure Appl. Math. 8, 371–376 (1955).
4. *Some non-selfadjoint operators.* Comm. Pure Appl. Math. 13, 609–639 (1960).

5. *Compact positive mappings in Lebesgue spaces.* Comm. Pure Appl. Math. 14, 693–705 (1961).
6. *Subdiagonalization of operators in Hilbert spaces with compact imaginary part.* Comm. Pure Appl. Math. 15, 159–172 (1962).
7. *On spectral operators in Hilbert space with compact imaginary part.* Comm. Pure Appl. Math. 15, 95–97 (1962).
8. *Some results on the spectra and spectral resolutions of a class of singular integral operators.* Comm. Pure Appl. Math. 15, 75–90 (1962).
9. *Some non-selfadjoint operators.* II. *A family of operators yielding to Friedrichs' method.* Comm. Pure Appl. Math. 14, 619–626 (1961).
10. *W* algebras.* Gordon and Breach, New York, 1967.

Schwartz, L.
6. *Théorie des distributions à valeurs vectorielles.* Ann. Inst. Fourier 7, 1–141 (1957).

Scroggs, J. E.
1. *Invariant subspaces of a normal operator.* Duke Math. J. 26, 95–111 (1959).

Sebastião e Silva, J.
4. *La définition de spectre d'un opérateur et les opérateurs à spectre élémentaire non borné.* Colloque sur l'Analyse Fonctionnelle (Louvain, 1960) pp. 47–50. Librairie Universitaire, Louvain, 1961.
5. *Sur le calcul symbolique d'opérateurs permutables, à spectre vide ou non borné.* Ann. Mat. Pure Appl. (4) 58, 219–275 (1962).

Seeley, R. T.
1. *The index of elliptic systems of singular integral operators.* J. Math. Anal. Appl. 7, 289–309 (1963).

Segal, I. E.
1. *Decompositions of operator algebras,* I, II. Memoirs Amer. Math. Soc. no. 9, 1951.

Sheth, I. H.
1. *On hyponormal operators.* Proc. Amer. Math. Soc. 17, 998–1000 (1966).

Shizuta, Y.
1. *Eigenfunction expansions associated with the operator* $-\Delta$ *in the exterior domain.* Proc. Japan Acad. 39, 656–660 (1963).

Sibuya, Y. (see Hukuhara, M.)

Sigalov, A. G.
1. *A new algorithm in the theory of perturbations of the continuous spectrum.* Doklady Akad. Nauk SSSR 158, 49–52 (1964). (Russian) Math. Rev. 29 #2659, 516 (1965).
2. *Integral perturbations.* Sibirsk Mat. Ž. 7, 373–408 (1966). (Russian) Math. Rev. 34 #3335, 602 (1967).

Sikorski, R.
1. *On multiplication of determinants in Banach spaces.* Bull. Acad. Polon. Sci. Cl. III. 1, 219–221 (1953).

2. *On Leżański's determinants of linear equations in Banach spaces.* Studia Math. 14 (1953), 24–48 (1954).
3. *On determinants of Leżański and Ruston.* Studia Math. 16, 99–112 (1957).
4. *Determinant systems.* Studia Math. 18, 161–186 (1959).
5. *On Leżański endomorphisms.* Studia Math. 18, 187–189 (1959).
6. *Remarks on Leżański's determinants.* Studia Math. 20, 145–161 (1961).
7. *On the Carleman determinants.* Studia Math. 20, 327–346 (1961).
8. *The determinant theory in Banach spaces.* Colloq. Math. 8, 141–198 (1961).
9. *Determinants in Banach spaces.* Studia Math. (Ser. Specjalna) Zeszyt. 1, 111–116 (1963).

Silberstein, J. P. O.
2. *Symmetrisable operators*, I–III.
I. J. Austral. Math. Soc. 2, 381–402 (1961/62).
II. ibid. 4, 15–30 (1964).
III. ibid. 4, 31–48 (1964).

Sills, W. H.
1. *On absolutely continuous functions and the well-bounded operator.* Pacific J. Math. 17, 349–366 (1966).

Silverman, R. J., and Yen, Ti
1. *Characteristic functionals.* Proc. Amer. Math. Soc. 10, 471–477 (1959).

Simon, A. B. (see Ionescu Tulcea, C.)

Simpson, J. E.
1. *On spectral measures and spectral operators.* Dissertation, Yale Univ., 1961.
2. *Nilpotency and spectral operators.* Pacific J. Math. 14, 665–672 (1964).
3. *On limits of scalar operators.* Trans. Amer. Math. Soc. 122, 163–176 (1966).

Sine, R. C.
1. *Spectral decomposition of a class of operators.* Pacific J. Math. 14, 333–352 (1964).

Singbal-Vedak, K.
1. *A note on semigroups of operators on a locally convex space.* Proc. Amer. Math. Soc. 16, 696–702 (1965).

Smart, D. R.
1. *Eigenfunction expansions in L^p and C.* Illinois J. Math. 3, 82–97 (1959).
2. *Conditionally convergent spectral expansions.* J. Austral. Math. Soc. 1, 319–333 (1959/60).
3. *Some examples of spectral operators.* Illinois J. Math. 11, 603–607 (1967).

Smith, K. T. (see Aronszajn, N.)

Smithies, F. (see also Bernau, S. J.)
1. *The Fredholm theory of integral equations.* Duke Math. J. 8, 107–130 (1941).

Šmul'jan, Ju. L. (see Brodskiĭ, M. S., and Kreĭn, M. G.)

Spitzer, F.
1. *A combinatorial lemma and its applications to probability.* Trans. Amer. Math. Soc. 82, 323–337 (1956).

Srinivasan, T. P. (see also Hasumi, M.)

1. *Simply invariant subspaces.* Bull. Amer. Math. Soc. 69, 706–709 (1963).
2. *Doubly invariant subspaces.* Pacific J. Math. 14, 701–707 (1964).

Stampfli, J. G. (see also McCarthy, C. A.)

1. *Roots of scalar operators.* Proc. Amer. Math. Soc. 13, 796–798 (1962).
2. *Hyponormal operators.* Pacific J. Math. 12, 1453–1458 (1962).
3. *Sums of projections.* Duke Math. J. 31, 455–461 (1964).
4. *Hyponormal operators and spectral density.* Trans. Amer. Math. Soc. 117, 469–476 (1965). Errata, ibid. 115 (sic), 550 (1965).
5. *Perturbations of the shift.* J. London Math. Soc. 40, 345–347 (1965).
6. *Extreme points of the numerical range of a hyponormal operator.* Michigan Math. J. 13, 87–89 (1966).
7. *Analytic extensions and spectral localization.* J. Math. Mech. 16, 287–296 (1966).
8. *Normality and the numerical range of an operator.* Bull. Amer. Math. Soc. 72, 1021–1022 (1966).
9. *Minimal range theorems for operators with thin spectra.* Pacific J. Math. 23, 601–612 (1967).
10. *A local spectral theory for operators.* J. Func. Anal. 4, 1–10 (1969).
11. *Adjoint abelian operators on Banach space.* Canadian J. Math. 21, 505–512 (1969).
12. *A local spectral theory for operators,* III; *Resolvents, spectral sets, and similarity* (unpublished).

Stankevič, I. V. (see also Gehtman, M. M.)

1. *On the perturbation theory of a continuous spectrum.* Doklady Akad. Nauk SSSR 144, 279–282 (1962). (Russian) Math. Rev. 27 #1834, 361 (1964).
2. *On the linear similarity between certain nonselfadjoint operators and selfadjoint ones and on the asymptotic behaviour of the nonstationary Schrödinger equation for $t \to \infty$.* Mat. Sbornik N. S. 69 (111), 161–207 (1966). (Russian) Math. Rev. 33 #3138, 542 (1967).

Stečenko, V. Ja. (see Bahtin, I. A., Esajan, A. R., and Zabreĭko, P. P.)

1. *An estimate for the spectrum of some classes of linear operators.* Doklady Akad. Nauk SSSR 157, 1054–1057 (1964). (Russian) Math. Rev. 29 #3898, 747 (1965).

Steckloff, W.

1. *Sur les expressions asymptotiques de certaines fonctions définies par des équations différentielles linéaires du deuxième ordre, et leurs applications au problème du développement d'une fonction arbitraire en séries procédant suivant les dites fonctions.* Comm. Soc. Math. Kharkow (2) 10 (2–6), 97–199 (1907–1909). Rev. Sem. Publ. Math. 21, 117 (1913).

Strang, W. G. (see Rota, G.-C.)

Štraus, A. V.

6. *On the multiplicity of the spectrum of a self-adjoint ordinary differential operators.* Doklady Akad. Nauk SSSR 155, 771–774 (1964). (Russian) Math. Rev. 28 #3346, 656 (1964).

Suciu, I. (see also Foiaş, C.)

1. *Dilatable spectral representations of a commutative Banach algebra.* Stud. Cerc. Mat. 16, 1211–1220 (1964).

Suzuki, Noboru

1. *On the spectral decomposition of dissipative operators.* Proc. Japan Acad. 42, 577–582 (1966).
2. *The algebraic structure of non self-adjoint operators.* Acta Sci. Math. Szeged 27, 173–184 (1966).

Švarc, A. S.

1. *On the homotopic topology of Banach spaces.* Doklady Akad. Nauk SSSR 154, 61–63 (1964). (Russian) Math. Rev. 28 #3309, 647 (1964).

Sz.-Nagy, B. (see also Riesz, F.)

3. *Spektraldarstellung linearer Transformationen des Hilbertschen Raumes.* Ergebnisse der Math., V 5, J. Springer, Berlin 1942. Reprinted Edwards Bros., Ann Arbor, Mich., 1947.
7. *On uniformly bounded linear transformations in Hilbert space.* Acta Sci. Math. Szeged 11, 152–157 (1947).
13. *On the stability of the index of unbounded linear transformations.* Acta Math. Acad. Sci. Hungar. 3, 49–52 (1952). (Russian summary).
16. *Contributions en Hongrie à la théorie spectrale des transformations linéaires.* Czech. Math. J. 6 (81), 166–176 (1956).
17. *Spectral sets and normal dilations of operators.* Proc. Internat. Congress Math. 1958, pp. 412–422. Cambridge Univ. Press, New York, 1960.
18. *On Schäffer's construction of unitary dilations.* Ann. Univ. Sci. Budapest. Eötvös Sect. Math. 3–4, 343–346 (1960/61).
19. *Bemerkungen zur vorstehenden Arbeit des Herrn S. Brehmer.* Acta Sci. Math. Szeged 22, 112–114 (1961).
20. *Un calcul fonctionnel pour les opérateurs linéaires de l'espace hilbertien et certaines de ses applications.* Studia Math. (Ser. Specjalna) Zeszyt. 1, 119–127 (1963).
21. *Sur les contractions de l'espace de Hilbert,* I, II.
    I. Acta Sci. Math. Szeged 15, 87–92 (1953).
    II. ibid. 18, 1–14 (1957).
22. *The "outer functions" and their role in functional calculus.* Proc. Internat. Congress Mathematicians (Stockholm, 1962), pp. 421–425. Inst. Mittag-Leffler, Djursholm, 1963.

Sz.-Nagy, B., and Foiaş, C.

1. *Sur les contractions de l'espace de Hilbert,* III–XII.
    III. Acta Sci. Math. Szeged 19, 26–45 (1958).
    IV. ibid. 21, 251–259 (1960).
    V. *Translations bilatérales.* ibid. 23, 106–129 (1962).
    VI. *Calcul fonctionnel.* ibid. 23, 130–167 (1962).
    VII. *Triangulations canoniques. Fonction minimum.* ibid. 25, 12–37 (1964).
    VIII. *Fonction caractéristiques. Modèles fonctionnels.* ibid. 25, 38–71 (1964).

IX. *Factorisations de la fonction caractéristique. Sous-espaces invariants.* ibid. 25, 283–316 (1964).
X. *Contractions similaires à des transformations unitaires.* ibid. 26, 79–91 (1965).
XI. *Tranformations unicellulaires.* ibid. 26, 301–324 (1965). Errata, ibid. 27, 265 (1966).
XII. ibid. 27, 27–33 (1966).
2. *Remark to the preceding paper of J. Feldman.* Acta Sci. Math. Szeged 23, 272–273 (1962).
3. *Modèles fonctionnels des contractions de l'espace de Hilbert. La fonction caractéristique.* C. R. Acad. Sci. Paris 256, 3236–3238 (1963).
4. *Propriétés des fonctions caractéristiques, modèles triangulaires et une classification de l'espace de Hilbert.* C. R. Acad. Sci. Paris 256, 3413–3415 (1963).
5. *Une caractérisation des sous-espaces invariants pour une contraction de l'espace de Hilbert.* C. R. Acad. Sci. Paris 258, 3426–3429 (1964).
6. *Quasi-similitude des opérateurs et sous-espaces invariants.* C. R. Acad. Sci. Paris, 261, 3938–3940 (1965).
7. *Décomposition spectrale des contractions presque unitaires.* C. R. Acad. Sci. Paris, 262, 440–442 (1966).
8. *Forme triangulaire d'une contraction et factorisation de la fonction caractéristique.* Acta Sci. Math. Szeged 28, 201–212 (1967).
9. *Echelles continues de sous-espaces invariants.* Acta Sci. Math. Szeged 28, 213–220 (1967).
10. *Analyse harmoniques des opérateurs de l'espace de Hilbert.* Akad. Kaidó, Budapest, 1967.
11. *Commutants de certains opérateurs.* Acta Sci. Math. Szeged 29, 1–17 (1968).
12. *On certain classes of power bounded operators in Hilbert space.* Acta Sci. Math. Szeged 27, 17–25 (1966).

Tamarkin, J. D.
2. *Sur quelques points de la théorie des équations différentielles linéaires ordinaires et sur la généralisation de la série de Fourier.* Rend. Circ. Mat. Palermo 24, 345–382 (1912).
3. *Some general problems of the theory of ordinary linear differential equations and expansions of an arbitrary function in a series of fundamental functions.* Math. Zeit. 27, 1–54 (1927).

Taylor, A. E. (see also Derr, J., Gindler, H. A., and Halberg, C. J. A.)
10. *Analysis in complex Banach spaces.* Bull. Amer. Math. Soc. 49, 652–669 (1943).
11. *Spectral theory of closed distributive operators.* Acta Math. 84, 189–224 (1951).
15. *Mittag-Leffler expansions and spectral theory.* Pacific J. Math. 10, 1049–1066 (1960).
16. *Spectral theory and Mittag-Leffler type expansions of the resolvent.* Proc. Int. Symposium on Linear Spaces, pp. 426–440. Jerusalem, 1960.

17. *The minimum modulus of a linear operator, and its use for estimates in spectral theory.* Studia Math. (Ser. Specjalna) Zeszyt 1, 131–132 (1963).

18. *Theorems on ascent, descent, nullity and defect of linear operators.* Math. Ann. 163, 18–49 (1966).

Teleman, S.

1. *On the relativization of set functions.* Rev. Roumaine Math. Pures Appl. 13, 683–689 (1968).

Thoe, D.

1. *Spectral theory for the wave equation with a potential term.* Arch. Rational Mech. Anal. 22, 364–406 (1966).

Thompson, A. C.

1. *A spectral theorem for positive operators.* J. London Math. Soc. 44, 485–495 (1969).

Thorp, E. O. (see Goldberg, S.)

Tillmann, H. G.

1. *Vector-valued distributions and the spectral theorem for selfadjoint operators in Hilbert space.* Bull. Amer. Math. Soc. 69, 67–71 (1963).

2. *Eine Erweiterung des Funktionalkalküls für lineare Operatoren.* Math. Ann. 151, 424–430 (1963).

3. *Darstellung vektorwertiger Distributionen durch holomorphe Funktionen.* Math. Ann. 151, 286–295 (1963).

Titchmarsh, E. C.

16. *Eigenfunction expansions associated with second-order differential equations.* Oxford Univ. Press, London, 1946.

Tôgô, S., and Shiraishi, R.

1. *Note on F-operators in locally convex spaces.* J. Sci. Hiroshima Univ. Ser. A-I Math. 29, 243–251 (1965).

Tomiuk, B. J. (see Bonsall, F. F.)

Trampus, A.

1. *A spectral mapping theorem for functions of two commuting linear operators.* Proc. Amer. Math. Soc. 14, 893–895 (1963).

Treves, F.

1. *Topological vector spaces, distributions and kernels.* Academic Press, New York, 1967.

Turner, R. E. L.

1. *Perturbation of compact spectral operators.* Comm. Pure Appl. Math. 18, 519–541 (1965).

2. *Perturbation of ordinary differential operators.* J. Math. Anal. Appl. 13, 447–457 (1966).

Tzafriri, L. (see also McCarthy, C. A.)

1. *The connection between normalizable and spectral operators.* Israel J. Math. 3, 75–80 (1965).

2. *On multiplicity theory for Boolean algebras of projections.* Israel J. Math. 4, 217-224 (1966).

3. *On perturbation theory for spectral operators.* Israel J. Math. 4, 62–64 (1966).
4. *Operators commuting with Boolean algebras of projections of finite multiplicity.* Pacific J. Math. 20, 571–587 (1967).
5. *Operators commuting with Boolean algebras of projections of infinite multiplicity.* Trans. Amer. Math. Soc. 128, 164–175 (1967).
6. *Quasi-similarity for spectral operators on Banach spaces.* Pacific J. Math. 25, 197–217 (1968).

Valickiĭ, Yu. N. (see Fišman, K. M.)

Vasilescu, F.-H.

1. *Spectral algebras of a generalized scalar operator.* Rev. Roumaine Math. Pures Appl. 10, 1241–1243 (1966).
2. *On an asymptotic behavior of operators.* Rev. Roumaine Math. Pures Appl. 12, 353–358 (1967).
3. *Spectral distance of two operators.* Rev. Roumaine Math. Pures Appl. 12, 733–736 (1967).
4. *On a class of scalar generalized operators.* Rev. Roumaine Math. Pures Appl. 12, 865–867 (1967).
5. *Some properties of the commutator of two operators.* J. Math. Anal. Appl. 23, 440–446 (1968).
6. *Asymptotic properties of the commutators of decomposable operators.* J. Math. Anal. Appl. (to appear).
7. *Residually decomposable operators in Banach spaces.* Tôhoku Math. J. 21, 504–522 (1969).
8. *Residual properties for closed operators on Fréchet spaces.* Illinois J. Math. (to appear).

Vidav, I.

2. *Eine metrische Kennzeichnung der selbstadjungierten Operatoren.* Math. Zeit. 66, 121–128 (1956).

Viziteĭ, V. N.

1. *Expansion in root vectors of a weakly perturbed normal operator.* Bul. Akad. Štiince RSS Moldoven. 1964, no. 6, 19–26. (Russian. Moldavian summary) Math. Rev. 34 #8199, 1507 (1967).
2. *On the decomposition with regard to the eigenvectors and root vectors of a bounded operator.* Bul. Akad. Štiince RSS Moldoven. 1965, no. 7, 33–39. (Russian. Moldavian summary) Math. Rev. 33 #6416, 1085 (1967).

Viziteĭ, V. N., and Markus, A. S.

3. *Convergence of multiple expansions of a bundle of operators in a system of eigenvectors and adjoined vectors.* Mat. Sbornik N. S. 66 (108), 287-320 (1965). (Russian). Math. Rev. 35 #801, 155 (1968).

Volk, V. Ja.

1. *On the spectral decomposition for a class of non-selfadjoint operators.* Doklady Akad. Nauk SSSR 152, 259–261 (1963). (Russian) Math. Rev. 27 #4085, 788 (1964).

Waelbroeck, L.

1. *Le calcul symbolique dans les algèbres commutatives.* J. Math. Pures Appl. (9) 33, 147–186 (1954).
2. *Étude spectrale des algèbres complètes.* Acad. Roy. Belg. Cl. Sci. Mém. Coll. in $-8°$ (2) 31, no. 7, 142 pp. (1960).
3. *Études spectrale de certaines algèbres complètes.* Colloque sur l'Analyse Fonctionnelle (Louvain, 1960), pp. 29–38. Librairie Universitaire, Louvain 1961.
4. *Le calcul symbolique lié à la croissance de la résolvente.* Rend. Sem. Math. Univ. Milano 34, 51–72 (1964).

Walsh, B. J. (see also Schaefer, H. H.)

1. *Banach algebras of scalar type elements.* Proc. Amer. Math. Soc. 16, 1167–1170 (1965).
2. *Structure of spectral measures on locally convex spaces.* Trans. Amer. Math. Soc. 120, 295–326 (1965).
3. *Spectral decomposition of quasi-Montel spaces.* Proc. Amer. Math. Soc. 17, 1267–1271 (1966).

Wecken, F. J.

2. *Unitärinvarianten sebstadjungierter Operatoren.* Math. Ann. 116, 422–455 (1939).

Weidmann, J.

1. *Ein Satz über nukleare Operatoren im Hilbertraum.* Math. Ann. 158, 69–78 (1965).
2. *Zur Spektraltheorie von Sturm-Liouville-Operatoren.* Math. Zeit. 98, 268–302 (1967).

Wermer, J.

1. *The existence of invariant subspaces.* Duke Math. J. 19, 615–622 (1952).
2. *Invariant subspaces of bounded operators.* Proc. XII Scand. Math. Congress Lund, 1953.
3. *Commuting spectral operators on Hilbert space.* Pacific J. Math. 4, 355–361 (1954).
4. *On invariant subspaces of normal operators.* Proc. Amer. Math. Soc. 3, 270–277 (1952).
7. *On a class of normed rings.* Arkiv. för Mat. 2, 537–551 (1953).

West, T. T. (see also Gillespie, T. A., and Olagunju, P. A.)

1. *The spectra of compact operators in Hilbert spaces.* Proc. Glasgow Math. Assoc. 7, 34–38 (1965).
2. *Operators with a single spectrum.* Proc. Edinburgh Math. Soc. (2) 15, 11–18 (1966).
3. *Riesz operators in Banach spaces.* Proc. London Math. Soc. (3) 16, 131–140 (1966).
4. *The decomposition of Riesz operators.* Proc. London Math. Soc. (3) 16, 737–752 (1966).

Weyl, H.

5. *Über gewöhnliche Differentialgleichungen mit Singularitäten und die zugehörigen Entwicklungen willkürlicher Funktionen.* Math. Ann. 68, 220–269 (1910).

Whitley, R. J.

1. *Strictly singular operators and their conjugates.* Trans. Amer. Math. Soc. 113, 252–261 (1964).
2. *The spectral theorem for a normal operator.* Amer. Math. Monthly 75, 856–861 (1968).

Wiener, N.

4. *The Fourier integral and certain of its applications.* Cambridge Univ. Press, 1933. Reprinted by Dover Pub., New York.
5. *Tauberian theorems.* Ann. of Math. (2) 33, 1–100 (1932).

Wiener, N., and Hopf, E.

1. *Über eine Klasse singulärer Integralgleichungen.* S.-B. Preuss. Akad. Wiss. Berlin Phys.-Math. Kl. 30/32, 696–706 (1931).

Wilder, C. E.

1. *Expansion problems of ordinary linear differential equations with auxiliary conditions at more than two points.* Trans. Amer. Math. Soc. 18, 415–442 (1917).
2. *Problems in the theory of ordinary linear differential equations with auxiliary conditions at more than two points.* Trans. Amer. Math. Soc. 19, 157–186 (1918).

Williams, J. P.

1. *Spectra of products and numerical ranges.* J. Math. Anal. Appl. 17, 214–220 (1967).

Williamson, J. H.

3. *Compact linear operators in linear topological spaces.* J. London Math. Soc. 29, 149–156 (1954).

Wolf, F.

2. *Simplicity of spectra in general operators.* (Abstract) Bull. Amer. Math. Soc. 60, 345 (1954).
3. *On majorants of subharmonic and analytic functions.* Bull. Amer. Math. Soc. 48, 925–932 (1942).
4. *Operators in Banach space which admit a generalized spectral decomposition.* Nederl. Akad. Wetensch. Proc. Ser. A 60, 302–311 (1957).
5. *Spectral decomposition of operators in a Banach space and the analytic character of their resolvent.* Seminars on Analytic Functions, Institute for Advanced Study, vol. 2, pp. 312–320 (1960).

Yen, T. (see Silverman, R. J.)

Yoshida, M. (see Nakamura, M.)

Yoshino, T. (see also Saitô, T.)

1. *On the spectrum of a hyponormal operator.* Tôhoku Math. J. (2) 17, 305–309 (1965). Errata ibid. (2) 19, 101 (1967). See Math. Rev. 32 #4548, 766 (1966).

2. *Spectral resolution of a hyponormal operator with the spectrum on a curve.* Tôhoku Math. J. (2) 19, 86–97 (1967).

Yosida, K.

13. *A perturbation theorem for semi-groups of linear operators.* Proc. Japan Acad. 41, 645–647 (1965).

14. *Functional analysis.* Springer-Verlag, Berlin, 1965.

Yood, B.

2. *Properties of linear transformations preserved under addition of a completely continuous transformation.* Duke Math. J. 18, 599–612 (1951).

Zaanen, A. C. (see also Luxemburg, W. A. J.)

10. *A note on perturbation theory.* Nieuw Arch. Wisk. (3) 7, 61–65 (1959).

Zabreĭko, P. P., Krasnosel'skiĭ, M. A., and Stečenko, V. Ja.

1. *Estimates of the spectral radius of positive linear operators.* Mat. Zametki 1, 461–468 (1967). (Russian) Math. Rev. 34 #8200, 1507 (1967).

Zinnes, I. I. (see Jauch, J. M.)

Zygmund, A.

1. *Trigonometrical Series.* Monografje Matematyczne, Warsaw, 1935. Reprinted Dover and Chelsea Pub. Co., New York.

Notation Index

Author Index

Subject Index

Section numbers are followed by page numbers in parenthesis.